NUTRITIONAL EVALUATION OF FOOD PROCESSING

SECOND EDITION

Other AVI Books

FOOD SCIENCE AND TECHNOLOGY
  AGRICULTURAL AND FOOD CHEMISTRY: PAST, PRESENT, FUTURE
    Teranishi
  CARBOHYDRATES AND HEALTH
    Hood, Wardrip and Bollenback
  CITRUS SCIENCE AND TECHNOLOGY
    Vols. 1 and 2 *Nagy, Shaw and Veldhuis*
  COMMERCIAL FRUIT PROCESSING
    Woodroof and Luh
  COMMERCIAL VEGETABLE PROCESSING
    Luh and Woodroof
  DIETARY NUTRIENT GUIDE
    Pennington
  DRUG-INDUCED NUTRITIONAL DEFICIENCIES
    Roe
  EGG SCIENCE AND TECHNOLOGY
    2nd Edition *Stadelman and Cotterill*
  ELEMENTARY FOOD SCIENCE
    Nicherson and Ronsivalli
  ELEMENTS OF FOOD TECHNOLOGY
    Desrosier
  ENCYCLOPEDIA OF FOOD SCIENCE
    Peterson and Johnson
  ENCYCLOPEDIA OF FOOD TECHNOLOGY
    Johnson and Peterson
  EVALUATION OF PROTEINS FOR HUMANS
    Bodwell
  FOOD ANALYSIS: THEORY AND PRACTICE
    Revised Edition *Pomeranz and Meloan*
  FOOD AND BEVERAGE MYCOLOGY
    Beuchat
  FOOD COLLOIDS
    Graham
  FOOD FOR THOUGHT
    2nd Edition *Labuza and Sloan*
  FOOD PROTEINS
    Whitaker and Tannenbaum
  FUNDAMENTALS OF FOOD FREEZING
    Desrosier and Tressler
  IMMUNOLOGICAL ASPECTS OF FOODS
    Catsimpoolas
  INTRODUCTORY FOOD CHEMISTRY
    Garard
  LABORATORY MANUAL IN FOOD PRESERVATION
    Fields
  PRINCIPLES OF FOOD CHEMISTRY
    deMan
  PROCESSED MEATS
    Kramlich, Pearson and Tauber
  TECHNOLOGY OF FOOD PRESERVATION
    4th Edition *Desrosier and Desrosier*

NUTRITIONAL EVALUATION OF FOOD PROCESSING
SECOND EDITION

Edited by **ROBERT S. HARRIS, Ph.D.**
*Professor Emeritus of Nutritional Biochemistry,
Massachusetts Institute of Technology,
Cambridge, Massachusetts*

and **ENDEL KARMAS, Ph.D.**
*Associate Professor of Food Science,
Rutgers University,
New Brunswick, New Jersey*

THE AVI PUBLISHING COMPANY, INC.
WESTPORT, CONNECTICUT

© *Copyright 1975 by*
THE AVI PUBLISHING COMPANY, INC.
Westport, Connecticut

Second Printing 1977

Copyright is not claimed in any portion of this work written by a United States Government employee as a part of his official duties.

All rights reserved. No part of this work covered by the copyright hereon may be reproduced or used in any form or by any means—graphic, electronic, or mechanical, including photocopying, recording, taping, or information storage and retrieval systems—without written permission of the publisher.

Library of Congress Catalog Card Number: 75-25422
ISBN-0-87055-189-2

Printed in the United States of America

Contributors

L. R. BAKER, Ph.D., Associate Professor, Horticulture, Department of Horticulture, Michigan State University, East Lansing, Michigan

PETER M. BLUESTEIN, Ph.D., Thomas J. Lipton, Inc., 800 Sylvan Avenue, Englewood Cliffs, New Jersey

BENJAMIN BORENSTEIN, Ph.D., Manager, Food Industry Technical Services, Roche Chemical Division, Hoffmann-LaRoche, Inc., Nutley, New Jersey

RICARDO BRESSANI, Ph.D., Head, Division of Agricultural and Food Chemistry, Institute of Nutrition of Central America and Panama, Guatemala City, Guatemala

T. C. BYERLY, Ph.D., Consultant to the Food Industry, 6-J Ridge Road, Greenbelt, Maryland

WILLIAM K. CALHOUN, Ph.D., Chief, Nutrition Division, Food Laboratory, U.S. Army Natick Laboratories, Natick, Massachusetts

WINIFRED M. CORT, Ph.D., Group Leader, Quality Control, Hoffmann-LaRoche, Inc., Nutley, New Jersey

HENRYK DAUN, Ph.D., Assistant Professor, Food Science, Department of Food Science, Rutgers University, New Brunswick, New Jersey

JOHN W. ERDMAN, JR., Ph.D., Department of Food Science, Rutgers University, New Brunswick, New Jersey

BERNARD FEINBERG, Consultant to the Food Industry, 1188 Keeler Ave., Berkeley, California

OWEN FENNEMA, Ph.D., Professor, Food Chemistry, Department of Food Science, University of Wisconsin, Madison, Wisconsin

VERNON L. FRAMPTON, Ph.D., Research Leader, Biochemistry and Biophysics, Southern Regional Research Center, Agricultural Research Service, U.S. Department of Agriculture, New Orleans, Louisiana

ROBERT S. HARRIS, Ph.D., Professor Emeritus, Nutritional Biochemistry, Massachusetts Institute of Technology, Cambridge, Massachusetts

NORMAN D. HEIDELBAUGH, V.M.D., M.P.H., Ph.D., U.S. Air Force Representative on the Department of Defense Food Program, U.S. Army Natick Laboratories, Natick, Massachusetts

GEORGE E. INGLETT, Ph.D., Chief, Cereal Properties Laboratory, Northern Regional Research Laboratory, Agricultural Research Service, U.S. Department of Agriculture, Peoria, Illinois

CONTRIBUTORS

IVAN D. JONES, Ph.D., Professor Emeritus, Food Science, North Carolina State University, Raleigh, North Carolina

EDWARD S. JOSEPHSON, Ph.D., Deputy Technical Director, Food Service Systems Program, U.S. Army Natick Laboratories, Natick, Massachusetts

MARCUS KAREL, Ph.D., Professor, Food Engineering and Deputy Department Head, Department of Nutrition and Food Science, Massachusetts Institute of Technology, Cambridge, Massachusetts

ENDEL KARMAS, Ph.D., Associate Professor, Food Science, Department of Food Science, Rutgers University, New Brunswick, New Jersey

JOHN. M. KROCHTA, Ph.D., Research Chemical Engineer, Western Regional Research Laboratory, Agricultural Research Service, U.S. Department of Agriculture, Berkeley, California

THEODORE P. LABUZA, Ph.D., Professor, Food Technology, Department of Food Science and Nutrition, University of Minnesota, St. Paul, Minnesota

PAUL A. LACHANCE, Ph.D., Professor, Nutritional Physiology, Department of Food Science, Rutgers University, New Brunswick, New Jersey

DARYL. B. LUND, Ph.D., Associate Professor, Food Science, Department of Food Science, University of Wisconsin, Madison, Wisconsin

D. D. MAKDANI, Ph.D., Department of Food Science and Human Nutrition, Michigan State University, East Lansing, Michigan

SAMUEL A. MATZ, Ph.D., Vice President, Research and Development, Ovaltine Products, Villa Park, Illinois

OLAF MICKELSEN, Ph.D., Professor, Food Science and Human Nutrition, Department of Food Science and Human Nutrition, Michigan State University, East Lansing, Michigan

ELDON E. RICE, Ph.D., Chief Nutritionist, Research and Development Center, Swift & Co., Oak Brook, Illinois

R. THIESSEN, JR., Corporate Research Manager, Nutrition Sciences, General Foods Corporation, White Plains, New York

MIRIAM H. THOMAS, M.S., Research Chemist, Nutrition Division, Food Laboratory, U.S. Army Natick Laboratories, Natick, Massachusetts

WALTER J. WOLF, Ph.D., Research Leader, Meal Products, Northern Regional Research Laboratory, Agricultural Research Service, U.S. Department of Agriculture, Peoria, Illinois

Preface to the Second Edition

The editors are indeed proud to present this Second Edition of *Nutritional Evaluation of Food Processing*. Thirty authors, who are competent in nutrition, food science, and food technology, collaborated in writing this volume. We are indebted to them for their scholarly contributions and for the time and effort which they have so unselfishly devoted to this task.

During the 15 years that have passed since the first edition was published there has been an accelerating interest in the effects of food handling and processing on the nutritional quality of prepared foods. Before 1960, farmers, food processors, and housewives were especially interested in food as a commodity and seemed only mildly interested in food quality, i.e., nutrient content. Some may find it difficult to believe that until recently the science of nutrition was seldom included in the curriculum of food scientists and food technologists; today basic and even advanced nutrition are important subjects.

An important breakthrough occurred during the 1960's when precise and rapid methods for estimating the vitamin, mineral, and amino acid contents of raw foods and food products were developed. By the use of these procedures it was soon proven that significant amounts of nutrients in processed (cooked and stored) foods were reduced, at times seriously, due to extraction into the cooking water or to destruction through chemical reactions.

The reader may be interested to learn that the first edition of *Nutritional Evaluation of Food Processing* was prompted by the results of three studies conducted at the Massachusetts Institute of Technology (Harris et al., Proc. Conf. Inst. Food Technologists, 1940; Harris et al., J. Lab. Clin. Med. 25, 838, 1940; Harris and Mosher, Food Res. 6, 387, 1940). These data led to the conclusion that a reference book concerned with the effects of food processing and food nutrients was urgently needed.

A remarkable change has taken place since then. Food scientists, food technologists, nutritional biochemists, clinical nutritionists, and dieticians are much more concerned about the beneficial and deleterious effects of food processing than ever before.

Genetically-improved food varieties are being developed by plant breeders to improve the quality of foods from plant and animal sources. Nutritionally-balanced diets are being provided for the nourishment of people of all ages, and especially those people with biomedical disorders who must avoid excess intakes or otherwise control their intake of specific nutrients or dietary factors. The federal government is now providing guidelines which will assist in the selection of more nutritious, less expensive, and more acceptable diets.

When the first edition was written, the subject was discussed in terms of commodities. In this second edition, it is discussed primarily in terms of food processing since modern food science is no longer taught in terms of food commodities.

It is our hope that this book will serve as a text and reference book and thus help readers in nutritional evaluation of food processing.

ROBERT S. HARRIS
ENDEL KARMAS

December, 1974

Preface to the First Edition

If food processing is defined to include all treatments of a foodstuff from the place of origin to the point of consumption, then more than 95% of our food is processed.

This book is concerned with the nutritional effects of the processing of foods as they proceed from garden to gullet. Most foodstuffs are not fully acceptable and must be trimmed and cooked to make them more palatable and nutritious. Some foods are contaminated with microorganisms and insects and must be treated to make them safe and acceptable. Most foodstuffs are not stable and must be milled, pasteurized, canned, refrigerated, frozen, dehydrated, and packaged so that they may be stored and transported to urban and suburban areas where most consumers work and live.

Approximately four dozen nutrients (amino acids, minerals, vitamins, calories) are required in human nutrition. These nutrients are present in a wide variety of foods. Because they are unequally distributed in various plant and animal tissues, and because they are unequally sensitive to temperature, light, air, etc., the losses in nutrients resulting from processing vary according to the type of food, process, time, and nutrient involved.

Food processing is essential if a population is to be fed. In most cases food processing causes a reduction of the nutritional value of a food. As a result of advances in the science of food and nutrition, the adverse effects of processing which formerly were unconscious are now becoming conscious. As we study the consequences of our actions, we should correct them. As we introduce new methods of processing, we should try to be aware of their consequences. The main purpose of this book is to evaluate the known effects of processing upon the nutritional values of foods, and to indicate how certain processing procedures may be altered to minimize losses in nutritional value.

It is with deep regret that I must record the death of the co-editor, Mr. von Loesecke, while this book was in preparation. It is a tragedy that the life of this able scientist and genuine friend has ended so prematurely.

I take this occasion to give abundant thanks to the authors who contributed to this book.

ROBERT S. HARRIS

February, 1960

Contents

CHAPTER   PAGE

Section I. Introduction

1. GENERAL DISCUSSION ON THE STABILITY OF NUTRIENTS, *Robert S. Harris* 1
2. THE MAJOR FOOD GROUPS AND THEIR NUTRIENT CONTENT, *Endel Karmas* 5
3. NUTRITIONAL ASPECTS OF FOOD PROCESSING METHODS, *Endel Karmas* 11
4. STABILITY OF NUTRIENTS IN FOODS, *Benjamin Borenstein* 16

Section II. Nutrients in Raw Foods

5. GENETIC MANIPULATION TO IMPROVE NUTRITIONAL QUALITY OF VEGETABLES, *L. R. Baker* 19
6. EFFECTS OF AGRICULTURAL PRACTICES ON THE COMPOSITION OF FOODS
   Part 1. EFFECTS OF AGRICULTURAL PRACTICES ON FOODS OF PLANT ORIGIN, *Robert S. Harris* 33
   Part 2. EFFECTS OF AGRICULTURAL PRACTICES ON FOODS OF ANIMAL ORIGIN, *T. C. Byerly*   58
7. EFFECTS OF HARVESTING AND HANDLING ON THE COMPOSITION OF FOODS
   Part 1. EFFECTS OF HARVESTING AND HANDLING ON FRUITS AND VEGETABLES, *John M. Krochta and Bernard Feinberg* 98
   Part 2. EFFECTS OF HANDLING AND STORAGE ON SEEDS, *Vernon L. Frampton* 118
   Part 3. EFFECTS OF POSTMORTEM HANDLING, *Eldon E. Rice* 125
8. EFFECTS OF REFINING OPERATIONS ON THE COMPOSITION OF FOODS
   Part 1. EFFECTS OF REFINING OPERATIONS ON CEREALS, *George E. Inglett* 139
   Part 2. EFFECTS OF REFINING OPERATIONS ON LEGUMES, *Walter J. Wolf* 158

Part 3. EFFECTS OF PROCESSING ON THE NUTRITIVE QUALITY OF OILSEED MEALS, *Vernon L. Frampton* 187

Section III. Effects of Commercial Processing and Storage on Nutrients

9. EFFECTS OF HEAT PROCESSING ON NUTRIENTS
   Part 1. EFFECTS OF BLANCHING, PASTEURIZATION, AND STERILIZATION ON NUTRIENTS, *Daryl B. Lund* 205
   Part 2. EFFECTS OF BAKING ON NUTRIENTS, *Samuel A. Matz* 240
10. EFFECTS OF FREEZE-PRESERVATION ON NUTRIENTS, *Owen Fennema* 244
11. EFFECTS OF MOISTURE REMOVAL ON NUTRIENTS, *Peter M. Bluestein and Theodore P. Labuza* 289
12. EFFECTS OF PROCESSING BY FERMENTATION ON NUTRIENTS, *Ivan D. Jones* 324
13. EFFECTS OF PROCESSING BY ADDITIVES ON NUTRIENTS
    Part 1. EFFECTS OF SALTING, CURING, AND SMOKING ON NUTRIENTS OF FLESH FOODS, *Henryk Daun* 355
    Part 2. EFFECTS OF HIGH-SUGAR PROCESSING ON NUTRIENTS, *R. Thiessen, Jr.* 382
    Part 3. EFFECTS OF TREATMENT WITH CHEMICAL ADDITIVES, *Winifred M. Cort* 383
14. EFFECTS OF TREATMENT OF FOODS WITH IONIZING RADIATION, *Edward S. Josephson, Miriam H. Thomas, and William K. Calhoun* 393
15. EFFECTS OF PACKAGING ON NUTRIENTS, *Marcus Karel and Norman D. Heidelbaugh* 412

Section IV. Effects of Preparation and Service of Food on Nutrients

16. EFFECTS OF FOOD PREPARATION PROCEDURES ON NUTRIENT RETENTION WITH EMPHASIS UPON FOOD SERVICE PRACTICES, *Paul A. Lachance* 463
17. EFFECTS OF HOME FOOD PREPARATION PRACTICES ON NUTRIENT CONTENT OF FOODS, *Paul A. Lachance and John W. Erdman, Jr.* 529

Section V. Nutrification and Nutrient Metabolism

18. NUTRIFICATION OF FOODS
   Part 1. ADDITION OF AMINO ACIDS TO FOODS, *Ricardo Bressani* 568
   Part 2. ADDITION OF VITAMINS AND MINERALS TO FOODS, *Benjamin Borenstein* 612
19. FACTORS AFFECTING NUTRIENT METABOLISM, *Olaf Mickelsen and D. D. Makdani* 621
INDEX 655

SECTION I

Introduction

CHAPTER 1

Robert S. Harris | General Discussion on the Stability of Nutrients

Nutrients are destroyed when foods are processed largely because they are sensitive to the pH of the solvent, to oxygen, light and heat, or combinations of these. Trace elements (especially copper and iron) and enzymes may catalyze these effects.

In Table 1.1 are tabulated the relative stabilities of the vitamins and amino acids under these various conditions. Vitamin A is stable under an inert atmosphere but rapidly loses activity when heated in the presence of oxygen, especially at higher temperatures. It is completely destroyed when oxidized or dehydrogenated. It is more sensitive to ultraviolet than to other wavelengths of light.

Ascorbic acid is fairly stable in acid solution and decomposes in light, and this decomposition is greatly accelerated in the presence of alkalies, oxygen, copper, and iron.

A 50% loss in biotin occurs when it is boiled for 6 hr in 30% hydrochloric acid or for 17 hr in 1 Normal potassium hydroxide, yet it is relatively stable in air and oxygen or when exposed to ultraviolet light. It is inactivated by agents which oxidize the sulfur atom, and by strong acids and alkalies.

Essential fatty acids isomerize when heated in alkali and are sensitive to light, temperature, and oxygen. When oxidized, they become inactive biologically and may even be toxic.

The stability of vitamin D is influenced by the solvent in which it is dissolved, but it is stable when crystals are stored in amber glass bottles. Generally, it is stable to heat, acids, and oxygen. It is slowly destroyed in foods and feeds which are slightly alkaline, especially in the presence of air and light.

The folic acid group is stable during boiling at pH 8 for 30 min, yet large losses occur during autoclaving in acids and alkalies. This destruction is accelerated by oxygen and light.

Inositol is stable during refluxing in strong hydrochloric acid or potassium hydroxide. It occurs in plants mainly in the form of phytic acid salts, and as plant and animal phosphoinositides. These complexes are broken down by phosphatases and similar enzymes. The free inositol has the highest biological value.

Vitamin K is stable to heat and reducing agents, and is labile to alcoholic alkali, oxidizing agents, strong acids, and light.

Niacin amide is partially hydrolyzed by acid and alkali, yet the resulting niacin has the same biological activity. Niacin is generally stable to air, light, heat, acids, and alkalies.

Pantothenic acid is most stable at pH 5.5–7.0, is rapidly hydrolyzed under stronger acid or alkaline conditions, and is labile to dry heat, hot acid, or hot alkalies.

p-Amino benzoic acid is only slightly destroyed by autoclaving in 6N sulfuric acid for 1 hr, is fairly stable in mild alkali, but is unstable in strong alkali.

Vitamin B-12 (cobalamin, etc.,) is stable to heat in neutral solution if pure, but is destroyed when heated in alkaline or acid media in crude preparations, as in foodstuffs. Choline is strongly alkaline and is slightly unstable in solutions in the presence of oxygen.

The vitamin B-6 group contains pyridoxine, pyridoxal, and pyridoxamine. Pyridoxine is stable to heat, strong alkali or acid but is sensitive to light, especially ultraviolet light, and when in alkaline solutions. Pyridoxal and pyridoxamine are rapidly destroyed by exposure to air, heat, and light. All three are sensitive to ultraviolet light when in neutral or alkaline solution. Pyridoxamine in foods is sensitive to processing.

Riboflavin is very sensitive to light, and the rate of destruction increases as the pH and temperature increase. Thus, the riboflavin of milk is rapidly lost (50% in 2 hr) on exposure to sunlight, and the resulting derivative (lumiflavin) in turn destroys the ascorbic acid in milk. It is stable to heat if in dry form or in an acid medium.

Thiamin suffers no destruction when boiled in acid for several hours, yet the loss approaches 100% when boiled at pH 9 for 20 min. It is unstable in air, especially at higher pH values, and is destroyed by autoclaving, sulfites, and alkalies.

The tocopherols are stable to vigorous boiling in acid in the absence of oxygen and are stable to visible light. They are unstable at room temperature in the presence of oxygen, alkalies, ferric salts, and when exposed to ultraviolet light. Considerable loss of tocopherols occurs in the oxidation of fats and in deep-fat frying due primarily to destruction by chemically active fatty acid derivatives formed in the fats during heating and oxidation. The esters of tocopherols are more stable than the free phenols.

Amino acids racemize in alkaline solutions, and the biological value

GENERAL DISCUSSION ON STABILITY OF NUTRIENTS

TABLE 1.1

STABILITY OF NUTRIENTS

Nutrient	Effect of pH			Air or Oxygen	Light	Heat	Max Cooking Losses
	Neutral pH 7	Acid <pH 7	Alkaline >pH 7				
Vitamins							%
Vitamin A	S	U	S	U	U	U	40
Ascorbic acid (C)	U	S	U	U	U	U	100
Biotin	S	S	S	S	S	U	60
Carotene (pro-A)	S	U	S	U	U	U	30
Choline	S	S	S	U	S	S	5
Cobalamin (B-12)	S	S	S	U	U	S	10
Vitamin D	S		U	U	U	U	40
Folic acid	U	U	S	U	U	U	100
Inositol	S	S	S	S	S	U	95
Vitamin K	S	U	U	S	U	S	5
Niacin (PP)	S	S	S	S	S	S	75
Pantothenic acid	S	U	U	S	S	U	50
p-Amino benzoic acid	S	S	S	U	S	S	5
Pyridoxine (B-6)	S	S	S	S	U	U	40
Riboflavin (B-2)	S	S	U	S	U	U	75
Thiamin (B-1)	U	S	U	U	S	U	80
Tocopherol (E)	S	S	S	U	U	U	55
Essential amino acids							
Isoleucine	S	S	S	S	S	S	10
Leucine	S	S	S	S	S	S	10
Lysine	S	S	S	S	S	U	40
Methionine	S	S	S	S	S	S	10
Phenylalanine	S	S	S	S	S	S	5
Threonine	S	U	U	S	S	U	20
Tryptophan	S	U	S	S	U	S	15
Valine	S	S	S	S	S	S	10
Essential fatty acids	S	S	U	U	U	S	10
Mineral salts	S	S	S	S	S	S	3

S = stable (no important destruction).
U = unstable (significant destruction).

of some is reduced as a result. Arginine, cystine, threonine and cysteine are partially destroyed, whereas glutamine and asparagine are deaminized by alkalies. In acid solution, tryptophan is rather readily destroyed, cysteine is partly converted to cystine, serine and threonine are partly destroyed. Phenylalanine and threonine are partially destroyed by ultraviolet light. All amino acids in foods, and especially lysine, threonine, and methionine, are sensitive to treatment with dry heat and radiations. Thus, in the roasting and toasting of cereals, legumes, and prepared dry mixtures of foodstuffs a significant reduction of the biological values of their proteins may occur.

Mineral salts are not significantly affected by these chemical and physical treatments. Some may be oxidized to higher valences by exposure to oxygen, but there is no convincing evidence that their nutritional value is affected.

In Table 1.1 are also given the limits of losses of these nutrients when the average food is cooked. More complete data on the losses of several of these nutrients in specific foods during cooking are presented in Chap. 16 and 17.

CHAPTER 2

Endel Karmas

The Major Food Groups and Their Nutrient Content

Nutrients are the building blocks of the human body. Nutrients are needed for growth, to maintain and repair the body tissues, to regulate body processes, and to furnish energy for the body's functions.

The nutrients that must be supplied in the daily food to keep man in good health belong to the groups of proteins, fats, and carbohydrates, the macronutrients; and vitamins and minerals, the micronutrients. Water is also an essential part of good nutrition.

More than 50 essential nutrients have been identified, and the identification of other nutrients is not yet complete. All these essential nutrients must be present in appropriate quantities to have balanced nutrition. Thus, the nutrient composition of a food is described in terms of its content of proteins, amino acids, fats, fatty acids, carbohydrates, vitamins, mineral salts, and water.

Man acquires his essential nutrients from foods obtained from the plant and animal world. The biochemistry of plants, animals, and man have much in common. Therefore, man requires much the same nutritional building blocks as those contained in plants and animals.

Figure 2.1 illustrates the biochemistry cycle of natural foods. Growing foods and harvesting foods belong to the realm of the agricultural sciences and technology. The sun's energy combines carbon dioxide, water, and soil nutrients to produce the so-called first-stage foods as related to harvesting. These are the foods of plant origin: various vegetables, fruits, seeds, tubers, etc. Foods of animal origin are obtained mostly from herbivorous animals. For instance, domestic animals providing red meat to man are mostly herbivores. This is the second stage of man's foods. And finally, the third-stage foods, such as eggs and milk, are produced by the animals.

Natural foods from all three stages may be manufactured into other processed products within that stage. For example, a texturized soy protein still belongs to the vegetable proteins, and cheese is a milk product. It is interesting to note that proteins increase in nutritive value as the amino acids from the first-stage proteins are reassembled to the second- and further to the third-stage proteins (compare the data from National Academy of Sciences—National Research Council 1963).

6 NUTRITIONAL EVALUATION OF FOOD PROCESSING

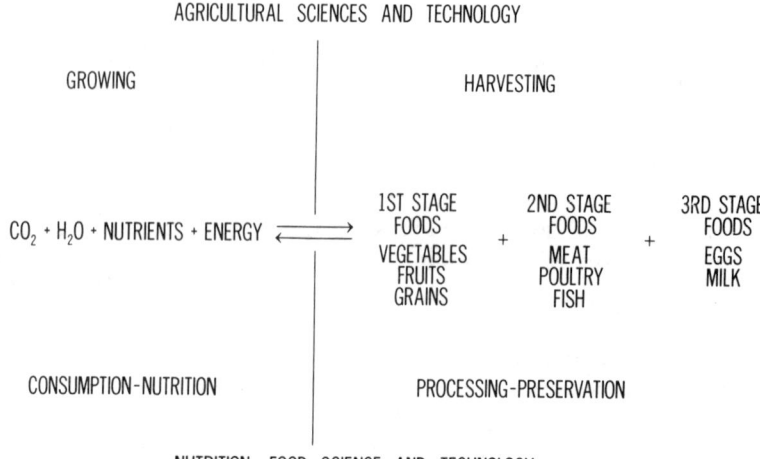

FIG. 2.1. THE RELATIONSHIPS BETWEEN THE VARIOUS LIFE SCIENCES AND THE CHEMISTRY CYCLE OF NATURAL FOODS

Natural foods are biological systems which spoil rapidly after harvest. Since man's need for food and his food harvests usually do not occur simultaneously, foods have to be preserved through processing. At times of hunger, man consumes his preserved food and starts the reverse of photosynthesis to release free energy and to obtain the essential nutrients for his biochemical needs. According to Nobel-laureate Albert Szent-Györgyi (1966), one of the basic principles of life that free energy can be preserved and stored in a foodstuff molecule and utilized when necessary.

Figure 2.1 also emphasizes the important relationship between the nutrient value of foods and food processing. Various foods serve as carriers for nutrients. Changes in nutrient levels of foods with respect to processing times and intensity are the discrete variables. The fields of nutrition as well as food science and technology deal with changes in these variables. This volume in particular concerns itself with nutritional evaluation of food processing.

Detailed data on food consumption in the United States are available from government sources (U.S. Dept. of Agr. 1974). Table 2.1 summarizes the per capita consumption of major food commodities in 1974. Foods of plant origin and foods of animal origin share about equally in American nutrition, totaling 85% of all the foods by weight. The total average per capita consumption of food in 1974 was estimated 1362 lb, which is equal to about 9 times the weight of an average man.

The available nutrient composition of natural foods in the diet in

THE MAJOR FOOD GROUPS AND THEIR NUTRIENT CONTENT

TABLE 2.1

PER CAPITA CONSUMPTION OF MAJOR FOOD COMMODITIES IN
THE UNITED STATES IN 1974

Food Groups	Lb	Consumption	%
Foods of plant origin			
Vegetables		308	23
Fresh	227		
Canned	56		
Frozen	11		
Dried	14		
Fruits		151	11
Fresh	99		
Canned	37		
Frozen	12		
Dried	3		
Cereal products		129	9
Foods of animal origin			
Meats and meat products		188	14
Beef	117		
Pork	67		
Veal and lamb	4		
Poultry and poultry products		50	4
Chicken	41		
Turkey	9		
Eggs		36	2
Fishery products		13	1
Dairy products		285	21
Milk	253		
Cheese	14		
Ice cream	18		
Other foods			
Sugar and syrup		130	10
Fats and oils		54	4
Beverage products		18	1

Source: Data adopted from the U.S. Dept. Of Agr. (1974).

the United States has been studied quite thoroughly (U.S. Dept. of Agr. 1974). The data on the contribution of major food groups to nutrient supplies available for civilian consumption in 1974 are tabulated in Table 2.2. An essential part of the food energy, protein, iron, vitamin A, niacin, vitamin B-6, and vitamin B-12 comes from meat, poultry, and fish products. Dairy products provide most of the calcium, phosphorus, magnesium, and riboflavin. Vegetable and fruit groups are rich in vitamin C or ascorbic acid. Cereal products contribute considerable amounts of carbohydrates, hence calories, and thiamin, iron, magnesium, and niacin.

Table 2.3 gives information about the quantity of nutrients available for civilian consumption per capita per day (U.S. Dept. of Agr.

NUTRITIONAL EVALUATION OF FOOD PROCESSING

TABLE 2.2

CONTRIBUTION OF MAJOR FOOD GROUPS TO NUTRIENT SUPPLIES AVAILABLE FOR CIVILIAN CONSUMPTION, 1974[1]

Food Groups	Food Energy (%)	Protein (%)	Fat (%)	Carbohydrate (%)	Calcium (%)	Phosphorus (%)	Iron (%)	Magnesium (%)	Vitamin A Value (%)	Thiamin (%)	Riboflavin (%)	Niacin (%)	Vitamin B-6 (%)	Vitamin B-12 (%)	Ascorbic Acid (%)
Meat (including pork fat cuts), poultry and fish	20.2	41.5	34.7	0.1	3.5	26.2	29.1	13.7	21.5	28.1	24.5	45.6	45.8	69.7	1 0
Eggs	2.0	5.1	2.9	0.1	2.3	5.4	5.1	1.3	5.8	2.2	5.1	0.1	1.9	8.3	0
Dairy products, excluding butter	11.1	22.5	12.4	6.6	75.7	36.1	2.3	21.6	12.9	9.0	41.0	1.6	10.2	20.5	4.0
Fats and oils, including butter	17.8	0.1	42.6	[2]	0.4	0.2	0	0.4	8.1	0	0	0	0	0	0
Citrus fruits	0.9	0.5	0.1	1.9	0.9	0.7	0.8	2.2	1.5	2.8	0.5	0.9	1.2	0	26.3
Other fruits	2.2	0.6	0.2	4.7	1.2	1.1	3.3	3.9	5.5	1.8	1.5	1.7	5.5	0	11.4
Potatoes and sweet potatoes	2.7	2.4	0.1	5.3	0.9	3.9	4.4	7.1	5.3	6.2	1.7	7.1	11.2	0	18.0
Dark green and deep yellow vegetables	0.3	0.5	[2]	0.5	1.6	0.7	1.6	2.1	21.2	0.9	1.1	0.7	1.7	0	8.3
Other vegetables, including tomatoes	2.5	3.3	0.4	4.7	4.9	5.0	9.0	10.4	15.5	6.9	4.5	6.1	9.2	0	27.6
Dry beans and peas, nuts, soy flour	3.2	5.4	4.0	2.2	2.8	6.2	6.4	11.7	[2]	5.7	2.0	7.6	4.3	0	[2]
Flour and cereal products	19.2	17.8	1.3	34.8	3.3	12.5	28.2	17.9	0.4	36.3	17.4	24.0	8.9	1.5	0
Sugars and other sweeteners	17.3	[2]	0	38.4	1.5	0.3	7.4	0.2	0	0.1	[2]	[2]	0.1	0	[2]
Miscellaneous[3]	0.7	0.4	1.2	0.6	1.0	1.8	2.4	7.6	2.3	0.1	0.7	4.8	[2]	0	3.5
Total[4]	100.0	100.0	100.0	100.0	100.0	100.0	100.0	100.0	100.0	100.0	100.0	100.0	100.0	100.0	100.0

Source: U.S. Dept. of Agr. (1974).
[1] Preliminary data.
[2] Less than 0.05%.
[3] Coffee and chocolate liquor equivalent of cocoa beans and fortification of products not assigned to a specific group.
[4] Components may not add to total due to rounding.

THE MAJOR FOOD GROUPS AND THEIR NUTRIENT CONTENT

TABLE 2.3

NUTRIENTS AVAILABLE FOR CIVILIAN CONSUMPTION, PER CAPITA PER DAY[1], AS COMPARED WITH THE DAILY RECOMMENDED DIETARY ALLOWANCES[2]

Nutrient	Unit	Quantity of Nutrients per Capita per Day	RDA, for Males, 23-50 Yr	RDA, for Females, 23-50 Yr
Food energy	Cal	3350	2700	2000
Protein	gm	101	56	46
Fat	gm	158		
Carbohydrate	gm	388		
Calcium	mg	950	800	800
Phosphorus	mg	1540	800	800
Iron	mg	18.3	10	18
Magnesium	mg	348	350	300
Vitamin A value	IU	8200	5000	4000
Thiamin	mg	1.94	1.4	1.0
Riboflavin	mg	2.33	1.6	1.2
Niacin	mg	23.4	18	13
Vitamin B-6	mg	2.28	2.0	2.0
Vitamin B-12	µg	9.7	3.0	3.0
Ascorbic acid	mg	119	45	45

[1] U.S. Dept. of Agr. (1974).
[2] National Academy of Sciences—National Research Council (1973).

1974). In these estimated data, no adjustment has been made for loss or destruction of nutrients during processing or preparation of food. The data include iron, thiamin, riboflavin, and niacin added to cereal products; vitamin A value added to margarine, milk, and milk extenders; and vitamin C added to fruit juices and drinks, flavored beverages and dessert powders, milk extenders, and cereals.

The nutrients available to the American population are compared with the Recommended Dietary Allowances (Table 2.3). Based on the present knowledge of nutrition, the Food and Nutrition Board of the National Academy of Sciences—National Research Council (1973) has established man's daily requirements for the various nutrients and published a list of Recommended Dietary Allowances (RDA). These allowances provide a safety margin above the minimum requirements for food energy and the various macro- and micronutrients. They are based on the amount of nutrients found in food as consumed and not on the levels before processing. The U.S. Recommended Daily Allowances (Food and Drug Administration 1973) are in accord with the ones recommended by the National Academy of Sciences—National Research Council.

The data indicate that the United States population gets considerably more food energy than specified by the RDA values in Table 2.3. The quantities of protein, ascorbic acid, and vitamin B-12 are 2-3 times the RDA. The apparent available quantities of the other nutrients well meet the given RDA values.

Although nutrients are extremely important to keep the body in

10 NUTRITIONAL EVALUATION OF FOOD PROCESSING

good health, one has to keep in mind that man obtains his nutrients through food, not directly. Therefore, it is essential to process foods so that they are palatable, safe, and, at the same time, high in nutrients. The nutrient content of many raw as well as processed foods is listed, e.g., in the USDA Agriculture Handbook 8 (Watt and Merrill 1963).

BIBLIOGRAPHY

FOOD AND DRUG ADMINISTRATION. 1973. Part 1. Regulations for the Enforcement of the Food, Drug, and Cosmetic Act and the Fair Packaging and Labeling Act. Nutritional Labeling. Federal Register *38-FR-6951*, Mar. 14.

NATIONAL ACADEMY OF SCIENCES—NATIONAL RESEARCH COUNCIL. 1963. Evaluation of Protein Quality. Food Nutr. Board, Natl. Acad. Sci.—Natl. Res. Council Publ. *1100*.

NATIONAL ACADEMY OF SCIENCES—NATIONAL RESEARCH COUNCIL. 1973. Recommended Dietary Allowances, 8th Edition. Food Nutr. Board, Natl. Acad. Sci.—Natl. Res. Council, Washington, D.C.

SZENT-GYÖRGYI, A. 1966. The strategy of life. Intern. Sci. Technol. No. 6, 48.

U.S. DEPT. OF AGR. 1974. National Food Situation. USDA Circ. *150*, Nov. 1974.

WATT, B. K., and MERRILL, A. L. 1963. Composition of Foods—Raw, Processed, Prepared. USDA Agr. Handbook *8*.

CHAPTER 3

Endel Karmas | Nutritional Aspects of Food Processing Methods

The title of this volume, *Nutritional Evaluation of Food Processing* refers to two discrete variables. These are: the change in nutrient level, the dependent variable, as a function of processing, the independent variable. Foods serve as carriers or media for the nutrients and influence their stability in different ways. The nutrient deterioration in foods is gradual and depends on the severity of the processing method. Various food processing methods either accelerate or retard the changes in nutrient levels. This relationship is illustrated schematically in Fig. 3.1.

Foods are processed for three reasons: (1) to preserve, package, and store foods (e.g., canning); (2) to manufacture desirable food products, including nutrification of foods (e.g., baking); and (3) to prepare foods for serving. Nutritional changes due to processing are numerous and they take place before, during and after processing.

All raw foods are perishable commodities. From the time of harvest or slaughter, the raw plant and animal tissues undergo gradual deterioration by various biological forces. The rate of the deterioration process may be very fast or relatively slow. One of the primary factors of food deterioration is the content of biologically active water in the tissue. Raw foods with a high biologically active water content, such as leafy vegetables and meat, deteriorate in only a few days, whereas dry seeds, containing only structural water, can be stored for years.

The major causes of food deterioration are microbial growth, enzyme action, and chemical changes, the former being by far the greatest cause of food loss. These actions and reactions take place fastest at high water activities as well as favorable temperature, pH, and other environmental factors. The principles of food preservation are based on the manipulation of these environmental factors. For instance, microorganisms require an optimum temperature for growth. Higher temperatures are injurious, while lower temperatures greatly retard their metabolism.

There are only six basic principles of food processing for preservation. These are:

1. Moisture removal—drying, dehydration, and concentration.
2. Heat treatment—blanching, pasteurization, and sterilization.

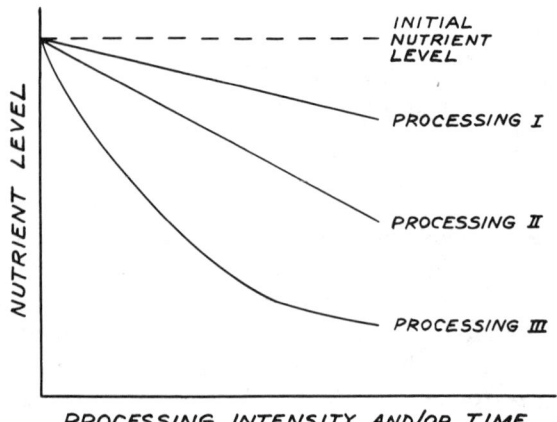

FIG. 3.1. SCHEMATIC ILLUSTRATION OF CHANGE IN NUTRIENT LEVEL AS A FUNCTION OF FOOD PROCESSING INTENSITY AND TIME

3. Low-temperature treatment—refrigeration and freezing.
4. Acidity control—fermentation and acidic additives.
5. Various chemical processing additives.
6. Irradiation.

Since all processed foods have to be stored until they are consumed between the two harvests, proper food packaging is an important coprocessing aspect to the basic food processing methods.

The metabolism of microorganisms requires plenty of free water. Removal of biologically active water through drying or dehydration stops the growth of microorganisms. It also reduces the rate of enzyme activity and chemical reactions. Rancidity of the lipid constituents of foods is reduced if the protective structural water is left intact. The effect of water removal on nutritional changes of dehydrated foods is relatively small if the dehydration temperature is kept moderate and the food is adequately packaged. Freeze-dehydration offers decisive nutrient preservation advantages over dehydration at elevated temperatures.

The principal effect of heat treatment is the denaturation of proteins, i.e., inactivation of microbial and other enzymes. Pasteurization frees the food from human pathogens and most of the vegetative microorganisms, whereas sterilization, by definition, means the destruction of all viable microorganisms. Heat sterilization, the most effective process of food preservation, has a severe effect on heat-labile nutrients, particularly vitamins, and reduces, mainly through the Maillard reaction, the nutritional quality of proteins.

NUTRITIONAL ASPECTS OF FOOD PROCESSING METHODS

Low-temperature preservation, particularly freeze-preservation, on the other hand, is the most harmless method of food preservation in many respects. Low temperature inhibits microbial growth and slows down the rate of chemical and enzyme reactions. The activity of meat enzymes is essentially stopped in commercial frozen storage, while foods of plant origin have to be blanched before freezing in order to avoid undesirable quality changes. Vitamin losses are minimal as compared with the other methods of food preservation. A main cause of quality deterioration is due to rancidity of animal fats. Losses in overall quality occur mainly through unfavorable freezing, storage, and thawing conditions.

Spoilage of low-acid foods is relatively rapid. The growth of food spoilage organisms is greatly inhibited in a highly acidic environment. One of the food preservation methods is to lower the pH of certain foods by anaerobic fermentation action on carbohydrates producing lactic acid. Acidity of some foods may also be increased by acidic additives, such as vinegar or citric acid, producing the same spoilage inhibitory effect. Loss of nutrients through fermentation is small. In some cases the nutrient level may even be increased, particularly through microbial vitamin and protein synthesis.

Chemical additives can substantially contribute to the preservation of foods by providing inhibitory environment for microbial growth as well as enzymes and chemical reactions. Such processing may involve curing agents and smoking of flesh foods, high-sugar preservation of fruits and vegetables, and treatment with various inhibitory chemical additives. The effect of these methods on nutrients is variable, but generally small.

Irradiation, the so-called cold-pasteurization or cold-sterilization process, is the most recent method of food preservation. Presently it does not have any practical importance. FDA categorizes it as "food additive" because high-energy irradiation produces, via highly reactive free radicals, new substances in irradiated foods. The free radical mechanisms are not deadly only to the microorganisms, they are also very detrimental to the nutrients, particularly to the vitamins. Another disadvantage of this method is the considerable detrimental flavor change. Sterilization doses do not inactivate the enzymes, therefore the enzymes have to be heat-inactivated.

Agriculture Handbook No. 8 (Watt and Merrill 1963), the most voluminous reference source of nutritional composition of foods, gives only the initial and final levels for a particular nutrient in a raw and a processed food, respectively. However, very little is known about the rate of change in nutrient levels as a function of processing intensity and time.

For instance, thiamin is known to be extremely sensitive to irradiation-induced reactions. Figure 3.2 (Karmas et al. 1962) illustrates a study on thiamin stability in meat irradiated at various intensities. The initial thiamin content of 0.96 mg per 100 gm of pork decreased to 12% of the initial level in fresh as well as freeze-dehydrated-rehydrated samples. About ⅓ of the initial thiamin content was destroyed by the freeze-dehydration process. However, thiamin was stable in a medium of low water activity when subjected to irradiation.

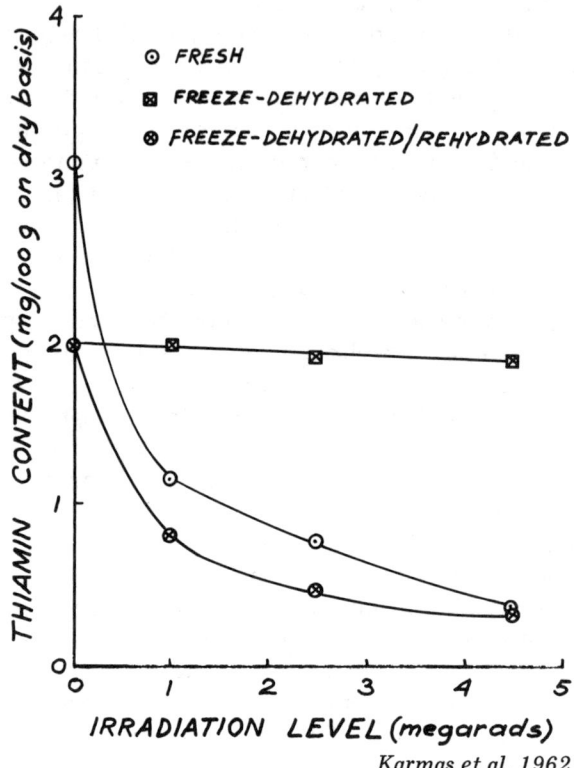

Karmas et al. 1962

FIG. 3.2. THIAMIN CONTENT OF IRRADIATED PORK

On an overall basis, the food processing techniques in greatest use today do not result in major losses in the nutritive value of foods (Institute of Food Technologists 1974). The more sophisticated food processing methods now being developed by advanced technology will retain an even higher percentage of nutrients. Factors to be considered in efforts to increase the retention of nutritional values must include further improvement in food processing technology, the various aspects of food storage and distribution, institutional food systems, and, last but not least, the home preparation of food.

BIBLIOGRAPHY

INSTITUTE OF FOOD TECHNOLOGISTS. 1974. The effects of food processing on nutritional values. A scientific status report by the IFT Expert Panel and the Committee on Public Information. Food Technol. 28, No. 10, 77-80.

KARMAS, E., THOMPSON, J. E., and PERYAM, D. B. 1962. Thiamine retention in freeze-dehydrated irradiated pork. Food Technol. 16, No. 3, 107.

WATT, B. K., and MERRILL, A. L. 1963. Composition of Foods—Raw, Processed, Prepared. USDA Agriculture Handbook 8.

CHAPTER 4

Benjamin Borenstein | Stability of Nutrients in Foods

The interest in nutrient stability during food processing has traditionally concentrated on vitamins on the not unreasonable assumption that many of the vitamins can be seriously depleted by leaching, heat degradation, light, and oxidation.

The attention to this field has not resulted in a body of literature enabling ready predictability of vitamin stability in new products or processes. This failure is due to a number of factors. Little of the published work on vitamin degradation supplies kinetics data. Much of the work emphasizes only a few of the vitamins, particularly vitamin C, and only a few studies are available on pantothenic acid, folic acid, B-12, E, and biotin. The analytical methodology for some of the vitamins at food concentrations is poor. For example, to determine 400 units (10 mcg) of vitamin D per kg of food with an accuracy of 10% requires a method accurate to 1 ppb. Such a method does not exist, nor do analogously accurate methods exist for folic acid, biotin, or E.

The significance of pH on heat lability has been ignored in many studies. The significance of water activity on vitamin stability has only recently been realized and much more work is necessary in this area.

Furthermore, vitamins naturally occurring in foods may be present in several coenzyme forms with different stability profiles. For example, vitamin B-6 occurs as pyridoxine, pyridoxamine and pyridoxal. The latter has poor heat stability compared to pyridoxine, which is the form usually discussed in stability studies and reviews. Similarly, thiamin stability has been studied carefully in model systems but little information is available about thiamin pyrophosphate, which occurs widely in foods.

The significance of packaging on vitamin stability has been largely ignored in the literature—head space, oxygen content, specific light transmission spectrum of glass; these can all be highly significant variables.

The optimum pH for at least one vitamin, C, is dependent on concentration and, in addition, the concentration of the metal catalysts copper and iron is important. Vitamin interactions can also occur but are usually not a problem at food concentrations.

It would be desirable to formulate a table for all the vitamins demonstrating their complete stability profile, but the problems discussed above and the interrelationship of variables, such as effect of pH on light stability, makes generalizations hazardous.

Thiamin is readily degraded in neutral and alkaline solutions even at low temperatures. Sulfite used in food processing can split thiamin into its pyrimidine and thiazole constituents.

Riboflavin is heat-stable in acid solution and in the presence of mild oxidizing agents, but is very sensitive to light at neutral and alkaline pH. In neutral solution it is moderately heat-stable. Reducing agents, such as ascorbic, plus light cause rapid degradation in clear glass bottled milk, for example.

Niacin is perhaps the most stable vitamin. It has excellent heat and light stability in the entire pH range of foods. It should be remembered, however, that niacin leaches readily in blanching and washing operations. It can be enzymatically degraded in ageing meats.

Pyridoxine is stable to heat in acidic and alkaline solution, but is light sensitive at pH $\geqslant 6.0$. Pyridoxal, a major form in milk and other foods, is heat-labile.

Pantothenic acid is stable at pH 5-7, but is more heat-sensitive at pH 3-4.

The stability of folic acid is effected by the conjugate form present in foods. The monoglutamate is moderately heat-stable in acid solution and at neutrality. The tri- and hepta-glutamate conjugates are heat-unstable. Folate is stabilized by ascorbate and destruction is catalyzed by copper.

Cobalamin has moderate to good heat stability at pH 4-5. Retorting at high pH causes rapid destruction, as does light. Ferrous salts catalyze destruction.

Ascorbic acid is a moderately strong reducing agent, redox potential +0.127 v at pH 5, hence, it is readily oxidized during food processing and storage. Copper and iron are effective oxidation catalysts at levels as low as 0.5 ppm. At food concentrations stability is superior at pH 3-4.5 than at pH 6-7. Ascorbic acid is more stable in concentrated solutions under anaerobic conditions, and the pH optimum is closer to neutrality. It reacts readily with anthocyanin pigments.

Vitamin A is readily oxidized and is extremely light-sensitive. In the vegetable kingdom vitamin A occurs only as provitamin carotenoids. These compounds have similar properties to vitamin A but are somewhat more stable. This may be due to the location of carotenoids in oxygen protected sites in foods, e.g., in colloidal dispersion in lipoid media or in protein complexes.

Vitamin D is sensitive to both oxygen and light. Vitamin K compounds are light-sensitive. Vitamin K-1, which occurs in green plants, is the predominant form in foodstuffs.

Alpha-tocopherol, the most important tocopherol with respect to vitamin E activity, is an effective fat phase antioxidant and, hence, can oxidize readily. Oxidation is catalyzed by very low levels of copper.

While it is generally agreed that food processing does not have major effects on mineral salts, there is little substantiating data. The recent literature concentrates on iron, but deals primarily with the bioavailability of added iron compounds rather than changes in native iron in foods caused by processing and storage.

Data on leaching, oxidation, and complexing of cations during processing and storage of foods would be desirable.

Cis-trans isomerization of fats during food processing can decrease biological value of the unsaturated fatty acids. Oxidation, which can be light- and/or metal-catalyzed, decreases biological value of fats.

Heating corn oil ($200°C$ for 24 hr) can decrease bioabsorption to as low as 31%.

The content of amino acids in a protein need not quantitatively reflect its nutritive value since a limiting step in protein utilization is protein digestibility. Processing can both increase and decrease digestibility of proteins. Heat-induced protein denaturation can increase the ease with which the protein is hydrolyzed by intestinal track proteases, but heat can also degrade protein quality by degradation and by blocking the epsilon-amino group of lysine in intact protein preventing hydrolysis by trypsin.

SECTION II

Nutrients in Raw Foods

CHAPTER 5

L. R. Baker | Genetic Manipulation to Improve Nutritional Quality of Vegetables

In the early 1970's, increasing attention was directed to the nutritive content of food for human consumption. Governmental response took the form of "nutrient labeling" for processed foods (Federal Register 1973) which was then proposed for fresh fruits and vegetables (as of 1974). To further encourage interest in the composition of foods for human consumption, the Food and Drug Administration (FDA) suggested that new varieties of plants developed for human consumption be classified on the Generally Recognized As Safe (GRAS) list before distribution to the public (Federal Register 1970, 1971). Precious little information is available as regards the effects of genetics and environment on the nutrient composition of foods for human consumption. *Agriculture Handbook 8* (Watt and Merrill 1963) is the only available source of information on the nutrient composition of the many plant species consumed by humans. The values contained in this valuable reference represent average values with deviations from these averages of more than 100% commonly observed. The concern for nutrient content has caused plant breeders to consider the improvement of nutrient content through genetic manipulation. The search for genetic variation has revealed that nutrient content might be enhanced, but the cost would be enormous in terms of assay techniques necessary to select from large populations of plants. Furthermore, much of this methodology requires development; i.e., such technology is unavailable. Each nutrient must be approached individually with its own heritable system. However, genetic improvement of nutrient content may be the least expensive means of producing more of the essential nutrients for human consumption. Strangely enough, mankind has probably expended more effort on improving the nutritive content of plants for animal consumption than for himself.

The feasibility of approaching the increase of nutrient content of plants through genetic methods is reasonable. Historically, food sufficiency has been uppermost to man as witnessed by the so-called "Green Revolution" most recently. Perhaps the quality of food can now be advanced dramatically without sacrificing quantity. The classical genetics research of Beadle and Tatum (1941) which established the "one-gene-one-enzyme" theory strongly suggested that plants could be manipulated by genetic means to serve the nutritional needs of man. Biochemical genetics research has well established biosynthetic pathways under genetic control which lead to "end-products." Obviously, such end-products might also include nutrients essential to the growth and development of man. The biosynthetic pathway for carotene production in the tomato fruit is a classic example (Porter and Lincoln 1950).

Plant breeders have enhanced the so-called "cosmetic values" of fruits and vegetables significantly over the years; e.g., tomatoes are more red and carrots more orange. This may seem nebulous but most of these improvements increase nutritive value as well as appearance; e.g., more red tomatoes and more orange carrots contain more provitamin A. So, generally, such so-called cosmetic value improvements are valid. It is the object of this chapter to review certain examples of nutrient improvements in vegetables through genetic manipulation for the benefit of mankind.

It must be noted that the environment greatly influences the nutrient content of plants. Interactions between genotype and environment often occur as well. The effect of agricultural practices on nutrient content of foods was thoroughly reviewed recently (Senti 1972). Horticultural symposia on the roles which genetics and plant breeding (Stevens 1970C; Tomes 1972; Stevens 1973) and other factors play in nutrient quality and quantity have been published. A symposium sponsored by the American Medical Association in 1973 focused attention on the possibility to improve the nutritive content of food for human consumption. In this symposium, the influence of cultural practices (Saloman 1974) and of genetics (Stevens 1974; Gabelman 1974; Kehr 1974) on nutrient content of major food crops was reviewed. These reviews emphasized that the nutrient content of fruits and vegetables could be increased (1) if methodology was developed to accommodate large numbers of samples; (2) if support of research to develop basic knowledge in this area was provided; and (3) if an interdisciplinary effort between geneticists, plant breeders, food scientists, biochemists, and others was encouraged. The challenge was made to plant breeders to genetically "lock-in" increased nutrient contents and values (Kehr 1973). Only the future will determine whether or not the advice of these scientists was timely and/or heeded by others.

GENETIC MANIPULATION TO IMPROVE VEGETABLES

Tomato (*Lycopersicon esculentum* Mill.)

In vegetable crops, perhaps more research effort has been expended on tomato than any other species. Therefore, proportionate attention will be directed to tomato. Much of this information may serve as a model for research on genetic manipulation to enhance the nutritive content of other vegetables.

The early work (Porter and Lincoln 1950) on carotene biosynthesis, which combined genetic mutants and biochemistry, might be viewed as the stimulus for subsequent nutrition research on this important vegetable species. Through the use of different loci which affected pigment content, biochemists established the biosynthetic pathway for carotene biosynthesis (Porter and Anderson 1962). The nutritional importance of this research resides with the provitamin A content of tomato represented as β-carotene in the fruit. The normal carotene concentration of tomato fruits is approx 88 µg per gm fresh weight (see Table 5.1).

TABLE 5.1

AVERAGE PIGMENT CONTENT OF TOMATO FRUITS OF THE COMMERCIAL "RUTGERS" VARIETY COMPARED TO MUTANT STRAINS

Genotype	Total Carotenes (µg/Gm Fresh Wt)	Lycopene (%)	β-Carotene (%)
Rutgers *RR TT bb*	88	97	3
Jubilee *RR tt bb*	100	tr	tr
Low total carotene *rr TT bb*	2	39	61
Beta-orange *RR TT BB*	84	7	93

Source: After Tomes *et al.* (1953).

It is now established that the carotenes found in mature fruits of red-fruited tomatoes are lycopene, α-carotene, β-carotene, and traces of certain other pigments and related polyenes. Lycopene (no vitamin A activity) comprises approx 95% of the pigment content and is responsible for the red color. Most of the remaining pigment is β-carotene. Different researchers (Lincoln and Porter 1950; Tomes *et al.* 1953, 1958) have demonstrated genetic control of this pigment content. Red (*RR TT bb*), yellow (*rr TT bb*), and beta-orange (*RR TT BB*) are described as quantitative shifts in lycopene and β-carotene. The fourth phenotype, Jubilee orange (*RR tt bb*), represents a change in the biosynthetic pathway as zeta-carotene and prolycopene replace lycopene and β-carotene resulting in little coloration of the fruit. These 3 loci are independently inherited with the 3 genes R, T, and B exhibiting complete dominance. The gene action of R seems to occur early in the biosynthetic pathway via production of a necessary precursor in quantity. The gene T converts pigment of the abnormal "Jubilee" system into the normal lycopene and β-carotene system. In the presence of both R and T, the gene B determined the relative proportions of lycopene and β-carotene. Other mutant genes affecting pigmentation have been identified and characterized. The gene hp (high pigment) was found to increase all the pigments of the fruit without changing the ratio of their total composition (Thompson 1955; Baker and Tomes 1964). This provided yet another gene to increase the β-carotene, and hence provitamin A, content of tomato fruit.

After the report of Tomes et al. (1953), it is apparent that the provitamin A content of tomato fruits could be increased some 30 times through genetic manipulation. However, the color of the fruits was changed from red to orange due to the increase in β-carotene content at the expense of lycopene. Such a variety was developed and offered to the public (Tomes and Quackenbush 1958). This variety with many times higher provitamin A content was similar to other commercial varieties except for the orange color. Consumer acceptance of this new variety never developed because of the color change. Seed of this variety named "Caro-Red" is still available through certain seed houses, but with little or no demand for it.

A comprehensive research effort by Stevens has developed considerable knowledge regarding the genetics of flavor, quality, and nutrient content of tomato. His initial effort related to the chemical constituents of flavor and their inheritance. Heritable differences for concentrations of 2-isobutylthiadiazole, methyl salicylate and eugenol were established (Stevens 1970A). The concentration of 2-isobutylthiadiazole was determined by a single gene with additive effects. The concentrations of methyl salicylate and eugenol were each determined by single genes, which were closely linked with dominance for low concentration. In another study by Stevens (1970B), the correlation of polyene-carotenes with a number of volatile constituents commonly found in tomato was firmly established. The volatile constituents were proposed as derivatives from polyenecarotene components as products of oxidation and enzymatic action. Stevens and Long (1971) were concerned with the organic acid content which is associated with total acidity. Some 40 to 90% of the organic acid content of tomato fruits is citrate; whereas malate is usually some 5 to 60% of the citrate concentration depending on the cultivar studied. Malate concentrations varied more than 10 times, dependent on the genotype. A single locus was suggested to condition the concentration of malate with low concentration dominant to high. Several alleles of this locus were thought to exist which accounted for different concentration levels. Earlier work (Lower and Thompson 1967) had suggested that titratable acidity was under quantitative genetic control. Stevens and Long (1971) agreed with this in general, but suggested that the individual components of total acidity might be simply inherited as in the case of malate. Stevens (1972A) then elucidated the relationships between different components thought to influence quality (flavor). He established that 2-isobutylthiazole concentration and the solids/acid ratio were primary factors conditioning flavor differences. His earlier work had already demonstrated genetic control of 2-isobutylthiazole (Stevens 1970A). The increase in solids was closely associated with increase in reducing sugars. Titratable acid and total acid were not closely associated. He established that differential buffering action was responsible for most of this variation which was accounted for by differential phosphate content in the fruit. Because citrate accounts for most of the acid (40-90%) contained in tomato fruit, it was closely associated with flavor. Because of the relatively low concentration (2-50%) of malate, it was not highly associated with flavor. In subsequent work (Stevens 1972B), he demonstrated the inheritance of both citrate and malate concentration to be controlled by single genes. There was dominance for high citrate and for low malate concentration. A survey of divergent tomato accessions indicated multiple alleles for each of these two loci which condition different degrees of concentration. Through genetic manipulation the concentration of both citrate and malate could be increased 2-3 times. Finally, Stevens and Paulson (1973) report that phosphorus concentration of the fruit was probably controlled by two major genes, but with a significant genotype-environment interaction; i.e., the concentration of phosphorus for a given genotype varied depending on test plot location.

In summary, Stevens demonstrated genetic control of the components thought important to tomato fruit quality. Flavor differences were associated primarily with the concentration of 2-isobutylthiazole and the solids/acid ratio. Each of these three constituents was demonstrated to be under genetic influence. This seems a classic model to approach other nutritional aspects of fruits and vegetables.

Carrot (*Daucus carota* L.)

This vegetable is one of the most popular root crops. It is consumed in both the fresh and cooked product. The bright orange color of carrot roots is due to the high content of β- and α-carotene.

Unlike tomato fruits, carrot roots contain little or no lycopene (red pigment). Because of this deep orange color, carrot roots are the best source of provitamin A among vegetables. They contain approx 10 times the concentration of provitamin A as compared to tomato fruits.

The nutrient content and quality may be affected markedly by such factors as variety, geographic location of growth, temperature, moisture, stage of maturity, etc. Bradley and Smittle (1965) studied the effects of variety, planting date, and maturity on the color, soluble solids, and total solids of the roots. There were wide differences in both color and solids due to different varieties. Orange color (β- and α-carotene) became much more intense with increasing maturity. They also observed high temperatures for 3-6 weeks prior to harvest resulted in reduced orange color whereas cool temperature enhanced color. Varietal differences also accounted for differences in soluble and total solids concentration. In a more elaborate investigation of 3 typical, commericial varieties (Bradley et al. 1967), the average content of β-carotene was 4.8 and 8.0 mg per 100 gm fresh weight for the low and high variety, respectively. The concentration of α-carotene was 3.3 and 5.6 mg per 100 gm fresh weight; somewhat less than the β-carotene. The same variety was low for both carotenes. Furthermore, the ratio of β:α-carotene was important in the color of both fresh and processed carrot roots with higher concentration of β- preferred to that of α-carotene. Again, averaging across all varieties, the best-colored carrot roots were grown under relatively low temperature conditions ($50°$-$65°F$) for the 3-6-week period prior to harvest. In later work (Bradley and Dyck 1968), it was suggested that carotene synthesis shifts more to β-carotene and less to α-carotene with both mature roots and cool growing temperatures. This results in approximately the same total carotene content but with the β:α ratio skewed more in favor of β-carotene, hence a more deep orange color.

Because of the importance of the orange color (provitamin A) in carrot roots, considerable research on the carotene and polyene content has occurred. Using the model system for carotenoid biosynthesis in tomato fruits, Gabelman and co-workers have elucidated the genetic control of carotene biosynthesis in carrot roots. The biosynthetic pathway has not been established. In a series of elaborate genetic experiments five genes were determined to control carotenoid biosynthesis. The concentration of carotene for these different genotypes ranged from less than 1 to more than 200 μg per gm fresh weight (Gabelman 1974). Initial work by LaFerriere and Gabelman (1968) demonstrated the dominance of white color over

GENETIC MANIPULATION TO IMPROVE VEGETABLES 25

yellow to orange colors conditioned by a single gene Y. The difference between yellow and orange color was suggested more complex. The genetic difference between yellow and orange color was later (Imam and Gabelman 1968) ascribed to two major genes which were independently inherited, but exhibited epistasis. In the presence of yy: lemon was dominant to light orange and light orange was dominant to orange. Both studies indicated that core and cortex color of the roots were independent of each other; i.e., could have a lemon core with an orange cortex. Subsequent research by Kust (1970) indicated that five different genes might account for the color of the core and cortex of the root with core color independent of cortex color. In a comprehensive study (Umiel and Gabelman 1972) between a red colored (high lycopene) Japanese variety and an orange colored U.S. line, two independent genes were proposed to condition orange versus red colored roots (see Table 5.2). Orange color was

TABLE 5.2

EFFECT OF GENOTYPE ON TOTAL CAROTENOID CONCENTRATION
(μg PER GM FRESH WEIGHT) OF CARROT ROOTS

Color Phenotype	Genotype	Total Carotenoids
White	Y—epistatic to all other loci	<1
Lemon	$yy\ Y_1$—epistatic to other loci	30
Light orange	$yy\ y_1y_1\ Y_2$—epistatic to other loci	60
Uniform orange	$yy\ y_1y_1\ y_2y_2\ IoIo\ 00$	90–110
Uniform intense orange	$yy\ y_1y_1\ y_2y_2\ IoIo\ 00\ RR$	—
Uniform intense red (no α-carotene)	$yy\ y_1y_1\ y_2y_2\ IoIo\ oo\ RR$	67–91
Uniform intense orange (no lycopene)	$yy\ y_1y_1\ y_2y_2\ IoIo\ 00\ rr$	—
Uniform intense orange (only β-carotene)	$yy\ y_1y_1\ y_2y_2\ IoIo\ oo\ rr$	—
White (hypothesized)	$yy\ y_1y_1\ y_2y_2\ ioio\ oo\ rr$	<1

Source: Data from Umiel and Gabelman (1972); genotypes as proposed by Gabelman (1974).

dominant to red; epistasis of the dominant O allele over the dominant R allele was noted. The authors noted the similarity in carotenoid content of red carrot roots with red tomato fruits. They further suggested that the pathway for carotenoid biosynthesis in carrot was similar to that of tomato (Porter and Anderson 1962). After reviewing this research, it is apparent that carrot varieties could be developed with extraordinary high concentrations of carotenes, and hence provitamin A. Because of the higher vitamin A conversion from β-carotene than from α-carotene, it would be advantageous to increase preferentially the concentration of β-carotene. Noteably for all genotypes which have been reported, the concentration of α-carotene never exceeded the concentration of β-carotene.

Little attention has been given to the genetic improvement of other nutritional factors aside from the carotenoids. An objectionable bitter principle was isolated and characterized (Dodson et al. 1956) from carrot roots. This was identified as an isocoumarin by Sondheimer (1957). Other research (Carlton et al. 1962) then demonstrated the production of this bitter compound could be induced through the catalytic action of ethylene gas (such as evolved by ripening fruit) with a requirement for oxygen or aerobic conditions. However, there has been no effort to determine if the ability of the root to produce isocoumarin might be under genetic control.

Carlton and Peterson (1963) evaluated the genetic variation for dry matter and sugar content. They found a range of 7.2 to 13.3% dry matter (of fresh weight) with more variation between roots of a given variety than between averages for varieties. The same was determined for both total and reducing sugars. The range was 1.9 to 6.4% total sugars (fresh weight) and 0.1 to 2.0% reducing sugars (fresh weight). They suggested that through selection and inbreeding, carrot lines might be established with significantly higher sugar concentration. They also observed reducing sugars should be increased whereas total sugars should be held constant as higher total sugar concentration resulted in brittle roots which cracked and split easily during growth and harvesting. Further research on the genetic improvement of the sugar content of carrots is unknown to date.

Concern by the public, and hence government agencies, that plant breeding programs might unintentionally alter the nutritive composition of food crops caused researchers to ascertain the validity of such a situation. Researchers at Michigan State University were among the first to develop new hybrid varieties of carrots. The development of hybrids required much genetic manipulation to develop parent lines with desirable horticultural characteristics. Following extensive determinations of vitamins A, B, and C, sugar, solids, N, and mineral content of hybrids compared to established varieties, it was apparent that the nutritional profile had not been altered (Kraut et al. 1975). There was little or no change in proximate and vitamin composition between established varieties and new hybrid varieties. The content of minerals seemed to vary somewhat but with no major deviations from the check varieties (see Table 5.3).

TABLE 5.3
PROXIMATE VITAMIN AND MINERAL COMPOSITION OF HYBRID COMPARED TO ESTABLISHED VARIETIES OF CARROT ROOTS

Variety	Total Solids (%)	Soluble Solids (%)	Proximate and Vitamin Composition (% Fresh Wt)				Carotenes		Thiamin	Riboflavin
			Reducing Sugars (%)	Total Acidity (%)	Ascorbic Acid	Total	B			
						Mg/100 Gm				
Hybrids										
Spartansweet	12.4	2.5	1.8	6.1	8.2	18	17		0.10	0.07
Spartan Fancy	12.7	8.6	1.8	6.1	9.4	20	19		0.11	0.07
Spartan Delite	12.7	8.6	1.5	6.0	10.0	19	17		0.10	0.07
Spartan Bonus	12.3	8.6	2.1	6.3	9.2	17	16		0.09	0.06
Check varieties										
Gold Pak	12.6	8.3	1.9	5.3	9.5	19	18		0.10	0.07
Danvers	11.3	7.8	2.6	5.5	7.1	16	14		0.07	0.06

Variety	Nitrogen and Mineral Composition (% Dry Wt)									
	N	K	P	Ca	Mg	Mn	Fe	Cu	Zn	B
			Gm/100 Gm					Mg/100 Gm		
Hybrids										
Spartansweet	1.39	2.86	0.43	0.23	0.12	0.56	3.5	0.46	1.9	2.0
Spartan Fancy	1.37	2.56	0.39	0.24	0.11	0.29	3.1	0.45	0.7	1.9
Spartan Delite	1.33	2.75	0.37	0.23	0.12	0.48	3.4	0.47	1.1	2.1
Spartan Bonus	1.35	2.66	0.36	0.27	0.13	0.74	3.3	0.39	0.8	2.1
Check varieties										
Gold Pak	1.69	2.99	0.46	0.26	0.14	0.75	3.6	0.68	1.4	2.3
Danvers	1.27	2.83	0.34	0.28	0.12	0.66	3.5	0.40	1.0	2.0

Source: Data after Kraut *et al.* (1975).

Beans (*Phaseolus vulgaris* L.) and Peas (*Pisum sativum* L.)

Most research on these two vegetables related to effects of environmental and production practices on nutrient content. However, the initial research efforts of Stevens and co-workers was on the garden green bean (snap bean). They (Stevens et al. 1967) characterized the difference in flavor between three commercial varieties of processed snap beans. The primary varietal differences in flavor were related to the presence of the volatile compounds cis-hex-3-en-1-ol, oct-1-en-3-ol, linalool, α-terpincol, pyridine, and furfural. They (Stevens and Fraizer 1967) then demonstrated a single gene difference for high vs. low concentration of oct-1-en-3-ol. The differences in concentration of linalool were also simply inherited (single gene) but with additive gene action. Table 5.4 gives the concentration of these 2 constituents for 3 varieties; note the association with flavor.

TABLE 5.4

INFLUENCE OF VARIETY ON TWO PRIMARY FLAVOR COMPONENTS OF PROCESSED SNAP BEAN QUALITY

	Mean Concn (Ppb)	
Variety	Oct-1-en-3-ol	Linalool
Desirable green bean aroma		
Blue Lake	160	7
Distinct perfumy aroma		
G-50	15	39
Strong bean flavor		
Romano	240	40
F_1 crosses		
Blue Lake X G-50	145	22
Blue Lake X Romano	121	23

Source: Data from Stevens and Fraizer (1967).

In an attempt to improve the nutritive value of dry beans or navy beans (*Phaseolus vulgaris* L.), Kelly (1971) assumed that the limiting factor was protein quality. The protein quality might be improved through genetic selection for higher methionine levels in the mature seed. In an extensive effort, the dry seeds of 3600 bean cultivars, varieties, lines, and strains were evaluated for methionine content when grown under the same conditions. Of this collection, 83 were found to contain 33% more methionine than is commonly found in dry beans (see Table 5.5). This difference in methionine concentration was consistent across 3 yr of testing indicating genetic control. Commercial varieties were uniform for methionine content within a variety which was indicative of variety purity. Most commercial

GENETIC MANIPULATION TO IMPROVE VEGETABLES 29

TABLE 5.5

VARIETAL (GENETIC) DIFFERENCES FOR METHIONINE
CONTENT OF MATURE SEEDS OF THE COMMON BEAN,
PHASEOLUS VULGARIS

Variety	Percentage of Check Variety	
	Total Methionine	Available Methionine
Check Sanilac	(2.44 mg/gm$N$) 100	(0.98 mg/gm$N$) 100
High concn.		
P.I. 282693	160	145
Blue Lake	148	219
P.I. 302542	121	200
Low concn.		
P.I. 172032	63	64

Source: Data from Kelly (1971).

varieties were similar for methionine content (approx 2.4 mg per gm N). The authors suggested that improvement of the nutritional value of beans could be rapidly advanced through genetic selection; i.e., ample genetic variation was available for selection to increase methionine content.

Other research has established genetic differences in the protein quality of peas. Utilizing a rat bioassay, Bajaj et al. (1971A), demonstrated a varietal range of 18 to 78% of that of casein in the inherent ability of mature seed from 28 lines to support the growth of rats. This was not closely correlated with PER and NIE values. They could establish no correlation between protein quality and protein quantity of the different lines. Subsequent research (Bajaj et al. 1971B) strongly indicated that albumin N content was responsible for the differences in protein quality. With a range in albumin N from 0.53 to 1.06%, a correlation coefficient of R = 0.95 with PER was determined. They concluded that development of pea varieties with improved protein quality via more albumin was possible.

Corn (*Zea mays* L.)

Although this chapter deals with vegetable species, the elegant research with corn cannot be overlooked. It is commonly known that the difference between starchy field corn and sugary sweet corn is conditioned by a single recessive gene, su, which reduces the rate of conversion from sugar to starch. Creech and collaborators have carefully elucidated the biosynthetic pathway for carbohydrate synthesis (Creech 1968). A total of seven different genes which affect the sugar concentration of sweet corn have been discovered and used to develop "super-sweet" sweet corn (Wann et al. 1971).

Perhaps the most exciting breakthrough in genetic research to improve nutritive content relates to "high lysine" corn. The collaborative research of the geneticist, O. E. Nelson, and biochemist, E. T. Mertz to improve the protein quality of corn is now classic. They discovered (Mertz et al. 1964; Nelson et al. 1965) two mutant genes, *opaque-2* and *floury-2*, which altered the quality of protein in corn. *Opaque-2* increased the lysine, tryptophan, histidine, arginine, aspartate, and glycine content of the protein. *Floury-2* produced a similar pattern of amino acids with methionine increased also (Nelson 1968). *Opaque-2* corn provides protein of sufficient quality to support good growth and development in feeding tests from laboratory rats to human children without supplemental protein in the diet. Obviously, this discovery has stimulated a search for genes affecting protein quality in many other plant species. It is used as a model system the world-over in an effort to improve the protein quality of plants.

It must be readily apparent that genetic manipulation to improve the nutritive quality and content of economic plants has an exciting beginning. Breakthroughs in other nutrients and plant species will occur by using the biochemical-genetic systems for provitamin A biosynthesis, protein quality, and flavor components as models. The capacity to lock in nutritive improvements by genetic means is extremely attractive. However, much research is necessary before knowledge can be made available to enable plant breeders to develop varieties more nutritious than those currently in use.

BIBLIOGRAPHY

BAJAJ, S., MICKELSEN, O., BAKER, L. R., and MARKARIAN, D. 1971A. The quality of protein in various strains of peas. Brit. J. Nutr. *25*, 207-212.

BAJAJ, S. et al. 1971B. Prediction of protein efficiency ratio of peas from their albumin content. Crop Sci. *11*, 813-815.

BAKER, L. R., and TOMES, M. L. 1964. Carotenoids and chlorophylls in two tomato mutants and their hybrid. Proc. Am. Soc. Hort. Sci. *85*, 507-513.

BEADLE, G. W., and TATUM, E. L. 1941. Genetic control of biochemical reactions in neurospora. Proc. Natl. Acad. Sci. *27*, 499-506.

BRADLEY, G., and DYCK, R. L. 1968. Carrot color and carotenoids as affected by variety and growing conditions. Proc. Am. Soc. Hort. Sci. *93*, 402-407.

BRADLEY, G., and SMITTLE, D. 1965. Carrot quality as affected by variety, planting, and harvest dates. Proc. Am. Soc. Hort. Sci. *86*, 397-405.

BRADLEY, G., SMITTLE, D., KATTAN, A. A., and SISTRUNK, W. A. 1967. Planting date, irrigation, harvest sequence and varietal effects on carrot yields and quality. Proc. Am. Soc. Hort. Sci. *90*, 223-234.

CARLTON, B. C., and PETERSON, C. E. 1963. Breeding carrots for sugar and dry matter content. Proc. Am. Soc. Hort. Sci. *82*, 332-340.

CARLTON, B. C., PETERSON, C. E., and TOLBERT, N. E. 1961. Effects of ethylene and oxygen on production of a bitter compound of carrot roots. Plant Physiol. *36*, 550-552.
CREECH, R. G. 1968. Carbohydrate synthesis in maize. Advan. Agron. *20*, 275-322.
DODSON, A. *et al.* 1956. Occurrence of a bitter principle in carrots. Science *124*, 984-985.
FEDERAL REGISTER. 1970. Food Additives. Federal Register *35*, Dec. 8, F. R. 18623.
FEDERAL REGISTER. 1971. Food & Drugs. Federal Register *36*, June 25, 12093.
FEDERAL REGISTER. 1973. Nutrient Labeling. Federal Register *38*, Mar. 14, F. R. 6951.
GABELMAN, W. H. 1974. The prospects of genetic engineering to improve nutritional values. *In* Nutritional Qualities of Fresh Fruits and Vegetables. Futura Publishing Co., Mount Kisco, N.Y.
IMAM, M. K., and GABELMAN, W. H. 1968. Inheritance of carotenoids in carrots, *Daucus carota* L. Proc. Am. Soc. Hort. Sci. *93*, 419-428.
KEHR, A. E. 1973. Naturally-occurring toxicants and nutritive value in food crops: The challenge to plant breeders. HortScience *8*, 4-5.
KEHR, A. E. 1974. Genetic engineering to remove undesirable compounds and unattractive characteristics. *In* Nutritional Qualities of Fresh Fruits and Vegetables. Futura Publishing Co., Mount Kisco, N.Y.
KELLY, J. F. 1971. Genetic variation in the methionine levels of mature seeds of common bean (*Phaseolus vulgaris* L.). J. Am. Soc. Hort. Sci. *96*, 561-563.
KRAUT, C. W., BAKER, L. R., and BEDFORD, C. L. 1975. New carrot hybrids: Nutrient composition compared to established open-pollinated hybrids. J. Am. Soc. Hort. Sci. (Submitted for publication.)
KUST, A. F. 1970. Inheritance and Differential Formation of Color and Associated Pigments in Xylem and Phloem of Carrot, *Daucus carota*. Ph.D. Thesis, Univ. Wisconsin, Madison.
LaFERRIERE, L., and GABELMAN, W. H. 1968. Inheritance of color, total carotenoids, alpha-carotene and beta-carotene in carrots, *Daucus carota* L. Proc. Am. Soc. Hort. Sci. *93*, 408-418.
LINCOLN, R. E., and PORTER, J. W. 1950. Inheritance of β-carotene in tomatoes. Genetics *35*, 206-211.
LOWER, R. L., and THOMPSON, A. E. 1967. Inheritance of acidity and solids of small-fruited tomatoes. Proc. Am. Soc. Hort. Sci. *91*, 486-494.
MERTZ, E. T., BATES, L. S., and NELSON, O. E. 1964. Mutant gene that changes protein composition and increases lysine content of maize endosperm. Science *145*, 279-280.
NELSON, O. E. 1968. The modification by mutation of protein quality in maize. *In* New Approaches to Breeding for Improved Plant Protein. Intern. At. Energy Agency, Vienna.
NELSON, O. E., MERTZ, E. T., and BATES, L. S. 1965. Second mutant gene affecting the amino acid pattern of maize endosperm proteins. Science *150*, 1469-1470.
PORTER, J. W., and ANDERSON, D. G. 1962. The biosynthesis of carotenes. Arch. Biochem. Biophys. *97*, 520-528.
PORTER, J. W., and LINCOLN, R. E. 1950. I. Lycopersicon selections containing a high content of carotenes and colorless polyenes. II. The mechanism of carotene biosynthesis. Arch. Biochem. *27*, 390-403.
SALOMAN, M. 1974. Influence of agronomic practice on nutritional values. *In* Nutritional Qualities of Fresh Fruits and Vegetables. Futura Publishing Co., Mount Kisco, N.Y.
SENTI, F. R. 1972. Effects of cultural practices on composition of food crops. *In* Influence of Agricultural Practices on Nutrient Composition. USDA ERRL Publ. *3786*.

SONDHEIMER, E. 1957. The isolation and identification of 3-methyl-6-methoxy-8-hydroxy-3, 4-dihydroisocoumarin from carrots. J. Am. Chem. Soc. 79, 5036-5039.
STEVENS, M. A. 1970A. Inheritance and flavor contribution of 2-isobutylthiazole, methyl salicylate and eugenol in tomatoes. J. Am. Soc. Hort. Sci. 95, 9-13.
STEVENS, M. A. 1970B. Relationship between polyene-carotene content and volatile compound composition of tomatoes. J. Am. Soc. Hort. Sci. 95, 461-464.
STEVENS, M. A. 1970C. Vegetable flavor. HortScience 5, 95-98.
STEVENS, M. A. 1972A. Relationships between components contributing to quality variation among tomato lines. J. Am. Soc. Hort. Sci. 97, 70-73.
STEVENS, M. A. 1972B. Citrate and malate concentrations in tomato fruits: Genetic control and maturational effects. J. Am. Soc. Hort. Sci. 97, 655-658.
STEVENS, M. A. 1973. The influence of multiple quality requirements on the plant breeder. HortScience 8, 110-112.
STEVENS, M. A. 1974. Varietal influence on nutritional value. In Nutritional Qualities of Fresh Fruits and Vegetables. Futura Publishing Co., Mount Kisco, N.Y.
STEVENS, M. A., and FRAZIER, W. A. 1967. Inheritance of oct-1-en-3-ol and linalool in canned snap beans (Phaseolus vulgaris L.). Proc. Am. Soc. Hort. Sci. 91, 274-285.
STEVENS, M. A., LINDSAY, R. C., LIBBEY, L. M., and FRAZIER, W. A. 1967. Volatile components of canned snap beans (Phaseolus vulgaris L.). Proc. Am. Soc. Hort. Sci. 91, 833-845.
STEVENS, M. A., and LONG, M. A. 1971. Inheritance of malate in tomatoes. J. Am. Soc. Hort. Sci. 96, 120-122.
STEVENS, M. A., and PAULSON, K. N. 1973. Phosphorus concentration: Inheritance and maturity effects. J. Am. Soc. Hort. Sci. 98, 607-610.
THOMPSON, A. E. 1955. Inheritance of high total carotenoid pigments in tomato fruits. Science 121, 896-897.
TOMES, M. L. 1972. Breeding for improved nutritional value. HortScience 7, 154-156.
TOMES, M. L., and QUACKENBUSH, F. W. 1958. Caro-red, a new provitamin rich tomato. Econ. Botany 12, 256-260.
TOMES, M. L., QUACKENBUSH, F. W., and KARGL, T. E. 1958. Synthesis of beta-carotene in the tomato fruit. Botan. Gaz. 119, 250-253.
TOMES, M. L., QUACKENBUSH, F. W., NELSON, O. E., and NORTH, B. 1953. The inheritance of carotenoid pigment systems in the tomato. Genetics 38, 117-127.
UMIEL, N., and GABELMAN, W. H. 1972. Inheritance of root color and carotenoid synthesis in carrot, Daucus carota L.: Orange vs. red. J. Am. Soc. Hort. Sci. 97, 453-460.
WANN, E. V., BROWN, G. B., and HILLS, W. A. 1971. Genetic modifications of sweet corn quality. J. Am. Soc. Hort. Sci. 96, 441-444.
WATT, B. K., and MERRILL, A. L. 1963. Composition of Foods—Raw, Processed, Prepared. USDA Agricultural Handbook 8.

CHAPTER 6

Effects of Agricultural Practices on the Composition of Foods

Robert S. Harris

PART 1
Effects of Agricultural Practices on Foods of Plant Origin

The nutrient content of freshly harvested edible plants varies. Usually this variation is less than severalfold, but on occasion it may exceed 20-fold. These variations result from the interplay of a number of factors, chiefly genetics, sunlight, reliable rainfall, topography, soils, location, season, fertilization of soils, and maturity. The composition of the same strain of edible plant grown in different parts of the world is often so different that one cannot safely use a general table of composition (e.g. FAO 1954) as a source of precise information as to the nutritional value of a food used in the diets of people in any one area. This is especially true of edible plants that have not been standardized genetically by crossing and back-crossing, in plants that have been growing randomly during long periods of time, and especially in those growing close to the area of their origin. Thus, it is usually advisable to analyze fresh samples of indigenous foods rather than to rely on pooled data taken from tables of food analysis. This explains why programs of food analysis have been established in recent years in many of the "underdeveloped" nations of the world. The data resulting from these programs have strengthened the conclusion that the composition of edible plants of the same species are variable, often extremely so.

The many varieties of edible plants in our food environment differ greatly in nutrient composition and nutrient stability. Varieties closely related genetically are somewhat similar while varieties unrelated are usually quite different in nutrient content.

A remarkable collaborative study was carried out by the Southern Cooperative group during 1943-1953, in which food scientists in eight agricultural experiment stations studied the effects of environmental factors upon the nutrient content of selected vegetables grown at each of these stations. Each food sample was analyzed with respect to calcium, phosphorus, iron, ascorbic acid, carotene, thiamin and riboflavin. The data from this ambitious project were published by Wade *et al.* 1945; and Sheets *et al.* 1954.

Extreme care was taken in the design of experiments, in standardizing analytical methods and in the analysis and interpretation of results. To our knowledge this was the first large-scale collaborative study of the composition of foods. This study has since served as a model for the many food analyses that have been conducted in recent years. It was so thoroughly and carefully done that I am using the result in this discussion of the effects of agricultural practices on the composition of foods of plant origin.

Knowledge concerning the occurrence and physiological functions of vitamins and minerals has expanded greatly during the past 25 yr, but it was only recently that investigators appreciated the important role of genetics in determining the nutrient content of edible plants. In the past most plant breeders were rightfully concerned with the color, flavor, and texture of foods, and in the production of more bushels per acre. It is now realized that the food scientist and nutritional biochemist can assist the plant breeder in producing foods that are nutritionally superior as well.

Fundamental studies by Beadle, Tatum, and others have revealed that a close relationship exists between genetics and biochemistry. Plant enzymes are required for the synthesis of growth substances which may later serve as vitamins for those who consume these plants. Although the ability to synthesize these compounds is controlled by genetic factors, the quantity of these factors present in plant tissues are influenced also by environmental factors (light, heat, water, soil composition).

Minute quantities of these nutrients may be measured by chemical and biological methods. Collaborative programs involving plant breeders and nutritional biochemists have already produced significant improvements in the nutritive values of staple foods, truck garden vegetables, and fruits. Different genetic factors may be responsible for the development of desirable size, color, or nutrient content in an edible plant. The desired genes may be introduced into the hereditary pattern of the cultivated strains by crossing and backcrossing. It may not be possible to combine all the good qualities in a plant in every case. When this occurs, it is necessary to make a choice between nutritive value and color, or size, etc. It may even be necessary to educate the people to accept a slightly different food of higher nutritive value.

Considerable progress has been made in improving the nutrient content of fruits and vegetables by genetic means. The ascorbic acid content of tomatoes has been doubled to 50 mg per 100 gm and, as a result, these strains of tomato contain as much ascorbic acid as citrus (Lincoln et al. 1949). The carotene content of tomatoes has been similarly increased to equal that of carrots.

Through the ages farmers, sometimes knowingly but often unknowingly, have been developing a corn which more nearly fits their purposes. Thus, the corn of today is quite unlike that of 50, 500, or 5000 yr ago. In one investigation it was noted that 50 generations of selection have produced corn containing a maximum of 19.5%, and a minimum of 4.9% protein. However, corns with high protein content often contained proportionally less lysine and tryptophan (two of the essential amino acids in which corn protein is especially deficient), and the biological values were thus lower (Mitchell et al. 1952). This high-protein corn was also richer in leucine, which is already present in excessive amounts in corn protein. Though high-protein corn may be more imbalanced in amino acid content than ordinary corn, it is likely that a protein of higher nutritional quality can be developed from these varieties of corn.

The niacin content of corn has been doubled, and it was observed that this effect was quite independent of the amount of protein in the corn (Richey and Dawson 1951). This development is especially fortunate because corn is pellagragenic, and niacin is specific in the prevention of this deficiency disease.

Most of the studies of the effect of genetics upon the nutrient content of foods have been concerned with ascorbic acid, carotene, and thiamin. Space permits mention of only a few reports which are typical of the rapidly expanding literature on this subject.

Ascorbic Acid.—Though ascorbic acid occurs in all plant tissues, the concentration varies greatly. Although the quantity of ascorbic acid in each plant strain is controlled by hereditary factors, it is affected also by temperature, light intensity, and moisture content.

Maclinn et al. (1937) reported a 3-fold variation in the ascorbic acid content of 98 strains of tomatoes in Massachusetts, while other researchers found a 5-fold variation in a group of accessions from the U.S. Bureau of Plant Industry. Anderson et al. (1954) observed a range between 1.8 and 29.3 mg ascorbic acid per 100 gm in 240 samples of fresh-pressed tomato juice.

Burrell et al. (1940) found a $3\frac{1}{2}$-fold range in the ascorbic acid content of 31 strains of cabbage. Walker and Foster (1946) concluded that the ascorbic acid values in cabbage could be raised considerably by controlled crossing and selection.

Wade et al. (1945) reported evidence of transgressive segregation of the ascorbic acid content of snap beans. Plans of the F_2 and F_4 generations contained more ascorbic acid than either parent. This type of segregation increases the possibility of values higher than predictions based on the content of parent stock.

Munsell et al. (1946) reported a range of 7 to 131 mg of ascorbic acid per 100 gm in 28 varieties of mangoes. Bell et al. (1942) found a 35-fold range among 7 varieties of Muscadine grapes. Zielinski (1948) found a range from 0.3 to 7.1 mg per 100 gm of 35 varieties of Oregon peaches.

Perhaps the most extreme example of the effect of genetics on ascorbic acid content is the Barbados cherry or acerola (*Malpighia punicifolia* l.). Cravioto et al. (1945) reported an ascorbic acid content of 8.4 mg % in acerola collected in Mexico; Aseñjo and de Guzman (1946) reported 1707 to 2963 mg % in acerola sampled in Puerto Rico; other researchers found 1100 mg % in acerola produced in Venezuela; and others reported a range of values between 577 and 3309 mg % ascorbic acid. This 300-fold difference in the ascorbic acid content of different samples of acerola is mainly due to genetic factors. It has been suggested that not all samples were valid samples of *Malpighia punicifolia*.

Carotene.—Lincoln (1943) developed a commercial tomato high in carotene content, and has produced some strains with nine times as much carotene as the highest commercial variety. Later, the Rutgers variety was back-crossed and a 12-fold increase in beta-carotene was obtained with no effect on the size, but an increase in the depth of red coloration. The role of specific genes in the synthesis of carotenoids in tomatoes has also been studied.

Different varieties of sweet potato have been variously reported to contain between nil and 7.2 mg % (Murthy and Swaminathan 1954) and between 0.1 and 4.0 mg % (Ezell and Wilcox 1948) of carotene, on a dry basis.

The beta-carotene content of 32 selections of Guatemalan corns varied between nil and 0.177 mg % (Bressani et al. 1954A). Sweet corn inbreds showed an 8-fold variation, whereas garden peas showed a 4-fold variation in carotene content (Scott and Belkengren 1944).

Niacin.—It was reported in 1944 that, on an air-dried weight basis, 28 strains of white field corn ranged between 12.7 and 29.4 mg %, 30 strains of yellow field corn ranged between 11.3 and 36.3 mg %, 46 strains of sweet corn ranged between 18.2 and 62.1 mg %, and 7 strains of popcorn ranged between 7.9 and 21.6 mg %. Preliminary data obtained for inbred lines, and for hybrids between them, suggest that genetic factors control the ability of the corn plant to store niacin. Hunt et al. (1947) observed a range from 1.43 to 2.95 mg % in the niacin content of 9 well-known double-cross corn hybrids. Richey and Dawson (1951) studied the possibilities of breeding corn with higher content of niacin, found a $3\frac{1}{2}$-fold range in this vitamin among 24 inbred lines, and concluded that corn hybrids with 5 mg % niacin could be developed.

Bressani et al. (1954B) reported that the niacin content of 10 strains of bean ranged narrowly between 2.14 and 2.53 mg %. Knox et al. (1944) observed a 3- to 4-fold variation in the niacin content of 29 varieties of sorghum.

Amino Acids.—Bressani et al. (1954B) found that the tryptophan content of 10 strains of beans varied between 0.13 and 0.19 mg %. Doty et al. (1946) studied the arginine, cystine, histidine, tryptophan, and tyrosine contents of 28 strains of corn grown from 28 single-cross hybrids, and concluded that a wider range in composition could probably be found by working with a larger number of inbred lines, and that it may be possible to develop a hybrid with a high proportion of essential amino acids.

Other Compounds.—In some cases it may be desirable to develop strains that are low in certain components. For example, strains of sorghum were developed that do not have an objectionable purple pigment, and, as a result, the white starch produced from them is superior. Forage crops, especially sorghums and Sudan grasses, often contain highly toxic compounds that hydrolyze to hydrocyanic acid in the rumen of cattle and cause illness and death. These toxic substances may be bred out.

The amounts of phytic acid in cereal grains such as wheat and corn are under genetic control. Phytic acid is of nutritional interest because it interferes with the uptake of minerals, especially calcium, from the intestinal tract (Harris 1955). Harris and Bunker (1935) reported that the content of phytic acid phosphorus in 40 strains of corn ranged between 0.14 and 0.32%, and that 45 to 85% of the total phosphorus is phytic phosphorus. Similarly, Young and Greaves (1940) studied 21 strains of wheat and reported that 57 to 94% of the total phosphorus is phytic phosphorus. It is evident that low-phytate varieties of cereals can be developed by genetic selection.

Horowitz and Winter (1957) reported that 4 sources of safflower oil (*Carthamus tinctorius*) contained between 11 and 68% linoleic acid (an essential fatty acid), according to genetic origin.

Summary.—The nutrient content of edible plants is largely under genetic control. In some instances, the differences between strains of the same plant are large, and occasionally these differences are very great. It is easier to increase the nutrient content of edible plants by genetic means than by any other approach.

ROLE OF ENVIRONMENTAL FACTORS

Amount and Intensity of Light

Ascorbic Acid.—Light affects the composition of leafy vegetables and fruits, especially their ascorbic acid content. Working with turnip greens, researchers found 28.2 mg % ascorbic acid after exposure to 200 ft-candles and 235.5 mg % after exposure to 5000 ft-candles of light. Reder et al. (1943) studied 3 crops of turnip greens and found 191 mg % ascorbic acid in samples grown in sunshine during 49% of the growing season, and 128 mg % in those receiving the least sunshine. Somers and Kelly (1957) reported that turnip plants grown in full sunlight contained more ascorbic acid than those grown in the shade.

Tomato plants grown outdoors in sand culture had 25.8 mg % ascorbic acid, whereas those grown under the same conditions but shaded with cloth, which permitted 25% as much light on bright days, had only 15.5 mg % ascorbic acid (Hamner et al. 1945). Strawberries showed a 36% reduction in ascorbic acid when exposed to only 57% as much light during development and ripening (Robinson 1949).

Variations in the ascorbic content of plants during the diurnal cycle have been reported. Snap beans, broccoli, cauliflower, kale, spinach, and Swiss chard contained slightly more ascorbic acid in the late afternoon than in the early morning, but these changes were caused by changes in moisture content (Platenius 1945). Barley, wheat, broad beans, and cabbage showed increases in ascorbic acid according to the amount of sunlight, with a maximum in the early afternoon (Sugawara 1941B).

Fifty pairs of tests using nine varieties of apples showed that the side exposed to sun was higher in ascorbic acid (Murphy 1939). Similar effects were noted with grapefruit (Harding and Thomas 1942), tangerines (Smith et al. 1945), and with apples (Sugawara 1941A). The removal of leaves to permit direct illumination of tomato fruit resulted in a significant increase in the ascorbic acid of tomatoes (Somers 1950). These investigators conducted extensive field studies with tomato fruits in 8 areas of 4 States. They found no correlation between the relative illumination 1, 2, and 3 weeks before harvest, and the ascorbic acid of tomato fruits. Kohman and Porter (1940) reported a rapid loss of ascorbic acid when the tomato plants were kept in the laboratory overnight, and a rapid recovery when the plants were exposed to direct sunlight. Thus, there is a striking relationship between solar irradiation and the ascorbic acid content of plants.

EFFECTS OF AGRICULTURAL PRACTICES ON COMPOSITION 39

Light is not essential for the synthesis of ascorbic acid in plants (Reid 1941A; Mapson et al. 1949). There is a mechanism in plants which converts sucrose, hexose, or other precursors into ascorbic acid. Light, temperature, and carbon dioxide affect the accumulation of ascorbic acid in plants (Somers 1950). A precursor of ascorbic acid is produced by photosynthesis, and this is then converted to ascorbic acid within the plant biologically. Because the loss of ascorbic acid is more rapid in detached than in intact leaves, it is likely that this loss is caused more by metabolic activity than by oxidation. Variations in light intensity can change the rate of formation of precursors, yet have no effect on conversion of these precursors to ascorbic acid or on the amount employed in the metabolic processes of the plant. Variations in temperature can change the metabolic activity or the rate of production of precursors, but appear to have no important effect upon the amount of ascorbic acid synthesized from precursors. This may explain why the literature on the effects of light upon ascorbic acid is often contradictory. Light seems to be the only environmental factor that influences the ascorbic acid content of leafy vegetables and fruits.

Carotene.—Observations on the effect of light upon the synthesis of carotene in plants have been confused with the effects of temperature and rainfall. The observation that the carotene content is lower in summer than in spring and fall has been interpreted to be caused by differences in light intensity. Using monochromatic light, a maximum of carotene synthesis in red kidney bean leaves was found when exposed to green light, and a minimum in leaves exposed to red light. Denison (1951) found decreasing amounts of carotene in tomatoes maturing in the dark, and increasing amounts in those maturing in the light. Turnip plants grown in full sunshine contained less carotene than those grown in the shade (Somers and Kelly 1957). This response to illumination was reversible.

Riboflavin.—Little has been published on the effects of light intensity on the riboflavin content of plants. Gustafson (1948) reported that bean, corn, pea, potato, and spinach plants grown under lowest light intensity had the least riboflavin, but the effect was not as striking as with thiamin. Turnip greens harvested in the morning have been reported to contain more riboflavin than greens harvested at other times (Reder et al. 1951).

Thiamin.—Light stimulates thiamin synthesis in plants. Bonner and Greene (1938) noted large increases in thiamin in decotylectomized pea seeds grown in the light as compared with those grown in the dark. Bonner and Bonner (1948) noted a similar effect with tomato seedlings. Hoffer et al. (1946) found no increase in thiamin in wheat seedlings grown in the dark during 18 days, and rapid synthesis when exposed to light. Gustafson (1948) showed that darkness decreased, and light increased, the thiamin content of seedlings and mature plants.

The synthesis of thiamin generally occurs in leaves and is then translocated to roots, tubers, and seeds. The concentration in leaves generally increases until after the plant matures. The consequent decrease appears to be due to transfer of the thiamin to seed, kernel, and root.

Fat.—Ultraviolet radiations (2900–3100 A) appear to favor the synthesis of lipids in plants (Tottingham and Moore 1931). Plants grown under Vitaglass showed significantly higher fat content than those grown under window glass.

Calcium.—Light does not seem to play a direct role in the uptake and metabolism of mineral elements by plants. Light exerts an indirect effect by increasing the temperature of plants and accelerating certain photochemical processes. A 5°C rise in the temperature of a leaf produced the same acceleration of evaporation as a 30–40% fall in relative humidity (Curtis 1936). Increased transpiration caused an increase in mineral absorption (Freeland 1937). The shading of calcium-deficient plants caused a large loss in combined calcium and an increase in free calcium in tomato plants (Nightingale et al. 1931). Other workers noted as much as 30% variation in the ash content of tomato leaves during a 24-hr period. Arens (1939) observed that the calcium content of tree leaves is higher at night than by day, and suggested that the calcium is removed by combination with sugar during the day, and returned during the night. After reaching full size, the leaves of many plants show large diurnal changes in carbohydrate, but no important changes in certain minerals (Phillis and Mason 1942).

Iron.—Since light stimulates plant growth, it increases the iron demand. Working with iron-deficient water cultures, a greater chlorotic effect on wheat was observed in sunlight than in shade.

Summary.—Although the literature on the subject is confused, it is evident that the amount and intensity of light have an effect upon the composition of plants, especially on the content of ascorbic acid.

Temperature

The temperature optimum for the most rapid rate of growth of each species of edible plant is not usually optimum for the synthesis and storage of nutrients in its tissues. Furthermore, the temperature condition promoting the greatest storage of one nutrient is often different from that favoring the greatest storage of another nutrient.

EFFECTS OF AGRICULTURAL PRACTICES ON COMPOSITION 41

Ascorbic Acid.—In general, the storage of ascorbic acid in turnip greens held in the dark was decreased as the temperature was increased from 10°-30°C (Somers et al. 1948). Aberg (1949) studied the loss of ascorbic acid when attached and detached leaves of parsley, spinach, lettuce, tomato, and kale were placed in the dark at 5° and 15°C. All leaves lost ascorbic acid when placed in the darkness, but the loss was less rapid at 5°C than at 15°C. Warm weather increases the ascorbic acid in strawberries.

Carotene.—Color develops more rapidly in carrots held at 60°-70°F than in those grown at higher or lower temperatures. Moster et al. (1952) noted equal and opposite changes in zeaxanthin and beta-carotene according to temperature, and concluded that these two pigments may be formed from a common precursor or are interconvertible.

Riboflavin.—Gustafson (1948) noted a higher riboflavin content in beans and soybeans grown at 28°-30°C, and in broccoli, cabbage, lupine, spinach, and wheat grown at 10°-15°C, than at other temperatures. Tomatoes grown at either of these temperatures had the same riboflavin content.

Thiamin.—Gustafson (1950) has noted species differences in the thiamin content of plants in response to temperature. Highest contents were found in broccoli, cabbage, and lupine held at 10°-15°C. No differences were noted in clover, peas, spinach, and wheat grown at either 10°-15°C or at 28°-30°C.

Summary.—The optimum temperature for the most rapid growth varies according to plant species, and it is not necessarily the optimum temperature for the highest nutrient content of plant tissues.

Season of the Year

It has been observed that the composition of edible plants produced from the same seed and soil are different in summer than in early or late season. It is likely that this is the result of differences in temperature, length of day, light intensity, and light spectrum, as well as of other minor factors.

Ascorbic Acid.—Increases in ascorbic acid content as the growing season progresses have been observed in beans (Wade and Kanapaux 1943; Heinze et al. 1944), cabbages (Poole et al. 1944; Smith and Walker 1946), and spinach (Sugawara 1941B). However, inconsistent results which might be attributed to season were found by Hansen (1945), who reported that climate changes had no effect upon the high ascorbic acid values of broccoli, collards, and kale, and the low values of chard and lettuce. Other workers found no significant changes in lima beans in relation to season. Cabbages harvested early in season had five times as much ascorbic acid as those harvested late (Pyke 1942). A similar difference was noted in onions (Becker 1951), in kale (Platenius 1945), and in grapefruit and oranges (Metcalfe et al. 1940).

Carotene.—Carrots maturing in fall appear to be richer in carotene than those maturing in winter (Booth and Dark 1949; Hansen 1945B; Janes 1949). Carrots planted early develop more carotenoid pigment than those sown late. The carotene content of broccoli, collards, and kale tended to decrease during fall and winter, then increased in early spring (Hansen 1945). Chard showed no seasonal trend. Winter-harvested lettuce was lower in carotene. Midsummer snap beans had more carotene than those harvested in the fall (Hibbard and Flynn 1945). Hothouse tomatoes were much lower in carotene than garden tomatoes harvested in summer (Ellis and Hamner 1943). Sweet potatoes from the same field averaged more than twice as much carotene from one season to the next (Ezell and Wilcox 1948).

Riboflavin.—The season of the year was found to have an effect on the riboflavin content of wheat, corn, and oats (Hunt et al. 1950), but not on the riboflavin content in snap beans (Hayden et al. 1948) or lima beans.

Thiamin.—The thiamin content of several varieties of snap beans was measured in spring and fall crops at 4 harvest dates during 3 yr (Hayden et al. 1948). In each year, the spring harvest was richer in thiamin and total solids than the fall harvest.

Magnesium.—Workers have reported that the nutritive elements in plants generally reach maximum concentrations in the early adult plant. Austin (1930) found that soybean forage harvested at 35, 73, and 110 days after seeding, contained 0.772, 0.776, and 0.659% magnesium, dry weight basis.

Summary.—Some vegetables and fruits show highest nutrient content when harvested in the spring, others are highest in the summer, still others are highest in the late season, and some do not differ in nutrient content during the season. The literature on this subject is limited, and the experimental design has been satisfactory in only a few investigations. Thus, it is not yet possible to draw any general conclusions as to the relation of season to nutrient content.

Effect of Location

A number of studies have demonstrated that the location where a plant is grown can significantly affect the nutrient content of its tissues.

EFFECTS OF AGRICULTURAL PRACTICES ON COMPOSITION 43

Ascorbic Acid.—Differences in ascorbic acid content according to location have been noted in the following plants: lima beans, snap beans (Janes 1944), cabbage (Janes 1944), tomatoes (Hamner et al. 1945; LoCoco 1945), turnip greens (Reder et al. 1943), and potatoes (Karikka et al. 1944). Differences as large as 36% were noted in lima beans at 6 locations. A 200% variation was observed in turnip greens by Reder et al. (1943), an influence which was about 14 times greater than that resulting from fertilizer treatment.

Carotene.—Peas were grown in 3 southeastern states during 2 yr, and comparatively large differences in carotene between locations were noted. A similar influence of location has been noted in the carotene content of turnip greens (Bernstein et al. 1945), but not in fruits (Ellis and Hamner 1943). Location affected the carotene content of 5 varieties of sweet potatoes, but had no effect upon 2 other varieties (Reder et al. 1943).

Riboflavin.—Several varieties of peas were studied in 3 locations during 2 yr. The riboflavin content was consistently lower in both years at one location at each stage of maturity tested, and especially at the snap pod stage, when the values of 0.87, 2.40, and 1.33 mg % in the first year, and 0.73, 1.11, and 1.07 mg % in the second year.

Lima beans were harvested at 3 stages of maturity, and the riboflavin was found to vary between 2.1 and 3.6 mg %. The same investigators later reported on 1 variety of lima bean grown for 4 yr at 4 locations. The riboflavin was consistently high at one location and low at the second.

Barley, oats, and wheat were studied during 2 yr; only small differences in riboflavin content according to soil areas in Alberta were found with oats and wheat, but not with barley. The same cereals grown in Manitoba showed no relation between soil zone and riboflavin content.

Thiamin.—Several varieties of pinto greens were grown at the same location during 3 yr, and were found to contain 0.84, 0.90, and 0.93 mg % thiamin, dry basis (Gough and Lantz 1950). Cowpeas were grown in 3 locations during 2 yr and harvested at 3 stages of maturity. Location had little effect, but the pods contained 1.08 and 1.65 mg % according to season. Lima beans were grown at 4 locations during 4 yr and showed no differences in thiamin content at 3 locations, and a difference (6.6 and 12.2 mg %, dry basis) at the fourth location.

Many studies have been reviewed indicating that environment may influence the thiamin content of cereals. Location was more important than genetics in determining the thiamin content (0.52-

0.72 mg %) of 30 wheat varieties grown throughout the United States. Canadian Hard Spring wheat showed variations in thiamin with location (Whiteside and Jackson 1943), but this was not observed with commercial Western Canadian wheats (Johannson and Rich 1942; Hoffer *et al.* 1944). Though no significant effect of soil zone was found by the latter authors, the thiamin content varied from 0.22 to 0.80 mg % in 1940 and 0.29 to 0.63 mg % in 1941. Some effect was noted in later years (Spencer and Galgan 1949; Robinson *et al.* 1948).

It is difficult to distinguish the effect of soil from that of weather (Robinson *et al.* 1950). An effect of location on the thiamin content was observed with peas, lima beans, green beans, and pinto beans but not with 7 varieties of snap beans grown in 2 areas in Michigan (Kelly *et al.* 1940).

Calcium and Phosphorus.—The mean calcium and phosphorus contents of 736 samples of cowpeas from 23 experiments in 15 locations were 0.078% and 0.520% (dry basis), respectively, and the mean iron content of 576 samples from 18 experiments was 0.065. The only significant correlation was a negative one between iron and calcium (Reder *et al.* 1943).

Amino Acids.—It has been reported that location had a highly significant effect upon all amino acids in 8 varieties of dried beans on a milligram percentage basis, and on 7 amino acids on a percentage of protein basis. In one location, there was a marked increase in arginine and a decrease in the other amino acids as compared with another location.

Summary.—The location where a plant is grown may have an effect upon its nutrient content, but generally this effect is small.

ROLE OF SOIL FERTILITY AND FERTILIZATION

The principal effect of improvement of soils is to increase the yield rather than to enhance the nutritional quality of plants grown on these soils (Beeson 1949; Maynard 1950, 1956). Minor effects on the composition of plants have been reported. The malnutrition developing in herbivorous animals fed crops which had been grown in soils deficient in trace elements is evidence that soils can affect the trace element content of plants. However, endemic goiter is the only evidence of a direct relationship between soil deficiency and malnutrition that has been established in human populations. Possibly this is because the diets of human beings are composed of a variety of foods produced in various areas, whereas the rations of animals often contain feedstuffs produced in a restricted area.

It is not to be expected that the addition of fertilizers to soils already replete with the nutrients required for the growth and development of plant crops will have any important effect, except possibly to reduce the yield.

The confusion in the literature relating to the effects of fertilizers upon nutrient content is caused in large part by the complex interrelationships which exist between the elements of the soil. The addition of one element may affect the availability of another, and the addition of combinations of elements may produce a variety of effects upon the plants grown on these soils. Because it is difficult to design an experiment, and even more difficult to interpret the results, there is still no clear evidence that organic fertilizers are superior to inorganic fertilizers in the production of foods of higher nutritive value.

Chelating compounds form complexes with mineral elements and thus promote their absorption by plants. Iron and zinc chelates are easily ionized and more readily available. Stewart and Leonard (1952) were able to cure chlorosis in citrus by the addition of iron chelates to the sandy soil on which they were growing. McIntire *et al.* (1954) increased the selective absorption and retention of K and Na, and decreased the uptake of Ca and Mg by the use of soil conditioners.

The growth of plants may be retarded if the soil contains excesses of Cu, Mn, B, Zn, and Mo, yet the content of certain nutrients may be increased. Excess boron caused a 60% increase in the thiamin and niacin content, and excess manganese caused a 35% increase in the riboflavin and niacin content, of turnip leaves. Excess copper caused a 60% increase in the ascorbic acid content of tomatoes (Lyon and Beeson 1948). However, this is not a practical method for increasing the nutritive quality of foods for the yield of nutrients per acre was greatly reduced.

Using a wide variety of soils, Reder *et al.* (1943) reported that the addition of K caused a decrease in the ascorbic acid of turnip greens. Varying concentrations of Ca, K, Mg, and other elements had no significant effect on the ascorbic acid and carotene contents of tomatoes and turnip greens. Bernstein *et al.* (1945) noted a correlation between the carotene content of plants and the ability of a fertilizer to cause chlorosis. Treatments which cause an increase in ascorbic acid content may cause a decrease in carotene content. There is no valid report in the literature to indicate that the thiamin content of plants is affected by inorganic fertilizers.

An 8-yr study was conducted in Michigan in which corn, wheat, oats, soybean, and hay were grown in adjacent plots of fertilized and depleted soil of the same type. No differences were noted in the macro- or micromineral contents of the grain crops, and only timothy hay was favorably affected by the fertilization (Duncan *et al.* 1955).

Greenhouse experiments were conducted with this same depleted soil after treatment with Ca and Mg carbonate, monocalcium phosphate, and K chloride at different levels. Though some beneficial and some deleterious changes in composition were produced, none was striking. These results support the conclusion from the field study that a completely balanced fertilizer does not affect plant composition. Duncan *et al.* (1955) studied the composition of soybeans, corn, wheat, oats, and brown hay grown on fertilized and unfertilized soil in the same plot areas. The small differences observed could not be attributed to fertilization. When Arnon *et al.* (1947) grew a variety of grass in soil and by water culture, the ascorbic acid content as measured by guinea pig assay, was the same in both crop samples.

Effects of Nitrogen Fertilization

Nitrogen represents the most common nutrient deficiency of soils. Plants grown in deficient soils show delayed growth, yellowed leaves, poor yields, and occasionally low protein content.

Protein.—Attempts to increase the protein content of plants by liberal fertilization with nitrogen have met with some success. A corn having 7.8% protein was raised to 10.4% protein during a single season of nitrogen fertilization (Hamilton *et al.* 1951). However, this high-protein corn contained a higher proportion of zein and the lysine was reduced from 3.0 to 0.99%. Thus, the nutritional value of the protein in the high-protein corn was definitely lower.

Mitchell *et al.* (1952) found that for rats the biological value of the protein increased considerably as the protein content of the corn decreased. Gundhardt and McGinnis (1957) have observed that high fertilization produced a high-protein wheat containing a lower percentage of lysine.

Reussner and Thiessen (1957) produced 2 corn varieties with 13.0 and 11.5% protein, instead of the 9.3% protein of regular corn. The growth and protein efficiency of these corns were equally as good as those of the regular corn.

Ascorbic Acid.—The literature on the effect of nitrogen fertilization on ascorbic acid in plants is conflicting. Other researchers found less ascorbic acid in kale, lettuce, parsley, tomato leaves, and wheat produced with suboptimal nitrogen supply, and a slight increase when the supply was adequate. Nitrogen fertilizer increased

EFFECTS OF AGRICULTURAL PRACTICES ON COMPOSITION 47

the ascorbic acid content of cabbage (Burrell et al. 1940), of collards (Sheets et al. 1954), and of spinach and Swiss chard (Wittwer et al. 1945), but decreased the ascorbic acid in grapefruit (Jones et al. 1945), in cantaloupe, potatoes, and sweet pepper (Finch et al. 1945), in sweet potatoes (Speirs et al. 1951) in apples (Murneek and Wittwer 1948), and in tomatoes (Somers et al. 1951).

The effect of nitrogen treatment on ascorbic acid in vegetables and fruits was not consistent; usually it caused a decrease. This effect may be caused by the increased shade resulting from denser foliage, or by increased acid metabolism in plants.

Carotene.—Nitrogen-deficient soils have been reported to cause a decrease in carotene in spinach (Ijdo 1936), and an increase in carotene in pasture plants (Virtanen and von Hansen 1932), in sweet potatoes, in collards (Sheets et al. 1954), and in carrots (Freeman and Harris 1951). On the contrary, other workers found no effect on carotene in carrots as a result of nitrogen fertilization.

Thiamin.—Though the effect of nitrogen fertilization upon the thiamin content of plants had been studied many times before 1942, all results except two were negative. Nitrogen fertilization of barley produced an increase in the thiamin content of the leaves of two corn hybrids (Burkholder and McVeigh 1940).

Iron.—Nitrogen fertilization caused a serious decrease in the Fe contents of plants, according to reports from researchers.

Summary.—It is evident that nitrogen fertilization of soils had a marked effect upon crop yields, and a mild effect upon nutrient composition.

Effects of Calcium Fertilization

The amount of Ca and Mg in the soil has an important effect upon soil pH. The roots of plants grown on Ca-deficient soils are short and bulbous, they are less able to absorb and assimilate nitrates, and as a result sugars accumulate in the tissues. Soils are seldom deficient in Ca, however, and Ca is used in agriculture primarily as a means for increasing the availability of other elements (N, Fe, P, Mn, etc.).

Sims and Volk (1947) contend that the Ca content of plants generally varies with the soils in which they are grown. Ca affects the pH of the soil. Tomatoes and lettuce have been observed to absorb Ca at constant rates between pH 5 and 9, and less at pH 4 to 5 (Arnon et al. 1942; Arnon and Johnson 1942). de Turk (1941) found that a decrease in soil acidity from pH 5 to 7 caused a 20-fold increase of Ca in clover plants. It is to be expected, therefore, that the Ca content of certain plants will be affected by the pH of the soil.

Nevertheless, the results from fertilization with calcium are contradictory and inconclusive. Some researchers noted no effect of treatment with limestone on the Ca content of kale, whereas others noted a marked increase. Sheets et al. (1954) observed no effect on the composition of collards following lime fertilization. Hester (1948) observed a Ca increase from 0.74 to 1.30% when okra was grown on lime-treated (3000 lb per acre) sandy loam. Dexter et al. (1950) studied the effect of lime treatment of depleted farm soil and reported that the plant growth was more luxuriant, but the nutritive content was the same. Sheets et al. (1944) conducted an extensive study on gypsum (Ca) fertilization of soils, and observed a very slight increase in the Ca content of turnip.

Effects of Phosphorus Fertilization

Adequate fertilization with phosphates is important in agriculture because plants grown on deficient soils become stunted, discolored, or misshapen, and the crop yields may be poor. The application of phosphate fertilizer may cause a distinct increase in the P content of plants (Matrone et al. 1954). Variable results were obtained by others (1955) who reported that the addition of phosphates to the soil increased the N, P, crude fiber, and ash content of oats, and had no effect on the N, Ca, and crude fiber content of alfalfa, yet increased the content of P and Fe.

Effects of Fertilization with Other Elements

The addition of simple iron salts to soils has little effect upon the Fe content of plants. Because of the pronounced antagonism between Mn and Fe, high fertilization with Mn produces iron chlorosis in plants.

Since magnesium is an integral part of the chlorophyll molecule, and the development of green leaves is impaired when plants are grown on Mg-deficient soils, the crop yields are decreased. Hester (1948) produced tomatoes on normal and on Mg-deficient soils with the following results: yield per acre 10.8 versus 7.6 tons; ascorbic acid 32.6 versus 26.5 mg %; sugars 4.82 versus 2.84% and Mg 4.4 versus 1.4%.

By the addition of sulfates to sand cultures, Sheldon et al. (1951) increased the sulfur and methionine contents of alfalfa and soybeans. Similar results were observed when alfalfa was raised on highly weathered soils after treatment with sulfate.

The calcium uptake of certain plants is increased by adding boron to the soil (Brenchley and Warington 1927) and decreased by fertilization with potash, magnesium, or sulfur as reported by others.

EFFECTS OF AGRICULTURAL PRACTICES ON COMPOSITION 49

When the calcium supply is low, the calcium uptake by the plant may be retarded by manganese.

Summary

The literature relating to the effects of soils and fertilizers upon the nutrient content of plants is voluminous and generally contradictory. The few carefully controlled experiments indicate that fertilizers increase the crop yield, often dramatically. On the other hand, the effect on nutrient content is minor and usually insignificant.

EFFECT OF SIZE

Small cabbages are richer in ascorbic acid than larger cabbages, and the core is especially high in the factor (Branion et al. 1948). Potatoes from larger plants are richer in ascorbic acid than those produced from small plants but the content of the skin of this tuber is essentially the same as the flesh.

The ascorbic acid of peaches is highest under the skin and lowest around the pit (Schroeder et al. 1943). Vitamins and minerals are present in larger amounts in the bran and scutellum layers of seeds than in the endosperm portion. Small fruits and seeds in which nutrients are concentrated near the surface will tend to be richer in these nutrients because the surface area per unit of weight is greater.

Nineteen varieties of apples were studied and it was found that although the peel contained 4.7 times more ascorbic acid than the flesh, a significant correlation between size and ascorbic acid content was found in only 3 varieties. Hallworth and Lewis (1944) reported that tomatoes weighing 30 gm or less contained more ascorbic acid than larger tomatoes, and they plotted the relationship between weight, surface area, and ascorbic acid content.

Murphy (1941), studied 16 varieties of fresh Maine onions and found that small onions contained 32 to 141% more ascorbic acid than large onions of the same variety.

Spiers et al. (1951) noted that small turnip leaves had more thiamin and riboflavin than ascorbic acid, and less carotene than larger leaves.

The ascorbic acid content of fresh lima beans increased with size. Those that tended to sink when placed in 15% brine contained less moisture and less ascorbic acid than those which floated (Thompson and Mahoney 1944).

The distribution of ascorbic acid in tomatoes parallels that of ascorbic acid oxidase. It is highest in the skin, lower in the pulp, and lowest in the juice. The skin contains 3 times as much ascorbic acid as the flesh or juice, and 10 times as much as the seeds (Wokes and

Organ 1943). It has been reported that, although the pulp of tomatoes contains twice as much ascorbic acid as the juice, it represents only 7-20% of the total ascorbic acid in the fruit.

EFFECT OF MATURITY

It appears that some edible plants are preferred if eaten while still immature (snap beans, greens, spinach, etc.), others when semimature (lima beans, peas, cob corn, etc.), fully mature (banana, berries, fruits, potato, cereals, dry beans, and peas, etc.), and others if eaten when postmature (certain fruits, broccoli, cauliflower, etc.). Some plants reach their highest nutritive value while still immature, others when mature, and still others when over-ripe. Furthermore, the stage at which a nutrient is at a maximum differs depending upon the nutrient and the plant species.

Ascorbic Acid

Increases in ascorbic acid content with ripeness have been reported for the following vegetable foods: asparagus, pepper, paprika, pimento, rhubarb, peas, tomatoes, turnip greens, acerola, currants, gooseberries, grapefruit, apples, peaches, and mangoes.

Decreases in ascorbic acid content with maturity have been reported for the following food plants: banana, kohlrabi, peas, citrus, grapefruit, mangoes, cantaloupe, and oranges.

Carotene

The carotene content of carrots reached a maximum at 100-140 days from seedling (Werner 1941; Pepkowitz et al. 1944). Flynn et al. (1946) and Hibbard and Flynn (1945) reported that the content of snap beans decreased after they had passed the very immature stage. During ripening to the red stage, peppers increased the carotene content even as much as 30-fold (Pepkowitz et al. 1944). Carotene decreases as snap beans mature (Flynn et al. 1946).

Most ripe fruits are richer in carotene than the unripe fruit, especially mangoes and oranges. Many have reported increases in carotene in tomatoes during ripening but the increase is less in picked than in vine-ripened tomatoes (Ellis and Hamner 1943).

Riboflavin

Decreases have been reported in riboflavin with increasing maturity of the seeds of green lima beans. Maturation had no effect on the content of peas but the amount of riboflavin in snap beans decreased after the bean had passed the immature stage (Hibbard and Flynn

1945). The content of barley and wheat decreased and the content of oats increased as the seed ripened (Robinson et al. 1948).

Thiamin

The thiamin content of peas increased with maturation until tenderometer readings exceeded 100, then stayed fairly constant. The thiamin in wheat and oat kernels increased until the seed was completely mature (Robinson et al. 1948). Thiamin in potatoes increased continuously as they matured.

Phytate

The phytate content of grain increases as the cereal matures (Giri 1938). Hence, immature grains are generally better sources of available phosphorus than mature grains.

BIBLIOGRAPHY

ÅBERG, B. 1945. Changes in ascorbic acid content of darkened leaves as influenced by temperature, sucrose applications, and severing from the plant. Ann. Roy. Agr. Coll. Sweden *13*, 239.
ÅBERG, B. 1946. Effects of light and temperature on the ascorbic acid content of green plants. Kgl. Lantbruks-Hogskol Ann. *13*, 239-273.
ÅBERG, B. 1948. Effects of nitrogen fertilization on the ascorbic acid content of green plants. Physiol. Plantarum *1*, 290-329.
ÅBERG, B. 1949. Changes in ascorbic acid content of darkened leaves as influenced by temperature, sucrose applications and severing from the plant. Physiol. Plantarum *2*, 164-183.
ÅBERG, B. 1953. The effect of different sugars upon the ascorbic acid content of detached leaves. Kgl. Lantbruks-Hogskol Ann. *20*, 129-138.
ANDERSON, E. E., FAGERSON, I. S., HAYES, K. M., and FELLERS, C. R. 1954. Ascorbic acid and sodium chloride content of commercially canned tomato juice. J. Am. Dietet. Assoc. *30*, 1250-1253.
ARENS, K. 1939. The day-night variation in the calcium content of leaves. Jahrb. Wiss. Botan. *88*, 169-175.
ARNON, D. I., FRATZKE, W. E., and JOHNSON, C. M. 1942. Hydrogen ion concentration in relation to absorption of inorganic nutrients by higher plants. Plant Physiol. *17*, 515-539.
ARNON, D. I., and JOHNSON, C. M. 1942. Influence of hydrogen ion concentration on the growth of higher plants under controlled conditions. Plant Physiol. *1*, 525-539.
ARNON, D. I., SIMMS, H. D., and MORGAN, A. F. 1947. The nutritive value of plants grown with and without soil. Soil Sci. *63*, 129-133.
ASEÑIO, C. F., and DE GUZMAN, F. 1946. The high ascorbic acid content of the West Indian cherry. Science *103*, 219.
AUSTIN, R. H. 1930. Effects of soil type and fertilizer treatment on composition of soy bean plant. J. Am. Soc. Agron. *22*, 135-156.
BANERJEE, B. N., and RAMASARMA, G. B. 1938. Vitamin A content of mangoes. Agr. Live-Stock India *8*, 253-258.
BASU, N. M., RAY, G. K., and DE, N. K. 1947. Relation of carotene, vitamin C, total acidity, pH and sugar content of various varieties of mangoes. J. Indian Chem. Soc. *24*, 355-357.
BECKER, W. 1951. Antiscorbutic activity of onions. Deut. Med. Wochschr. *76*, 615.

BECKLEY, V. A., and NOTLEY, V. E. 1943. The ascorbic acid content of sweet peppers. J. Soc. Chem. Ind. (London) *62*, 14-16.
BEESON, K. C. 1941. The mineral composition of crops with particular reference to the soils in which they were grown. USDA Misc. Publ. *369*.
BEESON, K. C. 1946. The effect of mineral supply on the mineral concentration and nutritional quality of plants. Botan. Rev. *12*, 424-455.
BEESON, K. C. 1949. The soil factor in human nutrition problems. Nutr. Rev. *7*, 353-355.
BELL, T. A., YARBROUGH, M., CLEGG, R. E., and SATTERFIELD, G. H. 1942. Ascorbic acid content of seven varieties of Muscadine grapes. Food Res. *7*, 144-147.
BERNSTEIN, L., HAMNER, K. C., and PARKS, R. Q. 1945. The influence of mineral nutrition, soil fertility and climate on carotene content of turnip greens. Plant Physiol. *20*, 540-572.
BONNER, J., and BONNER, H. 1948. The B vitamins as plant hormones. Vitamins Hormones *VI*, 225.
BONNER, J., and GREENE, J. 1938. Vitamin B_1 as a phytohormone. Botan. Gaz. *100*, 226-237.
BRANION, H. D., ROBERTS, J. S., CAMERON, C. R., and McCREADY, A. M. 1948. The ascorbic acid content of cabbage. J. Am. Dietet. Assoc. *24*, 101-104.
BRENCHLEY, W. E., and WARINGTON, K. 1927. The role of boron in the growth of plants. Ann. Botany London *41*, 167-187.
BRESSANI, R., CAMPOS, A. A., SQUIBB, R. L., and SCRIMSHAW, N. S. 1954A. Nutritive value of Central American beans. 4. The carotene content of thirty-two selections of Guatemalen corn. Food Res. *18*, 618-624.
BRESSANI, R., MARCUCCI, E., ROBLES, C. E., and SCRIMSHAW, N. S. 1954B. Nutritive value of Central American beans. 1. Variation in the nitrogen, tryptophan and niacin content of ten Guatemalen black beans (*Phaseolus vulgaris* L.) and the retention of niacin after cooking. Food Res. *19*, 263-268.
BURKHOLDER, P. R., and McVEIGH, I. 1940. Studies of thiamin in green plants with the Phycomyces Assay method. Am. J. Botany *27*, 853-861.
BURRELL, R. C., BROWN, H. D., and EBRIGHT, V. R. 1940. Ascorbic acid content of cabbage as influenced by variety, season, and soil fertility. Food Res. *5*, 247-252.
CRAVIOTO, R. *et al.* 1945. Composition of typical Mexican foods. J. Nutr. *29*, 317-329.
CURTIS, O. F. 1936. Leaf temperatures and the cooling of leaves by radiation. Plant Physiol. *11*, 343-363.
DENISON, E. L. 1951. Carotenoid content of tomato fruits as influenced by environment and variety. Iowa State Coll. J. Sci. *25*, 549-564.
DE TURK, E. E. 1941. Plant nutrient deficiency symptoms: Physiological basis. Ind. Eng. Chem. *33*, 648-653.
DEXTER, S. T. *et al.* 1950. Nutritive values of crops and cow's milk as affected by soil fertility. Mich Agr. Expt. Sta. Quart. Bull. *32*.
DOTY, D. M., BERGDOLL, M. S., NASH, H. A., and BRUNSON, A. M. 1946. Amino acids in grain from several single cross hybrids. Cereal Chem. *23*, 199-209.
DUNCAN, C. W. *et al.* 1955. The chemical composition and nutritive value of yellow dent corn grain. Mich. Agr. Expt. Sta. Quart. Bull. *33*.
ELLIS, G. H., and HAMNER, K. C. 1943. The carotene content of tomatoes as influenced by various factors. J. Nutr. *25*, 539-553.
EZELL, B. D., and WILCOX, M. S. 1948. Effect of variety and storage on carotene pigments in sweet potatoes. Food Res. *13*, 203-212.
EZELL, B. D., and WILCOX, M. S. 1958. Variation in carotene content of sweet potatoes. Agr. Food Chem. *6*, 61-65.

FINCH, A. H., JONES, W. W., and VAN HORN, C. W. 1945. The influence of nitrogen nutrition upon the ascorbic acid content of several vegetable crops. Proc. Am. Soc. Hort. 46, 314-318.
FINCH, L. M., HIBBARD, A. D., HOGAN, A. G., and MURNEEK, A. E. 1946. Effect of maturity on nutrients of snap beans. J. Am. Dietet. Assoc. 22, 415-419.
FAO. 1954. Food Composition Tables, Minerals and Vitamins for International Use. Food and Agriculture Organization, Rome.
FREELAND, R. O. 1937. Effect of transpiration upon the food absorption of mineral salts. Am. J. Botany 24, 373-374.
FREEMAN, J. A., and HARRIS, G. H. 1951. Effect of nitrogen, phosphorus, potassium and chlorine on the carotene content of the carrot. Sci. Agr. 31, 207-211.
GIRI, K. V. 1938. The availability of phosphorus from Indian foodstuffs. Indian J. Med. Res. 25, 869-877.
GOUGH, H. W., and LANTZ, E. M. 1950. Relations of variety and locality to niacin, thiamine, and riboflavin content of dried beans grown in three years. Food Res. 15, 308-312.
GUNDHARDT, H., and McGINNIS, J. 1957. Effect of nitrogen fertilization on amino acids in whole wheat. J. Nutr. 61, 167-176.
GUSTAFSON, F. G. 1948. Influence of light intensity upon the concentration of thiamine and riboflavin in plants. Plant Physiol. 23, 373-387.
GUSTAFSON, F. G. 1950. Influence of temperature on the vitamin content of green plants. Plant Physiol. 25, 150-157.
HALLWORTH, E. G., and LEWIS, V. M. 1944. Variation of ascorbic acid in tomatoes. Nature 154, 431-432.
HAMILTON, T. S., HAMILTON, B. C., JOHNSON, B. C., and MITCHELL, H. H. 1951. The dependence of the physical and chemical composition of the corn kernel on soil fertility and cropping systems. Cereal Chem. 28, 163-176.
HAMNER, K. C., BERNSTEIN, L., and MAYNARD, L. A. 1945. Effect of light intensity, day length, temperature and environmental factors on ascorbic acid content of tomatoes. J. Nutr. 29, 85-97.
HANSEN, E. 1945. Seasonal variations in the mineral content of certain green vegetable crops. Proc. Am. Soc. Hort. Sci. 46, 299-304.
HARDING, P. L. 1944. Seasonal changes in the ascorbic acid concentration of Florida grapefruit. Proc. Am. Soc. Hort. Sci. 45, 72-76.
HARDING, P. L., and FISHER, D. F. 1945. Seasonal changes in Florida grapefruit. USDA Tech. Bull. 886.
HARDING, P. L., and THOMAS, E. E. 1942. Relation of ascorbic acid concentration in juice of Florida grapefruit to variety, rootstock, and position of fruit on trees. J. Agr. Res. 64, 57-61.
HARDING, P. L., WINSTON, J. R., and FISHER, D. F. 1940. Seasonal changes in Florida oranges. USDA Tech. Bull. 753.
HARRIS, R. S. 1955. Phytic acid and its importance in human nutrition. Nutr. Rev. 13, 257-259.
HARRIS, R. S., and BUNKER, J. W. M. 1935. The phytin phosphorus of the corn component of a rachitogenic diet. J. Nutr. 9, 301-309.
HAYDEN, F. R., HEINZE, P. H., and WADE, B. L. 1948. Vitamin content of snap beans grown in South Carolina. Food Res. 13, 143-161.
HEINZE, P. H., et al. 1944. Ascorbic acid content of 39 varieties of snap beans. Food Res. 9, 19-26.
HESTER, J. B. 1948. Soil treatment for better foods. Food Technol. 2, 297-302.
HIBBARD, A. D., and FLYNN, L. M. 1945. Effect of maturity on the vitamin content of green snap beans. Proc. Am. Soc. Hort. Sci. 46, 350-354.
HOFFER, A., ALCOOK, A. W., and GEDDES, W. F. 1944. The effect of varia-

tions in Canadian spring wheat on the thiamine and ash of long extraction flours. Cereal Chem. *21*, 210-222.
HOFFER, A., ALCOOK, A. W., and GEDDES, W. F. 1946. The distribution of thiamine in wheat seedlings at different stages of germination in the dark. Cereal Chem. *23*, 76-83.
HOROWITZ, B., and WINTER, B. 1957. A new safflower oil with a low iodine value. Nature *179*, 582-583.
HUNT, C. H., DITZLER, L., and BETHKE, R. M. 1947. Niacin and pantothenic acid content of corn hybrids. Cereal Chem. *24*, 355-363.
HUNT, C. H., RODRIGUEZ, L. D., and BETHKE, R. M. 1950. The environmental and agronomical factors influencing the thiamine, riboflavin, niacin and pantothenic acid content of wheat, corn, and oats. Cereal Chem. *27*, 79-96.
IJDO, J. B. H. 1936. The influence of fertilizers on the carotene and vitamin content of plants. Biochem. J. *30*, 2307.
JANES, B. E. 1944. The relative effect of variety and environment in determining the variations of percent dry weight, ascorbic acid and carotene content of cabbage and beans. Proc. Am. Soc. Hort. Sci. *45*, 387-395.
JOHANNSON, H., and RICH, C. E. 1942. The variability in the thiamine content of Western Canadian wheat, corn and oats. Cereal Chem. *19*, 308-313.
JONES, W. W., VAN HORN, C. W., and FINCH, A. H. 1945. A note on ascorbic acid: nitrogen relationships in grapefruit. Ariz. Agr. Expt. Sta. Tech. Bull. *106*.
KARIKKA, K. J., DUDGEON, L. T., and HAUCK, H. M. 1944. Influence of variety, location, fertilizer and storage on the ascorbic acid content of potatoes grown in New York State. J. Agr. Res. *68*, 49-63.
KELLY, E., DIETRICH, K. S., and PORTER, T. 1940. Vitamin B_1 content of 8 varieties of beans grown in 2 locations in Michigan. Food Res. *5*, 253-262.
KNOX, G., HELLER, V. G., and SIELINGER, J. B. 1944. Riboflavin, niacin and pantothenic acid contents of grain sorghums. Food Res. *9*, 89.
KOHMAN, E. F., and PORTER, D. R. 1940. Solar rays and vitamin C. Science *92*, 561.
LINCOLN, R. E. 1943. Provitamin A and vitamin C in the genus *lycopersicon*. Botan. Gaz. *105*, 113-115.
LINCOLN, R. E., KOHLER, G. W., SILVER, W., and PORTER, J. W. 1949. Breeding for increased ascorbic acid content in tomatoes. Botan. Gaz. *111*, 343-353.
LoCOCO, G. 1945. Composition of Northern California tomatoes. Food Res. *10*, 114-121.
LYON, C. B., and BEESON, K. C. 1948. Influence of toxic concentrations of micro-nutrient elements in the nutrient medium on vitamin content of turnips and tomatoes. Botan. Gaz. *109*, 506-520.
MACLINN, W. A., FELLERS, C. R., and BUCK, R. E. 1937. Tomato variety and strain differences in ascorbic acid (vitamin C) content. Proc. Am. Hort. Soc. *34*, 543-552.
MAPSON, L. W., CRUICKSHANK, E. M., and CHEN, T. 1949. Factors affecting synthesis of ascorbic acid in cress seedlings. Biochem. J. *45*, 171-179.
MATRONE, G. et al. 1954. Effect of phosphate fertilization and dietary mineral supplements on the nutritive value of soybean forage. J. Nutr. *52*, 127-136.
MAYNARD, L. A. 1950. Soils and health. Council on Food and Nutrition. J. Am. Med. Assoc. *143*, 807.
MAYNARD, L. A. 1954. Animal species that feed mankind: The role of nutrition. Science *120*, 164-166.
MAYNARD, L. A. 1956. Effect of fertilizers on the nutritional value of foods. J. Am. Med. Assoc. *161*, 1478-1480.

McINTIRE, W. H., WINTERBERG, S. H., STERGER, A. J., and CLEMENTS, L. B. 1954. Chemical effects of a soil conditioner upon plant composition and uptake. J. Agr. Food Chem. 2, 463-468.
METCALFE, E., REHM, P., and WINTERS, J. 1940. Variations in ascorbic acid content of grapefruit and oranges from the Rio Grande valley of Texas. Food Res. 5, 233-240.
MILLER, E. V., and WINSTON, J. W. 1938. Seasonal changes in the carotenoid pigments in the juice of Florida oranges. Proc. Am. Hort. Soc. 38, 219-221.
MITCHELL, H. H., HAMILTON, T. S., and BEADLES, J. R. 1952. The relationship between the protein content of corn and the nutritional value of the protein. J. Nutr. 48, 461-476.
MOSELEY, M. A., JR., and SATTERFIELD, G. H. 1940. Ascorbic acid (vitamin C) content of 6 varieties of cantaloups. J. Home Econ. 32, 104-107.
MOSTER, J. B., QUACKENBUSH, F. W., and PORTER, J. W. 1952. The carotenoids of corn seedlings. Arch. Biochem. 38, 297-303.
MUNSELL, H. E. et al. 1946. Ascorbic acid content of mango in relation to variety. Food Res. 11, 95-98.
MURNEEK, A. E., and WITTWER, S. H. 1948. Some factors affecting ascorbic acid content of apples. Proc. Am. Soc. Hort. Sci. 51, 97-102.
MURPHY, E. F. 1939. Vitamin C and light. Proc. Am. Soc. Hort. Sci. 36, 498-499.
MURPHY, E. F. 1941. Ascorbic acid content of onions and observations on its distribution. Food Res. 6, 581-594.
MURTHY, H. B. N., and SWAMINATHAN, M. 1954. Nutritive value of different varieties of sweet potato. Current Sci. (India) 23, 14.
NIGHTINGALE, G. T., ADDOMS, R. M., ROBBINS, W. R., and SCHERMERHORN, L. G. 1931. Effects of calcium deficiency on nitrate absorption and on metabolism in tomato. Plant Physiol. 6, 605-630.
PEPKOWITZ, L. P., LARSON, R. E., GARDNER, J., and OWENS, G. 1944. The carotene and ascorbic acid concentrations of vegetable varieties. Plant Physiol. 19, 615-626.
PHILLIS, E., and MASON, I. G. 1942. On diurnal variations in the mineral content of the leaf of the cotton plant. Ann. Botany London (N.S.) 6, 437-442.
PLATENIUS, H. 1945. Diurnal and seasonal changes in the ascorbic acid content of some vegetables. Plant Physiol. 20, 98-105.
POOLE, C. F., GRIMBALL, P. C., and KANAPAUX, M. S. 1944. Factors affecting the ascorbic acid content of cabbage fines. J. Agr. Res. 68, 325-329.
PYKE, M. 1942. Effects of shredding and grating on vitamin C content of raw vegetables. J. Soc. Chem. Ind. 61, 149.
REDER, R., ASCHAM, L., and EHEART, M. S. 1943. Effect of fertilizer and environment on the ascorbic acid content of turnip greens. J. Agr. Res. 66, 375-388.
REDER, R. et al. 1951. The effects of maturity, N fertilization, storage and cooking on the ascorbic acid content of two varieties of turnip greens. Southern Coop. Ser. Bull. 10.
REID, M. E. 1938. The effect of light on the accumulation of ascorbic acid in young cowpea plants. Am. J. Botany 25, 701-711.
REID, M. E. 1941A. Metabolism of ascorbic acid in cowpea plants. Bull. Torrey Botan. Club 68, 359-371.
REID, M. E. 1941B. Relation of temperature to ascorbic acid content of cowpea plants. Bull. Torrey Botan. Club 68, 519-530.
REUSSNER, G., JR., and THIESSEN, R., JR. 1957. Studies on the protein quality of high-oil, high-protein corn. J. Nutr. 62, 575-584.
RICHEY, F. D., and DAWSON, R. F. 1951. Experiments on the inheritance of niacin in corn maize. Plant Physiol. 26, 475-493.
ROBINSON, A. D., LYND, L. E., and MILES, B. J. 1948. The distribution of

thiamine and riboflavin in wheat, oats and barley at successive stages of plant growth. Can. J. Res. *26B*, 711-717.
ROBINSON, W. B. 1949. The effect of sunlight on the ascorbic acid content of strawberries. J. Agr. Res. *78*, 257-262.
SCHROEDER, G. M., SATTERFIELD, G. H., and HOLMES, A. D. 1943. Influence of variety, size and degree of ripeness upon ascorbic acid content of peaches. J. Nutr. *25*, 503-509.
SCOTT, G. C., and BELKENGREN, R. O. 1944. Importance of breeding peas and corn for nutritional quality. Food Res. *9*, 371-376.
SHEETS, O. A. 1945. Vitamin losses in quantity cooking. Mississippi Expt. Sta. Record *92*.
SHEETS, O. A. *et al.* 1944. Effect of fertilizer soil composition and certain climatological conditions on the calcium and phosphorus content of turnip greens. J. Agr. Res. *68*, 145-190.
SHEETS, O. A. *et al.* 1954. The nutritive value of collards. I. Effects of nitrogen level, variety, maturity and cooking on vitamin and mineral content, and of storage on vitamin contents of collards. Southern Coop. Ser. Bull. *39*.
SHELDON, V. L., BLUE, W. G., and ALBRECHT, W. A. 1951. Biosynthesis of amino acids according to soil fertility. Tryptophan in forage crops. Plant Soil *3*, 33-40.
SIMS, G. T., and VOLK, G. M. 1947. Composition of Florida grown vegetables. I. Mineral composition of vegetables grown in Florida as affected by treatment, soil type, and locality. Florida Agr. Expt. Sta. Bull. *438*.
SMITH, F. G., and WALKER, J. C. 1946. Relation of environmental and hereditary factors to ascorbic acid in cabbage. Am. J. Botan. *33*, 120-129.
SMITH, M. C., CALDWELL, E., and FARRANKOP, H. 1945. Tangerines— their ascorbic acid content and factors affecting it. Ariz. Univ. Agr. Expt. Sta. Mimeo Rept. *72*.
SOMERS, G. F. 1950. Further studies on relationship between illumination and ascorbic acid content of tomato fruits. J. Nutr. *40*, 133-143.
SOMERS, G. F., HAMNER, K. C., and KELLEY, W. C. 1948. Changes in ascorbic acid content of turnip leaf discs as influenced by light, temperature and carbon dioxide concentration. Arch. Biochem. *18*, 59-67.
SOMERS, G. F., and KELLEY, W. C. 1957. Influence of shading upon changes in the ascorbic acid and carotene content of turnip greens as compared with changes in fresh weight, dry weight and nitrogen fractions. J. Nutr. *62*, 39-60.
SOMERS, G. F., KELLEY, W. C., and HAMNER, K. C. 1951. Influence of nitrate supply upon the ascorbic acid content of tomatoes. Am. J. Botany *38*, 472-475.
SPEIRS, M., WARREN, E. L., WILSON, F. H., and GREENLEAF, V. G. 1946. Influence of the degree of icing of vegetables on the retention of their quality and ascorbic acid content. Ice Refrig. *111*, No. 1, 19-22.
SPEIRS, M. *et al.* 1951. Effect of leaf size on moisture, ascorbic acid, carotene, thiamine and riboflavin content in turnip greens. Southern Coop. Ser. Bull. *10*.
SPENCER, E. Y., and GALGAN, M. W. 1949. Relations of thiamine content of Saskatchewan wheat to protein content, variety, and soil zone. Can. J. Res. *27*, 450-456.
STEWART, L., and LEONARD, C. D. 1952. Chelates as sources of iron for plants growing in the field. Science *116*, 564-566.
SUGAWARA, T. 1941A. The ascorbic acid content of various apples. Japan. J. Botany *11*, 327-341.
SUGAWARA, T. 1941B. Studies on the formation of ascorbic acid (vitamin C) in plants. Japan. J. Botany *2*, 344-356.
THOMPSON, A. H., and MAHONEY, C. H. 1944. Some ascorbic acid and moisture determinations on fresh and dehydrated lima beans. Proc. Am. Soc. Hort. Sci. *44*, 448-452.

EFFECTS OF AGRICULTURAL PRACTICES ON COMPOSITION

THORNTON, N. C. 1943. Carbon dioxide storage. XIV. The influence of carbon dioxide, oxygen, and ethylene on the vitamin C content of ripening bananas. Contrib. Boyce Thompson Inst. *13*, 201-220.

THORNTON, N. C. 1946. Factors influencing vitamin C content of asparagus, banana, and seedlings of garden pea during growth and in storage. Contrib. Boyce Thompson Inst. *14*, 295-304.

TOTTINGHAM, W. E., and MOORE, J. E. 1931. Some phases of plant development under Vitaglass. J. Agr. Res. *43*, 133-163.

VIRTANEN, A. I., and VON HAUSEN, S. 1932. The vitamin content of plants. Naturwissenschaften *20*, 905.

WADE, B. L., HEINZE, P. H., KANAPAUX, M. S., and GAETJENS, C. F. 1945. Inheritance of ascorbic acid content of snap beans. J. Agr. Res. *70*, 170-174.

WADE, B. L., and KANAPAUX, M. S. 1943. Ascorbic acid content of strains of snap beans. J. Agr. Res. *66*, 313-324.

WALKER, J. C., and FOSTER, R. E. 1946. The inheritance of ascorbic acid in cabbage. Am. J. Botany *33*, 758-761.

WATT, B. K., and MERRILL, A. L. 1950. Composition of Foods—Raw, Processed, Prepared. USDA Handbook *8*.

WERNER, H. O. 1941. Dry matter, sugar, and carotene of morphological portions of carrots through the growing and storage season. Proc. Am. Soc. Hort. Sci. *38*, 267-272.

WHITESIDE, A. G. O., and JACKSON, S. H. 1943. The thiamine content of Canadian hard red spring wheat varieties. Cereal Chem. *20*, 542-551.

WITTWER, S. H., SCHROEDER, R. A., and ALBRECHT, W. A. 1945. Vegetable crops in relation to soil fertility. II. Vitamin C and nitrogen fertilizer. Soil Sci. *59*, 329-336.

WOKES, F., and ORGAN, J. G. 1943. Oxidizing enzymes and vitamin C in tomatoes. Biochem. J. *37*, 259-265.

YOUNG, S. M., and GREAVES, J. E. 1940. Influence of variety and treatment on phytin content of wheat. Food Res. *5*, 103-108.

ZIELINSKI, Q. 1948. Ascorbic acid content of 33 peach varieties in relation to genetic and environmental factors. Proc. Am. Soc. Hort. Sci. *52*, 143-148.

ZILVA, S. S., KIDD, F., and WEST, C. 1938. Ascorbic acid in the metabolism of apple fruit. New Phytologist *37*, 345-357.

T. C. Byerly

PART 2
Effects of Agricultural Practices on Foods of Animal Origin

ROLE OF GENETICS

Genetic Differences among Livestock Species

Many of the important traits prized or despised in food products of animal origin are heritable. Table 6.1 gives data for the proximate composition of edible flesh of several livestock and poultry species. The data in Table 6.1 reflect average values for adequately fed, moderately fat animals and young chickens and turkeys.

Composition of the flesh and the vitamin content are in part characteristic of species and in part are determined by age, sex, and nutrition, e.g., proportion of fatty acids in fat of the several species. Fats of ruminants tend to be more saturated than those of nonruminants, but nonruminant fat saturations vary widely with relative saturation of their feed fat. Differences in meat flavor are affected by lipid factors (Wasserman and Talley 1968). Booren et al. (1973) associated flavor differences in beef and antelope with fat content of monocarbonyl—0.70 μM per gm vs 1.41 μM per gm.

Genetic factors markedly affect the distribution of flesh and fat, the age at which fat tends to accumulate under usual conditions of feeding, the proportion of flesh and bone, and the color of the fat. Many studies and hundreds of years of breeding experience have established the substantial heritability of the distribution of flesh and fat which, in animal husbandry terms, constitute "type" or "conformation." The rounded body of the beef breed animal contrasts markedly with the more angular and often leggier body of animals of the dairy breeds. However, animals of the large dairy breeds, e.g., Holstein and Brown Swiss, gain rapidly in the feed lot and generally produce acceptable carcasses.

Ziegler et al. (1971) reported data for steers of several beef breeds and for Holstein steers slaughtered at weights of about 470 kg. Data for cutability and quality grade of carcasses and for tenderness and acceptability of meat are shown in Table 6.2.

Cutability scores reflect yield of lean meat. Charolais scored highest among the four beef breeds and the dairy breed Holsteins next. In quality grade, Angus, Hereford, and Shorthorn were "choice," Charolais was "good," and Holstein "standard." The Tenderness

TABLE 6.1
PROXIMATE COMPOSITION OF EDIBLE PORTION OF CARCASS

Species	Water %	Protein %	Fat %	Ash %	Thiamin Mg %	Riboflavin Mg %	Niacin Mg %	Pantothenic Acid[1] Mg %	B-6[1] Mg %	B-12[2] Mg %
Beef	55	16.3	28	0.8	0.7	1.5	39	0.62	0.435	0.00180
Lamb	56	15.7	28	0.8	1.4	2.0	45	0.55	0.275	0.00215
Pork	42	11.9	45	0.6	5.8	1.4	31	0.79	0.450	0.00070
Chicken	71	20.2	7.2	1.1	0.8	1.6	102	0.90	0.500	0.00045
Turkey	58	20.1	20.2	1.0	0.9	1.4	80	0.80	—	—
Goose	51	16.4	32.0	0.9	1.0	2.4	56	—	—	—
Duck	54	16.0	29	1.0	1.0	2.4	56	—	—	—
Rabbit	59	30.3	7.8	1.4	1.0	1.2	—	0.78	0.440	—

Source: Watt and Merrill (1950), Abritton (1954), Ashbrook (1951), Orr (1969).

[1] Lean only.
[2] Cooked.

TABLE 6.2

CARCASS AND MEAT QUALITY FROM STEERS AND
HEIFERS OF DIFFERENT BREEDS

Breed	Cutability (%)	Quality (Grade)	Tenderness (Score)	Acceptability (Score)
Angus	46.9	Choice	7.0	7.0
Charolais	50.2	Good	5.8	6.0
Hereford	47.2	Choice	7.0	6.9
Shorthorn	45.3	Choice	7.1	6.9
Holstein	49.4	Standard	6.4	6.1

and Acceptability scores were on a 1 to 10 hedonic scale, 1 being least and 10 most acceptable. The scores for tenderness and acceptability rated all the meats as acceptable. The Angus, Hereford, and Shorthorn meats were moderately superior to Charolais and Holstein.

Bond et al. (1972) reported data for Holstein, Milking Shorthorn, Jersey, and Hereford carcasses from steers fed concentrate rations to a slaughter weight of 400-500 kg. The Holsteins grew faster and were leaner than the others. Hereford meat was slightly more tender than meat from the carcasses of the other breeds.

Warwick (1968) reported the following estimates of heritability of some important beef traits, without respect to breed, approximated from many research estimates.

Traits	Heritability %
Weaning conformation score	25-30
Carcass grade	30-40
Rib eye (longissimus dorsi) area	30-40
Fat thickness over rib eye	25-45
Tenderness of lean	40-70
Mature cow weight	50-70

Cundiff et al. (1969) reported estimated heritabilities of carcass yield of 0.43 ± 0.19% and of primal cuts of 0.34% among carcasses from cattle of Hereford, Angus, and Shorthorn breeds and crosses among those breeds.

Genetic Factors Affecting Lamb Quality

Sheep vary widely in shape, fleshing, and distribution of flesh and fat. U.S. range sheep are largely of fine-wool breeding such as the Rambouillet. Range ewes are often crossed with mutton breeds such as Suffolk or Hampshire to produce slaughter lambs.

EFFECTS OF AGRICULTURAL PRACTICES ON COMPOSITION 61

Botkin et al. (1969) reported heritability estimates for several traits in Rambouillet, Columbia, and Corriedale breeds and all possible crosses among them. Based on half-sib correlation, average daily gain had an estimated heritability of 0.24 ± 0.09%; loin eye area, 0.34 ± 0.11%; retail cut weight 0.40 ± 0.12%; and percentage of fat 0.54 ± 0.13%; all sufficiently high to make improvement in these traits through selective breeding likely. Hazel and Terrill (1949) reported heritability estimates of 13 and 16% for body type and condition scores in Rambouillet lambs at 124 days of age.

Dickerson et al. (1972) reported data for carcass traits in Suffolk, Hampshire, Dorset, Rambouillet, Targbee, Corriedale, and Coarse Wool lambs fed in a feedlot from weaning to slaughter at 26 weeks. The Suffolks clearly excelled in quantity of lean meat produced. The mutton breeds—Suffolk, Hampshire, and Dorset—clearly excelled in conformation score, and they slightly excelled the others in percentage of boneless lean cuts. As McMeekan (1940) pointed out, even if a long leg contains the same amount of meat as a thick, blocky leg, it will dry out more in cooking and yield less attractive slices.

Palsson (1940) reported carcass traits for lambs of several breeds and crosses in Scotland. Suffolk-sired lambs reported by Palsson excelled in lean yield as did Suffolks in the data reported by Dickerson et al. (1972).

Genetic Factors Affecting the Composition of Pork

In the United States our fast-growing, corn-fed pigs have historically tended to be fat when brought to market. Each pig has traditionally produced enough meat for 2 people and enough lard for 3 people per year. Lard production per hog slaughtered fell from 32.8 lb in 1955 to 20.5 lb in 1971. Consumption per capita during the same period fell from 10.7 lb to 4.7 lb (USDA 1972). Despite the increase in leanness of slaughter pigs, the lard surplus over domestic food use exceeded a billion pounds in both years.

Meatiness of slaughter pigs has improved substantially as indicated in an Economic Research Service Report (USDA 1969). All breeds and crosses can be improved in meatiness through selective breeding. The U.S. Swine Breed Record Associations (Anon. 1970) recognize selection standards which include for carcasses adjusted to 220 lb slaughter weight less than 1.5 in. of backfat, more than average cross-sectional area of the loin eye (longissimus dorsi) and more than 50% of carcass weight in the lean cuts—ham, loin, and shoulder.

U.S. grades for swine carcasses recognize differences in meatiness. Grade 1 is expected to yield more than 53% of carcass weight as lean cuts, grades 2 and 3 yield less, while grade 4 is expected to yield less than 47% carcass weight (USDA 1970).

There is still substantial opportunity for genetic improvement in meatiness of U.S. pigs.

Economic production of pigs requires that they go to market weight in less than six months from birth. Rate of gain is substantially heritable. Like meatiness, it is also affected by diet. Dickerson (1947) analyzed the researches in the experimental stations cooperating in the Regional Swine Breeding Laboratory as to carcass composition, growth rate, and carcass yield. He found that, in general, the faster gaining pigs tended to be fatter.

Trimmed cuts from fatter carcasses may be of excellent quality. Hiner et al. (1965) reported that Duroc carcasses had significantly more intramuscular fat, and more tender and juicier loin than Yorkshires from pigs used in studies of inheritance of fatness.

Pork muscle consists of two kinds of fibers: (1) red, small diameter, myoglobin-rich, and slow contracting; and (2) white, large diameter, myoglobin-poor, and quick contracting. Fiber diameter may be associated with tenderness, the smaller fibers being more tender.

Strickland and Goldspink (1973) reported that fiber diameters of Pietrain muscles were significantly greater than Large White muscles. Hendricks et al. (1971) reported that muscle fibers from Poland China carcasses were larger than those from Hampshire carcasses.

Heritabilities for several swine traits were summarized by Altman and Dittmer (1962). Composite heritability estimates were reported for: loin area, 72%; percentage of lean cuts, 43%; thickness of backfat for Durocs, 49%; and conformation of Poland China pigs, 20%.

Genetic Factors Affecting the Composition of Poultry Meat

Genetic differences among poultry species in carcass traits are striking. Thus, young ducks and geese tend to be fat, other young poultry tend to be lean. Breast flesh of chickens and turkeys is light in color in comparison with that of geese and ducks. The yield of edible flesh from turkeys as compared to percentage of liveweight is greater than that in chickens (Broadbent and Bean 1952).

Within species, fleshing may vary widely due to genetic differences. The change in breast fleshing of our present "Broadbreasted" turkeys compared with wild turkeys and the domestic varieties of 30 yr ago has been dramatic. Heritability estimates for breast width of Broadbreasted turkeys reported by Altman and Dittmer (1962) averaged about 33%.

Our present broiler stocks tend to yield somewhat more flesh than those of 30 yr ago, too. Infusions of germ plasm from the Cornish breed have been general. Most of the Leghorn-type egg-laying stock males are killed at hatching time due to growth rate and carcass inferiority. Goodman (1973) reported heritability estimates of more than 0.5, or 50% for broiler body weight. Mass selection for this trait has made possible the production of broilers of more than 3 lb. liveweight in less than 9 weeks.

Hayes and Marquis (1973) reported the yield of the several component parts of carcasses of current 8-week-old broilers. Males yielded 73.48% and females 75.23% edible flesh of raw carcass weight. In comparison to data reported by Jull et al. (1943) 30 yr ago, breast yield is about 3% greater. Giblet yield is about 1.5% less, perhaps reflecting changed energy content of the diet and its density.

Genetic Factors Affecting the Composition of Eggs

Direct genetic factors on the proximate composition of the edible portion of the eggs of one species are relatively small, though indirect effects reflected in differences in the weight of the whole egg and of its parts are substantial. Heritability of egg size in chickens is important (Farnsworth and Nordskog 1955). The edible portion of genetically small chicken eggs contains relatively more yolk, therefore more fat and total solids, than genetically large eggs from hens of similar age under the same environment.

Heritable variations in vitamin content of eggs exist, and Maw (1954) has established a population of chickens which are genetically deficient in their capacity to store riboflavin in their eggs.

Genetic Factors Affecting the Composition of Milk

The percentage of fat in milk is substantially heritable. Variation in nonfat solids in cows' milk is closely related to fat content, but there is some indication that variation in protein content may be independently heritable.

Milk from various breeds varies in content of fat and solids; variations within each of the breeds are at least as great as the variations among the average values for the breeds given in Table 6.3.

The color of Guernsey and Jersey milk is deeper than that of the other breeds when all are on the same diet. About 90% of the pigment in milk is beta-carotene (Cary 1939). Color of milk is not necessarily a reflection of its vitamin A potency. Holsteins, Brown Swiss, Ayrshires, and Shorthorns produce milk of about the same total vitamin A potency as Guernseys and Jerseys.

Altman and Dittmer (1962) reported heritability estimates from

TABLE 6.3

AVERAGE COMPOSITION OF MILK OF SIX BREEDS OF COWS

Breed	Water (%)	Total Solids (%)	Fat (%)	Protein (%)	Lactose (%)	Ash (%)
Guernsey[1]	85.13	14.87	5.19	4.02	4.91	0.74
Jersey[1]	85.31	14.69	5.18	3.86	4.94	0.70
Ayrshire[1]	86.89	13.11	4.14	3.58	4.69	0.68
Holstein[1]	87.50	12.50	3.55	3.42	4.86	0.68
Shorthorn[1]	87.43	12.57	3.62	3.32	4.89	0.73
Brown Swiss[2]	86.59	13.41	4.01	3.61	5.04	0.73

[1] Wright et al. (1939).
[2] Espe (1946).

the literature for: butterfat percentage in milk, 52-86%; nonfat solids, 35-97%; and protein percentage in milk, 48-53%.

Edwards et al. (1973) reported that variation in the proportions of the different fatty acids in milk fat from 1- and 2-egg Ayrshire twin females indicate a high degree of genetic control.

AGE, SEX, AND SEASON

Beef

As cattle increase in age, flesh color darkens, flavor increases, and ultimately most muscles decrease in tenderness. Mature bull beef carcasses have heavy necks. Under conditions of ad libitum feeding, mature nonlactating cows become fat; indeed, all sexes tend to become fatter as they grow older with ad libitum feeding. Seasonally, some grass-fat young cattle come to slaughter in the summer and early fall. Their carcasses seldom exceed the Good grade, and their fat is likely to be tinged with the yellow from the carotenoid pigments in the grass they have eaten.

With the exception of cows culled from breeding and milk herds, most of our beef comes from the feedlot. Such beef shows minimal seasonal change in carcass composition and consists largely of 12-24 month old steers and heifers fed to Good and Choice grades. There is a growing interest in feeding young bulls for slaughter. Their rate of gain in feedlots is generally greater than that for steers and heifers. Their flesh tends to be heavier than steer or heifer flesh under comparable feeding conditions.

Data for Holstein veal carcasses reported by Brekke and Wellington (1969) are given in Table 6.4. The percentage of boneless meat from the carcass and from all selected wholesale cuts increased significantly with increasing carcass weight.

EFFECTS OF AGRICULTURAL PRACTICES ON COMPOSITION 65

TABLE 6.4

CARCASS YIELD OF HOLSTEIN CALVES IN RELATION TO AGE AND LIVEWEIGHT

Age (Days)	Liveweight (Kg)	Carcass Wt (Kg)	Dressing (%)	Edible Flesh (Kg)	Edible Flesh Carcass Wt (%)	Percent of Edible Flesh Protein (%)	Fat (%)	Ash (%)
4	44.1	25.9	58	16.8	64.4	19.1	3.0	1.0
70	89.5	54.5	63	39.7	72.2	16.8	11.8	0.8
110	131.2	82.9	63	62.7	76.0	17.4	16.5	1.0

Hooven et al. (1972) reported carcass data for 180-day-old Holstein, Milking Shorthorn, Hereford, and Jersey calves. Data for calves from a high level plane of nutrition are given in Table 6.5. Dressing percentages were lower than those for the veal calves described in Table 6.4. Edible flesh as a percentage of carcass weight was about the same for the older calf carcasses and for the 180-day calves.

TABLE 6.5

CARCASS TRAITS OF 180-DAY-OLD CALVES FED A HIGH PLANE OF NUTRITION

Breed	Slaughter Wt in Kg	Carcass Wt in Kg	Dressing	Edible	Lean	Fat
			Percentage of Carcass Wt			
Holstein	267	152.2	57	73	57.8	14.4
Milking Shorthorn	242	140.4	58	75	52.0	23.5
Hereford	204	110.2	54	76	58.1	18.0
Jersey	168	90.7	54	74	55.1	18.7

Data for composition of beef according to U.S. Grades are reported by Pecot et al. (1965). Not included in Table 6.6 are Canner and Cutter grades which are generally used for sausage making. These grades are generally applied to mature thin cows. The Commercial grade contains a high proportion of fat cows yielding carcasses suitable for cooking (USDA 1965). Prime, Choice, Good and Standard are restricted to beef from young cattle. The Commercial grade is reserved for cattle too mature to qualify as Good or Standard. Utility, Cutter and Canner grades may contain thin, poor conformation, young carcasses as well as carcasses from older cattle.

Bulls generally grow faster and produce leaner carcasses than steers or heifers under similar feeding conditions. Meat from young bulls,

66 NUTRITIONAL EVALUATION OF FOOD PROCESSING

TABLE 6.6

PHYSICAL AND CHEMICAL COMPOSITIONS OF BEEF CARCASSES BY GRADE

Grade	As Purchased Separable			Edible Portions Including Kidney and Kidney Fat			
	Lean (%)	Fat (%)	Bone (%)	Water (%)	Protein (%)	Fat (%)	Ash (%)
Prime	47	39	14	44.8	13.6	41	0.6
Choice	51	34	15	49.4	14.9	35	0.7
Good	56	28	16	54.7	16.5	28	0.8
Standard	60	22	18	60.1	18.0	21	0.9
Commercial	54	31	16	52.4	15.8	31	0.8
Utility	62	20	18	62.5	18.6	18	0.9

feedlot fed, may be less tender than that of steers. Glimp et al. (1971) reported data for bulls and steers with like nutritional history. Differences in acceptability of cooked steaks were not significant. Bull carcasses were leaner than steer carcasses.

Laflamme and Burgess (1973) reported that bull carcasses had less fat cover and higher yield of edible meat than steers. In two trials on several rations, bulls gained faster than steers and were leaner than steers.

Ziegler et al. (1971) reported that meat from young Holstein bulls was less acceptable than that from Holstein steers or from steers or heifers of the beef breeds. Meat from heifer carcasses of Angus, Hereford, or Charolais breeds was slightly less acceptable than that from steers of those breeds.

Age, Sex, and Seasonal Effects on the Composition of Lamb

Lambs start to market about Easter. The "new crop" lambs are generally marketed directly off the ewe with carcass weight not more than 20 kg and fat of no more than 30%. Similar but often older lambs are marketed through the summer and early fall. Lambs too small or too thin for slaughter go to the feedlots. They supply the domestically-produced proportion of the winter supply of lamb. They are older, heavier, and generally fatter than the "new crop" lambs marketed in spring and summer.

Paul et al. (1961) compared quality of leg, loin, and shoulder cuts from "new crop," 5½-months-old, and "old crop," 12-months-old, lambs. The lean meat of the new crop lambs was highest in moisture and lowest in fat. Old crop cuts had higher scores in marbling, juiciness, and flavor of lean.

Ram lambs are generally superior to ewe or wether lambs in rate of gain and cutability. Jacobs *et al.* (1972) summarized the findings of previous research and presented new information on heavy lambs (65–68 kg bodyweight). Average daily gain of rams was 15–30% greater than that of wethers. Cutability of heavy ram lamb carcasses exceeded that of heavy wether carcasses. Palatability and tenderness were acceptable.

Age, Sex, and Seasonal Differences in Pork

As pigs grow older and heavier they tend to become fatter. Historically, there has been a somewhat higher proportion of heavy hogs, especially old sows, among the hogs marketed during the summer than during other seasons. Currently, year-round farrowing is eliminating seasonal differences. Boar carcasses tend to be leaner than carcasses of sows and barrows. As boars mature sexually their flesh may develop a characteristic, unpleasant, "boar odor." Plimpton *et al.* (1971) reported that about 10% of carcasses from young boars of 70 kg bodyweight and 30% of carcasses from boars of 100 kg bodyweight had "boar odor."

Ellis and Hankins (1925) reported that as pigs grow older and fatter there is an increase in the percentage of total saturated fatty acids in their fat. There is a decrease in linoleic acid while oleic acid remains constant.

Age, Sex, and Seasonal Differences in the Composition of Poultry Meat

Young chickens were formerly seasonally abundant in summer and fall; young turkeys only in fall and winter. Old hens were sent to market at the end of the laying season during summer and fall. These seasonal changes have largely disappeared. Broiler chickens and young broiler and roaster turkeys are marketed throughout the year. "Spent" hens from laying flocks are also marketed throughout the year, usually for processing.

Poultry tend to become fatter as they grow older when fed ad libitum. Pullets fatten as they approach sexual maturity. The onset of egg layings diverts fat from deposits to egg yolk. When egg laying ceases, fat may be deposited rapidly. Capons are much fatter than cocks.

Seasonal Changes in the Composition of Eggs

Eggs from commercial egg-laying flocks, the bulk of our egg supplies, show minimal seasonal variations. Hot weather may cause some decrease in egg weight (Bennion and Warren 1933). Resultant seasonal variation in protein, fat, and water content of eggs may amount to about one percentage point each.

Eggs from farm flocks have darker yolk color as the grass greens in the spring.

Spring and summer sunshine assure high vitamin D content in eggs from hens allowed to range. Devaney et al. (1936) reported that February eggs from hens receiving direct sunlight but no added vitamins in their diet, contained 1.4 IU per gm egg yolk compared to 3.9 IU per gm egg yolk for June eggs from the same hens. Bethke et al. (1936) reported that eggs from hens behind glass, without dietary vitamin D, decreased in vitamin D content of 17 IU per yolk in December to 5 IU per yolk in May.

Initial pullet eggs often weigh about 44 gm with about 12 gm of yolk and 27 gm of albumin. Maximum weight of about 60 gm may be reached in 2-5 months after laying starts.

Seasonal Changes in Composition of Milk

The proportion of our milk supply from confined cows has increased rapidly during the past few years. Seasonal changes in their milk reflect changes in feed, such as decline of vitamin A content of stored hay.

Milk from pasture-fed cows may show substantial seasonal changes in vitamins A and D. Espe (1946) summarized several reports showing that winter milk under general farm conditions may contain only 5-8 IU vitamin D per quart while summer milk may contain 40 IU per quart. Such differences are largely due to the effects of ultraviolet in summer sunshine directly on the lactating animal. Steenbock et al. (1925) reported that direct irradiations of lactating goats previously depleted of vitamin D could produce a 3- to 6-fold increase in the vitamin C content in their milk.

Vitamin A content is dependent on the carotene and A content of the diet. Lush pasture is an excellent source of these precursors of milk vitamin A.

Hand and Sharp (1939) reported that summer commercial milk (that is, from cows on pasture) contained 20% more riboflavin than winter milk. Riboflavin varied from 0.60 to 3.42 mg per liter. They reported a negative correlation of 0.830 between milk production and riboflavin content.

Fat and nonfat milk solids are affected by ambient temperature. Morrison (1948) summarized relevant research reports which indicated a drop of about 0.2-0.3 in butterfat percentage for each $10°F$ increase in temperature from $30°$ to $80°F$ ambient temperature. Regan and Richardson (1938) reported a drop in nonfat milk solids of as much as one percentage point due to high environmental temperature.

ANTEMORTEM STRESS AND MEAT QUALITY

Stress due to excitement, fear, or exhaustion prior to slaughter may adversely affect meat quality. Extensive glycolysis immediately before slaughter may result in rapid postmortem cessation of glycolysis and rapid onset of rigor mortis. DeFremery and Pool (1960) described this process in relation to toughness in poultry meat. DeFremery and Pool also compiled data from the literature on the time of onset of rigor mortis, ultimate muscle pH, and initial content of adenosine triphosphate (ATP) for several species. These data are given in Table 6.7.

TABLE 6.7

TIME OF RIGOR MORTIS ONSET, ULTIMATE MUSCLE PH AND INITIAL ATP CONTENT FOR SEVERAL SPECIES

Species	Rigor Mortis Onset Hours Postmortem	Ultimate pH	Initial ATP Mg/Gm Fresh Tissue
Beef	4-10	5.4	3.1
Pig	—	5.4	—
Rabbit	1.5-4	5.9	4.1
Chicken	2-4.5	5.8-5.9	4.8

DeFremery and Pool (1960) stated that every treatment that resulted in more rapid loss of ATP, more rapid onset of rigor mortis, more rapid drop in pH, and more rapid loss of glycogen resulted in increased muscle toughness. Khan and Lentz (1973) reported similar findings for beef.

In pork, there is a substantial frequency of carcasses with pale, soft, exudative (PSE) muscle. Such PSE pork exudes large quantities of fluid when cut. Pomeroy (1968) ascribed the immediate cause to rapid postmortem fall in muscle pH to values below 6.0 during the first 45 min postmortem.

Hilton et al. (1972) reported marked improvement in water binding capacity and darker color of pork from pigs subcutaneously injected with 0.1-0.5 mg adrenaline 2-4 hr before slaughter.

MINERALS, DRUGS, AND PESTICIDES IN LIVESTOCK PRODUCTS

Phosphorus and Cobalt Deficiencies

Soils in substantial areas of the United States and the forage produced there is deficient in minerals essential in animal nutrition.

Ruminants grown on phosphorus- or cobalt-deficient forage are likely to be thin and poorly fleshed (Beeson 1941).

Copper, Molybdenum, Sulphate, and Manganese

The metabolism and retention of copper, molybdenum, and sulphate, and perhaps manganese, are interrelated. Dick (1956) reported that molybdenum content of sheep tissues rose very sharply when the molybdenum content of the diet was increased and dietary sulfate was low but not when sulfate was adequate.

Dietary molybdenum inhibits the accumulation of copper in the liver and this inhibition is increased by an increase in dietary sulphate, and decreased by an increase in dietary manganese. The manganese content of eggs is dependent on the manganese content of the diet (Lyons and Insko 1937).

Huber et al. (1971) reported data for liver and muscle Cu and Mo of lactating cows fed a basal ration containing 6 ppm Cu to which graded amounts from 53 to 300 ppm of Mo were added. Liver Cu decreased and liver Mo increased as level of Mo fed increased as shown in Table 6.8. Milk Cu and Mo content increased as Mo was increased in the diet.

TABLE 6.8

CONTENT OF COPPER AND MOLYBDENUM IN LIVER AND MILK OF COWS FED GRADED AMOUNTS OF MOLYBDENUM

Mo Added To Diet Ppm	Cu Ppm		Mo Ppm	
	Milk	Liver	Milk	Liver
0	0.10	136.0	0.03	3.6
53	1.11	25.1	1.03	10.4
173	1.31	24.0	1.83	32.4
200-300	—	17.6	—	22.7

Copper is used in the United Kingdom as a growth promotant in pig diets. Braude and Ryder (1973) reported that 250 mg of Cu per kilogram of diet gave best performance. Their study shows the Cu in livers of pigs that had been fed different levels of dietary Cu:

Cu Mg/Kg Diet	Control	150	200	250
Ppm Cu liver, dry wt	37.63	81.68	143.8	286.57

Murphy et al. (1972) reported Cu content of market milk in various areas of the United States. The range was 0.044–0.190 mg per liter and the mean, 0.086 mg per liter. They reported that Mn content ranged from 0.033–0.211 mg per liter with an average con-

EFFECTS OF AGRICULTURAL PRACTICES ON COMPOSITION

tent of 0.091 mg per liter. Murphy *et al.* reported significant but small geographic variations.

Zinc

Zinc is a dietary essential. Hoekstra *et al.* (1956) demonstrated an interrelationship between the zinc and calcium content of swine diets and the accumulation of Zn in their tissues.

Murphy *et al.* (1972) reported a range in Zn content of market milk from areas throughout the United States of 2.3–5.1 ppm with an average value of 3.28. There was no significant seasonal variation. Geographic variation was small. Neathey *et al.* (1973) reported apparently higher values for Zn in normal cows' milk. Experimental cows fed a control diet containing 39.5 ppm of Zn produced milk containing 4.22 ppm Zn. Cows fed a low Zn diet containing 16.6 ppm Zn produced milk containing 3.26 ppm Zn.

Turk (1965) reported that hens fed a sesame-based, zinc-deficient diet had 90 ppm in the femur and 17.7 ppm in egg yolk. Hens fed a casein-based diet with added Zn had 140 ppm Zn in their livers, 327 ppm in their femurs and 35.3 ppm in their egg yolks.

Iodine

Iodine is deficient in many areas but is generally compensated for by the use of iodized salt. The iodine content of eggs may be changed by varying dietary iodine (Table 6.9).

TABLE 6.9

IODINE CONTENT OF HEN'S EGGS AS AFFECTED BY DIET

Iodine Added to Diet (Mg/Head/Day)	Iodine in Shell-Free Fresh Egg (Ppm)
Basal diet	0.1
Basal diet plus 2 mg iodine	7.69
Basal diet plus 5 mg iodine	15.59

Source: Wilder *et al.* (1933).

Selenium

Allaway (1973) summarized the research information on selenium in the food chain. Homeothermic animals must have some selenium present and available in their diets. Hartley (1967) stated that 0.25 mcg per gm for liver and 1.0 mcg per gm for kidney may be considered satisfactory for normal lambs. He reported levels of 0.02 mcg per gm liver and 0.15 mcg per gm kidney for lambs severely depleted from Se-deficient areas in New Zealand.

McFarland et al. (1970) reported values of 0.48 and 0.80 ppm for chicken and turkey liver; 0.66 and 1.10 ppm for chicken and turkey kidney; and 0.18 and 0.19 ppm for chicken and turkey breast muscle for birds on diet containing 0.27 ppm Se. Amounts present in tissues of lambs, pigs, and poults are dependent on the form as well as the amount of selenium in the diet. Animals fed diets to which selenite-Se has been added contain lower levels of Se in their muscles than those fed diet containing equal amounts of "naturally occurring" plant Se.

Selenium is found in toxic amounts in forage and grain in some areas of western United States. The accumulation of selenium may be inhibited by arsenic, which is sometimes effective as an antidote. Moxon et al. (1944) reported liver contents of 5.4, 5.3, and 6.7 ppm of selenium in steers grazed on 45, 29 and 18 acres per head of seleniferous range, indicating that close grazing, which reduces selection, produces greater selenium accumulation than light or moderate grazing.

Arsenic

Certain arsenic compounds are widely used as growth promotants for pigs and poultry. At the tonic levels used, arsenic does not accumulate in edible tissues in amounts considered hazardous to human health. Liver residues are likely to be higher than those in muscle (Frost and Spruth 1956). During the second quarter of 1973 no samples of chicken or pig liver from animals slaughtered in USDA-inspected plants had arsenic levels in excess of 2 ppm, the established tolerance; 43 of 1001 chicken liver samples tested contained more than 1 but less than 2 ppm arsenic (USDA 1973).

Arsenic compounds are used as cotton desiccants and as herbicides in cotton fields. Peoples (1963) fed lactating cows arsenic acid at levels of 0, 0.5, 0.25, and 1.25 ppm of diet in order to determine whether or not arsenic acid residues in cotton seed meal might cause arsenic residues in milk. Maximum arsenic found in milk from these cows was 0.09 ppm. Liver arsenic residues found were:

| Diet arsenic ppm: | 0 | 0.05 | 0.25 | 1.25 |
| Liver arsenic ppm: | 0 | 0.25 | 0.50 | 1.30 |

Arsenic is seldom found in market cattle tissues. Spalding (1972) reported that the highest such residue found in samples from beef slaughtered in USDA-inspected plants was 0.22 ppm in beef liver.

Lead

Lead is a very common contaminant, not of feed purposefully fed but of the diet of cattle which seem to have an appetite for paint, old storage batteries, and other lead-containing materials.

Dinius et al. (1973) fed groups of calves a control diet that was supplemented with newsprint containing 10 ppm lead or with 10 or 100 ppm lead as lead chromate. Only in liver and kidney were lead values elevated. Control livers contained 0.6 ppm lead; livers from calves fed 100 ppm lead, 2.3 ppm.

Spalding (1972) reported values for lead in tissues of animals slaughtered in USDA-inspected plants. The average values were: liver 0.536 ppm, muscle 0.361, kidney 0.625.

Cadmium

Spalding (1972) reported cadmium residues in beef tissues as follows: liver 0.207 ppm, muscle 0.082 ppm, kidney 0.546 ppm.

Mercury

Mercury is generally present in insignificant amounts in foods of animal origin. Large marine fish are exceptional. Large swordfish may contain amounts in excess of the 0.5 ppm guidelines. Former use of mercurial fungicides as slimicides in pulp and paper manufacture and mercury leaked from chloralkali plants have caused excessive levels in freshwater fish. Seed treatment with mercurial fungicides for planting has led to mercury toxicity when such seeds were fed to animals or man (Nelson 1971).

Wright (1973) and Wright et al. (1973) fed cattle, sheep, and chickens the mercurial fungicides ethyl-mercury-paratoluene-sulfonanilide (Ceresan) and dicyan-diamide (Panogen). Accumulation of mercury in liver and kidney was much higher than in brain or muscle.

Goldwater (1964) reported 3-10 ng per gm Hg in whole milk and 4 ng per gm beef in U.S. samples. Spalding (1972) reported data for 2176 beef carcasses: 18% of livers, 26% of muscle, and 53% of kidney samples were positive for mercury. Mercury content of these samples was about 26 ng per gm.

Westöö (1969) reported mercury content of pork, beef and hens' eggs in Sweden from 1964-1966 while use of methyl-mercury as a fungicide treatment for cereal seeds was permitted, and from 1967-1968 after such use of methyl-mercury stopped:

Food Product	1964-1966 Avg Hg (Ng/Gm)	1967-1968 Avg Hg (Ng/Gm)
Pork chop	30	8
Beef filet	12	2
Hens' egg	29	9

Westöö (1969) reported that Hg content of 300 samples of Swedish pike ranged from 0.04 to 8400 ng per gm; 99% of the Hg found was methyl-mercury. Westöö (1973) reported Hg content of Swedish salmon from 1 to 7 yr of age. Mercury content increased with age from 62 ng per gm in the 1-yr-old to 251 ng per gm in the 7-yr-old fish. Regardless of age, 90-98% of the mercury found was methyl-mercury.

Fluorine

Fluorine is present in such feed ingredients as bone meal, rock phosphate, phosphatic limestone, and smelter effluents which may contaminate feed as well as some water sources. It is transferred in small amounts from these sources to the tissues, especially the bone. Poultry are much more tolerant than mammalian species and may accumulate it in bones in substantial amounts.

Haman et al. (1936) fed diets ranging from 0 to 1000 ppm added fluorine to laying hens. Their tissues accumulated fluorine in proportion to that fed. Bone fluorine in the controls was 613 ppm; for birds receiving 1000 ppm in diet fluorine was 8353 ppm. Liver and muscle of controls contained about 5 ppm fluorine while these tissues for birds fed 1000 ppm fluorine contained about 15 ppm.

Insecticide Residues

Insecticide residue in meat, milk, eggs, or poultry may result from direct applications to animals or indirectly through feed, forage, or treated premises. Chlorinated hydrocarbon insecticides accumulate in animal tissues and products. Their use has been regulated to prevent residues which might be hazardous, and use of such insecticides has been greatly reduced.

Rubin et al. (1947) reported that DDT accumulates in eggs of hens fed DDT in their diet. Marsden and Bird (1947) reported accumulations of DDT in flesh and especially in fat of turkeys fed DDT, reaching a level of 6235 ppm in fat of turkeys fed 1500 ppm in their diet.

Draper et al. (1952) reported DDT in eggs from hens fed 200 ppm DDT in their diet of about 56 ppm, while their depot fat contained about 2000 ppm.

Stadelman et al. (1965) and Stadelman (1973) reported data for lindane, heptachlor epoxide, Dieldrin, DDT, and other chlorinated hydrocarbon residues accumulated in fat and eggs from hens fed these compounds. They also reported data for the disappearance of such residues after treatment ceased. Stadelman et al. (1965) administered 10-15 ppm feed-equivalent of the insecticides orally by capsule for 5 successive days. Data for residues following treatment are shown in Table 6.10.

Level of calcium in the diet may affect storage of DDT in the egg.

EFFECTS OF AGRICULTURAL PRACTICES ON COMPOSITION

TABLE 6.10

RESIDUES OF LINDANE, HEPTACHLOR EPOXIDE, DIELDRIN, DDT, AND DDE RESULTING FROM ORAL DOSAGE OF 10-15 PPM FEED-EQUIVALENT FOR 5 DAYS

Weeks After Treatment	Ppm in Fat (F) and Yolk (Y)							
	Lindane		Heptachlor Epioxide		Dieldrin		DDT and DDE	
	F	Y	F	Y	F	Y	F	Y
1	0.7	0.5	10.2	1.1	3.6	0.7	9.6	5.1
10	0.1	0.1	1.1	0.4	3.9	1.6	1.6	0.5
26	0.0	0.0	0.3	0.2	1.0	0.3	0.7	0.2

Cecil et al. (1973) reported results of tests designed to resolve conflicting reports of DDT on shell thickness. They reported that DDT fed at 50 ppm had little effect on shell thickness of hens' eggs at either 3.5 or 1.5% level of calcium in the diet. DDT residues in egg yolks plateaued at about the dietary level in eggs from hens fed the 3.5% calcium diet and at 1.4 times the dietary level in eggs from hens fed the 1.5% calcium diet.

Lampert (1947) reported cows' milk with 44 ppm DDT from cows fed feed containing 2000 ppm.

Link et al. (1963) reported that cows fed 50 ppm heptachlor or Dieldrin or 100 ppm DDT for 12 weeks had residues of about 300 ppm heptachlor epoxide, 120 ppm Dieldrin, and 50 ppm DDT respectively in omental fat.

Spalding (1972) reported data for ten chlorinated hydrocarbon insecticides in stratified random samples of carcasses of meat and poultry slaughtered in USDA-inspected plants. DDT and its metabolites and Dieldrin were most frequently found, as shown in Table 6.11.

Verified sources of violative residues were chiefly either accidental feed contamination or single feed ingredients containing unusually high residues.

TABLE 6.11

RESIDUES OF DDT AND ITS METABOLITES AND OF DIELDRIN IN MEAT AND POULTRY, PPM FAT BASIS

Insecticide	Meat: No. of Samples			Poultry: No. of Samples		
	ND[1]	<0.5	>0.5	ND[1]	<0.5	>0.5
DDT and Metabolite	550	1608	245	138	1474	192
Dieldrin	1546	853	4	758	1043	3

[1] Not Detected.

EFFECTS OF DIET ON COMPOSITION OF ANIMAL PRODUCTS

Effects of Nutrition on Beef Composition

Plane of Nutrition.—Quantitative limitation of feed is a major source of variation in chemical composition of beef. Beef cattle are characteristically kept on maintenance or even submaintenance rations for considerable periods of their lives. In many areas of the world, periods of adequate or luxus feed intake are brief.

Although animals inadequately fed are lean and bony, the research of Winchester and Howe (1955) at Beltsville with identical twin calves indicated clearly that, for calves, long periods of severe limitation of caloric intake followed by adequate periods of full feeding resulted in carcasses equal to those of their continuously full-fed twins, slaughtered at the same weight. Dockerty et al. (1973) reported that 6 months interruption in gain in weight of Hereford steers at 227-kg liveweight followed by full feeding to slaughter weight of 454 kg, produced carcasses and meat fully equal to those from steers of the same origin full-fed from initial 227-kg weight to 454-kg slaughter weight.

One of the basic studies of the effect of plane of nutrition on composition of the beef animal is that of Moulton et al. (1922) who maintained 3 lots of steers from birth to 48 months of age on high, medium, and low planes of nutrition respectively. Protein, as percentage of empty body weight, fell from an initial value of 20% to a final value of 13% for the high-plane steers, from 21% to 18% for the medium-plane steers, but changed only from an initial 21% to a final 22% for the low-plane group.

Hiner and Bond (1971) reported data for Angus steers fed a diet consisting of 70% concentrate and 30% roughage ad libitum (full-fed) or restricted to amounts producing average daily gains of 0.34–0.50 kg per day (Table 6.12).

TABLE 6.12

SEPARABLE FAT AND LEAN IN CARCASSES OF ANGUS STEERS FULL-FED OR RESTRICTED IN FEED INTAKE FROM SIX MONTHS OF AGE

Feeding Level	Carcass Component	Slaughter Age In Months					
		6 (%)	12 (%)	18 (%)	24 (%)	30 (%)	36 (%)
Full-fed	Lean	76	63	66	59	60	58
	Fat	24	27	34	41	40	42
Restricted	Lean	83	76	72	77	77	66
	Fat	17	26	28	23	23	34

At 24 months of age the full-fed steers had reached a weight of 506 kg, the restricted steers only about 350 kg. The full-fed steer carcasses at that age contained 2.5 times as much weight of fat as the restricted steers but only 15% more protein. At 36 months, the full-fed steer carcasses and the restricted ones contained almost identical weights of lean. The carcasses from full-fed, 36-month steers contained almost 140 kg fat while the carcasses from restricted-fed steers contained only about 99 kg.

Vitamins in Feed and in Beef.—Vitamin A is stored in the liver, and the amount may vary enormously. Since feedlot cattle eat little roughage, they may undergo severe vitamin A depletion with resulting night blindness and anasarca and very low liver vitamin A. Conversely, grass-fed cattle may have livers with much higher vitamin A content than the average value. Davis (1956) has found from about 3 μg per gm vitamin A in livers from depleted beef cattle to about 1500 μg per gm in cases of carotenosis.

Beef fat may contain appreciable amounts of vitamin A (Table 6.13).

TABLE 6.13

VITAMIN A IN BEEF FAT IN RELATION TO CAROTENE INTAKE

Daily Carotene Intake μg/Cwt Liveweight	Vitamin A in Fat IU/100 Gm
1250	24
1360	54
2500	33
5000	96
Very high pasture	630

Source: Cabell et al. (1943).

Hormones and Composition of Beef.—Burroughs et al. (1955) summarized experiments at 9 agricultural experiment stations, including their own work at Iowa, showing an average increase in daily gain of ⅓ lb per head per day following the addition of small amounts of diethylstilbestrol (DES) in the cattle feed, a feed saving of 12%, without difference in carcass grade as compared with controls fed for the same time period. Feed and ear implant use of DES to increase rate of feedlot gain became general.

Feed use of DES and other growth-promoting substances is subject to control of the FDA. Since DES has been reported to produce cancer when administered to experimental animals, its use is subject to the Delaney Clause of the Food, Drug and Cosmetic Act which states that any amount of a known carcinogen in edible tissues of food animals cannot be permitted. The FDA (1972) announced in the Federal Register that feed use of DES must cease on January 1, 1973. On April 25, 1973, FDA (1973A, B) announced that all use of DES ear implants must end on April 29, 1973. These actions were based on the results of research by the USDA which identified 0.52 ppb of DES by radioactive tracer methods in the liver of a steer slaughtered 7 days after withdrawal of DES from its feed (FDA 1972A). Further USDA research identified 0.04-0.12 ppb DES in livers of steers slaughtered 120 days after implantation. DES had been used for two decades without a single known instance of harm to humans (FDA 1973).

Several compounds remain available for use as growth promotants. Melengestrol acetate (MGA) suppresses oestrous and increases the rate of feedlot gain of heifers (Ray et al. 1969; Hawkins et al. 1972). Zeranol (resorcyclic acid lactone) is an estrogen used as an implant for steers. Synovex S (progesterone and estradiol benzoate) is used as an implant for steers. Synovex H (testosterone propionate and estradiol benzoate) is used as implants for heifers.

Baker and Arthaud (1972) summarized data from 46 comparisons in 17 studies in the literature of hormone treatments on average daily gain (ADG) and carcass grade of young slaughter bulls. The comparisons included DES, Zeranol, Synovex S, combinations of Synovex S and H with DES and Zeranol. There were no consistent responses in ADG or carcass grade.

Burris et al. (1953) reported that calves treated with androgens had a higher percentage of round and a lower percentage of loin than control calves. Androgen-treated calves had a higher percentage of lean than controls.

Antibiotics and Composition of Beef.—Antibiotics are widely used as growth promotants and prophylactics for calves, especially calves of dairy breeds which are generally reared on milk surrogates. Such use is undergoing review to make sure that it does not contribute to development of pathogenic bacteria resistant to the antibiotics used therapeutically in human and veterinary medicine (Swann 1969). Of 482 cow carcasses examined, 46 had antibiotic levels above tolerance, and of 518 calf carcasses 48 had antibiotic levels above tolerance—streptomycin, penicillin, or other antibiotics (USDA 1973).

Nutritional Effects on Fatty Acid Composition of Beef Fat.—The fatty acid composition of beef fat can be modified somewhat by diet. Dryden and Marchello (1973) reported that the carcass fat of steers fed diets containing 6% safflower oil had carcass fat containing

EFFECTS OF AGRICULTURAL PRACTICES ON COMPOSITION 79

more linoleic acid (2.7-3.8% of tissue fats) than carcasses from steers fed animal fat-supplemented diets (1.3-2.0% linoleic acid in tissue fats).

Rumsey et al. (1972) reported data for total percentages of unsaturated fatty acids in depot fat of roughage-fed and concentrate-fed steers. The former had 40.8% of depot fat as unsaturated fatty acids while the depot fat from the concentrate-fed steers contained 34.0% unsaturated fatty acids. The difference was mostly in the oleic acid fractions. It may be possible to produce dramatic change in proportions of polyunsaturated fatty acid content of milk or depot fat by feeding certain formaldehyde-treated casein-vegetable oil mixtures (cf Plowman et al. 1972).

Effects of Nutrition on Lamb Composition

Plane of Nutrition.—Comments with respect to effects of nutrition on composition of beef apply equally to lamb. Barbella et al. (1936) reported roasted lamb legs from full-fed lambs to be better in flavor and tenderness than those from otherwise similar lambs which were held without gain by restricted feeding during the feeding period. Full-fed lambs had 3.4% intramuscular fat in the longissimus dorsi muscle compared to 1.8% in the same muscle from restricted-fed lambs.

Vitamin A.—Lamb liver is a rich source of vitamin A. Hoefer and Gallup (1947) fed lambs for 55-103 days on various levels of carrot oil, alfalfa meal, and fish-liver oil as sources of carotene or vitamin A. Table 6.14 shows standard equivalents (0.6 mg carotene = 1 IU of activity) for each feed source. Storage was grossly different depending on the three feed sources.

TABLE 6.14

VITAMIN A INTAKE (EQUIVALENT) FROM VARIOUS FEED SOURCES AND LEVEL OF VITAMIN A IN LAMBS

Carrot Oil		Alfalfa Meal		Fish-Liver Oil	
IU A Activity Intake/Kg Liveweight/Day	Liver A Gm/100Gm	IU A Activity Intake/Kg Liveweight/Day	Liver A Gm/100Gm	IU A Activity Wt/Day	Liver A Gm/100Gm
—	—	50	209	50	665
75	144	84	190	—	—
168	309	165	690	—	—
259	1294	262	4340	—	—
—	—	500	3322	500	19,743

Source: Hoefer and Gallup (1947).

Effects of Nutrition on Pork Composition

Plane of Nutrition.—Restricted feeding is generally practiced in Denmark, England, and often in Canada, to assure satisfactorily lean carcasses. Crampton et al. (1954) reported that rather mild restriction of feed intake improves carcass quality for 50- to 100-kg pigs.

A second way of restricting feed intake, is through the use of fibrous feeds to reduce energy intake. Crampton (1954) reported data from a trial of this sort with 50- to 100-kg pigs.

Hale and Southwell (1967) and Gilster and Wahlstrom (1973) reported data on the effect of different dietary protein levels of swine carcass compositions. Ham and loin as percentages of carcass weight are decreased on low protein diets; proportion of lean decreased and proportion of fat increased. Cross section of longissimus dorsi in carcasses from 115-kg pigs fed on 18% protein diet was 34.25 per sq cm, for carcasses from pigs fed a 12–10% protein diet cross section area of longissimus dorsi was 28.60 per sq cm. Palatability of pork chops was acceptable regardless of protein level in the swine diet.

Feed Fat and Pork Fat.—A former complaint of U.S. packers was that the carcasses of lean hogs were soft and flabby. It was generally conceded that a minimum of 1.5 in. of backfat was necessary in a cornfed hog to give a firm carcass. Recent experience indicates that 1.2 in. is enough in well-fleshed, meat-type hogs. It is none-the-less true, as Ellis and Hankins (1925) demonstrated, that hogs yield firmer carcasses as they grow older and heavier. Both old and young pigs store feed fat, but the older ones convert an increasing proportion of carbohydrates to hard fat. Feeds containing fats liquid at room temperature, such as soybeans and peanuts, produce a softer pork than feeds containing little fat.

Vitamins in Feed and Pork.—Pigs are principally dependent on the vitamins in their feed to meet their metabolic needs and for tissue storage. Pork is an important dietary source of thiamin, generally excelling other meats in this respect. Pence et al. (1945) reported the thiamin content of pork tissues from pigs maintained on a basal diet containing 1.2 mg and 5 mg per gm per head per day thiamin for periods of 5 to 155 days (Table 6.15).

Bratzler et al. (1950) reported a great increase in the tocopherol content of swine tissues by feeding tocopherol-supplemented rations.

Hormones and Composition of Pork.—The use of hormones in pork production has not been established as practical. The results with estrogens have shown little or no increase in the rate of gain, and little or no effect on composition. Pseudopregnancy and mammary development are undesirable effects. Beeson et al. (1955) reported

TABLE 6.15

EFFECT OF SUPPLEMENTARY FEED THIAMIN ON STORAGE
OF THIAMIN IN PORK TISSUES

		Thiamin Content of Tissue		
Treatment	Time (Days)	Shoulder (μg/Gm)	Loin (μg/Gm)	Liver (μg/Gm)
Basal	155	7.6	9.9	3.6
Basal + 5 mg thiamin/head/day	5	10.2	12.1	3.2
	15	11.7	15.5	5.2
	22	13.8	17.3	5.5
	35	18.4	24.3	6.8
	155	14.1	21.8	4.1

Source: Pence et al. (1945).

that the fat content of the edible portion of carcasses from control 100-kg liveweight was 55.1%; from pigs fed 2 mg per head per day DES, 53.7%; and from pigs fed 20 mg per head per day testosterone, 50.0%.

Effects of Antibiotics on Composition of Pork.—Clausen (1956) presented an excellent account of Danish researches, and a summary of those in other countries on the effects of antibiotics on carcass composition, and concluded that the only circumstances under which antibiotic feeding produced fatter carcasses was the ad lib feeding of low-protein diets supplemented with antibiotic. Ashton et al. (1955) reported data showing little effect of antibiotic on carcass composition with ad lib feeding and protein levels of 10–20%.

Broquist and Kohler (1954) were unable to detect antibiotic activity in the flesh of pigs fed 200 mg per kg body weight of chlortetracycline. They found 0.3 gm per kg in the tissues of pigs fed 2 gm per kg body weight. Jukes (1955) found 0.3 gm per kg in muscle of pigs fed 2 gm per kg, but none in the flesh of those fed 200 mg per kg.

There is general use of antibiotics as growth promotants in swine production in the United States. This usage is under current review (Swann 1969; FDA 1972A) in order to assure that such usage is compatible with therapeutic use of antibiotics.

Effects of Nutrition on Poultry Composition

Plane of Nutrition.—Poultry species are generally fed ad libitum. Caged laying hens thus fed often develop fatty livers. Ivy and Nesheim (1973) reported data from experiments designed to determine whether or not such livers resulted from overeating. They force-fed laying hens daily amounts of feed 10% greater than ad libitum intake of controls. Bodyweight of the force-fed hens increased 10% in 30 days. The liver fat content of the force-fed hens was 75.6% compared to 58.5% for the controls. Bragg *et al.* (1973) reported that hens receiving diets supplemented with 8% animal tallow or rapeseed oil developed fatty livers while hens receiving 8% supplements of soybean or sunflower seed oil did not.

There is an ancient European practice of force-feeding ("cramming") geese to produce fatty livers which are used in making paté de foie gras. Nir (1972) reported that geese force-fed from 12 to 16 weeks of age increased in body weight by about 77%; liver weight increased 6-fold. The livers varied in lipid content from 45 to 65%.

It has been reported (Anon. 1973) that surgical destruction of the satiety center in the brain causes geese to eat continually, thus obviating the need for force-feeding.

Calorie-to-Protein Ratio.—It is possible through adjustment of the energy and protein content of the diet to produce young chickens and turkeys of widely different carcass fat percentages. Fraps (1943) demonstrated that very young chickens with fat contents of from about 2% to more than 16% fat in carcass could be produced by varying the calorie-to-protein ratio as shown in Fig. 6.1.

Leong *et al.* (1955), Waibel (1955), Olson *et al.* (1972) and Boomgaardt and Baker (1973) have substantiated and extended our knowledge of the effects of energy and protein levels. Olson *et al.* reared

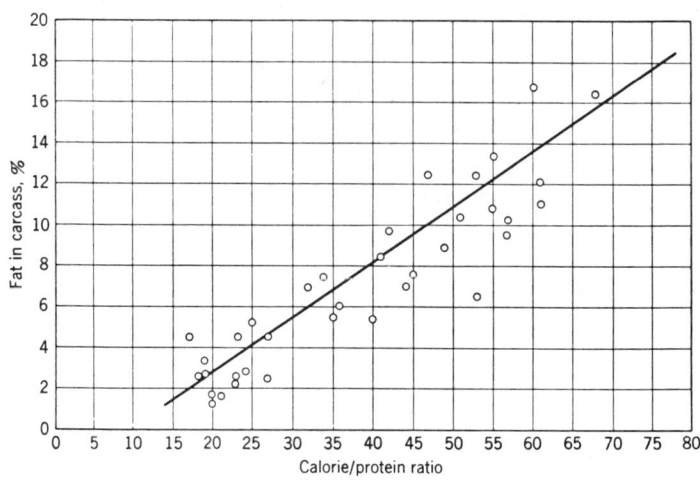

Fraps (1943)

FIG. 6.1. RELATION BETWEEN CALORIE/PROTEIN RATIO IN THE DIET AND PERCENTAGE OF FAT IN CARCASSES OF YOUNG CHICKENS

EFFECTS OF AGRICULTURAL PRACTICES ON COMPOSITION 83

groups of experimental chicks on 1 of 3 dietary levels of metabolizable energy and under ambient temperatures varying from 13 to 40.5°C. Carcass fat was increased and protein content decreased with increasing ambient temperature or dietary energy level. Thomas and Combs (1967) fed different groups of chicks with a wide range of metabolizable energy and protein. Data for carcass fat in the chicks so fed are given in Table 6.16.

TABLE 6.16

EFFECTS OF DIETARY METABOLIZABLE ENERGY AND PROTEIN CONTENT ON PERCENTAGE OF CARCASS FAT IN YOUNG CHICKENS

Metabolizable	Percentage of Dietary Protein		
Energy	25.6	16.0	6.4
KCal/Kg	Fat %	Fat %	Fat %
3647	9.3	13.9	27.0
2906	5.8	11.2	20.1
2165	4.1	6.5	15.9
1424	1.4	2.1	5.3

Dean (1973) reported that narrowing the calorie-protein ratio in duckling diets lowered the percentage of carcass fat. Eviscerated carcasses of market age ducklings fed a 16% protein diet contained 36.0% fat. Those of ducklings fed an isocaloric diet with 24% protein had 32.6% fat.

Nutrition and the Vitamin Content of Poultry Tissues.—With the exception of ascorbic acid, poultry require all the vitamins in their diet. Their body tissues reflect, but not always proportionately, the vitamin content of their diet.

Rubin and Bird (1941) reported that chicks store more vitamin A from alfalfa than from either crystalline carotene or vitamin A.

Denton et al. (1947) found some increased storage of niacin in breast and liver at higher levels of niacin intake.

Fatty Acids in Feed Fats and Carcass Fat.—Watts (1954) stated that polyunsaturated fatty acids in poultry carcass fat are derived from feed fat. Chu and Kummerow (1950) showed that rate of rancidification of skin fat of young chickens increases with increased content of linolenic acid. By feeding levels of linseed oil from 0 to 25% of the diet, Chu and Kummerow obtained skin fat percentages of linolenic acid from 1.9% in the controls to 26% in the skin of chickens fed 25% linseed oil. Klose et al. (1951) reported that turkey carcass fat from turkeys fed diets containing 5% linseed oil contained 17.7% linolenic acid while carcass fat from turkeys fed diets containing 5% coconut oil contained 1.5% linolenic acid. Carcasses of linseed oil-fed turkeys developed off-flavor and -odor during six months' storage while that from coconut oil-fed birds showed little change in storage.

Sim et al. (1973) reported data for fatty acid composition of liver and adipose tissue from hens fed a basal diet or that diet supplemented with animal tallow, soybean oil, sunflower seed oil, or rapeseed oil. Data for oleic and linoleic acid percentages in liver fat and adipose tissue fat are shown in Table 6.17. Liver fat from hens

TABLE 6.17

OLEIC AND LINOLEIC ACID PERCENTAGES IN LIVERS AND ADIPOSE TISSUE AND PERCENTAGES OF THOSE ACIDS IN DIETARY FAT

| Diet | Oleic Acid (%) | | | Linoleic Acid (%) | | |
| | | in Fat of | | | in Fat of | |
	Diet	Liver	Adipose	Diet	Liver	Adipose
Basal	36.4	51.2	54.5	23.2	8.7	11.7
Animal tallow[1]	47.0	53.6	52.0	2.5	5.1	10.8
Soybean oil[1]	29.0	24.3	39.5	51.2	26.5	32.4
Sunflower seed oil[1]	16.4	27.7	35.7	73.3	28.2	39.5
Rapeseed oil[1]	23.2	41.5	46.7	36.5	16.2	24.6

[1] As an 8% supplement to the basal diet.

fed the soybean oil and rapeseed oil supplemented diets contained about 2% linolenic acid. Liver fat from the rapeseed oil-fed birds contained 5.2% erucic acid.

Hormones and Poultry Carcass Composition.—Neck-implanted pellets of diethylstilbestrol, once widely used in the production of young meat chickens are no longer permitted. Residues in excess of 2 ppb were demonstrated in edible tissues (Morrison and Munro 1969). Other hormones are not widely used in poultry production.

Chickens respond to administered estrogen by increased deposition of fat, unlike cattle and sheep. Detwiler et al. (1950) reported that 12-week-old broilers, each implanted with a 12 mg pellet of diethylstilbestrol when 7 weeks old, contained 16.9% fat compared to 11.9% fat for their controls. Libby et al. (1955) reported that the thigh flesh of chickens which were given 0.5 mg DES per day for 28 days contained 8% fat compared to 3.9% for the controls.

Thiouracil, protamone, testosterone, and growth hormone have also been investigated singly and in various combinations with respect to their effect on carcass composition. Detwiler et al. (1950) showed that 0.2% thiouracil in the diet produced 18.5% fat in the carcasses of 12-week-old broilers compared to 11.9% in the controls and 16.9% in a diethylstilbestrol-treated lot. Libby et al. (1955) found little change in fat in the thigh flesh on response to administration of growth hormone, protamone, or testosterone. Hill (1953) reported 3 mg per 100 gm of thiouracil in the flesh of nonfasted chickens fed 0.2% thiouracil in their diet.

Antibiotics and Carcass Composition.—Antibiotics are widely used in poultry feed as growth promotants and prophylactics. Such use has been and continues to be under challenge as a possible source of antibiotic-resistant pathogens, including those such as *Salmonella*, transmissable to man. Further concern exists over the possibility of transfer of genetic resistance to other bacteria with impairment of therapeutic use of antibiotics (Swann 1969).

The Food and Drug Administration (FDA 1972B) has recently established new requirements for subtherapeutic and growth promotant use of antibiotics which are also used in clinical human medicine. Subtherapeutic and growth promotant use of such antimicrobial agents will depend on data establishing their safety and effectiveness under specific criteria. These criteria are based on guidelines contained in the Report of the FDA Task Force on the Use of Antibiotics in Animal Feeds (FDA 1972A).

Effects of Nutrition on Egg Composition

Limited data relating to the effects of nutrition on the proximate chemical analysis of egg yolk and egg white are available. Titus *et al.* (1933) reported small but significant differences in the amount of protein in the dry matter of eggs from hens fed different protein sources. Ingram *et al.* (1950) failed to demonstrate differences in the tryptophan, lysine, cystine, or methionine content in the few eggs they had obtained from hens fed diets deficient in amino acids.

The fatty acid composition of egg yolk fats may be substantially affected by dietary fat. Reiser (1950) compared the eggs from pullets on a synthetic fat-free diet with eggs from pullets receiving the same diet plus 4% cottonseed oil. Reiser concluded that the hen can synthesize 3, 4, and 5 but not 6 double-bond acids from the 2 double-bond linoleic acids (Table 6.18).

Sim *et al.* (1973) fed groups of hens a basal diet or the same diet supplemented with 1, 2, 4, or 8% of animal tallow, soybean oil, sunflower oil, or rapeseed oil. The fatty acid compositions of the fats fed and of the egg yolk fats produced are given in Table 6.18. Oleic (18:1) and linoleic (18:2) in the egg yolk reflected the relative content of those fatty acids in the diet. As more linoleic acid was deposited in the egg yolk fat, less oleic acid was retained. Data

TABLE 6.18

EFFECT OF DIETARY FAT SOURCE ON FATTY ACID COMPOSITION OF
EGG YOLK FATS (PERCENTAGE OF TOTAL FATTY ACIDS)

Fatty Acid	14:0	16:0	16:1	18:0	18:1	18:2	Total Polyenoic
Basal diet	1.5	28.5	3.5	6.9	36.4	23.2	23.2
Egg yolk	0.4	29.5	7.9	5.8	47.4	9.0	9.0
Animal tallow[1]	2.8	28.3	3.6	15.8	47.0	2.5	2.5
Egg yolk	0.4	27.0	5.7	2.5	57.8	6.6	6.6
Soybean oil[1]	—	10.9	—	3.7	29.0	51.2	56.4
Egg yolk	0.3	25.1	4.8	8.8	35.2	24.2	25.8
Sunflower oil[1]	—	6.8	—	3.5	16.4	73.3	73.3
Egg yolk	0.3	25.2	2.5	9.2	29.7	33.1	33.1
Rapeseed oil[1]	—	4.9	0.6	2.1	23.2	36.8	69.5
Egg yolk	0.4	19.8	5.9	4.9	44.4	22.8	24.6

[1] As an 8% supplement added to basal diet.

reported by Roland and Edwards (1972) for eggs from hens fed a stock diet or that diet supplemented with lard, linseed oil, corn oil, or menhaden oil, showed that egg yolk fat from the soybean oil groups contained 1.6% linolenic (18:3) and that from the rapeseed oil-fed groups contained 1.8% erucic acid (22:1).

Cholesterol content of egg yolk is substantially affected by diet. Weiss *et al.* (1967) reported that cholesterol was increased by feeding supplements of cholesterol and greatly augmented by feeding supplements of both cholesterol and safflower oil. Inclusion of 1% B-sitosterol in the feed may have inhibited cholesterol deposition moderately. Data are shown in Table 6.19.

Clarenburg *et al.* (1971) fed 1, 2, or 4% supplements of sitosterol with a standard diet. Data are given in Table 6.20. Sitosterol supple-

TABLE 6.19

CHOLESTEROL CONTENT OF EGG YOLK FROM HENS FED A LOW FAT DIET
OR SUPPLEMENTS OF CHOLESTEROL, SAFFLOWER OIL AND/OR SITOSTEROL

Diet	Cholesterol Mg/Gm Yolk
Low fat	12.5 ± 0.9
30% Safflower oil	15.6 ± 0.2
1% Cholesterol	19.9 ± 2.6
1% Cholesterol plus 29% safflower oil	30.2 ± 2.1
1% Cholesterol plus 1% B sitosterol	15.0 ± 0.3
1% Cholesterol plus 29% safflower oil plus 1% B sitosterol	25.7 ± 0.9

EFFECTS OF AGRICULTURAL PRACTICES ON COMPOSITION 87

TABLE 6.20

CHOLESTEROL AND SITOSTEROL IN EGGS FED A STANDARD DIET
OR THAT DIET SUPPLEMENTED WITH SITOSTEROL

Groups	Cholesterol Control Period Mg/Gm yolk	Cholesterol 15-35 Day Feeding Mg/Gm yolk	Sitosterol 15-35 Day Feeding Mg/Gm yolk
Standard diet	14.2	13.9	
1% Sitosterol	14.1	13.0	1.28
2% Sitosterol	14.3	9.2	2.03
4% Sitosterol	13.9	9.0	2.07

mentation of 2-4% resulted in a decrease of about 23% in egg yolk cholesterol and deposition of substantial amounts of sitosterol.

The major determining factor for egg yolk color is the amount of xanthophyll in the hen's diet (Hughes and Payne 1937). Littlefield *et al.* (1973) reported that egg yolks from hens receiving a white corn diet contained 2.3 mcg per gm per yolk while egg yolks from hens receiving a yellow corn diet contained 28.9 mcg per gm per yolk. Bartov and Bornstein (1967) fed diets containing yellow corn from different sources in diets resulting in widely different levels of xanthophyll, from about 17 mcg per gm of egg yolk produced to about 138 mcg per gm of egg yolk. The corresponding variation in the egg yolks was from about 7 mcg per gm yolk to 40 mcg per gm yolk B-carotene equivalent.

The vitamin A content of egg yolk is readily affected by diet. Bearse and Miller (1937) published data on the relationship between vitamin A in the feed and vitamin A in the egg yolk; the values in Fig. 6.2 were calculated from these data.

The vitamin D activity of egg yolk is substantial and varies widely with the vitamin D content of the feed, or, in range poultry with the season. Devaney *et al.* (1936) showed that the amount of vitamin D transferred from feed to egg is about 2% over a range of 1.4-14.4 IU per gm of feed; about the same efficiency as shown in milk secretion.

The thiamin content of the egg varies with the dietary content of this vitamin. It is possible to produce depleted eggs by use of special diets, but the usual ration produces eggs adequate in thiamin for the needs of the developing chick embryo. Scrimshaw *et al.* (1945) reported 1.05 mg per gm thiamin per egg from hens receiving a diet containing 5 mg per gm thiamin and Mayfield *et al.* (1955) reported values of 1-1.17 mg per gm thiamin.

The riboflavin and niacin content of eggs is markedly affected by

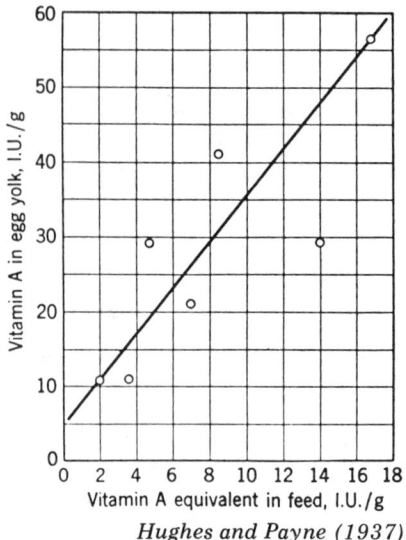

Hughes and Payne (1937)

FIG. 6.2. RELATION BETWEEN VITAMIN A IN FEED AND VITAMIN A IN EGG YOLK

the riboflavin and niacin content of the diet (Peterson et al. 1947; Bauernfind and Norris 1939; Snell et al. 1941).

Waibel (1955) found 360, 304 and 276 ng per gm of biotin in the yolks of eggs from hens receiving 200, 5, and 0 mg of penicillin per kilogram of feed respectively.

Effects of Nutrition on Milk Composition

Proximate composition of milk is affected by plane of nutrition, energy content of the diet, the composition and amount of feed fats and degree of comminution of the roughage.

Espe (1946) summarized research reports on the effect of feed withdrawal. Fasting sharply reduces the amount of milk secreted with a concomitant increase in the percentages of fat, total nitrogen, and chlorides, and a decrease in lactose.

Maynard et al. (1940) reported that cows receiving a diet containing 7% fat produced more milk than similar cows receiving a diet containing 4% fat. Porter et al. (1953) reported that inclusion of large amounts of soybean oil in the diet increased the oleic acid content of butterfat produced by cows fed such diets. Macleod et al. (1972) reported that cows receiving a low fat basal diet produced milk with normal butterfat content but with increased proportion of shorter-chain fatty acids in the butterfat. When soybean oil was added to the basal diet, stearic and oleic acid proportions were increased;

shorter-chain fatty acids decreased. Linoleic acid was increased. Cows receiving hydrogenated animal tallow produced fatty acid proportions similar to those produced by cows fed the basal. Storry et al. (1973) reported data for fatty acid proportions in butterfat from cows receiving a low fat basal diet or that diet with additions of 2, 4, 7, or 10% of animal tallow. There was no significant effect on total milkfat. Stearic and oleic acids increased in proportion to the amount of tallow in the diet. Myristic and palmitic acid proportions were reduced. Linoleic acid proportion was unaffected.

Dramatic changes in linoleic acid content of butterfat resulted when cows were fed diets containing a formaldehyde-treated safflower oil-casein particle (Scott et al. 1970; Plowman et al. 1972). The formaldehyde treatment prevented hydrogenation of the linoleic acid of the safflower oil in the rumen. Some of the dietary linoleic acid was thereafter transferred unchanged to the butterfat, increasing the proportion of linoleic acid content to 35% of total fatty acids. Proportions of C_8-C_{16} fatty acids were reduced; proportion of 18:0 and 18:1 acids was unchanged.

Powell (1938) showed that the fat percentage in milk may be reduced very sharply by feeding large amounts of grain with small amounts of ground hay. Porter et al. (1953) reported a decrease in fat percentage in the milk of cows fed ground or pelleted alfalfa hay. Hess and Young (1972) reported a diet of 3.5 kg of alfalfa hay plus flaked corn ad libitum depressed butterfat content of milk from initial 3.36% to 1.98%.

Vitamin A content of milk may be changed greatly by changing amounts of carotene or vitamin A in the diet. Espe (1946) stated that 10- to 20-fold increases can be obtained. Under general farm conditions late winter butter may have only 29,000 IU vitamin A per kilogram while levels of 300,000 IU per kilogram have been produced by high level feeding of vitamin A.

While Holstein and Brown Swiss cows tend to produce whiter milk than Jerseys and Guernseys, total vitamin A activity is about the same when the feed intake is similar (Table 6.21). There is wide variation in the efficiency with which feed carotene is utilized for the production of milk carotene and vitamin A.

The vitamin D content of milk as secreted varies directly with the amount of vitamin D fed, and the amount of direct sunlight to which the cow is exposed. Lampert (1947) reviewed the relationship between vitamin D in feed and vitamin D in milk, and reported that irradiated yeast or irradiated ergosterol raised the vitamin D content of milk from initial values of 4 to 60 IU per qt to values as high as 430 IU per qt. The efficiency of transfer is low.

TABLE 6.21

CAROTENE AND VITAMIN A CONTENT OF BUTTER FROM COWS OF DIFFERENT BREEDS ON VARIOUS FEEDS

Diet	Guernseys Carotene	Vitamin A	Holsteins Carotene	Vitamin A	Jerseys Carotene	Vitamin A	Brown Swiss Carotene	Vitamin A
			(μg/Gm Butter)					
Grain, corn silage, alfalfa, hay	6.0	2.9	3.5	4.0	3.7	3.1	3.4	5.8
Grain, alfalfa silage, timothy hay	9.5	8.5	5.4	10.0	7.1	7.7	5.0	9.7
Pasture	17.0	8.7	6.6	15.0	10.7	10.2	9.8	12.6

Source: Beeson (1934).

Antibiotics may be present in milk following their use in mastitis therapy (Randall 1956). The antibiotic content is very low, and generally occurs after therapeutic use rather than from feed use. Randall (1956) reported finding antibiotic activity in 55 of 474 qt examined, penicillin in all but 1 instance. The amount found varied from 0.0003 to 0.08 unit per milliliter. Even these amounts are sufficient to interfere with cheese making.

BIBLIOGRAPHY

ALBRITTON, E. C. 1954. Standard Values in Nutrition and Metabolism. W. B. Saunders Co., Philadelphia.
ALLAWAY, W. H. 1973. Selenium in the food chain. Cornell Vet. *63*, 151-170.
ALTMAN, P. L., and DITTMER, D. S. 1962. Growth. Federation Am. Soc. Exptl. Biol., Washington, D.C.
ANON. 1970. Certification revision: New Standards proposed. Natl. Hog Farmer, Jan., 34.
ANON. 1973. I.N.P.A. poultry research serves the industry and the consumer. World Poultry Sci. J. *29*, 175-180.
ASHBROOK, F. G. 1951. Crops in peace and war. In Agriculture Yearbook, U.S. Dept. Agr.
ASHTON, G. C. et al. 1955. Protein levels and carcass leanness. J. Animal Sci. *14*, 82-93.
BAKER, F. H., and ARTHAUD, V. H. 1972. Use of hormones or hormone active agents in production of slaughter bulls. J. Animal Sci. *35*, 752-754.
BARBELLA, N. G., HANKINS, O. G., and ALEXANDER, L. M. 1936. The influence of retarded growth in lambs on flavor and other characteristics of the meat. Proc. Am. Soc. Animal Production, 29th Ann. Mtg., 289-294.
BARTOV, I., and BORNSTEIN, S. 1967. Studies on egg yolk pigmentation. Poultry Sci. *46*, 796-805.
BAUERNFIND, J. C., and NORRIS, L. C. 1939. Effect of the level of dietary riboflavium upon the quantity stored in eggs. Poultry Sci. *18*, 400.
BEARSE, G. E., and MILLER, W. H. 1937. The effect of varying levels of vitamin A in the hen ration on the vitamin A content of the egg yolk, on hatchability and on chick livability. Poultry Sci. *16*, 39-43.

BEESON, K. C. 1941. The mineral composition of crops with particular reference to the soils in which they were grown. USDA Misc. Publ. *369*.
BEESON, W. M. 1934. Influence of breed and ration on the vitamin A and carotene content of milk. Proc. Am. Soc. Animal Production, 27th Ann. Mtg., 54-56.
BENNION, N., and WARREN, D. C. 1933. Temperature and its effect on egg size in the domestic fowl. Poultry Sci. *12*, 69-82.
BETHKE, R. M., RECORD, P. R., WILDER, O. H. M., and KICK, C. H. 1936. Effect of different sources of vitamin D on the laying bird. Poultry Sci. *15*, 336-344.
BOND, J. et al. 1972. Influence of breed and plane of nutrition on performance of dairy, dual-purpose and beef steers. J. Animal Sci. *34*, 1046-1053.
BOOMGAARDT, J., and BAKER, D. H. 1973. Tryptophan requirements of growing chicks. J. Animal Sci. *33*, 595-599.
BOOREN, A., FIELD, R. A., and KUNSMAN, J. E., JR. 1973. Carbonyl and fatty acid analyses of antelope and beef fat. J. Food Sci. *38*, 63-65.
BOTKIN, M. P. et al. 1969. Heritability of carcass traits in lambs. J. Animal Sci. *29*, 251-255.
BRAGG, D. B., SIM, J. S., and HODGSON, G. C. 1973. Influence of dietary energy source on performance and fatty liver syndrome in white leghorn laying hens. Poultry Sci. *52*, 736-740.
BRATZLER, J. W., LOOSLI, J. K., KRUKOWSKY, V. N., and MAYNARD, L. A. 1950. Effect of the dietary level of tocopherols on their metabolism in swine. J. Nutr. *42*, 50-70.
BRAUDE, R., and RYDER, K. 1973. Copper levels in diets for growing pigs. J. Agr. Sci. *80*, 489-493.
BREKKE, C. J., and WELLINGTON, G. H. 1969. Meat yield from Holstein veal calves. J. Animal Sci. *29*, 6-10.
BROADBENT, M., and BEAN, H. W. 1952. The yield of edible meat from turkeys, ducklings and different market classes of chickens. Poultry Sci. *31*, 447-450.
BROQUIST, H. P., and KOHLER, A. R. 1953. Studies of the antibiotic potency in the meat of animals fed chlortetracycline. *In* Antibiotics Annual. Medical Encyclopedia, New York.
BURRIS, M. J., BOGART, R., and OLIVER, A. W. 1953. Effect of male hormones on beef cattle. J. Animal Sci. *12*, 740-746.
BURROUGHS, W., et al. 1955. Gains and feed efficiency of cattle fed diethylstilbestrol. Iowa State Coll. Agr. Expt. Sta. A. H. Leaflet *201*.
CABELL, C. A., ELLIS, N. R., and MADSEN, L. L. 1943. Vitamin A activity of lean meat and fat from cattle fed various levels of carotene. Food Res. *8*, 496-501.
CARY, C. A. 1939. Food and life. *In* Agriculture Yearbook, U.S. Dept. Agr.
CECIL, H. C. et al. 1973. Changes in egg shell quality and pesticide content of laying hens and pullets fed DDT. Poultry Sci. *52*, 646-653.
CHU, T. K., and KUMMEROW, F. A. 1950. The deposition of linolenic acid in chickens fed linseed oil. Poultry Sci. *29*, 846-851.
CLARENBURG, R., CHUNG, I. A. K., and WAKEFIELD, L. N. 1971. Reducing egg cholesterol level by including emulsified sitosterol in standard chicken diet. J. Nutr. *101*, 289-297.
CLAUSEN, H. J. 1956. Influence of antibiotics on carcass quality of pigs. *In* Antibiotics in Agriculture. Natl. Acad. Sci.—Natl. Res. Council Publ. *397*.
CRAMPTON, E. W. 1954. Improvement of bacon carcass quality by fibrous feeds in finishing ration. J. Animal Sci. *13*, 327-331.
CRAMPTON, E. W., ASHTON, G. C., and LLOYD, L. E. 1954. Effect of restricting feed intake of hogs on carcass quality. J. Animal Sci. *13*, 321-326.
CUNDIFF, L. V., GREGORY, K. E., KOCH, R. M., and DICKERSON, G. E. 1969. Genetic variation in total and differential growth of carcass components in beef cattle. J. Animal Sci. *29*, 233-244.

DAVIS, R. E. 1956. Personal communication. 5609 Kennedy Street, East Riverdale, Maryland.
DEAN, W. E. 1973. Recent findings in duck nutrition. Proc. Cornell Nutr. Conf., Cornell Univ., Ithaca, N.Y. Oct. 31-Nov. 2, 1972.
DeFREMERY, D., and POOL, M. F. 1960. Biochemistry of chicken muscle as related to rigor mortis and tenderization. Food Res. 25, 73-87.
DENTON, C. A., KELLOGG, W. L., and BIRD, H. R. 1947. The effect of diet, age and sex on the nicotinic acid content of the tissues of chickens. Poultry Sci. 26, 299-303.
DETWILER, R. W., ANDREWS, F. N., and BOHREN, B. B. 1950. The influence of thiouracil and stilbestrol on broiler quality. Poultry Sci. 29, 513-519.
DEVANEY, G. M., MUNSELL, H. E., and TITUS, H. W. 1936. Further studies on the effect of sources of vitamin D in the diet of the chicken on storage of the antirachitic factor. Poultry Sci. 15, 149-153.
DICK, A. T. 1956. Molybdenum in animal nutrition. Soil Sci. 81, 229-236.
DICKERSON, G. E. 1947. Composition of hog carcasses as influenced by heritable differences in rate and economy of gain. Iowa Agr. Expt. Sta. Res. Bull. 354.
DICKERSON, G. E., GLIMP, H. A., TUMA, H. J., and GREGORY, K. E. 1972. Genetic resources for efficient meat production in sheep. J. Animal Sci. 34, 940-951.
DINIUS, D. A., BIMSFIELD, T. H., and WILLIAMS, E. E. 1973. Effect of subclinical lead intake on calves. J. Animal Sci. 37, 169-173.
DOCKERTY, T. R. et al. 1973. Carcass development in beef cattle subsequent to interrupted growth. J. Animal Sci. 36, 1057-1062.
DRAPER, C. E. et al. 1952. The transfer of DDT from the feed to eggs and body tissues of White Leghorn hens. Poultry Sci. 31, 388-393.
DRYDEN, F. B., and MARCHELLO, J. A. 1973. Influence of dietary fats on carcass lipid composition in the bovine. J. Animal Sci. 37, 33-39.
EDWARDS, R. A., KING, J. W. B., and YOUSEF, I. 1973. A note on the genetic variation in the fatty acid composition of cow's milk. Animal Prod. 16, 309-310.
ELLIS, N. R., and HANKINS, O. G. 1925. Soft pork studies. J. Biol. Chem. 66, 101-122.
ESPE, D. 1946. Secretion of Milk, 3rd Edition. Collegiate Press, Ames, Iowa.
FARNSWORTH, G., JR., and NORDSKOG, A. W. 1955. Breeding for egg quality. Poultry Sci. 34, 16-26.
FDA. 1972A. Diethylstilbestrol: Order denying a hearing and withdrawing approval. Federal Register 37, No. 151, 15717.
FDA. 1972B. Diethylstilbestrol: Revocation of all provisions for use in animal feed. Federal Register 37, No. 238, 26307.
FDA. 1973A. Diethylstilbestrol: Order denying a hearing and withdrawing approval of new animal drug applications for diethylstilbestrol implants. Federal Register 38, No. 81, 10185.
FDA. 1973B. Diethylstilbestrol. Health, Education, Welfare News Release 73-17, Apr. 25.
FRAPS, G. S. 1943. Relation of the protein fat, fat, and energy of the ration to the composition of chickens. Poultry Sci. 22, 421-424.
FROST, D. V., and SPRUTH, H. C. 1956. Arsenicals in feeds. In Symposium on Medicated Foods. Medical Encyclopedia, New York.
GILSTER, K. E., and WAHLSTROM, R. C. 1973. Protein levels for swine fed to heavy weight. J. Animal Sci. 36, 888-893.
GLIMP, H. A. et al. 1971. Effect of sex condition on growth and carcass traits of male Hereford and Angus cattle. J. Animal Sci. 33, 1242-1247.
GOLDWATER, L. J. 1964. Occupational exposure to mercury. J. Roy. Inst. Public Health Hyg., Dec.

GOODMAN, B. L. 1973. Heritabilities and correlations of body weight and dressing percentage in broilers. Poultry Sci. 52, 379-381.
HALE, O. W., and SOUTHWELL, B. L. 1967. Differences in swine performance and carcass characteristics because of dietary protein level, sex and breed. J. Animal Sci. 26, 341-344.
HAMAN, K., PHILLIPS, P. H., and HALPIN, J. G. 1936. The distribution and storage of fluorine in the tissues of the laying hen. Poultry Sci. 15, 154-157.
HAND, D. B., and SHARP, P. F. 1939. The riboflavin content of cow's milk. J. Dairy Sci. 22, 779-783.
HARTLEY, W. J. 1967. Levels of selenium in animal tissues and methods of selenium administration. In Symposium: Selenium in Biomedicine. O. H. Muth et al. (Editors). Avi Publishing Co., Westport, Conn.
HAWKINS, D. R., HANDERSON, H. E., and NEWLAND, H. W. 1972. Melengesterol acetate and feedlot performance. J. Animal Sci. 35, 1257-1262.
HAYES, P. L., and MARQUIS, W. B. 1973. Eviscerated yield, component parts, and meat, skin and bone ratios in the chicken broiler. Poultry Sci. 52, 718-722.
HAZEL, L. N., and TERRILL, C. E. 1946. Heritability of weanling traits in range sheep. J. Animal Sci. 5, 371-377.
HENDRICKS, H. B. et al. 1971. Relation of porcine muscle fiber type and size to post-mortem shortening. J. Animal Sci. 32, 57-61.
HESS, G. S., and YOUNG, J. W. 1972. Preventing and alleviating milk fat depression by feeding 1-3 butanediol to cows. J. Dairy Sci. 55, 1097-1105.
HILL, F. W. 1953. Thyroprotein and antithyroid drugs in poultry feeding. In Hormonal Relationships and Applications in the Production of Meat, Milk, and Eggs. Natl. Acad. Sci.—Natl. Res. Council Publ. 266.
HILTON, M. W., LAWRIE, R. A., RATCLIFFE, P. W., and WAYNE, N. 1972. Effect of preslaughter adrenalin injection on muscle. Metabolites and meat quality of pigs. J. Food Technol. 7, 443-453.
HINER, R. L., and BOND, J. 1971. Growth of muscle and fat in beef steers from 6 to 36 months of age. J. Animal Sci. 32, 225-232.
HINER, R. L., THORNTON, J. W., and ALSMEYER, R. H. 1965. The palability and quality of pork as influenced by breed and fatness. J. Food Sci. 30, 550-559.
HOEFER, J. A., and GALLUP, W. D. 1947. Maintaining vitamin A level in lambs. J. Animal Sci. 6, 325-333.
HOEKSTRA, W. G., LEWIS, P. K., JR., PHILLIPS, P. H., and GRUMMER, R. H. 1956. The relationship of parakeratosis, calcium and zinc to zinc content of swine body components. J. Animal Sci. 15, 752-764.
HOOVEN, N. W. et al. 1972. Influence of breed and plane of nutrition on the performance of dairy, dual-purpose and beef steers. J. Animal Sci. 34, 1037-1045.
HUBER, J. T., PRICE, N. O., and ENGEL, R. W. 1971. Response of lactating dairy cows to high levels of dietary molybdenum. J. Animal Sci. 32, 364-367.
HUGHES, J. S., and PAYNE, L. F. 1937. The relation of the carotenoid pigments of the feed to the carotenoid pigments of egg yolk. Poultry Sci. 16, 135-138.
INGRAM, G. R., CRAVENS, C. W., ELVEHJES, C. A., and HALPIN, J. G. 1950. Relation of tryptophan and lysine to egg production, hatchability and composition of the protein of hens' eggs. Poultry Sci. 29, 793-803.
IVY, C. A., and NESHEIM, M. C. 1973. Factors influencing the liver fat content of laying hens. Poultry Sci. 52, 281-291.
JACOBS, J. A. et al. 1972. Effects of weight and castration on lamb carcass composition and quality. J. Animal Sci. 35, 926-930.
JUKES, T. H. 1955. Antibiotics in Nutrition. Medical Encyclopedia, New York.

JULL, M. G., PHILLIPS, R. E., and WILLIAMS, C. S. 1943. Meat contributed by breast, humeri and legs of fryers in relation to shank length. U.S. Egg Poultry Mag. *49*, 364-365.

KHAN, A. W., and LENTZ, C. P. 1973. Influence of antemortem glycolypids and dephosphorylation on high energy phosphates on beef aging and tenderness. J. Food Sci. *38*, 56-58.

KLOSE, A. A., NECCHI, E. P., HANSON, H. L., and LINEWEAVER, H. 1951. The role of dietary fat in the quality of fresh and frozen storage turkeys. J. Am. Oil Chemists' Soc. *28*, 162-164.

LAFLAMME, L. F., and BURGESS, T. D. 1973. Effect of castration, ration and hormone implants on the performance of finishing cattle. J. Animal Sci. *36*, 762-766.

LAMPERT, L. M. 1947. Milk and Dairy Products. Chemical Publishing Co., New York.

LEONG, K. C., SUNDE, M. L., BIRD, H. R., and ELVEHJEM, C. A. 1955. Effect of energy: protein ratios in growth rate, efficiency, feathering and fat deposition in chickens. Poultry Sci. *34*, 1206-1207.

LIBBY, D. A., MEITES, J., and SCHAIBLE, P. J. 1955. Growth hormone effects in chickens. Poultry Sci. *34*, 1329-1331.

LINK, R. P., BRUCE, W. N., and DECKER, G. C. 1963. The effect of chlorinated hydrocarbon insecticides on dairy cattle. Ann. N.Y. Acad. Sci. *111*, 788-792.

LITTLEFIELD, L. H., BILETNER, J. T., and GOFF, O. E. 1973. The effect of feeding laying hens various levels of cow manure on the pigmentation of egg yolks. Poultry Sci. *52*, 179-181.

LYONS, M., and INSKO, W. W., JR. 1937. Chondrodystrophy in the chick embryo produced by manganese deficiency in the diet of the hen. Kentucky Agr. Expt. Sta. Bull. *371*.

MACLEOD, G. K., WOOD, A. S., and YAO, Y. T. 1972. Influence of dietary fat on rumen fatty acids, plasma lipids and milk composition in the cow. J. Dairy Sci. *55*, 446-453.

MARSDEN, S. J., and BIRD, H. R. 1947. Effects of DDT on growing turkeys. Poultry Sci. *26*, 3-6.

MAW, A. J. G. 1954. Inherited riboflavin deficiency in chicken eggs. Poultry Sci. *33*, 216-217.

MAYFIELD, H. L., ROEHM, R. R., and BEECKLER, A. F. 1955. Riboflavin and thiamine content of eggs. Poultry Sci. *34*, 1106-1111.

MAYNARD, L. A., LOOSLI, J. K., and McCAY, C. M. 1940. Further studies on the influence of fat intake on milk and fat secretion. Proc. Am. Soc. Animal Production, 33rd Ann. Mtg., Nov. 29-Dec. 1, 340-344.

McFARLAND, L. Z., WINGET, C. M., WILSON, W. O., and JOHNSON, C. M. 1970. The role of selenium in neural physiology of avian species. Poultry Sci. *49*, 216-221.

McMEEKAN, C. P. 1940. Growth and development in the pig with special reference to carcass characteristics. J. Agr. Sci. *30*, 276-343.

MORRISON, F. B. 1948. Feeds and Feeding. 21st Edition. Morrison Publishing Co., Ithaca, N.Y.

MORRISON, A. B., and MUNRO, I. C. 1969. The use of drugs in animal feeds. Natl. Acad. Sci.—Natl. Res. Council Publ. *1679*.

MOULTON, C. R., TROWBRIDGE, P. F., and HAIGH, L. D. 1922. Studies in animal nutrition. II. Changes in proportion of carcass and offal on different planes of nutrition. Missouri Agr. Expt. Sta. Bull. *54*.

MOXON, A. L., RHIAN, M. A., ANDERSON, H. D., and OLSON, O. E. 1944. Growth of steers on seleniferous range. J. Animal Sci. *3*, 299-309.

MURPHY, G. R., RHEA, U. S., and PEELER, J. T. 1972. Copper, iron, manganese, strontium and zinc content of market milk. J. Dairy Sci. *55*, 1666-1670.

NEATHEY, N. W., MILLER, W. J., BLACKMAN, D. M., and GERITY, R. P. 1973. Performance and milk zinc from low-zinc intake in Holstein cows. J. Dairy Sci. 56, 212-215.
NELSON, N. et al. 1971. Hazards of mercury. Environ. Res. 4, 1-69.
NIR, I. 1972. Modification of blood plasma concentration as related to the degree of hepatic steatosis in the forced-fed goose. Poultry Sci. 51, 2044-2049.
OLSON, D. W., SUNDE, M. L., and BIRD, H. R. 1972. The effect of temperature on metabolizable energy determination and utilization by the growing chick. Poultry Sci. 56, 1915-1922.
ORR, M. L. 1969. Pantothenic acid, vitamin B-6 and vitamin B-12 in foods. USDA Home Econ. Res. Rept. 36.
PALSSON, H. 1940. Meat quality in the sheep with special reference to Scottish breeds and crosses. J. Agr. Sci. 30, 1-82.
PAUL, P. C., TORTEN, J., and SPIERLOCK, G. M. 1961. Eating quality of lamb. II. Effect of preslaughter nutrition. Food Technol. 18, (10) 121-130.
PECOT, R. K., JAEGER, C. M., and WATT, B. K. 1965. Proximate composition of beef from carcass to cooked meat. USDA Home Econ. Res. Rept. 31.
PENCE, J. W., MILLER, R. C., DUTCHER, R. A., and ZIEGLER, P. T. 1945. Storage of thiamine in pork muscle. J. Animal Sci. 4, 141-145.
PEOPLES, S. A. 1963. Arsenic toxicity in cattle. Ann. N.Y. Acad. Sci. 111, 644-649.
PETERSON, C. F., LAMPMAN, C. E., and STAMBERG, O. E. 1947. Effect of riboflavin intake on egg production and riboflavin content of eggs. Poultry Sci. 26, 180-186.
PLIMPTON, R. F. et al. 1971. Palatability, composition and quality of beef, pork. J. Animal Sci. 32, 51-56.
PLOWMAN, R. D. et al. 1972. Milk fat with increased polyunsaturated fatty acids. J. Dairy Sci. 55, 204-210.
POMEROY, R. W. 1968. Identification of quality factors in meat. In Proceedings 2nd World Conference on Animal Production. Bruce Publishing Co., St. Paul.
PORTER, G. H. et al. 1953. Relative value for milk production of field-cured and field-baled, artificially dried chopped alfalfa when fed as the sole source of roughage to dairy cattle. J. Dairy Sci. 36, 1140-1149.
POWELL, E. B. 1938. One cause of fat variation in milk. Proc. Am. Soc. Animal Production, 31st Ann. Mtg., 40-47.
RANDALL, W. A. 1956. Antibiotic residues. In Antibiotics in Agriculture. Natl. Acad Sci.—Natl. Res. Council Publ. 397.
RAY, D. E., HALE, W. H., and MARCHELLO, J. A. 1969. Influence of season, sex, and hormonal growth stimulant on feedlot performance of beef cattle. J. Animal Sci. 29, 490-495.
REGAN, W. M., and RICHARDSON, M. A. 1938. Reaction of the dairy cow to changes in environmental temperature. J. Dairy Sci. 21, 73-79.
REISER, R. 1950. Fatty acid changes in egg yolk of hens on a fat free and a cotton seed oil ration. J. Nutr. 40, 429-440.
ROLAND, D. A., SR., and EDWARDS, M. A., JR. 1972. Effect of linoleic acid reserves on essential fatty acid deficiency of the chick. Poultry Sci. 51, 382-389.
RUBIN, M., and BIRD, H. R. 1941. Some experiments on the physiology of vitamin A storage in the chick. Poultry Sci. 20, 291-297.
RUBIN, M., BIRD, H. R., GREEN, N., and CARTER, R. H. 1947. Toxicity of DDT to laying hens. Poultry Sci. 26, 410-413.
RUMSEY, T. S., OLTJEN, R. R., BOVARD, K. P., and PRIODE, B. M. 1972. Depot fat composition in beef cattle. J. Animal Sci. 35, 1069-1075.
SCOTT, T. W., et al. 1970. Production of polyunsaturated milk fat in domestic ruminants. Australian J. Sci. 32, 291-293.

SCRIMSHAW, N. S., HUTT, F. B., SCRIMSHAW, M. W., and SULLIVAN, C. R. 1945. The effect of genetic variation in the fowl on the thiamine content of the egg. J. Nutr. *30*, 375-383.
SIM, J. S., BRAGG, D. B., and HODGSON, G. C. 1973. Effect of dietary animal tallow and vegetable oil on fatty acid composition of egg yolk, adipose tissue and liver of laying hens. Poultry Sci. *52*, 51-57.
SNELL, E. E., ALIME, E., COUCH, J. R., and PEARSON, P. B. 1941. The effect of diet on the panthothenic acid content of eggs. J. Nutr. *21*, 201-205.
SPALDING, J. F. 1972. Pesticide and heavy metal residues. In Proc. Meat Inst. Res. Conf., Am. Meat Inst. Found. Chicago, Mar. 23-24, 1972.
STADELMAN, W. J. 1973. Record of some chemical residues in poultry products. Bioscience *23*, 424-428.
STADELMAN, W. J. et al. 1965. Persistence of chlorinated hydrocarbon insecticide residues in chicken tissues and eggs. Poultry Sci. *44*, 435-437.
STEENBOCK, H., HART, E. B., HOPPERT, C. A., and BLACK, A. 1925. Fat soluble vitamin. XVI. The amount in milk and its increase by direct irradiation or by irradiation of the animal. J. Biol. Chem. *66*, 441-449.
STORRY, J. E., HALL, A. J., and JOHNSON, U. W. 1973. The effects of increasing amounts of dietary tallow on milk-fat secretion in the cow. J. Dairy Res. *40*, 293-299.
STRICKLAND, N. C., and GOLDSPINK, G. 1973. A possible indicator muscle for the fiber content and growth characteristics of porcine muscle. Animal Prod. *16*, 135-146.
SWANN, M. 1969. Report Joint Committee on Use of Antibiotics in Animal Husbandry and Veterinary Medicine. HM Stationery Office, London, England.
THOMAS, O. P., and COMBS, G. F. 1967. Relationship between serum protein level and body composition in the chick. J. Nutr. *91*, 468-472.
TITUS, H. W., BYERLY, T. C., and ELLIS, N. R. 1933. Effect of diet on egg composition. J. Nutr. *6*, 127-138.
TURK, D. E. 1965. Effect of diet on the tissue zinc distribution and reproduction in the fowl. Poultry Sci. *44*, 122-126.
USDA. 1965. Official United States Standards for Grades of Carcass Beef. U.S. Dept. Agr. SRA-CMS *99*.
USDA. 1969. Improvements in Grades of hogs slaughtered from 1960-61 to 1967-68. U.S. Dept Agr. ERS Marketing Res. Rept. *849*.
USDA. 1972. Meat and lard production and consumption in the United States, 1930-71. U.S. Dept. Agr., Agr. Statist., p. 418.
USDA. 1973. USDA reports on residues found in meat and poultry. U.S. Dept. Agr. Press Release *2475*, Aug. 10.
WAIBEL, P. E. 1955. Effect of dietary protein level and added tallow on growth and carcass composition of chicks. Poultry Sci. *34*, 1226.
WARWICK, E. J. 1968. Effective performance recording in beef cattle. In Proceedings 2nd World Conference on Animal Production. Bruce Publishing Co., St. Paul, Minn.
WASSERMAN, A., and TALLEY, E. 1968. Organoleptic identification of roasted beef, veal, lamb and pork as affected by fat. J. Food Sci. *33*, 219-223.
WATT, B. K., and MERRILL, A. L. 1950. Composition of Foods—Raw, Processed, Prepared. U.S. Dept. Agr. Handbook *8*.
WATTS, B. M. 1954. Oxidative rancidity and discoloration in meat. In Advances in Food Research, Vol. 50. Academic Press, New York.
WEISS, J. F., JOHNSON, R. M., and NABER, E. C. 1967. Effect of some dietary factors on cholesterol concentration in the egg and plasma of the hen. J. Nutr. *91*, 119-128.
WESTÖÖ, G. 1969. Mercury and methyl mercury level in some animal food products. Var Foda *7*, 137-154.

WESTÖÖ, G. 1973. Methyl mercury as percentage of total mercury in flesh and viscera of salmon and sea trout of various ages. Science *181*, 567-568.
WILDER, O. H. M., BETHKE, R. M., and RECORD, P. R. 1933. The iodine content of the hen's eggs as affected by the ration. J. Nutr. *6*, 407-412.
WINCHESTER, E. F., and HOWE, P. E. 1955. Relative effects of continuous and interrupted growth on beef steers. U.S. Dept. Agr. Tech. Bull. *1108*.
WRIGHT, F. C., PALMER, J. S., and RINER, J. C. 1973. Accumulation of mercury in tissues of cattle, sheep and chickens given the mercurial fungicide, Panogen 150, orally. J. Agr. Food Chem. *21*, 414-416.
WRIGHT, F. C., PALMER, J. S., and RINER, J. C. 1973. Retention of mercury in tissues of cattle and sheep given oral doses of a mercurial fungicide, Ceresan M. J. Agr. Food Chem. *21*, 614-618.
WRIGHT, P. A., DEYSHER, E. F., and CARY, C. A. 1939. Food and Life. *In* Agriculture Yearbook, U.S. Dept. Agr.
ZIEGLER, J. H., WILSON, I. I., and COLE, J. H. 1971. Comparison of certain carcass traits of several breeds and crosses of cattle. J. Animal Sci. *32*, 446-450.

CHAPTER 7

Effects of Harvesting and Handling on the Composition of Foods

John M. Krochta
and
Bernard Feinberg

PART 1
Effects of Harvesting and Handling on Fruits and Vegetables

This chapter is concerned with the changes that take place during the time interval between the separation of a fruit or vegetable from the parent plant or from the soil, and its arrival as a fresh plant product to the end consumer or its arrival at a food processing plant. When vegetables and fruits such as potatoes, peas, bananas, and apples are harvested and stored, chemical and biochemical changes continue to take place. Quality will usually show a gradual decline concurrently with transpiration, respiration, and a number of other biochemical and physical changes. Eventually, through the action of enzyme activity and/or spoilage microorganisms, the plant product reaches a point where it is not acceptable to the consumer or processor.

During the period of growth and maturation, fruits and vegetables are highly dependent on photosynthesis and absorption of water and minerals from the parent plant. However, once detached they become independent units in which respiratory processes and transpiration now play a proportionately major role (Eskin et al. 1971).

It may appear that portions of this chapter are more concerned with the economic loss of fruits and vegetables than with the loss of various nutrients. Kramer (1965) points out that discussions of quality related to economic loss are almost always concerned with factors of a primarily aesthetic nature, i.e., those that are evaluated by sight, touch, taste, or smell. However, this chapter will show the direct relationship between market quality and nutritional value. Harvesting and handling techniques are developed to ensure delivery of fruits and vegetables to consumers in as near fresh condition as possible. When this objective is met, the nutrients, as well as other quality factors, have been protected.

The main contribution of fruits and their processed products to the nutrition of mankind is undoubtedly their supply of L-ascorbic acid (vitamin C). Fruits, together with vegetables, are the main sources from which all the primates derive this vitamin. Very little definite information is available as to the role ascorbic acid plays in plant metabolism (Mapson 1970). Retention of vitamin C is often used as an indication of all nutrient retention (Tressler and Evers 1957). In his review of nutrient stability, Nelson (1972) cites the belief that other nutrients, being more stable, are less affected by any processing than vitamin C. Thus, discussions in this chapter will deal mainly with vitamin C.

Several fruits and vegetables are good sources of beta carotene (provitamin A) also. These include apricots, peaches, melons, cherries, carrots, leafy green vegetables, and sweet potatoes. Other fruits contain moderate amounts of pantothenic acid (apricots, gooseberries, figs) and some contain biotin.

Kelly (1972) points out that some major contributions of fruit and vegetable crops to the human diet—variety, flavor, color—are not easily evaluated. Consumers ignore the fact that apples are poor sources of vitamins A, B, and C, as well as protein, for the pleasure of eating a big, red, crunchy apple.

MATURITY

Just as a human progresses through various stages of growth, from birth to old age, a fruit or vegetable passes through similar phases. These are frequently marked by differences in terminology but the names suggested by Gortner et al. (1967)—development, prematuration, maturation, ripening, and senescence—are accepted by most plant physiologists. These stages are illustrated in Fig. 7.1. *Development* starts with the formation of the edible part: the setting of fruit, the emergence of a seedling, the swelling of a root, tuber, or bulb, or the elongation of a stalk or petiole. Development ceases with the termination of a desirable or natural enlargement, or with a change in the growth pattern of the edible part. Development occurs largely before harvest and includes prematuration and part of the maturation. *Prematuration* starts with development and lasts until the beginning of edible use. *Deterioration*, inevitably associated with the life span of a fresh fruit or vegetable, must be distinguished from senescence. *Senescence* applies only to normal physiological changes such as changes in flavor, composition, texture, color, or mode of growth. Deterioration, on the other hand, encompasses all aspects of quality loss: senescence; physiological disorders; diseases induced by fungi, bacteria, or viruses; wilting; freezing; or the aftermath of mechanical injury. Furthermore, deterioration can start any time during growth and continue beyond the end of usefulness of the

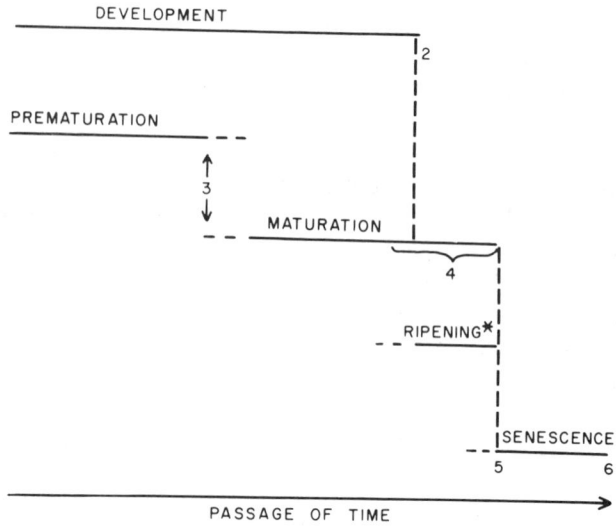

FIG. 7.1. STAGES DURING LIFE SPAN OF FRESH VEGETABLES AND FRUITS

Ryall and Lipton (1972)

(1) Initiation of edible part. (2) Termination of natural or desirable growth in size or type. (3) Start of period of usefulness, but too immature for most users. (4) Period of maximum usefulness. (5) Degradative changes become predominant. (6) End of usefulness for human consumption.

foodstuff for human consumption. The *ripening* of fruits may be defined as the sequence of changes in color, flavor, and texture which lead to the state at which the fruit is acceptable to eat. This does not necessarily mean that this is a fixed physiological state—it can and does vary from one type of fruit to another (Rhodes 1970).

Apples and pears will ripen while still attached to the tree. However, pears ripened on the tree tend to become "mealy." With varieties of apples, the best eating quality is often obtained by tree ripening. Ripening of apples on the tree is, however, seldom encountered in practice. Storm and bird damage would make this practice uneconomic in many apple growing countries. In addition, it has been found that ripe fruit has a short storage and shelf-life (Hulme and Rhodes 1971). Wilkinson (1970) points out that the stage at which a fruit is picked has a profound effect upon the possibility of eventual storage disorders. If the fruit is picked exceptionally early, development of its flavor, nutrients, etc., can be interrupted. On the other hand, a fruit which has become over-ripe on the plant has also lost food value. There are stages of development between these extremes which are associated with susceptibility to disorders. Scald is more likely to develop when apples are picked too early (Fidler 1959 as cited in Hulme 1970), while brown heart is more likely to develop if apples are harvested too late (Dewey 1962 as cited in Hulme 1970). Research on transport and storage has shown that the best results are obtained if the fruit is picked before the onset of ripening, and this has become general commercial practice. Most storage practices are, in fact, aimed at delaying or prolonging the first stages of ripening (the respiration climacteric). The above also applies to other fruits such as peaches, apricots, and pears. Pears for canning are usually held in cold storage, about $-0.6°C$ for Bartletts, and then ripened at about $18°C$ for 10–14 days. This practice results in uniform ripening. Bartlett pears picked at optimum maturity can be stored for about 3 months at $-1.1°$ to $-0.6°C$ (USDA 1968).

Ethylene gas (C_2H_4) is produced by most if not all plant materials, and may have important beneficial or detrimental effects on fruits and vegetables during postharvest handling. The compound induces ripening in fruits and senescence (loss of green color, shedding of leaves, etc.) in other plant tissues. For the gas to exert an effect a certain threshold concentration must accumulate in the internal atmosphere of the tissues, and temperature of produce must be above a minimum level. Neither threshold concentration nor minimum temperature requirements for ethylene activity are well defined. The rate of production and action of ethylene are temperature-dependent. Rapid cooling and good temperature management are desirable to limit the effects of the gas on ripening and senescence for most commodities. The maximum effect occurs when produce temperature ranges from $17°$ to $21°C$. Ethylene gas is widely used to initiate commercial ripening of bananas and melons, and for the degreening of citrus (Mitchell et al. 1972).

A large number of fruits exhibit a sudden sharp rise in respiratory activity following harvesting. This is commonly called the "climacteric" rise in respiration. Fruits which do not show this rise are classified as "nonclimacteric" (Eskin et al. 1971). With the exception of avocado, climacteric fruits will normally ripen on the tree; however, as mentioned above, they are usually harvested prior to the onset of the climacteric and stored under carefully controlled conditions to suppress the ripening process. When required for marketing or processing, ripening is then induced, thus eliminating losses which otherwise would have resulted had they been allowed to ripen and enter the deterioration processes following senescence. Nonclimacteric fruits, such as oranges or strawberries, are normally allowed to ripen on the parent plant prior to harvesting (Eskin et al. 1971).

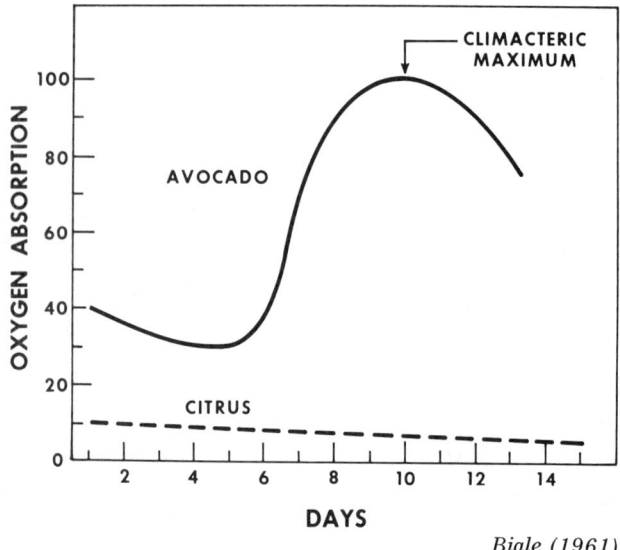

FIG. 7.2. PATTERNS OF RESPIRATION—CITRUS VERSUS AVOCADO

Biale (1961)

Bananas are never allowed to ripen on the tree. Fruit so ripened is subject to insect damage and rot, and the edible quality of tree-ripened fruit is inferior to that of fruit ripened off the tree. Bananas are therefore harvested green, and ripened under controlled temperature and humidity. The most characteristic change during ripening is a marked decrease in starch, with a corresponding increase in sugars (von Loesecke 1950). In the green fruit, sugars are present in small concentrations, i.e., about 1–2% of the fresh pulp. This increases to 15–20% in fully ripe fruit. The starch concentration goes

TABLE 7.1

CLASSIFICATION OF VARIOUS FRUITS ACCORDING TO THEIR RESPIRATORY ACTIVITY

Climacteric Fruits	Nonclimacteric Fruits
Apples	Orange
Pears	Grapefruit
Bananas	Lemon
Apricots	Pineapple
Avocado	Melon
Passion fruit	Strawberry

Source: Eskin et al. (1971).

EFFECTS OF HARVESTING AND HANDLING ON COMPOSITION 103

down concurrently, dropping from about 20% in the green fruit to 1-2% in the ripe. In one study, Harris and Poland (1939) found that ascorbic acid content doubled when bananas were ripened from the green peel-hard pulp stage to the yellow-flecked-with-brown stage.

TABLE 7.2

ASCORBIC ACID CONTENT OF COSTA RICA BANANAS AT VARIOUS STAGES OF RIPENESS

Stage of Ripeness	No. of Analysis	Range of Value (Mg/Gm)	Avg Value (Mg/Gm)	Change (%)
Green peel, hard pulp	10	0.058-0.049	0.053	—
Green peel, some yellow	15	0.066-0.053	0.058	9.4
Yellow more dominant in peel than green	15	0.144-0.038	0.063	18.9
Yellow peel, green tip	20	0.130-0.052	0.088	66.0
Completely yellow peel	20	0.132-0.056	0.091	71.7
Yellow, flecked with brown	18	0.143-0.089	0.111	109.4
Black peel, overripe	7	0.054-0.010	0.032	-39.6

Source: Harris and Poland (1939).

Tomatoes destined for the fresh market are usually harvested at a stage of development known as "mature green" and shipped to urban markets where they are ripened. It is difficult to ship mature ripe tomatoes by truck or train or to hold them for more than two days. Pantos and Markakis (1973), in a study of two tomato cultivars, concluded that table-ripe tomatoes contained $1/4 - 1/3$ more ascorbic acid when they were ripened on the vine than when they were picked green and ripened off of the vine. Results were similar for 4 different ripening temperatures, ranging from $13°$ to $21°C$. Vine-ripened tomatoes contained 25 mg per 100 gm ascorbic acid while, after 6 days of ripening mature-green tomatoes to the red-ripe stage, the level was 18-19 mg per 100 gm. Loss of ascorbic acid because of damage from machine harvesting, hauling, and canning may be compensated for by the fact that canning tomatoes are harvested as red-ripe. Levels of ascorbic acid in canned tomato juice frequently exceed 20 mg per 100 gm (see Table 7.3).

Comparison of studies on ascorbic acid loss in tomatoes should be reviewed with caution. There is a wide range of ascorbic acid levels in different tomato varieties, from 5 to 50 mg per 100 gm.

Ripening on the vine or plant does not always produce higher vitamin content. Singleton (1959 as cited in Hulme 1971) found that pineapples left on the plant underwent a normal ripening pattern

TABLE 7.3

VITAMIN C CONTENT OF TOMATO JUICE[1]

Year	% of Tomatoes Machine Harvested[2]	No. of Samples Examined	Ascorbic Acid (Mg/100 Ml)			% of Samples	
			Max	Min	Avg	Under 15 Mg per 100 Ml	20 Mg or Over per 100 Ml
1960		30	26.1	9.4	16.0	37	13
1961		29	22.2	5.5	13.5	69	7
1962		35	18.8	6.4	12.6	80	0
1963		31	20.2	6.4	14.2	58	3
1964	7	26	19.2	3.0	13.8	58	0
1965	25	26	22.5	5.0	14.2	62	12
1966	65	29	22.4	6.9	15.1	62	17
1967	85	32	31.4	5.2	16.0	41	22
1968	99	39	21.7	3.2	12.8	79	3
1969		34	27.5	5.8	15.4	35	9
1970		32	18.9	6.3	12.5	75	0
1971		38	25.3	4.9	12.5	66	8
1972		35	20.0	5.0	12.4	74	3

[1] Courtesy of Canners League of California.
[2] Source: Allewelt (1969).

with the ascorbic acid dropping from 12.4 mg per 100 gm to 9.1 mg per 100 gm over a 10-day period. Ascorbic acid in fruit stored for the same length of time (at an ambient temperature of 24°C) increased from 12.4 to 13.0 mg per 100 gm.

In vegetables, the climacteric rise in respiration, so well-exhibited in fruits such as the avocado and the apple, is not apparent. Thus, there is no clear-cut division between maturation and breakdown. Intensities and rates of respiration differ among the different vegetables. There are different stages in the history of starch-sugar interconversion at which the vegetable is acceptable in terms of quality for eating or processing. For instance, potatoes should be low in sugar content. Sugars, especially reducing sugars, are responsible for poor texture after cooking, a sweet taste, and participate in the initial steps of the Maillard reaction which eventually results in browning in such products as chips, French fries, and dehydrated potatoes.

By contrast, high quality in peas is associated with tenderness and a high sugar content. During maturation, sucrose and other metabolites are translocated to the developing pea seed (Eriksson and von Sydow 1964, cited in Eskin et al. 1971). As maturity progresses, the sugar concentration decreases in the pea seed, and is accompanied by a concomitant increase in starch content. Thus, it is the immature seeds which are sweet, low in starch, and possess the highest quality.

In order to minimize sugar loss, peas should be cooled rapidly after harvesting. It has been demonstrated that in peas, stored at room temperatures, there is a rapid fall in sugar concentration accompanied by some starch formation (Wager 1964, cited in Eskin et al. 1971).

As sweet corn matures, sugars may be converted to starch or respired to carbon dioxide and water. Therefore, at harvest this vegetable requires similar treatment to that of the peas.

CELLULAR ORGANIZATION

Jansen (1969) writes that plant products . . .

> are composed of living cells that are highly organized entities, containing large populations of many kinds of subcellular entities—for example, the nucleus, mitochondria, chloroplasts, Golgi bodies, membrane systems, and enzymes. In fact, the cell may be pictured as a community of several, separate organized systems, living symbiotically with one another. It is easy to see why the disturbance of one part of this interlinked system should turn organization into chaos. The mere bruising of living tissue is enough to produce disorder and ruin.

Many cell constituents must remain in their elaborately complex forms in order to play their parts in the life processes of the cell. One can well understand the concern of growers and processors with bruising that might result from mechanical harvesting, bulk transportation, bulk storage, etc. There are several recent reviews on these topics (USDA 1968; Hulme 1970; Ryall and Lipton 1972). As noted above, an important prerequisite for the maintenance of good quality is the avoidance of mechanical injury. Damage not only causes structural and physiological disorganization of the tissues but greatly facilitates the entry of microorganisms which cause spoilage. Injury most often occurs during handling operations such as harvesting, grading, cleaning, washing, and packing and again during unpacking and sale in the market. It can also occur during transportation as a result of faulty packing combined with the vibration and jerky movement of the vehicle (Duckworth 1966).

The contrast between the stability of ascorbic acid in fruits in which cellular disorganization has occurred and in those in which it has not is well illustrated by a study on processing rose hips. While rose hips are not marketed as a commercial fruit or vegetable, the study nevertheless demonstrates what happens to at least one nutritional component when the complex packaging of cell structures is disturbed. In this study, some of the rose hips were frozen (which disrupted the structural integrity) and then stored at various temperatures. Unfrozen hips were stored at $1°C$ for 174 days, after which they had lost none of their ascorbic acid. The hips which had been frozen lost 60% of their ascorbic acid when held for the same length of time at $-10°C$ (Mapson 1970).

There are at least four enzymes that occur in fruits which may be responsible for the oxidative destruction of ascorbic acid; these are ascorbic acid oxidase, phenolase, cytochrome oxidase, and peroxidase. Any of these may initiate the oxidative destruction of the vitamin. In the intact fruit these enzymic systems are controlled; only when cellular disorganization occurs, as a result of mechanical damage, rot, or senescence, do these oxidation activities become operative (Mapson 1970).

Carotenoid pigments can be oxidized under certain circumstances with a consequent loss of provitamin activity. For example, green leafy tissues contain a lipoxidase-like enzyme system which can destroy carotene rapidly if the tissues are damaged. However, the extent of destruction of carotene in intact living tissues is very small. In some products, e.g., carrots, tomatoes, and peaches, the synthesis of carotene can continue after harvest, leading to an actual increase in provitamin (Duckworth 1966).

Mechanized harvesting for many crops has increased in recent years as domestic migrant labor has shrunk and the "bracero" program, which permitted the temporary migration of Mexican labor, came to

Courtesy of USDA

FIG. 7.3. MECHANICAL CHERRY HARVESTER

Shaker arm with trunk clamp at left; canvas catching frame around tree.

an end in 1965. Two crops, cherries and tomatoes, that are now extensively harvested by machine, will be discussed in this section; although others, such as potatoes, carrots, peas, etc., have long been mechanically harvested.

The inertia shaker used for harvesting some fruits from trees can result in damage to fruits in many ways: fruit may strike each other or limbs before detachment, strike limbs while falling, strike other fruit as they fall onto the catching frame, and strike other fruit or parts of the conveying and filling equipment while being moved along into conveyors. The shake-catch method has not been satisfactory for fresh market fruits such as oranges, pears, plums, apples, or apricots (O'Brien and Kasmire 1972).

Bruising from mechanical harvesting may be compensated, at least in part, by harvesting in the cool hours of the night, quick application of precooling, and by harvesting a sufficient quantity to avoid delay in accumulating loads to be sent to the canner or freezer. Fruits and vegetables usually take much longer to get from the field to the retail fresh market than they do to get to the processor. This latter journey is frequently less than 1 hr.

Transportation equipment must be carefully considered to avoid bruising of fruits shipped to the processor or to market. O'Brien and Kasmire (1972) found severe in-transit damage in peaches hauled 100 miles on a truck with leaf spring suspension while little damage was evident on peaches hauled the same distance on a truck with airride suspension.

Tomatoes are the most important vegetable for processing in the United States. In 1971, processed tomatoes and tomato products (paste, catsup, etc.) were 38% of the total canned vegetable consumption (King et al. 1973). California alone supplied 78% of the 5.8 million tons of canning tomatoes harvested in 1972. In 1964, tomato varieties with small cores, tough skin, small size, uniform ripening, and easily detachable stems, as well as unique harvesting machines, became available. By 1968, more than 99% of California's canning tomatoes were machine-harvested.

Mechanical harvesting of tomatoes poses special problems. When a tomato fruit becomes fully ripe, the usual maturity for harvesting for processing, tissue disorganization becomes increasingly dominant. Cell walls become thin and the organized cytoplasmic units begin to disintegrate. The degradation of cellulose, as well as most of the pectic components, leads to a progressive loss of tissue cohesion (Hobson and Davies 1971). As a result, mature ripe tomatoes are a difficult vegetable to haul and to handle. Before the advent of mechanical harvesting, canning tomatoes were hand-picked, loaded into 40-lb lug boxes, and hauled directly to the cannery. About 1965, the use of a large bin holding about 1000 lb became common. Now, tomatoes are commonly hauled from field to cannery in trucks holding 20 tons or more. Hauling can take its toll. With fruits of ripe maturity, tests on 100-mile transportation after 20 and 44 hr showed an increased loss from damaged tomatoes of from 12 to 16% (O'Brien et al. 1972). Nevertheless, analyses of more than 500 samples of canned tomato juice packed from 1960 to 1972, shown in Table 7.3, showed no significant change in the level of ascorbic acid between hand-picked and machine-harvested tomatoes (National Canners Association 1973).

In 1956, agricultural engineers of USDA at Michigan State University began to build equipment for mechanizing the tart cherry harvest. Cherries are shaken from the trees onto catching frames and hauled to the processing plant cushioned in water trucks (Roberts 1968; Whittenberger and Labelle 1969). The quality of the canned or frozen produce will largely depend on how quickly the cherries are cooled during the "holding period," i.e., the time between harvesting and processing. The most practical cooling technique for handling machine-harvested cherries during the critical holding period is to use cold-water tanks, holding about 1200 lb of fruit. Cooling the cherries quickly and holding them in water at temperatures of 18°C or below helps to prevent scald, facilitates pitting, and increases finished product yield (Cargill 1967). Mechanical harvesting obviously can result in bruising because falling fruit hits tree limbs or the catching frame. This can be compensated by quick cooling and by careful handling.

TEMPERATURE AND HUMIDITY

Moisture loss or wilting after harvest can cause nutritive loss and result in unsalability. Moisture loss begins as soon as a fruit or vegetable is removed from the soil or separated from its parent plant. The rate of moisture loss depends on the extent and nature of the surface area of the product and upon its environment. Leafy vegetables, because of their extensive and rather permeable surfaces, wilt very rapidly in unfavorable environments. This is seldom a problem with products such as potatoes or avocados, which have smaller surface-to-volume ratios and less permeable skins (Ryall 1965).

Water is lost from produce as vapor. Most fruits and vegetables are composed of cells loosely bound together, with considerable intercellular space which is interconnected and leads to natural openings called lenticles or stomates (Fig. 7.4). Water from the cells vaporizes

EFFECTS OF HARVESTING AND HANDLING ON COMPOSITION 109

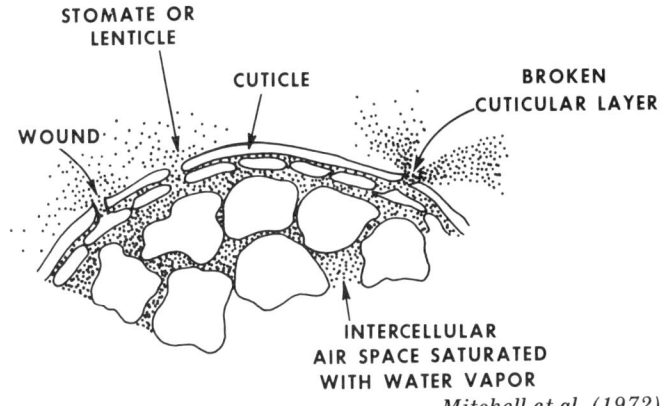

FIG. 7.4. SCHEMATIC OF COMMON ROUTES OF WATER LOSS IN FRESH PRODUCE

Cuticle is a natural waxy covering found on some fruits and vegetables.

into the intercellular spaces, and maintains an essentially saturated atmosphere within the product. Water vapor may then move to the outside atmosphere through the lenticles, through stems or stem scars through any injured area, or directly through the cuticle. Water vapor will leave produce in direct proportion to the difference between its internal concentration and that of the surrounding atmosphere.

Most fruits and vegetables are grown in districts far removed from the population centers and must be transported long distances to reach the consumer. Conditions surrounding the products during the period between harvest and the time they reach the consumer determine to a large degree what part of the at-harvest vitamin values are available to the consumer.

Fresh, green, leafy vegetables wilt (lose moisture) when transported or stored at high temperatures and/or low humidities. The same conditions result in loss of ascorbic acid, with temperature having the more severe effect. Ezell and Wilcox (1959) found that, with slight to moderate wilting, kale would lose about 40% of its ascorbic acid content in 3 weeks when stored at 0°C, in 4 days at 10°C, and in 2 days at 21°C. With rapid wilting, the same losses took place in, respectively, 9 days, 2 days, and 1 day. Snap beans were found to lose 40% of their ascorbic acid content in 10 days at 0°C, and in 6 days at 10°C or 21°C. Wilting had little effect on ascorbic acid in snap beans.

Carotene supplies approximately $3/5$ of the vitamin A values in the normal diet. Dark green, leafy vegetables are especially rich in carotene, but may lose it rapidly unless properly handled after harvest. Temperature and humidity are the primary factors in the preservation of these crops. Low humidity and/or rapid air movement often result in wilting and a less attractive product. Wilting plasmolysis may also hasten oxidation of cell constituents and thereby have an adverse effect on carotene content (Ezell and Wilcox 1962).

TABLE 7.4

RELATIVE EFFECTS OF TEMPERATURE AND
OF WILTING ON LOSS OF CAROTENE
DURING FIRST FOUR DAYS
OF STORAGE

Rate of Wilting	Loss of Carotene			
	32°F (%)	50°F (%)	70°F (%)	Avg (%)
Kale				
Slow	0.0	17.1	58.5	25.2
Moderate	0.0	18.9	65.9	28.3
Rapid	4.9	31.2	75.8	37.3
Avg	1.6	22.4	66.7	
Collards				
Slow	2.4	29.5	57.1	29.7
Moderate	7.5	32.4	64.2	34.7
Rapid	12.6	40.9	82.3	45.3
Avg	7.5	34.3	67.9	

Source: Ezell and Wilcox (1962).

After harvesting, fresh fruits and vegetables are thus subject to deterioration in appearance, flavor, and nutritive value. Refrigeration is the most commonly-used procedure for slowing this process and extending the shelf-life and quality of most fruits and vegetables. Temperature reduction slows the respiratory activities of plant products, reduces water loss, decreases the chances of new decay invasions, and inhibits or slows the growth of incipient infections (Ryall 1965). Spinach, for example, becomes unsalable 13 times as rapidly at 27°C as 2°C (Ryall and Lipton 1972). Tangerines stored for 8 weeks at 0°C lost little ascorbic acid but when held for the same length of time at 7°-9°C, they lost almost $1/2$ of their initial ascorbic acid (Bratley 1939). Harding (1954), cited in Hulme (1971), studied changes in fruit quality during the marketing period for oranges. Oranges were stored for 3-6 days at 10°C followed by 7 days at 21°C to simulate market conditions. Only a very slight loss of vitamin C was found. Hulme (1971) also cited the work of Kefford (1966), who concluded that the loss of vitamin C in fresh fruit is unlikely to exceed 10% under reasonable conditions of distribution and marketing. The level of ascorbic acid decreases rapidly in asparagus spears during postharvest storage; 50% of the original ascorbic acid content in one lot of asparagus was lost after 1 week storage at 0°C; but after 1 week at 21°C, 90% was lost (Scott and Kramer 1949, cited in Eskin et al. 1971). Riboflavin in vegetables is relatively stable during their storage prior to processing or marketing. Fresh carrots increased in riboflavin (not due to a change in moisture content) after three months in cold storage (Ingalls et al. 1950).

The potato is considered to be a good source of ascorbic acid. During storage, however, the ascorbic acid content of tubers decreases. The literature varies on the effects of temperature, some workers finding faster disappearance of ascorbic acid at low temperature, others at high (Smith 1967). In a study of 20 varieties of potatoes by Allison and Driver (1953), the decrease of ascorbic acid in the course of 8 months of storage was from 16 to 10 mg per 100 gm.

Sometimes vegetables are deliberately held at relatively high temperatures for a short period. For example, the skin of sweet potatoes is usually damaged during harvesting. If the roots are held at elevated temperature for a few days the skin will form a corky layer or "suberize" (heal) minor lesions. The sweet potato roots are "cured" by holding them at 29°C and 85-90% RH for 4-7 days. After curing, the sweet potatoes can be satisfactorily stored from 4 to 6 months at 13°-16°C. During curing the level of ascorbic acid diminishes slightly. Ascorbic acid content of sweet potatoes continues to decrease slowly during storage, but the carotene increases (Ezell et al. 1952).

Refrigeration does not always aid in extending the storage life of fruits and vegetables. Some commodities are subject to injury from cool storage temperature well above their freezing point. Although some temperature reduction is often desirable when shipping mature green tomatoes, sustained temperatures below 13°C will prevent proper ripening and make the tomatoes subject to decay. The same is true of pepper, eggplant, banana, and other crops.

At temperatures of 0° to 10°C, some fruits and vegetables become weakened because they are unable to carry on normal metabolic processes. Often, products that are chilled look sound when removed from low temperature storage, but symptoms of chilling—pitting or other skin blemishes, internal discoloration, or failure to ripen—

become evident in a few days at warmer temperatures. Sensitive fruits and vegetables that have been chilled may be particularly susceptible to decay. Alternaria rot is often severe on tomatoes, squash, peppers, or cantaloupe that have been chilled.

Chilling of certain fruits and vegetables can accelerate decay because the low temperature reduces the resistance of cells to pathogens. In addition, in some crops, such as sweet potatoes, low temperatures prevent wound healing, thus enhancing the chance of decay. Other symptoms of chilling injury are discoloration, pitting, and the loss of the ability to ripen normally, depending on the fruit or vegetable involved (Ryall and Lipton 1972). Chilling can not only cause the conspicuous injuries discussed above but it can also reduce the nutritive value of some crops. Carotene increases in sweet potatoes stored at 13°C or higher, but not when stored at 10°C, which is in the chilling range (Ezell et al. 1952). Chilling may also reduce the ascorbic acid in some vegetables, although the opposite is true in most. The influence of chilling on nutrition has received little attention in research, most likely because the visible symptoms of chilling

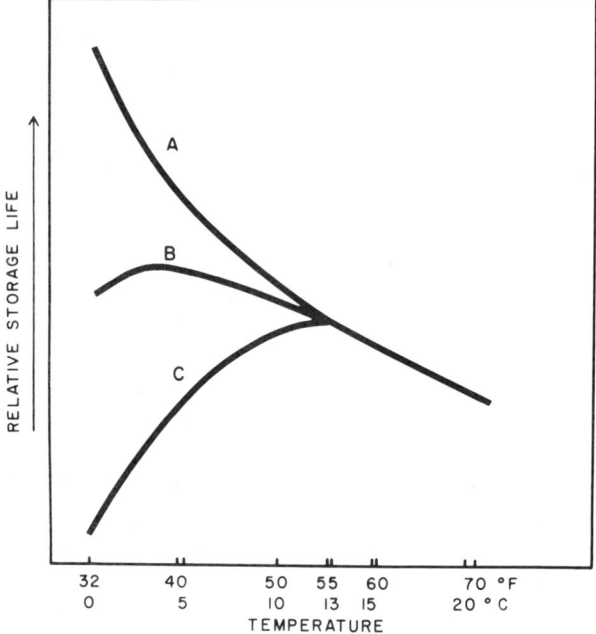

Ryall and Lipton (1972) after Tomkins (1966)

FIG. 7.5. CHANGE IN RELATIVE LENGTH OF STORAGE LIFE AT VARIOUS TEMPERATURES OF CROPS WITH (A) NO SENSITIVITY TO CHILLING INJURY, (B) SLIGHT SENSIBILITY, AND (C) HIGH SENSIBILITY TO CHILLING INJURY

EFFECTS OF HARVESTING AND HANDLING ON COMPOSITION 113

injury are more important economically than the invisible and possibly minor changes in nutritive value. The relationship between sensitivity to chilling injury, storage life, and temperature was well illustrated in Ryall and Lipton (1972) after Tomkins (1966). In Fig. 7.5, curve A represents crops, such as cabbage, that are not sensitive to chilling and whose storage life is longest when held just above their freezing point. Curve B applies to asparagus, whose storage life is longest at about 2°C. Curve C represents crops, such as sweet potatoes, which can be stored longest at about 13°C, and which are the most sensitive to chilling.

PRECOOLING

Usually, storage rooms designed for holding produce under refrigeration do not have either the refrigeration capacity or the air movement needed for rapid cooling. Thus, precooling for storage is generally a separate operation requiring special facilities and equipment.

Ryall and Lipton (1972)
FIG. 7.6. HYDROCOOLING

114 NUTRITIONAL EVALUATION OF FOOD PROCESSING

Precooling refers to the rapid removal of field heat before shipment or storage, and is essential for the more perishable horticultural crops. Precooling is accomplished commercially by several methods. Most involve rapid transfer of heat from the commodity to a cooling medium, such as air, water, or ice. From less than 30 min to more than 24 hr may be required for adequate precooling.

Hydrocooling is a popular precooling method wherein produce is flooded with or immersed in cold water (near 0°C). Asparagus, sweet corn, topped carrots, celery, and peaches are commodities often hydrocooled (USDA 1968).

Almost all lettuce shipped from California and Arizona is now vacuum cooled. Commodities to be vacuum cooled, such as lettuce, are introduced into large vacuum chambers and held at a pressure of 4.6 mm of mercury. Temperature is quickly reduced to about 0°C by the latent heat required to evaporate moisture. About 1% of the moisture is lost for every 5°C reduction in temperature, with an average loss of about 2.5–3%. Since this loss occurs about equally from all parts of the produce being cooled, wilting generally is not apparent.

Refrigeration retards respiration and lengthens storage life. For some products, respiration, ripening decay, and certain physiological

Ryall and Lipton (1972)
FIG. 7.7. VACUUM COOLING

disorders can be reduced still further by reducing the oxygen level or increasing the carbon dioxide content of the storage air, or both.

CONTROLLED ATMOSPHERE

Controlled atmosphere (CA) storage is a system for holding fresh fruit or vegetables in an atmosphere that differs substantially from normal air in its proportions of nitrogen, oxygen, or carbon dioxide. A commercial storage system using this method was developed in Great Britain more than 50 yr ago. No one mixture of gases is suitable under all circumstances. The storage temperature, storage period, and the commodity to be stored are all interdependent (Ryall and Lipton 1972).

CA storage is used mostly with apples. Desirable conditions for CA storage of most apple varieties range from $-0.5°$ to $4°C$, and oxygen concentration from 1 to 3%, and carbon dioxide concentrations from 1 to 6%. For some varieties of apples, especially those subject to the physiological diseases "core flush" or "brown heart," an atmosphere of low oxygen (2.5-3.0%) with no carbon dioxide is optimum for storage (Hulme and Rhodes 1971). As with other cellular disorganization, brown heart has been associated with disappearance of ascorbic acid (Bogdanski 1960 as cited in Hulme 1970).

FUMIGANTS

The use of fumigants is another type of atmosphere modification for a longer storage life. The only fumigant used extensively for decay control is sulfur dioxide. This is universally used for quality protection of the vinifera-type table grapes produced in California, Arizona, and several foreign countries. The fruit is initially treated for about 30 min with 1% sulfur dioxide as soon after harvest as possible. Fruit subsequently placed in cold storage is refumigated with lower concentrations at intervals of 7-10 days. Sulfur dioxide in the refrigerated air during storage reduces the gray mold rot *Botrytis cinerea*. Fumigation with sulfur dioxide also reduces the rate of respiration in grapes. Emperor grapes containing only 22 ppm of sulfur dioxide had their respiration rate reduced 82% in storage at $28°C$ (Winkler 1962). Other types of grapes and most other fruits and vegetables will not tolerate sulfur dioxide in sufficient concentration to be useful (Ryall 1965).

In sound fresh grapes, B vitamins are reasonably stable in cold storage, whereas in cool storage the initial amount may decrease as much as 10 to 30% in a month. Thiamin in grapes is destroyed by sulfur dioxide in proportion to the amount of free sulfur dioxide present.

Ascorbic acid is present in grapes in minute, though measurable, quantities (1 to 12 mg per 100 gm) (Winkler 1962).

CONCLUSION

Providing quality fruits and vegetables to the consumer for fresh consumption or to the food processor for preservation is a challenge. Most fruits and vegetables are highly perishable and deteriorate rapidly after harvest. Without proper handling and treatment the quality and nutritional value of these crops are reduced or lost. Mechanical harvesting and bulk transportation for processing crops have helped to keep preserved food prices in check. At the same time, there has been no indication of a drop in processed fruit or vegetable quality or nutritional value because of these practices. Major changes in the growing areas for market-fresh fruits and vegetables have occurred in recent years. The result has been a significant alteration in distribution patterns, with products required to travel greater distances. Procedures have been developed with the goal to provide delivery of these fruits and vegetables to the consumer with as near garden-fresh quality and nutritional value as possible.

BIBLIOGRAPHY

ALLEWELT, W. F., JR. 1969. A tribute to Prof. G. C. (Jack) Hanna. Natl. Canners Assoc. Inform. Letter *2197*, 95-96.
ALLISON, R. M., and DRIVER, C. M. 1963. The effect of variety, storage, and locality on the ascorbic acid content of the potato tuber. J. Sci. Food Agr. *4*, 386-396.
BIALE, J. B. 1961. Postharvest physiology and chemistry. *In* The Orange. W. B. Sinclair (Editor). Univ. Calif., Div. Agr. Sci.
BRATLEY, C. O. 1939. Loss of ascorbic acid (vitamin C) from tangerines during storage on the market. Proc. Am. Soc. Hort. Sci. *37*, 526-528.
CARGILL, B. F. 1967. Cooling stations. Mich. State Univ. Ext. Bull. *659*.
DUCKWORTH, R. B. 1966. Fruits and Vegetables. Pergamon Press, London, England.
ERIKSSON, C., and VON SYDOW, E. 1964. Postharvest metabolism of green peas (*Pisum sativum*) with special reference to glutamic acid and related compounds. J. Food Sci. *29*, 59-64.
ESKIN, N. A. M., HENDERSON, H. N., and TOWNSEND, R. J. 1971. Biochemistry of Foods. Academic Press, New York.
EZELL, B. D., and WILCOX, M. S. 1959. Loss of Vitamin C in fresh vegetables as related to wilting and temperature. J. Agr. Food Chem. *7*, 507-509.
EZELL, B. D., and WILCOX, M. S. 1962. Loss of carotene in fresh vegetables as related to wilting and temperature. J. Agr. Food Chem. *10*, 124-126.
EZELL, B. D., WILCOX, M. S., and CROWDER, J. N. 1952. Pre- and postharvest changes in carotene, total carotenoids, and ascorbic acid content of sweet potatoes. Plant Physiol. *27*, 355-369.
GORTNER, W. A., DULL, C. G., and KRAUSS, B. H. 1967. Fruit development, maturation, ripening, and senescence: a biochemical basis for horticultural terminology. HortScience *2*, 141-144.

HARDING, P. L. 1954. Effects of simulated transit and marketing periods on quality of Florida oranges. Food Technol. 8, 311-312.
HARRIS, P. L., and POLAND, G. L. 1939. Variations in ascorbic acid content of bananas. Food Res. 4, 317-327.
HOBSON, G. E., and DAVIES, J. N. 1971. The tomato. In The Biochemistry of Fruits and Their Products, Vol. 2. A. C. Hulme (Editor). Academic Press, New York.
HULME, A. C. (Editor). 1970. The Biochemistry of Fruits and Their Products, Vol. 1. Academic Press, New York.
HULME, A. C. (Editor). 1971. The Biochemistry of Fruits and Their Products, Vol. 2. Academic Press, New York.
HULME, A. C., and RHODES, M. J. C. 1971. Pome Fruits. In The Biochemistry of Fruits and Their Products, Vol. 2. A. C. Hulme (Editor). Academic Press, New York.
INGALLS, R. et al. 1950. The nutritive value of canned foods. III. Changes in riboflavin content of vegetables during storage prior to canning. Food Technol. 4, 258-263.
JANSEN, E. F. 1969. Quality-related chemical and physical changes in frozen foods. In Quality and Stability of Frozen Foods. W. B. Van Arsdel, M. J. Copley, and R. L. Olson (Editors). Wiley-Interscience, New York.
KELLY, J. F. 1972. Horticultural crops as sources of proteins and amino acids. HortScience 7, 149-151.
KING, G. A., JESSE, E. V., and FRENCH, B. C. 1973. Economic trends in the processing tomato industry. Univ. Calif. Agr. Ext.-Agr. Exp. Sta. Inform. Ser. Agr. Econ. 73-74.
KRAMER, A. 1965. Evaluation of quality of fruits and vegetables. In Food Quality. G. W. Irving, and S. W. Hoover (Editors). Am. Assoc. Advan. Sci. Publ. 77.
MAPSON, L. W. 1970. Vitamins in fruits. In The Biochemistry of Fruits and Their Products, Vol. 1. A. C. Hulme (Editor). Academic Press, New York.
MITCHELL, F. G., GUILLOU, R., and PARSONS, R. A. 1972. Commercial cooling of fruits and vegetables. Univ. Calif. Div. Agr. Sci. Manual 43.
NATIONAL CANNERS ASSOCIATION. 1973. Report of Technical Data on Cans Examined from the 1972 Fruit and Vegetable Pack. Canners Leagues of California, Berkeley, Calif.
NELSON, P. E. 1972. Processing effects on the nutritional components of horticultural crops. HortScience 7, 151-153.
O'BRIEN, M., and KASMIRE, R. F. 1972. Engineering developments and problems at production source of fresh product. Trans. ASAE 15, 566-568.
O'BRIEN, M., LEONARD, S. J., MARSH, G., and OLSON, N. A. 1972. Processing losses of tomatoes as affected by harvesting and handling procedures. Trans. ASAE 15, 563-565, 568.
PANTOS, C. E., and MARKAKIS, P. 1973. Ascorbic acid content of artificially ripened tomatoes. J. Food Sci. 38, 550.
RHODES, M. J. C. 1970. The climacteric and ripening of fruits. In The Biochemistry of Fruits and Their Products, Vol. 1. A. C. Hulme (Editor). Academic Press, New York.
ROBERTS, N. R. 1968. Tree shaker saves our cherry pies. USDA 1968 Book of Agriculture.
RYALL, A. L. 1965. Protecting the quality of fruits and vegetables after harvest. In Food Quality. G. W. Irving, and S. W. Hoover (Editors). Am. Assoc. Advan. Sci. Publ. 77.
RYALL, A. L., and LIPTON, W. J. 1972. Handling, Transportation, and Storage of Fruits and Vegetables. Avi Publishing Co., Westport, Conn.
SCOTT, L. E., and KRAMER, A. 1949. Physiological changes in asparagus after harvest. Proc. Am. Soc. Hort. Sci. 54, 357-366.
SIMMONDS, N. W. 1966. Bananas. Longmans, Green, & Co., London, England.

118 NUTRITIONAL EVALUATION OF FOOD PROCESSING

SMITH, O. 1975. Effect of transit and storage conditions on potatoes. *In* Potato Processing, 3rd Edition. W. F. Talburt, and O. Smith (Editors). Avi Publishing Co., Westport, Conn.
TOMKINS, R. C. 1966. The choice of conditions for the storage of fruits and vegetables. Ditton Lab. Mem. *91.*
TRESSLER, D. K., and EVERS, C. F. 1957. The Freezing Preservation of Foods, Vol. 1. Avi Publishing Co., Westport, Conn.
USDA. 1968. The Commercial Storage of Fruits, Vegetables, and Florist and Nursery Stock. USDA Agr. Handbook *66.*
VON LOESECKE, H. W. 1950. Bananas. Interscience Publishers, New York.
WAGER, H. G. 1964. Physiological studies of the storage of green peas. J. Sci. Food Agr. *15,* 245-252.
WHITTENBERGER, R. T. 1970. Buying and selling tart cherries by volume. Great Lakes Fruit Grower News *9,* 37-38.
WHITTENBERGER, R. T., and LABELLE, R. L. 1969. Effect of mechanization and handling on cherry quality. *In* Fruit and Vegetable Harvest Mechanization. Mich. State Univ.
WILKINSON, B. G. 1970. Physiological disorders of fruit after harvesting. *In* The Biochemistry of Fruits and Their Products, Vol. 1. A. C. Hulme (Editor). Academic Press, New York.
WINKLER, A. J. 1962. General Viticulture. University of California Press, Berkeley, Calif.

Vernon L. Frampton

PART 2
Effects of Handling and Storage on Seeds

Only a few species of seeds are used in the feeding of man and domestic animals. The useful seeds are put in storage at harvest time to serve as a reservoir of food to be used to tide over until the next harvest. It is self-evident that the economic stability of society depends upon an adequate storage of edible seeds.

A seed consists of an embryo containing reserve nutrients and a seed coat. It is excellently adapted to fulfill its biological functions, conservation of, and dissemination of its species. It is remarkably resistant to deteriorating factors and will survive storage from a few years to more than a century depending upon variety and storage conditions (Ewart 1908). The storage ability of seeds usually correlates with their life span. Short-lived species remain viable under

EFFECTS OF HARVESTING AND HANDLING ON COMPOSITION

good storage conditions for about 3 yr, while long-lived seeds may remain viable for up to 100 yr. The capacity of a seed to pass into a state of latent life while it is maturing on the plant is most remarkable, and this dormant seed is highly resistant to all kinds of unfavorable conditions. Some seeds are protected naturally, as in peanut shells, corn husks, or in paddy rice. In others the seed coat itself may be highly resistant to insect and fungal attacks (e.g., soybeans). Seeds with damaged seed coats are more susceptible to attack, but in some instances their constituents provide a metabolic defense against attacking insects (e.g., saponins in the soybean) (Gay et al. 1973).

Seeds should be kept in a viable condition until they are used or processed because rapid deterioration begins at the time of death of the seed. Seeds stored in bulk are not in a natural environment and deterioration can be very rapid when the conditions of storage are adverse (Barton 1941).

Some seeds entering into storage suffer damage during the preharvest period while they are still green and attached to the plant, due to sucking insects, termites, nematodes, larvae, and plant pathogens. Moreover, there is physical damage to the mature seed coat as a result of harvesting techniques. In some instances, for example with the peanut, immature as well as mature seeds are present when placed in storage. With the increased use of mechanical harvesters there has been a great tendency to harvest seed while it is still damp. These seeds are ready prey for insects and fungi, and should be reprocessed.

Over the years man has determined the environmental conditions that should be controlled in order to avoid deterioration of seed in storage. Unfortunately, the lessons taught by history are often ignored and then there is disaster. This is particularly true in regions of the world where agricultural methods are still primitive. Substantial portions of the harvest are lost to insects, fungi, and rodents. Losses in stored grain in the United States due to insects and fungi are estimated to be as high as 4.5% of the total crop (Barton 1961). Production of edible seeds in the United States (counting the major cereals and oilseeds only) was estimated to be in excess of 230 million tons in 1972. Thus, estimates of total loss of our food grains to insects and fungi approaches the staggering total of about 10 million tons per year.

Among the factors which must be controlled if deterioration of stored seed is to be avoided are the moisture (both of the seed itself and the ambient atmosphere), the temperature of storage (since the rate of growth of insects and fungi are temperature-dependent), the maturity of seed going into storage, the destruction by insects, mites, fungi, and rodents (Milner 1950; Simpson 1953) and the handling of the seed during harvest (to reduce the mechanical damage to the seed coat (Hall 1963).

Of course, the nutritional quality of seeds in storage is adversely affected by (a) proliferation of fungi in the seed and the production of toxic fungus metabolites; (b) the growth and proliferation of insects and mites which contaminate the seed with insect parts and insect frass; (c) the insect secretions and metabolites that give foul tastes and odors and discolor the seed; and (d) contamination by rodents of the seed with feces and urine, and the introduction of harmful fungi and pathogenic bacteria (Slater and Ando 1970). Moreover, there are physical changes in the seed itself unrelated to the infestations or contaminations cited above which alter the solubility and digestibility of the seed protein (Jones and Gersdorff 1938, 1941). These latter alterations in seed properties probably cannot be controlled entirely.

Viable seeds must respire at a low rate to produce energy for life processes (Osley 1948). The measurable respiration in healthy stored seeds appears to be almost entirely in the pericarp and is due to the presence of microorganisms under the surface of the case of the grain (the so-called subepidermal fungi). The respiration observed in seed stored in bulk is attributable to saprophytic fungi growing on dead seed parts, on plant debris and seed fragments, and to insects and mites that infect the seed. High rates of respiration that occur in stored seeds are associated with high moisture contents of the seeds (Karon and Altschul 1946).

Actually, most of the difficulties encountered in the storage of seeds are attributable to unfavorable moisture levels in the seed and in the ambient air spaces. Frequently, damp spots in the grain are generated through the translocation of moisture through the air spaces to cold surfaces where condensation occurs. At times moisture pockets are created when damp seeds are placed in a dry bulk of seed, or are produced because of inadequate protection of the seed from precipitation, be it snow or rain.

Localized heating occurs in these damp regions because moist conditions favor the growth and reproduction of fungi, bacteria, yeasts, insects, and mites. These localized regions are referred to as "hot spots." Temperatures in the hot spots may increase to the point where the grain becomes charred.

Although it has been known from antiquity that moldy seed may be toxic to man and animal, serious attention to the tremendous capacity of saprophytic fungi to produce metabolites of astonishing toxicity developed only recently. Several well-documented incidents during the past two decades relating to Turkey disease in England and the "Yellow Rice" toxicity in Japan, serve as dramatic demonstrations of the hazards of eating or feeding moldy seeds or products produced from them.

Many species of fungi can grow on stored seed and the physiological symptoms produced in animals which injest these seeds or seed products are many and varied—none of them are good.

The mycological aspects of storage of edible seeds are extensive. The fungal population growing in contaminated seed during storage is dynamic, and a variety of factors determine the composition of the population at any one point in time. The particular fungus species that becomes active under a given set of environmental conditions depends on the types of competing microorganisms and the resistance of the host to invasion. The spores of many genera are present simultaneously on weathered and nonweathered seed at the time of harvest. Sinha and Wallace (1965) studied the microorganisms sporulating on wheat harvested in Canada when the grain was kept on moistened filter paper in petri dishes; they identified species of *Alternaria, Cladosporium, Hormodendrum, Trichothecium, Rhizopus, Chaetonnium, Absidia, Aspergillus, Penicillium* and *Streptomyces*, together with species of bacteria and pink and white yeasts. Kurata et al. (1947) reported that milled rice that had been treated with silver nitrate harbored species of *Alternaria, Phyllostica, Cladosporium, Epicoccum, Helminthosporium* and *Penicillium*.

Fortunately for man, most of the fungus species found on seed at the time of harvest are innocuous because the spores do not germinate and the fungi do not grow on harvested seed in storage. However, the many species of fungi, generally referred to as storage fungi, present superficially on the seed at the time of harvest, present a distinct hazard because their spores germinate and the fungi grow on stored seed. Microbial activity in these dangerous species is generally initiated on seed with a moisture content in the range of 13-19%.

The ecological succession of microorganisms on contaminated seed in storage frequently overlaps. For example, a succession of populations was reported by Sinha and Wallace (1965). A hot spot developed in the stored grain four months after it was stored. The heating was initiated in the winter time (storage temperature was in the range of -5° to 8°C) primarily by the activity of the low temperature fungi *Penicillium cyclopium* and *P. funiculosum*. The moisture in the pocket increased to 18.5-20%, while the temperature finally reached 64°C and then subsided. A succession of fungus populations developed in the hot spot in the following order: *P. cyclopium, P. funiculosum, Aspergillus flavus, A. glaucus, A. versicolor*, followed by

TABLE 7.5

POTENTIAL OF MIXTURES OF TOXIC SUBSTANCES THAT MAY BE PRESENT IN SEED INVADED BY STORAGE FUNGI

Metabolite	Physiological Symptoms in Animals Ingesting the Contaminated Seed or Seed Product	Storage Fungus
Aflatoxins	Liver involvement.	*Aspergillus flavus*
Ascladiol		*A. flavus*
Aspergillic acid		*A. sojae* X - 1
Aspertoxin		*A. flavus*
Citrevirdin		*Penicillium ochrosalmoneum*
		P. citrinum
		P. oralicum
		P. commune
Citrinin	Liver involvement.	*A. terreus*
		A. candidus
		P. citrinum
		P. implicatum
Crotocin	Hemorrhage, anemia.	*Fusarium tricinctum*
		F. sporotrichioides
		Cephalosporium species
Crotocol		*F. tricinctum*
Diacetoxyscirpenol		*F. tricinctum*
		Paecilomyces varioti
Flavoskyrin	Liver involvement.	*P. islandicum*
Rubroskyrin		*P. islandicum*
Luteoskyrin		*P. islandicum*
Fusarenon	Degeneration and necrosis of mucosal epithelium of small intestines. Destructive changes in lymph nodes, spleen, thymus and bone.	*F. nivale*
		F. graminearum
		F. kuhnii
		Gibberella fujikuroi
Islanditoxin	Cirrhosis.	*P. islandicum*
Leutoskyrin		*P. islandicum*
Maltorizine	Kills cows.	*A. oryzae* variety *microsporus*
		F. graminearum
Nivalenol	Inhibits protein synthesis.	*F. nivale*
Ochratoxin		*A. ochraceus*
O-methylsterigmatocystin		*A. flavus*
Patulin	Carcinogen.	*A. clavatus*
		P. expansum
		P. viridicatum
		P. urticae
Roridin		*P. tricinctum*

TABLE 7.5 (Continued)

Metabolite	Physiological Symptoms in Animals Ingesting the Contaminated Seed or Seed Product	Storage Fungus
Rugulosin	Vomiting.	P. wortmanni P. tardum G. zeae A. oryzae variety microsporus P. brunneum Epicoccum nigrum F. moniliforme P. wortmanni
Sterimatocystin Trichodermin		A. flavus Trichoderma ligonorum
Trichothecin		F. tricinctum
Zearalenone	Estrogenic effect in cattle and hogs.	G. zeae

species of *Tariboshi*, *Abisidia* and *Streptomyces*. *A. flavus* and *A. glaucus* appeared only after the optimum temperature of 64°C had subsided.

This teaches the lesson that in the dealing with stored seed, one must expect to find residual mycelia and fungal metabolities of those species of fungi which were successively active in the hot spot previously and had become inactive for any one of several reasons. It will, in general, be difficult to identify all of the species that grew in the succession of populations, and it will be equally difficult to know which, or how many, of the toxic fungus metabolites may be present.

While major emphasis in the current literature has been placed on aflatoxins produced by *A. flavus*, it is to be expected that other toxic fungus metabolites, in addition to aflatoxins, will be present in seed that has been contaminated.

Thus, any lot of stored seed or seed product (such as oilseed meals) which show evidence of microbial growth should be considered as dangerous, and it should not be used for feed. It is suggested, also, that the absence of aflatoxins in such a seed specimen does not ensure the absence of other toxic fungus metabolites. The presence of aflatoxins is a warning that other highly-toxic fungal metabolites are likely present also, possibly in substantial concentrations.

A succession of populations of fungi and other microorganisms, harbored on stored seed under conditions that are conducive to microbial activity, is almost certain. The specific microbial populations that follow each other in succession (with some overlapping) may be different in each instance of spoilage, and the identity of each succeeding population cannot be anticipated. It is also certain that mixtures of fungal metabolites will be present, whether or not the fungi and other microorganisms remain viable. Research on storage is very active, and the list of fungi developing during storage, and the identification of the toxic metabolites which they produce, is growing rapidly. In Table 7.5 are presented a sampling of the potential of mixtures of toxic substances that may be present in seed which has been invaded by storage fungi (Slater and Ando 1970).

BIBLIOGRAPHY

BARTON, L. V. 1941. Relation of certain air temperatures and humidities to the viability of seeds in storage. Contrib. Boyce Thompson Inst. *12*, 85.

BARTON, L. V. 1961. Seed Preservation and Longevity. Leonard Hill, London, England.

EWART, A. J. 1908. On the longevity of seeds. Proc. Roy. Soc. Victoria *21*, 1.

GAY, E. J., MUSON, H. L., and PEARMAN, G. G. 1973. Damage to stored soybeans by the cowpea weevil *Callosobruchus maculatus*. J. Georgia Soc. Entomol. *8*, 164.

HALL, D. W. 1963. Some essential considerations of the storage of food grain (cereals, legumes, oilseeds) in tropical Africa. FAO, United Nations Informal Working Bull. *24*.

JONES, D. B. and GERSDORFF, E. F. 1938. Changes that occur in the proteins of soybean meal as a result of storage. J. Am. Chem. Soc. *60*, 723.

JONES, D. B., and GERSDORFF, E. F. 1941. The effect of storage on the protein of wheat; white flour and whole wheat flour. Cereal Chem. *18*, 417.

KARON, M. L., and ALTSCHUL, A. M. 1946. Respiration of cottonseed. Plant Physiol. *21*, 506.

KURATA, H., OGASAWARA, K., and FRAMPTON, V. L. 1957. Microflora of milled rice. Cereal Chem. *34*, 47.

MILNER, M. 1950. Biological processes in stored soybeans. *In* Soybean Products. K. S. Markley (Editor). Interscience, London, England.

OSLEY, T. A. 1948. The Scientific Principles of Grain Storage. Northern Publishing Co., Liverpool, England.

SIMPSON, D. M. 1953. Cottonseed storage in various gases under controlled temperatures and humidities. Univ. Tenn. Agr. Expt. Sta. Bull. *228*.

SINHA, R. N., and WALLACE, H. A. H. 1965. Ecology of a fungus-induced hot spot in stored grain. Can. J. Plant Sci. *45*, 48.

SLATER, J. A., and ANDO, H. 1970. Coordinators. Proc. 1st. U.S.-Japan Conf. Toxic Microorganisms, Mycotoxins and Botulism. U.S. Dept. Interior. (Library of Congress Card. *77-604719*)

Eldon E. Rice

PART 3
Effects of Postmortem Handling

Foods of animal origin are quite varied in nature, ranging from fluids, e.g., milk, to solid materials, such as meat, poultry, fish, and eggs. All of them—with the possible exception of shell eggs—are subject to prompt bacterial attack, to oxidative changes in fatty portions exposed to air, and to other flavor and color changes unless refrigeration or some other method of preservation is used.

Except for milk and eggs, commercial processing of foods of animal origin requires delivery of the live animal to a plant designed for its efficient conversion to food items. Unless the production site is near the processing unit, decreases in gross weight of the animals being moved can be expected, due to loss of moisture and solids in excreta and moisture through respiration. The magnitude of these losses, conventionally called "shrink," varies considerably, depending upon the species, age, and condition of the animal, weather, distance traveled, and the type of transport.

Decentralization of the livestock and poultry processing industry has decreased the distances between the sites of production and processing so that shipment takes only a few hours, in contrast to the several days of a few years ago. Although varying considerably, ordinarily shrink in gross weight of livestock will be 1-4%. As an example, 4 truckloads of 1100 turkeys each, shipped about 120 miles, showed shrinks of 1.6, 3.0, 2.8, and 2.6% in transit. Due to equipment failure, the first 2 lots had to be held on the trucks at a processing plant for about 18 hr after arrival, incurring additional 2.4 and 2.7% shrinks. Losses for broilers are similar. Normal feedlot to processing plant shrinks are 1-2% for swine and 2-4% for cattle. Unless the livestock is subjected to exceptionally long transit time without water there is little loss of nutritive value, the decrease in weight being primarily moisture and solid waste.

The low-temperature storage needed to preserve organoleptic quality and to retard bacterial spoilage also promotes retention of the nutritive qualities of the fresh animal products and minimizes losses other than those due to mechanical factors such as breakage, spillage, or drip. Because of the small losses of nutrients, the processor of these foods ordinarily directs his attention to preservation of optimum fresh color, odor, and flavor. Packaging is also directed toward this objective.

Perhaps the most obvious change that occurs during the processing of meat animals is the decrease in weight caused by removal of inedible materials such as feathers or hide, blood, organs (many of which are not used for human food), and the contents of the intestinal tract and bladder. The yield of salable product varies from species to species, and to a certain extent from animal to animal

TABLE 7.6

TYPICAL YIELDS OF SALABLE PRODUCT FROM PROCESSING OF SEVERAL ANIMAL SPECIES

Kind of Product	Live Wt	Yield[1]
	lb	%
Beef		
Prime steer	1100	64
Choice steer	1100	62
Good steer	1100	59
Standard steer	1100	58
Commercial	1100	57
Commercial cow	950	52
Utility cow	950	49
Cutter cow	950	47
Canner cow	950	48
Pork[2]		
	183	80
	218	81
	289	83
Lamb[3]		
	80	44
	90	44
Chicken[4]		
Broilers	3	64
Roasters	5	67
Hens	5.5	67
Turkeys[4]		
Fryers	7	72
Hens and toms	11	74
Hens and toms	18	77
Hens and toms	27	78
Ducks[4]	6	69
Geese[4]	14	72

[1] Salable carcass weight as percentage of live weight. This does not include edible organs such as the heart, liver, and brain except for chickens where the giblets are included. The values for pork include the heads.
[2] Hankins and Titus (1939).
[3] Hankins (1947).
[4] USDA Production & Marketing Administration (1952).

within a species. The age and grade of animal and the nature of the shipping and handling also influence yield. The data in Table 7.6 illustrate the typical values.

Even within a single type of product, such as beef, there are so many factors influencing composition that nutritional evaluations are difficult. Breed, strain, age, sex, type, and amount of feed and individual variations permit marked differences in compositions between animals (and even animals in the same live grade); there are equally large differences from cut to cut within a single animal; e.g., beef round is quite lean and rib roast carries much more fat. The vitamin content has also been shown to differ from one muscle to another within the same animal (Rice et al. 1945). Sometimes even the section of the country or season of the year may influence composition. Because of the variability of the raw materials, the following discussions are unavoidably general in nature.

RELATION OF GRADE TO COMPOSITION

To group foods of animal origin into classes of somewhat similar properties, attempts are being made by the food industry to grade the various products according to standards. The grade of the animal carcass and, frequently, the subsequent grade of the specific product as sold at retail depend on many factors: age, weight, conformation (build or shape of carcass, determining the meatiness and the relative size of the various cuts), finish (degree, color, and uniformity of fatness), the quality of the flesh (tenderness, flavor, amounts and strength of connective tissue, color and size of muscle fibers), and size and character of the bones, yield of usable meat, etc.

Official standards detail the characteristics of each type of meat and are revised from time to time. Those currently in effect for beef require seven pages in the fine print of the Federal Register of April 7, 1965 (pp. 4456-4462). The American Meat Science Association (1972) has suggested numerical grades based on the USDA standards for quality (palatability factors) and cuttability (yield of retail cuts).

For the most part, the product in each grade is of comparable quality and of similar composition; occasionally, the physical appearance upon which most grading must of necessity be based does not reflect an inherent difference in an item. Thus, variations in the tenderness of cuts of meat within a given grade are common and unavoidable. Similarly, colors of egg yolk and flavors of milk may not be reflected by grade. In general, however, grades are as reliable an estimate as can be arrived at regarding overall quality and composition, despite the fact that composition may differ within wide limits within any grade. There is much overlapping in composition from one grade to another, since grades are intended to reflect other properties important to the marketing of food. This is well illustrated by data assembled by Doty (1956), showing the relative amounts of fat and lean in several grades of beef carcasses (see Table 7.7).

TABLE 7.7

CARCASS COMPOSITION RANGE OF DIFFERENT GRADES AND WEIGHTS

Beef	Separable Lean (%)	Separable Fat (%)	Separable Bone (%)
Light prime	46-59	31-43	11-16
Heavy prime	45-57	31-43	10-14
Light good	54-64	21-30	13-20
Heavy good	47-64	20-42	11-20
Commercial cow	46-58	26-44	11-18

Source: Doty (1956).

Meats are divided into several grades, each characteristic of certain general qualities. For example, beef may be prime, choice, good, standard, commercial, utility, cutter, or canner.

Prime beef originates from well-fattened, beef-type cattle. Carcasses are of good conformation and have uniform fat covering. Liberal quantities of fat are interspersed between the muscle fibers, but recent trends have led to a decrease in the amount of fat covering. Choice beef lacks a little in conformation and does not carry the degree of finish found in prime beef. Beef that grades "good" shows beef type characteristics but lacks uniformity in fat covering or is actually deficient in fat covering and fat between the muscle fibers. Standard and commercial beef do not have the thickness and meatiness of the top grades. The standard grade shows soft bones and not much fatness. The commercial grade is from the older cattle and may show some fat covering and some marbling. Utility beef is largely from older animals. It is angular in shape and lacks meatiness and fatness. Canner and cutter beef has little fat, poor conformation, and lacks meatiness. No grade of beef is likely to be tender unless it has been aged.

On the average, the higher grade beef carcass will be more tender a few days after it is dressed out than the next lower grade. However, grade is no guarantee of tenderness, and equal aging gives more added tenderness to lower commercial, standard, and good grades of beef than to prime and choice beef.

Pork, lamb, veal, poultry, and fish are graded similarly. Since this book is concerned more with nutrition than with other factors of quality, these grades are only shown along with brief descriptions in Table 7.8.

There is a paucity of data concerning the nutrient composition of meats of different grades. The great variability within grades makes such data difficult to accumulate, and of little value insofar as giving information relative to the probable composition of any particular cut. For example, 6 beef rib roasts (9th to 11th rib, inclusive) from U.S. good carcasses gave protein and fat analyses of the edible portion as follows: for protein—12.6, 15.3, 11.8, 11.4, 14.6 and 12.9%; for fat—45.0, 46.0, 40.5, 47.5, 36.7, and 44.5%. These large differences were caused partly by differences in trimming and cutting, and partly by different degrees of fatness of carcasses.

So many considerations other than composition enter into a determination of grade, that this alone is an inadequate indicator of nutritional quality. Beef that carries enough fat to justify a prime classification may actually grade choice or good, depending upon other factors, such as conformation of the cut, color, tenderness, age, etc. Thus, grade alone, although relating to composition, is far from a reliable guide with respect to any one sample. Frequently, descriptions such as fat, medium fat, lean, and very lean are more useful from a strictly nutritional point of view. These characteristics have been approximately related to grade and composition in Table 7.9.

Fish are even more variable than mammals according to Stansby (1954). He states that the composition of fish frequently differs more from fish to fish within a species than the average for one species differs from that for another species. Fat content, particularly, depends upon the season of the year, the geographical area, feed available, age and sex of the fish, and on other unidentifiable reasons. For example, Stansby refers to the oil content of mackerel, which varies by 30-fold, and to the vitamin A content of shark liver, which varies by more than a 1000-fold.

The average value for the fat content of fish in a single species may differ from the average for fish in another species by a 100-fold. Despite these large variations, which make average values of little significance in estimating the composition of any particular fish, certain species commonly contain more fat than others and are referred to as "fat" fish in comparison with other species that are "lean." Salmon, shad, herring, cuttlefish, and mackerel are fat fish. Cod, haddock, cusk, pollock, and flounder are examples of lean fish.

TABLE 7.8

GRADES USED IN DESCRIBING PRODUCTS OF ANIMAL ORIGIN

Product[1]	Grade	Description
Beef	Prime	Young, well-fattened, beef-type cattle.
	Choice	Well-fed, beef-type cattle; less fat than prime.
	Good	Relatively lean but good conformation.
	Commercial	Lean but lacking in meatiness.
	Utility	Very lean and poor in conformation.
Veal	Prime	Light colored, blocky, well-finished, firm.
	Choice	Lacks a little in finish compared to prime.
	Good	Less meaty, good color, less fat.
	Commercial	Lacks meatiness, darker in color.
	Utility	Poor color, conformation, and finish.
Pork[2]	Choice No. 3	Average back fat thickness 2.0 in. or more.
	Choice No. 2	Average back fat thickness 1.7–2.0 in.
	Choice No. 1	Average back fat thickness 1.4–1.7 in.
	Medium	Average back fat thickness 1.0–1.4 in.
	Cull	Average back fat thickness less than 1.0 in.
Lamb[3]	Choice or No. 1	Compact, well- and evenly-fleshed; smooth covering of fat.
	Good or No. 2	Moderately compact, less well-covered with fat; fat is evenly distributed.
	Medium or No. 3	Rangy or unsymmetrical, little fat, and poor fat distribution; muscles thin.
	Common or No. 4	Poor conformation, thin flesh, and little fat.
	Cull or No. 5	Very poor conformation, very little fat, small proportion of flesh to bone.
Chicken[4]	A quality	Typical conformation; well-fleshed, well covered with fat over entire carcass; very few pinfeathers, and no cuts, tears, or broken bones.
	B quality	Slightly crooked breastbones or back, and/or misshapen legs and wings; fairly well-fleshed and sufficient fat to cover breast and legs; few pinfeathers, no more than one broken bone, and only moderate bruises or discolorations.
	C quality	Abnormal conformation, thin and little fat covering; may have pinfeathers, broken bones and/or bruises.
Turkey[5]	A, B, and C	Approximately same as for chickens.
Shell eggs	AA	Clear, unbroken, practically normal shell; clear, firm, white; well-centered yolk, outline only slightly defined, and an air cell $1/8$ in. or less in depth.
	A	Clean, unbroken, practically normal shell; clear, reasonably firm white; yolk fairly well-centered; air cell no more than $1/4$ in. in depth.

TABLE 7.8 (Continued)

Product[1]	Grade	Description
	B	Clean, may show slight stains but no adhering dirt; unbroken shell—may be slightly irregular; clear, weak white; yolk may be off center, enlarged, and flattened; yolk may show slight defects; air cell not over $3/8$ in. in depth.
	C	Clean, may show slight to moderate stains but no adhering dirt; unbroken shell—may be abnormal; clear white, but watery and may have spots; yolk may be off center, enlarged, flattened, and may have visible defects.
Raw milk (Certified)	Grade A	Clean milk from healthy, properly housed and handled cows, having a bacterial count not exceeding 200,000/ml.
	Grade B	Same requirements as A, but may have up to 1,000,000 bacteria/ml.
	Grade C	Milk not meeting B requirements.
Pasteurized milk	Grade A	Raw grade A milk, pasteurized sufficiently to pass a phosphatase test and to maintain a bacterial count under 30,000/ml until delivered and which is properly packaged.
	Grade B	Pasteurized milk that has bacterial counts not exceeding 50,000/ml before delivery.
	Grade C	Pasteurized milk not meeting B standards.

[1] Some types of meat have several broad classifications; thus pork carcasses may come from barrows or gilts, the preferred types, or from sows, boars, or stags. Mutton or lamb may originate from lambs, yearlings, or mature sheep. Each type may be graded according to its quality.
[2] Grades for 165–209 lb carcasses that are 30–32.9 in. in length.
[3] Grade descriptions are for lambs; but, in general, apply to all sheep.
[4] Chicken may be classified as fryer or broiler, roaster, capon, stag, hen or fowl, or cock.
[5] Turkey is classified as fryer (usually under 16 weeks of age); young hens and young toms, or old hens and old toms.

Poultry, like meats and fish, may vary through a wide range of composition. Here again, the differences are primarily caused by the quantity of fat present. Each type of poultry, e.g., broilers or roasters, has certain desirable levels of fat, but these will vary considerably.

In general, the younger the bird, the less fat it carries; for example, chicken broilers have less than 10% fat, roasters have 12–18% fat, and fat hens have 20–30% fat. Naturally, chickens of any age may be thin, depending upon their feeding and management control.

The vitamin content of meats, fish, poultry, milk, and eggs depends more upon the species, breed, and the feeding and handling of the animal than upon grade, and correlations between vitamin content

TABLE 7.9

RELATION OF PROXIMATE COMPOSITION TO GRADES OF ANIMAL PRODUCTS (EDIBLE PORTIONS)

Product	Grade	Description	Water (%)	Protein (%)	Fat (%)	Ash (%)	Thiamin (Mg/100 Gm)	Riboflavin (Mg/100 Gm)	Niacin (Mg/100 Gm)
Beef sides	Prime	Very fat	47	13.7	39	0.7	0.06	0.12	3.3
	Choice	Fat	55	16.3	28	0.8	0.07	0.15	3.9
	Good	Medium	60	17.5	22	0.9	0.08	0.16	4.2
	Commercial } Utility	Lean	66	18.8	14	1.0	0.08	0.17	4.5
Pork sides	Choice No. 3 } Choice No. 2	Fat	35	9.8	55	0.5	0.48	0.12	2.6
	Choice No. 1	Medium	42	11.9	45	0.6	0.58	0.14	3.1
	Medium } Cull	Lean	50	14.1	35	0.8	0.69	0.16	3.7
Veal sides	Prime } Choice	Fat	65	18.5	16	0.9	0.14	0.25	6.2
	Good	Medium	68	19.1	12	1.0	0.14	0.25	6.4
	Commercial } Utility	Lean	71	19.7	8	1.0	0.14	0.25	6.6
Lamb sides	Choice } Good	Fat	46	13.0	40	0.7	0.12	0.16	3.8

EFFECTS OF HARVESTING AND HANDLING ON COMPOSITION

Poultry									
Chicken	Medium Commercial	Medium	56	15.7	28	0.8	0.14	0.20	4.5
	Cull	Lean	66	17.1	15	0.9	0.15	0.21	4.9
	Broilers	Medium fat	71	20.2	7	1.1	0.08	0.16	10.2
	Roasters	Medium fat	66	20.2	13	1.0	0.08	0.16	8.0
	Hens	Medium fat	56	18.0	25	1.1	0.08	0.16	8.0
Turkey	Roasters	Medium fat	58	20.1	20	1.0	0.09	0.14	8.0
Fish	Fat fish	Herring	67	18.3	13	2.7	0.02	0.15	3.4
		Mackerel	68	18.7	12	1.2	0.15	0.35	8.4
		Salmon	63	17.4	17	1.0	0.10	0.23	7.2
	Intermediate lean fish	Halibut	75	18.6	5	1.0	0.07	0.06	9.2
		Sunfish	76	19.2	4	1.3	0.05	0.05	9.1
	Lean fish	Cod	83	14.9	0.4	1.2	0.06	0.09	2.2
		Flounder	83	14.9	0.5	1.3	0.06	0.05	1.7
		Haddock	81	18.2	0.1	1.4	0.05	0.08	2.4
Shell eggs	Any grade		74	12.8	12	1.0	0.10	0.29	0.1
Raw milk (Certified)	Any grade		87	3.5	4	0.7	0.04	0.17	0.1

Source: Data adapted from Table 21, Watt and Merrill (1950).

and grade are more apt to be reflections of the relative amounts of fat and lean tissue than of grade. Vitamins of the B complex are associated with the lean tissues and occur most liberally in carcasses which have relatively little fat. This phenomenon is reflected in the values for the B complex vitamins listed in Table 7.9.

EFFECTS OF HOLDING RAW MATERIALS FOR PROCESSING OR SALE

The perishable nature of foods of animal origin necessitates refrigeration during storage and shipment. This minimizes nutrient changes during handling, and, for most practical purposes, the nutrient content of the product as purchased may be considered equivalent to that of the product as it is produced, except for trimming and other mechanical losses.

Surface dehydration along with very small losses of tissue fluid from cut surfaces reduces the weight of meats held under refrigeration. In addition to weight loss, this dehydration or "cooler shrink" dulls and hardens meat surface, thereby reducing salability. Ordinarily, the actual weight losses are not large, ranging from 1 to 5%. However, when considered for the billions of pounds of meat handled annually, even these small decreases become very significant.

Beef carcasses shrink approximately $2\frac{1}{2}\%$ during the first 48 hr of refrigeration, according to records of the American Meat Institute (1945), and there is a $\frac{1}{4}$–1% additional shrink in the next 24 hr. Poultry undergoes similar shrinkage during chilling, but it may lose much more if held under refrigeration for several days.

Many factors influence the amount of cooler shrink, principal ones being the rate of air movement and the humidity of the air. High humidities, while reducing shrink, increase the tendency for bacterial and mold growth.

Although shrink is a very important economic consideration for the food handler, it does not markedly affect nutrient content since removal of moisture only concentrates the solids of the food and does not change their total quantity.

There have been few reported investigations of nutrient changes during storage and transport of meats and milk, possibly because of the obvious lack of change in the proximate compositions other than the slight concentrations caused by dehydration. Modern packaging is designed to minimize these losses. Workers with these products are familiar with their lack of obvious chemical changes under usual handling conditions, and attention is directed more often to problems relating to color, flavor, odor, or tenderness.

That the vitamins in pork are stable during storage that is compatible with organoleptic acceptability and bacteriological safety has

EFFECTS OF HARVESTING AND HANDLING ON COMPOSITION 135

been demonstrated by Rice et al. (1946) who reported the data summarized in Table 7.10. Except for decreases in niacin in unground pork, only slight losses occurred before the sample spoiled. Niacin was stable in ground pork, but has been shown to decrease in whole tissues of other meats (Rice et al. 1948). Samples which were heavily contaminated with bacteria increased in riboflavin and pantothenic acid. Similar storage losses have not been reported for other flesh foods, but they may be assumed to be low. Assays of fresh meats of all types may be repeated after frozen storage without detectable changes in values. Nutrient retentions during storage are discussed in more detail in subsequent chapters.

TABLE 7.10

STABILITY OF B COMPLEX VITAMINS IN PORK

Storage Conditions Days	Temp (°C)	Percentage of Initial Value after Storage			
		Thiamin (%)	Riboflavin (%)	Niacin (%)	Pantothenate (%)
Ground pork loin					
14	4	93	104	97	105
28[1]	4	87	154	101	128
56[1]	4	89	141	105	146
Cured pork					
14	4	95	105	90	116
56	4	91	120	93	102
Whole pork loin					
11	4	97	100	72	112
Sterile whole pork loin					
14	25	92	94		

Source: Adapted from Rice et al. (1946, 1948).
[1] Sample badly decomposed by bacterial spoilage.

There does not seem to be significant proteolysis during aging and storage (Wierbicki et al. 1955), and amino acid stability may be assumed. Ginger et al. (1954) have shown that there is no change in the arginine, lysine, leucine, glutamic acid, and tyrosine content of beef loins during two weeks of aging. The changes that do occur during postmortem storage greatly influence the tenderness of the meat and its structure (Dutson et al. 1974). Proper injection of enzymes immediately prior to dispatch of the animal can significantly improve the tenderness of the cooked product (Robinson and Goeser 1962; Kang and Rice 1970). Premortem addition of enzymes does not greatly increase proteolysis until the meat is heated during cooking.

Fish, particularly, are subject to rapid spoilage due to both bacterial and autolytic changes. Tressler and Lemon (1952) indicate the first changes in iced fish to be autolytic, leading to increases in the amounts of free amino acids and polypeptids. Later, secondary decomposition to amino acid, indole, hydrogen sulfide, methylamine, etc., takes place. Organoleptic quality is so perishable that most attention has been diverted toward its preservation. It is doubtful if there are significant losses of nutrients before spoilage sufficient to make the product unacceptable.

Milk, also, because of its ready attack by bacteria is cooled and handled as a "fresh" item, even after pasteurization. Except for a very rapid loss of ascorbic acid, the nutrients of milk are stable during normal periods of storage and sale. If milk is exposed to sunlight or to strong light, riboflavin is also destroyed. As much as 60% may be lost during several hours' storage in glass bottles in sunlight, according to Holmes and Jones (1944).

Evans and co-workers (1949 through 1953) have made extensive studies of the retention of nutrients in shell eggs during storage periods up to 12 months at $0°C$. Evaporation of moisture from shell eggs occurs in an almost linear relationship with time of storage and will account for an approximate 5% weight loss during 12 months' storage at $0°C$ under usual storage conditions. At higher temperatures or at low humidities, greater losses occur. There are also some losses of carbon dioxide, ammonia, hydrogen sulfide, and other gases, so that the actual solids content decreases.

Part of the data of Evans et al. has been summarized in Table 7.11, which indicates that storage at $0°C$ results in slow loss of protein, riboflavin, niacin, vitamin B-6, and pantothenic acid, but not of choline or biotin. In many cases, the loss of moisture (14% for these eggs) offsets the loss of vitamin, so that the concentration remained constant or even increased, despite the fact that a loss on a per-egg basis had occurred. Folic acid and vitamin B-6 were the only two vitamins showing marked decrease during storage. Evans et al. also showed that changes in the relative concentration of the nutrients in yolk and white changed, usually by a diffusion of moisture into the yolk and of vitamins into the white.

The relative concentrations of arginine, lysine, histidine, methionine, cystine, tryptophan, phenylalanine, proline, and tyrosine in egg substance did not change, but there were 12–18% decreases on a per-egg basis, reflecting a loss of protein (to ammonia and other nitrogenous components). These changes, along with other less well-defined chemical changes, account for the gradual decrease in egg quality during long storage.

In general, it may be observed that the principal change which occurs in foods of animal origin during their processing to fresh foods is

EFFECTS OF HARVESTING AND HANDLING ON COMPOSITION

TABLE 7.11

RETENTION OF B COMPLEX VITAMINS IN EGGS DURING STORAGE AT 0°C

Vitamin		Storage Periods, Months			
		0	3	6	12
Riboflavin	µg/gm	3.49	3.32	2.93	3.07
	µg/egg	170	163	146	147
	% retained	—	96	86	87
Niacin	µg/gm	0.66	0.60	0.54[1]	0.67
	µg/egg	31.3	28.9	25.8[1]	32.1
	% retained	—	92	82[1]	102
Pantothenic	µg/gm	12.5	11.7	11.7	11.8
acid	µg/egg	608	570	578	560
	% retained	—	94	95	92
Biotin	µg/gm	10.76	11.63	10.36[1]	10.42
	µg/egg	225	244	220[1]	228
	% retained	—	108	96[1]	97
Folic acid	µg/gm	94	93	80	74
	µg/egg	4.59	4.59	3.84	3.37
	% retained	—	100	84	74
Vitamin B-6	µg/gm	2.52	2.06	1.78	1.34
	µg/egg	124	99	83	60
	% retained	—	80	67	48
Choline	µg/gm	7.5	—	—[1]	8.7
	µg/egg	252	255	260[1]	251
	% retained	—	101	103[1]	100
		Fresh eggs:	33.6 gm		
		12 months' old:	28.9 gm		

Source: Adapted from Evans et al. (1949 to 1953).
[1] 7 months' storage.

that of loss in weight caused by the discard of inedible portions and by moisture losses during chilling and temporary storage. Nutrient losses are slight, permitting emphasis upon retention of desirable color, flavor, odor, and other organoleptic qualities.

BIBLIOGRAPHY

AMERICAN MEAT INSTITUTE. 1945. Beef, Veal, and Lamb Operations, 4th Edition. J. W. Edwards, Publisher, Ann Arbor, Mich.

AMERICAN MEAT SCIENCE ASSOCIATION. 1972. Recommended Procedure for Beef Carcass Evaluation and Carcass Contents. Am. Meat Sci. Assoc., Chicago.

DOTY, D. M. 1956. Laboratory characteristics of graded beef carcasses. Proc. 9th Ann. Reciprocal Meat Conf., Chicago.

DUTSON, D. R., PEARSON, A. M., and MERKEL, R. A. 1974. Ultrastructural postmortem changes in normal and low quality porcine muscle fibers. J. Food Sci. 39, 32–37.

EVANS, R. J., BUTTS, HELEN A., and DAVIDSON, J. A. 1951. The niacin content of fresh and stored shell eggs. J. Poultry Sci. *30*, 132-135.
EVANS, R. J., BUTTS, HELEN A., and DAVIDSON, J. A. 1952B. The riboflavin content of fresh and stored shell eggs. J. Poultry Sci. *31*, 269-273.
EVANS, R. J., and DAVIDSON, J. A. 1951. The choline content of fresh and stored shell eggs. J. Poultry Sci. *30*, 29-33.
EVANS, R. J., DAVIDSON, J. A., BANDEMER, S. L. and BUTTS, HELEN A. 1949A. The amino acid content of fresh and stored shell eggs. 2. Arginine, histidine, lysine, methionine, aptine, tyrosine, phenylalanine, and proline. J. Poultry Sci. *28*, 697-702.
EVANS, R. J., DAVIDSON, J. A., BAUER, DORIS, and BUTTS, HELEN A. 1953A. The biotin content of fresh and stored shell eggs. J. Poultry Sci. *32*, 680-683.
EVANS, R. J., DAVIDSON, J. A., BAUER, DORIS, and BUTTS, HELEN A. 1953B. Folic acid in fresh and stored shell eggs. J. Agr. Food Chem. *1*, 170-172.
EVANS, R. J., DAVIDSON, J. A., and BUTTS, HELEN A. 1949B. Changes in egg proteins occurring during cold storage of shell eggs. J. Poultry Sci. *28*, 206-214.
EVANS, R. J., DAVIDSON, J. A., and BUTTS, HELEN A. 1952A. The pantothenic acid content of fresh and stored shell eggs. J. Poultry Sci. *31*, 777-780.
GINGER, I. D., WACHTER, J. P., DOTY, D. M., and SCHWEIGERT, B. S. 1954. Effect of aging and cooking on the distribution of certain amino acids and nitrogen in beef muscle. Food Res. *19*, 410-416.
HANKINS, O. G. 1947. Estimation of the composition of lamb carcasses and cuts. USDA Agr. Bull. *944*.
HANKINS, O. G., and TITUS, H. W. 1939. Growth, fattening and meat production. In USDA Yearbook of Agriculture.
HOLMES, A. D., and JONES, C. P. 1944. Effect of sunshine upon the ascorbic acid and riboflavin content of milk. J. Nutr. *29*, 201-209.
KANG, C. K., and RICE, E. E. 1970. Degradation of various meat fractions by tenderizing enzymes. J. Food Sci. *35*, 563-565.
RICE, E. E., DALY, M. E., BEUK, J. F., and ROBINSON, H. E. 1945. The distribution and comparative content of certain B-complex vitamins in pork muscular tissues. Arch. Biochem. *7*, 239-246.
RICE, E. E., FRIED, J. F., and HESS, W. R. 1946. Storage and microbial action upon vitamins of the B complex in pork. Food Res. *11*, 305-312.
RICE, E. E., SQUIRES, E. M., and FRIED, J. F. 1948. Effect of storage and microbial action on vitamin content of pork. II. Food Res. *13*, 195-202.
ROBINSON, H. E., and GOESER, P. A. 1962. Enzymatic tenderization of meat. J. Home Econ. *54*, 195-200.
STANSBY, M. E. 1954. Composition of certain species of fresh-water fish. 1. Introduction: The determination of the variation of composition of fish. Food Res. *19*, 231-234.
TRESSLER, D. K., and LEMON, J. McW. 1952. Marine Products of Commerce, 2nd Edition. Reinhold Publishing Corp., New York.
U.S. DEPT. OF AGR. 1952. Poultry—Approximate Weights and Processing Shrinkages. USDA Production Marketing Admin. Leaflet.
U.S. DEPT. OF AGR. 1965. Applications of Standards for Grades of carcass beef. Federal Register *30*, 53.102, 4456-4462.
WATT, B. K., and MERRILL, A. L. 1950. Composition of Foods—Raw, Processed, Prepared. USDA Agr. Handbook *8*.
WIERBICKI, E., KUNKLE, L. E., CAHILL, V. R., and DEATHERAGE, F. E. 1955. Postmortem changes in meat and their possible relation to tenderness together with some comparisons of meat from heifers, bulls, steers, and diethylstilbestrol-treated bulls and steers. Food Technol. *10*, 80-86.

CHAPTER 8

Effects of Refining Operations on the Composition of Foods

George E. Inglett

PART 1
Effects of Refining Operations on Cereals

Cereals are seed grains grown worldwide principally as a food source. The major cereals are wheat, corn (maize), rice, oats, barley, rye, grain sorghum, and millet. Although mystery shrouds the prehistoric development of these grains, civilization could not exist today without early man's careful perpetuation of these seeds.

Morris and Sears (1967) refer to the growing of wheat for food as early as 10,000–8000 B.C. Early civilizations of the Western Hemisphere (Aztecs, Mayas, and Incas) were based principally on corn as the dominant food staple. Mature Indian corn was found in pre-Inca graves dating back to 3000 B.C. (Inglett 1970A). The major emphasis on cereal production since prehistoric times was to increase yields to provide adequate grain for food and feed. Only in the 20th Century have plant scientists become interested in breeding cereal crops for nutritional improvement (Inglett 1972A).

CEREAL COMPOSITION

The average chemical composition of wheat, corn, rice, oats, barley, rye, grain sorghum, and millet is given in Table 8.1. All cereal grains contain starch as the principal component, which is indicated by the high nitrogen-free extract values in Table 8.1. The second highest component of cereal grains is protein. Both content and nutritive value of cereal protein vary widely and depend particularly on seed heredity and environment during cultivation and harvest. An indication of protein composition differences can be seen by the amino acid patterns of the different cereal proteins (Table 8.2). Improvement in cereal protein quality and quantity has been a major thrust in plant breeding since the discovery by Mertz et al. (1964) that *opaque-2* corn (high lysine) had a protein composition that provides considerably better nutrition than ordinary corn.

TABLE 8.1
AVERAGE COMPOSITION OF CEREAL GRAINS

Property	Wheat[1]	Corn[2]	Rice[3]	Oats[4]	Barley[5]	Rye	Sorghum	Millet
Moisture, %	12.5	13.8	12.0	8.3	11.1	11.0	11.0	11.8
Calories/100 gm	330	348	360	390	349	334	332	327
Protein, %	12.3	8.9	7.5	14.2	8.2	12.1	11.0	9.9
Fat, %	1.8	3.9	1.9	7.4	1.0	1.7	3.3	2.9
N-free extract, %	71.7	72.2	77.4	68.2	78.8	73.4	73.0	72.9
Fiber, %	2.3	2.0	0.9	1.2	0.5	2.0	1.7	3.2
Ash, %	1.7	1.2	1.2	1.9	0.9	1.8	1.7	2.5
Thiamin, mg/100 gm	0.52	0.37	0.34	0.60	0.12	0.43	0.38	0.73
Riboflavin, mg/100 gm	0.12	0.12	0.05	0.14	0.05	0.22	0.15	0.38
Niacin, mg/100 gm	4.3	2.2	4.7	1.0	3.1	1.6	3.9	2.3

Source: USDA (1963).
[1] Whole grain, hard red winter.
[2] Field, whole grain, raw.
[3] Raw, brown.
[4] Oatmeal, dry.
[5] Pearled, light.

EFFECTS OF REFINING ON COMPOSITION OF FOODS 141

TABLE 8.2

AMINO ACID CONTENT OF CEREALS (PERCENTAGE OF AMINO ACID IN THE PROTEIN)

Amino Acid	Wheat	Corn (Field)	Rice (Brown)	Barley	Oats	Rye	Sorghum	Millet
Tryptophan	1.2	0.6	1.1	1.2	1.3	1.1	1.1	2.2
Threonine	2.9	4.0	3.9	3.4	3.3	3.7	3.6	4.0
Isoleucine	4.3	4.6	4.7	4.3	5.2	4.3	5.4	5.6
Leucine	6.7	13.0	8.6	7.0	7.5	6.7	16.1	15.3
Lysine	2.8	2.9	4.0	3.4	3.7	4.1	2.7	3.4
Methionine	1.3	1.9	1.8	1.4	1.5	1.6	1.7	2.4
Cystine	2.2	1.3	1.4	2.0	2.2	2.0	1.7	1.3
Phenylalanine	4.9	4.5	5.0	5.2	5.3	4.7	5.0	4.4
Tyrosine	3.7	6.1	4.6	3.6	3.7	3.2	2.7	—
Valine	4.6	5.1	7.0	5.0	5.9	5.2	5.7	6.0
Arginine	4.8	3.5	5.8	5.2	6.6	4.9	3.8	4.6
Histidine	2.0	2.1	1.7	1.9	1.8	2.3	1.9	2.1
Alanine	3.5	9.9	3.6	4.6	6.1	—	—	—
Aspartic acid	5.5	12.4	4.7	5.6	4.1	—	—	—
Glutamic acid	31.2	17.6	13.7	22.4	20.1	21.3	21.9	—
Glycine	6.1	3.4	6.8	4.6	4.6	—	—	—
Proline	10.4	8.4	4.8	9.0	5.7	—	—	—
Serine	4.6	5.6	5.1	4.6	4.0	4.1	5.1	—
Total protein, %	14.0	10.0	7.5	12.8	14.2	12.1	11.0	11.4

Source: Orr and Watt (1968); Juliano *et al.* (1964).

CONSUMPTION OF CEREAL PRODUCTS

Modern cereal processing is practiced worldwide to provide many nutritious food products or ingredients for food processors. Wheat, for example, is processed in large industrial mills that produce purified flour used principally for leavened white bread. In contrast in some developing countries, cereals are frequently eaten after boiling, parching, or grinding to a flour. Many times ground wheat kernels are made into unleavened bread by simple and sometimes primitive procedures (Pomeranz and Shellenberger 1971).

Corn is processed in modern factories to produce an abundant number of principal products—meal, starch, syrups, and dextrose. These primary products are used in the preparation of other foods which people eat without even realizing that they are consuming corn ingredients. An estimated 100 million people eat corn as the major component in unleavened bread or as porridge (Inglett 1970B).

Besides wheat and corn, rice, oats, barley, and rye are also milled today to produce cereal food products. This chapter covers these six cereals in particular. An estimated 159 lb per capita of these cereal

142 NUTRITIONAL EVALUATION OF FOOD PROCESSING

TABLE 8.3

CEREAL CONSUMPTION IN THE
UNITED STATES FOR 1971

Cereal	Consumption[1] Million Bu	Per Capita, Lb
Wheat	515	
Flour		110
Breakfast cereal		2.9
Corn	226	
Breakfast cereal		2.3
Meal		7.4
Syrup		16.2
Sugar		5.2
Starch		1.9
Rice, milled	15.4[2]	
Products		7.6
Oats	44	
Products		3.2
Barley	8	
Products		1.2
Rye	5.3	
Products		1.2
Total		159.1

Source: USDA (1972).
[1] Preliminary
[2] Million cwt.

products are consumed annually in the United States (Table 8.3). Wheat flour is the major cereal product consumed annually, at least 110 lb per capita.

WHEAT

Milling

Milling of good clean wheat (*Triticum aestivum*) is controlled by modern processes to give as much quality flour, farina, and germ as practical. Products from hard wheat are: farina, patent flour, first clear flour, second clear flour, germ, shorts, and bran. Since flour products from milling are primarily for food applications, the wheat is cleaned before conditioning (or tempering) and grinding. Modern flour milling requires an intricate process involving many grinding and sifting steps. Similar flour milling processes for soft and durum wheats provide products for many foods that differ from those incorporating hard wheat products. Additives, such as maturing agents, bleaching agents, self-rising ingredients, are frequently blended into wheat flours at the mill. Flour being shipped to the baker is not enriched at the mill, but at the bakery. Family flour,

EFFECTS OF REFINING ON COMPOSITION OF FOODS 143

packaged at the mill, is enriched at the mill (Anderson and Inglett 1974).

Nutrient Composition of Wheat Flour

Wheat flour, the principal refined product of wheat milling, is the major ingredient in almost all breads, rolls, chapaties, crackers, cookies, biscuits, cakes, doughnuts, muffins, pancakes, waffles, noodles, macaroni, and spaghetti. Flour composition and functionality vary greatly depending upon the milled wheat's heredity and the environmental conditions of its culture and harvest (Inglett 1974). Nutritional value of wheat foods depends largely on the chemical composition of the refined flour used in preparation. The average percentage composition of hard wheat milled products is given for illustrative purposes in Table 8.4.

TABLE 8.4

AVERAGE PERCENTAGE COMPOSITION OF HARD WHEAT PRODUCTS

Constituent	Wheat	Farina	Patent Flour	First Clear Flour	Second Clear Flour	Germ	Shorts	Bran
Moisture	12.0	14.2	13.9	13.4	12.4	10.5	13.5	14.1
Ash	1.8	0.4	0.4	0.7	1.2	4.0	4.1	6.0
Protein	12.0	10.3	11.0	12.7	13.5	30.0	16.0	14.5
Crude fiber	2.5	—	—	—	—	2.0	5.5	10.0
Fat	2.1	0.8	0.9	1.3	1.3	10.0	4.5	3.3

Source: Inglett (1974).

Vitamin and mineral compositions vary directly with the degree of flour extraction. These relationships have been extensively reviewed by Dimler (1960A). Vitamin concentrations in wheat and wheat food products from later data are recorded in Tables 8.5 and 8.6 (Toepfer et al. 1972).

The protein content of wheat is usually around 12%, but heredity and environment exert a strong influence on that level. Since the discovery of high-lysine corn (Mertz et al. 1964), some research emphasis has been placed on increasing the protein quality and quantity of wheat (Johnson et al. 1972). In the World Wheat Collection, 15,000 hexaploid and tetraploid wheats have been screened for protein and lysine contents. Mean protein value was nearly 13%, ranging between 7 and 22%. The lysine content of wheat protein varied between 2.2 and 4.2%, with a mean value of approximately 3.0%. Successful breeding of wheat for better protein quality and quantity

TABLE 8.5

VITAMIN CONCENTRATION IN WHEAT AND WHEAT PRODUCTS

Nutrient	Hard Red Wheat (Dry Wt) (γ/Gm)	Concentration of Wheat-Grain Nutrient in Product Bread			Durum Wheat (Dry Wt) (γ/Gm)	Concentration of Wheat-Grain Nutrient in Product	
		Flour (%)	Conventional Dough-Mix (%)	Continuous Dough-Mix (%)		Semolina (%)	Macaroni (%)
Thiamin	5.7	23	30	30	6.7	48	48
Riboflavin	1.2	34	161	122	1.1	88	86
Niacin							
Total	74	28	39	36	111	35	40
Free	36	44	64	61	47	47	45
Vitamin B-6	3.5	15	15	12	4.3	28	25
Pyridoxine	2.6	11	5	6	3.3	24	20
Pyridoxal	0.5	28	41	26	0.6	31	31
Pyridoxamine	0.4	22	55	35	0.4	53	58
Tocopherols	58	11	2	11	58	43	5
α-T	13	2	1	2	10	30	2
β-T	7	7	3	4	5	31	4
γ-T[1]	—	—	—	1.9	—	—	—
δ-T[1]	—	—	—	3.6	—	—	—
α-T-3	5	30	50	0	7	37	3
β-T-3	33	17	3	3	36	48	6

Source: Toepfer et al. (1972).
[1] Micrograms per gram.

EFFECTS OF REFINING ON COMPOSITION OF FOODS

TABLE 8.6

VITAMIN CONCENTRATION IN WHEAT AND WHEAT PRODUCTS

		Concentration of Wheat-Grain Nutrient in Product					
Nutrient	Soft Wheat (Dry Wt) (γ/Gm)	Patent Flour (%)	Cake (%)	Straight-Grade Flour (%)	Crackers (%)	Cut-Off Flour (%)	Crackers (%)
Thiamin	5.4	23	10	39	20	76	39
Riboflavin	1.1	29	119	40	50	55	72
Niacin							
Total	72	14	5	17	12	19	14
Free	38	15	7	21	17	21	19
Vitamin B-6	3.3	10	7	14	10	25	21
Pyridoxine	2.5	7	3	10	5	22	12
Pyridoxal	0.5	22	30	30	35	39	50
Pyridoxamine	0.3	18	11	21	14	32	50
Tocopherols	54	6	185	—	—	59	35
α-T	12	3	94	—	—	59	31
β-T	7	5	—	—	—	66	38
γ-T[1]	—	—	66	—	—	—	1.0
δ-T[1]	—	—	21	—	—	—	0.3
α-T-3	5	14	—	—	—	34	18
β-T-3	30	6	6	—	—	61	34

Source: Toepfer et al. (1972).
[1] Micrograms per gram.

can have an important impact on the nutrient content of wheat food products.

CORN

Milling

Corn (*Zea mays Linnaeus*) is processed to give food ingredients, industrial products, feeds, and alcoholic beverages. Approximately 360 million bushels are used annually by wet corn processors, corn dry-millers, and fermentation processors. Wet millers produce starch; modified starch products, including dextrose and syrups; feed products; and oil (Anderson 1970). The corn dry-milling process and the effects of dry-milling on nutrient composition have been reviewed by Dimler (1960B). In modern corn dry-milling plants a degerming system is the principal process employed. In this system the corn kernel is separated into hull, germ, and endosperm fractions which vary in particle size and fat content. Endosperm products—grits, meal, and flour—are the primary products used by the food processors and in consumer markets (Brekke 1970; Inglett 1970B).

Nutrient Composition of Corn Dry-Milled Products

A modern corn dry miller using a conventional degerming system can produce a wide variety of products. Typical yields and analyses

146 NUTRITIONAL EVALUATION OF FOOD PROCESSING

TABLE 8.7

TYPICAL YIELDS AND ANALYSES FOR PRODUCTS A DEGERMING-TYPE
DRY CORN MILL MIGHT PRODUCE

Products	Yield (%)	Typical Particle Size Range[1]	Moisture, % Wet Basis	Fat, % Dry Basis	Crude Fiber, % Dry Basis	Ash, % Dry Basis	Crude Protein, % Dry Basis
Corn	100		15.5	4.5	2.5	1.3	9.0
Primary Products							
Cereal flaking (hominy) grits	12	-3.5 + 6	14.0	0.7	0.4	0.4	8.4
Coarse grits	15	-10 + 14	13.0	0.7	0.5	0.4	8.4
Regular grits	23	-14 + 28	13.0	0.8	0.5	0.5	8.0
Coarse meal	3	-28 + 50	12.0	1.2	0.5	0.6	7.6
Dusted meal	3	-50 + 75	12.0	1.0	0.5	0.6	7.5
Flour[2]	4	-75 + Pan	12.0	2.0	0.7	0.7	6.6
Oil	1						
Hominy feed	35		13.0	6.3	5.4	3.3	12.5
Shrinkage	4						
Alternative Products							
Brewers' grits	30	-12 + 30	13.0	0.7	0.5	0.5	8.3
100% Meal	10	-28 + Pan	12.0	1.5	0.6	0.6	7.2
Fine meal	7	-50 + Pan	12.0	1.6	0.6	0.7	7.0
Germ fraction[3]	10	-3.5 + 20	15.0	18.0	4.6	4.7	14.9

Source: Brekke (1970).
[1] U.S. standard sieve.
[2] Break flour.
[3] Yield is distributed between corn oil and hominy feed.

of his products are reported for illustrative purposes in Table 8.7 (Brekke 1970). Human diets are improved by fortifying corn dry-milled products with vitamins and minerals. Enriched corn grits contain thiamin, riboflavin, niacin, and iron according to Federal Standards of Identity (Table 8.8). Calcium and vitamin D may be added as optional ingredients. These nutrients may be blended with such carriers as wheat or corn starches containing anticaking agents. The nutrients are added in powder form to the milled products (Brockington 1970).

The discovery of the superior protein quality of *opaque-2* corn (Mertz et al. 1964) makes possible the production of dry-milled products with improved nutritional value. Conventional dry-milling of *opaque-2* corn (high-lysine) gave products of acceptable fat content using ordinary dry-milling equipment (Brekke et al. 1971). However, the prime product spectrum of the high-lysine corn produced no flaking grits while a considerable amount of table grits, meal, and flour resulted. *Opaque-2* corn is a floury variety, so the lack of large grit particles after milling is not surprising. Corn breed-

TABLE 8.8

NUTRIENTS IN DRY-MILLED CORN PRODUCTS[1]

Corn Products	Thiamin		Riboflavin		Niacin		Iron	
	Unenriched	Enriched	Unenriched	Enriched	Unenriched	Enriched	Unenriched	Enriched
Corn grits, degermed	0.59	2.0	0.18	1.2	5.4	16.0	4.5	13.0
Corn meal								
Whole ground	1.72	2.0	0.50	1.2	9.1	16.0	10.9	13.0
Bolted	1.36	2.0	0.36	1.2	8.6	16.0	8.2	13.0
Degermed	0.64	2.0	0.23	1.2	4.5	16.0	5.0	13.0

[1] Milligram per pound.

148 NUTRITIONAL EVALUATION OF FOOD PROCESSING

ing programs are underway to develop high-lysine corn varieties with a greater horny (flinty) character.

A review of compositional and nutritional values of corn germ, a product of current degerming dry-milling operations, indicated that this fraction would provide a good quality protein source when properly handled (Inglett 1972B). Corn germ flour prepared from a commercial dry-milled fraction appears to be a promising fortifying ingredient for the food industry (Blessin et al. 1972, 1973).

RICE

Rice (*Oryza sativa* L.) is one of the leading food crops of the world. In the United States, rice products are consumed at a yearly rate of 7.6 lb per capita (Table 8.3). However in Asiatic countries, the annual per capita consumption is believed to be higher than 200 lb, accounting for 70-80% of the daily caloric intake (Witte 1970).

Arkansas, California, Louisiana, and Texas produce more than 97% of the rice grown in the United States. The annual U.S. production is approximately 4 million metric tons.

Milling

Rice milling processes have been reported by many workers, the latest by Witte (1970, 1972). Four basic operations are performed by all rice mills: (1) removal of foreign matter from rough rice; (2) removal of hulls; (3) removal of bran; and (4) sizing of milled rice. The primary purpose of rice milling is to obtain the maximum yield of unbroken grain. A broken kernel has about half the commercial value of unbroken kernels.

Solvent extractive rice milling is a relatively new process with reported higher yields of rice, fewer broken kernels, and two by-products—an edible defatted bran and a crude dewaxed oil (Hunnell and Nowlin 1972). The basic steps involved in this process are: (1) pretreatment of brown rice before milling, (2) milling the pretreated brown rice in the presence of rice oil/hexane miscella, and (3) separation and recovery of the defatted bran, crude oil, and hexane.

The major constituent of milled rice is starch, and it is most concentrated in the endosperm portion of the kernel. Protein is the second most abundant constituent of rice grain and is unique among the cereal proteins because it contains at least 80% glutelin (alkali-soluble protein). Glutelin has the closest amino acid composition to milled rice protein probably because it is the major protein fraction (Table 8.9). The protein content of rice of any variety can vary

EFFECTS OF REFINING ON COMPOSITION OF FOODS

TABLE 8.9

LEVELS OF ESSENTIAL AMINO ACIDS OF PROTEIN FRACTIONS
AND PROTEIN OF MILLED RICE (gm/16.8 gm N)

Amino Acid	Protein Fraction				Milled Rice Protein, %
	Albumin	Globulin	Prolamin	Glutelin	
Isoleucine	4.05	3.03	4.68	5.27	4.13
Leucine	7.89	6.56	11.3	8.19	8.24
Lysine	4.92	2.56	0.51	3.47	3.80
Methionine	2.54	2.27	0.50	2.61	3.37
Methionine + cystine	5.40	2.27	0.80	4.09	4.97
Phenylalanine	2.97	3.32	6.26	5.42	6.02
Threonine	4.65	4.55	2.86	3.92	4.34
Tryptophan	1.88	1.34	0.94	1.16	1.21
Valine	8.72	6.18	6.97	7.31	7.21

Source: Juliano (1972).

considerably even when grown at the same location (Cagampang et al. 1966; Tecson et al. 1971). For example, protein content of the high-protein variety BPI-76-1 may range from 8 to 14% and the low protein variety, Intan, from 5 to 11% protein (at 12% moisture). The effect of differences in protein contents of milled rice on their nutritive quality is shown in Table 8.10 (Juliano 1972). Although protein quality tended to decrease as protein contents increased, the decrease in quality was less than proportional to the increase in protein content.

TABLE 8.10

SUMMARY OF PROTEIN QUALITY INDICES FOR FOUR MILLED RICE SAMPLES
AND CASEIN BASED ON WEIGHT GAIN IN WHITE RATS

Protein Source	Protein Content[1] at 12% Moisture	PER[2]	NPR[3]	N Growth Protein		N Growth Relative	
				Index	Quality[4]	Index	Quality[4]
Intan	5.68	2.56	3.71	3.49	80	2.37	47
IR8	7.32	2.20	3.36	3.25	75	2.30	46
IR8	9.73	1.94	3.07	3.04	71	2.17	43
BPI-76-1[5]	14.3	1.50	2.57	2.47	57	2.12	42
Casein	86.2	2.20	3.36	3.23	75	3.78	75

Source: Juliano (1972).
[1] $N \times 5.92$ for rice protein and $N \times 6.25$ for casein.
[2] Protein efficiency ratio at 5% protein.
[3] Net protein ratio at 5% protein.
[4] Based on a value of 75 for casein.
[5] Corrected for differences in casein values between the two feeding experiments. BPI-76-1 was tested later than the other rice samples.

Nutrient Composition of Milled Rice

The composition of milled rice will vary depending on the variety, its agronomic conditions during growth, and extent of milling. The milled-rice kernel is not homogeneous in composition, but will vary by layers. The outer layer and the amount removed during milling are most important in determining the nutrient composition of the rice. The compositional differences of rice are illustrated by data given in Table 8.11. The chemical composition of the outer layer and the endosperm of milled rice is compared with the parent milled kernel. These data are based mainly on short-grain Balilla rice originally milled to 10% (bran removed by weight of brown rice). The

TABLE 8.11

CHEMICAL COMPOSITION OF OUTER LAYER, ENDOSPERM, AND ENTIRE KERNEL OF MILLED RICE[1]

Constituent	Unit	Outer Layer[2]	Endosperm	Entire Kernel[3]
Starch[4]	%	61.86	92.00	90.68
Amylose[4]	%	16.12	29.85	29.46
Reducing sugars	gm maltose/100 gm rice	0.50	0.07	0.12
Nonreducing sugars	gm sucrose/100 gm rice	2.42	0.11	0.26
Total sugars	%	2.92	0.18	0.38
Fiber	%	1.47	0.22	0.28
Total N	gm N/100 gm rice	2.53	1.27	1.39
Nonprotein N	gm N/100 gm rice	0.04	0.02	0.02
Protein N	gm N/100 gm rice	2.49	1.25	1.37
Albumin	gm ($N \times 5.95$)/100 gm rice	1.75	0.29	0.30
Globulin	gm ($N \times 5.95$)/100 gm rice	1.12	0.60	0.67
Prolamin	gm ($N \times 5.95$)/100 gm rice	0.72	0.22	0.25
Glutelin	gm ($N \times 5.95$)/100 gm rice	7.93	5.05	5.25
Insoluble fraction	gm ($N \times 5.95$)/100 gm rice	3.28	1.48	1.69
Free amino N	mg/100 gm	25.11	2.55	3.40
Total lipids[5]	%	4.44	0.45	0.66
Free fatty acids	%	1.34	0.15	0.21
Neutral fats	%	2.53	0.26	0.38
Phospholipids	%	0.57	0.04	0.07
Ash	%	6.10	0.45	0.72
Calcium[4]	%	0.36	0.05	0.02
Iron[6]	%	0.03	—	0.00
Phosphorus[4]	%	1.02	0.10	0.14

[1] Dry basis data adopted from Barber (1972).
[2] Five percent by weight of entire kernel, unless otherwise specified.
[3] Approximately 10% milling, unless otherwise specified.
[4] Commercially milled rice; outer layer 4.4%.
[5] Chloroform:methanol (2:1) extractable lipids.
[6] Commercially milled rice; outer layer 4.27%.

outer layer, accounting for 5% by weight of the milled kernel, was obtained by tangential-abrasive milling that left mainly the endosperm portion.

Vitamin contents of rice and its by-products are summarized in Table 8.12 (Houston and Kohler 1970). The refining of rice to give

TABLE 8.12

VITAMIN CONTENTS OF RICE AND ITS BY-PRODUCTS (mg/100 gm)

Vitamin	Brown Rice	Milled Rice	Rice Bran	Rice Polish	Rice Germ
Thiamin	0.34	0.07	2.26	1.84	6.5
Riboflavin	0.05	0.03	0.25	0.18	0.5
Niacin	4.7	1.6	29.8	28.2	3.3
Pyridoxine	1.03	0.45	2.5	2.0	1.6
Pantothenic acid	1.5	0.75	2.8	3.3	3.0
Folic acid	0.02	0.02	0.15	0.19	0.43
Inositol	119	10	463	454	373
Choline	112	59	170	102	300
Biotin	0.01	0.01	0.06	0.06	0.06

Source: Houston and Kohler (1970).

the milled product causes significant losses of essential vitamins. The enrichment of breakfast cereals based on milled rice appears justified.

OATS

Milling

Oats rank as the third most important grain grown in the United States—next to wheat and corn. The groat or fruit represents about 75% of the kernel weight and is tightly held within the chaff or hull. The fat content of oat groats will average about 7%, which is distributed throughout the kernel with a slight concentration in the germ and outer layers. The protein content will average 16–17% with only a slight concentration in the germ and outer layers (Salisbury and Wichser 1971).

The initial step in oat milling is cleaning to remove foreign materials, such as sticks, corn, seeds, soybeans, barley, wheat, and dust. The cleaned oats are dried on pan driers normally 10–12 ft in diameter and placed one above the other in stacks of 7 to 14. As the oats gradually pass down the stack, normally 3–4% moisture is removed. The temperature of the oats seldom exceeds 200°F, but it is sufficient to cause a slight roasted flavor considered desirable.

Rotary steam tube driers are sometimes used by smaller millers, and many European plants use charcoal-fired kilns. In some mills that dehull oats without drying or conditioning, the groats are heated separately to develop the desired toasted flavor. Besides flavor development, heating inactivates the lipolytic or fat-splitting enzymes sufficiently to prevent the development of undesirable flavors during processing.

Oats after drying and cooling are ready for the huller which separates the hulls from the groats. An impact huller produces the best yields and requires less horsepower than earlier stone hullers did. Sizing the grain before hulling assists oat and groat separation after hulling. Disc machines are generally used for this separation in the United States.

In a large system the final step is the separation of free groats used for producing old-fashioned flakes. Free groats are separated by all machines, and all other oats pass to the cutting operation. Cutting converts the groats into uniform pieces with a minimum of fine granules or flour. Cutting is done with rotary granulators giving 2-4 pieces per groat. Cut groats are separated from the uncut groats, oats, and long hulls by a cylinder separator or disc machine. The cut material is heated with live steam at atmospheric pressure just before flaking. The steam-heated groats or cut-groats are flaked on rolls that are adjusted to produce flakes of uniform thickness or density measurement. Oat flour is made by grinding steam-heated groats. For a white and lower fiber flour, high-fiber fractions must be removed from the groats.

Nutrient Composition of Oat Products

The composition of oats, like other cereal grains, is greatly influenced by variety (heredity) and agronomic conditions during culture and harvest. Milled products will vary some between processors because of differences in milling operations. Typical compositions for groats, rolled oats, and oat flour are listed in Table 8.13.

The nutritional quality of oat protein is good; rolled oats have a protein efficiency ratio (PER) of 2.2 compared to casein that has a PER of 2.5. Feeding studies of 7 pure oat varieties—Garland, Clintland, Bonkee, Newton, Beedee, Lodi, and Newaha, which had PER values between 2.25 and 2.38, indicated that the small variations in amino acids observed in these oat samples were not great enough to influence growth response (Clark and Potter 1972; Hischke et al. 1968). The protein content of oat groats being used for food has a value between 11 and 15%. However, some oats found in the Near East have protein contents varying between 14 and 25% in the groats.

TABLE 8.13
COMPOSITION OF GROATS, ROLLED OATS, AND OAT FLOUR (%, AS-IS BASIS)

Fraction	Moisture	Protein (N × 6.25)	Crude Fat	Crude Fiber	Ash	Nitrogen-Free Extract	Calcium	Phosphorus	Iron
Whole groats									
Groats to rolls	7.0	15.6	8.0	1.5	1.8	66.1	0.0625	0.4456	0.0054
Fines from flaking	10.0	18.6	10.0	2.0	3.0	56.4	0.0748	0.7078	0.0065
Package-grade regular rolled oats	10.0	15.7	7.8	1.5	1.8	63.2	0.0576	0.4362	0.0053
Cut groats									
Groats to rolls	7.0	15.5	7.8	1.5	1.8	66.4	0.0613	0.4522	0.0046
Fines from flaking	10.0	11.0	7.5	1.2	1.1	69.2	0.0380	0.2687	0.0030
Package-grade quick-cooking rolled oats	10.0	15.4	7.7	1.5	1.8	63.4	0.0638	0.4544	0.0047
Whole oat flour	6.4	16.7	6.0	1.1	1.7	68.1	—	—	—

Source: Salisbury and Wichser (1971).

Practically no research had been done on individual oat protein fractions until the pioneering work by Wu *et al.* (1972) at the USDA Northern Regional Research Laboratory. Protein concentrates have been prepared from oat flours of ordinary and high-protein varieties by air classification (Wu and Stringfellow 1973) and by wet-milling procedures (Wu *et al.* 1973; Cluskey *et al.* 1973). These protein products appear to have some promise for food applications.

Studies on selected B vitamin contents of oats and oat products have been reviewed by Geddes (1960) and are summarized in Table 8.14.

TABLE 8.14

SELECTED B VITAMIN CONTENTS OF OATS AND OAT PRODUCTS FROM SAME MILLING (mg/100 gm)

Product	Thiamin	Riboflavin	Niacin	Pyridoxine
Dry milled oats	0.65	0.14	1.15	0.20
Finished groats	0.77	0.14	0.97	0.12
Hulls	0.15	0.16	1.04	—
Oat shorts	0.44	0.35	1.62	—
Oat flour, chips, and meal	0.78	0.17	1.25	—

Source: Geddes (1960).

BARLEY

Barley is used for human food in the form of parched grain, pearled grain for soups, flour for flat bread, and ground grain for porridge. Barley flour is milled generally by conventional roller-milling (Pomeranz *et al.* 1971). Air classification can be used to separate barley flours into high-protein and low-protein fractions.

Conventional roller-milling of barley gives four major products: flour, tailings flour, shorts, and bran. The flour contains primarily the starchy endosperm; the shorts and tailings flour, a mixture of aleurone and pericarp with some germ and endosperm; and the bran, hulls and pericarp. On a dry-matter basis, barley contains 63–65% starch, 1–2% sucrose, 1% other sugars, 1–1.5% soluble gums, 8–10% hemicellulose, 4–5% cellulose, 2–3% lipids, 8–11% protein ($N \times 6.25$), 2–2.5% ash, and 5–6% other substances. The protein content of milled products can vary widely. The yield, protein contents, and amino acid composition of roller-milled barley to 65% extraction barley flour are contained in Table 8.15 (Robbins and Pomeranz 1972).

A screening program of the World Barley Collection for genetic

TABLE 8.15

YIELD, PROTEIN CONTENTS, AND AMINO ACID COMPOSITION
OF ROLLER-MILLED BARLEY PRODUCTS

Assay	Whole Kernel	Flour, 65% Extraction	Tailings Flour	Shorts	Bran
Yield, %	100.0	65.0	17.7	11.9	5.4
Protein[1]	9.3	9.8	11.3	8.8	3.1
Amino acids[2]					
Lysine	4.2	4.1	4.1	4.8	5.0
Histidine	2.4	2.4	2.4	2.1	1.4
Ammonia	3.1	3.1	3.0	2.9	3.5
Arginine	5.3	5.5	5.7	5.9	4.6
Aspartic acid	7.4	7.1	7.5	8.2	8.6
Threonine	3.6	3.6	3.6	3.8	4.2
Serine	4.1	4.0	4.1	4.2	4.7
Glutamic acid	22.6	23.3	22.9	21.2	20.6
Proline	11.4	10.1	9.6	9.2	9.9
Cystine/2	1.1	1.4	1.3	1.1	0.3
Glycine	4.5	4.3	4.7	5.1	5.0
Alanine	4.6	4.4	4.7	5.1	5.0
Valine	5.3	5.2	5.3	5.5	6.1
Methionine	2.5	2.7	2.5	2.5	2.3
Isoleucine	3.6	3.7	3.6	3.7	3.7
Leucine	6.8	7.0	6.8	6.9	7.5
Tyrosine	2.7	3.2	3.0	2.9	2.5
Phenylalanine	4.9	5.0	5.2	5.0	5.1

Source: Robbins and Pomeranz (1972).
[1] $N \times 6.25$, %.
[2] Grams of amino acid per 100 gm recovered.

varieties having high lysine and high protein was successful, and the most promising variety, CI 3947 (Hagberg and Karlsson 1969) was later called Hiproly. The opportunities offered by this improved barley variety and its properties have been reviewed by Munck (1972).

RYE

In the United States, about 5.6 million bushels of rye are processed in a year for food products. The principal products are rye flours and meals for bread, crackers, and snack foods. Rye bread is favored by the U.S. population with a recent European background. Cracker and biscuit manufacturers frequently use about 10% white rye flour in their flour blend. Meat processors may use rye flours as fillers in ground meat products.

Rye is milled after it has been cleaned and tempered (Shaw 1970). Tempered rye should have from 14.5 to 15.0% moisture for optimum milling. The milling process involves passing rye grain over

156 NUTRITIONAL EVALUATION OF FOOD PROCESSING

reduction rolls. More than 60% of the total flour is produced during the first 2 breaks, the sizings, and the first 2 reductions. Basically, two grades of rye flour are made—white rye flour or patent and dark rye flour or clear. About 80% of the milled flour is white rye flour. Many grades of rye flour are possible by combining the white and dark rye flour streams in varying percentages. Ash specifications on rye flours are one criteria for selling the flours. Color and ash contents are fairly closely related. White flours range from 0.6 to 0.7% ash, whereas the dark rye flours will range from 2.2 to 3.0% ash. Little attention is given to the protein content of rye flours.

BIBLIOGRAPHY

ANDERSON, R. A. 1970. Corn wet milling industry. In Corn: Culture, Processing, Products. G. E. Inglett (Editor). Avi Publishing Co., Westport, Conn.
ANDERSON, R. A., and INGLETT, G. E. 1974. Flour milling. In Wheat: Production and Utilization. G. E. Inglett (Editor). Avi Publishing Co., Westport, Conn.
BARBER, S. 1972. Milled rice and changes during aging. In Rice Chemistry and Technology. D. F. Houston (Editor). American Association of Cereal Chemists, St. Paul, Minn.
BLESSIN, C. W. et al. 1972. Defatted germ flour—food ingredient from corn. Food Prod. Develop. 6, No. 3, 34-35.
BLESSIN, C. W. et al. 1973. Composition of three food products containing defatted corn germ flour. J. Food Sci. 38, 602-606.
BREKKE, O. L. 1970. Corn dry milling industry. In Corn: Culture, Processing, Products. G. E. Inglett (Editor). Avi Publishing Co., Westport, Conn.
BREKKE, O. L., GRIFFIN, E. L., JR., and BROOKS, P. 1971. Dry-milling of opaque-2 (high-lysine) corn. Cereal Chem. 48, 499-511.
BROCKINGTON, S. F. 1970. Corn dry milled products. In Corn: Culture, Processing, Products. G. E. Inglett (Editor). Avi Publishing Co., Westport, Conn.
CAGAMPANG, G. B. et al. 1966. Studies on the extraction and composition of rice proteins. Cereal Chem. 43, 145-155.
CLARK, W. L., and POTTER, G. C. 1972. The compositional and nutritional properties of protein in selected oat varieties. In Symposium: Seed Proteins. G. E. Inglett (Editor). Avi Publishing Co., Westport, Conn.
CLUSKEY, J. E. et al. 1973. Oat protein concentrates from a wet-milling process: Preparation. Cereal Chem. 50, 475-481.
DIMLER, R. J. 1960A. Effects of commercial processing of cereals on nutrient content. A. Milling. 1. Wheat. In Nutritional Evaluation of Food Processing. R. S. Harris, and H. von Loesecke (Editors). John Wiley & Sons, New York. Reprinted in 1971 by Avi Publishing Co., Westport, Conn.
DIMLER, R. J. 1960B. Effects of commercial processing of cereals on nutrient content. A. Milling. 3. Corn. In Nutritional Evaluation of Food Processing. R. S. Harris, and H. von Loesecke (Editors). John Wiley & Sons, New York. Reprinted in 1971 by Avi Publishing Co., Westport, Conn.
GEDDES, W. F. 1960. Oats and oat products. In Nutritional Evaluation of Food Processing. R. S. Harris, and H. von Loesecke (Editors). John Wiley & Sons, New York.
HAGBERG, A., and KARLSSON, K. E. 1969. Breeding for high protein con-

EFFECTS OF REFINING ON COMPOSITION OF FOODS 157

tent and quality in barley. Symposium: New approaches to breeding for improved plant protein. Intern. Atomic Energy Agency, Vienna (1968).
HISCHKE, H. H., JR., POTTER, G. C., and GRAHAM, W. R. 1968. The nutritive value of oat protein. I. Varietal differences as measured by amino acid analysis and rat growth response. Cereal Chem. 45, 374–378.
HOUSTON, D. F., and KOHLER, G. O. 1970. Nutritional properties of rice. Food and Nutrition Board, Natl. Res. Council—Natl. Acad. Sci., Washington, D.C.
HUNNELL, J. W., and NOWLIN, J. F. 1972. Solvent extractive rice milling. In Rice Chemistry and Technology. D. F. Houston (Editor). American Association of Cereal Chemists, St. Paul, Minn.
INGLETT, G. E. 1970A. Corn in perspective. In Corn: Culture, Processing, Products. G. E. Inglett (Editor). Avi Publishing Co., Westport, Conn.
INGLETT, G. E. 1970B. Food uses of corn around the world. In Corn: Culture, Processing, Products. G. E. Inglett (Editor). Avi Publishing Co., Westport, Conn.
INGLETT, G. E. 1972A. Seed proteins in perspective. In Symposium: Seed Proteins. G. E. Inglett (Editor). Avi Publishing Co., Westport, Conn.
INGLETT, G. E. 1972B. Corn proteins related to grain processing and nutritional value of products. In Symposium: Seed Proteins. G. E. Inglett (Editor). Avi Publishing Co., Westport, Conn.
INGLETT, G. E. 1974. Wheat in perspective. In Wheat: Production and Utilization. G. E. Inglett (Editor). Avi Publishing Co., Westport, Conn.
JULIANO, B. O. 1972. Studies on protein quality and quantity of rice. In Symposium: Seed Proteins. G. E. Inglett (Editor). Avi Publishing Co., Westport, Conn.
JULIANO, B. O., BAUTISTA, G. M., LUGAY, J. C., and REYES, A. C. 1964. Studies on physicochemical properties of rice. J. Agr. Food Chem. 12, 131–138.
JOHNSON, V. A., MATTERN, P. J., and SCHMIDT, J. W. 1972. Genetic studies of wheat proteins. In Symposium: Seed Proteins. G. E. Inglett (Editor). Avi Publishing Co., Westport, Conn.
MERTZ, E. T., BATES, L. S., and NELSON, O. E. 1964. Mutant gene that changes protein composition and increases lysine content of maize endosperm. Science 145, 279–280.
MORRIS, R., and SEARS, E. R. 1967. The cytogenetics of wheat and its relatives. In Wheat and Wheat Improvement. K. S. Quisenberry, and L. P. Reitz (Editors). Agronomy Monograph 13. American Society of Agronomy, Madison, Wisconsin.
MUNCK, L. 1972. Barley seed proteins. In Symposium: Seed Proteins. G. E. Inglett (Editor). Avi Publishing Co., Westport, Conn.
ORR, M. L., and WATT, B. K. 1968. Amino acid content of foods. USDA Home Econ. Res. Rept. 4.
POMERANZ, Y., KE, H., and WARD, A. B. 1971. Composition and utilization of milled barley products. I. Gross composition of roller-milled and air-separated fractions. Cereal Chem. 48, 47–58.
POMERANZ, Y., and SHELLENBERGER, J. A. 1971. Bread Science and Technology. Avi Publishing Co., Westport, Conn.
ROBBINS, G. S., and POMERANZ, Y. 1972. Composition and utilization of milled barley products. III. Amino acid composition. Cereal Chem. 49, 240–246.
SALISBURY, D. K., and WICHSER, W. R. 1971. Oat milling—systems and products. Assoc. Operative Millers Bull. pp. 3242–3247.
SHAW, M. 1970. Rye milling in the United States. Assoc. Operative Millers Bull. pp. 3203–3207.
TECSON, E. M. S. et al. 1971. Studies on the extraction and composition of rice endosperm glutelin and prolamin. Cereal Chem 48, 168–181.

TOEPFER, E. W. et al. 1972. Nutrient composition of selected wheats and wheat products. XI. Summary. Cereal Chem. 49, 173-186.
USDA. 1963. Composition of foods. USDA Agr. Res. Serv., Agricultural Handbook 8.
USDA. 1972. Agricultural Statistics. USDA, U.S. Government Printing Office.
WITTE, G. C., JR. 1970. Rice milling in the United States. Assoc. Operative Millers Bull. pp. 3147-3159.
WITTE, G. C., JR. 1972. Conventional rice milling in the United States. In Rice Chemistry and Technology. D. F. Houston (Editor). American Association of Cereal Chemists, St. Paul, Minn.
WU, Y. V. et al. 1972. Oats and their dry-milled fractions: Protein isolation and properties of four varieties. J. Agr. Food Chem. 20, 757-761.
WU, Y. V., and STRINGFELLOW, A. C. 1973. Protein concentrates from oat flours by air classification of normal and high-protein varieties. Cereal Chem. 50, 489-496.
WU, Y. V. et al. 1973. Oat protein concentrates from a wet-milling process: Composition and properties. Cereal Chem. 50, 481-488.

Walter J. Wolf

PART 2
Effects of Refining Operations on Legumes[1]

In this subchapter, the term legumes is restricted to soybeans, lentils, peas, and beans. All are members of the Leguminosae family but soybeans are atypical of the group because of their high oil content and lack of starch. Indeed, soybeans are often classified as oilseeds.

SOYBEANS

Soybeans are available in a number of varieties that have been selected for composition, growth at a given latitude, disease resistance, and other agronomic characteristics. Consequently, soybeans

---

[1] Contribution from the Northern Regional Research Laboratory, Agricultural Research Service, U.S. Department of Agriculture, Peoria, Illinois. The mention of firm names or trade products does not imply that they are endorsed or recommended by the Department of Agriculture over other firms or similar products not mentioned.

EFFECTS OF REFINING ON COMPOSITION OF FOODS 159

will vary in composition, and data selected may reflect varietal and environmental factors as well as effects of processing. For a detailed review of composition, processing, nutritional properties, and food uses of soybeans see Smith and Circle (1972).

Whole Soybeans

Seed Structure.—Anatomically soybeans consist of three parts: seed coat (hull), cotyledons, and hypocotyl. Proximate compositions of whole soybeans and the three parts are given in Table 8.16.

TABLE 8.16

PROXIMATE COMPOSITIONS OF SOYBEANS AND SEED PARTS[1]

Fraction	Protein ($N \times 6.25$) (%)	Fat (%)	Carbohydrate (%)	Ash (%)
Whole bean (100%)	40	21	34	5
Seed coat (8%)	9	1	86	4
Cotyledon (90%)	43	23	29	5
Hypocotyl (2%)	41	11	44	4

Source: Kawamura (1967).
[1] Moisture-free basis.

Because of their high fiber content the hulls are usually removed when soybeans are processed into products for human consumption. The cotyledons are the major seed part and have a subcellular structure characteristic of many oilseeds. The bulk of the proteins are stored in particles called protein bodies or aleurone grains and the oil is deposited in spherical sites called spherosomes (Fig. 8.1). The protein bodies vary from 2 to 20 μ in diameter while the spherosomes are only about 0.2-0.5 μ in diameter.

Composition.—Kawamura (1967) reported proximate compositions for 6 U.S. and 3 Japanese varieties. Data for the U.S. varieties are shown in Table 8.17.

Mineral contents of soybeans are summarized in Table 8.18. The major mineral is potassium followed by phosphorus. The phosphorus is distributed as follows: phytin 77%; phosphatides 14%; inorganic phosphorus 5%; and residual phosphorus (possibly nucleic acids) 4% (Earle and Milner 1938).

Amino acid composition of soybeans is listed in Table 8.19 (column for defatted, undehulled meal). Soybeans are a good source of essential amino acids with one exception: content of the sulfur

160 NUTRITIONAL EVALUATION OF FOOD PROCESSING

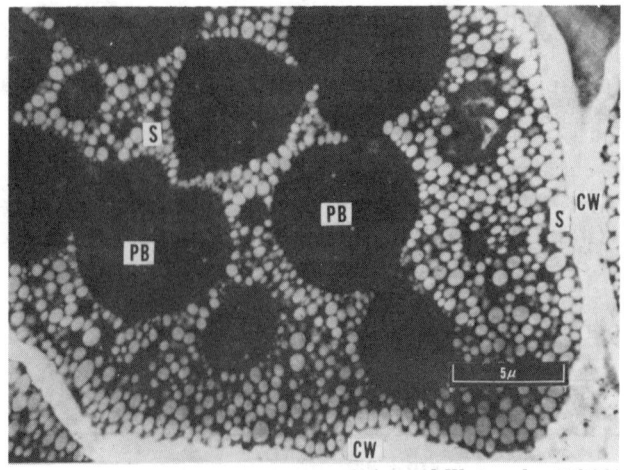

Saio and Watanabe (1968)

FIG. 8.1. ELECTRON MICROGRAPH OF A SECTION OF MATURE SOYBEAN COTYLEDON

Seed was soaked in water overnight, fixed with asmium tetroxide, and stained with uranyl acetate and lead citrate. Protein bodies (PB), spherosomes (S), and cell wall (CW) are labeled.

amino acids, methionine and cystine, is low and these are the first limiting amino acids.

The second largest constituent of nutritional importance in soybeans is the oil which constitutes about ⅕ of the seed. Fatty acid composition of soybean oil varies with crop year, growing location, and variety (Collins and Sedgwick 1959). Data for the major fatty acids in a commercial soybean oil are given in Table 8.20. Ranges for these fatty acids plus minor ones are also included in the table.

TABLE 8.17

PROXIMATE COMPOSITION FOR U.S. VARIETIES OF SOYBEANS[1]

Variety	Ash (%)	Oil (%)	Protein (%)	Nitrogen-Free Extract + Fiber (%)
Chippewa	4.8	20.6	42.9	31.7
Merit	5.5	21.9	40.8	31.9
Harosoy	4.7	22.0	38.6	34.8
Hawkeye	4.7	22.8	41.1	31.4
Hampton	5.0	23.6	40.1	31.3
Lee	4.8	22.8	42.0	30.5

Source: Kawamura (1967).

[1] Moisture-free basis.

EFFECTS OF REFINING ON COMPOSITION OF FOODS 161

TABLE 8.18

MINERAL CONTENTS OF SOYBEANS AND DERIVED PRODUCTS[1]

Mineral	Soybeans[2]	Defatted Flour[3]	Protein Concentrate[3]	Protein Isolate[3,4]
	(%)	(%)	(%)	(%)
Ash	4.60	6.4	5.1	4.2
Potassium	1.83	2.7	2.1	0.30
Phosphorus	0.78	0.73	0.70	0.77
Magnesium	0.31	0.26	0.25	0.07
Sodium	0.24	0.01	<0.005	1.1
Calcium	0.24	0.17	0.22	0.30
Sulfur	0.24	0.35	0.42	0.33
Chlorine	0.03	0.08	0.11	0.24
	Ppm	Ppm	Ppm	Ppm
Iron	80	65	100	167
Manganese	28	25	30	25
Boron	19	40	25	22
Zinc	18	73	46	110
Copper	12	14	16	14
Barium	8	6.5	3.5	5.7
Silicon	—	140	150	7
Molybdenum	—	3.9	4.5	3.8
Iodine	—	0.09	0.17	0.10
Aluminum	—	7.7	7.7	18
Strontium	—	0.85	0.85	2.3
Chromium	—	<1.5	<1.5	<1.5

[1] Moisture-free basis.
[2] Beeson (1941).
[3] Tech. Serv. Bull., Central Soya Co., Chicago.
[4] Sodium proteinate.

Approximately ⅓ of soybeans consists of soluble sugars and insoluble polysaccharides. Kawamura (1967) determined the soluble sugars in 6 U.S. and 3 Japanese soybean varieties. Average values for the nine varieties studied by Kawamura are in Table 8.21. Hymowitz et al. (1972A, B) examined a large number of soybean lines for sugar contents; mean values for 18 varieties are included in Table 8.21. Raffinose and stachyose have become of practical importance because of their contribution to flatulence when soybeans are ingested (Rackis et al. 1970). Soaking soybeans in water overnight results in a 3-4% loss of solids (Albrecht et al. 1967). These solids contain the oligosaccharides plus nitrogenous constituents. Leaching out of these components also occurs during boiling. For example, boiling soybeans for 20 min at a 1:10 bean-to-water ratio removes 33% of the oligosaccharides and 1% of the protein (Ku 1973).

The polysaccharide fraction of soybeans is much more complex

TABLE 8.19

AMINO ACID COMPOSITION OF SOYBEAN MEAL FRACTIONS

Amino Acid	Undehulled, Defatted Meal[1]	Dehulled, Defatted Meal[2]	Hulls[2]	Gm Amino Acid/16 Gm N Protein Concentrate[3]	Whey Protein[2]	Residue[2]	Acid-Precipitated Protein[2]	Protein Isolate[3]
Arginine	6.75	8.42	4.38	7.53	6.64	7.44	9.00	7.8
Histidine	2.72	2.55	2.54	2.65	3.25	2.70	2.83	2.5
Lysine	6.13	6.86	7.13	6.28	8.66	6.14	5.72	6.1
Tyrosine	3.04	3.90	4.66	3.93	4.67	3.30	4.64	3.7
Tryptophan	1.78	1.28	—	1.50	1.28	—	1.01	1.4
Phenylalanine	4.61	5.01	3.21	5.20	4.46	5.24	5.94	5.4
Cystine	2.06	1.58	1.66	1.58	1.82	0.71	1.00	1.0
Methionine	1.29	1.56	0.82	1.35	1.92	1.63	1.33	1.1
Serine	4.64	5.57	7.02	5.65	7.62	5.97	5.77	5.5
Threonine	3.54	4.31	3.66	4.16	6.18	4.67	3.76	3.7
Leucine	7.04	7.72	5.93	7.80	7.74	8.91	7.91	7.7
Isoleucine	4.74	5.10	3.80	4.76	5.06	6.02	5.03	4.9
Valine	5.33	5.38	4.55	4.91	6.19	6.37	5.18	4.8
Glutamic acid	18.08	21.00	8.66	19.84	15.64	17.76	23.40	20.5
Aspartic acid	11.33	12.01	10.05	12.00	14.08	12.39	12.87	11.9
Glycine	4.21	4.52	11.05	4.37	5.74	5.21	4.56	4.0
Alanine	4.18	4.51	3.98	4.35	6.16	5.73	4.48	3.9
Proline	5.14	6.28	5.76	5.17	6.66	5.35	6.55	5.3
Hydroxyproline	0	0	7.57	—	—	—	—	—
Ammonia	1.68	2.05	1.55	1.92	1.53	2.61	2.20	2.0

[1] Tkachuk and Irvine (1969).
[2] Rackis et al. (1961).
[3] Commercial product; Tech. Serv. Bull, Central Soya Co., Chicago.

EFFECTS OF REFINING ON COMPOSITION OF FOODS 163

TABLE 8.20

FATTY ACID COMPOSITION OF SOYBEAN OIL[1]

Fatty Acid	Commercial Soybean Oil[2]	Ranges[3]
C<14	—	<0.1
C14:0	—	<0.5
C16:0	10.2	7.0–12
C16:1	—	<0.5
C18:0	4.6	2.0–5.5
C18:1	25.5	19–30
C18:2	51.8	48–58
C18:3	7.8	4–10
C20:0	—	<1.0
C20:1	—	<1.0
C22:0	—	<0.5

[1] Expressed as wt % of methyl esters as analyzed by gas liquid chromatography.
[2] Degummed, alkali refined (Cowan et al. 1970).
[3] Ranges typical of commercial soybean oil (FAO/WHO Codex Alimentarius Commission 1974).

than the oligosaccharide portion and is still poorly understood. The hulls are predominantly polysaccharides whereas only about 14–17% of defatted cotyledon meal is estimated to be polysaccharides. Hulls contain galactomannans (9–11%), acidic polysaccharides that resemble pectic polysaccharides from other sources (10–12%), xylan (9–10%), cellulose (~40%), proteins and peptides (~11%); the remainder is probably mainly lignin (Aspinall et al. 1967). The cotyledon polysaccharides originate primarily from the cell wall whose composition is estimated to be 30% pectic polysaccharides, 50% hemicelluloses (arabinogalactans), and 20% cellulose (Kikuchi et al. 1971). When soybeans are cooked, an arabinogalactan and two acidic polysaccharides dissolve in the cook water (Kikuchi 1972).

TABLE 8.21

SUGARS IN SOYBEANS AND SOYBEAN FRACTIONS

Sugar	Whole Soybeans[1] (%)	Whole Soybeans[2] (%)	Hulls[1]	Defatted Hypocotyls[1]	Defatted Cotyledons[1]
Arabinose	0.002	—	0.02	—	—
Glucose[3]	0.006	—	0.05	—	—
Sucrose	5.0	5.96	0.60	7.0	6.6
Raffinose	1.1	0.75	0.13	1.9	1.4
Stachyose	3.8	2.65	0.41	7.7	5.3

[1] Kawamura (1967).
[2] Hymowitz et al. (1972B).
[3] May include trace of fructose.

Vitamin contents of soybeans and various derived products are summarized in Table 8.22. Liener (1972) recently reviewed these constituents and pointed out that variations in analytical values probably reflect differences between analytical techniques as well as the influence of processing conditions.

Other minor constituents of soybeans include 1.05% nucleic acids (DiCarlo et al. 1955). Phenolic acids including syringic, vanillic, ferulic, gentistic, salicylic, p-coumaric, and p-hydroxybenzoic acids were isolated from hexane-defatted flour by Arai et al. (1966). Syringic acid was present in highest concentration.

Food Usage.—Very small amounts of whole soybeans are consumed as foods in the United States. Usage is limited to canned soybeans in tomato sauce, canned green soybeans, and extraction of whole beans to make soy milk-based infant formulas and beverages. Consumption of these products is restricted mainly to vegetarians and infants allergic to cow's milk. Roasted soybeans have been available for several years as snack items. Roasting consists of a rapid dehydration of water-soaked beans followed by a partial pyrolysis. Some destruction of lysine, cystine, and tryptophan occurs during roasting. As a result of roasting, protein efficiency ratio increases from 0.6 for raw beans up to 1.7 and then decreases again as roasting is continued (Badenhop and Hackler 1971). A soy-based product simulating lightly roasted peanuts was introduced in 1971. It is prepared by soaking and leaching soybeans followed by removal of the seed coat and an undisclosed dehydration process that minimizes shrinkage of the expanded cotyledons. The product analyzes 48% protein, 26% fat, 2% moisture, 4% ash, 4% fiber, and 16% carbohydrates. The high protein and fat contents as compared to whole soybeans reflect removal of the seed coat and some of the soluble sugars during soaking.

The largest use of whole soybeans is in the Orient where they are converted into soy milk, tofu, miso, and tempeh. These are discussed briefly in a later section (Oriental Soybean Foods).

Full-Fat Soy Flour

Processing.—A minimum of processing is involved in preparation of full-fat flour but the seed structure is disrupted in the process. In commerce, beans are steamed, dried, cracked between corrugated rolls, dehulled by screening and aspiration, and finally ground. The initial steaming inactivates antinutritional factors plus lipoxygenases and other enzymes that may catalyze degradation of lipids if allowed to remain active when the seed structure is disrupted (Mustakas et al. 1969; Nelson et al. 1971).

TABLE 8.22

VITAMIN CONTENT OF SOYBEANS AND DERIVED FOOD PRODUCTS[1]

Soybean Product	β-Carotene (μg/Gm)	Thiamin (μg/Gm)	Riboflavin (μg/Gm)	Niacin (μg/Gm)	Pantothenic Acid (μg/Gm)	Pyridoxine (μg/Gm)	Biotin (μg/Gm)	Folic Acid (μg/Gm)	Inositol (Mg/Gm)	Choline (Mg/Gm)	Ascorbic Acid (Mg/Gm)
Immature bean	2–7	6.4	3.5		12	3.5	0.5	1.3		3.0–3.3	0.2
Mature bean	0.2–2.4	11.0–17.5	2.3	20.0–25.9	12	6.4	0.6	2.3	1.9–2.6	3.4	0.2
Sprouts		11.9–21.9	4.8–7.0	29.9–48.0	18.8–34.4	14.1–17.7	1.1–1.7	3.7	2.5–3.9		0.4
Defatted flour[3]	<0.01		2.7	21.1	22.8	6.2	0.5	2.6	1.8	4.3	0.1
Protein concentrate[3]	<0.01		1.8	11.6	2.1	2.6	0.5	3.6	1.2	0.7	0.06
Protein isolate[3]	<0.01		1.5		5.7	1.6	0.1	0.9	1.5	1.2	0.1
Curd (tofu)		3.9	3.7	6							
Milk[4]	7.50	0.8	1.1	5.5							
Miso		1.3	1.4	2.5							21.6

Source: Liener (1972) except as indicated otherwise.

[1] Where a range of values is shown, average value is very closely given by taking the average of the two extreme values.
[2] Weakley and McKinney (1958) report that mature soybeans contain no ascorbic acid; values reported by earlier workers may be in error.
[3] Tech. Serv. Bull., Central Soya Co., Chicago.
[4] Expressed as mg/liter with the exception of β-carotene which is expressed in terms of IU of vitamin A per liter.

Extrusion cooking of soybeans as an alternative process for preparing full-fat flour was reported by Mustakas et al. (1970). Beans were dehulled by conventional means, heated, raised to 18% moisture, extruded (250°–290°F), cooled, and ground. A second experimental method for making full-fat soy flour consists of soaking beans in water, boiling, drying, dehulling, and grinding (Albrecht et al. 1967); alternatively, the boiled beans (hulls remain in product) may be ground into a paste and then drum-dried to yield a full-fat "flake" (Shemer et al. 1973).

Composition.—The major fractionation that occurs in the manufacture of full-fat flour is the separation of the hulls and cotyledons. The hulls are low in protein content (Table 8.16) but are unique in their amino acid content because they contain hydroxyproline (Table 8.19). Full-fat soy flour closely resembles the composition of whole soybeans because only the hulls are removed during processing. Proximate analyses for commercial full-fat flours are similar to values for experimentally cooked soy flour (Table 8.23).

Nutritional Properties.—Under optimum conditions a high nutritional value is obtained for full-fat flours (Table 8.23); little destruction of available lysine or thiamin occurs and loss of tocopherols is only about 15% (Mustakas et al. 1970). Heat treatment during processing inactivates antinutritional factors such as trypsin inhibitor that are responsible for poor growth of young animals when fed raw

TABLE 8.23

PROXIMATE ANALYSES AND BIOCHEMICAL PARAMETERS OF SOY FLOURS, PROTEIN CONCENTRATES, AND PROTEIN ISOLATES

	Full-Fat Flour[1]	Toasted Defatted Flour[2]	Protein Concentrate[2,3]	Protein Isolate[2]
Moisture, %	3.4	6.5	8.0	4.8
Protein ($N \times 6.25$), %	41.0	53.0	65.3	92.0
Crude fat, %	22.5	1.0	0.3	—
Crude fiber, %	1.7	3.0	2.9	0.25
Ash, %	5.1	6.0	4.7	4.0
PER[4]	2.15	2.3	2.3	1.1–1.2
Inactivation of trypsin inhibitors, %	89	—	—	—
Urease activity, pH change	0.1	—	—	—
Nitrogen solubility index	16	15–25	5	75

[1] Experimental sample (Mustakas et al. 1970).
[2] Tech. Serv. Manual, Central Soya Co., Chicago.
[3] Prepared by extraction with aqueous alcohol.
[4] Corrected to casein = 2.50.

EFFECTS OF REFINING ON COMPOSITION OF FOODS   167

soybean products. The effects of heat treatment on the nutritive properties of soy proteins are discussed in greater detail in a following section (Defatted Grits and Flours). All full-fat flours in the United States are heat-treated whereas in Great Britain unheated full-fat flour is used as a lipoxygenase source to bleach pigments in bread baking (Pringle 1974). Heating during baking inactivates the antinutritional factors in the raw soy flour.

Processing Soybeans into Oil and Defatted Flakes

Commercial processing of soybeans into oil and defatted flakes is outlined in Fig. 8.2. In the first steps, soybeans are cleaned,

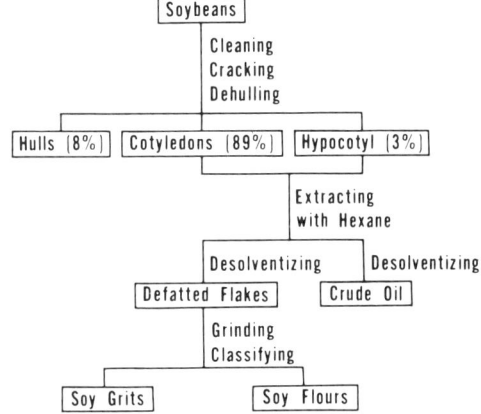

FIG. 8.2. OUTLINE OF COMMERCIAL PROCESSING OF SOYBEANS INTO OIL AND DEFATTED GRITS AND FLOURS

cracked, and dehulled. The dehulled material is then conditioned, flaked, and extracted with hexane to remove the oil. After defatting, the flakes are desolventized. For feed use the flakes are cooked in a desolventizer-toaster to remove the hexane efficiently as well as to inactivate antinutritional factors. For edible purposes, other methods are used to desolventize and to control the amount of heat treatment given to the flakes (Becker 1971). The final step is grinding and classifying into grits and flours.

Edible Soybean Oil Products.—The oil-laden hexane from the extraction step is filtered to remove fines. The hexane is then stripped from the crude oil in film evaporators and stripping columns. The next step is degumming in which about 1% moisture is added to the oil to hydrate the phosphatides. The hydrated phosphatides are centrifuged out and further refined into lecithin which is used in a wide variety of processed foods (Wolf and Cowan 1971). The degummed oil is the starting material for shortening, margarine, plus salad and cooking oils (Fig. 8.3). The first step is alkali refining to remove free fatty acids. Bleaching follows although it may be omitted for some salad oil uses. The bleached oil is hydrogenated,

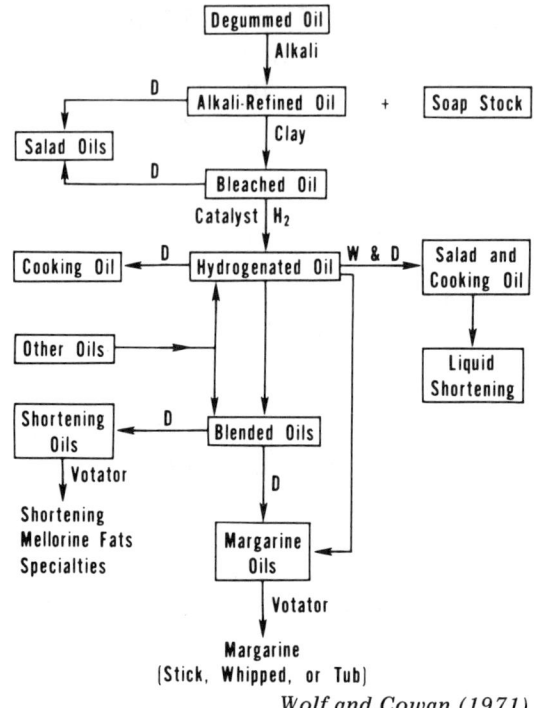

Wolf and Cowan (1971)

FIG. 8.3. OUTLINE OF CONVERSION OF SOYBEAN OIL INTO EDIBLE PRODUCTS, WHERE D STANDS FOR DEODORIZATION AND W FOR WINTERIZATION

Votator is the name of the chilling machine used to control crystal structure of the products.

winterized, and deodorized to yield salad and cooking oils and liquid shortenings. The hydrogenated oils are also converted into shortenings and margarines but are often blended with other oils to give desired physical and chemical properties. Further details about the oil refining process and products can be found in Wolf and Cowan (1971).

Defatted Grits and Flours.—These products are differentiated on the basis of particle size. Grits have particle sizes larger than 100 mesh whereas flours are 100 mesh or finer. Grits are further classified as coarse (10-20 mesh), medium (20-50 mesh) and fine (50-80 mesh). Grits and flours are available with varying degrees of moist heat treatment. Heat treatment inactivates enzymes, improves flavor and nutritional quality but increases color and decreases water dispersibility of the proteins. Degree of heat treatment is measured by determining the Nitrogen Solubility Index (NSI) which is defined as the percentage of the total nitrogenous constituents (primarily proteins) that is dispersible in water under specified conditions. A similar method called the Protein Dispersibility Index (PDI) is also used (American Oil Chemists' Society 1970). The following table indicates the relationship between moist heat treatment and NSI:

Amount of Heat	NSI
Minimum	85-90
Light	40-60
Moderate	20-40
Fully toasted	10-20

Flakes prepared with a minimum of heat treatment retain their enzyme activities and have high protein dispersibilities. Such flakes are preferred as starting materials for protein isolates, for certain types of protein concentrates, and for enzyme-active soy flours used in bread making.

When raw soybean meal is fed to rats and other experimental animals, they grow poorly. About 40% of the poor growth induced by raw soybean meal is attributed to trypsin inhibitors but the other antigrowth factors are still unidentified (Kakade et al. 1973). Accompanying the poor growth is an enlargement of the pancreas (hypertrophy) which is believed to be caused by trypsin inhibitors. The mechanism by which trypsin inhibitors retard growth and bring about pancreatic hypertrophy is still uncertain (Rackis 1972).

The antinutritional factors in raw soybean meal are readily inactivated by steaming. Raw meal has a protein efficiency ratio (PER) of about 1.3 but after 15 min of steaming the protein efficiency increases to 1.9-2.0 depending on the initial moisture content (Fig. 8.4). Simultaneously, trypsin inhibitor activity drops rapidly and reaches a residual activity of less than 5% in the same time period. Fully toasted flakes and grits are often undesirable in food applications because of insolubility of the proteins and the dark color of the toasted products. Consequently, less heated products are used

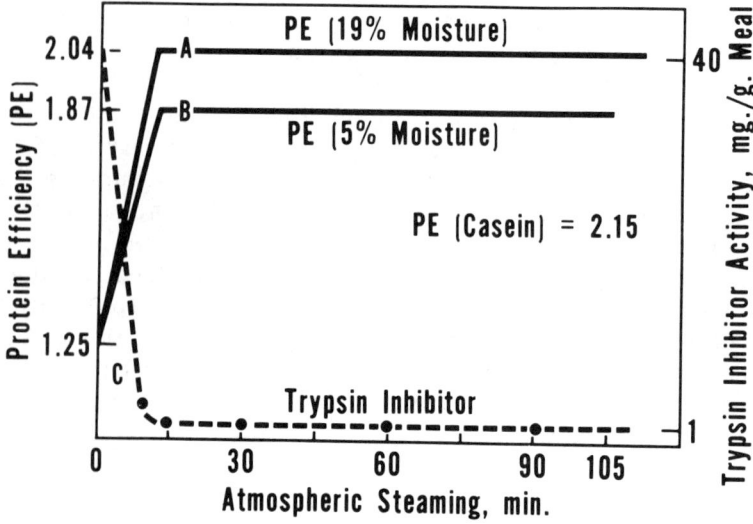

Rackis (1972)

FIG. 8.4. EFFECT OF AUTOCLAVING ON PROTEIN EFFICIENCY (PE) AND TRYPSIN INHIBITOR ACTIVITY OF RAW SOYBEAN MEAL

Conditions: Live steam at atmospheric pressure, 100°C. Curve A, meal with initial moisture of 19%; curve B, meal with initial moisture of 5%; curve C, trypsin inhibitor activity.

but subsequent heat treatment of the food is relied upon to completely inactivate the residual antinutritional factors.

As a result of removing the hulls (largely carbohydrates) and oil during processing, defatted flours and grits have higher protein contents than full-fat flour (Table 8.23). Commercial defatted flours contain a minimum of 50% protein but typically will analyze 52–53%. Mineral content for a commercial defatted flour is given in Table 8.18. Amino acid composition is not altered appreciably during processing of soybeans into defatted meal (Table 8.19); removal of the amino acids of the hulls has little effect on overall composition. Sugars are not soluble in hexane and as a result the sugar content of defatted meal (defatted cotyledons in Table 8.21) is higher than for whole soybeans. Vitamin analyses for defatted flour are included in Table 8.22.

Soy Protein Concentrates

By removing the soluble sugars and other low-molecular-weight components of defatted soy flakes, the protein content is raised to about 70% and the resulting product is called a protein concentrate. Three established processes are used to prepare protein concentrates

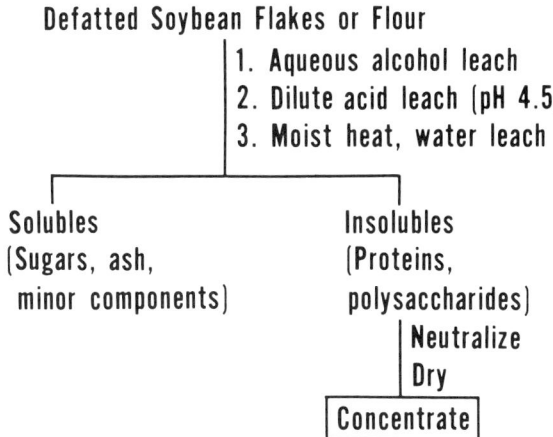

FIG. 8.5. OUTLINE OF PROCESSES FOR PREPARING SOY PROTEIN CONCENTRATES

Initial extraction can be made by one of three alternate methods as described in the text.

as outlined in Fig. 8.5. In the first method aqueous alcohol is used to dissolve the sugars and nonprotein constituents; the proteins and polysaccharides are insoluble in alcohol and constitute the protein concentrate after the solvent is removed (Mustakas et al. 1962). In the second process defatted flakes or flour are extracted with dilute acid at pH 4.5. At this pH the globulin protein fraction is insoluble and after neutralizing and drying, the protein and insoluble polysaccharides make up the second type of protein concentrate (Sair 1959). In the third process the soy flakes or flours are toasted to heat-denature the proteins. Subsequent washing with water then removes the sugars leaving the denatured proteins and the polysaccharides (McAnelly 1964). A fourth process recently described involves adding alcohol to hexane-wet flakes after they leave the extractor. The hexane-alcohol mixture is then separated to remove residual lipids not extracted by hexane alone. Subsequent washing with aqueous alcohol yields a protein concentrate of improved flavor (Hayes and Simms 1973).

Proximate analyses for a protein concentrate prepared by aqueous alcohol extraction are given in Table 8.23; the other types of protein concentrates are similar in composition. Although alike in composition, the three types of concentrates differ significantly in solubility of the proteins. The acid-leached protein concentrates may contain about 70% water-soluble protein whereas the other 2 protein concentrates have protein solubilities of only 3–5% as a result of protein denaturation by alcohol and heat. Mineral and vitamin contents of the alcohol-extracted protein concentrate are listed in Tables 8.18 and 8.22, respectively. Amino acid composition of protein concentrate made by leaching defatted flakes with aqueous alcohol is given in Table 8.19. The first limiting amino acid is methionine and the composition of the concentrate approximates that of defatted meal. Longenecker et al. (1964) found the protein efficiency ratio (PER) for a protein concentrate (protein contents of two other samples labeled protein concentrates in this study indicate that they are soy flours rather than concentrates) to be 1.86 and on heating it increased to 2.02 as compared to a casein control of 3.00. A later report by Meyer (1966), however, gave the following results for three concentrates:

	PER	
	Methionine Supplementation	
Concentrate	None	0.15%
A	2.29	3.00
B	2.16	2.86
C	2.36	3.06
Casein	2.50	

Heat treatment was unnecessary and was attributed to improved manufacturing processes. When 0.15% methionine was added to the diet, all concentrates out-performed the casein control. The nutritive value of one type of protein concentrate (made by moist heating and water leaching) was also demonstrated in its use as a protein supplement for wheat flour in bread making (Wilding et al. 1968):

Concentrate:Wheat Protein Ratio	PER
0:100	1.1
25:75	1.8
50:50	2.5
75:25	2.6
100:0	2.4
Casein	2.8

Soy Protein Isolates

Isolates are the most refined form of soy protein available. They are processed one step beyond the protein concentrates by removing the water-insoluble polysaccharides as well as the water-soluble sugars and other minor constituents (Fig. 8.6). Defatted flakes or flour with a high protein solubility [desolventized with a minimum of heat treatment as described by Becker (1971)] are extracted with

EFFECTS OF REFINING ON COMPOSITION OF FOODS 173

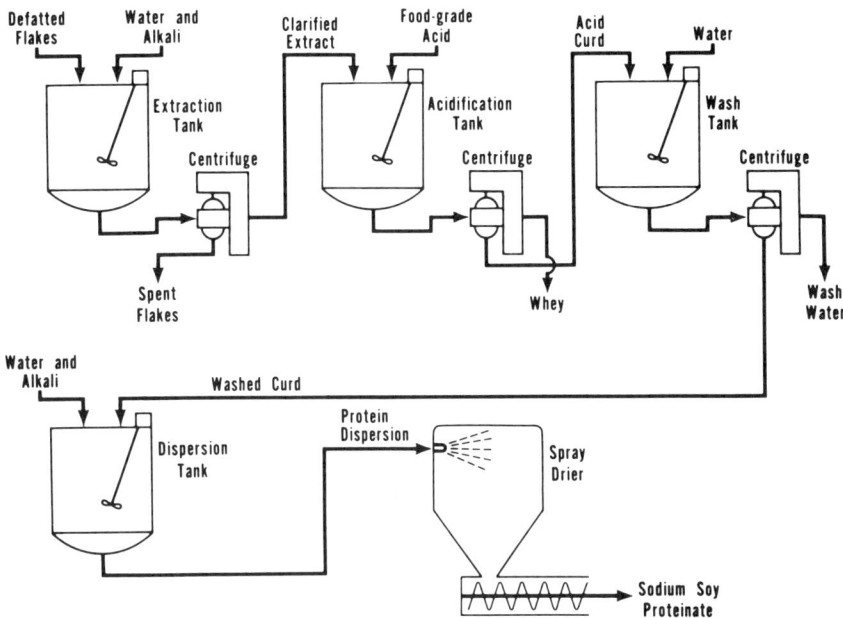

FIG. 8.6. OUTLINE OF PROCESS FOR COMMERCIAL PRODUCTION OF SOYBEAN PROTEIN ISOLATES

dilute alkali (pH 7-9) at 50°-55°C. The spent flake residue (water-insoluble polysaccharides plus residual protein) is then removed by filtration or centrifugation. The resulting extract which contains the dissolved proteins plus the soluble sugars is then adjusted to about pH 4.5. This is the approximate isoelectric point for the bulk of the proteins and the pH region of minimum solubility. Consequently, the proteins precipitate. After centrifuging to remove the whey (contains soluble sugars, some proteins, peptides, salts, and minor constituents) the resulting curd is washed to more thoroughly remove soluble compounds. After washing, the protein curd can be concentrated to 15-30% solids and dried to yield the isoelectric (water-insoluble) form of the protein. More frequently, however, the washed curd is resolubilized by neutralizing to about pH 7 and then spray dried. This latter process yields the sodium, potassium, or calcium proteinates depending on the alkali used for neutralization. The proteinates are usually the preferred form because they are water-dispersible and therefore easier to incorporate into food products.

Proximate analyses for a commercial isolate are listed in Table 8.23 which also includes analyses for the less refined forms of soy protein. Protein contents ($N \times 6.25$) on a dry basis typically are 95% or higher. Mineral and vitamin analyses for this same commercial isolate are found in Tables 8.18 and 8.22, respectively. Amino acid composition data in Table 8.19 for an experimental protein isolate (acid-precipitated protein) and the two by-products—whey and residue—show the distribution of amino acids when an isolate is prepared from defatted meal. Data for a commercial protein isolate are also given in Table 8.19. The proteins left in the whey and residue fractions are higher in sulfur amino acids than the isolate. Consequently, the nutritive value of the isolate is lower than for defatted meal as demonstrated with humans (Longenecker 1963), rats (Meyer 1966), and pigs (Berry et al. 1966).

Some workers report that heat treatment improves the nutritive value of soy isolates whereas others do not observe any beneficial effects as a result of heating (Liener 1972). Isolates vary in PER depending on their source. In a study by Longenecker et al. (1964) one isolate had a PER 45% that of the casein standard and on heating increased to only 49% whereas another had a PER 44% of the standard but rose to 76% after heating. Differences in the extent of removing the antinutritional factors existing in raw meal may account for this variability.

Another factor that can adversely affect nutritional quality of isolates is the pH during extraction and subsequent processing. In alkaline solutions, cystine and serine residues of soy isolates may be converted to dehydroalanine which, in turn, can react with the ϵ-amino groups of lysine to form lysinoalanine:

$$\begin{array}{cc} | & | \\ NH & NH \\ | & | \\ C{=}CH_2 \;+\; H_2N(CH_2)_4 & CH \\ | & | \\ CO & CO \\ | & | \end{array}$$

$$\downarrow$$

$$\begin{array}{cc} | & | \\ NH & NH \\ | & | \\ CHCH_2NH(CH_2)_4 & CH \\ | & | \\ CO & CO \\ | & | \end{array}$$

$$\downarrow \text{Hydrolysis}$$

$$\begin{array}{cc} NH_2 & NH_2 \\ | & | \\ CHCH_2NH(CH_2)_4 & CH \\ | & | \\ COOH & COOH \end{array}$$

Lysinoalanine

EFFECTS OF REFINING ON COMPOSITION OF FOODS 175

DeGroot and Slump (1969) found that soybean meal or isolated protein held at pH 12.2 and 40°C for 4 hr had respective contents of lysinoalanine of 0.57 and 0.80 gm per 16 gm N and significant decreases in their net protein utilization values. Cystine and lysine contents of the proteins decreased during such alkali treatment; serine also decreased when alkali treatment was conducted at 60°-80°C. Lysinoalanine was not formed when isolate was treated at pH 7-9 but was detected after 4 hr boiling at pH 10 at 40°C. When the alkali-treated proteins were fed to rats, histological examination showed no evidence of tissue change except for nephrocalcinosis in females. This was overcome by supplementary calcium in the diet.

A recent study by Woodard and Short (1973), however, reports that feeding alkali-treated soy protein to rats causes cytomegalia in the straight portion of the proximal tubule of the kidney. They also isolated lysinoalanine from a hydrolyzate of the alkali-treated protein. They did not feed authentic lysinoalanine in a control diet to establish that this compound is, in fact, responsible for the effects noted. In agreement with DeGroot and Slump, they found that the commercial, edible soy isolate produced no unusual effects.

Oriental Soybean Foods

Soybeans have been an important source of protein in China, Japan, and other Oriental countries for many centuries. A variety of foods has been developed and they can be conveniently classified as nonfermented and fermented foods. Some of these products are listed in Table 8.24.

TABLE 8.24

COMPOSITION OF ORIENTAL SOYBEAN FOODS[1]

Food	Description	Moisture (%)	Protein (%)	Fat (%)	Carbohydrate (%)	Ash (%)
Nonfermented						
Tofu	Calcium precipitated oil-protein complex	88.0	6.7	3.5	1.9	0.6
Aburage	Fried tofu	44.0	20.4	31.4	—	1.4
Kori-tofu	Dried tofu	10.4	58.8	26.4	7.0	2.6
Yuba	Oil-protein film from heated soy milk	8.7	57.6	24.1	11.9	3.0
Kinako	Roasted soybean powder	5.0	42.1	19.2	29.5	5.0
Fermented						
Miso	Paste of soybeans fermented with or without rice or barley	50.0	12.6	3.4	19.4	12.8
Natto	Soybeans fermented with *Bacillus* natto	58.5	16.5	10.1	10.1	2.6

[1] From Smith and Circle (1972) Chap. 10 and 11.

These products are not well known in Western countries except in Oriental communities. One food not listed in Table 8.24 that is well known outside of the Orient is soy sauce (called shoyu in Japan). The preferred form in Japan is made by fermenting soybeans (often defatted meal), wheat, and salt until extensive hydrolysis of the protein and carbohydrates occurs and a bright reddish brown liquor is obtained which is then separated from the insoluble residue. A related product is made by acid hydrolysis of defatted soybean meal or other plant protein materials and this is the form most people in this country are familiar with. However, in 1973, Kikkoman Shoyu Co. of Japan completed a plant in the United States to produce fermented soy sauce (Ziemba 1974). A good fermented soy sauce contains about 18% sodium chloride plus a complex mixture of amino acids, peptides, peptones, sugars, and other constituents. A detailed discussion of Oriental soybean foods can be found elsewhere (Smith and Circle 1972).

PEAS AND BEANS

Five groups of legume seeds classified as peas and beans are of major importance for food uses: chick peas (*Cicer arietinum* also called garbanzo, Bengal gram, chenata or chana); peas (*Pisum sativum*, var. *arvense* Poir., field or smooth pea, and *P. sativum* L., or wrinkled pea); broad beans (*Vicia faba* also called horse or field bean); lentils (*Lens esculenta*) and beans (*Phaseolus vulgaris*, *P. lunatus*, *P. aureus*, and *P. mungo*). Deschamps (1958) has written an excellent review on these legumes (often referred to as pulses) which should be consulted for more detailed information. Only mature peas and beans are considered here. Other legumes such as certain *Lathyrus* species that contain toxic compounds have also been excluded; these are reviewed elsewhere (Sarma and Padmanaban 1969).

Structure

Dry chick peas have the most unusual shape of these pulses. The seed resembles the head of a pullet which is believed to be the origin of the name. Deschamps (1958) shows a picture of the seed and describes some of its other characteristics. Transmission electron microscopy of cotyledon cells of peas revealed protein bodies and starch granules plus the usual cellular organelles—nuclei, mitochondria, golgi bodies, and chloroplasts (Varner and Schidlovsky 1963). A study of structure of *Vicia faba* beans with the scanning electron microscopy was published recently (McEwen *et al.* 1974). Rockland and Jones (1974) have examined the cellular structure of lima beans and how it is affected by cooking.

EFFECTS OF REFINING ON COMPOSITION OF FOODS

Composition

Proximate analyses of various peas and beans are given in Table 8.25. Protein contents range from about 20 to 30%. Fat contents tend to be low as compared to soybeans and vary from 1 to 5%. Ash (3-5%) and fiber (2-10%) make up the other minor fractions. Carbohydrates are the major constituent of peas and beans and fall into the range of 55 to 66%.

TABLE 8.25

COMPOSITION OF PEAS AND BEANS[1]

Legume	Protein[2] (%)	Fat (%)	Ash (%)	Fiber (%)	Carbohydrate (%)	Reference
Chick pea (Cicer arietinum)	20.6	5.4	2.8	10.3	61	Verma et al. (1964)
Pea (Pisum sativum L.)	29.8	1.8	3.1	6.8	59	Zarkadas et al. (1965)
Broad bean (Vicia faba)	31.4	1.5	4.0	8.0	55	Eden (1968)
Lentil (Lens esculenta)	24	1.4	3.5	4.2	66	Kahn and Baker (1957)
Large lima (Phaseolus lunatus)	24.3	3.7	5.1	8.4	59	Maneepun et al. (1974)
Pink Rajmah (Phaseolus vulgaris)	27.8	1.6	3.7	1.7	65	Pant and Tulsiani (1969)
Urd (Phaseolus mungo)	28.8	1.7	3.4	3.2	63	Pant and Tulsiani (1969)

[1] Moisture-free basis.
[2] Kjeldahl $N \times 6.25$.

Essential amino acid contents of some of the peas and beans are listed in Table 8.26. Most of these legumes are good sources of lysine but are low in methionine. The pea sample listed in Table 8.26 contained no methionine although earlier workers report 0.3-1.3 gm per 16 gm N (Deschamps 1958).

Mineral and vitamin contents of peas and beans are summarized in Table 8.27. All of the legumes are high in phosphorus content and this element occurs in a variety of forms. For example, Verma and Lal (1966) found the following distribution of phosphorus in Bengal gram (Cicer arietinum): Acid-soluble 74%; inorganic 11%; phytate 45%; phosphatide 16%; and unaccounted (mainly nucleic acid) 10%. In general, the legumes (including soybeans, Table 8.22) are poor sources of carotene and ascorbic acid.

Fatty acid analysis of the oils obtained from peas and beans (Table 8.28) shows that the unsaturated fatty acids predominate but composition varies greatly. For example, linolenic acid varies from less than 2% in lentils to 51% of the fatty acids in kidney beans.

The major carbohydrate in peas and beans is starch but numerous other sugar constituents are also present. For example, Pritchard et

TABLE 8.26

ESSENTIAL AMINO ACID CONTENT OF PEAS AND BEANS

	Chick Pea[1] (Cicer arietinum)	Pea[2] (Pisum sativum)	Broad Bean[3] (Vicia faba)	Lentil[3] (Lens esculenta)	Navy Bean[4] (Phaseolus vulgaris)	Lima Bean[5] (Phaseolus lunatus)	Mung Bean[3] (Phaseolus mungo)
Protein content, %[6]	17.5	29.8	24.7	24.1	24.0[7]	24.3	26.0
Amino Acids				Gm/16 Gm N			
Arginine	7.98	5.33	7.45	8.45	6.7	4.34	5.15
Histidine	2.57	1.49	3.26	3.81	2.1	2.75	3.28
Isoleucine	4.53	4.81	6.16	6.30	4.6	3.57	5.80
Leucine	7.63	9.70	10.29	10.9	7.2	7.48	10.5
Lysine	7.72	6.98	6.68	7.96	7.3	6.56	6.80
Methionine	1.16	0	0.63	0.70	1.1	0.92	1.12
Phenylalanine	6.46	4.29	5.37	6.25	5.9	4.68	6.30
Threonine	3.86[3]	3.43	4.41	4.47	3.4	3.43	3.85
Tryptophan	1.78[3]	—[8]	1.01	1.22	—[8]	—[8]	1.40
Valine	4.63	4.36	5.48	5.42	5.1	3.81	5.57

[1] Shehata and Fryer (1970).
[2] Zarkadas et al. (1965).
[3] Kahn and Baker (1957).
[4] Bandemer and Evans (1963).
[5] Maneepun et al. (1974).
[6] Moisture-free basis.
[7] As-is basis on air-dried sample.
[8] Not determined.

TABLE 8.27

MINERAL AND VITAMIN CONTENT OF PEAS AND BEANS[1]

Legume	Phosphorus	Calcium	Iron	Carotene	Thiamin	Riboflavin	Niacin	Ascorbic Acid
				Mg/100 Gm Whole Seed (Dry Basis)				
Chick pea (*Cicer arietinum*)	438	115	9.7	0.07	0.81	0.19	1.68	0.72
Pea (*Pisum sativum*)	320	80	8.4	0.02	1.01	0.20	2.56	0.00
Broad bean (*Vicia faba*)	482	54	8.0	0.09	1.00	0.34	2.52	5.38
Lentil (*Lens esculenta*)	354	82	8.3	0.08	0.77	0.22	2.20	3.99
White bean (*Phaseolus vulgaris*)	448	173	7.3	0.01	0.12	0.20	2.01	3.42

[1] From Cravioto et al. (1951).

180 NUTRITIONAL EVALUATION OF FOOD PROCESSING

TABLE 8.28

FATTY ACID COMPOSITION OF OILS FROM PEAS AND BEANS

	Chick Pea[1] (Cicer arietinum) (%)	Pea[2] (Pisum sativum) (%)	Broad Bean[3] (Vicia faba) (%)	Lentil[1] (Lens esculenta) (%)	Lima[4] (Phaseolus lunatus) (%)	Kidney[4] (Phaseolus vulgaris) (%)
Total lipids	4.30	1.78	2.32	1.42	1.02	1.9
Fatty acids in glycerides			% of Total Fatty Acids			
Myristic	0.3	0.3	0.2	1.1	0.4	Tr
n-Pentadecanoic	0	0	0	0.6	0.3	Tr
Palmitic	12.7	13.1	20.7	23.2	22.5	13.4
Stearic	1.5	3.5	2.4	4.6	3.3	0.7
Oleic	19.3	26.1	30.6	36.0	8.0	8.3
Linoleic	62.9	49.9	41.0	20.6	43.7	26.9
Linolenic	3.3	7.0	5.0	1.6	21.9	50.6
Arachidic	Tr	3.0	0	2.3	0	0
Behenic	0	0	0	0	0	0

[1] Baker et al. (1961).
[2] Zarkadas et al. (1965).
[3] Takayama et al. (1965).
[4] Korytnyk and Metzler (1963). Average values for two crop years of lima beans.

al. (1973) did a comprehensive study of the carbohydrates of broad beans (Vicia faba) and isolated the following:

Fraction	Gm/100 Gm Dry Matter
Starch	27.7
Hemicellulose	7.6
Water-soluble polysaccharide	6.6
Cellulose	4.1
Glucose	0.4
Fructose	1.4
Sucrose	2.2
Raffinose Stachyose	~6

Comprehensive quantitative studies such as this are rare except for the oligosaccharides in several of the legumes (Table 8.29). As in soybeans, raffinose and stachyose in peas and beans have recently become of practical concern because of their contribution to flatulence when legumes are eaten (Calloway et al. 1971).

Nutritional Properties

Several raw legumes, especially beans, contain heat-labile growth inhibitors. Black and red kidney beans eaten raw do not support growth of rats but when the beans are soaked in water and auto-

EFFECTS OF REFINING ON COMPOSITION OF FOODS

TABLE 8.29

OLIGOSACCHARIDE CONTENT OF PEAS AND BEANS

Sugar	Chick Pea[1] (Cicer arietinum)	Broad Bean[2] (Vicia faba)	Lentil[1] (Lens esculenta)	Black Gram[1] (Phaseolus mungo)	Mung[3] (Phaseolus aureus)	California White Bean[4] (Phaseolus vulgaris)
			Gm/100 Gm Dry Matter			
Sucrose	2.4	2.2	2.1	1.6	0.8	ND[5]
Raffinose	1.0		0.6	0.5	0.6	0.4
Stachyose	2.5	~6	2.2	1.8	0.8	3.7
Verbascose	4.2	—	3.0	3.7	0.2	—

[1] Nigam and Giri (1961).
[2] Pritchard et al. (1973).
[3] Kawamura (1963).
[4] Becker et al. (1974).
[5] Not determined.

claved they give a growth rate equal to 60% of the control diet (Jaffee and Vega Lette 1968). Beneficial effects of heat on the nutritive value have also been reported for lima beans but results with lentils have been contradictory (Liener 1958). Bengal gram (*Cicer arietinum*), mung beans, and peas are reported to show no improvement on heating (Liener 1958; Honavar et al. 1962). Raw, ground broad beans (*Vicia faba*) gave a lower protein efficiency ratio than heated soybean meal when fed to male rats and heating decreased the nutritive value of the beans (Nitsan 1971). Legumes are low in sulfur amino acids and methionine is limiting in most of them (Jaffe 1949). In addition, tryphophan and threonine may also be limiting in Bengal gram and lentils (Esh and Som 1955).

Food Uses

The use of beans of one type or another for food is very widespread. In countries where animal protein is scarce and expensive, peas and beans supply large quantities of protein as well as carbohydrate to man's diet. Nearly all of the legumes can be prepared in the home by soaking in water, removing the outer skin and then cooking for an hour or more to soften them. The beans may also be mashed, cooked, and finally dried to form the grams of India. Another method of preparation in India is grinding into powders or flours called "dhals." The majority of these legumes are consumed directly in the home (Dean 1958; Deschamps 1958).

Processing.—In contrast to soybeans, which in the United States are almost always refined into oil plus defatted flakes and grits, peas and beans are not processed commercially to concentrate their constituents. Rather, they are canned, baked, and dehydrated by food processors, but their consumption in the United States is declining.

Research is therefore being done to make these products more convenient to use by the housewife.

A disadvantage of dried legumes for home use is the long time required for soaking and cooking. One approach to shorten cooking is to vacuum infiltrate lima beans and other legumes with a hydration solution containing sodium chloride, sodium tripolyphosphate, sodium bicarbonate, and sodium carbonate followed by soaking, draining, rinsing, and drying. Cooking times for most beans processed this way varied from 25-35 min as compared to 30-75 min for untreated legumes. The quick-cooking process did not impair nutritional value of lima bean protein (Rockland and Metzler 1967).

A simpler process to hasten cooking of beans consists of mechanical peeling to remove the seed coat. Peeled beans require 26-36% less cooking time than unpeeled soaked beans. Peeling has no effect on the protein efficiency ratio of the beans and causes no loss of the water-soluble vitamins—thiamin, niacin, pyridoxine, and folic acid (Kon et al. 1973).

Instant bean powders have also been developed to make bean products more convenient for home and commercial use. In one process the beans are soaked in water overnight and then cooked for 1 hr at $210°F$ with constant stirring. Next, the cooked beans are disintegrated and the resulting slurry is then drum dried into flakes. In a second process the beans are ground coarsely and then slurried in dilute hydrochloric acid to give a final pH of 3.5. The acid treatment inhibits lipoxygenase and thereby helps prevent formation of off-flavors (Kon et al. 1970). The slurry is passed through a disintegrator, cooked for 15 min in the acid solution, neutralized to pH 6.0-7.0 and cooked for another 45 min. The final step is drum drying of the cooked slurry (Miller et al. 1973).

Conversion of peas and beans on a large commercial scale into products analogous to those of soybeans—flours, concentrates, and isolates—appears to be still in the development stage. In India, Multipurpose Food was developed as a protein supplement consisting of 75% low-fat peanut flour and 25% chick pea flour. Chick pea flours have also been blended with corn or wheat flours to prepare nutritionally balanced protein foods (Parpia 1969). Chick pea-wheat flour blends have been evaluated for baking Egyptian bread (Shehata and Fryer 1970). Up to 20% chick pea flour had no significant effects on the dough properties but crumb color was darker than for all-wheat bread. Bread fortified with 20% chick pea flour had a significantly better protein efficiency ratio (1.56) than all-wheat bread (0.96).

Bell and Youngs (1970) prepared a protein concentrate (60% pro-

EFFECTS OF REFINING ON COMPOSITION OF FOODS 183

tein) from field peas by stirring pea flour in lime water at pH 9.3 and centrifuging off the starch. The protein-rich supernatant was drum dried to yield the protein concentrate. This type of product differs from soy protein concentrate in the nature of the carbohydrates present. In soy protein concentrate the carbohydrates are water-insoluble polysaccharides whereas in the pea concentrate the insoluble polysaccharides (starch) have been removed and the sugars have been retained. The pea concentrate was deficient in methionine.

Deschamps (1958) has reviewed some of the technical problems of making isolates from chick peas. A major difficulty is that when protein is isolated from undefatted seed the oil remains with the protein fraction and oxidizes readily. A typical undefatted protein has the following composition: 60% protein; 28% oil; 3% ash; and 9% nitrogen-free extract. Yields from the seed are: 25% undefatted protein; 38% starch; 15% residue; and 22% soluble sugars, salts, and minor constituents. Countercurrent extraction of the protein with a lipid solvent before drying yielded an isolate containing 95% protein, 4% oil, and 1% impurities. Oxidation of the polyunsaturated fatty acids by lipoxygenase is a likely source of difficulty when the oil is not removed from the seed before proceeding with the wet processing involved in making isolates.

Pilot scale studies on preparation of protein isolates from broad beans (*Vicia faba*) have been reported (Guinat 1969; Flink and Christiansen 1973). For example, Guinat extracted broad bean flour with $NaHSO_3$ solution at pH 7.4 and screened off the insoluble residue. Next the resulting protein extract and starch suspension were separated in a hydrocyclone to yield the starch and the clarified extract. The protein isolate was then obtained by isoelectric precipitation at pH 4.5 and centrifugation. Flink and Christiansen (1973) prepared their extract with dilute alkali at pH 8-10 and precipitated the protein at pH 3.5.

Protein isolates from *Vicia faba* have been prepared commercially in England by Rank, Hovis, McDougal, Ltd., for spinning into a fiber used in a meat analog. Although the process is technically feasible, soy protein isolates are now used instead because of uncertainty of sufficient supplies of field beans at economic prices (Smith 1974).

BIBLIOGRAPHY

ALBRECHT, W. J. et al. 1967. A simple method for making full-fat soy flour. Cereal Sci. Today *12*, 81-83.
AMERICAN OIL CHEMISTS' SOCIETY. 1970. Official and Tentative Methods, 3rd Edition. Methods Ba 10-65 and Ba 11-65.
ARAI, S. et al. 1966. Studies on flavor components in soybean. II. Phenolic acids in defatted soybean flour. Agr. Biol. Chem. (Tokyo) *30*, 364-369.

ASPINALL, G. O., BEGBIE, R., and McKAY, J. E. 1967. Polysaccharide components of soybeans. Cereal Sci. Today *12*, 223, 226-228, 260-261.
BADENHOP, A. F., and HACKLER, L. R. 1971. Protein quality of dry roasted soybeans: amino acid composition and protein efficiency ratio. J. Food Sci. *36*, 1-4.
BAKER, B. E. *et al.* 1961. Protein and lipid constitution of some Pakastani pulses. J. Sci. Food Agr. *12*, 205-207.
BANDEMER, S. L., and EVANS, R. J. 1963. The amino acid composition of some seeds. J. Agr. Food Chem. *11*, 134-137.
BECKER, K. W. 1971. Processing of oilseeds to meal and protein flakes. J. Am. Oil Chemists' Soc. *48*, 299-304.
BECKER, R. *et al.* 1974. Conditions for the autolysis of alpha-galactosides and phytic acid in California small white beans. J. Food Sci. *39*, 766-769.
BEESON, K. C. 1941. The mineral composition of crops with special reference to the soils in which they were grown. USDA Misc. Publ. *369*.
BELL, J. M., and YOUNGS, C. G. 1970. Studies with mice on the nutritional value of pea protein concentrate. Can. J. Animal Sci. *50*, 219-226.
BERRY, T. H. *et al.* 1966. Responses of the growing pig to alterations in the amino acid pattern of isolated soybean protein. J. Animal Sci. *25*, 722-728.
CALLOWAY, D. H., HICKEY, C. A., and MURPHY, E. L. 1971. Reduction of intestinal gas-forming properties of legumes by traditional and experimental food processing methods. J. Food Sci. *36*, 251-255.
COLLINS, F. I., and SEDGWICK, V. E. 1959. Fatty acid composition of several varieties of soybeans. J. Am. Oil Chemists' Soc. *36*, 641-644.
COWAN, J. C. *et al.* 1970. Flavor evaluation of copper-hydrogenated soybean oils. J. Am. Oil Chemists' Soc. *47*, 470-474.
CRAVIOTO, R. O., MASSIEU H., G., and GUZMAN G., J. 1951. Composition of Mexican foods. Ciencia (Mexico) *11*, No. 5-6, 129-155.
DEAN, R. F. A. 1958. Use of processed plant proteins as human food. *In* Processed Plant Protein Foodstuffs. A. M. Altschul (Editor). Academic Press, New York.
DEGROOT, A. P., and SLUMP, P. 1969. Effects of severe alkali treatment of proteins on amino acid composition and nutritive value. J. Nutr. *98*, 45-56.
DESCHAMPS, I. 1958. Peas and beans. *In* Processed Plant Protein Foodstuffs. A. M. Altschul (Editor). Academic Press, New York.
DiCARLO, F. J., SCHULTZ, A. S., and KENT, A. M. 1955. Soybean nucleic acid. Arch. Biochem. Biophys. *55*, 253-256.
EARLE, F. R., and MILNER, R. T. 1938. The occurrence of phosphorus in soybeans. Oil Soap *15*, 41-42.
EDEN, A. 1968. A survey of the analytical composition of field beans (*Vicia faba* L.). J. Agr. Sci. (Cambridge) *70*, 299-301.
ESH, G. C., and SOM, J. M. 1955. Studies on the nutritive value of plant proteins. I. Pulse proteins—their improvement by amino acid supplementation. Proc. Natl. Inst. Sci. (India) *21B*, No. 2, 68-73.
FAO/WHO CODEX ALIMENTARIUS COMMISSION. 1974. Report of the seventh session of codex committee on fats and oils. London, March 25-29.
FLINK, J., and CHRISTIANSEN, I. 1973. Production of a protein isolate from *Vicia faba*. Lebensm.-Wiss. Technol. *6*, 102-106.
GUINAT, E. 1969. Separation of starch and proteins from pulverized plant materials. Ger. Offen. 1,911,107, C.A. *72*, 45263 (1970).
HAYES, L. P., and SIMMS, R. P. 1973. Defatted soybean fractionation by solvent extraction. U.S. Pat. 3,734, 901. May 22.
HONAVAR, P. M., SHIH, C.-V., and LIENER, I. E. 1962. Inhibition of the growth of rats by purified hemagglutinin fractions isolated from *Phaseolus vulgaris*. J. Nutr. *77*, 109-114.
HYMOWITZ, T., COLLINS, F. I., PANCZNER, J., and WALKER, W. M. 1972A.

Relationship between the content of oil, protein, and sugar in soybean seed. Agron. J. *64*, 613-616.
HYMOWITZ, T. *et al.* 1972B. Stability of sugar content in soybean strains. Commun. Soil Sci. Plant Anal. *3*, 367-373.
JAFFE, W. G. 1949. Limiting essential amino acids of some legume seeds. Proc. Soc. Exptl. Biol. Med. *71*, 398-399.
JAFFE, W. G., and VEGA LETTE, C. L. 1968. Heat-labile growth-inhibiting factors in beans (*Phaseolus vulgaris*). J. Nutr. *94*, 203-210.
KAHN, N. A., and BAKER, B. E. 1957. The amino-acid composition of some Pakistani pulses. J. Sci. Food Agr. *8*, 301-305.
KAKADE, M. L., HOFFA, D. E., and LIENER, I. E. 1973. Contribution of trypsin inhibitors to the deleterious effects of unheated soybeans fed to rats. J. Nutr. *103*, 1772-1778.
KAWAMURA, S. 1963. The oligosaccharides of some Japanese legumes. II. Higher oligosaccharides in mung beans. Kagawa Univ. Fac. Tech. Bull. *15*, 29-33.
KAWAMURA, S. 1967. Quantitative paper chromatography of sugars of the cotyledon, hull, and hypocotyl of soybeans of selected varieties. Kagawa Univ. Fac. Tech. Bull. *18*, 117-131.
KIKUCHI, T. 1972. Food-chemical studies on soybean polysaccharides. Part III. The polysaccharides from soybeans obtained by cooking. J. Agr. Chem. Soc. Japan *46*, 405-409.
KIKUCHI, T. *et al.* 1971. Food-chemical studies on soybean polysaccharides. Part I. Chemical and physical properties of soybean cell wall polysaccharides and their changes during cooking. J. Agr. Chem. Soc. Japan *45*, 228-234.
KON, S. *et al.* 1970. pH adjustment control of oxidative off-flavors during grinding of raw legume seeds. J. Food Sci. *35*, 343-345.
KON, S. *et al.* 1973. Split peeled beans: preparation and some properties. J. Food Sci. *38*, 496-498.
KORYTNYK, W., and METZLER, E. A. 1963. Composition of lipids of lima beans and certain other beans. J. Sci. Food Agr. *14*, 841-844.
KU, S. 1973. Extraction of oligosaccharides in processing whole soybeans. Diss. Abstr. Intern. B *34*, 709-710.
LIENER, I. E. 1958. Effect of heat on plant proteins. In Processed Plant Protein Foodstuffs. A. M. Altschul (Editor). Academic Press, New York.
LIENER, I. E. 1972. Nutritional value of food protein products. In Soybeans: Chemistry and Technology, Vol. 1, Proteins. A. K. Smith and S. J. Circle (Editors). Avi Publishing Co., Westport, Conn.
LONGENECKER, J. B. 1963. Utilization of dietary protein. In Newer Methods of Nutritional Biochemistry, Vol. 1. A. A. Albanese (Editor). Academic Press, New York.
LONGENECKER, J. B., MARTIN, W. H., and SARETT, H. P. 1964. Protein quality improvement: improvement in the protein efficiency of soybean concentrates and isolates by heat treatment. J. Agr. Food Chem. *12*, 411-412.
MANEEPUN, S., LUH, B. S., and RUCKER, R. B. 1974. Amino acid composition and biological quality of lima bean protein. J. Food Sci. *39*, 171-174.
McANELLY, J. K. 1964. Method for producing a soybean protein product and the resulting product. U.S. Pat. 3,142,571. July 28.
McEWEN, T. J., DRONZEK, B. L., and BUSHUK, W. 1974. A scanning electron microscope study of Fababean seed. Cereal Chem. *51*, 750-757.
MEYER, E. W. 1966. Soy protein concentrates and isolates. Proc. Intern. Conf. Soybean Protein Foods. USDA Agr. Res. Serv. *71-35*.
MILLER, C. F., GUADAGNI, D. G., and KON, S. 1973. Vitamin retention in bean products: cooked, canned and instant bean powders. J. Food Sci. *38*, 493-495.
MUSTAKAS, G. C., KIRK, L. D., and GRIFFIN, E. L., JR. 1962. Flash de-

solventizing defatted soybean meals washed with aqueous alcohols to yield a high-protein product. J. Am. Oil Chemists' Soc. *39*, 222-226.
MUSTAKAS, G. C. et al. 1969. Lipoxidase deactivation to improve stability, odor and flavor of full-fat soy flours. J. Am. Oil Chemists' Soc. *46*, 623-626.
MUSTAKAS, G. C. et al. 1970. Extruder-processing to improve nutritional quality, flavor and keeping quality of full-fat soy flour. Food Technol. (Chicago) *24*, 1290-1296.
NELSON, A. I., WEI, L. S., and STEINBERG, M. P. 1971. Food products from whole soybeans. Soybean Dig. *31*, No. 3, 32-34.
NIGAM, V. N., and GIRI, K. V. 1961. Sugars in pulses. Can. J. Biochem. Physiol. *39*, 1847-1853.
NITSAN, Z. 1971. Vicia faba beans vs. soyabean meal as a source of protein. J. Sci. Food Agr. *22*, 252-255.
PANT, R., and TULSIANI, D. R. P. 1969. Solubility, amino acid composition, and biological evaluation of proteins isolated from leguminous seeds. J. Agr. Food Chem. *17*, 361-366.
PARPIA, H. A. B. 1969. Protein foods of India based on cereals, legumes, and oilseed meals. *In* Protein-Enriched Cereal Foods for World Needs. M. Milner (Editor). American Association of Cereal Chemists, St. Paul, Minn.
PRINGLE, W. 1974. Full-fat soy flour. J. Am. Oil Chemists' Soc. *51*, 74A-76A.
PRITCHARD, P. J., DRYBURGH, E. A., and WILSON, B. J. 1973. Carbohydrates of spring and winter field beans (*Vicia faba* L.). J. Sci. Food Agr. *24*, 663-668.
RACKIS, J. J. 1972. Biologically active components. *In* Soybeans: Chemistry and Technology, Vol. 1, Proteins. A. K. Smith and S. J. Circle (Editors). Avi Publishing Co., Westport, Conn.
RACKIS, J. J. et al. 1961. Amino acids in soybean hulls and oil meal fractions. J. Agr. Food Chem. *9*, 409-412.
RACKIS, J. J. et al. 1970. Soybean factors relating to gas production by intestinal bacteria. J. Food Sci. *35*, 634-639.
ROCKLAND, L. B., and JONES, F. T. 1974. Scanning electron microscope studies on dry beans. Effects of cooking on the cellular structure of cotyledons in rehydrated large lima beans. J. Food Sci. *39*, 342-346.
ROCKLAND, L. B., and METZLER, E. A. 1967. Quick-cooking lima and other dry beans. Food Technol. (Chicago) *21*, 344-348.
SAIO, K., and WATANABE, T. 1968. Observation of soybean foods under electron microscope. J. Food Sci. Technol. (Tokyo) *15*, 290-296.
SAIR, L. 1959. Proteinaceous soy composition and method of preparing. U.S. Pat. 2,881,076. Apr. 7.
SARMA, P. S., and PADMANABAN, G. 1969. Lathyrogens. *In* Toxic Constituents of Plant Foodstuffs. I. E. Liener (Editor). Academic Press, New York.
SHEHATA, N. A., and FRYER, B. A. 1970. Effect on protein quality of supplementing wheat flour with chickpea flour. Cereal Chem. *47*, 663-670.
SHEMER, M., WEI, L. S., and PERKINS, E. G. 1973. Nutritional and chemical studies of three processed soybean foods. J. Food Sci. *38*, 112-115.
SMITH, A. K., and CIRCLE, S. J. 1972. Soybeans: Chemistry and Technology, Vol. 1, Proteins. Avi Publishing Co., Westport, Conn.
SMITH, P. R. 1974. Personal communication. The British Food Manufacturing Industries Research Association, Leatherhead. Surrey, England.
TAKAYAMA, K. K., MUNETA, P., and WIESE, A. C. 1965. Lipid composition of dry beans and its correlation with cooking time. J. Agr. Food Chem. *13*, 269-272.
TKACHUK, R., and IRVINE, G. N. 1969. Amino acid composition of cereals and oilseed meals. Cereal Chem. *46*, 206-218.

VARNER, J. E., and SCHIDLOVSKY, G. 1963. Intracellular distribution of proteins in pea cotyledons. Plant Physiol. *38*, 139-144.
VERMA, S. C., and LAL, B. M. 1966. Physiology of Bengal gram seed. II. Changes in phosphorus compounds during ripening of the seed. J. Sci. Food Agr. *17*, 43-46.
VERMA, S. C., LAL, B. M., and PARKASH, V. 1964. Changes in the chemical composition of the seed parts during ripening of bengal gram (*Cicer arietinum* L.) seed. J. Sci. Food Agr. *15*, 25-31.
WEAKLEY, F. B., and McKINNEY, L. L. 1958. A modified indophenol-xylene extraction method for the determination of ascorbic acid in soybeans. J. Am. Oil Chemists' Soc. *35*, 281-284.
WILDING, M. D., ALDEN, D. E., and RICE, E. E. 1968. Nutritive value and dietary properties of soy protein concentrates. Cereal Chem. *45*, 254-259.
WOLF, W. J., and COWAN, J. C. 1971. Soybeans as a Food Source. CRC Press, Cleveland, Ohio.
WOODARD, J. C., and SHORT, D. D. 1973. Toxicity of alkali-treated soy protein in rats. J. Nutr. *103*, 569-574.
ZARKADAS, C. G., HENNEBERRY, G. D., and BAKER, B. E. 1965. The constitution of leguminous seeds. V. Field peas (*Pisum sativum* L.). J. Sci. Food Agr. *16*, 734-738.
ZIEMBA, J. V. 1974. Authentic soy sauce brewed in Wisconsin. Food Eng. *46*, No. 2, 58-60.

PART 3

Vernon L. Frampton

Effects of Processing on the Nutritive Quality of Oilseed Meals

Whereas a substantial part of the peanut and soybean crop is used for food on a worldwide basis, only an insignificant portion of the oilseed meals that are produced as a by-product of the vegetable oil industry is used as human food. Almost all of the oilseed meals, with the exception of the toxic castor and tung oil meals, are used in animal feeding where they serve as a supplemental source of protein.

Apparently, the food use of oilseed meals, or proteins derived from them, is being expanded in the United States, where protein nutrition of the adult human being is assuming a role of primary importance. These substances serve as extenders of meat products or as substitutes for milk and egg proteins in food products.

The nutritive value of oilseed meals is greatly influenced by the processing conditions used in oil extraction. Moreover, as with most vegetable materials, many of the oilseeds contain physiologically active constituents, and the physiological activity often survives the processing operations used in rendering and recovering the oil.

The mere presence in a food, or feed, of toxic substances does not make the food unsafe—it is their presence in excessive amounts that is important. We tend to rely on centuries of human experience to assure us of the safety of the foods we eat. However, until we understand completely the etiology of all the so-called metabolic and degenerative diseases we should keep an open mind about the potentialities for harm, or good, inherent in the composition of even our most common and important foods. While there is concern about chemical additives, intentional or accidental, we must add the problem of determining the safe level for each physiologically active plant constituent. This is a challenge to biological scientists that is only now beginning to gain acceptance as a respectable scientific activity.

PHYSIOLOGICALLY ACTIVE SUBSTANCES IN OILSEED AND OILSEED MEALS

Trypsin Inhibitors

The concensus among earlier research workers was that the inhibition of trypsin activity in the digestive tract of young laboratory animals receiving raw soybean meals accounted for the observed depression in growth (Ham *et al.* 1945). This explanation has stood the test of time (Birk and Bondi 1955). The quantity of precipitate obtained on adding trichloracetic acid to the intestinal contents of young chicks on a raw soybean meal diet is far greater than that obtained with chicks receiving toasted soybean meal. The quantitative data for the presence of peptides precipitable with trichloracetic acid in significant quantities was correlated negatively with the rate of growth. However, it is not the trypsin inhibitor per se that is important, for indeed the addition of soybean trypsin inhibitor to toasted soybean meal used in rations for small laboratory animals had only a slight effect of the growth of the young animals (Gertler *et al.* 1967; Brochere *et al.* 1948; Alumot and Nitsan 1961; Bieloria and Bondi 1963).

It remains, however, that the inactivation of the trypsin inhibitor by the action of heat is correlated in a simple way with the improved nutritive quality of soybean meals. Actually, the observation that improved nutritive value is obtained when heat is used is common to many leguminous seeds (as well as other seeds) (Osborne and Mendel

1917; Orru and Demel 1941; Sure and Read 1921; Jones and Fink 1921; Fink and Jones 1921).

The presence or absence of trypsin inhibitors in seeds apparently has but little influence on the improvement in the nutritive quality that is observed when mild heat is applied. Seeds containing trypsin inhibitors which are improved in nutritive quality on being heated include the lima bean, common bean, soybean, the blackeyed pea. Seeds containing no trypsin inhibitors but which are also improved in nutritive quality include the jack bean, lentil, and the horse bean. Seeds that contain trypsin inhibitors and which are not improved in their nutritive quality are the peanut, partridge pea, chick-pea, sweet pea, lespedeza, and the mung bean. Seeds that do not contain trypsin inhibitors and which are not improved in their nutritive quality include guar, the common pea, and common vetch (Brochere and Ackerman 1950).

Evidently the improvement in the nutritive quality of leguminous seeds by the application of heat involves more than the denaturation of trypsin inhibitors. Apparently, one of the effects of trypsin inhibitors includes that of inducing enlargement of the pancreas and changes in the proteolytic systems of the exocrine pancrease.

Richetogenic Agents

Richetogenic agents are reported to be present in raw soybean meals (Thompson *et al.* 1968; Carlson *et al.* 1964A). These agents are inactivated when the meals are heated.

Anticoagulants

An anticoagulant is reported to be present in raw soybean meals (Balloun and Johnson 1953; Ferguson 1942; Grob 1943; Tagon and Soulier 1946; Rush and Clifton 1951). The blood clotting time was increased as much as 50% in chicks receiving raw soybean meal. Although the growth of the chicks was retarded, the retardation was independent of the effect on clotting time.

Hemostatic Agents

The retardation of the clotting time observed with raw soybean meal is the reverse of that observed from peanut constituents (Frampton *et al.* 1966A; Jackson *et al.* 1966). The bleeding time of transected arterioles in the cheek pouch of hamsters ingesting an alcohol extract of defatted peanuts was shortened to about $3/4$ that observed for the control animals.

Saponins

Beneficial effects that accrue from the heating of oilseeds may very well include those that arise from the hydrolysis of saponins (Potter and Kummerow 1954; Peterson 1950). Apparently the sapogenins that are produced on the hydrolysis of saponins do not inhibit the growth of chicks or induce bloat in ruminants; the saponins apparently retard the growth of chicks and also induce bloat in ruminant animals. The inhibition by saponins of normal growth in cattle is counteracted by cholesterol and phytosterol.

Goitrogenic Agents

Enlargement of the thyroid gland is one of the oldest complaints of man; the affliction was known in ancient Egypt and pre-Christian China. Many articles have been written concerning this disease; many theories of goiter etiology have been advanced through the years. It is now known that iodine plays a very important role in the control of thyroid function, but other substances that are ingested are also important. Goitrogenic substances are frequently found in food, and an awareness of this fact was reached in the third decade of this century (Chesney et al. 1928) when it was discovered that cabbage induced enlargement of thyroids in rabbits.

In the light of modern knowledge, all goitrogens cause thyroid hypertrophy by diminishing the supply of thyroid hormone available to the body by reducing the supply of iodine by interfering with the organic binding of the iodide ion which does reach the thyroid. The inhibiting break of circulating thyroxin on pituitary thyrotrophin production is reduced and more thyrotrophin is poured into the blood stream and the thyroid is stimulated to hypertrophy, hyperplasia, and increased function. It is unlikely that a naturally-occurring goitrogen will cause thyroid hyperfunction per se; it is probable that the stimulus is on the pituitary production of thyrotrophin. Actually, feeding rape seed to rats (Kennedy and Purves 1941) resulted in the abnormal increase in size of the thyroid during the first three weeks, growth depression, hypertrophy of the adrenal cortex, delayed ovary development, and histological changes in the pituitary.

Goitrogenic activity has been reported for cabbage and mustard seed, in addition to rape (Hercus and Purves 1936). The active agent was identified as L-5-vinyl-2-thiooxazolidone (Astwood et al. 1949), which arises following the enzymic hydrolysis of the glucoside sinigrin, a plant constituent.

Among the earliest reports relating goitrogenicity to oilseeds are those of Blum (1937) and McCarrison (1933). Other glycosides, in addition to sinigrin, which are reported to be goitrogenic are arachidoside, anacardioside, and catechin (Srinivasan et al. 1957; Moudgal et al. 1957, 1958). The goitrogenic moiety in these glycosides has not been characterized.

These several goitrogenic glycosides are not oil soluble and are not removed from the marc when the seeds are processed for oil. The only process for the preparation of meals that contain these glycosides seems to be that of aqueous extraction of the glycosides before they have been hydrolyzed by enzyme action. One technique is to destroy the enzymes by heat before enzyme action can occur. In the case of rapeseed, the present processing methods do not produce meals that can be fed without restriction to any domestic animal.

Thiocyanates

Abnormalities of the blood and bone marrow in cattle seem to be attributable to thiocyanates in rapeseed (Backgren and Johnson 1969). Thiocyanates appear as glycoside moieties.

Allergens

Oilseed proteins constitute the most highly allergenic food groups. Seeds of mustard, flax, and cotton are common potent allergens, and may produce in sensitive persons violent symptoms following their ingestion. These proteins find their way into bakery goods.

Hemagglutinins

Hemagglutinins have been most frequently detected in leguminous seeds. It is presumed by some research workers that a part of the enhanced nutritive quality in soybean meal obtained when the beans are cooked prior to oil extraction is attributable to denaturation (and inactivation) of soybean hemagglutinins (Liener and Pallansch 1952; Jaffe 1969). The addition of soybean hemagglutinin to autoclaved soybean meal depressed the growth of rats to about 75% that of the control animals, although the preparation was not lethal to the rats when administered via stomach tube at a rate of 500 mg per kg of body weight (Liener and Wada 1953). On the other hand, the hemagglutinin from the kidney bean was lethal (Honavar et al. 1962).

Hemagglutinins are digested by pepsin and trypsin, but they differ greatly in their susceptibility to digestion.

The four hemagglutinins reported to be present in soybeans are glycoproteins (Lis et al. 1966, 1969) which contain mannose and glucose amine.

These glycoproteins are concentrated in the whey fraction in the course of fractionation of soybean proteins.

Flatus Factors

The appearance of flatulence seems to depend on the food ingested and the microbial populations in the intestinal tract. Malodorous gasses, including ammonia, amines, hydrogen sulfide, and mercaptans are voided in the flatus, but in addition to the egestion of rectal gas the most common complaints include nausea, cramps, diarrhea, and pain. The view is advanced (Rackis 1972) that flatus is composed chiefly of gas produced on the fermentation of water-soluble carbohydrates in the intestinal tract. Raffinose and stachyose, components of soybean meals, are considered to be of importance in flatus production (Gall 1968; Steggerda 1968; Levitt and Ingelfinger 1968; Calloway and Murphy 1968; Rackis 1970).

Cyclopropenoid Fatty Acids

Phelps et al. (1965) reviewed the biological properties of cyclopropene fatty acids, while Carter and Frampton (1964) prepared a review of the chemistry of the cyclopropene compounds. The fatty acids ω-(2-n-octylcycloprop-1-enyl) octanoic and ω-(2-n-octylcycloprop-1-enyl) heptanoic acids occur in plants belonging to the families *Sterculiaceae, Malvaceae, Bombaceae*, and *Symplocaceae*. Of immediate interest are the seeds of *Gossypium hirsutum, G. barbadense, G. arboreum, G. herbaceum*, the four commercial cottons, and *Hibiscus esculentus* (okra). These two acids appear as components of the triglycerides, but are of interest here because they appear in the residue of oil left in the meal.

Particular attention has been given to eggs produced by hens ingesting these two acids; when such eggs are kept in cold storage one finds that the albumins become red (Sherwood 1928, 1931; Evans et al. 1957), the pH value of the yolk is increased (Thompson 1930; Schaible et al. 1946; Frampton et al. 1962), and the yolk (when cottonseed meal is used) becomes discolored. Commercial candlers place these eggs in the lowest grades, or discard them as rots because of the dark yolk color. Hatchability of fertile eggs is reduced (Schneider et al. 1961), and there is an alteration in the component fatty acids of the yolk fat (Evans et al. 1960, 1962; Frampton et al. 1966B). While papers concerned with egg yolk fat carry the implication that the increase in the ratio of stearic to oleic acid that is observed in the abnormal eggs is specific to the cyclopropenoid fatty acids, evidence has been reported (Frampton et al. 1973) that the ratio can also be affected by the level of choline in the ration, and therefore the effect is not specific to cyclopropenoid fatty acids.

The chief effect on eggs seems to be one of alteration of the permeability of the vitelline membrane—increase in the yolk pH is attributed to the migration of albumins into the yolk and the development of the red color in the albumin is attributed to a migration of iron to the albumin where a red complex of iron and conalbumin is formed (Schaible et al. 1946; Schaible and Bandemer 1946).

Growth retardation of pullets and rats was observed when *Sterculia foetida* seed oil (which contains sterculic acid) was ingested (Schneider et al. 1962; Schneider 1962). Whether comparable results may be obtained with cottonseed meals containing a residuum of crude cottonseed oil remains to be determined.

Gossypol

Gossypol, 1,1',6,6',7,7'-hexahydroxy-3,3'-dimethyl-5,5'-diisopropyl-2,2'binaphthyl-8,8'-dialdehyde, is a yellow pigment found in the cotton plant which was first purified by Marchlewski (1899) and subsequently synthesized by Edwards (1958). This compound is found in most of the species of *Gossypium* (El-Nockrashy et al. 1969) and it appears, together with other very similar compounds (Boatner 1948), in specialized structures in the cotton plant.

Gossypol is not oil soluble and most of the gossypol of the cottonseed remains as a residue in the meal when the seed is processed for oil. It is a highly reactive compound and the great bulk of the material is found in the meal in chemical combination with other seed constituents when the seed is processed in the conventional way. Apparently a small portion remains free.

Consensus ascribes the adverse effects, observed on feeding conventional cottonseed meals to nonruminant animals, to the "free" gossypol in the meals. According to the concept, the chemically-bound gossypol is essentially innocuous.

Gossypol has the capacity to combine with the epsilon-amino groups of lysine in proteins (Conkerton and Frampton 1959). Aside from this property its specific physiological role has not been identified. The evidence seems reliable that it is responsible for the dark green-brown pH sensitive pigment found in the yolks of eggs produced from hens on cottonseed meal rations. Ceroid pigments that appear in the organs of animals ingesting cottonseed meals are also ascribed to the influence of gossypol.

Gossypol accumulates in the liver (Buitrago et al. 1970) where it is presumed to be detoxified—the detoxification pathway has not been discovered. Apparently, the elimination is via the bile duct.

Cottonseed meals that have been prepared through the use of petroleum ether, whether through an initial light pressing of the cooked oilseed flakes or through the action of the solvent alone, are toxic to swine. Actually, the mortality in any one feeding test may be as high

194 NUTRITIONAL EVALUATION OF FOOD PROCESSING

as 100% by the time the shoats reach market weight. The most obvious symptom is a curious abdomenal spasm, generally referred to as "the thumps." The lethal effect is attributed by some to the "free" gossypol that may be present in the meals, but the cause of toxicity and death is not actually known.

The repression of the growth of broilers is also attributed by some to the "free" gossypol present in the rations. Data developed in carefully-conducted tests with broilers, however, can support different conclusions, one of which is that the poor response of broilers to cottonseed meal rations can more properly be interpreted in terms of lysine starvation. Cottonseed meals, which are deficient in lysine, are further impaired in protein quality through the combination of gossypol with the free epsilon-amino groups of protein lysine. Such lysine is not available to the broilers. The analysis of variance of data (Table 8.30) obtained in a broiler feeding test with cottonseed meals of commerce serving as the protein supplement is instructive (Frampton 1965).

TABLE 8.30

ANALYSIS OF VARIANCE OF THE REGRESSION OF WEIGHT GAIN ON TOTAL GOSSYPOL + AVAILABLE LYSINE

Source of Variance	Sum of Squares	Degrees of Freedom	Mean Square	F Ratio
Total	21,855	7		
Due to total gossypol + available lysine	21,171	2	10,586	77
Due to total gossypol after allowing for available lysine	558	1	551	4
Due to available lysine after allowing for total gossypol	14,322	1	14,322	104
Due to total gossypol alone	6,849	1	6,849	50
Due to available lysine alone	20,620	1	20,620	151
About the regression	684	5	137	

All regressions in Table 8.30 are significant, with the exception of the regression on total gossypol after the effect of available lysine is accounted for. Evidently the correlation between weight gain of broilers and total gossypol of the cottonseed meals used in the rations is spurious. Simple correlations can be, and frequently are, misleading.

It may be noted that the correlation between the weight gain and the "free" gossypol of these rations is essentially zero. Even in feed-

EFFECTS OF REFINING ON COMPOSITION OF FOODS 195

ing tests (Frampton 1965) with broilers where the "free" gossypol ranged from 0.026 to 0.313%, the correlation between weight gain and "free" gossypol was only 0.08 for 7 degrees of freedom.

Cyaniferous Glycosides

Cyaniferous glycosides are very widespread in the plant kingdom, although the concentrations in most plants are low. Of the several glycosides that occur in seeds, the only one of importance to this discussion is linamarin, found in flaxseed. The recommendations are that flaxseed meal used for feed should be prepared from cooked flakes where the glycoside splitting enzymes of the seed are inactivated. Apparently the glycoside itself is not toxic, but difficulties arise when enzymic-liberated cyanide is present in the meal.

INFLUENCE OF HEAT ON NUTRITIVE QUALITY OF OILSEED MEALS

Effect on Lysine

A good deal is known about the effects of heat on the protein quality of oilseed meals. The denaturation of physiologically active proteins in the soybean has been cited. In general, a mild heating of oilseed flakes in the processing of oilseeds is beneficial; severe heating is deleterious due, chiefly, to a reduction in the level of lysine available to nonruminant animals, as well as to an actual destruction of lysine in the meals. This is important because most oilseed meals are deficient in available lysine, and any reduction in the level of available lysine will be reflected in a reduced rate of growth of young nonruminant animals receiving such meals in their rations. The epsilon-amino groups of lysine in cottonseed may be bound to the carbonyl group of gossypol (Conkerton and Frampton 1959) or they may be bound to fatty aldehydes that are present in the meals (El-Nockrashy and Frampton 1972). In addition to the influence of reducing sugars on lysine is the evidence (El-Nockrashy and Frampton 1967) that nonreducing sugars may also reduce the level of lysine in oilseed meals.

The total lysine, at least in cottonseed meals, is available to nonruminants when the seed is processed in the absence of cooking procedures to render the oil, as, for example, in the use of the acetone-hexane-water azeotrope with raw flakes. An illustration of the differences in levels of available lysine found in cottonseed meals due to the differences in processing methods is found in Table 8.31 (Frampton 1965).

The superiority of these azeotrope-extracted cottonseed meals for nonruminants, as compared with conventional commercial cotton-

TABLE 8.31

AVAILABLE LYSINE LEVELS IN COTTONSEED MEALS
PRODUCED BY DIFFERENT PROCESSING METHODS

Screw Press (Commercial)	Prepress Solvent (Commercial)	Hexane Extraction (Commercial)	Azeotrope Extraction (Pilot Plant)
	(Grams/16 gm of N)		
2.59	3.14	3.30	4.23
3.07	3.21	3.31	4.18
2.70	3.38	3.20	4.15
3.04	3.70	2.82	4.18
3.22	3.60	3.67	4.26
3.42	3.17	3.64	4.25
2.22	3.52	3.75	4.25
3.11	3.36	3.12	4.28

seed meals, has been established (Vix *et al.* 1969). In the work published by Barns *et al.* (1966), both of the cottonseed meals supplied by USDA were azeotrope-extracted meals prepared in the pilot plant at the New Orleans Regional Research Laboratory.

Methionine

It is well known that the nutritive quality of protein from soybean meal can be markedly enhanced by supplementation with methionine or by mild cooking of the meal (Hayward and Hafner 1941). This effect might be, in part, a result of an increased availability of methionine to digestion, as revealed in *in vitro* experiments (Melnick *et al.* 1946), although the fraction of the total methionine that is not absorbed is the same for raw as for cooked soybeans. According to Liener and Rose (1953) the poor growth rate of rats on raw soybean flour may be attributed entirely to an impairment in the availability of methionine.

Other Amino Acids

With the exception of a slight reduction in the level of arginine induced through the use of heat in processing oilseed, there does not seem to be any significant effect (nutrition-wise) on the other amino acids of oilseed meal proteins.

Influence on Vitamin Requirement

Raw soybean flour is deficient in B-12, and, moreover, it contains a factor which increases the requirement of B-12 for the rat (Edelstein and Guggenheim 1969; Edelstein 1970). Growth of young animals on raw soybean flour is poor. However, supplementation of

EFFECTS OF REFINING ON COMPOSITION OF FOODS 197

rations containing raw soybean flour with B-12 resulted in a substantial increase in growth rate in the test animals. Supplementation of rations containing cooked soybean flour (which supported good growth) did not result in any significant increase in growth rate.

Results reported by Fisher et al. (1969) indicate an antivitamin E activity in soybean proteins for the chick, and the effect is related to the processing used in preparing the proteins. About ½ of the chicks on a diet containing 30–35% soybean protein died in 4 weeks due to encephalomalacia or exudative diathesis. Deaths were prevented when casein replaced soybean proteins, when corn oil was omitted from the diets, when alpha-tocopherol (50 mg per k of ration) was added, or when the antioxidant ethoxyquin was added.

Symptoms of richitis appeared in turkey poults on raw soybean meal (Thompson et al. 1968; Carlson et al. 1964A). The intensity of the symptoms was reduced, but not eliminated, when the rations were supplemented with 4–10 times the recommended levels of cholecalciferol, or if the proteins were first autoclaved.

Flaxseed is considered to be a poor source of proteins for the chick. A considerable improvement in the response of chicks may be obtained if the meal is extracted with water or autoclaved. Dry heat was not effective in improvement of the growth response. Improvement could also be obtained by supplementing the ration with pyridoxine (MacGregor and McGinnis 1948; Kratzer et al. 1954; Klosterman et al. 1967). Klosterman reported the isolation of a pyridoxine antagonist (1-amino proline) which occurs naturally as a peptide with glutamic acid. 1-amino proline was reported to be much more toxic than the peptide, and its toxicity was reported to be counteracted by the administration of pyridoxine.

INFLUENCE OF SOLVENTS IN PROCESSING OILSEEDS

The economic structure in the United States has made the use of the hydraulic press obsolete and is driving the screw press to obsolescence as well. Actually, there is a tendency to consolidate oilseed processing into larger and fewer plants where an organic solvent is used to remove the oil from the seed. The solvent of choice at the present time is a relatively low boiling petroleum naphtha.

Two aspects of importance are suggested in the processing of oilseeds: (1) the separation from the meal of ingredients that may have protective action and (2) the concentration of deleterious materials, such as solvent residues and toxic materials of microbiological origin.

The more efficient operations where a low boiling petroleum naphtha is used include a cooking step that serves to render the oil.

In some instances, such as with cottonseed, the cooking is necessary, for otherwise very finely-divided cottonseed fragments become colloidally dispersed in the miscella and create technical problems that have not been resolved.

Rather extensive attention has been given to other solvents that may be used in processing oilseeds. Included is water. The process with water is a very ancient one—seeds are comminuted in the presence of water and the oil is recovered from the emulsion that rises to the top of the liquid mixture. There does not seem to be very much information about the nutritive quality of the proteins recovered from the aqueous layer.

There is rather an extended literature on the use of alcohols. The use of alcohols has not become industrial, however. Attention is attracted to alcohol extraction of petroleum naphtha-extracted soybean meal (on a laboratory scale) because of the possibility of improving the quality of the meal for human food by removing adverse flavor factors (Beckel and Smith 1944; Beckel et al. 1948; Moser et al. 1967). While there is a reduction of the intensity of the beany and bitter flavors obtained on the extraction with ethyl alcohol, a completely bland preparation apparently has not been produced.

The use of mixtures of petroleum naphtha and several alcohols in the extraction of cottonseed and soybean flakes was reported to result in a more rapid and more extensive oil extraction (Ayer and Scott 1952) than was obtained with petroleum naphtha alone. These authors did not determine the quality of the meals obtained through the use of these mixtures. There probably is an advantage in the use of these mixtures in the production of superior meals; there is a disadvantage in the use of these solvent mixtures because of the glycerol-alcohol ester exchange that will occur to modify the properties of the oils.

The hazard of using highly flammable petroleum naphtha in solvent extraction of oilseeds has lead to the examination of the safer chlorinated hydrocarbons as solvents that may be used. Disaster resulted when, for example, trichloroethylene was used in the extraction of soybeans in that severe cases of toxicity occurred among cattle receiving such meals (Pritchard et al. 1952, 1956).

A number of the lower boiling chlorinated hydrocarbons have been considered in the extraction of oilseeds. It is suggested, however, that the use of chlorinated solvents be studied carefully before their use is recommended, since there is a potential for toxicity problems. For example, it was established (Morrison and Munro 1965; Munro and Morrison 1967A,B) that meals prepared on extracting cod fillets with 1,2-dichloroethane are toxic to rats.

EFFECTS OF REFINING ON COMPOSITION OF FOODS 199

With the recognition that cooking of oilseed flakes in the process of oil removal reduces the nutritive value of the resultant meals, attention has been given to the development of processes that avoid cooking. Some attention was given to ethyl methyl ketone as a potential solvent for use with cottonseed; but a better solvent appears to be any one of several mixtures of commercial hexane, acetone, and water (King and Frampton 1961; Frampton 1961); there is advantage in using the azeotrope (boiling point 49°C) since solvent recovery is simplified.

Comparisons are made in Table 8.32 of data obtained with this azeotrope and commercial hexane.

TABLE 8.32

COMPARISON OF RESULTS OBTAINED WITH THE ACETONE-HEXANE-WATER AZEOTROPE AND COMMERCIAL HEXANE ON THE EXTRACTION OF COTTONSEED

Comparison	Hexane Extraction	Azeotrope Extraction
Yield of neutral oil per ton of kernels, lb	494	512
Gossypol in meal, %	1.3	0.1-0.4
Residual oil in meal, %	1	0.1
Abnormalities in stored shell eggs produced by hens receiving cottonseed meals (6 months), %	100	0.0
Mortalities in swine receiving cottonseed meals (to market weight at 200 lb), %	100	0.0
Time required for complete oil extraction (pilot plant), min	90	3

Some comments about the comparisons should be offered. The abnormalities that appear in eggs consist of enlarged and discolored yolks and discolored albumins. The yolks assume a brown-green color due to a pigment that is pH-sensitive (Frampton et al. 1961) and is ostensibly a derivative of gossypol; the pH of the yolk increases abnormally and there is a migration of albumin into the yolk sack. Moreover, the albumins become red or pink in color.

Cottonseed meals prepared on extracting cottonseed with commercial hexane are lethal to swine. The animals develop a curious abdominal spasm (referred to as the "thumps") and death comes suddenly, presumably because of cardiac failure. So far as is known, the azeotrope-extracted meals have not killed any pigs.

It has been established that one of the adverse effects which result from cooking oilseed flakes is a reduction of the level of available lysine below that normally present in the seed proteins. The evidence is that there is no reduction in the level of available lysine on the extraction of raw oilseed flakes. One might then expect very little difference in the nutritive quality of the meals (excluding those prepared from gossypol-containing cottonseed) prepared with different solvents. There is, however, a substantial difference in the growth response of young laboratory animals when the meals prepared from gossypol-free seed with the azeotrope are compared with those prepared with commercial hexane (Johnston and Watts 1965); data for chicks show a daily gain of 8.3 gm per day for the azeotrope meal and 7.3 gm for the hexane meal in tests with protein-depleted birds. The conclusion reached in this research was that the effects due to azeotrope extraction are physical rather than chemical.

This report is in keeping with the observation (Hensarling *et al.* 1970) that the intracellular cytoplasm is completely disrupted by this solvent system; and this effect appears to be virtually instantaneous (Frampton *et al.* 1967; Frampton and Pepperman 1967).

BIBLIOGRAPHY

ALUMOT, E., and NITSAN, S. 1961 The influence of soybean antitrypsin on the intestinal proteolysis of the chick. J. Nutr. *73*, 73.

ASTWOOD, E. B., GREER, M. A., and ETTLINGER, M. G. 1949. L-5-vinyl-2-thiooxazolidone, an antithyroid compound from yellow turnip and *Brassica* seeds. J. Biol Chem. *181*, 121.

AYER, A. L., and SCOTT, C. R. 1952. A study of extraction rates for cottonseed and soybean flakes using various hexane-alcohol mixtures. J. Am. Oil Chemists' Soc. *29*, 1913.

BACKGREN, A. W., and JONSSON, G. 1969. Blood and bone marrow studies on cattle feeding on *Brassica* species. Acta Vet. Scand. *10*, 309.

BALLOUN, S. C., and JOHNSON, E. L. 1953. Anticoagulant properties of unheated soybean meal in chick diets. Arch. Biochem. *42*, 355.

BARNES, R. H., POND, W. G., KWONG, E., and REID, I. 1966. Effect of severe protein-calorie malnutrition on the baby pig upon relative utilization of different dietary proteins. J. Nutr. *89*, 355.

BECKEL, A. C., BELTER, P. A., and SMITH, A. K. 1948. Solvent effects on the products of soybean oil extraction. J. Am. Oil Chemists' Soc. *25*, 7.

BECKEL, A. C., and SMITH, A. K. 1944. Alcohol extraction improves soya flour flavor and color. Food Eng. *16*, 616, 664.

BIELORIA, R., and BONDI A. 1963. Relationship between antitryptic factors of some plant protein feeds and products of proteolysis precipitable by trichloracetic acid. J. Sci. Food Agri. *14*, 124.

BIRK, Y., and BONDI, A. 1955. The action of proteolytic enzymes on protein feeds. II. Intermediary products precipitated by trichloracetic acid and phosphotungstic acid from peptic and pancreatic digests. J. Sci. Food Agri. *6*, 549.

BLUM, H. 1937. Physiology of the thyroid gland. Endokrinologie *19*, 19.

BOATNER, C. H. 1948. Pigments of the cottonseed. In Cottonseed and Cottonseed Products. A. E. Bailey (Editor). Interscience Publishers, New York.

BROCHERE, R., and ACKERMAN, C. W. 1950. The nutritive value of legume seeds. X. Effect of autoclaving on the trypsin inhibitor for 17 species. J. Nutr. *41*, 339.

BROCHERE, R., ACKERMAN, C. W., and MUSSEHL, F. E. 1948. Trypsin inhibitor. VIII. Growth inhibiting properties of soybean trypsin inhibitor. Arch. Biochem *19*, 317.
BUITRAGO, J. A., CLAWSON, A. J., and SMITH, F. H. 1970. Effect of dietary iron on gossypol accumulation in and elimination from procine liver. J. Animal Sci. *31*, 554.
CALLOWAY, D. H., and MURPHY, E. L. 1968. The use of expired air to measure intestinal gas formation. Ann. N.Y. Acad. Sci. *150*, 82.
CARLSON, C. W., McGINNIS, J., and JENSEN, L. S. 1964B. Antirichitic effects of soybean preparations. J. Nutr. *82*, 366.
CARLSON, C. W., SAXENA, L. S., JENSEN, L. S., and McGINNIS, J. 1964A. Rachitogenic activity of soybean fractions. J. Nutr. *82*, 507.
CARTER, F. L., and FRAMPTON, V. L. 1964. Review of the chemistry of cyclopropeneoid compounds. Chem Rev. *64*, 497.
CHESNEY, A. M., CLAWSON, T. A., and WEBSTER, B. 1928. Endemic goiter in rabbits. I. Incidence and characteristics. John Hopkins Hospital Bull. *43*.
CONKERTON, E. J., and FRAMPTON, V. L. 1959. Reactions of gossypol with free epsilon amino groups of lysine in proteins. Arch. Biochem. Biophys. *81*, 130.
EDELSTEIN, S. 1970. Cause of the increased requirement for vitamin B-12 in rats subsisting on unheated soybean flour. J. Nutr. *100*, 1377.
EDELSTEIN, S., and GUGGENHEIM, K. 1969. Effect of raw soybean flour on vitamin B-12 requirements of the rat. Israel J. Med. Sci. *5*, 415.
EDWARDS, J. D., JR. 1958. Total synthesis of gossypol. J. Am. Chem. Soc. *80*, 3798.
EL-NOCKRASHY, A. S., and FRAMPTON, V. L. 1967. Destruction of lysine by nonreducing sugars. Biochem. Biophys. Res. Commun. *78*, 675.
EL-NOCKRASHY, A. S., and FRAMPTON, V. L. 1972. Destruction of available lysine by fatty aldehydes in petroleum ether defatted cottonseed meals. Grasas Aceites (Seville, Spain) *23*, 126.
EL-NOCKRASHY, A. S., SIMMONS, J. G., and FRAMPTON, V. L. 1969. A chemical survey of seeds of the genus *Gossypium*. Phytochemistry *8*, 1949.
EVANS, R. J., BANDEMER, S. L., ANDERSON, M., and DAVIDSON, J. A. 1962. Fatty acid and lipid distribution in egg yolk from hens fed cottonseed oil or *Sterculia foetida* seeds. J. Nutr. *73*, 282.
EVANS, R. J., BANDEMER, S. L., and DAVIDSON, J. A. 1960. Fatty acid distribution in lipids from eggs produced by hens fed cottonseed oil and cottonseed fatty acid fractions. Poultry Sci. *39*, 1199.
EVANS, R. J., BANDEMER, S. L., DAVIDSON, J. A., and SCHAIBLE, P. J. 1957. The occurrence of pink whites and salmon-colored yolks in stored shell eggs from hens fed crude cottonseed oil and cottonseed meal. Poultry Sci. *36*, 798.
FERGUSON, J. H. 1942. Crystalline trypsin-inhibitor and blood clotting. Proc. Soc. Exptl. Biol. Med. *51*, 373.
FINK, A. J., and JONES, C. O. 1921. Studies on the nutritive value of the lima bean, *P. lunatus*. Am. J. Physiol. *56*, 205.
FISHER, H., GRIMINGER, P., and PUDOWSKI, P. 1969. Antivitamin E activity of isolated soybean protein for the chick. Z. Ernaehrungsweiss *9*, 271.
FRAMPTON, V. L. 1961. Azeotrope extraction of cottonseed. Oil Mill Gaz. Aug. 1961, 1199.
FRAMPTON, V. L. 1965. Cottonseed proteins; their status in nonruminant feeding. Cereal Sci. Today *10*, 557.
FRAMPTON, V. L., CARTER, F. L., PICCOLO, B., and HEYWANG, B. W. 1962. Cottonseed constituents and discoloration of stored shell eggs. J. Agr. Food Chem. *10*, 46.
FRAMPTON, V. L., LEE, L. S., MORRIS, N. J., and BOUDREAUX, H. B. 1966A. Influence of peanut extract on bleeding time. Thromb. Diath. Haemorrhag. *16*, 265.

FRAMPTON, V. L., and PEPPERMAN, A. B., JR. 1967. On the extraction of oil from raw comminuted cottonseed kernels with the acetone-hexane-water azeotrope. J. Am. Oil Chemists' Soc. 44, 455.
FRAMPTON, V. L., PEPPERMAN, A. B., JR., SIMMONS, J., and KING, W. H. 1967. Countercurrent extraction of raw cottonseed flakes with the acetone-hexane-water azeotrope. Agr. Food Chem. 15, 790.
FRAMPTON, V. L., PICCOLO, B., and HEYWANG, B. W. 1961. Discolorations in stored shell eggs produced by hens fed cottonseed meal. Agr. Food. Chem. 9, 59.
FRAMPTON, V. L. et al. 1966B. Some physiological properties of Halphen positive cottonseed oils. Poultry Sci. 45, 527.
FRAMPTON, V. L. et al. 1973. Influence of ingested fat on the fatty acid composition of lipids in rat tissue. Grasas Aceites (Seville, Spain) 24, 85.
GALL, L. S. 1968. The role of intestinal flora in gas formation. Ann. N.Y. Acad. Sci. 150, 27.
GERTLER, A., BIRK, Y., and BONDI, A. 1967. A comparative study of the nutritional and physiological significance of pure soybean trypsin inhibitor and of ethanol-extracted soybean meal in chicks and rats. J. Nutr. 91, 353.
GROB, D. 1943. The antiproteolytic activity of serum. I. The nature and experimental variation of the antiproteolytic activity of serum. J. Gen. Physiol. 26, 243.
HAM, W., STANDSTEDT, F. E., and MUSEHL E. F. 1945. The proteolytic inhibiting substance in the extract from unheated soybean meal and its effect upon the growth of the chick. J. Biol. Chem. 161, 635.
HAYWARD, J. W., and HAFNER, F. H. 1941. The supplementary effect of cystine and methionine upon the protein of raw and cooked soybeans as determined with chicks and rats. Poultry Sci. 20, 139.
HENSARLING, T. P., YATSU, L. Y., and JACKS, T. J. 1970. Extraction of lipids from cottonseed tissue. II. Ultrastructure effects of lipid extraction. J. Am. Oil Chemists' Soc. 47, 224.
HERCUS, C. E., and PURVES, A. D. 1936. Studies on endemic and experimental goiter. J. Hyg. 36, 182.
HONAVAR, P. M., SHIH, C. V., and LIENER, I. E. 1962. The inhibition of the growth of rats by purified hemagglutinin fractions isolated from *Phaseolus vulgaris*. J. Nutr. 77, 109.
JACKSON, B., OWEN, W., and BOUDREAUX, H. B. 1966. Pharmacological and chemical characteristics of peanut extracts effective in hemostasis. Thromb. Diath. Haemorrhagica 16, 256.
JAFFE, W. G. 1969. Hemagglutinin. In Toxic Constituents of Plant Food Stuffs. I. E. Liener (Editor). Academic Press, New York.
JOHNSTON, C., and WATTS, A. B. 1965. The characterization of a growth inhibitor of glandless cottonseed. Poultry Sci. 44, 652.
JONES, C. O., and FINK, A. J. 1921. Studies in nutrition. VII. The nutritive value of the adsuki bean, *P. angularis*. Am. J. Physiol. 56, 208.
KENNEDY, T. H., and PURVES, A. D. 1941. Studies on experimental goiter. I. Effect of *Brassica* seed diets on the rat. Brit. J. Exptl. Pathol. 22, 241.
KING, W. H., and FRAMPTON, V. L. 1961. Properties of oil extracted from cottonseed with acetone-hexane-water solvent mixtures. J. Am. Oil Chemists' Soc. 38, 497.
KLOSTERMAN, H. J., LAUMOUREUX, N. L., and PATSONS, J. L. 1967. Isolation, characterization and synthesis of linatine, a vitamin B-6 antagonist from flax seed, *Linum usilatissimum*. Biochem. 6, 170.
KRATZER, F. H., WILLIAMS, D. E., MARSHALL, B., and DAVIS, P. N. 1954. Some properties of the chick growth inhibitor in linseed oil meal. J. Nutr. 52, 555.
LEVITT, D. M., and INGELFINGER, F. J. 1968. Hydrogen and methane production in man. Ann. N.Y. Acad. Sci. 150, 75.

LIENER, I. E., and PALLANSCH, M. J. 1952. Purification of a toxic substance from defatted soybean flour. J. Biol. Chem. *197*, 29.
LIENER, I. E., and ROSE, J. E. 1953. Soyin, a toxic protein from the soybean. IV. Immunochemical properties. Proc. Soc. Exptl. Biol. Med. *83*, 539.
LIENER, I. E., and WADA, S. 1953. Liver xanthine oxidase activity in relation to available methionine from soybean protein. Proc. Soc. Exptl. Biol. Med. *82*, 484.
LIS, H., FRIDMAN, C., SHARON, N., and KATCHALSKI, E. 1966. Multiple hemagglutinins in soybean. Arch. Biochem. Biophys. *117*, 301.
LIS, H., SHARON, N., and KATCHALSKI, E. 1969. Identification of the carbohydrate protein linkage group in soybean hemagglutinin. Biochem. Biophys. Acta. *192*, 364.
McCARRISON, R. 1933. The goitrogen action of soybeans and ground nuts. Indian J. Med. Res. *21*, 179.
MacGREGOR, H. I., and McGINNIS, J. 1948. Toxicity of linseed meal for chicks. Poultry Sci. *27*, 141.
MARCHLEWSKI, L. 1899. Gossypol, a constituent of cottonseed. J. Prak. Chem. *60*, 84.
MELNICK, D., OSER, B. L., WEISS, S. 1946. Rate of enzymic digestion of protein as a factor in nutrition. Science *103*, 326.
MORRISON, A. B., and MUNRO, I. C. 1965. Factors influencing the nutritive value of fish flour. IV. Reaction between 1,2-dichloroethane and protein. Can. J. Biochem. *43*, 33.
MOSER, H. A. et al. 1967. Sensory evaluation of soy flour. Cereal Sci. Today *12*, 269, 298, 314.
MOUDGAL, R. N., RAGHUPATHY, E., and SARMA, P. S. 1958. Studies on goitrogenic agents in food. III. Goitrogenic action of some glycosides isolated from edible nuts. J. Nutr. *66*, 291.
MOUDGAL, R. N., SRINIVASAN, V., and SARMA, P. S. 1957. Studies on goitrogenic agents in food. II. Goitrogenic action of arachidoside. J. Nutr. *61*, 97.
MUNRO, I. C., and MORRISON, A. B. 1967A. Factors influencing the nutritive value of fish flour. V. Chlorochlone chloride, a toxic material in samples extracted with 1,2-dichloroethane. Can. J. Biochem. *45*, 1049.
MUNRO, I. C., and MORRISON, A. B. 1967B. Toxicity of 1,2-dichloroethane extracted fish protein concentrate. Can. J. Biochem. *45*, 1779.
ORRU, A., and DEMEL, V. C. 1941. Physiological and anatomical-pathological observations on rats fed seeds of *Canavalia ensiformis*. Quaderni Nutriz. *7*, 237.
OSBORNE, T. B., and MENDEL, L. B. 1917. The use of soybeans as food. J. Biol. Chem. *32*, 369.
PETERSON, O. W. 1950. Effect of sterols on the growth of chicks fed high alfalfa diets or a diet containing *Quillaja* saponin. J. Nutr. *42*, 597.
PHELPS, R. A., SHENSTONE, F. S., KEMMERER, A. R., and EVANS, R. J. 1965. A review of cyclopropeneoid compounds; biological effects of some derivatives. Poultry Sci. *44*, 358.
POTTER, G. C., and KUMMEROW, F. A. 1954. Chemical similarity and biological activity of saponins from alfalfa and soybeans. Science *120*, 224.
PRITCHARD, W. R., REHFELD, C. E., and SAUTTER, H. H. 1952. Aplastic anemia of cattle associated with ingestion of trichlorethylene-extracted soybean oil meal. I. Clinical and laboratory investigation of field cases. J. Am. Vet. Med. Assoc. *121*, 1.
PRITCHARD, W. R. et al. 1956. Studies on trichlorethylene extracted feeds. I. Experimental production of acute aplastic anemia in young heifers. J. Am. Vet. Res. *17*, 425.
RACKIS, J. J. 1970. Soybean factors relating gas production by intestinal bacteria. J. Food Sci. *35*, 634.

RACKIS, J. J. 1972. Biologically active compounds. *In* Soybeans: Chemistry and Technology, Vol. 1. A. K. Smith, and S. J. Circle (Editors). Avi Publishing Co. Westport, Conn.
RUSH, B., and CLIFTON, E. C. 1951. Control of proteolytic activity in serum; effect of soy bean inhibitor *in vivo* in the mouse. Am. J. Physiol. *166*, 458.
SCHAIBLE, P. J., and BANDEMER, S. L. 1946. Composition of fresh and stored eggs from hens fed cottonseed and noncottonseed rations. V. Cause of discoloration. Poultry Sci. *25*, 456.
SCHAIBLE, P. J., BANDEMER, S. L., and DAVIDSON, J. A. 1946. Composition of fresh and storage eggs from hens fed cottonseed and noncottonseed rations. I. General observations. Poultry Sci. *25*, 440.
SCHNEIDER, D. L. 1962. Some physiological and biochemical effects of *Sterculia foetida* oil on animal systems. Ph.D. Thesis. Library Univ. Arizona.
SCHNEIDER, D. L., KURNICK, A. A., VAVICH, M. G., and KEMMERER, A. R. 1962. Delay of sexual maturity in chickens by *Sterculia foetida* oil. J. Nutr. *77*, 403.
SCHNEIDER, D. L., VAVICH, M. G., KURNICK, A. A., and KEMMERER, A. R. 1961. Effect of *Sterculia foetida* oil on mortality of the chick embryo. Poultry Sci. *40*, 1644.
SHERWOOD, R. M. 1928. The effect of various rations on the storage quality of eggs. Texas Agr. Expt. Sta. Bull. *376*.
SHERWOOD, R. M. 1931. The effect of cottonseed meal and other feeds on the storage quality of egg. Texas Agr. Expt. Sta. Bull. *429*.
SRINIVASAN, V., MOUDGAL, R. N., and SARMA, P. S. 1957. Studies on goitrogenic agents in food. I. Goitrogenic action of ground nuts. J. Nutr. *61*, 87.
STEGGERDA, F. R. 1968. Gastrointestinal gas following food consumption. Ann. N.Y. Acad. Sci. *150*, 57.
SURE, B., and READ, J. W. 1921. Biological analysis of seed of the Georgia velvet bean *Sitzolobium deeringianum*. J. Agr. Res. *22*, 5.
TAGNON, H. J., and SOULIER, J. P. 1946. Anticoagulant activity of the trypsin inhibitor from soya bean flour. Proc. Soc. Exptl. Biol. Med. *61*, 440.
THOMPSON, O. J., CARLSON, C. W., PALMER, I. S., and OLSON, O. E. 1968. Destruction of rachitogenic activity of isolated soybean by autoclaving as demonstrated with turkey poults. J. Nutr. *94*, 227.
THOMPSON, R. B. 1930. Mystery shown egg quality betterment. Rept. Oklahoma A&M Coll. Agr. Expt. Sta. *1926-30*.
VIX, H. L. E., DUPUY, H. R., and LAMBU, M. G. 1969. Critical evaluation of the use of acetone in solvent extraction process. *In* Conf. Protein Rich Food Prod. Oil Seeds, New Orleans. USDA Agr. Res. Serv. *72-71*.

SECTION III

Effects of Commercial Processing and Storage on Nutrients

CHAPTER 9

Effects of Heat Processing on Nutrients

Daryl B. Lund

PART 1
Effects of Blanching, Pasteurization, and Sterilization on Nutrients

Heat processing is one of the most important methods developed by man for extending the storage life of foodstuffs. Because of this extended storage life, foods which are abundantly available only during relatively short harvesting periods are made available throughout the year. There is no doubt that this has increased the availability of nutrients to the consumer. However, heat processing also has a detrimental effect on nutrients since thermal degradation of nutrients can and does occur. Therefore, thermal processing makes it possible to extend and increase availability of a foodstuff to the consumer but the foodstuff may have a lower nutrient content (compared to the fresh foodstuff). The challenge to the food processing industry is to minimize the loss of nutrients during thermal processing while providing an adequate process to ensure an extended storage life.

Several processes involving the use of heat are currently applied to foodstuffs. For some, the primary objective is to increase the palatability of the food. An example is cooking which includes baking, broiling, roasting, boiling, frying, and stewing. For other thermal processes, the objectives are to increase storage life of the foodstuff and to minimize food-borne diseases. Blanching, pasteurization, and sterilization are examples of these processes.

Reduction in nutrient content of the foodstuff as a result of a thermal process depends on the severity of the process. In this chapter effects of blanching, pasteurization, and sterilization on nutrients shall be discussed. The approach will be to define each of these heat processes, mention general reviews currently available, discuss the effects of heat on nutrients in general, examine the interaction between current technology for accomplishing the objectives of each process and nutrient retention, and review the effect of storage on nutrients in foodstuffs that have received these heat processes.

DEFINITION OF BLANCHING, PASTEURIZATION, AND STERILIZATION

Blanching is a heat process frequently applied to tissue systems prior to freezing, drying, or canning. The objectives of the blanching process depend on the subsequent treatment of the foodstuffs. For example, blanching prior to freezing or drying is used primarily to inactivate enzymes which would contribute to undesirable changes in color, flavor, or nutritive value during storage. Blanching prior to canning serves several different functions including wilting the tissue to facilitate packing, removing tissue gases prior to container filling, increasing the temperature of the tissue prior to container closing, and inactivating or activating enzymes. Although the objectives of the blanching process are dependent on the subsequent treatment, a criterion frequently used for evaluating the adequacy of the blanching operation, regardless of subsequent treatment, is enzyme inactivation. Generally, if enzymes are inactivated, the heat treatment was sufficient to accomplish the objectives of blanching prior to canning. In the special case of blanching to activate the enzyme pectin methyl esterase, van Buren et al. (1960) and Kaczmarzyk et al. (1963) showed that blanching water temperature must be greater than 150°F but less than 180°F. An important concept about blanching is that microbial destruction is not a primary objective of the process.

Pasteurization is a heat process designed to inactivate part but not all of the vegetative microorganisms present in the food. Since the food is not sterile, pasteurization, like blanching, must also be used in conjunction with other preservation techniques such as fermentation (e.g., pickles), refrigeration (e.g., milk), maintenance of anaerobic conditions (e.g., beer), or must be used on products such as high acid fruit juices where the environment is not particularly suited for growth of spoilage and health-hazard microorganisms. The basis of the process may be a spoilage microorganism (e.g., yeast in beer, yeast and molds in high acid fruit juices) or a health-hazard organism (e.g., *Coxiella burnetti*, the rickettsia organism responsible for Q fever in milk).

Sterile is a term which refers to a condition in which no viable microorganisms are present, a viable organism being one that is able to reproduce under conditions optimum for its growth. Sterilization then is a term applied to any process which produces a sterile condition in the food. Some microorganisms and their spores are extremely heat resistant and generally it is not practical to render a food sterile by heat processing. To do so would alter the organoleptic and nutritive value of the food to the point that it would be unacceptable. Therefore, the "sterilization" process used in heat processing foods is also used in conjunction with other preservation techniques, namely packaging and control of storage temperature. The requirement for these techniques is that the remaining dormant microorganisms or their spores will not grow in the environment of the food under the conditions of storage. Foods which have been thermally processed and meet this requirement are said to be "commercially sterile."

GENERAL REVIEWS ON THE EFFECTS OF PROCESSING ON NUTRIENTS

Several review articles which discuss the effect of heat processing on nutrients have been published, and their contents are summarized in Table 9.1. With the consumer demand for more nutrition information, particularly as it applies to processed foods, it is understandable that there is great interest and increased publication on this subject. Several of these reviews treat only one particular product, process, or nutrient (e.g., Dugan 1968; Goldblith 1971; Gortner 1972; Hartman and Dryden 1965), while others are quite general (e.g., Bender 1966; Chichester 1973; Mapson 1956). Most, however, point out that thermal processing can in some instances enhance the nutritive value of food. Examples of this are given elsewhere in this book.

EFFECT OF HEAT ON NUTRIENTS

Although several review articles have been published on the effects of heat processing on nutrients, few authors have attempted to summarize the kinetic data which can be used to describe the time/temperature effect on nutrients. Labuza (1972) took this approach in an excellent review article on the effects of dehydration and subsequent storage on nutrient content of dehydrated foods. But as Labuza pointed out, not much data exist. The same situation is true for the thermal destruction of nutrients under conditions normally found in blanching, pasteurization, and commercial sterilization processes. Many authors have reported the percentage loss of a nutrient in a food product that was given a particular treatment. But these data

TABLE 9.1
GENERAL REVIEW ON THE EFFECT OF PROCESSING ON NUTRIENTS

Reference	Products[1]								Processes[2]							Nutrients[3]					Other	No. of References[4]
	V	F	M	D	Ce	Fi	E	C	P	B	C	D	Fr	F	R	Pa	V	M	P	C	L		
Ashton (1972)	X			X							X	X				X	X	X	X	X	X		42
Barratt (1973)	X		X	X	X	X											X	X	X	X			0
Bender (1966)	X		X	X	X	X					X	X	X	X			X	X	X	X			132
Bender (1972)		X	X	X	X						X								X	X			76
Berk (1970)																	X					Heat, fractionation	13
Burger and Walters (1973)			X	X		X					X	X	X	X			X		X				55
Burton et al. (1970)				X														X				Ultra-high temperature	9
Cain (1967)	X	X							X	X				X		$X_{\cdot}$		X					45
Cameron et al. (1955)	X	X	X					X	X	X	X	X	X				X		X				88
Chichester (1973)	X	X	X	X	X				X	X	X	X	X	X			X	X	X	X	X		30
Clifcorn (1948)	X	X	X					X		X	X						X		X	X			90
Dugan (1968)																					X	Heat, oxidation	68
Ford et al. (3-M) (1969)				X									X			X	X					Ultra-high temperature	24
Goldblith (1971)							X		X									X	X				21
Gortner (1972)																	X	X					0
Harris and Von Loesecke (1960)	X	X	X	X	X	X	X	X	X	X	X	X	X	X	X	X	X	X	X	X	X		1212
Hartman and Dryden (1965)				X												X	X						
Hein and Hutchings (1971)									X	X							X						27

EFFECTS OF HEAT PROCESSING ON NUTRIENTS

Reference	Products	Processes	Nutrients	Comments	No. of refs[4]
Henshall (1973)	X X	X X X X X	X X X		21
Hollingsworth (1970)		X	X X		19
Hollingsworth and Martin (1972)	X	X X	X X X X	X X		175
Holmquist et al. (1954)	X	X			5
Labuza (1972)	X X X		X	X X X		79
Lang (1970)	X X X X	X X	X X X X X	X X X		408
Lawrie (1968)	X	X		X X	Cooking	90
Lee (1958)			X	X X		63
Lund et al. (1973)			X X	X X		23
Mapson (1956)	X	X X	X X X X X	X X X		51
Orr (1969)			X		Composition Table for B-6, B-12, PA[5]	107
Osner and Johnson (1968)				X	Protein degradation	62
Rolls and Porter (1973)		X	X	X		29
Sabrey (1968)		X		X	Heat, extraction	5
Schroeder (1971)	X X X X X	X	X	X X		26
Thompson (1969)		X	X	X	Ultra-high temperature Composition Tables	60
Watt and Murphy (1970)				X		27
Woodham (1973)		X	X	X	Texturizing	46

[1] Products: V = vegetable, F = fruit, M = meat, D = dairy, Ce = cereal, Fi = fish, E = eggs, C = canned products.
[2] Processes: P = preparation, B = blanching, C = canning, D = dehydration, Fr = freezing, F = fermentation, R = radiation, Pa = pasteurization.
[3] Nutrients: V = vitamins, M = minerals, P = protein, C = carbohydrate, L = lipid.
[4] Number of references cited in the review.
[5] PA—pantothenic acid.

usually have not been complete enough to allow estimation of the kinetic parameters that can be used to predict or calculate the response of the nutrient to the heat treatment. For example, after analyzing the data compiled by Orr (1969), Schroeder (1971) concluded that vitamin B-6 and pantothenic acid losses could be as high as 91% in canned foods and that the Recommended Daily Allowance (RDA) for these 2 nutrients probably could not be obtained from a menu of refined, processed, and canned foods. Yet the parameters which are needed to predict the susceptibility of these two nutrients to thermal processing are not known. The necessary parameters cannot be obtained from the data presented by Schroeder or Orr due to the wide variation in processing conditions used in the canning industry. Therefore, optimization of the process for retention of these two nutrients cannot yet be determined.

Several kinetic parameters have been used to describe the effect of time/temperature treatments on the rate and extent of nutrient destruction. Basically two parameters are needed: (1) the rate of nutrient destruction at a reference temperature and (2) the dependence of the rate of destruction on temperature. In most chemical and engineering applications these two parameters have been the reference reaction rate constant (k_r) at temperature T_r and the Arrhenius activation energy (E_a). For the food processing industry, these two parameters are expressed as the time to reduce the concentration of the component by 90% (D_r) at a reference temperature T_r and the degrees Fahrenheit temperature change necessary to cause a 10-fold change in D value (z value). Finally, the biological scientists have used the reference reaction rate constant (k_r) and Q_{10} value (reaction rate at $(T + 10)°C$/reaction rate at $T°C$). Determination and use of these values and concepts are covered elsewhere (e.g., Aiba et al. 1965; Blakeborough 1968; Pflug and Schmidt 1968; Stumbo 1973) and will not be reviewed here. Suffice it to say that these two parameters are necessary to adequately describe and predict the effect of heat on nutrients.

Table 9.2 is a compilation of data from existing literature where the authors have determined E_a and D_r (or their corresponding values in other terms) or have provided sufficient data to allow calculation of these parameters. The parameters E_a and D_r for a particular nutrient or component are dependent on several variables: (1) pH, (2) oxidation-reduction potential, and (3) medium composition (including presence of catalytic factors such as heavy metals). For each study, information is given on the component that was evaluated, the medium, pH, and temperature range over which the param-

TABLE 9.2

KINETIC PARAMETERS FOR THE THERMAL DEGRADATION OF FOOD COMPONENTS

Reference	Component	Medium	pH	Temp Range (°F)	E_a (Kcal/Mole)	D_{121} [1]	Other
Bendix et al. (1951)	Thiamin	Whole peas	Nat[2]	220–270	21.2	164 min	
Feliciotti and Esselen (1957)	Thiamin	Carrot purée	5.9	228–300	27	158 min	
	do	Green bean	5.8	do	do	145 do	
	do	Pea	6.6	do	do	163 do	
	do	Spinach	6.5	do	do	134 do	
	do	Beef heart	6.1	do	do	115 do	
	do	Beef liver	6.1	do	do	124 do	
	do	Lamb	6.2	do	do	120 do	
	do	Pork	6.2	do	do	157 do	
Mulley et al. (1974)	Thiamin	Phosphate buffer	6.0	250–280	29.4	156.8 min	
	do	Pea purée	Nat[2]	do	27.5	246.9 do	
	do	Beef purée	do	do	27.4	254.2 do	
	do	Peas-in-brine purée	—	—	—	—	
Gillespy (1962)	Thiamin	—	do	do	27.0	226.7 do	
	Riboflavin	—	—	—	20.0	—	
Garrett (1956)	B-1 · HCl	—	—	—	23.0	—	
	d-Pantothenic acid	Liquid multi-vitamin prep	3.2	39–158	26	1.35 days	
	C	do	do	do	21	4.46 do	
	B-12	do	do	do	23.1	1.12 do	
	Folic acid	do	do	do	23.1	1.94 do	
Garrett (1956)	A	Vitamin prep	3.2	39–158	16.8	1.95 days	
			do		14.6	12.4 do	
Davidek et al. (1972)	Inosinic acid (IMP)	Buffer soln	3	140–208	34.0	—	
	(IMP)	do	4	do	30.4	—	
		do	5	do	28.1	—	
Gupte et al. (1964)	Chlorophyll a	Spinach	6.5	260–300	15.5	13.0 min	
	Chlorophyll b	do	5.5	260–300	7.5	14.7 do	

212 NUTRITIONAL EVALUATION OF FOOD PROCESSING

TABLE 9.2 (Continued)

Reference	Component	Medium	pH	Temp Range (°F)	E_a (Kcal-Mole)	D_{121}[1]	Other
Gold and Weckel (1959)	Chlorophyll do	Pea purée do	Nat[2] do	240–280 do	16.1 12.6	14.0 min 13.9 do	Blanched Unblanched
Lenz and Lund (1974)	Chlorophyll do	Pea purée Spinach purée	6.5 6.5	175–280 do	22 19	113 min 166 do	
Mackinney and Joslyn (1941)	Chlorophyll a Chlorophyll b	Buffered soln do	— —	32–122 do	7.5 9.0	— —	
Dietrich et al. (1959)	Chlorophyll	Green beans	Nat[2]	190–212	12.0	10.1 min	
Timbers (1971)	Color	Peas Asparagus Green beans	Nat[2] do do	175–300 do do	15.0 14.0 15.0	25.0 min 17.0 do 21.0 do	
	Quality by taste panel	Whole kernel corn Whole peas Whole green beans	do do do	do do do	19.5 22.5 22.0	6.0 do 2.3 do 4.0 do	Time for taste panel to judge 4.0/9.0 compared to frozen control.
Herrmann (1970)	Chlorophyll a Chlorophyll b Maillard reaction Nonenzymic browning B-1	Spinach Spinach Apple juice Apple juice Pork	Nat[2] do do do do	212–266 do 100–266 do ? –250	12.5 10.0 27.0 20.7 19.5	34.1 min 48.3 do 4.52 hr 4.75 hr 6.03 hr	
Ponting et al. (1960)	Anthocyanin	Grape juice Boysenberry juice Strawberry juice	Nat[2] do do	68–250 do do	28.0 20.0 19.0	17.8 min 102.5 do 110.3 do	
Tanchev and Joncheva (1973)	Cyanidin-3-rutinoside Peonidin-3-rutinoside	Citrate buffer Plum juice Citrate buffer Plum juice	4.5 4.5 4.5 4.5	170–225 do do do	23.7 23.1 26.2 22.1	28.5 min 41.6 do 21.6 do 27.7 do	

EFFECTS OF HEAT PROCESSING ON NUTRIENTS

Reference	Component	Substrate	pH	Temp (°C)	z-value	D-value	Notes
Ramakrishnan and Francis (1973)	Carotensids	Paprika	Nat[2]	125-150	34.0	0.038 min	
von Elbe et al. (1974)	Betanin	Buffer	5.0	122-212	12.5	19.5 min	
	do	Beet juice	do	do	10.0	46.6 do	
Burton (1963)	Browning	Goat's milk	6.5-6.6	200-250	27.0	1.08 min	Homogenized
	do	do	do	do	do	0.91 do	Unhomogenized
Williams and Nelson (1974)	Methylmethionine sulfonium bromide → DMS	Sodium-citrate buffer	6.0	178-212	34.8	8.4 min	
		Sweet corn	6.9	do	31.6	4.5 do	
		Tomato	4.4	do	27.2	23.2 do	
Taira et al. (1966)	Lysine	Soybean meal	—	—	30.0	13.1 hr	
Mansfield (1974)	Texture	Peas	Nat[2]	212-260	19.5	1.4 min	Time to acceptable product.
	Overall cook quality	Peas		170-200	19.5	2.5 do	
	do	Beets		210-240	34.0	2.0 do	
	do	Whole kernel corn		180-210	16.0	2.4 do	
	do	Broccoli		212-250	13.0	4.4 do	
	do	Squash		182-240	25.0	1.5 do	
	do	Carrots		176-240	38.0	1.4 do	
	do	Green beans		182-240	41.0	1.0 do	
	do	Potato		161-240	27.5	1.2 do	
Hackler et al. (1965)	Trypsin inhibitor	Soybean milk	—	200-250	18.5	13.3 min	
Adams and Yawger (1961)	Peroxidase	Whole peas	Nat[2]	230-280	16.0	3.0 min	
Licciardello et al. (1967)	C. botulinum toxin-type E	Growth Media	6.2	125-135	26.4	3.8 min	Temp-dependent activation energy.
		do	do	135-140	57.4	@60°C	
		do	do	140-145	153.5		
Read and Bradshaw (1966)	Staphylococcus enterotoxin B	Milk	6.4-6.6	210-260	25.9	9.4 min	
Stumbo (1973)	C. botulinum spores (type A and B)	Variety	>4.5	220-?	64-82	0.1-0.2 min	
	B. stearothermophilus	do	do	do	53-82	4.0-5.0 min	

[1] D-value at 121°C.
[2] Indicates natural pH of the system.

eters were determined. In addition to presenting data on nutrients, parameters for factors such as color (pigments and nonenzymatic browning), texture, flavor precursors, enzymes, microbial toxins, and microbial vegetative and spore cells are given. These values are included in order to examine how each of the three processes (blanching, pasteurization, and commercial sterilization) may be optimized with respect to nutrient retention.

It should be pointed out that for each of the studies reported in Table 9.2, the component under study obeyed first-order reaction kinetics, and the D_r value can be directly related to the first-order rate constant by $k_r = 2.303/D_r$.

Several conclusions can be drawn from Table 9.2. Most important is the observation that there is a scarcity of kinetic data on many nutrients. In fact, the only nutrient that has been extensively studied is thiamin and even for thiamin the studies have been limited to relatively few foodstuffs. For thiamin, the activation energy appears to be nearly independent of medium, pH, and composition indicating that the mechanism of thermal degradation is the same in all of the media. The D_{121}-value, however, is strongly dependent on composition of the medium and pH. For example, vitamin B-1 degradation was determined in pea purée at pH 6.6 and the resulting D_{121}-value was 163 min (Feliciotti and Esselen 1957). However, the thiamin destruction data for a liquid multivitamin preparation at pH 3.2 gave a D_{121}-value of 1.35 days (Garrett 1956). Thus thiamin destruction at pH 6.6 is nearly 12 times faster than at pH 3.2. Only a few studies have dealt with the stability of other vitamins in foods.

It is commonly assumed that vitamin C (ascorbic acid) is very heat labile; however, few kinetic data are available for the determination of the activation energy and reference D-value. Garrett (1956) conducted one of the most complete studies on other nutrients. However, that study was done on a liquid multivitamin preparation of low pH. The application of those data to foodstuffs should be verified.

The data reported on chlorophyll degradation illustrate the variability of results obtained in different laboratories. In most of these studies, heating techniques were used that required a correction for the heating and cooling lag periods. Failure to apply this correction to the data may account for part of the data variability. For chlorophyll degradation, the activation energy is between 10 and 25 Kcal per mole. Also chlorophyll b appears to be more stable than chlorophyll a (Gupte et al. 1964; Herrmann 1970). For other pigments and the browning reaction, the activation energy is approximately the same as that for vitamin destruction.

Other quality attributes such as texture and flavor exhibit temperature dependencies for thermal degradation that are similar to that for nutrient destruction. For example, Timbers (1971) reported activation energies between 19.5 and 22.5 Kcal per mole for texture and flavor attributes in selected canned vegetables. The attributes were determined by taste panel. Williams and Nelson (1974) reported that dimethyl sulfide production from methyl-methionine sulfonium bromide in sweet corn and tomato gave an activation energy of 31.6 and 27.2 Kcal per mole, respectively. And finally, Mansfield (1974) stated that the activation energy for texture in a variety of vegetables ranged from 13 to 41 Kcal per mole.

The activation energy for destruction of heat-resistant enzymes is also comparable to that for nutrients. Adams and Yawger (1961) reported an activation energy of 16 Kcal per mole for heat-resistant peroxidase from peas.

Finally, for comparative purposes, the range of activation energy for the spores of two microorganisms used for the basis of thermal process calculations should be considered. *C. botulinum* and *B. stearothermophilus* have reported activation energies between 53 and 82 Kcal per mole (Stumbo 1973). The important fact is that these activation energies are considerably higher than those of nutrients. This difference has been used to optimize processes for nutrient retention. The fact that the activation energy is larger for microbial destruction than for nutrients means that for a given increase in processing temperature, the rate of microorganism destruction will increase more than the rate of nutrient destruction. Thus, as the temperature is raised, times yielding the same degree of microbial lethality will result in greater retention of nutrients.

OPTIMIZATION OF THERMAL PROCESSES FOR NUTRIENT RETENTION

One of the newest developments in the thermal processing of foods has been an attempt to optimize the thermal process for nutrient retention. This has resulted primarily from the use of computers and the increasing public concern over the nutrient content of processed foods. To determine optimum conditions for nutrient retention, equations describing the time/temperature history of a product must be coupled with the parameters describing the reaction kinetics for destruction of nutrients and other factors. This allows the process to be optimized with respect to nutrient retention while assuring that the objective of the process has been accomplished.

Optimization of the blanching process with respect to nutrient retention involves consideration of losses of nutrients in addition to losses by thermal degradation. For example, blanching in hot water can result in a considerable loss of nutrients due to leaching (Lee 1958). Similarly, losses due to oxidation can result during blanching in hot air. However, even if one only considers thermal degradation of nutrients for blanching optimization, it is difficult to predict an optimum process. This is true because the basis for the process (heat-resistant enzymes) and nutrient factors exhibit nearly the same temperature dependence. Therefore, blanching for a long time at a low temperature has no real advantage over blanching for a short time at a high temperature. If, however, significant leaching or oxidative losses could occur, then high temperature-short time (HTST) blanching would result in a greater retention of nutrients.

For pasteurization and commercial sterilization, there is an opportunity to optimize the process for nutrient retention. For foods or food fluids which are pasteurized, the HTST process results in maximum nutrient retention (Hartman and Dryden 1965). This can be predicted by comparing the activation energies of microorganisms to those of nutrients. An increase in process temperature (with an appropriate decrease in process time) will have a greater effect on increasing the rate of microbial destruction than it will on the rate of nutrient destruction. Consequently, HTST results in greater nutrient retention.

For commercial sterilization, optimization of the thermal process is not as straightforward. For commercial sterilization either out-of-container (aseptic thermal processing) or in-container by convection heating, high temperature-short time processes will result in maximum retention of nutrients and quality factors (Anon. 1969; Clifcorn et al. 1950; Everson et al. 1964A, B; Feaster et al. 1949; Jackson and Benjamin 1948). As in pasteurization treatments, this is due to the difference in temperature response of the rate of microbial destruction compared to the rate of destruction of nutrients and quality factors. This is used to advantage in aseptic canning units where temperatures up to 350°F can be employed. In food systems where natural enzymes may be present, however, there are limitations on the maximum temperature that may be used. That maximum occurs when the thermal process may impart sufficient lethality for microorganisms but insufficient lethality for enzymes. This is a consequence of the difference in the response of microbial and enzymic rates of degradation to temperature (Farkas et al. 1956).

At relatively low thermal processing temperatures, the destruction rate for enzymes is greater than that for microorganisms, but as process temperature is increased the destruction rate for microorganisms increases faster than that for enzymes. Hence there exists some temperature at which the destruction rate for the heat-resistant enzyme is equal to the destruction rate for the microorganism used as the basis of the process. Above that temperature, inactivation of the enzyme must be used as the basis of the process since the destruction rate of the enzyme is less than that of the microorganism. If this is not considered in processing products containing natural heat-resistant enzymes, product quality can deteriorate during storage because of residual enzymic activity (Anon. 1969). The temperature range where the destruction rate of enzymes equals that for microorganisms is generally $270°$–$290°F$. Therefore, for products containing heat-resistant enzymes, processes above this crossover temperature must be based on enzyme inactivation. Under these circumstances, process optimization for nutrient retention is difficult to predict since the rate of destruction for nutrients and quality factors exhibit a temperature dependence similar in magnitude to that of heat-resistant enzymes.

In addition to considering enzyme activity as a basis of the process, Mansfield (1962) suggested that HTST processes over approximately $260°F$ may have to be based on product quality considerations. In particular, the desired degree of cooking may not be attained under these processing conditions. From Table 9.2 it can be seen that the texture (degree of cook) for many vegetable products exhibits a dependence on temperature similar to that for heat-resistant enzymes. Therefore, HTST processes may result in adequate microbial lethality but poor consumer preference because of a too firm texture. This is particularly important to the consumer since "heat and serve" items appear more desirable than "cook and serve."

For products which heat primarily by convection and contain particulates, two important assumptions are made: (1) surfaces of pieces in the brine or fluid are at the temperature of the surrounding fluid and (2) particulates are sterile in the interior. Thus, if the thermal process is based on the slowest heating point in the container, the lethality at all points in the container and at the surface of the particulates will be adequate for commercial sterilization. However, in foods containing particulates that may not be sterile in the center (e.g., foods containing fabricated pieces such as meatballs, stuffed noodles, etc.), the basis of the process should be the temperature at the center of a particulate (the slowest heating point). Under these conditions and provided it can be assumed that the nutrients are located in the particulates, designing the process for optimizing nutrient retention is basically the same as that for conduction-heating foods.

Optimization of thermal processes for conduction-heating foods is much more difficult than for the previously discussed situations. The difficulty in optimizing the process lies in the fact that each point in the cross-section of the container or particulate receives a different thermal process and these thermal histories may or may not be equivalent in microbial and nutrient destruction. It had previously been thought that the center point received the least lethality. However, recent investigations have shown that the location of the least-lethal point (critical lethal point) is dependent on the geometry of the container and the boundary conditions of the process (Teixeira et al. 1969A). Although this will not be discussed here, suffice it to say that the overall lethality is the mass-average (or volume-average) lethality obtained by integrating the effect of the heat treatment at every point in the container over the volume of the container.

Several investigators have developed methods for calculating the average destruction of nutrients and microorganisms in foods which heat by conduction (Ball and Olson 1957; Cohen and Wall 1971; Hayakawa 1969; Jen et al. 1971; Lund and Lenz 1973; Manson et al. 1970; Stumbo 1973; Teixeira et al. 1969B). Most of these methods require major computational effort and some require a computer. However, for conduction-heating foods, the general considerations for optimizing a thermal process for nutrient retention can be illustrated by considering calculations presented in the study by Teixeira et al. (1969B).

Figure 9.1 shows the percentage of retention of a component (with the D_r and z-value indicated on the curves) as a function of processes of equivalent microbial lethality. It can be seen that the optimum retention of a low z-value nutrient is obtained at a low temperature-long time process; whereas a high temperature-short time process is optimum for a nutrient with a high z-value. The curve for $z = 45°F$, $D = 188$ min is that for thiamin destruction in green bean purée (Feliciotti and Esselen 1957). It can be seen that the optimum process for thiamin retention is 90 min at $248°F$, very close to existing processes for green bean purée. More significantly, as the temperature of the process is increased, thiamin retention decreases sharply. Thus in-container HTST is not the best thermal process for nutrient retention in conduction-heating foods, and, moreover, each process must be individually optimized.

In conclusion, the optimization of a thermal process for nutrient retention is dependent on the relative temperature dependence of the rate of destruction of the basis of the operation (enzyme or microorganism) to the rate of destruction of nutrients. In Table 9.3 the

EFFECTS OF HEAT PROCESSING ON NUTRIENTS 219

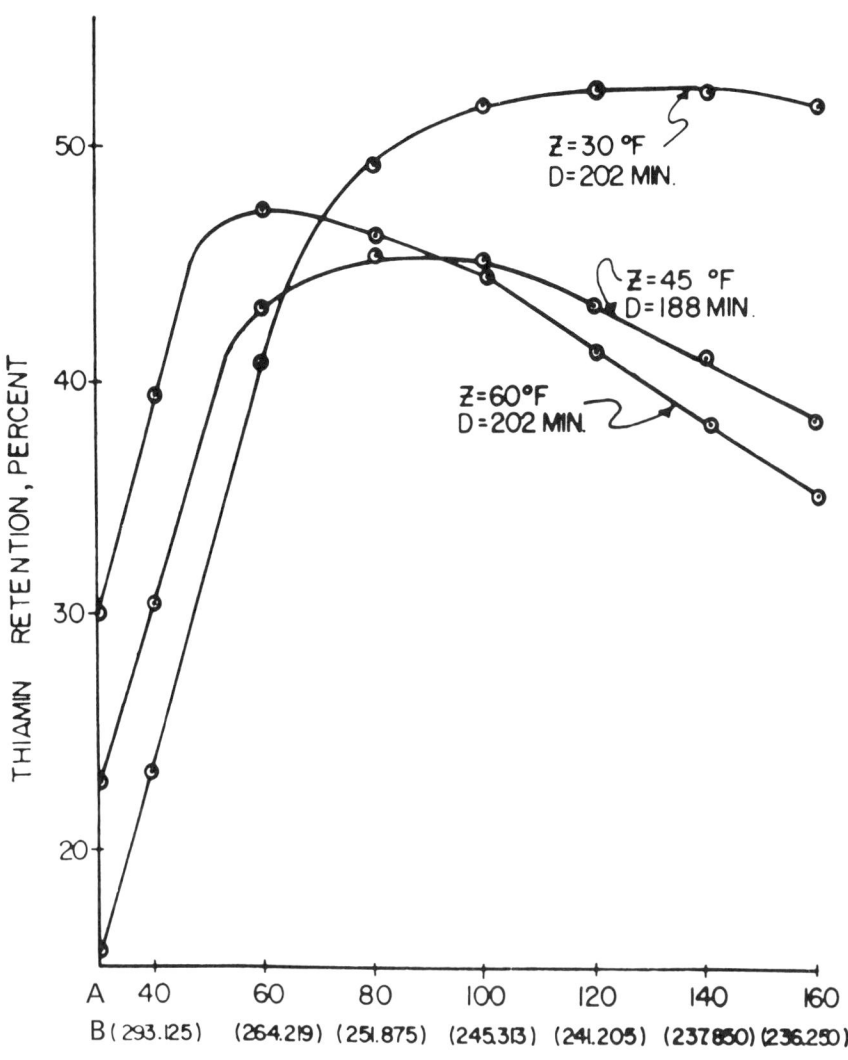

A - PROCESS TIME, MIN.
B - CORRESPONDING RETORT TEMPERATURE °F

Teixeira et al. (1969B)

FIG. 9.1. MULTINUTRIENT OPTIMIZATION

Percentage of nutrient retention versus process time with corresponding retort temperature.

220 NUTRITIONAL EVALUATION OF FOOD PROCESSING

TABLE 9.3

OPTIMIZATION OF THERMAL PROCESSES FOR NUTRIENT RETENTION

Process	Method of Optimization
Blanching	Based on considerations other than thermal losses (e.g., leaching losses, oxidative degradation, damage to product).
Pasteurization	HTST if heat-resistant enzymes are not present.
Commercial Sterilization	Convection-heating foods and aseptic processing: HTST until heat-resistant enzymes become important. Conduction-heating foods: not HTST, necessarily. Difficult but not impossible calculation.

methods by which blanching, pasteurization, and commercial sterilization can be optimized with respect to nutrient retention are summarized.

EFFECT OF BLANCHING METHODS ON NUTRIENTS

For the blanching process, the effect of various methods of accomplishing the objectives of blanching on nutrients can be assessed by considering thermal, leaching, and oxidative losses. For data published prior to 1958, Lee (1958) presented an excellent review of the blanching process. Feaster (1960B) also considered nutrient losses in the blanching operation. Table 9.4 is a supplement to those reviews, and illustrates the effect of various methods of blanching on nutrient losses.

The two traditional methods of blanching use either hot water or steam as the heat transfer medium. Many systems have been designed to contact product with the heating medium for the time required to achieve a "blanched" condition. Since a process designed with either of these heating mediums would accomplish the desired objectives of the blanching operation and since there would not appear to be an advantage for an HTST process from the standpoint of thermal degradation of nutrients (see previous discussion on blanching optimization), the primary difference between these two processes with respect to nutrient retention is the extent of leaching. As expected, for water blanching the loss of water-soluble vitamins increases with contact time, and fat-soluble vitamins are relatively unaffected (Table 9.4); (Guerrant et al. 1947). Factors expected to affect losses during water blanching would be those factors affecting mass transfer: (1) surface area, (2) concentration of solutes in the hot water, and (3) agitation of the water.

Steam blanching results in greater retention of water-soluble nutrients than water blanching (Table 9.4) (Raab et al. 1973; Dietrich and Neumann 1965; Holmquist et al. 1954; Korobkina et al. 1969; Schwerdtfeger 1971). A new steam-blanching method called individual quick blanch (IQB) designed to reduce blanching effluent, has recently been developed (Lazar et al. 1971). Bomben et al. (1973) (Table 9.4) indicate that there may be a slight improvement in ascorbic acid retention with IQB as compared to conventional steam blanching. The slight improvement may be the result of the fact that in IQB each individual particle receives nearly the same heat treatment. With conventional steam blanching, the particles on the periphery of the bed are generally overblanched while particles in the center of the bed are just adequately blanched.

Microwave heating has also been applied for blanching food products. Since it can be assumed that microwave energy has no direct enhancing effect on degradation of food components other than through temperature elevation (Lopez and Baganis 1971), microwave blanching should result in nutrient retentions at least equal to that achieved during steam blanching and better than that achieved during water blanching. Dietrich et al. (1970) compared microwave, steam, and water blanching and verified that microwave blanching resulted in better ascorbic acid retention in brussel sprouts; however, the best product was achieved with combination processes involving microwave and water-blanching procedures. The microwave treatment gave rapid heat input into the product and a holding period in hot water following microwave treatment allowed thermal equilibration in the brussel sprouts. Although microwave blanching is inviting from a nutrient retention consideration, the cost per unit of product is generally exhorbitant (Huxsoll et al. 1970). Other efforts have been reported with respect to combination blanching processes involving microwave heating and hot gas treatments (Jeppson 1968, 1969); however, no data are available on the retention of nutrients.

Hot gas blanching also has been developed primarily to reduce effluent generated during the blanching operation (Ralls et al. 1972). Although temperatures up to 250°F are used, product temperature would not be expected to exceed 212°F because of evaporation of surface moisture. Ralls et al. (1973) (Table 9.4) reported the content of selected nutrients in spinach after water or hot gas blanching. The authors concluded that there was no significant difference between the two blanching methods. Although no studies have been reported on the effect of hot air blanching on nutrients, one of the factors which could contribute significantly to nutrient loss is oxidation. Loss of nutrients during dehydration has recently been reviewed by Labuza (1972).

TABLE 9.4

EFFECTS OF BLANCHING ON NUTRIENTS

Reference	Product	Nutrient	Process[1]	% Loss	Comment
Bomben et al. (1973)	Green beans	Vitamin C	S 2.5 min	8 mg/100 gm	No initial content available; content after blanching.
			IQB	11 mg/100 gm	
			IQB predry	7 mg/100 gm	
	Lima beans	Vitamin C	S 3.0 min	16 mg/100 gm	
			IQB	24 mg/100 gm	
			IQB predry	22 mg/100 gm	
	Brussel sprouts	Vitamin C	S	47 mg/100 gm	
			IQB	46 mg/100 gm	
			IQB predry	43 mg/100 gm	
	Peas	Vitamin C	S	21 mg/100 gm	
			IQB	18 mg/100 gm	
			IQB predry	20 mg/100 gm	
Dietrich and Neumann (1965)	Brussel sprouts	Vitamin C	W 9 min/190°F	24	
			S 11 min/190°F	16	
			W 6 min/200°F	16	
			S 7 min/200°F	16	
			W 5 min/212°F	19	
			S 6 min/212°F	20	
Dietrich et al. (1970)	Brussel sprouts	Vitamin C	W 6 min/212°F	43	
			M 1 min + W 4 min/212°F	29	
			M 3 min + W 2 min/212°F	35	
Guerrant et al. (1947)	Peas	Ascorbic acid	W 3 min/200°F	33	Also did other temp/time combinations on other vegetables: green beans, lima beans, spinach.
			W 6 min/200°F	46	
			W 9 min/200°F	58	
	Peas	Riboflavin	W 3 min/200°F	30	
			W 6 min/200°F	30	
			W 9 min/200°F	50	
	Peas	Thiamin	W 3 min/200°F	16	
			W 6 min/200°F	16	
			W 9 min/200°F	34	

EFFECTS OF HEAT PROCESSING ON NUTRIENTS 223

Reference	Food	Nutrient	Process	Result
	Peas	Carotene	W 3 min/200°F	2
			W 6 min/200°F	0
			W 9 min/200°F	0
	Lima beans	Niacin	W 2 min/200°F	32
			W 4 min/200°F	32
			W 6 min/200°F	37
Holmquist et al. (1954)	Peas	Vitamin C	S	12.3
			W	25.8
Korobkina et al. (1969)	Mussels	Niacin	S	57.6
			W	71.8
	Mussels	B-6	S	42.4
			W	67.5
	Mussels	Co	S	50.6
			W	49.8
	Mussels	Mn	S	17.7
			W	36.5
Raab et al. (1973)	Lima beans	B-6	W 10 min/212°F	21
			S 10 min/212°F	14
Ralls et al. (1973)	Spinach	Carotene	W	5.4 mg/100 gm
			Hot gas	3.9 mg/100 gm
	Spinach	Riboflavin	W	0.12 mg/100 gm
			Hot gas	0.10 mg/100 gm
	Spinach	Vitamin C	W	20.8 mg/100 gm
			Hot gas	34.2 mg/100 gm
	Spinach	Ca, Mg, P	W	No differences
			Hot gas	
Schwerdtfeger (1971)	Peas	Amino acid	W 212°F	25
			S	13
	Spinach	Protein	W 212°F	none
	Peas	Protein	S	none
	Spinach	Amino acid	W 212°F	80
			S	60

Evaluation after canning and processing. Initial level not reported.

[1] Process to adequate blanch. Generally determined by peroxidase inactivation. W = Water; S = Steam; IQB = Individual Quick Blanch; Hot gas = hot gas blanching; M = microwave.

224 NUTRITIONAL EVALUATION OF FOOD PROCESSING

Superheated steam also has been used to blanch and partially dry vegetables (Lazar 1972). Although no data were reported on the effect of this process on nutrients, based on the fact that an enzyme was used to assess blanching efficacy, it is likely that this treatment would have no more effect on nutrients than hot gas blanching.

In conclusion, it appears that the blanching operation can significantly reduce the nutrient content of foods, the extent being dependent on the blanching method and the product. Variation of nutrient losses between blanching methods can be rationalized on the basis of losses by leaching and oxidative degradation.

STORAGE OF BLANCHED FOODS

As previously pointed out, blanching is a thermal operation applied to foods which will subsequently receive an additional treatment. For those foods that are frozen or dehydrated, see the appropriate sections in this book (Chap. 10 and 11, respectively) on the effect of subsequent storage on nutrients. Those foods receiving an additional thermal process will be covered later.

TABLE 9.5

EXAMPLES OF PASTEURIZATION TREATMENTS USED FOR FOOD PRODUCTS

Temp Range (°C)	Product Pasteurized
60-65	Milk (holding process), milk for butter manufacture, egg, ice cream mix, smoked hams (meat temperature), carbonated beverages.
65-70	Ready-to-eat smoked meats (meat temperature), pickled sausages ([U.S.A.] meat temperature), canned hams (U.S.A.), wine (low temperature pasteurization), nonalcoholic fruit drinks.
70-75	Dill pickles, piccalilli, milk (flash process), carbonated fruit juices, mortadella sausage (pork and tongue).
75-80	Apple juice (holding process), grape juice, bread and butter pickles, cream for butter manufacture, raspberries, strawberries, bilberries, etc., in syrup in cans or jars.
80-85	Jamaica pickle, wine (U.S.A.), preserved and pickled vegetables, vegetables in oil, ice cream mix (flash process), desiccated coconut (other temperatures have been suggested).
85-90	Apple juice (flash process), canned olives, citrus juices, peeled tomatoes (pH 4.1).
90-95	Marroni sciroppati (chestnuts in syrup), tomato purée, citrus juices (flash process), prosciutti salati inscatolati (packaged ham), tomato juice, peeled tomatoes (pH 4.5), jam.
95-100	Wine (flash process), fruit purée, fruit juices, canned fruits (internal can temperature), canned mortadella sausage (pork and tongue).

Source: Shapton et al. (1971).

EFFECTS OF HEAT PROCESSING ON NUTRIENTS

EFFECT OF PASTEURIZATION METHODS ON NUTRIENTS

Some foods which receive pasteurization treatments are listed in Table 9.5 (Shapton et al. 1971). Examination of Table 9.5 reveals that most of the products which are pasteurized have a low pH either because the natural pH of the system is low or the product has been fermented to produce an acid environment. Since most of the heat-labile nutrients are relatively stable in acid conditions, nutrient losses in those products are relatively minor.

Although thermal losses during pasteurization may be small, oxidative losses can be high. Thus, pasteurization of food fluids such as fruit juices, beer, wine, etc., is generally accomplished in indirect heat exchangers (such as the plate or double-tube heat exchanger) rather than open-film type pasteurizers (Heid 1960). Oftentimes, fluids are deaerated prior to pasteurization.

The most important nonacid food fluid is milk. The effect of pasteurization treatments on nutrients in milk has received considerable attention. Vitamins in milk and milk products were extensively reviewed by Hartman and Dryden (1965) in one of the best reviews published on the effects of processing on nutrients in milk. Table 9.6 from Thompson (1969) summarizes the effect of pasteurization and sterilization on nutrients in milk.

TABLE 9.6

LOSS OF NUTRIENTS IN MILK DURING PROCESSING

Nutrient	Pasteurized		Sterilized	
	HTST (%)	Holder (%)	UHT (%)	In Bottle (%)
Protein	0	0	Whey proteins denatured	
Fat	0	0	Some loss of polyunsaturated fatty acids	
Sugar	0	0	0	Slight loss of nutritive value
Minerals	0	0	0	0
Vitamin A				
Vitamin D				
Riboflavin				
Vitamin B-6	0	0	0	0
Pantothenic acid				
Biotin				
Nicotinic acid				
Thiamin	10	10	10	35
Vitamin C	10	20	10	50
Folic acid	0	0	10	50
Vitamin B-12	0	10	20	30

Source: Thompson (1969).

226 NUTRITIONAL EVALUATION OF FOOD PROCESSING

As indicated in our earlier discussion, the HTST process results in greater nutrient retention for those nutrients affected by the pasteurization treatment (primarily thiamin, vitamin C, and vitamin B-12). Milk and milk products can be considered as primary sources for these nutrients, especially for the younger age groups (Hartman and Dryden 1965), and, therefore, these losses are very important nutritionally. However, this is a perfect example of the need to provide a heat treatment even though there are adverse nutritional consequences.

In addition to pasteurization, Thompson (1969) also reviewed sterilization of milk. The data included in Table 9.6 indicate that ultra-high-temperature (UHT) processing results in significantly greater retention of the heat-labile nutrients. In UHT processing, temperatures up to 300°F are used for very short periods (on the order of seconds).

STORAGE OF PASTEURIZED FOODS

Little information has been published on the storage stability of nutrients in high acid, pasteurized products. However, those nutrients that are more sensitive to high temperature are generally the same ones that are of concern during storage. It would be reasonable that the lower the storage temperature the slower the rate of nutrient degradation. Usually in these kinds of products, proper packaging is paramount for extending the storage life since oxidative losses and light-catalyzed (both visible and ultraviolet) losses can be the major mechanism of loss.

In contrast to other pasteurized products, storage of pasteurized milk has received extensive consideration (Hartman and Dryden 1965). Low storage temperature and the relatively short storage time minimize the loss of nutrients in milk. However, some nutrient destruction does occur and is catalyzed primarily by visible and ultraviolet light. Therefore, packaging considerations are of primary importance (Karel 1960; Singh et al. 1974). Packaging as a means of maintaining nutrients in foods is covered in another section of this book.

EFFECT OF COMMERCIAL STERILIZATION METHODS ON NUTRIENTS

The various methods available for commercial sterilization of food have been reviewed recently by Brody (1971). Since the destruction of nutrients during the thermal process is dependent on: (1) time/temperature treatment used as the basis of the process, and (2) rate of heat transfer into the product, commercial developments have

EFFECTS OF HEAT PROCESSING ON NUTRIENTS 227

focused primarily on increasing the rate of heat transfer into the product (Gutterson 1972). Hence, agitated retorts such as the orbitort, steritort, flame-sterilizer, and hydrostatic cooker have been developed.

In addition to increasing the rate of heat transfer, however, there also has been a gradual shift to higher processing temperatures. As pointed out in the discussion on optimizing nutrient retention in commercial sterilization, a high temperature/short time process results in greater nutrient retention in those products heating primarily by convection. Ammerman (1957) presented an excellent study on the effects of heat treatments of equal microbial lethality on selected food constituents including nutrients, colors, proteins, and flavor compounds. Figure 9.2 from Ammerman (1957) illus-

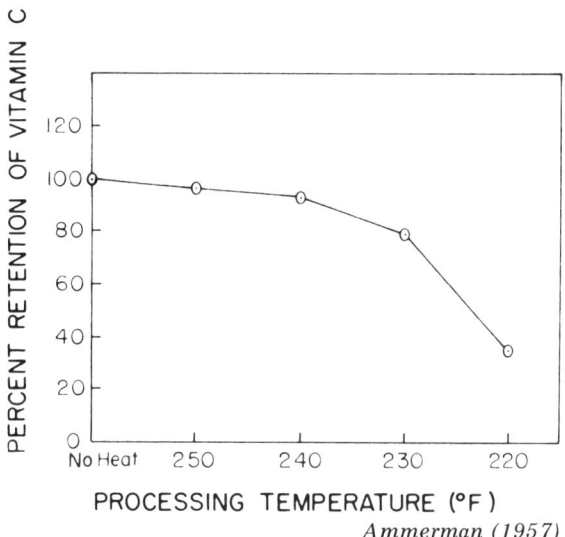

Ammerman (1957)

FIG. 9.2. THE EFFECT OF EQUIVALENT LETHAL HEAT TREATMENTS AT THE INDICATED TEMPERATURE ON THE RETENTION OF VITAMIN C IN TOMATO JUICE

trates that retention of vitamin C in tomato juice is improved when processing is conducted at a high-temperature short-time condition. For natural products containing enzymes the limitation of the benefit of HTST processing, as pointed out earlier, occurs when the basis of the process shifts from microbes to enzymes (about 270°–290°F).

The use of high-temperature short-time processes is particularly adaptable to aseptic processing. In this system, processing tempera-

228 NUTRITIONAL EVALUATION OF FOOD PROCESSING

tures in excess of 300°F are used for very short periods (order of seconds). Under these conditions nutrient retention may be greatly enhanced. In an evaluation of HTST aseptic processing, Everson *et al.* (1964A, B) found that thiamin retention was significantly greater in HTST products than in conventionally canned and retorted products (Table 9.7). For pyridoxine, the benefit of HTST was not as

TABLE 9.7

EFFECT OF ASEPTIC AND CONVENTIONAL THERMAL PROCESSING METHODS ON NUTRIENT LOSSES

Product	Thiamin Loss (%)		Pyridoxine Loss (%)	
	HTST	Conventional	HTST	Conventional
Strained lima beans	15.8	40.3	9.5	10.1
Strained beef	9.2	21.6	4.1	2.9
Tomato juice concentrate	0	2.8	0	0

Source: Everson *et al.* (1964A, B)

evident, probably indicating that thermal destruction of pyridoxine is not as temperature-dependent as that of thiamin. HTST aseptic canning also results in a significant improvement in organoleptic qualities (Anon. 1970). Currently there is activity in developing aseptic processing equipment for handling food particulates.

As pointed out earlier, most of the reports on the effect of thermal processing on nutrients only contain information on the content of a specific nutrient after the thermal process and gives the percentage retention or loss of the nutrient. In light of the fact that there are numerous processing methods and time/temperature possibilities for accomplishing commercial sterilization, it is not appropriate to assume that the nutrient losses reported in the literature represent the average or norm for the industry. For this reason, data of this type are of limited value. However, these reports can be used, as Schroeder (1971) used them, to point out a critical lack of particular nutrients in our processed food supply.

USDA Handbook No. 8 (Watt and Merrill 1963), Orr (1969), and Mitchell *et al.* (1968) report nutrient content of processed foods. Some of the data that they assembled were used to calculate percentage loss of nutrients in selected vegetables during canning and are presented in Table 9.8. Nutrient losses range from 0 to 91% depending on the nutrient and product. These losses represent the sum of the losses during the entire canning process and, as pointed out in Table 9.4, blanching losses can be quite large. However, the impor-

TABLE 9.8

LOSSES OF NUTRIENTS IN THE CANNING PROCESS

Product	Biotin[1] (%)	Folacin[1] (%)	B-6[2] (%)	Pantothenic[2] Acid (%)	A[3] (%)	Thiamin[3] (%)	Riboflavin[3] (%)	Niacin[3] (%)	C[3] (%)
Asparagus	0	75.2	64.0	—	43.3	66.7	55.0	46.6	54.5
Lima beans	—	61.8	47.1	72.3	55.2	83.3	66.7	64.2	75.9
Green beans	—	57.1	50.0	60.5	51.7	62.5	63.6	40.0	78.9
Beets	—	80.0	9.1	33.3	50.0	66.7	60.0	75.0	70.0
Carrots	40.0	58.8	80.0	53.6	9.1	66.7	60.0	33.3	75.0
Corn	63.3	72.5	0	59.2	32.5	80.0	58.3	47.1	58.3
Coropeas	—	36.6	90.6	84.8	83.8	79.1	61.5	68.8	89.7
Mushrooms	54.4	83.8	—	54.5	—	80.0	45.6	52.3	33.3
Green peas	77.7	58.8	68.8	80.0	29.7	74.2	64.3	69.0	66.7
Spinach	66.7	34.7	75.0	78.3	32.1	80.0	50.0	50.0	72.5
Tomatoes	55.0	53.75	—	30.3	0	16.7	25.0	0	26.1

[1] Mitchell et al. (1968).
[2] Orr (1969).
[3] Watt and Merrill (1963).

tant observation is that nutrient losses appear to be quite significant in the canning process.

Studies with more complete data will be forthcoming as nutrient labeling expands. More complete data on the kinetics of destruction and temperature dependence of rate constants for nutrients will be needed to allow optimization of processes and to predict the effect of a change in process temperature on nutrient retention.

STORAGE OF COMMERCIALLY STERILE FOODS

A common misconception among consumers and many food industry personnel is that commercially sterile products remain unchanged during storage. This is not the case. Organoleptic and nutrient changes do occur during storage, the extent of the changes being dependent on the time and temperature of storage, the packaging system, and the product characteristics. Several extensive studies have been conducted on the storage stability of nutrients in canned foods (Cameron et al. 1949; Goresline et al. 1955; Cecil and Woodroff 1962; Ball et al. 1963).

Suprisingly, few authors have considered the kinetic approach in assessing the effect of storage conditions on nutrients. Labuza (1972) used this approach in describing nutrient stability in stored dehydrated foods, and Wanninger (1972) developed a mathematical model for predicting the stability of ascorbic acid in stored foods as a function of temperature, time, moisture content, and oxygen content. This approach also would be applicable to storage of commercially sterile products. The dependence of reaction rates on temperature should be similar in magnitude to those reported in

TABLE 9.9

RETENTION OF VITAMINS IN CANNED FOODS DURING STORAGE

Product	Storage Conditions °F	Months	% Retention Ascorbic Acid	Carotene	Niacin	Riboflavin	Thiamin
Apricots	50	12	96	94	—	—	—
	65	12	93	85	—	—	—
	80	12	85	83	—	—	—
Apricots	50	24	94	91	—	—	—
	65	24	90	84	—	—	—
	80	24	56	76	—	—	—
Asparagus, green	50	12	97	97	89	92	89
	65	12	94	88	85	87	79
	80	12	89	85	84	83	66
Asparagus, green	50	24	93	88	93	81	85
	65	24	91	84	91	77	72
	80	24	86	76	87	72	54
Asparagus, white	50	12	96	—	96	—	82
	65	12	94	—	94	—	74
	80	12	87	—	97	—	62
Asparagus, white	50	24	90	—	96	—	72
	65	24	87	—	98	—	65
	80	24	82	—	97	—	52
Beans, green	50	12	92	—	83	72	92
	65	12	90	—	81	69	86
	80	12	85	—	80	62	78
Beans, green	50	24	88	—	86	62	82
	65	24	81	—	86	57	80
	80	24	74	—	86	42	67
Beans, lima	50	12	100	—	101	95	88
	65	12	98	—	100	91	82
	80	12	95	—	99	88	74
Beans, lima	50	24	86	—	99	75	87
	65	24	83	—	97	75	76
	80	24	78	—	100	70	66
Carrots	50	12	—	94	—	—	—
	65	12	—	97	—	—	—
	80	12	—	93	—	—	—
Carrots	50	24	—	90	—	—	—
	65	24	—	95	—	—	—
	80	24	—	91	—	—	—
Corn, white	50	12	98	—	82	—	97
	65	12	92	—	85	—	85
	80	12	86	—	88	—	78
Corn, white	50	24	90	—	84	—	94
	65	24	88	—	86	—	89
	80	24	78	—	88	—	71

EFFECTS OF HEAT PROCESSING ON NUTRIENTS 231

TABLE 9.9 (*Continued*)

Product	Storage Conditions °F	Months	% Retention Ascorbic Acid	Carotene	Niacin	Riboflavin	Thiamin
Corn, yellow	50	12	98	85	89	84	90
	65	12	94	87	89	80	86
	80	12	89	84	91	78	74
Corn, yellow	50	24	92	69	91	71	89
	65	24	89	72	90	68	76
	80	24	81	87	96	61	60
Grapefruit juice	50	12	95	—	—	—	99
	65	12	91	—	—	—	100
	80	12	75	—	—	—	93
Grapefruit juice	50	24	94	—	—	—	99
	65	24	82	—	—	—	94
	80	24	57	—	—	—	84
Grapefruit segments	50	12	94	—	—	—	—
	65	12	91	—	—	—	—
	80	12	73	—	—	—	—
Grapefruit segments	50	24	87	—	—	—	—
	65	24	77	—	—	—	—
	80	24	46	—	—	—	—
Orange juice	50	12	97	—	—	—	100
	65	12	92	—	—	—	98
	80	12	77	—	—	—	89
Orange juice	50	24	95	—	—	—	101
	65	24	80	—	—	—	94
	80	24	50	—	—	—	83
Peaches	50	12	98	95	101	—	92
	65	12	85	90	102	—	90
	80	12	72	86	101	—	81
Peaches	50	24	98	75	100	—	88
	65	24	80	64	98	—	100
	80	24	53	63	99	—	86
Peas, Alaska	50	12	91	97	82	91	91
	65	12	89	95	77	84	86
	80	12	84	91	82	82	75
Peas, Alaska	50	24	90	95	99	80	89
	65	24	88	93	87	73	85
	80	24	81	89	85	68	68
Peas, sweet	50	12	94	98	95	93	93
	65	12	92	92	87	89	88
	80	12	88	91	90	84	73
Peas, sweet	50	24	92	94	96	88	91
	65	24	89	90	95	84	85
	80	24	81	90	95	81	72
Pineapple juice	50	12	110	—	—	—	93
	65	12	108	—	—	—	93
	80	12	93	—	—	—	87

232 NUTRITIONAL EVALUATION OF FOOD PROCESSING

TABLE 9.9 (Continued)

Product	Storage Conditions °F	Months	% Retention Ascorbic Acid	Carotene	Niacin	Riboflavin	Thiamin
Pineapple juice	50	24	108	—	—	—	103
	65	24	100	—	—	—	100
	80	24	79	—	—	—	93
Pineapple, sliced	50	12	100	—	—	—	97
	65	12	95	—	—	—	96
	80	12	74	—	—	—	89
Pineapple, sliced	50	24	83	—	—	—	102
	65	24	78	—	—	—	103
	80	24	53	—	—	—	89
Plums, purple (prunes)	50	12	—	102	95	84	—
	65	12	—	100	93	82	—
	80	12	—	97	103	78	—
Plums, purple (prunes)	50	24	—	90	86	84	—
	65	24	—	98	91	82	—
	80	24	—	86	95	76	—
Spinach	50	12	93	91	100	92	96
	65	12	91	90	103	89	89
	80	12	86	84	99	85	76
Spinach	50	24	90	80	96	82	90
	65	24	88	80	100	80	82
	80	24	81	81	101	69	71
Tomatoes	50	12	95	94	91	94	94
	65	12	94	98	93	95	93
	80	12	82	95	93	91	82
Tomatoes	50	24	89	75	88	96	91
	65	24	87	75	88	98	87
	80	24	70	74	85	97	70
Tomato juice	50	12	100	98	99	88	95
	65	12	97	100	99	84	93
	80	12	86	99	99	83	85
Tomato juice	50	24	102	94	92	92	103
	65	24	92	97	91	94	94
	80	24	74	98	90	94	77

Source: Cameron et al. (1955).

Table 9.2. Since the activation energy is weakly temperature-dependent, the temperature range over which it was determined must be given and extrapolation over broad temperature ranges should be avoided. The concept, however, is applicable at storage temperatures.

Cameron et al. (1955) reviewed thermal processing and storage of foods and compiled Table 9.9 from existing literature. It can be seen

EFFECTS OF HEAT PROCESSING ON NUTRIENTS 233

TABLE 9.10

MAXIMUM STORAGE TEMPERATURES FOR CANNED FOODS TO ASSURE NOT MORE THAN 10% LOSS OF A SELECTED VITAMIN

Canned Product	Ascorbic Acid (Vitamin C)			Thiamin (Vitamin B-1)			Carotene (Vitamin A Precursor)			Niacin			Riboflavin (Vitamin B-2)		
	12	18	24	12	18	24	12	18	24	12	18	24	12	18	24
Apricots	76	68	60	>32	—	—	70	62	54	—	—	—	—	—	—
Asparagus	74	65	58	45	38	32	72	58	45	2	2	2	57	50	45
Beans, green	60	45	32	57	45	32	—	—	—	2	2	2	57	50	45
Beans, lima	74	60	45	>32	—	—	—	—	—	3	3	3	65	40	32
Carrots	—	—	—	—	—	—	80+	80+	80+	—	—	—	—	—	—
Corn, sweet	71	60	50	64	58	53	58	66	75	4	4	4	70	55	48
Frankfurters and beans	—	—	—	56	47	41	—	—	—	—	—	—	—	—	—
Grapefruit segments	60	45	32	—	—	—	—	—	—	—	—	—	—	—	—
Peaches	65	50	40	76	80	84	70	40	32	3	3	3	—	—	—
Peas	74	66	52	62	56	50	72	58	45	2	2	2	55[6]	50[6]	45[6]
Plums	—	—	—	—	—	—	80+	80+	80+	4	4	4	—	—	—
Pineapple slices	68	54	32	80	80	80	—	—	—	—	—	—	—	—	—
Spinach	68	60	52	63	57	49	46[1]	41[1]	38[1]	3	3	3	63[4]	46[4]	32[4]
Tomatoes	68	60	45	70	60	45	—	—	—	5	5	5	—	—	—

Source: Kramer (1974).
[1] 10% loss in 17 months—no temperature effect.
[2] 20% loss in 12 months, 15% in 24 months, no temperature effect.
[3] No loss.
[4] 10% loss—no time or temperature effect.
[5] 8% loss in 12 months, 13% in 24 months, no temperature effect.
[6] 20% loss, no time or temperature effect.

that low temperature storage results in an improvement in nutrient retention. The 50° and 65°F storage temperatures would require refrigerated warehousing and would be economically feasible only if the consumer is willing to pay for the increased cost.

With the increased consumer awareness of nutrition, and with the advent of nutritional labeling requirements, it may become economically and legally advantageous to select storage temperatures that will result in a stated shelf-life with a specified level of nutrient.

Kramer (1974) adapted data from Feaster (1960A) and Cecil and Woodroof (1962) to predict the maximum storage temperatures that could be used for a variety of canned foods to assure that no more than 10% of a specified nutrient was lost during a stated storage period (Table 9.10). It can be seen that the maximum storage temperature is dependent upon the nutrient under consideration since each nutrient has a characteristic activation energy. It is also evident that most of the temperatures are below ambient temperature and therefore the products would require refrigerated storage.

In conclusion, it is evident that there is a significant loss of nutrients during canning and that these losses increase during storage. Altering processing and storage conditions to maximize nutrient retention is an important and necessary direction for the food processing industry.

SUMMARY

The fact that application of thermal energy to foods reduces the nutritive value of some components cannot be contested. However, it is necessary to evaluate that consequence in view of the fact that application of thermal processes results in decreased food wastage through spoilage and decreased food-borne diseases. To have food available only at the season of harvest will not nourish the people in the kind of industrial nation we have become. It is the responsibility of the preserved-food industry to produce the most nutritious food supply it is capable of. This one area, alteration of thermal processes and storage conditions for maximizing nutrient retention, requires much more diligence than it has received in the past. Although the tools for determining the best conditions within the economic constraints placed on the system are available, the appropriate kind of data, in many cases, are lacking. This need will be filled as we continue to recognize the importance of using nutrient retention as one of the bases for establishing optimum thermal processes and storage conditions.

BIBLIOGRAPHY

ADAMS, H. W., and YAWGER, E. S. 1961. Enzyme inactivation and color of processed peas. Food Technol. *15*, 314-317.
AIBA, S., HUMPHREY, A. E., and MILLIS, N. F. 1965. Biochemical Engineering. Academic Press, New York.
AMMERMAN, G. R. 1957. The effect of equal lethal heat treatments at various times and temperatures upon selected food components. PhD Thesis. Purdue Univ., W. Lafayette, Ind.
ANON. 1969. Aseptic Processing, 2nd Edition. Cherry-Burrell Corp., Tech. Dig. *Cb-201*.
ANON. 1970. Physico chemical differences of puréed vegetables packed by the aseptic and retort processes. United States Steel *ADUSS 33-4406-01*.
ASHTON, W. M. 1972. The components of milk, their nutritive value and the effects of processing. Part I. Part II. Dairy Ind. *37*, 535-536, 538; *37*, 602-606, 611.
BALL, C. O., JOFFE, F. M., STIER, E. F., and HAYAKAWA, K. 1963. The role of temperature in retaining quality in canned foods. ASHRAE J. *5*, No. 6, 93-108, 144, 146.
BALL, C. O., and OLSON, F. C. W. 1957. Sterilization in Food Technology. McGraw-Hill Book Co., New York.
BARRATT, B. 1973. Nutrition 1: The building blocks. Nutrition 2: Effects of Processing. Food Can. *33*, No. 1, 13-16, *33*, No. 2, 28-31.
BENDER, A. E. 1966. Nutritional effects of food processing. J. Food Technol. *1*, 261-289.
BENDER, A. E. 1972. Processing damage to protein food. A review. J. Food Technol. *7*, 239-250.
BENDIX, G. H., HEBERLEIN, D. G., PTAK, L. R., and CLIFCORN, L. E. 1951. Thiamine destruction in peas, corn, lima beans, and tomato juice from 104.5° to 132°C (220°-270°F). J. Food Sci. *16*, 494-503.
BERK, Z. 1970. Processing and storage damage to nutritional value of foods. Proc. 3rd Intern. Congr. Food Sci. Technol. 189-191.
BLAKEBOROUGH, N. 1968. Preservation of biological materials especially by heat treatment. *In* Biochemical and Biological Engineering Science, Vol. 2. N. Blakeborough (Editor). Academic Press, New York.
BOMBEN, J. L. *et al.* 1973. Pilot plant evaluation of individual quick blanching (IQB) for vegetables. J. Food Sci. *38*, 590-594.
BRODY, A. L. 1971. Food canning in rigid and flexible packages. Critical Rev. Food Technol. *2*, 187-244.
BURGER, I. H., and WALTERS, C. L. 1973. The effect of processing on the nutritive value of flesh foods. Proc. Nutr. Soc. *32*, 1-8.
BURTON, H. 1963. A note on the effect of heat on the colour of goat's milk. J. Dairy Res. *30*, 217-222.
BURTON, H. *et al.* 1970. Comparison of milks processed by the direct and indirect methods of ultra-high-temperature sterilization. IV. The vitamin composition of milks processed by different processes. J. Dairy Res. *37*, 529-533.
CAIN, R. F. 1967. Water soluble vitamins: Changes during processing and storage of fruits and vegetables. Food Technol. *21*, 998-1007.
CAMERON, E. J., PILCHER, R. W., and CLIFCORN, L. E. 1949. Nutrient retention during canned food production. Am. J. Public Health *39*, 756-763.
CAMERON, E. J. *et al.* 1955. Retention of nutrients during canning. National Canners Association, Washington, D.C.
CECIL, S. R., and WOODROOF, J. G. 1962. Long-term storage of military rations. Quartermaster Food and Container Institute for the Armed Forces, Chicago.

CHICHESTER, C. O. 1973. Nutrition in food processing. World Rev. Nutr. Dietet. *16*, 318-333.
CLIFCORN, L. E. 1948. Factors influencing the vitamin content of canned foods. Food Res. *1*, 39-104.
CLIFCORN, L. E., PETERSON, G. T., BOYD, J. M., and O'NEIL, J. H. 1950. A new principle for agitating in processing of canned foods. Food Technol. *4*, 450-457.
COHEN, J. S., and WALL, M. A. 1971. A method of calculating average sterilizing value in cylindrical containers. Trans. ASAE *3*, 329-333.
DAVIDEK, J., VELISEK, J., and JANICEK, G. 1972. Stability of inosinic acid, inosine and hypoxanthine in aqueous solutions. J. Food Sci. *37*, 789-790.
DIETRICH, W. C., HUXSOLL, C. C., and GUADAGNI, D. G. 1970. Comparison of microwave, conventional and combination blanching of brussels sprouts for frozen storage. Food Technol. *24*, 613-617.
DIETRICH, W. C., and NEUMANN, H. J. 1965. Blanching brussels sprouts. Food Technol. *19*, 1174-1177.
DIETRICH, W. C. et al. 1959. Time-temperature tolerance of frozen foods. XVIII. Effect of blanching conditions on color stability of frozen beans. Food Technol. *13*, 258-261.
DUGAN, L. R., JR. 1968. Processing and other stress effects on the nutritive value of lipids. World Rev. Nutr. Dietet. *9*, 181-205.
EVERSON, G. J. et al. 1964A. Aseptic canning of foods. II. Thiamine retention as influenced by processing method, storage time and temperature and type of container. Food Technol. *18*, 84-86.
EVERSON, G. J. et al. 1964B. Aseptic canning of foods. III. Pyridoxine retention as influenced by processing method, storage time and temperature and type of container. Food Technol. *18*, 87-88.
FARKAS, D. F., GOLDBLITH, S. A., and PROCTOR, B. E. 1956. Stopping storage off-flavor by curbing peroxidase. Food Eng. *28*, No. 1, 52-53.
FEASTER, J. F. 1960A. Effects of commercial storage on the nutrient content of processed foods. A. Foods of plant origin. 1. Fruits and vegetables. *In* Nutritional Evaluation of Food Processing. R. S. Harris and H. von Loesecke (Editors). John Wiley & Sons, New York. Reprinted in 1971 by Avi Publishing Co., Westport, Conn.
FEASTER, J. F. 1960B. A. Washing, trimming, and blanching. *In* Nutritional Evaluation of Food Processing. R. S. Harris and H. von Loesecke (Editors). John Wiley & Sons, New York. Reprinted in 1971 by Avi Publishing Co., Westport, Conn.
FEASTER, J. F., TOMPKINS, M. D., and IVES, M. 1949. Retention of vitamins in low acid canned foods. Food Ind. *20*, 14-17, 150, 152, 154.
FELICIOTTI, E., and ESSELEN, W. B. 1957. Thermal destruction rates of thiamine in puréed meats and vegetables. Food Technol. *11*, 77-84.
FORD, J. E. et al. 1969. Effects of ultra-high-temperature (UHT) processing and of subsequent storage on the vitamin content of milk. J. Dairy Res. *36*, 447-454.
GARRETT, E. R. 1956. Prediction of stability in pharmaceutical preparations. II. Vitamin stability in liquid multivitamin preparations. J. Am. Pharm. Assoc. *45*, 171-178.
GILLESPY, T. G. 1962. Principles of heat sterilization. *In* Recent Advances of Food Sciences, Vol. II, Processing. J. Hawthorn and J. M. Leitch (Editors). Butterworths, London, England.
GOLD, H. J., and WECKEL, K. G. 1959. Degradation of chlorophyll to pheophytin during sterilization of canned green peas by heat. Food Technol. *13*, 281-286.
GOLDBLITH, S. A. 1971. Thermal processing of foods: a review. World Rev. Nutr. Dietet. *13*, 165-193.

GORESLINE, H. E., LEINEN, N. J., and MRAK, E. M. (Editors) 1955. Establishing optimum conditions for storage and handling of semiperishable subsistence items. Office of the Quartermaster General, Washington, D.C.
GORTNER, W. A. 1972. The impact of food technology on nutrient supplies. Food Technol. Australia 24, 504-517.
GUERRANT, N. B. et al. 1947. Effect of duration and temperature of blanch on vitamin retention by certain vegetables. Ind. Eng. Chem. 39, 1000-1007.
GUPTE, S. M., EL-BISI, H. M., and FRANCIS, F. J. 1964. Kinetics of thermal degradation of chlorophyll in spinach purée. J. Food Sci. 29, 379-382.
GUTTERSON, M. 1972. Food Canning Techniques. Noyes Data Corporation, Park Ridge, N.J.
HACKLER, L. R. et al. 1965. Effect of heat treatment on nutritive value of soymilk protein fed to weanling rats. J. Food Sci. 31, 723-728.
HARRIS, R. S., and VON LOESECKE, H. (Editors) 1960. Nutritional Evaluation of Food Processing. John Wiley & Sons, New York. Reprinted in 1971 by Avi Publishing Co., Westport, Conn.
HARTMAN, A. M., and DRYDEN, L. P. 1965. Vitamins in milk and milk products. American Dairy Science Association, Champaign, Illinois.
HAYAKAWA, K. 1969. New parameters for calculating mass average sterilizing value to estimate nutrients in thermally conductive foods. Can. Inst. Food Technol. 2, 165-172.
HEID, J. L. 1960. C. Pasteurization, sterilization and storage. In Nutritional Evaluation of Food Processing. R. S. Harris, and H. von Loesecke (Editors). John Wiley & Sons, New York. Reprinted in 1971 by Avi Publishing Co., Westport, Conn.
HEIN, R. E., and HUTCHINGS, I. J. 1971. Influence of processing on vitamin-mineral content and biological availability in processed foods. Council Foods Nutrition, AMA and AMA-Food Liaison Comm., New Orleans.
HENSHALL, J. D. 1973. Fruit and vegetable products. Proc. Nutr. Soc. 32, 17-22.
HERRMANN, J. 1970. Calculation of the chemical and sensory alterations in food during heating and storage processes. Ernährungsforschung 15, 279-299.
HOLLINGSWORTH, D. F. 1970. Effects of some new production and processing methods on nutritive values. J. Am. Dietet. Assoc. 57, 246-249.
HOLLINGSWORTH, D. F., and MARTIN, P. E. 1972. Some aspects of the effects of different methods of production and of processing on the nutritive value of foods. World Rev. Nutr. Dietet. 15, 1-34.
HOLMQUIST, J. W. et al. 1954. Steam blanching of peas. Food Technol. 8, 437-445.
HUXSOLL, C. C., DIETRICH, W. C., and MORGAN, A. I., JR. 1970. Comparison of microwave with steam or water blanching of corn-on-the-cob. 1. Characteristics of equipment and heat penetration. Food Technol. 24, 290-292.
JACKSON, J. M., and BENJAMIN, A. A. 1948. Sterilization of foods. Ind. Eng. Chem. 40, 2241-2246.
JEN, Y., MANSON, J. E., STUMBO, C. R., and ZAHRADNIK, J. W. 1971. A procedure for estimating sterilization of and quality factor degradation in thermally processed foods. J. Food Sci. 36, 692-698.
JEPPSON, M. R. 1968. Treating food products with microwave energy and hot gas of decreasing humidity. U.S. Pat. 3,409,447, Nov. 5.
JEPPSON, M. R. 1969. Apparatus for treating food products and the like with microwave energy. U.S. Pat. 3,478,900, Nov. 18.
KACZMARZYK, L. M., FENNEMA, O., POWRIE, W. D. 1963. Changes produced in Wisconsin green snap beans by blanching. Food Technol. 17, 943-946.
KAREL, M. 1960. Effects of packaging on maintenance of nutrients in food

products. *In* Nutritional Evaluation of Food Processing. R. S. Harris and H. von Loesecke (Editors). John Wiley & Sons, New York. Reprinted in 1971 by Avi Publishing Co., Westport, Conn.

KOROBKINA, G. S., DANILOVA, E. N., and KALININA, N. N. 1969. Effect of processing on the food value of mussels. Vopr. Pitaniya *28*, No. 5, 85-86. (Russian). From Nutr. Abstr. *40*, No. 2, No. 2384.

KRAMER, A. 1974. Storage retention of nutrients. Food Technol. *28*, 50-58.

LABUZA, T. P. 1972. Nutrient losses during drying and storage of dehydrated foods. Critical Rev. Food Technol. *3*, 217-240.

LANG, K. 1970. Influence of cooking on foodstuffs. World Rev. Nutr. Dietet. *12*, 266-317.

LAWRIE, R. A. 1968. Chemical changes in meat due to processing. A review. J. Sci. Food Agr. *19*, 233-240.

LAZAR, M. E. 1972. Blanching and partial drying of foods with super-heated steam. J. Food Sci. *37*, 163-166.

LAZAR, M. E., LUND, D. B., and DIETRICH, W. C. 1971. IQB—A new concept in blanching. Food Technol. *25*, 684-686.

LEE, F. A. 1958. The blanching process. Advan. Food Res. *8*, 63-109.

LENZ, M. K., and LUND, D. B. 1974. Personal communication. Department of Food Science, University of Wisconsin, Madison.

LICCIARDELLO, J. J., RIBICH, C. A., NICKERSON, J. T. R., and GOLDBLITH, S. A. 1967. Kinetics of thermal inactivations of type E *Clostridium botulinum* toxin. Appl. Microbiol. *15*, 344-349.

LOPEZ, A., and BAGANIS, N. A. 1971. Effect of radio-frequency energy at 60 MHz on food enzyme activity. J. Food Sci. *36*, 911-914.

LUND, D. B., and LENZ, M. K. 1973. Lethality calculations for heat/cool and heat/hold/cool processes by the L-Fourier number method. Presented at the 33rd Ann. Meeting Inst. Food Technologists, Minneapolis.

LUND, D. B. *et al.* 1973. Symposium: Effects of processing, storage, and handling on nutrient retention in foods. Food Technol. *27*, No. 1, 16-38, 51.

MACKINNEY, G., and JOSLYN, M. A. 1941. Chlorophyll-pheophytin: Temperature coefficients of the rate of pheophytin formation. J. Am. Chem. Soc. *63*, 2530-2531.

MANSFIELD, T. 1962. High temperature, short time sterilization. First Intern. Congr. Food Sci. Technol., London.

MANSFIELD, T. 1974. A brief study of cooking. FMC Corporation. (Personal Communication).

MANSON, J. E., ZAHRADNIK, J. W., and STUMBO, C. R. 1970. Evaluation of lethality and nutrient retention of conduction—heating foods in rectangular containers. Food Technol. *24*, 1297-1301.

MAPSON, L. W. 1956. Effect of processing on the vitamin content of foods. Brit. Med. Bull. *12*, 73-77.

MITCHELL, A. S., RYNBERGEN, H. J., ANDERSON, L., and DIBBLE, M. V. 1968. Cooper's Nutrition in Health and Disease, 15th Edition. J. B. Lippincott Co., Philadelphia.

MULLEY, E. A., STUMBO, C. R., and HUNTING, W. M. 1974. Kinetics of thiamine degradation by heat: A new method for studying reaction rates in model systems and food products. Presented at 34th Ann. Meeting Inst. Food Technologists, New Orleans.

ORR, M. 1969. Pantothenic acid, vitamin B_6 and vitamin B_{12} in foods. USDA Agr. Res. Serv. Home Econ. Res. Rept. *36*.

OSNER, R. C., and JOHNSON, R. M. 1968. Nutritional changes in protein during heat processing. J. Food Technol. *3*, 81-86.

PFLUG, I. J., and SCHMIDT, C. F. 1968. Thermal destruction of microorganisms. *In* Disinfection, Sterilization, and Preservation. C. A. Lawrence, and S. S. Block (Editors). Lea & Febiger, Philadelphia.

PONTING, J. D., SANSHUCK, D. W., and BREKKE, J. E. 1960. Color mea-

EFFECTS OF HEAT PROCESSING ON NUTRIENTS 239

surement and deterioriation in grape and berry juices and concentrates. J. Food Sci. 25, 471-478.

RAAB, C. A., LUH, B. S., and SCHWEIGERT, B. S. 1973. Effects of heat processing on the retention of vitamin B_6 in lima beans. J. Food Sci. 38, 544-545.

RALLS, J. W., MAAGDENBERG, H. J., YACOUB, N. L., and MERCER, W. A. 1972. Reduced waste generation by alternate vegetable blanching system. Proc. 3rd Natl. Symp. Food Processing Wastes, New Orleans.

RALLS, J. W. et al. 1973. In-plant, continuous hot-gas blanching of spinach. J. Food Sci. 38, 192-194.

RAMAKRISHNAN, T. V., and FRANCIS, F. J. 1973. Color degradation in paprika. J. Food Sci. 38, 25-28.

READ, R. B., JR., and BRADSHAW, J. G. 1966. Staphylococcal enterotoxin B thermal inactivation in milk. J. Dairy Sci. 49, 202-203.

ROLLS, B. A., and PORTER, J. W. G. 1973. Some effects of processing and storage on the nutritive value of milk and milk products. Proc. Nutr. Soc. 32, 9-15.

SABRY, Z. I. 1968. The nutritional consequences of developments in food processing. Can. J. Public Health 59, 471-474.

SCHROEDER, H. A. 1971. Losses of vitamins and trace minerals resulting from processing and preservation of foods. Am. J. Clinical Nutr. 24, 562-573.

SCHWERDTFEGER, E. 1971. Changes in the content of nutritive components during preparation and processing of vegetables. 3. Protein and amino acids. Qualitas Plant. Mater. Vegetabiles 21, 97-110, from Nutr. Abstr. 42, No. 3, No. 5190.

SHAPTON, D. A., LOVELOCK, D. W., and LAURITA-LONGO, R. 1971. The evaluation of sterilization and pasteurization processes from measurements in degrees celcius (°C). J. Appl. Bacteriol. 34, 491-500.

SINGH, R. P., HELDMAN, D. R., and KIRK, J. R. 1974. Kinetic analysis of light-induced vitamin loss in liquid foods. Presented at the 34th Ann. Meeting Inst. Food Technologists, New Orleans.

STUMBO, C. R. 1973. Thermobacteriology in Food Processing, 2nd Edition. Academic Press, New York.

TAIRA, H., TAIRA, H., and SUKURAI, Y. 1966. Studies on amino acid contents of processed soybean. Part 8. Effect of heating on total lysine and available lysine in defatted soybean flour. Japan. J. Nutr. Food. 18, 359.

TAN, C. T., and FRANCIS, F. J. 1962. Effects of processing temperature on pigments and color of spinach. J. Food Sci. 27, 232-241.

TANCHEV, S. S., and JONCHEVA, N. 1973. Kinetics of the thermal degradation of cyanidin-3-rutinoside and peonidin-3-rutinoside. Z. Lebensm. Unters.-Forsch. 153, 37-41.

TEIXEIRA, A. A., DIXON, J. R., ZAHRADNIK, J. W., and ZINSMEISTER, G. E. 1969A. Computer determination of spore survival distributions in thermally-processed conduction-heated foods. Food Technol. 23, 352-354.

TEIXEIRA, A. A., DIXON, J. R., ZAHRADNIK, J. W., and ZINSMEISTER, G. E. 1969B. Computer optimization of nutrient retention in the thermal processing of conduction-heating foods. Food Technol. 23, 845-850.

THOMPSON, S. Y. 1969. Nutritional aspects of UHT products. In Ultra-high temperature Processing of Dairy Products. Society of Dairy Technology, London, England.

TIMBERS, G. E. 1971. Some aspects of quality degradation during the processing and storage of canned foods. Ph.D. Thesis. Rutgers, The State University, New Brunswick, N.J.

VAN BUREN, J. P. et al. 1960. Influence of blanching conditions on sloughing, splitting, and firmness of canned snap beans. Food Technol. 14, No. 5, 233-236.

VON ELBE, J. H., MAING, I.-Y., and AMUNDSON, C. H. 1974. Color stability of betanin. J. Food Sci. *39*, 334-337.
WANNINGER, L. A., JR. 1972. Mathematical model predicts stability of ascorbic acid in food products. Food Technol. *26*, No. 6, 42-45.
WATT, B. K., and MERRILL, A. L. 1963. Composition of foods. Handbook *8*, Consumer Food Econ. Res. Div., ARS, USDA, Washington, D.C.
WATT, B. K., and MURPHY, E. W. 1970. Tables of food composition—scope and needed research. Food Technol. *24*, 674-684.
WILLIAMS, M. P., and NELSON, P. E. 1974. Kinetics of the thermal degradation of methyl methionine sulfonium ions in citrate buffers and in sweet corn and tomato serum. J. Food Sci. *39*, 457-460.
WOODHAM, A. A. 1973. The effect of processing on the nutritive value of vegetable-protein concentrates. Proc. Nutr. Soc. *32*, 23-29.

Samuel A. Matz

PART 2
Effects of Baking on Nutrients

Destruction of nutrients in bakery foods is related primarily to the temperature and duration of the oven exposure during the baking process and to the pH of the dough or batter. It does not appear that significant amounts of vitamins or proteins are lost in the mixing, fermentation, and make-up phases. In fact, content of some of the vitamins may actually increase slightly during fermentation because of synthesis by yeast cells.

The interior of most foods of this class does not get much above the boiling point of water at any time during the oven treatment. Exceptions are certain cookies (e.g., macaroons), meringues, dry-baked specialties such as zwieback, and pie crusts. The crusts of nearly all bakery foods will reach temperatures considerably in excess of $212°F$ toward the end of the baking cycle, but crusts constitute but a small fraction of the total weight of the product, so that the reactions occurring in them do not have a great effect on the overall composition of the piece.

MINERALS

Although it is not to be expected that the total content of an element will change solely as the result of the baking process, the availability of certain mineral nutrients may very well do so. Phytins present in the bran of wheat can firmly complex calcium and probably other cations, making them unavailable for human nutrition. When it is necessary to use bread made from whole wheat flour as a substantial part of the diet, calcium supplementation (as by adding chalk to bread) has been recommended as a means of off-setting this effect.

Ranhotra (1972) indicates that considerable hydrolysis of phytic acid leading to release of inorganic phosphorus and reduced binding of calcium and iron can occur during the fermentation steps of bread making.

VITAMINS

Thiamin has been studied far more than any other vitamin, so far as retention during baking is concerned. In the mildly acid environment of most fermented goods (bread, rolls, etc.), thiamin suffers only small losses, and the amount retained appears to be related to the intensity of the heat treatment. For example, Zaehringer and Personius (1949) found that bread baked until the crusts were pale, medium, or dark colored retained 83, 80, and 78%, respectively, of the thiamin in the dough. Furthermore, greatest retention occurs in dough pieces which bake in the shortest period of time, so that bread loses more thiamin than do rolls. Double baking, as in the production of zwieback, results in an additional loss, 14-24% according to Meckel and Anderson (1944).

If the pH rises much above 6, nearly all the thiamin can be lost (Harris and Levenberg 1960). Such conditions prevail in most chemically-leavened foods such as muffins, cakes, corn bread, and some doughnuts. As a result of experiments in which the soda content of plain muffins was varied, Briant and Klosterman (1950) showed there was a negative correlation between pH of the finished product and thiamin retention. At the highest level of sodium bicarbonate addition (highest pH) only 6% of the vitamin was retained, as compared to 83% at the lowest level of soda. Similar results were obtained by Barackman (1942) for biscuits.

Farrer (1955) reviewed earlier studies on this subject, and quoted a 1942 report by Escudero and de Alvarez Herrero claiming that potassium bromate (a common bread dough improver) causes a rapid destruction of thiamin.

It would be interesting to know the fate of thiamin in pretzels, which actually undergo a dip in fairly strong alkali solution before they are baked, but there seems to be no published information on this subject.

Keagy and Stokstad (1973) studied folacin stability during flour storage and bread processing. Natural folacin, as determined using *L. casei*, ascorbate, and hog-kidney conjugase, nearly doubled during fermentation, but baking losses resulted in a final concentration close to that expected on the basis of the ingredient contributions.

Most investigators agree that very little riboflavin is lost in home or industrial processing of baked foods. For example, Thomas *et al.* (1952) found essentially no loss in corn bread at pH 5.4, but a loss of 11% at pH 6.6. In doughnuts processed by deep-fat frying, however, 23% loss of riboflavin was reported (Everson and Smith 1945).

Niacin is well retained during baking, with losses usually under 5% (Harris and Levenberg 1960). Its exceptional stability does not seem to be much affected by the pH of the product, within the normal range.

PROTEIN AND AMINO ACIDS

The principal concern of most investigators has been the effect of baking on the lysine content in the finished products, since this amino acid is limiting in most cereal flours. The extent of destruction of the free amino acid appears to be strongly related to the amount of reducing sugars present.

Rosenberg and Rohdenburg (1951) reported average losses of about 15% of lysine (determined by microbiological assay) in bread baking, with a range of 9.5 to 23.8%. Added lysine (as the isolated amino acid) was lost at about the same rate.

Jansen *et al.* (1964A, B) found a nutritional loss of 15% of the lysine monohydrochloride which had been added to water bread, as a result of baking 30 min in an oven held at 450°F. Increasing the baking times increased the loss of native and supplemental lysine. Adding moderate amounts of nonfat dry milk to the formula greatly increased loss of lysine, probably due to the intensified Maillard reaction resulting from the higher concentration of reducing sugars (i.e., lactose).

BIBLIOGRAPHY

BARACKMAN, R. A. 1942. Thiamine retention in self-rising flour biscuits. Cereal Chem. *19*, 121-128.
BRIANT, A. M., and KLOSTERMAN, A. M. 1950. Influence of ingredients on thiamine and riboflavin retention and quality of plain muffins. Trans. Am. Assoc. Cereal Chem. *8*, 69-77.
EVERSON, G. J., and SMITH, A. H. 1945. Retention of thiamine, riboflavin, and niacin in deep-fat cooking. Science *101*, 338-339.
FARRER, K. T. H. 1955. The thermal destruction of vitamin B_1 in foods. Advan. Food Res. *6*, 257-311.

EFFECTS OF HEAT PROCESSING ON NUTRIENTS 243

HARRIS, R. S., and LEVENBERG, R. K. 1960. B. Effects of home preparation on nutrient content of foods of plant origin. *In* Nutritional Evaluation of Food Processing. R. S. Harris, and H. von Loesecke (Editors). John Wiley & Sons, New York. Reprinted in 1971 by Avi Publishing Co., Westport, Conn.

JANSEN, G. R., EHLE, S. R., and HAUSE, N. L. 1964A. Studies of the nutritive loss of supplemental lysine in baking. I. Loss in a standard white bread containing 4% nonfat dry milk. Food Technol. *18*, 109-113.

JANSEN, G. R., EHLE, S. R., and HAUSE, N. L. 1964B. Studies on the nutritive loss of supplemental lysine in baking. II. Loss in water bread and in breads supplemented with moderate amounts of nonfat dry milk. Food Technol. *18*, 114-117.

KEAGY, P. M., and STOKSTAD, E. L. R. 1973. Folacin stability during flour storage and bread processing. 58th Ann. Meeting, AACC, St. Louis. Nov. 4-8, 1973.

MECKEL, R. B., and ANDERSON, G. 1944. Thiamine retention in the commercial production of Zwieback. Cereal Chem. *21*, 280-283.

RANHOTRA, G. S. 1972. Hydrolysis during breadmaking of phytic acid in wheat protein concentrate. J. Food Sci. *37*, 12-13.

ROSENBERG, H. R., and ROHDENBURG, E. L. 1951. The fortification of bread with lysine. I. The loss of lysine during baking. J. Nutr. *45*, 593-598.

THOMAS, K., PACE, J. K., and WHITEACRE, J. 1952. Effects of enrichment on the thiamine, riboflavin, and niacin of corn meal and grits as prepared for eating. Texas Agr. Expt. Sta. Bull. *753*.

ZAEHRINGER, M. V., and PERSONIUS, C. J. 1949. Thiamine retention in bread and rolls baked to different degrees. Cereal Chem. *26*, 384-392.

CHAPTER 10

Owen Fennema

Effects of Freeze-Preservation on Nutrients

The handling, storage and preservation of food often involves changes in nutritive value, most of which are undesirable. The freezing process (prefreezing treatments, freezing, frozen storage, and thawing), if properly conducted, is generally regarded as the best method of long-term food preservation when judged on the basis of retention of sensory attributes and nutrients. The freezing process is, however, not perfect, as is apparent from the fact that substantial amounts of the more labile nutrients can be lost. Vitamin losses during freezing preservation vary greatly depending on the food, the package, and the conditions of processing and storage. Losses of nutrients can result from physical separation (e.g., peeling and trimming during the prefreezing period, or exudate loss during thawing), leaching (especially during blanching), or chemical degradation. The seriousness of these losses depends on the nutrient (whether it is abundant or meager in the average diet), and the particular food item (does this food generally supply a major or a minor amount of the nutrient in question?).

The approach taken in this chapter is to summarize nutrient losses that occur in important classes of foods during various stages of the freezing process. Although most of the important literature has been reviewed, no claim is made as to total coverage. Because of the emphasis in the literature, attention is given mainly to vitamins, especially vitamin C in fruits and vegetables, and B-vitamins in animal products. (Values of vitamin C reported in this paper refer to reduced L-ascorbic acid unless otherwise stated.) Vitamin C and thiamin (B-1) have been studied most extensively since they are water-soluble (cannot be stored in the body and are subject to leaching during processing), are highly susceptible to chemical degradation, are present in many foods, and are sometimes deficient in the diet. If these vitamins are well retained during processing it generally can be assumed that all other nutrients also are well retained.

When considering the various data presented in this chapter, it is important to note that the reference value used for calculating nutrient losses differs depending on the phase of the freezing process that is under consideration. For example, blanching losses are based on the difference in nutrient content between blanched and *fresh*

EFFECTS OF FREEZE-PRESERVATION ON NUTRIENTS 245

products, freezing losses in vegetables are based on the difference in nutrient content between blanched-frozen-unstored products and *blanched-unfrozen* products, and storage losses are based on the difference in nutrient content between frozen-stored and *frozen-unstored* products, etc. This is mentioned so that improper conclusions will be avoided. For example, if a given product loses 20% of its vitamin C content during blanching and another 20% of its vitamin C content during frozen storage, the *absolute amounts* of vitamin C lost during these two phases will not be the same since the vitamin C content of the fresh product (large value) is used as the reference value for computing blanching losses, and the vitamin C content of the blanched-frozen-unstored product (a lower value) is used as the reference value for computing storage losses.

It will be seen that nutrient losses reported for a given product differ greatly from one investigator to another, and in some instances the results are contradictory. This should not be regarded as unusual since nutrient losses depend on many factors (product, handling, processing, packaging, etc.) and when any two studies are compared, it is rare to find that all factors have been controlled in a similar fashion.

One final introductory comment is worthy of mention. The reader will soon note, especially with animal tissues, that apparent increases in the amounts of some vitamins occasionally occur during the freezing process (indicated by + signs in the tables). In some instances this can be attributed to analytical errors or to biological variability, but there are instances where the increase is too large to be accounted for in this manner. In such instances, it has been suggested that the freezing process releases bound, biologically inactive forms of the vitamin, or converts inactive precursors to the active vitamin.

VEGETABLES

Loss of Nutrients During Blanching and Cooling

Shown in Table 10.1 are mean losses of vitamins C and B-1 that occur during water blanching and water cooling of vegetables. Losses of vitamin C often amount to about 10 to 50%, and losses of vitamin B-1 are often about 9 to 60% depending on the product and the conditions. Water blanching also causes approximately a 30% (14–53) loss of niacin from lima beans (Cook *et al.* 1961; Guerrant *et al.* 1947; Guerrant and O'Hara 1953), a 19% (14–24) loss of riboflavin from green peas (Guerrant and O'Hara 1953; Guerrant *et al.* 1947; Lee and Whitcombe 1945; Van Duyne *et al.* 1950), a 14% (11–17) loss of riboflavin from green beans (Guerrant and Dutcher 1948; Lee and Whitcombe 1945; Phillips and Fenton 1945; Van Duyne *et al.*

TABLE 10.1

LOSS OF VITAMINS C AND B-1 FROM VEGETABLES
DURING BLANCHING AND COOLING[1]

Product	Loss of Vitamins (Mean % and Range)		References
	Vitamin C	Vitamin B-1	
Asparagus	10 (6-15)		[2]
Beans, green	23 (12-42)	9 (0-14)	[3]
Beans, lima	24 (19-40)	36 (20-67)	[4]
Broccoli	36 (12-50)		[5]
Brussels sprouts	22 (21-25)		[6]
Cauliflower	20 (18-25)		[7]
Peas, green	21 (1-35)	11 (3-23)	[8]
Spinach	50 (40-76)	60 (41-80)	[9]

[1] Vitamin content after cooling or after freezing, as compared to the vitamin content of the fresh product. The effect of freezing has been shown to be negligible. Almost all of the data relates to water blanching and water cooling.
[2] Gordon and Noble (1959), Noble and Gordon (1964).
[3] Bedford and Hard (1950), Dawson et al. (1949), Farrell and Fellers (1942), Fisher and Van Duyne (1952), Gordon and Noble (1959), Guerrant and Dutcher (1948), Guerrant et al. (1947), Hartzler and Guerrant (1952), Lee and Whitcombe (1945), Melnick et al. (1944), Noble and Gordon (1964), Phillips and Fenton (1945), Proctor and Goldblith (1948), Retzer et al. (1945).
[4] Cook et al. (1961), Guerrant and O'Hara (1953), Guerrant et al. (1947, 1953), Gustafson and Cooke (1952), Tressler et al. (1937).
[5] Batchelder et al. (1947), Eheart (1967), Fisher and Van Duyne (1952), Gordon and Noble (1959), Hartzler and Guerrant (1952), Noble and Gordon (1964), Proctor and Goldblith (1948).
[6] Dietrich and Neumann (1965), Gordon and Noble (1959), Noble and Gordon (1964).
[7] Gordon and Noble (1959), Noble and Gordon (1964), Retzer et al. (1945).
[8] Barnes and Tressler (1943), Batchelder et al. (1947), Bedford and Hard (1950), Feaster et al. (1949), Guerrant and O'Hara (1953), Guerrant et al. (1947, 1953), Hartzler and Guerrant (1952), Holmquist et al. (1954), Jenkins and Tressler (1938), Lamb et al. (1948), Lee and Whitcombe (1945), Moyer and Tressler (1943), Proctor and Goldblith (1948), Tressler et al. (1936), Van Duyne et al. (1950).
[9] Bedford and Hard (1950), Fisher and Van Duyne (1952), Guerrant et al. (1947), Hartzler and Guerrant (1952), Proctor and Goldblith (1948), Sweeney and Marsh (1971), Tressler et al. (1936), von Kamienski (1972).

1950), and no significant loss of carotenes from broccoli, green peas, green beans, corn, brussels sprouts, spinach, squash, collards, kale, beet greens, endive, carrots, and sweet potatoes (Guerrant and O'Hara 1953; Guerrant et al. 1947; Stimson and Tressler, 1939; Sweeney and Marsh 1971; Zimmerman et al. 1940).

Losses of water-soluble vitamins occur primarily by leaching rather than by chemical degradation. This is borne out by the behavior of broccoli and spinach. Both of these products have large surface-to-mass ratios that favor leaching, and both exhibit very large losses of vitamin C during water blanching.

Several investigators have compared water and steam blanching with respect to losses of nutrients in vegetables (Batchelder et al. 1947; Dietrich and Neumann 1965; Dietrich et al. 1959; Guerrant et al. 1947; Holmquist et al. 1954; Moyer and Stotz 1945; Noble and

EFFECTS OF FREEZE-PRESERVATION ON NUTRIENTS   247

Gordon 1964; Proctor and Goldblith 1948; Retzer et al. 1945; Wedler 1971). Substantially smaller losses of water-soluble vitamins and most minerals generally occur during steam blanching than during water blanching. This is particularly true when products with large surface-to-volume ratios are being blanched, and when steam blanching is followed by a method of cooling that does not involve liquid water (e.g., air cooling). Furthermore, microwave blanching reportedly results in substantially smaller losses of nutrients than steam blanching (Moyer and Stotz 1945; Proctor and Goldblith 1948; Samuels and Wiegand 1948).

One purpose of blanching is to render the product more resistant to vitamin losses during frozen storage (the major effect apparently is to inactivate enzymes that otherwise would catalyze degradation of vitamins). The advantage of blanching with respect to retarding losses of vitamins C, B-1, B-2, and carotene from vegetables during frozen storage is clearly shown in Table 10.2.

One final point should be considered with respect to blanching. If nutritional differences result from different blanching techniques, do these differences persist during subsequent phases of the freezing

TABLE 10.2

EFFECT OF BLANCHING ON LOSS OF VITAMINS FROM VEGETABLES DURING FROZEN STORAGE

Product	Storage Conditions	Vitamin Loss				References
		C (%)	B-1 (%)	B-2 (%)	Carotene (%)	
Beans, green						
Unblanched	9 months at	35			A[1]	Bedford and Hard (1950)
Blanched	-19°C	10			≪ A	
Unblanched	1 yr at	91	74	39		Farrell and Fellers (1942)
Blanched	-20°C	47	22	3		
Beans, lima						
Unblanched	6-24 months				B	Zscheile et al. (1943)
Blanched	at -20°C				B-10	
Peas, green						
Unblanched	6-24 months				C	Batchelder et al. (1947)
Blanched	at -20°C				C-50	Zscheile et al. (1943)
Unblanched	9 months at				40	Batchelder et al. (1947)
Blanched	-19°C				14	
Unblanched	5 months at	D				Jenkins and Tressler (1938)
Blanched	-18°C	≪ D				
Spinach						
Unblanched	9 months at	54			E	Bedford and Hard (1950)
Blanched	-19°C	24			E	
Unblanched	6-24 months				F	Zscheile et al. (1943)
Blanched	at -20°C				F-28	

[1] When absolute loss values were unavailable, it was necessary to express losses between blanched and unblanched products on a comparative basis. For example, if the loss of vitamin C in a given unblanched product is set equal to A then the loss in the comparable blanched product can be expressed in terms of A.

process? Unfortunately, only a few studies have dealt with this question and the results are contradictory (Fisher and Van Duyne 1952; Retzer et al. 1945). Additional information on blanching is available elsewhere in this volume.

Although blanching can decrease significantly the amount of water-soluble vitamins in vegetables, it is not the only prefreezing treatment of concern. Substantial losses of vitamins can occur if vegetables are not promptly frozen following harvest. This important matter is thoroughly discussed in another chapter in this volume.

Loss of Nutrients During Freezing

The losses of vitamins that occur in vegetables during freezing (effect of freezing alone) are shown in Table 10.3. The values reported are percentage losses based on comparable unfrozen products. In addition to the data in Table 10.3, freezing was reported to cause no loss of certain nutrients in the following products: B-6 in potatoes (Secomska et al. 1973), pantothenic acid in kale (Holmes et al.

TABLE 10.3

LOSS OF VITAMINS FROM VEGETABLES DURING FREEZING

Product	C (%)	B-1 (%)	B-2 (%)	Niacin (%)	Carotene (%)	References
Asparagus	6 (ns)[1]					[2]
Beans, green	3 (0-10)	3 (0-7)	~0		8	[3]
Beans, lima	14 (0-28)	5 (0-9)	6		~0	[4]
Broccoli	6 (5-7)				~0	[5]
Brussels sprouts	5 (ns)				~0	[6]
Cauliflower	3 (0-6)					[7]
Kale	~0			~0	~0	[8]
Peas, green	9 (0-18)	8 (1-15)	2 (0-4)	16	~0	[9]
Potatoes		~0	~0	~0		[10]
Spinach	4 (0-8)		~0		~0	[11]
Grand means	6	4	2	4	1	

[1] Not significant.
[2] Gordon and Noble (1959).
[3] Fisher and Van Duyne (1952), Gordon and Noble (1959), Lee et al. (1946), Phillips and Fenton (1945), Retzer et al. (1945), Van Duyne et al. (1950).
[4] Cook et al. (1961), Guerrant and O'Hara (1953), Guerrant et al. (1953).
[5] Fisher and Van Duyne (1952), Gordon and Noble (1959), Sweeney and Marsh (1971).
[6] Gordon and Noble (1959), Sweeney and Marsh (1971).
[7] Gordon and Noble (1959), Retzer et al. (1945).
[8] Holmes et al. (1945), Sweeney and Marsh (1971).
[9] Guerrant and O'Hara (1953), Guerrant et al. (1953), Lee et al. (1946), Van Duyne et al. (1950).
[10] Secomska et al. (1973).
[11] Fisher and Van Duyne (1952), Noble and Gordon (1964), Sweeney and Marsh (1971), Tinklin and Filinger (1956), Van Duyne et al. (1950).

1945), and carotene in collards, beet greens, endive, carrots, squash, and sweet potatoes (Sweeney and Marsh 1971). In judging these data, it should be noted that: (1) loss values of less than 5% probably are not significant, and (2) in many instances the period studied was somewhat longer than that needed to complete freezing, e.g., some data were collected 24 hr after the conclusion of freezing, and other data relate to the period beginning after blanching and cooling, and ending when the final storage temperature had been attained. In the latter instance, this could, particularly in commercial situations, involve a significant holding period between postblanch cooling and the start of freezing (grading, packaging, handling). With these factors in mind, it must be concluded from the data presented, that losses of vitamins during freezing generally are negligible.

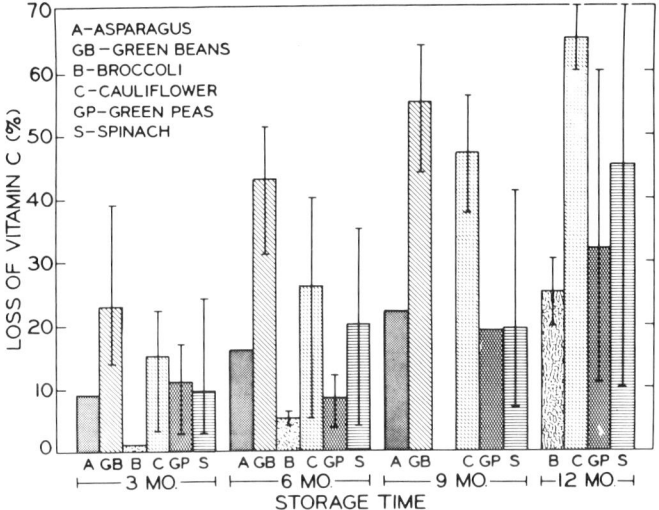

FIG. 10.1. CUMULATIVE LOSSES (AFTER COMPLETION OF FREEZING) OF VITAMIN C FROM VEGETABLES DURING STORAGE AT $-18°C$ (MEANS PLUS RANGES)

References: Asparagus (Gordon and Noble 1959; Guerrant 1957; Jenkins et al. 1940 [reduced ascorbic acid plus dehydroascorbic acid]); green beans (Derse and Teply 1958; Dietrich et al. 1959; Pierce et al. 1955; Retzer et al. 1945); broccoli (Fisher and Van Duyne 1952; Gordon and Noble 1959; Guerrant 1957); cauliflower (Gordon and Noble 1959; Guerrant 1957; Jenkins et al. 1940 [reduced ascorbic acid plus dehydroascorbic acid]; Retzer et al. 1945); green peas (Bedford and Hard 1950; Derse and Teply 1958; Guerrant and O'Hara 1953; Guerrant et al. 1953; Volz et al. 1949; Weitz et al. 1970); spinach (Bedford and Hard 1950; Dietrich et al. 1960; Fisher and Van Duyne 1952; Guerrant 1957; Jurics 1970; Tinklin and Filinger 1956; Volz et al. 1949; Weits et al. 1970).

Loss of Nutrients During Frozen Storage

Shown in Fig. 10.1 and Tables 10.4 and 10.5 are losses of various vitamins and minerals that occur from vegetables during frozen storage. Percentage loss values are based on the difference in nutrient content between blanched-frozen-stored product and blanched-frozen-unstored product. Cumulative losses of vitamin C during storage of frozen vegetables (Fig. 10.1) vary with the product and increase with time at $-18°C$. Percentage losses of vitamin C are large in green beans and cauliflower and small in broccoli and peas. The fact that losses of vitamin C from broccoli are very large during blanching and relatively small during frozen storage again indicates that losses during blanching occur primarily by the mechanism of leaching rather than by chemical degradation.

Losses of minerals and vitamins (other than vitamin C) from 6 vegetables during storage for 12 months at $-18°C$ are listed in Table 10.4.

TABLE 10.4

APPROXIMATE LOSSES OF VITAMINS AND MINERALS FROM VEGETABLES DURING FROZEN STORAGE

Product	Loss of Nutrients During 12 Months at $-18°C$									
	B-1 (%)	B-2 (%)	Niacin (%)	B-6 (%)	Vitamin K (%)	Folic Acid (%)	Pantothenic Acid (%)	Carotene (%)	Fe (%)	Other[1] Minerals (%)
Beans, green[2]	0-32	0	0	0-21	0	6	53	0-23	18	0
Beans, lima[3]		45	26	0						
Broccoli[4]						6		0		
Cabbage[5]						0	0			
Peas, green[6]	0-16	0-8	0-8	7			29	0-4	20	0
Spinach[7]			0		42					

[1] P, Mg, Ca, K, Na.
[2] Barnes and Tressler (1943), Dawson et al. (1949), Derse and Teply (1958), Lee et al. (1946), Richardson et al. (1961A,B), Tinklin and Filinger (1956).
[3] Guerrant and O'Hara (1953), Richardson et al. (1961A).
[4] Martin et al. (1960), Richardson et al. (1961B).
[5] Richardson et al. (1961A,B).
[6] Barnes and Tressler (1943), Derse and Teply (1958), Guerrant and O'Hara (1953), Lee et al. (1946), Stimson and Tressler (1939), Van Duyne et al. (1950).
[7] Richardson et al. (1961B), Van Duyne et al. (1950).

Iron is the only mineral that appears to undergo a significant loss and this is probably an analytical inaccuracy rather than a real loss. Vitamin losses vary greatly depending on the product. Losses of vitamins from green beans generally are greater than from peas, and losses of B-2 and niacin from lima beans are quite large compared to losses of these vitamins from other vegetables. Vitamins B-1 and pantothenic acid appear to be the least stable of the vitamins listed in Table 10.4, and losses of these vitamins are similar to the losses of vitamin C reported in Fig. 10.1.

TABLE 10.5

COMPARATIVE LOSSES OF VITAMINS IN VEGETABLES HELD AT $-18°C$ AND $-7°C$

Product	Storage Time, Months (Value of X)	Ratio of % Loss at Constant $-18°C$ for X Months to % Loss at $-18°C$ for X-1 Months Plus 1 Month at $-7°C$						
		C	β-carotene	Folic Acid	Niacin	Pantothenic Acid	B-2	B-1	B-6
Beans, green, French style	3	39:66	20:37	0:0	0:0	51:68	0:0	0:0	0:0
	6	44:69	53:57	0:61	0:0	16:16	9:0	0:12	12:22
	12	52:76	23:27	6:0	0:0	53:57	0:9	0:0	21:21
Peas, green	3	17:48	15:15	12:37	0:0	27:53	0:8	0:0	0:0
	6	12:52	15:17	50:75	0:0	0:4	8:17	0:0	9:0
	12	11:64	4:10	0:6	0:1	29:40	0:0	3:0	7:9

Source: Derse and Teply (1958).

In Table 10.5, losses of nutrients in green beans and peas during storage at a constant temperature of −18°C are compared to losses during storage at the same time-temperature condition except that temperature during the last month of storage was −7°C. Exposure to the −7°C condition almost invariably resulted in greater losses of vitamin C, β-carotene, folic acid, and pantothenic acid, but had no significant effect on losses of niacin, riboflavin (B-2), thiamin (B-1), B-6 and minerals (not shown). It is also evident that losses of vitamin C, β-carotene, pantothenic acid and B-6 are greater in green beans at any given time than in peas. Vitamin C and pantothenic acid are by far the most labile vitamins in both green beans and peas.

The effect of storage temperature on loss of vitamin C from four frozen vegetables is shown in Fig. 10.2. Supplementary information

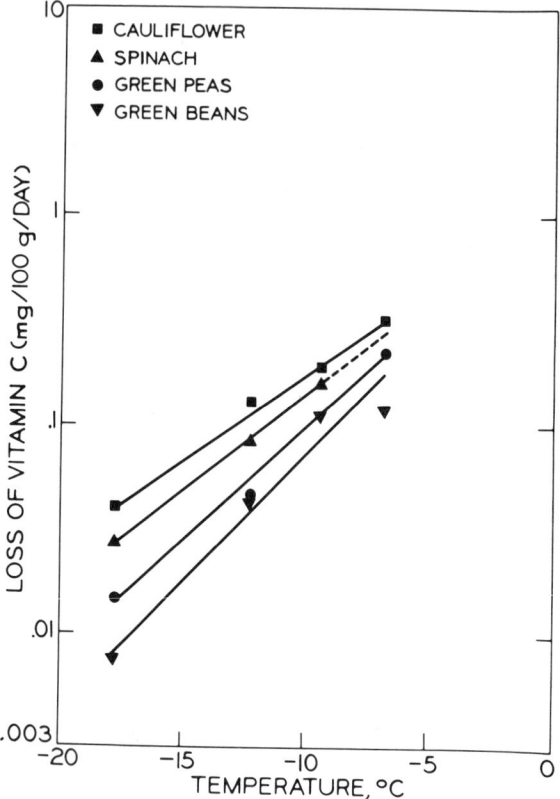

FIG. 10.2. TEMPERATURE DEPENDENCE OF VITAMIN C DEGRADATION IN FROZEN VEGETABLES

See Table 10.6 for conditions and references.

EFFECTS OF FREEZE-PRESERVATION ON NUTRIENTS 253

TABLE 10.6

SUPPLEMENTARY INFORMATION FOR THE PRODUCTS IN FIG. 10.2

Product	Approximate Maximum Storage Period Over Which Rates of Vitamin C Degradation are Valid[1] Months at:				References
	$-7°C$	$-9°C$	$-12°C$	$-18°C$	
Beans, green	1.5	3	4	6	2
Cauliflower	2	3	4	6	3
Peas, green	2		12	12+	4
Spinach	4	6	6	8	5

[1] Periods vary greatly depending on the conditions.
[2] Bennett et al. (1954), Dietrich et al. (1957, 1959), Jenkins et al. (1940 [data include both reduced ascorbic acid and dehydroascorbic acid]), Jurics (1970).
[3] Bennett et al. (1954), Dietrich et al. (1957, 1962), Jenkins et al. (1940 [data include both reduced ascorbic acid and dehydroascorbic acid]).
[4] Dietrich et al. (1957), Guerrant et al. (1953), Lindquist et al. (1950).
[5] Bennett et al. (1954), Dietrich et al. (1960), Jurics (1970).

for the products in Fig. 10.2 is given in Table 10.6. Each line in Fig. 10.2 is a mean of at least 3 studies and most of the studies involved at least 3 temperatures. Below $-18°C$ the data sometimes departed from a linear relationship, and when this occurred, loss values generally were larger than would be expected from extrapolation of the lines presented. Considerable variability is associated with each mean in Fig. 10.2, so little significance should be attached to the relative positions of the products. The lines, however, fit the mean values quite well, indicating that the slopes probably are reasonably accurate. If this is true then it is apparent that changes in temperature affect the rate of vitamin C degradation to a similar degree in all products, i.e., a $10°C$ rise in temperature within the range $-18°C$ to $-7°C$ causes the rate of vitamin C degradation to accelerate in all products by a factor of 6-20X (Q_{10} = 6-20). This is an uncommonly large dependence on temperature. Thus, low storage temperatures ($-18°C$ or lower) will preserve vitamin C content of vegetables far better than higher subfreezing temperatures. Comparable information is not available for nutrients other than vitamin C.

A few studies have dealt with nutrient losses during storage of products at fluctuating storage temperatures as compared to comparable constant temperatures. With respect to frozen green peas (Boggs et al. 1960; Gortner et al. 1948), frozen cauliflower (Dietrich et al. 1962), and green beans (Gortner et al. 1948), losses of vitamin C do not differ significantly between the two conditions.

Loss of Nutrients During Thawing

Only a few studies of frozen vegetables have been conducted in a manner so that the effect of thawing on nutrient value can be determined (Fenton and Tressler 1938; Jenkins and Tressler 1938; Holmes et al. 1945; Phillips and Fenton 1945). Results of these studies indicate that proper thawing, as such, has a very small and probably insignificant effect on the nutrient content of foods.

Loss of Nutrients During the Entire Freezing Process

Shown in Table 10.7 are losses of vitamin C from 10 important vegetables during the entire freezing process (based on differences

TABLE 10.7

LOSS OF VITAMIN C FROM VARIOUS VEGETABLES DURING THE ENTIRE FREEZING PROCESS

Product	Typical Amount of Vitamin C in Fresh Product[1] (Mg/100Gm)	Loss of Vitamin C During 6-12 Months at $-18°C$ (Mean % and Range)
Asparagus[2]	33	10 (12-13)
Beans, green[3]	19	45 (30-68)
Beans, lima[4]	29	51 (39-64)
Broccoli[5]	113	49 (35-68)
Cauliflower[6]	78	50 (40-60)
Peas, green[7]	27	43 (32-67)
Spinach[8]	51	65 (54-80)

[1] From USDA Handbook 8 (Watt and Merrill 1963).
[2] Batchelder et al. (1947), Gordon and Noble (1959).
[3] Bedford and Hard (1950), Dawson et al. (1949), Fisher and Van Duyne (1952), Gordon and Noble (1959), Jurics (1970), Retzer et al. (1945).
[4] Guerrant and O'Hara (1953), Guerrant et al. (1953).
[5] Batchelder et al. (1947), Fisher and Van Duyne (1952), Gordon and Noble (1959).
[6] Gordon and Noble (1959), Retzer et al. (1945).
[7] Batchelder et al. (1947), Bedford and Hard (1950), Guerrant and O'Hara (1953), Guerrant et al. (1953).
[8] Bedford and Hard (1950), Fisher and Van Duyne (1952), Von Kamienski (1969).

of vitamin C contents of fresh products and products that were blanched, frozen and stored for 6-12 months at $-18°C$). Losses differ among products and among different lots and varieties of the same type of product. With the exception of asparagus, losses of vitamin C average about 50% during the freezing process. This is unfortunate since all of the vegetables with the exception of green beans contain substantial amounts of vitamin C in the fresh state.

The losses in Table 10.7 can be directly compared to those in Table 10.1 (for blanching) since losses in both instances are calculated on the basis of the fresh product. This comparison shows that blanching is responsible for a major share of the loss of water-soluble nutrients that occurs in vegetables during the entire freezing process. Additional information on losses of nutrients during the entire freezing process can be computed from data in USDA Handbook 8 (Watt and Merrill 1963). Absolute values for a broad range of nutrients in frozen vegetables "as purchased" are available in the reference just cited and in a report by Burger et al. (1956).

Loss of Nutrients During Cooking of Previously Frozen Vegetables

This topic is appropriate for coverage since there is no assurance that cooking will cause the same loss of nutrients from frozen and fresh vegetables. Shown in Table 10.8 are losses of vitamin C that occur in various frozen vegetables during cooking. These losses are calculated on the basis of comparable uncooked products. Since large differences are apparent with respect to conditions of frozen storage and cooking, it is not feasible to condense the data any further than has been done. It is evident that vegetables usually lose substantial amounts of vitamin C during cooking, and that time of frozen storage prior to cooking usually is unimportant; but when it is, storage usually is detrimental (green beans, green peas). It is also evident from the amounts of vitamin C that are recoverable in the cooking water, that leaching is a major mechanism of loss. The independent effects of cooking time and amount of cooking water cannot be determined from these data.

Losses of minerals and vitamins (other than vitamin C) from various frozen vegetables during cooking are listed in Tables 10.9 and 10.10. When losses of minerals are averaged over all vegetables, it can be seen that losses of iron are by far the greatest and losses of phosphorus and calcium are the least. With the exception of iron and in some instances phosphorus, most of the mineral losses can be accounted for in the cooking water. One would expect 100% recovery of minerals in the cooking water. According to Teply and Derse (1958), "These exceptions may have been due in part to package-to-package variation, but were probably due mainly to large errors, percentagewise, in determining some of the components present at low levels."

When the vitamin losses shown in Table 10.10 are averaged over all vegetables, losses of B-6 (29% loss), pantothenic acid (27% loss), and folic acid (22% loss) are the greatest; losses of niacin (12% loss) and ribloflavin (11% loss) are intermediate; and losses of β-carotene are least ($<2\%$). Losses of vitamin C from frozen vegetables (Table 10.8) during cooking generally exceed any of the mean values in Table 10.10.

TABLE 10.8

LOSS OF VITAMIN C FROM PREVIOUSLY FROZEN VEGETABLES DURING COOKING

Product	Conditions of Frozen Storage Prior to Cooking	Total Cooking Time in Boiling Water (Min)	Cooking Water ÷ Product (by Wt)	Loss of Vitamin C (%)	Amt of Lost Vitamin C Recoverable in Cooking Water (%)	References
Asparagus	No storage	10–15	2.33	49		Gordon and Noble (1959)
	6 months −18°C	10–15	2.33	51		Gordon and Noble (1959)
	APP[1]	8	0.2	9[2]	100[2]	Teply and Derse (1958)
Beans, green	No storage	15	?	13		Dawson et al. (1949)
	10 months −18°C	15	?	47		Dawson et al. (1949)
	No storage	10–15	3.7	36		Gordon and Noble (1959)
	6 months −18°C	10–15	3.7	25		Gordon and Noble (1959)
	<3 weeks −18°C	25	0.17	27		Phillips and Fenton (1945)
	<3 weeks −18°C	25	0.8	48		Phillips and Fenton (1945)
	No storage	10	2.0	32		Lee et al. (1946)
	6 months −21°C	10	2.0	31		Lee et al. (1946)
	APR[3]	11–22	0.46	33		Sweeney et al. (1961)
	APP[1]	9	0.4	49[2]	10[2]	Teply and Derse (1958)
Beans, lima, baby	APP[1]	9	0.4	35[2]	49[2]	Teply and Derse (1958)
Beans, lima, Fordhook	APP[1]	13–14	0.3	25[2]	60[2]	Teply and Derse (1958)
	APR[3]			28		Sweeney et al. (1961)

EFFECTS OF FREEZE-PRESERVATION ON NUTRIENTS

Broccoli	?	11	0.5	34²	Guerrant et al. (1953)
	No storage	8–10	5.5	35	Gordon and Noble (1959)
	6 months –18°C	8–10	5.5	33	Gordon and Noble (1959)
	APR³	5–10	0.4–0.8	23–34	Martin et al. (1960)
	APR³	12	0.3	25	Sweeney et al. (1961)
				17	Eheart (1970)
Brussels sprouts	No storage	9–10	5	31	Gordon and Noble (1959)
	6 months –18°C	9–10	5	22	Gordon and Noble (1959)
	APR³	13–16	0.3	26	Sweeney et al. (1961)
	APP¹	10	0.2	7²	Teply and Derse (1958)
Cauliflower	No storage	8–10	3.7	24	Gordon and Noble (1959)
	6 months –18°C	8–10	3.7	22	Gordon and Noble (1959)
	APP¹	7	0.2	27²	Teply and Derse (1958)
Corn, cut	?	7	0.3	25	Barnes and Tressler (1943)
	APP¹	6	0.2	55²	Teply and Derse (1958)
Corn, on cob	3–12 weeks –34°C	?	?	Slight	Payne (1967)
Peas, green	No storage	9	1.64	17	Lee et al. (1946)
	6 months –18°C	9	1.64	52	Lee et al. (1946)
	APR³	10–12	0.2	32	Sweeney et al. (1961)
	APP¹	8	0.2	17²	Teply and Derse (1958)
Spinach, chopped	APR³	9–10	0.3	42	Sweeney et al. (1961)
	APP¹	11	0.17	34²	Teply and Derse (1958)

Values in the column with 75–80, ns, 70², 37², 100, 13², 53², 21² appear aligned as additional data:

Vegetable	Condition	col a	col b	col c	extra	Reference
Broccoli APR³ (Sweeney)		12	0.3	25	75–80	Sweeney et al. (1961)
Brussels sprouts 6 months		9–10	5	22	ns	Gordon and Noble (1959)
Brussels sprouts APP¹		10	0.2	7²	70²	Teply and Derse (1958)
Cauliflower APP¹		7	0.2	27²	37²	Teply and Derse (1958)
Corn, cut APP¹		6	0.2	55²	100, 13²	Teply and Derse (1958)
Peas APP¹		8	0.2	17²	53²	Teply and Derse (1958)
Spinach APP¹		11	0.17	34²	21²	Teply and Derse (1958)

¹ As purchased frozen from processor.
² Total ascorbic acid (reduced ascorbic acid plus dehydroascorbic acid).
³ As purchased frozen in retail market.

TABLE 10.9

LOSS OF MINERALS FROM FROZEN VEGETABLES DURING COOKING[1]

Product	Total Cooking Time[2] (Min)	Cooking Water ÷ Product (by Wt)	Ca (%) Loss	Ca RCW[3]	Fe (%) Loss	Fe RCW[3]	P (%) Loss	P RCW[3]	Mg (%) Loss	Mg RCW[3]	K (%) Loss	K RCW[3]	Na (%) Loss	Na RCW[3]
Asparagus spears	8	0.2	10	60	13	70	5	60	17	53	8	100	0	
Beans, green, cut	11	0.2	6	67	+		2	100	10	70	3	100	11	100
Beans, lima, baby	9	0.4	5	100	12	50	10	40	8	62	13	77	10	100
Beans, lima, Fordhook	9	0.4	10	100	1	100	0		13	100	7	100	3	100
Broccoli, chopped	10	0.2	4	100	31	13	0		6	100	7	100	17	76
Broccoli, spears	8	0.2	4	100	38	11	8	50	9	100	14	71	25	36
Brussels sprouts	10	0.2	3	100	+		2	100	4	100	4	100	11	55
Cauliflower	7	0.2	8	87	11	45	+		9	100	1	100	0	
Corn, cut	6	0.2	12	42	37	11	9	67	9	100	10	90	12	100
Peas, green	8	0.2	6	83	21	19	0		8	87	10	100	13	85
Spinach, chopped	11	0.17	+		9	11	+		14	21	4	67	5	100
Spinach, leaf	11	0.17	0		12	?	0		5	80	3	?	6	?

[1] Losses based on comparable uncooked product. Product purchased frozen from processor.
[2] Boiling or near boiling water.
[3] Approximate amount of lost nutrient that is recoverable in cooking water. From Teply and Derse (1958).

EFFECTS OF FREEZE-PRESERVATION ON NUTRIENTS 259

TABLE 10.10

LOSS OF VITAMINS FROM PREVIOUSLY FROZEN VEGETABLES DURING COOKING[1]

Product	Total Cooking Time[2] (Min)	Cooking Water ÷ Product (by Wt)	β-carotene (%)		Folic Acid (%)		Niacin (%)		Pantothenic Acid (%)		Riboflavin, B-2 (%)		Thiamin, B-1 (%)		B-6 (%)	
			Loss	RCW[3]	Loss	RCW[3]	Loss	RCW[3]	Loss	RCW[3]	Loss	RCW[3]	Loss	RCW[3]	Loss	RCW[3]
Asparagus spears	8	0.2	+		26	62	19	53	21	38	15	60	10	70	42	14
Beans, green, cut	11	0.2	+		19	84	26	35	35	20	14	50	0		27	26
Beans, lima, baby	9	0.4	1	0	23	70	23	26	11	100	17	59	15	100	26	23
Beans, lima, Fordhook	9	0.4	+		15	100	23	50	11	100	19	74	27	37	30	23
Broccoli, chopped	10	0.2	+		26	42	9	100	14	100	11	91	+		6	100
Broccoli, spears	8	0.2	1	0	25	100	14	100	32	34	22	54	14	79	26	38
Brussels sprouts	10	0.2	+		12	67	8	62	28	19	8	50	16	50	19	26
Cauliflower	7	0.2	0		50	28	15	87	31	32	16	69	26	35	36	22
Corn, cut	6	0.2	11	0	56	12	0		40	15	7	100	20	35	39	15
Peas, green	8	0.2	15	7	+		0		34	29	+		12	67	33	52
Spinach, chopped	11	0.17	+		3	100	4	100	26	23	+		22	36	39	10
Spinach, leaf	11	0.17	0		11	55	+		36	14	0		18	28	26	46

[1] Losses based on comparable uncooked product. Product purchased frozen from processor.
[2] Boiling or near boiling water.
[3] Approximate amount of lost nutrient that is recoverable in cooking water. From Teply and Derse (1958).

260 NUTRITIONAL EVALUATION OF FOOD PROCESSING

Large but variable amounts of the water-soluble vitamins are recoverable in the cooking water, again indicating that leaching is a major mechanism by which water-soluble vitamins are lost. Vitamin losses during cooking generally vary greatly from product to product, and no doubt depend on factors such as product pH, surface-to-mass ratio, and perhaps on the amount of the original vitamin content that was lost during earlier stages of processing (von Kamienski 1972).

The vitamin data in Table 10.10 are in reasonably good agreement with other comparable studies dealing with cooking losses of β-carotene in peas and green beans (Lee et al. 1946), thiamin losses in green peas (Barnes and Tressler 1943; Lee et al. 1946; Martin et al. 1960) and riboflavin losses in green beans (Lee et al. 1946; Phillips and Fenton 1945; Van Duyne et al. 1950).

Lee et al. (1946) reported that riboflavin losses during cooking of frozen green peas range from 0 to 39% which is contrary to the results in Table 10.10. Studies by Dawson et al. (1949), Lee et al. (1946), and Phillips and Fenton (1945) indicated that thiamin losses during cooking of frozen green beans range from 10 to 35% which also is contrary to the result in Table 10.10, but is probably a more typical finding.

The adverse effects of increased cooking time and increased cook-

TABLE 10.11

LOSS OF VITAMIN C (REDUCED ASCORBIC ACID PLUS DEHYDROASCORBIC ACID) DURING COOKING OF PREVIOUSLY FROZEN LEAF SPINACH

Wt of Cooking Water ÷ Product Wt	Loss (%)	Percentage of Loss Recoverable in Cooking Water[1] (%)
"Underdone" spinach		
0.17	30	40
0.33	42	31
0.67	49	49
Done to "optimum flavor"		
0.17	35	17
0.33	55	24
0.67	65	28
"Overdone"		
0.17	45	4
0.33	60	12
0.67	66	20

[1] Percentage of the amount represented in column to the left. From Teply and Derse (1958).

EFFECTS OF FREEZE-PRESERVATION ON NUTRIENTS   261

ing water on losses of vitamin C from frozen spinach are clearly indicated in Table 10.11.

The results of a study by Sweeney et al. (1961) indicate that losses of vitamin C during cooking of six commonly frozen vegetables do not differ between product obtained directly from the processor and product obtained from the retail market.

According to van der Meer et al. (1973), loss of vitamin C in packaged potatoes, spinach, green peas, and brussels sprouts is the same regardless of whether they are cooked in a microwave oven, in a convection oven, or in hot water.

Comparative Losses of Vitamins from Vegetables During Various Methods of Preparation and Preservation

A point of considerable importance is how do losses of nutrients from vegetables during the freezing process compare to losses during other common methods of preparation and preservation. Data with respect to this matter are compiled in Table 10.12. The mean values

TABLE 10.12

COMPARATIVE LOSSES OF VITAMINS FROM VEGETABLES DURING CANNING AND FREEZE-PROCESSING

Method of Preservation and Preparation	No. of Vegetables Examined		Loss of Vitamins as Compared to Values of Fresh-Cooked Products[1]				
			A (%)	B-1 (%)	B-2 (%)	Niacin (%)	C (%)
Frozen, cooked (boiled), drained	10[2]	Mean	12	20	24	24	26
		Range	0–50	0–61	0–45	0–56	0–78
Canned, drained solids	7[3]	Mean	10	67	42	49	51
		Range	0–32	56–83	14–50	31–65	28–67

Source: Data from USDA Handbook 8 (Watt and Merrill 1963). Based on edible portions of vegetables "as purchased."

[1] Boiled and drained.
[2] Asparagus, lima beans, green beans, broccoli spears, brussels sprouts, cauliflower, whole kernel corn, green peas (immature), potatoes (frozen, mashed, heated), and leaf spinach.
[3] Same as above except broccoli, brussels sprouts, and cauliflower are not included. Values for potatoes include solids and liquid.

should be accepted with some caution since large variances can occur depending on the product and the conditions.

It is apparent that 10 common vegetables when frozen, stored, and cooked, contain, on the average, about 10% less vitamin A, and about 25% less of niacin and vitamins B-1, B-2, and C than when fresh-cooked. It is also evident that seven common vegetables when canned, lose on the average, about three times more vitamin B-1, about twice as much of niacin and vitamins B-2 and C, and about the same amount of vitamin A as when frozen-cooked. Thus, water-soluble vitamins are usually present at lower concentrations in frozen-stored-cooked vegetables than in fresh-cooked vegetables, and these same vitamins are present at lower concentrations in canned vegetables than in frozen-stored-cooked vegetables. The concentrations of water-soluble vitamins in frozen-stored-cooked vegetables could be raised significantly if blanching and cooling were conducted in the absence of liquid water.

FRUITS

Loss of Nutrients During Prefreezing Treatments

Prefreezing treatments that have adverse effects on the nutrient content of fruits usually involve storage for excessive periods or storage at a temperature that is too high. For example, Loeffler (1946) found that raspberries, when allowed to stand at ambient temperatures for 24 or 48 hr prior to freezing, lost, respectively, 17% and 30% of their vitamin C content. On the other hand, storage of strawberries for 2-4 days at $0°-11°C$ prior to freezing apparently has only a slight effect on vitamin C content (Scott and Schrader 1947). Additional information on this subject is available elsewhere in this book.

Loss of Nutrients During Freezing

Unfortunately, data on this subject are almost nonexistent. Two studies on raspberries and black currants indicate that losses of vitamin C during freezing are insignificant (Loeffler 1946; Sulc 1973).

Loss of Nutrients During Frozen Storage

Losses of vitamin C from fruits during storage at $-18°C$ are indicated in Fig. 10.3. Losses of vitamin C from muskmelon can be large, particularly if the melons are not packed in syrup. The variety of muskmelon also has a large influence on loss of vitamin C during frozen storage (Wolfe et al. 1949). According to Guadagni et al. (1957C), raspberries packed in syrup lose only small amounts of vitamin C during frozen storage. Based on several studies, losses of vitamin C from frozen peaches (Bennett et al. 1954; DuBois and Colvin 1945; Guadagni et al. 1957A; Guerrant 1957) and strawberries (Bennett et al. 1954; Crivelli et al. 1969; Derse and Teply 1958; Guadagni et al. 1961; Guerrant 1957; Pierce et al. 1955) generally are moderate during frozen storage. The oxygen barrier properties of

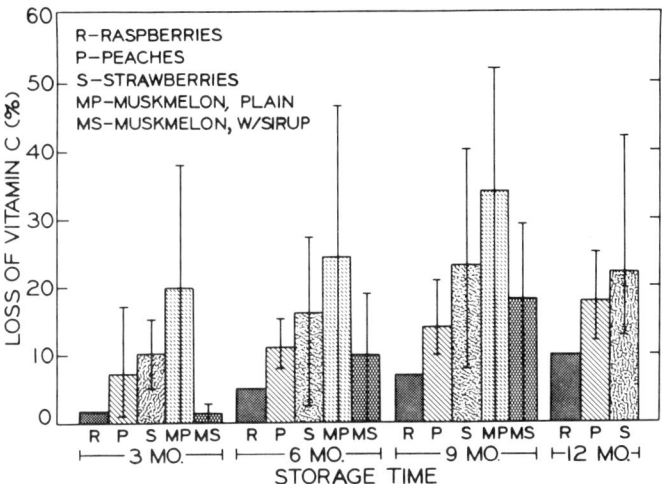

FIG. 10.3. CUMULATIVE LOSSES (AFTER COMPLETION OF FREEZING) OF VITAMIN C FROM FRUITS DURING STORAGE AT −18°C (MEANS PLUS RANGES)

References and conditions: Muskmelon—two varieties, sliced, packaged in cellophane-lined pint containers (Wolfe et al. 1949). Peaches—sliced, in syrup, various packages (Bennett et al. 1954; DuBois and Colvin 1945; Guadagni et al. 1957A; Guerrant 1957). Raspberries—whole, 3+1 part 50% sucrose syrup, retail composite cartons (Guadagni et al. 1957C). Strawberries—whole or sliced, sugared or plain, various packages (Bennett et al. 1954; Crivelli et al. 1969; Derse and Teply 1958; Guadagni et al. 1961; Guerrant 1957; Pierce et al. 1955).

the package have a profound influence on losses of vitamin C during frozen storage, as is clearly indicated in Fig. 10.4 and 10.5.

Losses of vitamin C from citrus juice concentrates during storage for 9-12 months at −18°C usually are less than 5% (Derse and Teply 1958; Huggart et al. 1954; McColloch et al. 1957; Wolfe et al. 1949). Plain or syruped boysenberries, regardless of the kind of container, apparently lose less than 1% vitamin C during storage for 3 months at −18°C (Guadagni et al. 1960).

Few data are available concerning losses of minerals and vitamins other than vitamin C from fruits during frozen storage.

The effect of storage temperature on losses of vitamin C from four frozen fruits and fruit juices is shown in Fig. 10.6. Supplementary information for the products in Fig. 10.6 is given in Table 10.13. Separation of the strawberry data into two groups (I and II) was done simply because the two groups behaved differently. The varieties used could account for the differences observed (Crivelli et al. 1969).

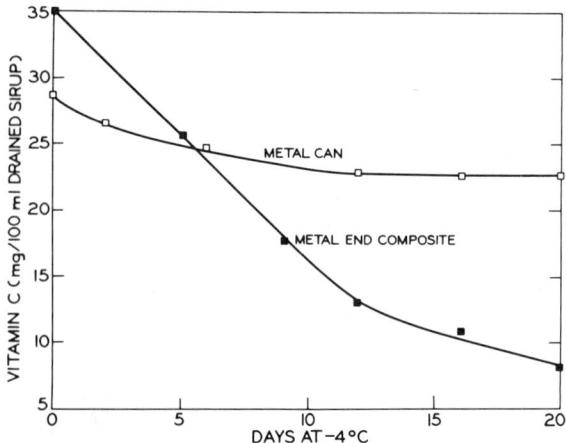

FIG. 10.4. EFFECT OF CONTAINER TYPE ON LOSSES OF VITAMIN C FROM FROZEN PEACHES

Sliced peaches in 50% sucrose syrup containing 0.1% ascorbic acid. Enameled metal cans or 12-oz composite-type containers.
Adapted from Guadagni and Nimmo (1957) courtesy of Institute of Food Technologists.

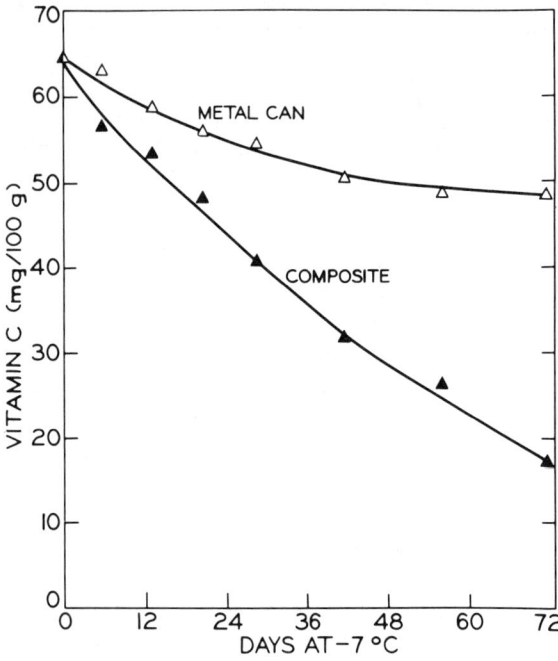

FIG. 10.5. EFFECT OF CONTAINER TYPE ON LOSSES OF VITAMIN C FROM FROZEN STRAWBERRIES

Sliced strawberries, 4+1 sugar, enameled metal cans or composite-type containers.
Adapted from Guadagni et al. (1957B) courtesy of Institute of Food Technologists.

EFFECTS OF FREEZE-PRESERVATION ON NUTRIENTS 265

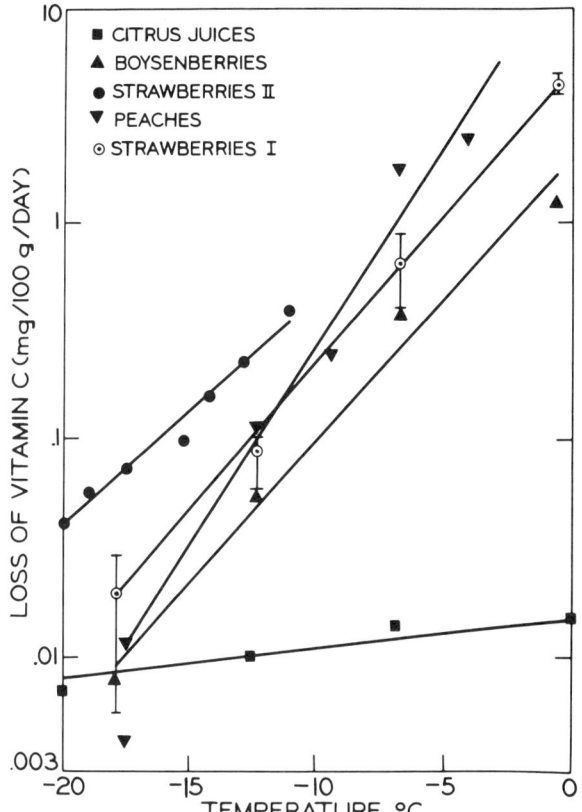

FIG. 10.6. TEMPERATURE DEPENDENCE OF VITAMIN C DEGRADATION IN FROZEN FRUITS

See Table 10.13 for conditions and references.

With the exception of the data for boysenberries, each line in Fig. 10.6 represents at least 2 studies, and most of the studies involved at least 3 temperatures. All data points have been plotted for peaches and strawberries II, whereas points for strawberries I, boysenberries (plain and syruped products) and citrus juices are means. The ranges of values for strawberries I are indicated on Fig. 10.6. The data ranges for boysenberries and citrus juices are much smaller than those for strawberries I.

Rates of vitamin C loss in peaches, strawberries I and boysenberries are affected similarly by temperature. For these products, a 10°C rise in temperature in the range of -18°C to -7°C causes the rate of vitamin C degradation to accelerate by a factor of 30 to 70X (Q_{10} = 30–70). This is an uncommonly large dependence on temperature. For strawberries II, the factor is 10X and for citrus juice concen-

TABLE 10.13

SUPPLEMENTARY INFORMATION FOR THE PRODUCTS IN FIG. 10.6

Product	Conditions	Approximate Maximum Storage Period Over Which Rates of Vitamin C Degradation are Valid Months at:				References
		$-18°C$	$-12°C$	$-7°C$	$-1°C$	
Strawberries I	Sliced or whole, 4 + 1 sugar, 30-lb tins or composite retail cartons.	12	6	1	<1 week	1
Strawberries II	Sliced or whole, sugared or plain, plastic packages.	10	2	—	—	2
Peaches	Sliced, in syrup, composite or plastic cartons.	12	6	2 week	—	3
Boysenberries	Whole, plain or in syrup, packaged in cardboard cartons, or composite retail containers.	>4	>4	1 week	<1 week	4
Citrus juices	Grapefruit, orange, tangerine, 42° Brix; type of package unknown.	>12	>12	>12	>12	5

[1] Bennett et al. (1954), Guadagni et al. (1957B, 1961).
[2] Crivelli et al. (1969), Pierce et al. (1955).
[3] Bennett et al. (1954), Guadagni et al. (1957A).
[4] Guadagni et al. (1960).
[5] Huggart et al. (1954).

trates, 1.5X. Thus, low storage temperatures ($-18°C$ or lower) will preserve vitamin C contents of fruits (not applicable to citrus juice concentrates) much better than high subfreezing temperatures. Comparable information is unavailable for nutrients other than vitamin C.

A few studies have dealt with the effects of fluctuating storage temperatures versus comparable constant temperatures on nutrient losses from fruits. With respect to strawberries and raspberries, losses of vitamin C do not differ significantly between the two conditions (Gortner et al. 1948; Guadagni and Nimmo 1958).

Loss of Nutrients During and Following Thawing

Information concerning losses of nutrients from fruits during thawing is almost nonexistent and what little there is deals only with vitamin C. As would be expected, losses of vitamin C from citrus juice concentrates during thawing are insignificant (Huggart et al. 1954). According to Bauernfeind et al. (1946), the vitamin C contents of packaged peaches, apricots, nectarines, and fruit salad (all in syrup), are not affected significantly by the method of thawing (thawing times ranged from 20 min to 19 hr). Thus it is probably reasonable to assume that properly conducted thawing has almost no detrimental effect on the nutrient contents of fruits provided the syrup and thaw-exudate are consumed.

EFFECTS OF FREEZE-PRESERVATION ON NUTRIENTS 267

Holding periods following thawing can result in significant losses of vitamin C. For example, thawed raspberries-in-syrup can lose 15% vitamin C when held 1 day at 20°C (Loeffler 1946), and thawed peaches-in-syrup can lose 13% vitamin C when held 2 hr at room temperature (Strachan and Moyls 1949).

Loss of Nutrients During the Entire Freezing Process

Shown in Table 10.14 are losses of vitamin C from several fruits and fruit juices during the entire freezing process (based on differences in the vitamin C contents of fresh products and products that were frozen and stored for various times at −18°C). Losses vary

TABLE 10.14

LOSS OF VITAMIN C FROM FRUITS DURING THE ENTIRE FREEZING PROCESS

Product	Storage Time at −18°C (Months)	Loss of Vitamin C (%)	References
Strawberries			
44 brands "as purchased"	?	45 (9-85)[1]	Fagerson et al. (1954)
17 varieties, sliced, sugared, in metal cans	5	17 (0-44)	Scott and Schrader (1947)
Puree, 5 + 1 or 3 + 1 sugar	6	16	Wolfe et al. (1949)
Whole, no syrup or sugar, in polyethylene bags	10	34	Crivelli et al. (1969)
Partially sliced, 6 + 1 sugar, polyethylene boxes	10	42	Pierce et al. (1955)
Citrus products			
Orange juice conc, 42° Brix	9	1	Marshall et al. (1955)
Orange juice, unconcentrated	6	32	Tingleff and Miller (1960)
Orange segments	6	31	Tingleff and Miller (1960)
Grapefruit juice conc, 42° Brix	9	5	Marshall et al. (1955)
Grapefruit sections:			
Plain	9	4	Wolfe et al. (1949)
With syrup	9	4	Wolfe et al. (1949)
Apricots in syrup	5	19	Crow and Scoular (1955)
Apricots in syrup + added vitamin C	5	22	Crow and Scoular (1955)
Acerola juice, natural or diluted, + added sugar, pH 3.3	8	16	Fitting and Miller (1960)
Cantaloupes			
In syrup	5-9	9-44	Crow and Scoular (1955), Wolfe et al. (1949)
In syrup + added vitamin C	5	23	Crow and Scoular (1955)
Plain	9	65-85	Wolfe et al. (1949)
Cherries, sweet, pitted, in syrup, with or without added vitamin C and citric acid	10	19 (11-28)	Strachan and Moyls (1949)
Peaches			
Sliced, in syrup, with added vitamin C, 12 varieties	8	23 (12-40)	Crow and Scoular (1955), Strachan and Moyls (1949)
Sliced, in syrup, 12 varieties, moistureproof containers	8	69 (38-82)	Strachan and Moyls (1949)
Sliced, in syrup, in glass jars	5	29	Crow and Scoular (1955)

[1] Based on estimated initial concentration of 60 mg vitamin C per 100 gm of product.

greatly depending on the type of product, the variety (Crivelli *et al.* 1969), whether or not syrup is present, the solids content of the juices, and the type of package. Most of the conventional fruits, when processed in a recommended manner, lose less than 30% of their original vitamin C contents during the entire freezing process, and concentrated citrus juices lose less than 5%. Most of these losses occur during frozen storage. The small losses of vitamin C from citrus juice concentrates are probably attributable to the low pH values and low oxygen contents of these products (Thompson and Fennema 1971).

Some fruits are fortified with vitamin C (150–350 mg vitamin C per pound of finished pack) (Bauernfeind and Pinkert 1970) prior to freezing so that enzyme-catalyzed oxidative browning will be restricted. It has been shown that about 80% of the vitamin C added to peaches, apricots, nectarines, and cataloupes is retained after 8 months at $-18°C$ (Bauernfeind *et al.* 1946; Crow and Scoular 1955).

Information on the absolute nutritive composition of frozen fruits "as purchased" can be obtained from USDA Handbook 8 (Watt and Merrill 1963) and in papers by Burger *et al.* (1956) and Lowenberg and Wilson (1959).

Comparative Losses of Vitamins During Various Methods of Preparation and Preservation

Comparative losses of vitamins from fruits during freeze-processing, canning, and drying are shown in Table 10.15. The mean values should be accepted with some caution since large variances can occur depending on the product and the conditions.

It is apparent that 8 common fruits, when frozen and stored, contain, on the average, about $1/3$ less of vitamins A and B-1, and about $1/5$ less of niacin and vitamins B-2 and C than when fresh. Furthermore, 8 common fruits, when canned, lose, on the average, about 2–3 times more niacin and vitamins B-1, B-2, and C, and about the same amount of vitamin A as when frozen and stored. Four common fruits, when dried, lose, on the average, about $1/6$ as much vitamin A, significantly less of niacin and vitamin B-2, and about twice as much of vitamins B-1 and C as when frozen-stored.

Thus, water-soluble vitamins are usually present at lower concentrations in frozen-stored fruits than in comparable fresh fruits, and are usually present at lower concentrations in canned fruits than in frozen-stored fruits. Based on limited data, niacin and vitamins A and B-2 appear to be preserved less well, and vitamins B-1 and C better by freeze-processing than by drying.

TABLE 10.15

COMPARATIVE LOSSES OF VITAMINS FROM FRUITS
DURING VARIOUS METHODS OF PRESERVATION

Method of Preservation	Number of Fruits Examined		Loss of Vitamins as Compared to Values of Fresh Products				
			A (%)	B-1 (%)	B-2 (%)	Niacin (%)	C (%)
Frozen, not thawed	8[1]	Mean	37	29	17	16	18
		Range	0-78	0-66	0-67	0-33	0-50
Canned, solids and liquid	8[2]	Mean	39	47	57	42	56
		Range	0-68	22-67	33-83	25-60	11-86
Dried, uncooked	4[3]	Mean	6	55	0	0	39
		Range	0-18	11-90	—	—	0-65

Source: USDA Handbook 8 (Watt and Merrill 1963). Values based on edible portions of fruits "as purchased."

[1] Apples, apricots, blueberries, cherries (red, sour), orange juice concentrate (calculated on rediluted basis), peaches, raspberries and strawberries.
[2] Same as above, except orange juice was single-strength originally.
[3] Apples (sulfured), apricots (sulfured), orange juice (calculated on rediluted basis), and peaches (sulfured).

ANIMAL TISSUES

Loss of Nutrients During Prefreezing Treatments

A major prefreezing treatment for animal tissues is aging. In one study, 0.5-lb samples of longissimus dorsi and semimembranous muscles from beef were aged 21 days at 1°C (Meyer et al. 1963). This resulted in no loss of thiamin and riboflavin, but a 35% loss of niacin. In another study, pork roasts were aged 7 days at -1°C and this had no consistent effect on the amounts of thiamin, riboflavin, pantothenic acid, and nicotinic acid (Westerman et al. 1955). If these data are typical, it is likely that short-term aging at low nonfreezing temperatures has little adverse effect on the nutritive value of animal tissues.

Loss of Nutrients During Freezing

Losses of B-vitamins in various animal tissues during freezing are listed in Table 10.16. With the exception of pork, losses of the five B-vitamins are insignificant. Since freezing as such causes no significant alteration in the nutrient contents of vegetables, fruits, and most animal tissues, the losses encountered during freezing pork are indeed puzzling. The losses for pork may be inflated somewhat since both freezing and thawing are involved, but even after compensating for the approximate effects of thawing, the values are still substantial.

270 NUTRITIONAL EVALUATION OF FOOD PROCESSING

TABLE 10.16

LOSS OF B-VITAMINS FROM VARIOUS ANIMAL TISSUES DURING FREEZING[1]

Vitamin	Product	Loss (%)	References
Thiamin	Beef steak	ns[2]	Lee et al. (1950)
	Beef liver slices	ns	Kotschevar (1955)
	Beef stew	ns	Kahn and Livingston (1970)
	Chicken a la king	ns	Kahn and Livingston (1970)
	Shrimp newburg	ns	Kahn and Livingston (1970)
	Oysters	ns	Fieger (1956)
	Pork chops	17-20[3]	Lee et al. (1954), Lehrer et al. (1951)
Riboflavin	Beef steak	ns	Lee et al. (1950)
	Beef liver slices	ns	Kotschevar (1955)
	Oysters	ns	Fieger (1956)
	Pork chops	+16-25[3]	Lee et al. (1954), Lehrer et al. (1951)
Niacin	Beef steak	ns	Lee et al. (1950)
	Beef liver slices	ns	Kotschevar (1955)
	Pork chops	6-18[3]	Lee et al. (1954), Lehrer et al. (1951)
Pantothenic acid	Beef steak	ns	Lee et al. (1950)
	Oysters	ns	Fieger (1956)
	Pork chops	18[3]	Lee et al. (1954)
Pyridoxine	Beef steak	ns	Lee et al. (1950)
	Oysters	ns	Fieger (1956)
	Pork chops	22[3]	Lee et al. (1954)

[1] Effect of freezing alone unless stated otherwise.
[2] Not significant.
[3] Includes effect of thawing, but no frozen storage.

Furthermore, these values represent the work of two groups, so it is not possible to simply disregard the findings. Thus more detailed consideration of these results is justified. Lee et al. (1954) reported that pork, during freezing, lost sizeable amounts of thiamin, riboflavin, pantothenic acid, and pyridoxine, and a small amount of niacin, whereas during 6 months' storage at −18°C they reported *increases* in the amounts of thiamin and niacin, *no change* in the amounts of riboflavin and pantothenic acid, and a further decrease in pyridoxine. Lehrer et al. (1951) reported that pork during freezing lost sizeable amounts of thiamin and niacin, whereas during storage for 6 months at −18°C the amount of thiamin did not change and niacin decreased further. Although it is possible that large losses of vitamins could occur during freezing (acceleration of reaction rates by freeze-concentration) (Fennema et al. 1973) and be followed by no further losses or even increases during extended frozen storage, this pattern is not normal in frozen foods. These results with pork, therefore, should be viewed with scepticism until they are verified by other workers.

Loss of B-vitamins during freezing muscles of bovine (Lee et al. 1950), ovine (Lehrer et al. 1952), and porcine (Lee et al. 1954; Lehrer et al. 1951) animals is not affected significantly by freezing rate. However, freezing rate can influence the amounts of exudate during thawing and cooking (discussed later) and the extent of structural damage in poultry (damage assessed by the amount of thaw-exudate) and cod (damage assessed by the concentration of DNA-phosphorus in the expressible juice from thawed cod). Furthermore, structural damage in poultry and cod is not linearly related to freezing rate (Crigler and Dawson 1968; Love 1958).

One final point with respect to freezing rate is that oxidative changes in beef and pork apparently occur more slowly following slow freezing than following rapid freezing (Nestorov et al. 1969). If this result is verified, it will have nutritional significance, since several vitamins can be inactivated by oxidation.

Only one study was found that dealt with nutrient retention in meat during storage under fluctuating temperatures as compared to a comparable constant temperature. This study, which involved frozen pork loin roasts stored for 12 months, showed that thiamin loss did not differ significantly between the two conditions (Gortner et al. 1948).

A study by Lane (1966) is also worth mentioning. He found that frozen fish, once it leaves the warehouse for distribution to retail outlets, is generally exposed to temperatures that average well above −18°C. This situation probably applies to most frozen foods and is highly undesirable from the standpoint of vitamin retention.

Loss of Nutrients During Frozen Storage

Shown in Table 10.17 are limited data with respect to losses of B-vitamins in several animal tissues during 6 months' storage at −18°C. Losses of pyridoxine appear to be the greatest and losses of niacin and pantothenic acid the least. B-vitamins are apparently more unstable in pork and oysters than in beef and lamb. Unavailability of data prevents this subject from being covered as thoroughly as was done for fruits and vegetables.

Loss of Nutrients During Thawing

Shown in Table 10.18 are thiamin losses that occur during thawing of a few animal tissues. Loss of thiamin from beef liver is substantial, but much of this loss is probably accountable for in the thaw-exudate. Losses of thiamin in the other products are relatively small and probably border on being within the range of experimental error. If these limited results are typical of all vitamins and all animal prod-

TABLE 10.17

LOSS OF B-VITAMINS FROM ANIMAL TISSUES DURING FROZEN STORAGE

Product	Loss of Vitamins During 6 Months Storage at $-18°C$ [1]					
	B-1 (%)	B-2 (%)	Niacin (%)	Pantothenic Acid (%)	Pyridoxine (%)	Ref
Beef steaks[2]	0	<1	+	<10	22	3
Pork, chops and roasts	+ to 18	0–37	+ to 5	0–8	18[2]	4
Lamb chops	+		+			5
Oysters[2]	33	19	3	17	59	6

[1] Independent effect of storage except some values include effect of thawing. Plus sign (+) indicates an apparent increase in vitamin content.
[2] Data are of limited value since only one study is involved.
[3] Kotschevar (1955).
[4] Lee et al. (1954), Lehrer et al. (1951), Westerman et al. (1955).
[5] Lehrer et al. (1952).
[6] Fieger (1956).

ucts, then nutrient losses during thawing (aside from losses in the thaw-exudate) appear to be small.

Several studies have been conducted to determine the effect of thawing method on nutrient losses. Losses of B-vitamins from unpackaged beef steak are generally greater when thawing is conducted in water rather than in air at various temperatures (Westerman et al. 1949). Leaching is undoubtedly an important mechanism of loss when meat is thawed directly in water. When heat transfer methods of thawing (air or water) are applied to beef steak or poultry, the slowest method of thawing (in refrigerator) generally results in losses of B-vitamins that are less, or no greater than, those which occur

TABLE 10.18

LOSS OF THIAMIN FROM ANIMAL TISSUES DURING THAWING

Product	Loss of Thiamin During Thawing (%)[1]	References
Beef liver, sliced[2]	22[3]	Kotschevar (1955)
Beef stew	1–11	Kahn and Livingston (1970)
Chicken a la king	2–9	Kahn and Livingston (1970)
Shrimp newburg	4–10	Kahn and Livingston (1970)

[1] Various methods.
[2] Apparent values for riboflavin and niacin increased during thawing.
[3] Probably mostly thaw-exudate.

EFFECTS OF FREEZE-PRESERVATION ON NUTRIENTS 273

during more rapid thawing (Westerman et al. 1949; Singh and Essary 1971). With respect to beef stew, chicken a la king and shrimp newburg, thiamin losses are least during microwave thawing, slightly greater during infrared thawing, and greatest during thawing in boiling water (packaged) (Kahn and Livingston 1970). Except when unpackaged products are thawed in water (undesirable), the method of thawing appears to have a small effect on losses of B-vitamins from animal tissues.

Loss of Nutrients in the Thaw-Exudate

This topic is appropriate for discussion since thaw-exudate contains a substantial amount of nutrients, as is indicated in Table 10.19.

TABLE 10.19

COMPOSITION OF THAW-EXUDATE FROM MAMMALIAN MEAT

Component	Calves Liver[1]	Beef[2] (% W/W)	Beef[3] (Mg/100Gm)	Pork[4] (Mg/100Gm)	Lamb[5] (Mg/100Gm)
Total solids	20% w/w (66%)[6]	13.8	—	—	—
Soluble nitrogen	—	2.24	—	—	—
Ash	—	1.16	—	—	—
Chloride	—	0.032	—	—	—
Thiamin	73 mg/100gm (116%)[6]	—	0.14 (233%)[7]	0.2 (100%)[7]	7 (150%)[7]
Riboflavin	5 mg/100gm (50%)[6]	—	1.0 (98%)[7]	0.2 (100%)[7]	4 (68%)[7]
Niacin	46 mg/100gm (116%)[6]	—	0.16 (94%)[7]	0.2 (66%)[7]	8.6 (166%)[7]

[1] Sliced fresh, frozen slowly and stored 60 days at −20°C (Kotschevar 1955).
[2] Pieces of psoas and longissimus dorsi muscles measuring 6 X 3 X 1 cm (the shortest dimension being along the fibers) were cut from frozen sirloin immediately after removal from storage, placed on glass grids inside air-tight dishes and thawed for 48 hr at 10°C. Thaw-exudate was collected under the samples (Howard et al. 1960).
[3] Means for blade roasts, rib roasts, short ribs, and rib steaks. Samples were slowly frozen and stored 60 days at −20°C (Kotschevar 1955).
[4] Means for tenderloin, loin roast, chops. Samples were slowly frozen and stored 60 days at −20°C (Kotschevar 1955).
[5] Means for leg roasts and stew meat. Samples were slowly frozen and stored for 60 days at −20°C (Kotschevar 1955).
[6] Expressed as percentage of the concentration in the product (after exudate was lost).
[7] Expressed as percentage of the concentration in the fresh product.

It is apparent that the concentrations of thiamin, riboflavin, and niacin present in thaw-exudate from the products listed are roughly equal to those existing in the parent products.

Actual percentage losses of vitamins and amino acids that can occur in beef and pork via the mechanism of thaw-exudate are illustrated in Table 10.20. Although losses of most nutrients generally are not large, an awareness should be created about them so that thaw-exudate will be utilized by the consumer and so that the processor will exercise whatever control he can to minimize the amount of thaw-exudate.

TABLE 10.20

NUTRIENTS LOST FROM BEEF AND PORK IN THAW-EXUDATE

Nutrient	Percentage of Original Nutrient Lost in Thaw-exudate From:	
	Beef steak[1] (%)	Pork chops[2] (%)
Thiamin	12	9
Riboflavin	10	4
Niacin	15	10.7
B-6 (pyrimidine)	9	8.7
Pantothenic acid	33	6.9
Folic acid	8	—
B-12	—	5.1
Isoleucine	—	11.1
Leucine	—	9.7
Lysine	—	8.6
Methionine	—	7.6
Tryptophan	—	7.1

[1] Frozen at −18°C, stored at this temperature until needed for analysis, then thawed 14–15 hr in air at 26°C (Pearson et al. 1951).
[2] Frozen at −18°C, stored at this temperature until needed for analysis, then thawed 18–20 hr in air at 2°C (Pearson et al. 1959).

The amount of thaw-exudate from animal tissue can range from less than 1% to more than 30%. Factors that influence the amount of thaw-exudate can be categorized as follows: (1) type of product, (2) natural variations within a given product, and (3) processing variables.

Type of Product.—Differences in the amount of thaw-exudate can occur depending on the type of product being frozen. Amounts of thaw-exudate obtained from beef (Bennett et al. 1954; Cook et al. 1926; Kotschevar 1955; Ramsbottom and Koonz 1939, 1940, 1941; Pearson and Miller 1950; Westerman et al. 1949) and pork muscle (Kotschevar 1955; Pearson et al. 1959) often range from about 1 to 10%, amounts from poultry (Crigler and Dawson 1968; Khan and Lentz 1965; Khan and van den Berg 1967) are generally less than 5%, and amounts from fish are highly variable, but values of 5–10% are common for salmon, halibut, and fresh-water species (Botta et al. 1973; Manohar et al. 1973). Other product characteristics also have a bearing on the amount of thaw-exudate. For example, thaw-exudate from some salt-water fish is greater than that from fresh-water fish (Botta et al. 1973; Manohar et al. 1973), and glandular meats generally exhibit greater thaw-exudate than muscle meats (Kotschevar 1955).

Properties of a Given Product.—The pH of meat has a profound influence on the amount of thaw-exudate. A high ultimate pH of 6.4 will minimize thaw-exudate in pork, mutton, and beef, and thaw-exudate will increase as the pH is lowered to 5-5.2 (Ramsbottom and Koonz 1940; Sair and Cook 1938).

State of rigor at the time of freezing also can have a profound effect on the amount of thaw-exudate. For tissues which are sensitive to this effect (some fish, whale muscle, poultry, mammalian muscle), a processing sequence involving rapid freezing prerigor, storage at a low temperature, and rapid thawing can result in very large amounts of thaw-exudate (Bendall 1972; Khan and Lentz 1965; Manohar et al. 1973; Tanaka and Tanaka 1956A, 1957).

Processing Variables.—Aging of beef prior to freezing has been found to decrease the amount of thaw-exudate (Ramsbottom and Koonz 1940).

Small cuts of animal tissue (large surface-to-volume ratio) are much more likely to exhibit large amounts of thaw-exudate than large cuts of animal tissue (Ramsbottom and Koonz 1939, 1941).

Rapid freezing often results in less thaw-exudate than slow freezing, particularly with small cuts of animal tissue (Ramsbottom and Koonz 1939, 1941; Cook et al. 1926; Khan and van den Berg 1967). With large cuts of beef, freezing rate generally has no influence on the amount of thaw-exudate (Ramsbottom and Koonz 1941). With poultry, freezing rate and the amount of thaw-exudate are not linearly related (Fig. 10.7), and this behavior may well extend to small cuts of other kinds of animal tissue (Love 1958). When animal tissues are frozen prerigor, slow freezing generally results in less thaw-exudate than rapid freezing (Tanaka and Tanaka 1957). Slow freezing causes the product to remain at high subfreezing temperatures for a relatively long period, thus allowing glycolysis to continue and thereby lessening the possibility of thaw-rigor (a large amount of thaw-exudate is often associated with thaw-rigor).

The effect of storage temperature on the amount of thaw-exudate apparently varies with the product and with other processing variables. In two studies, large and small cuts of beef were stored for 4-12 months at temperatures ranging from $-12°$ to $-29°C$ and no difference was noted in the amount of thaw-exudate as a function of storage temperature (Bennett et al. 1954; Ramsbottom and Koonz 1941). However, with cod, the amount of thaw-exudate apparently increases as the storage temperature is increased over the range $-29°$ to $-7°C$ (Miyauchi 1962).

The amount of thaw-exudate generally increases with increased

276 NUTRITIONAL EVALUATION OF FOOD PROCESSING

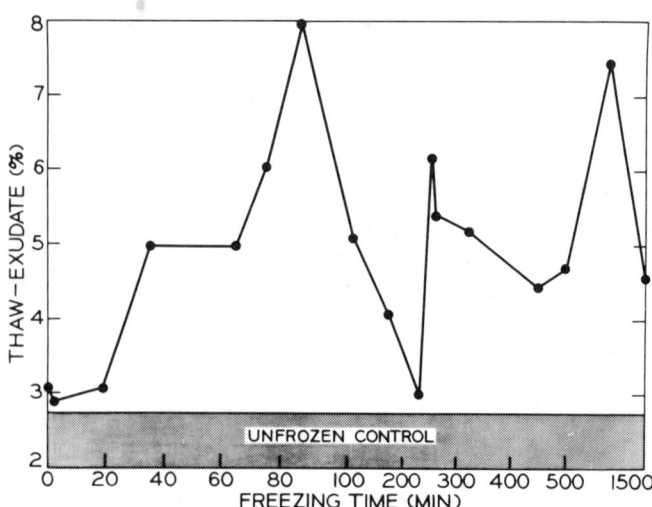

FIG. 10.7. EFFECT OF FREEZING TIME (RATE) ON THE PERCENTAGE OF THAW-EXUDATE FROM POULTRY

Adapted from Crigler and Dawson (1968) courtesy of Institute of Food Technologists.

storage time, especially when small cuts of animal tissue are involved (Ramsbottom and Koonz 1941; Pearson and Miller 1950). The opposite effect can sometimes occur when animal muscles are frozen prerigor (Tanaka and Tanaka 1956B). When this occurs, prolonged storage apparently allows glycolysis to proceed sufficiently so that thaw-rigor is less likely.

Thawing rate may or may not influence the amount of thaw-exudate. In two studies involving pork and beef steaks, thawing rate was shown to have little or no affect on the amount of thaw-exudate (Vail et al. 1943; Westerman et al. 1949). However, the amount of thaw-exudate from animal tissues frozen prerigor often will be less if thawing is slow rather than rapid (Tanaka and Tanaka 1957). The reason for this was mentioned earlier when freezing rates and storage times were discussed.

In a study involving thawing of relatively large blocks of postrigor whale meat, very rapid dielectric thawing was shown to result in much less thaw-exudate than very rapid thawing by resistance heating or by slow thawing in air (Tanaka and Tanaka 1956A).

Chemical treatments also can influence the amount of thaw-exudate. For example, if fish, prior to freezing, are dipped in a solution of trisodium polyphosphate, the amount of thaw-exudate will be reduced (Manohar et al. 1973).

TABLE 10.21

LOSS OF VITAMINS FROM ANIMAL TISSUES DURING THE ENTIRE FREEZING PROCESS[1]

Product	Storage Conditions	B-1 (%)	B-2 (%)	Loss of B-Vitamins Niacin (%)	Pantothenic Acid (%)	Pyridoxine (%)	References
Beef liver, sliced	60 days −20°C	32	35	+			3
Beef steak, l. dorsi	6 months −18°C	8	9	0	8	24	4
	10–12 months −18°C	2	43	4			4
Beef (l. dorsi and semimembranous)	3 yr −18°C, unaged	+34	+9	20			5
	3 yr −18°C, aged	3	+8	+1			5
Pork loins[2]	1 yr −18°C	11	+44	+14	+6		6
Poultry, fowls, whole	8 months −18°C						
	light meat	12	3	+10			7
	dark meat	42	11	0			7
Poultry, less than 84 days old	8–12 months −18°C						
	breast	16	0	9			8
	leg	5	33	22			8
Turkeys, 27–36 weeks old	3 months −23°C						
	breast	0	8	0			9
	leg	18	0	0			9
Oysters[2]	6 months −18°C	22	0	35	+	46	10

[1] May or may not include thawing.
[2] Data are highly variable.
[3] Kotschevar (1955).
[4] Lee et al. (1950).
[5] Meyer et al. (1963).
[6] Westerman et al. (1952).
[7] Millares and Fellers (1949).
[8] Morgan et al. (1949).
[9] Cook et al. (1949).
[10] Fieger (1956).

Loss of Nutrients During the Entire Freezing Process

Losses of B-vitamins from animal tissues during the entire freezing process are shown in Table 10.21. It is evident that losses are quite variable. Changes in thiamin content during the freezing process range from +34 to −42%, changes in riboflavin content range from +44 to −43%, changes in niacin content range from +14 to −35%, and, based on limited data, changes in pantothenic acid content range from +6 to −8% and changes in pyridoxine content range from −24 to −46%.

Variability of vitamin losses within a given type of product (e.g., beef) appears to be fully as large as variability among different types of products.

The phases of the freezing process that cause the greatest losses of B-vitamins appear to be frozen storage and thawing (thaw-exudate). With pork, it is also possible that substantial amounts of B-vitamins are lost during freezing, but further work is needed to confirm or reject this notion.

TABLE 10.22

LOSS OF B-VITAMINS DURING COOKING
OF PREVIOUSLY FROZEN BEEF AND PORK

Product	Conditions of Frozen Storage Prior to Cooking	Loss of B-Vitamins[1]					References
		B-1 (%)	B-2 (%)	Niacin (%)	Pantothenic Acid (%)	Pyridoxine (%)	
Beef steak (l. dorsi)	No storage	40	5	6	+2	14	2
	6–10 months −18°C	35	3	15	15	+16	2
Pork chops (l. dorsi)	No storage	15	0	+			3
	6 months −18°C	19	0	14	+	8	3

[1] Losses based on thawed, uncooked product that had been stored for the same time and temperature.
[2] Lee et al. (1950).
[3] Lee et al. (1954).

Loss of Nutrients During Cooking of Previously Frozen Animal Tissues

Shown in Table 10.22 are the results of two studies dealing with losses of B-vitamins during cooking of previously frozen beef and pork. The losses indicated presumably are total losses, i.e., they include both cook-exudate and chemical degradation of vitamins within the meat. Losses of most B-vitamins, with the exception of thiamin, are relatively small.

EFFECTS OF FREEZE-PRESERVATION ON NUTRIENTS

Comparative Losses of B-vitamins from Fresh-Cooked and Frozen-Stored-Cooked Animal Tissues

The results of two studies pertinent to this topic are summarized in Table 10.23. Frozen-stored-cooked lamb chops, in one study, contained as much thiamin and more niacin than the fresh-cooked product. Frozen-stored-cooked pork chops contained slightly less thiamin and riboflavin and considerably less niacin than the fresh-cooked product.

TABLE 10.23

COMPARATIVE NUTRITIVE VALUES OF FRESH-COOKED AND FROZEN-STORED-COOKED MEATS

	Vitamin Content (μg/Gm)						
	Fresh-Cooked			Cooked[1] after 6 Months at $-22°C$			
Product	B-1	B-2	Niacin	B-1	B-2	Niacin	References
Lamb chops	4		82	4 $(0)^2$		134 $(+63)^2$	3
Pork chops	15	3.5	207	14 $(-7)^2$	3.1 $(-11)^2$	113 $(-45)^2$	4

[1] Cooked directly from the frozen state.
[2] Percentage change as compared to cooked, unfrozen product.
[3] Lehrer *et al.* (1952).
[4] Lehrer *et al.* (1951).

DAIRY PRODUCTS AND MARGARINE

During freezing and 24 months' storage at $-10°C$, butter loses only a small amount of its original vitamin D content, and butter and margarine lose about 17% of their vitamin A contents (Chick and Roscoe 1926; Deuel and Greenberg 1953). According to Holmes *et al.* (1946), ice cream frozen and stored for 7 months at $-23°C$ loses 5% of its original riboflavin, 16% of its original carotenes and 100% of its vitamin C (actually, freshly-made ice cream is totally devoid of vitamin C). According to Lawrence *et al.* (1946), frozen milk stored for 19 weeks at $-14°C$ loses no biotin or nicotinic acid.

MISCELLANEOUS OBSERVATIONS

Broccoli that has been contaminated with malathion (a pesticide) losses 45-77% of the malathion during blanching, freezing, and storage for 6 months at $-9°C$ (Kilgore and Windham 1970).

Almost no information is available with respect to the effect of the freezing process on the nutritive value of proteins. The most relevant study is one conducted by de Groot (1963) on freeze-dried beef, chicken, fish, fish patties, and corn. He found that cooking, and freeze drying caused no noticeable change in protein quality (based on feeding studies with rats) as compared to the fresh-cooked products. Thus, it is likely that the freezing process has no significant detrimental effects on proteins.

DISCUSSION

Several points of interest are apparent when losses of vitamins during freeze-processing of vegetables, fruits, and animal tissues are compared (Table 10.24). During the freezing process, vitamin losses in vegetables are caused primarily by blanching and prolonged (6-12 months) frozen storage, in fruits by prolonged frozen storage and thawing (if the syrup and thaw-exudate are not consumed), and in animal tissues by prolonged frozen storage and by thawing (thaw-exudate).

Moderate-to-large amounts of water-soluble vitamins are lost from vegetables during blanching, from animal tissues during thawing (thaw-exudate), and presumably from fruits by leaching into the syrup and during thawing (thaw-exudate). Almost all of these losses could be avoided if: (1) vegetables were blanched and cooled by means that do not involve liquid water, and (2) the thaw-exudate from animal tissues and the syrup and thaw-exudate from fruits were consumed.

With vegetables and animal tissues, significant losses of water-soluble nutrients can occur during cooking (Table 10.24). With vegetables, these losses can be minimized by using minimal cooking times and minimal amounts of cooking water, and with animal tissues by consuming the cooking-exudate, e.g., in the form of gravy.

The temperature-dependence of vitamin C degradation in frozen fruits and vegetables is also worthy of comment. The rates of vitamin C degradation in several fruits and vegetables at $-18°C$ are similar (Fig. 1.2 and 1.6). Furthermore, a rise in temperature causes uncommonly large increases in the rates of vitamin C degradation in several vegetables (Q_{10} = 6-20 for green beans, cauliflower, green peas, and spinach) and in several fruits (Q_{10} = 30-70 for peaches, boysenberries, and some strawberries). This large temperature-dependence is no doubt caused by the fact that a high concentration of solutes exists in the unfrozen phases of frozen foods and that this concentration (and accordingly, oxygen solubility, viscosity, and molecular mobility) changes markedly with a change in temperature

TABLE 10.24

SUMMARY OF VITAMIN LOSSES FROM VARIOUS CLASSES OF FOODS DURING FREEZE-PROCESSING AND COOKING

Processing Step	Vitamin Losses		
	Vegetables	Fruits	Animal Tissues
Prefreezing treatments	10–44% loss of vitamin C during blanching. Substantial losses can occur during storage if time-temp conditions are abusive.	Slight if properly handled prior to freezing.	Slight if properly handled prior to freezing.
Freezing	Slight.	Slight (2 studies).	Insignificant, except perhaps for B-vitamins in pork.
Frozen storage	Substantial losses of vitamin C and pantothenic acid. Moderate losses of vitamins B-1 and B-2. Losses of vitamin C are highly temp dependent ($Q_{10} = 6$–20).	Substantial losses of vitamin C except in citrus juice conc. No data for other vitamins. Losses of vitamin C are highly temp dependent ($Q_{10} = 30$–70).	Data limited. Substantial losses of pyridoxine from beef, pork, oysters. Moderate losses of vitamins B-1 and B-2 in pork. Substantial losses of B-vitamins in oysters.
Thawing	Slight.	Little data; probably slight, except perhaps for loss of vitamins in the syrup and thaw-exudate.	Moderate losses of B-vitamins and amino acids in thaw-exudate.
Total freezing process	Many common vegetables lose approx 50% of their original vitamin C contents.	Losses of vitamin C are usually less than 30%, except in citrus juice conc where losses of vitamin C are less than 5%.	Losses of B-vitamins are highly variable. Losses of B-1 usually <25%; B-2 often <15%; niacin often <10%; pyridoxine about 25–50% (few data).
Cooking of frozen food	Vitamin C 10–50%; β-carotene slight; folic acid 0–50%; niacin 0–25%; B-1 0–25%; B-6 20–40%; B-2 0–20%; pantothenic acid 10–40%; leaching is an important factor.	Not applicable for most fruits.	B-1 15–40%; B-2 0–5%; niacin 5–15%; pantothenic acid 0–15%; pyridoxine 0–15%; substantial losses in cooking exudate.

(Fennema et al. 1973; Pincock and Kiovsky 1966; Thompson and Fennema 1971). From the standpoint of retention of vitamin C, the desirability of storing fruits and vegetables at temperatures of $-18°C$ or lower should be abundantly clear.

When vegetables are exposed to a freezing process and then cooked, their contents of common vitamins (A, B-1, B-2, niacin, and C) almost invariably will be significantly lower (often 25% ± 20) than those existing in comparable fresh-cooked products. Likewise, when fruits are exposed to a freezing process their contents of common vitamins (A, B-1, B-2, niacin, and C) almost invariably will be significantly lower (often 20% ± 15) than those existing in comparable fresh products.

Canning of vegetables and fruits often results in 2-3 times greater losses of niacin and vitamins B-1, B-2, and C than occur in vegetables during freezing, storing, and cooking, or in fruits during freezing and frozen storage.

Data gathered with respect to animal tissues are insufficient to warrant comparisons similar to those made with fruits and vegetables.

BIBLIOGRAPHY

BARNES, B., and TRESSLER, D. K. 1943. Thiamin content of fresh and frozen peas and corn before and after cooking. Food Res. *8*, 420-427.

BATCHELDER, E. L., KIRKPATRICK, M. E., STEIN, K. E. and MARRON, I. M. 1947. Effect of scalding method on the quality of three home-frozen vegetables. J. Home Econ. *39*, 282-286.

BAUERNFEIND, J. C., JAHNS, F. W., SMITH, E. G. and SIEMERS, G. F. 1946. Vitamin C stability in frozen fruit processed with crystalline L-ascorbic acid. Fruit Prod. J. *25*, 324-330, 347.

BAUERNFEIND, J. C. and PINKERT, D. M. 1970. Food processing with added ascorbic acid. Advan. Food Res. *18*, 219-315.

BEDFORD, C. L., and HARD, M. M. 1950. The effect of cooling method on the ascorbic acid and carotene content of spinach, peas, and snap beans preserved by freezing. Proc. Am. Soc. Hort. Sci. *55*, 403-409.

BENDALL, J. R. 1972. Postmortem changes in muscle. In The Structure and Function of Muscle, Vol. 2, Part 2. G. H. Bourne (Editor). Academic Press, New York.

BENNETT, G. et al. 1954. Some factors affecting the quality of frozen foods. II. Penn. State Coll. Agr. Expt. Sta. Bull. *580*.

BOGGS, M. M. et al. 1960. Time-temperature tolerance of frozen foods. XXI. Frozen peas. Food Technol. *14*, 181-185.

BOTTA, J. R., RICHARDS, J. F., and TOMLINSON, N. 1973. Flesh pH, color, thaw drip, and mineral concentration of pacific halibut (*Hippoglossus stenolepis*) and Chinook salmon (*Oncorhynchus tshawytscha*) frozen at sea. J. Fisheries Res. Board Can. *30*, 71-77.

BURGER, M. et al. 1956. Vitamin, mineral, and proximate composition of frozen fruits, juices, and vegetables. J. Agr. Food Chem. *4*, 418-425.

CHICK, H., and ROSCOE, M. H. 1926. LXXXIV. Influence of diet and sunlight upon the amount of vitamin A and vitamin D in the milk afforded by a cow. Biochem. J. *20*, 632-649.

COOK, B. B., GUNNING, B., and UCHIMOTO, D. 1961. Variations in nutritive value of frozen green baby lima beans as a result of methods of processing and cooking. J. Agr. Food Chem. 9, 316-321.
COOK, B. B., MORGAN, A. F., and SMITH, M. B. 1949. Thiamin, riboflavin, and niacin content of turkey tissues as affected by storage and cooking. Food Res. 14, 449-458.
COOK, G. A., LOVE, E. F. J., VICKERY, J. R., and YOUNG, W. J. 1926. Studies on the refrigeration of meat. I. Investigations into the refrigeration of beef. Australian J. Exptl. Biol. Med. Sci. 3, 15-31.
CRIGLER, J. C., and DAWSON, L. E. 1968. Cell disruption in broiler breast muscle related to freezing time. J. Food Sci. 33, 248-250.
CRIVELLI, G., ROSATI, P., and MONZINI, A. 1969. Chemical stability of frozen strawberries during storage. In Frozen Foods. International Institute of Refrigeration, Paris, France.
CROW, L. S., and SCOULAR, F. I. 1955. Effects of antioxidant ascorbic acid upon the ascorbic acid content of certain frozen fruits. J. Home Econ. 47, 259-260.
DAWSON, E. H., REYNOLDS, H., and TOEPFER, E. W. 1949. Home-canned versus home-frozen snap beans. J. Home Econ. 41, 572-574.
DE GROOT, A. P. 1963. The influence of dehydration of foods on the digestibility and the biological value of the proteins. Food Technol. 17, 339-343.
DERSE, P. H., and TEPLY, L. J. 1958. Effect of storage conditions on nutrients in frozen green beans, peas, orange juice, and strawberries. J. Agr. Food Chem. 6, 309-312.
DEUEL, H. J., JR., and GREENBERG, S. M. 1953. A comparison of the retention of vitamin A in margarines and in butters based upon bioassays. Food Res. 18, 497-503.
DIETRICH, W. C. et al. 1957. The time-temperature tolerance of frozen foods. IV. Objective tests to measure adverse changes in frozen vegetables. Food Technol. 11, 109-113.
DIETRICH, W. C. et al. 1959. Time-temperature tolerance of frozen foods. XVI. Quality retention of frozen green snap beans in retail packages. Food Technol. 13, 136-145.
DIETRICH, W. C., BOGGS, M. M., NUTTING, M-D. F., and WEINSTEIN, N. E. 1960. Time-temperature tolerance of frozen foods. XXIII. Quality changes in frozen spinach. Food Technol. 14, 522-527.
DIETRICH, W. C., and NEUMANN, H. J. 1965. Blanching Brussels sprouts. Food Technol. 19, 1174-1177.
DIETRICH, W. C., NUTTING, M-D. F., BOGGS, M. M., and WEINSTEIN, N. E. 1962. Time-temperature tolerance of frozen foods. XXIV. Quality changes in cauliflower. Food Technol. 16, 123-128.
DuBOIS, C. W., and COLVIN, D. L. 1945. Loss of added vitamin C in the storage of peaches. Fruit Prod. J. and Am. Food Mfgr. 25, 101-103.
EHEART, M. S. 1967. Effect of microwave vs. water-blanching on nutrients in broccoli. J. Am. Dietet. Assoc. 50, 207-211.
EHEART, M. S. 1970. Effect of storage and other variables on composition of frozen broccoli. Food Technol. 24, 1009-1011.
FAGERSON, I. S., ANDERSON, E. E., HAYES, K. M., and FELLERS, C. R. 1954. Vitamin C and frozen strawberries. Quick Frozen Foods 16, No. 9, 84-85.
FARRELL, K. T., and FELLERS, C. R. 1942. Vitamin content of green snap beans. Influence of freezing, canning, and dehydration on the content of thiamin, riboflavin, and ascorbic acid. Food Res. 7, 171-177.
FEASTER, J. F., MUDRA, A. E., IVES, M., and TOMPKINS, M. D. 1949. Effect of blanching time on vitamin retention in canned peas. Canner 108, No. 1, 27-30.
FENNEMA, O., POWRIE, W. D., and MARTH, E. H. 1973. Low-Temperature Preservation of Foods and Living Matter. Marcel Dekker, New York.

FENTON, F., and TRESSLER, D. K. 1938. Losses of vitamin C during commercial freezing, defrosting, and cooking of frosted peas. Food Res. 3, 409-416.
FIEGER, E. A. 1956. Vitamin content of fresh, frozen oysters. Quick Frozen Foods 19, No. 4, 152, 155.
FISHER, W. B., and VAN DUYNE, F. O. 1952. Effect of variations in blanching on quality of frozen broccoli, snap beans, and spinach. Food Res. 17, 315-325.
FITTING, K. O., and MILLER, C. D. 1960. The stability of ascorbic acid in frozen and bottled acerola juice alone and combined with other fruit juices. Food Res. 25, 203-210.
GORDON, J., and NOBLE, I. 1959. Effects of blanching, freezing, freezing-storage, and cooking on ascorbic acid retention in vegetables. J. Home Econ. 51, 867-870.
GORTNER, W. A., FENTON, F., VOLZ, F. E., and GLEIM, E. 1948. Effect of fluctuating storage temperatures on quality of frozen foods. Ind. Eng. Chem. 40, 1423-1426.
GUADAGNI, D. G., DOWNES, N. J., SANSHUCK, D. W., and SHINODA, S. 1961. Effect of temperature on stability of commercially frozen bulk pack fruits—strawberries, raspberries, and blackberries. Food Technol. 15, 207-209.
GUADAGNI, D. G., EREMIA, K. M., KELLY, S. H., and HARRIS, J. 1960. Time-temperature tolerance of frozen foods. XX. Boysenberries. Food Technol. 14, 148-150.
GUADAGNI, D. G., and NIMMO, C. C. 1957. The time-temperature tolerance of frozen foods. III. Effectiveness of vacuum, oxygen removal, and mild heat in controlling browning in frozen peaches. Food Technol. 11, 43-47.
GUADAGNI, D. G., and NIMMO, C. C. 1958. Time-temperature tolerance of frozen foods. XIII. Effect of regularly fluctuating temperatures in retail packages of frozen strawberries and raspberries. Food Technol. 12, 306-310.
GUADAGNI, D. G., NIMMO, C. C., and JANSEN, E. F. 1957A. The time-temperature tolerance of frozen foods. II. Retail packages of frozen peaches. Food Technol. 11, 33-42.
GUADAGNI, D. G., NIMMO, C. C., and JANSEN, E. F. 1957B. Time-temperature tolerance of frozen foods. VI. Retail packages of frozen strawberries. Food Technol. 11, 389-397.
GUADAGNI, D. G., NIMMO, C. C., and JANSEN, E. F. 1957C. Time-temperature tolerance of frozen foods. X. Retail packs of frozen red raspberries. Food Technol. 11, 633-637.
GUERRANT, N. B. 1957. Changes in light reflectance and ascorbic acid content of foods during frozen storage. J. Agr. Food Chem. 5, 207-212.
GUERRANT, N. B., and DUTCHER, R. A. 1948. Further observations concerning the relationship of temperature of blanching to ascorbic acid retention in green beans. Arch. Biochem. 18, 353-359.
GUERRANT, N. B., and O'HARA, M. B. 1953. Vitamin retention in peas and lima beans after blanching, freezing, processing in tin and in glass, after storage and after cooking. Food Technol. 7, 473-477.
GUERRANT, N. B. et al. 1947. Effect of duration and temperature of blanch on vitamin retention by certain vegetables. Ind. Eng. Chem. 39, 1000-1007.
GUERRANT, N. B. et al. 1953. Some factors affecting the quality of frozen foods. Penn. State Coll. Agr. Expt. Sta. Bull. 565.
GUSTAFSON, F. G., and COOKE, A. R. 1952. Oxidation of ascorbic acid to dehydro-ascorbic acid at low temperatures. Science 116, 234.
HARTZLER, E. R., and GUERRANT, N. B. 1952. Effect of blanching and of frozen storage of vegetables on ascorbic acid retention and on the concomitant activity of certain enzymes. Food Res. 17, 15-23.
HOLMES, A. D. et al. 1945. Vitamin content of field-frozen kale. Am. J. Diseases Children 70, 298-300.

HOLMES, A. D., KUZMESKI, J. W., and CANAVAN, F. T. 1946. Stability of vitamins in stored ice cream. J. Am. Dietet. Assoc. 22, 670-672.
HOLMQUIST, J. W., CLIFCORN, L. E., HEBERLEIN, D. G., and SCHMIDT, C. F. 1954. Steam blanching of peas. Food Technol. 8, 437-445.
HOWARD, A., LAWRIE, R. A., and LEE, C. A. 1960. Studies on beef quality. VIII. Some observations on the nature of drip. CSIRO, Melbourne, Australia.
HUGGART, R. L., HARMAN, D. A., and MOORE, E. L. 1954. Ascorbic acid retention in frozen concentrated citrus juices. J. Am. Dietet. Assoc. 30, 682-684.
JENKINS, R. R., and TRESSLER, D. K. 1938. Vitamin C content of vegetables. VIII. Frozen peas. Food Res. 3, 133-140.
JENKINS, R. R., TRESSLER, D. K., MOYER, J., and McINTOSH, J. 1940. Storage of frozen vegetables. Vitamin C experiments. Refrig. Eng. 39, 381-382.
JURICS, E. W. 1970. Comparative investigations of the vitamin C contents of frozen and fresh vegetables in the raw and cooked states. Nahrung 14, 107-114. (German)
KAHN, L. N., and LIVINGSTON, G. E. 1970. Effect of heating methods on thiamin retention in fresh or frozen prepared food. J. Food Sci. 35, 349-351.
KHAN, A. W., and LENTZ, C. P. 1965. Influence of prerigor, rigor and postrigor freezing on drip losses and protein changes in chicken meat. J. Food Sci. 30, 787-790.
KHAN, A. W., and VAN DEN BERG, L. 1967. Biochemical and quality changes occurring during freezing of poultry meat. J. Food Sci. 32, 148-150.
KILGORE, L., and WINDHAM, F. 1970. Disappearance of Malathion residue in broccoli during cooking and freezing. J. Agr. Food Chem. 18, 162-163.
KOTSCHEVAR, L. H. 1955. B-vitamin retention in frozen meat. J. Am. Dietet. Assoc. 31, 589-596.
LAMB, F. C., LEWIS, L. D., and LEE, S. K. 1948. Effect of blanching on retention of ascorbic acid and thiamin in peas. Western Canner Packer 5, 60-62.
LANE, J. P. 1966. Time-temperature tolerance of frozen seafoods. Food Technol. 20, 549-553.
LAWRENCE, J. M., HERRINGTON, B. L., and MAYNARD, L. A. 1946. The nicotinic acid, biotin, and pantothenic acid content of cow's milk. J. Nutr. 32, 73-91.
LEE, F. A. et al. 1950. Effect of freezing rate on meat. Appearance, palatability, and vitamin content of beef. Food Res. 15, 8-15.
LEE, F. A. et al. 1954. Effect of rate of freezing on pork quality. J. Am. Dietet. Assoc. 30, 351-354.
LEE, F. A., GORTNER, W. A., and WHITCOMBE, J. 1946. Effect of freezing rate on vegetables. Ind. Eng. Chem. 38, 341-346.
LEE, F. A., and WHITCOMBE, J. 1945. Blanching of vegetables for freezing. Effect of different types of potable water on nutrients of peas and snap beans. Food Res. 10, 465-468.
LEHRER, W. P., JR., WIESE, A. C., HARVEY, W. R., and MOORE, P. R. 1951. Effect of frozen storage and subsequent cooking on the thiamin, riboflavin, and nicotinic acid content of pork chops. Food Res. 16, 485-491.
LEHRER, W. P., JR., WIESE, A. C., HARVEY, W. R., and MOORE, P. R. 1952. The stability of thiamin, riboflavin, and nicotinic acid of lamb chops during frozen storage and subsequent cooking. Food Res. 17, 24-30.
LINDQUIST, F. E., DIETRICH, W. C., and BOGGS, M. M. 1950. Effect of storage temperature on quality of frozen peas. Food Technol. 4, 5-9.
LOEFFLER, H. J. 1946. Retention of ascorbic acid in raspberries during freezing, frozen storage, puréeing, and manufacture into velva fruit. Food Res. 11, 507-515.
LOVE, R. M. 1958. Expressible fluid of fish fillets. VIII. Cell damage in slow freezing. J. Sci. Food Agr. 9, 257-262.

LOWENBERG, M. E., and WILSON, E. D. 1959. Nutrients in frozen foods. Natl. Assoc. Frozen Food Packers, Washington, D.C.

MANOHAR, S. V., RIGBY, D. L., and DUGAL, L. C. 1973. Effect of sodium tripolyphosphate on thaw drip and taste of fillets of some freshwater fish. J. Fisheries Res. Board Can. *30*, 685-688.

MARSHALL, J. R., HAYES, K. M., FELLERS, C. R., and DuBOIS, C. W. 1955. Stability of ascorbic acid in citrus concentrates during storage. Quick Frozen Foods *17*, No. 12, 50-52, 129.

MARTIN, M. E., SWEENEY, J. P., GILPIN, G. L., and CHAPMAN, V. J. 1960. Factors affecting the ascorbic acid and carotene content of broccoli. J. Agr. Food Chem. *8*, 387-390.

McCOLLOCH, R. J., RICE, R. G., BANDURSKI, M. B., and GENTILI, B. 1957. The time-temperature tolerance of frozen foods. VII. Frozen concentrated orange juice. Food Technol. *11*, 444-449.

MELNICK, D., HOCHBERG, M., and OSER, B. L. 1944. Comparative study of steam and hot water blanching. Food Res. *9*, 148-153.

MEYER, B., MYSINGER, M., and BUCKLEY, R. 1963. The effect of three years of freezer storage on the thiamin, riboflavin, and niacin content of ripened and unripened beef. J. Agr. Food Chem. *11*, 525-527.

MILLARES, R., and FELLERS, C. R. 1949. Vitamin and amino acid content of processed chicken meat products. Food Res. *14*, 131-143.

MIYAUCHI, D. 1962. Application of centrifugal method for measuring shrinkage during the thawing and heating of frozen cod fillets. Food Technol. *16*, No. 1, 70-72.

MORGAN, A. F. et al. 1949. Thiamin, riboflavin, and niacin content of chicken tissues, as affected by cooking and frozen storage. Food Res. *14*, 439-448.

MOYER, J. C., and STOTZ, E. 1945. Electronic blanching of vegetables. Science *102*, 68-69.

MOYER, J. C., and TRESSLER, D. K. 1943. Thiamin content of fresh and frozen vegetables. Food Res. *8*, 58-61.

NESTOROV, N. et al. 1969. Proc. Eur. Mktg. Meat Res. Workers *15*, 110. Cited by Burger, I. H., and Walters, C. L. 1973. The effect of processing on the nutritive value of flesh foods. Proc. Nutr. Soc. *32*, 1-8.

NOBLE, I., and GORDON, J. 1964. Effect of blanching method on ascorbic acid and color of frozen vegetables. J. Am. Dietet. Assoc. *44*, 120-123.

PAYNE, I. R. 1967. Ascorbic acid retention in frozen corn. J. Am. Dietet. Assoc. *51*, 344-348.

PEARSON, A. M., and MILLER, J. I. 1950. The influence of rate of freezing and length of freezer-storage upon the quality of beef of unknown origin. J. Animal Sci. *9*, 13-19.

PEARSON, A. M. et al. 1951. Vitamin losses in drip obtained upon defrosting frozen meat. Food Res. *16*, 85-87.

PEARSON, A. M., WEST, R. G., and LUECKE, R. W. 1959. The vitamin and amino acid content of drip obtained upon defrosting frozen pork. Food Res. *24*, 515-519.

PHILLIPS, M. G., and FENTON, F. 1945. Effects of home freezing and cooking on snap beans: thiamin, riboflavin, ascorbic acid. J. Home Econ. *37*, 164-170.

PIERCE, R. T., SHAW, M. D., HECK, J. G., and BENNETT, G. 1955. Small storage temperature differences can affect the quality of frozen strawberries and green beans. Refrig. Eng. *63*, No. 11, 52-57.

PINCOCK, R. E., and KIOVSKY, T. E. 1966. Kinetics of reactions in frozen solutions. J. Chem. Educ. *43*, 358-362.

PROCTOR, B. E., and GOLDBLITH, S. A. 1948. Radar energy for rapid food cooking and blanching, and its effect on vitamin content. Food Technol. *2*, 95-104.

RAMSBOTTOM, J. M., and KOONZ, C. H. 1939. Freezing temperature as related to drip of frozen-defrosted beef. Food Res. *4*, 425-431.

RAMSBOTTOM, J. M., and KOONZ, C. H. 1940. Relationship between time of freezing beef after slaughter and amount of drip. Food Res. 5, 423-429.
RAMSBOTTOM, J. M., and KOONZ, C. H. 1941. Freezer storage temperature as related to drip and to color in frozen-defrosted beef. Food Res. 6, 571-580.
RETZER, J. L., VAN DUYNE, F. O., CHASE, J. T., and SIMPSON, J. I. 1945. Effect of steam and hot-water blanching on ascorbic acid content of snap beans and cauliflower. Food Res. 10, 518-524.
RICHARDSON, L. R., WILKES, S., and RITCHEY, S. J. 1961A. Comparative vitamin B_6 activity of frozen, irradiated and heat-processed foods. J. Nutr. 73, 363-368.
RICHARDSON, L. R., WILKES, S., and RITCHEY, S. J. 1961B. Comparative vitamin K activity of frozen, irradiated and heat-processed foods. J. Nutr. 73, 369-373.
SAIR, L., and COOK, W. H. 1938. Relation of pH to drip formation in meat. Can. J. Res. 16D, 255-267.
SAMUELS, C. E., and WIEGAND, E. H. 1948. Radiofrequency blanching of cut corn and freestone peaches. Fruit Prod. J. and Am. Food Mfgr. 28, 43-44, 61.
SCOTT, L. E., and SCHRADER, A. L. 1947. Ascorbic acid content of strawberry varieties before and after processing by freezing. Proc. Am. Soc. Hort. Sci. 50, 251-253.
SECOMSKA, B., IWANSKA, W., and NADOLNA, I. 1973. Retention of some vitamins in cooked potatoes stored in the frozen state. 13th Intern. Congr. Refrig. 3, 311-316.
SINGH, S. P., and ESSARY, E. O. 1971. Vitamin content of broiler meat as affected by age, sex, thawing and cooking. Poultry Sci. 50, 1150-1155.
STIMSON, C. R., and TRESSLER, D. K. 1939. Carotene (vitamin A) content of fresh and frosted peas. Food Res. 4, 475-483.
STRACHAN, C. C., and MOYLS, A. W. 1949. Ascorbic, citric and dihydroxymaleic acids as antioxidants in frozen pack fruits. Food Technol. 3, 327-332.
SULC, S. 1973. The influence of different freezing methods, temperature, and time, on the preservation of important factors. 13th Intern. Congr. Refrig. 3, 447-454.
SWEENEY, J. P., CHAPMAN, V. J., MARTIN, M. E., and DAWSON, E. H. 1961. Quality of frozen vegetables purchased in selected retail markets. Food Technol. 15, 341-345.
SWEENEY, J. P., and MARSH, A. C. 1971. Effect of processing on provitamin A in vegetables. J. Am. Dietet. Assoc. 59, 238-243.
TANAKA, K., and TANAKA, T. 1956A. Defrosting of frozen whale meat. J. Tokyo Coll. Fisheries 42, 80-82.
TANAKA, K., and TANAKA, T. 1956B. Biochemical condition of whalemeat before or after freezing and cold storage of frozen meat. J. Tokyo Coll. Fisheries 42, 83-88.
TANAKA, K., and TANAKA, T. 1957. Drip from frozen whalemeat affected by freezing rate and air temperature in air defrosting. J. Tokyo Coll. Fisheries 43, 19-23.
TEPLY, L. J., and DERSE, P. H. 1958. Nutrients in cooked frozen vegetables. J. Am. Dietet. Assoc. 34, 836-840.
THOMPSON, L. U., and FENNEMA, O. 1971. Effect of freezing on oxidation of L-ascorbic acid. J. Agr. Food Chem. 19, 121-124.
TINGLEFF, A. J., and MILLER, E. V. 1960. Studies on ascorbic acid retention in frozen juice, segments, and whole oranges. Food Res. 25, 145-147.
TINKLIN, G. L., and FILINGER, G. A. 1956. Effects of different methods of blanching on the quality of home frozen spinach. Food Technol. 10, 198-201.
TRESSLER, D. K., MACK, G. L., and JENKINS, R. R. 1937. Vitamin C in vegetables. VII. Lima beans. Food Res. 2, 175-181.

TRESSLER, D. K., MACK, G. L., and KING, C. G. 1936. Factors influencing the vitamin C content of vegetables. Am. J. Public Health 26, 905-909.
VAIL, G. E., JEFFREY, M., FORNEY, H., and WILEY, C. 1943. Effect of method of thawing upon losses, shear, and press fluid of frozen beefsteaks and pork roasts. Food Res. 8, 337-342.
VAN DER MEER, M. A., LASSCHE, M. J. B., and PASHCHA, C. N. 1973. Nutritive value and sensory qualities of deep-frozen components of meals and the effect of three reheating methods on these properties. Voeding 34, 2-15. (Dutch)
VAN DUYNE, F. O., WOLFE, J. C., and OWEN, R. F. 1950. Retention of riboflavin in vegetables preserved by freezing. Food Res. 15, 53-61.
VOLZ, F. E., GORTNER, W. A., and DELWICHE, C. V. 1949. The effect of desiccation on frozen vegetables. Food Technol. 3, 307-313.
VON KAMIENSKI, E. S. 1969. Retention of vitamin C during the processing of frozen spinach. In Frozen Foods, International Institute of Refrigeration, Paris, France.
VON KAMIENSKI, E. S. 1972. Retention of vitamin C during the processing of frozen spinach. Sci. dell'Alimentazione 18, No. 7, 243-244.
WATT, B. K., and MERRILL, A. L. 1963. Composition of Foods. USDA Agr. Handbook 8.
WEDLER, A. 1971. Results of experiments on the change in contents of nutritional compounds through processing and preparation of vegetables. Qualitas Plant. Mater. Vegatibiles 21, No. 1-2, 79-95.
WEITS, J., VAN DER MEER, M. A., and LASSCHE, J. B. 1970. Nutritive value and organoleptic properties of three vegetables fresh and preserved in six different ways. Intern. Z. Vitaminforsch. 40, 648-658.
WESTERMAN, B. D., OLIVER, B., and MACKINTOSH, D. L. 1955. Influence of chilling rate and frozen storage on B-complex vitamin content of pork. J. Agr. Food Chem. 3, 603-605.
WESTERMAN, B. D., VAIL, G. E., TINKLIN, G. L., and SMITH, J. 1949. B-complex vitamins in meat. II. The influence of different methods of thawing frozen steaks upon their palatability and vitamin content. Food Technol. 3, 184-187.
WESTERMAN, B. D. et al. 1952. B-complex vitamins in meat. III. Influence of storage temperature and time on the vitamins in pork muscle. J. Am. Dietet. Assoc. 28, 49-52.
WOLFE, J. C., OWEN, R. F., CHARLES, V. R., and VAN DUYNE, F. O. 1949. Effect of freezing and freezer storage on the ascorbic acid content of muskmelon, grapefruit sections, and strawberry purée. Food Res. 14, 243-252.
ZIMMERMAN, W. I., TRESSLER, D. K., and MAYNARD, L. A. 1940. Determination of carotene in fresh and frozen vegetables. 1. Carotene content of green snap beans and sweet corn. Food Res. 5, 93-101.
ZSCHEILE, F. P., BEADLE, B. W., and KRAYBILL, H. R. 1943. Carotene content of fresh and frozen green vegetables. Food Res. 8, 299-313.

CHAPTER 11

Peter M. Bluestein
and
Theodore P. Labuza

Effects of Moisture Removal on Nutrients

Dehydrated foods and concentrated foods, both as ingredients for further processing and as consumer products, are major industrial products. Milk, eggs, fruits and fruit juices, vegetables, meat, and other items of nutritional importance can be found in the dehydrated form. Fruit juices and milk are the major products of nutritional significance which can be found in the concentrated form. Products produced by both of these processing operations have a wide variety of preliminary processes such as washing, peeling, blanching, and cooking which can affect the nutritional value and these are considered elsewhere in this book and have been reviewed by Bender (1966). Dehydrated foods, if stored under the proper conditions, will not spoil from microbial attack. Concentrated products are usually not stable to microbial attack and, therefore, concentration is often used with a further preservative process. The preservative processes used with a concentrated product, with the exception of drying, are not covered in this chapter.

Evaporation and most drying processes involve the addition of heat to the food and the removal of moisture as water vapor. In many cases the temperature of processing is above room temperature but below the temperatures used for sterilization. As will be shown later, there are a wide variety of processes available for producing dried or concentrated products. Each process has its own advantages when compared to other methods of production. For any single process there is a range of processing conditions which will affect the nutrient retention of the processed product. It would simplify this discussion greatly if a simple rule of thumb could be devised for all drying operations to predict the best processing conditions for drying and concentration. This is not possible because of the complexity of the changes in foods which occur during these processes. Some of these changes will be considered here. However, for more detailed discussions, reference works on drying and concentration should be consulted (Van Arsdel 1963; Van Arsdel and Copley 1964).

CHEMICAL KINETICS AND MOISTURE REMOVAL

Chemical Kinetics

It is worthwhile to review some aspects of chemical kinetics with emphasis on the phenomena which occur during processing and storage of dried and concentrated foods. Since the loss of nutritive value is usually the destruction of a single chemical compound, this loss can be described by a simple monomolecular reaction as in Equation 1:

$$A \xrightarrow{k} B \qquad (1)$$

where compound A reacts to form compound B with a reaction rate constant, k. If Equation 1 applies to the loss of nutrient A, then the rate of that loss can be described by Equation 2:

$$-\frac{d[A]}{dt} = k[A] \qquad (2)$$

where $d[A]/dt$ is the rate of loss of nutrient A, k is the reaction rate constant and $[A]$ is the concentration of A. The reaction rate constant is related to temperature by Equation 3:

$$k = k_o e^{[E_a/RT]} \qquad (3)$$

where k_o is a constant, E_a is the activation energy of the reaction, R is the gas constant and equal to 1.986 cal per gm-mole°K, and T is the absolute temperature, °K.

If the reaction rate constant is a true constant and the only change in the concentration of A is due to chemical reaction, the extent of the loss of A can be found by integrating Equation 2 and substituting the appropriate boundary conditions to obtain Equation 4:

$$-\ln \frac{[A]}{[A]_o} = kt \qquad (4)$$

where $[A]_o = [A]$ initially. Many food reactions can be described by Equation 4 when the limitations above are observed. This may be the case during storage when the temperature and moisture content of the product are held constant. However, during drying and most concentration processes neither the temperature or moisture content are held constant. To include the effects of variation in moisture content and temperature rigorously would complicate the discussion in this section and limit the usefulness of the equations derived.

Temperature

The temperature of a food during drying or concentrating varies over a wide range during the process and with different processing techniques. Temperatures usually found can range from $-20°F$ to above $212°F$ depending on the process and product. This range is much smaller when considering the processing alternatives for a single product. In considering a product dried or concentrated by one process, it is clear that higher temperatures experienced by the food result in increased rates of chemical reactions. This effect is a result of the change in the reaction rate constant with a change in temperature. Since nonenzymatic chemical reactions have activation energies between 10 to 40 Kcal per gm-mole, the reaction rate constant can be expected to increase from 2-fold to 15-fold for a $10°C$ increase (Labuza 1972). The activation energy for moisture removal is approximately 10 Kcal per gm-mole (King 1970). Because of this difference in activation energies for chemical deterioration and moisture removal, low temperature processing should produce products with the least amount of chemical deterioration. However, low temperature processing is usually more expensive because of the longer processing time. In addition, there is also the possibility of microbial growth during processing, at least between $4°$ to $40°C$. Therefore, methods which reduce the processing time without going much above the upper temperature for microbial growth in the food will result in the maximum nutrient retention. These methods would include better air flow patterns and increased surface to volume ratio.

Water

Water is distributed throughout dried and concentrated foods in many forms. Water may be found as a liquid containing solutes when the food is "wet" and associated with other constituents. The thermodynamic parameter which describes the state of water is the water activity, which, as a working definition, can be defined as the relative humidity in equilibrium with a food divided by 100. Water activity is related to the moisture content of a product by the moisture sorption isotherm, as shown in Fig. 11.1. As the water activity of a product decreases, as it does during drying and concentration, the state of water changes. Over most of the range of the water activity scale, water behaves as a solvent even though some literature classifies it as bound. In concentrated products, this aqueous solution is a liquid but in dried and "semimoist" foods this solution may be found in capillaries or held by swollen protein or polysaccharide gels. As the water activity decreases, the predominating form of water shifts to water hydrating hydrophilic constituents. However, water in this region still acts as an aqueous phase down to the BET (Brunauer-Emmett-Tetler) monolayer (Brunauer *et al.*

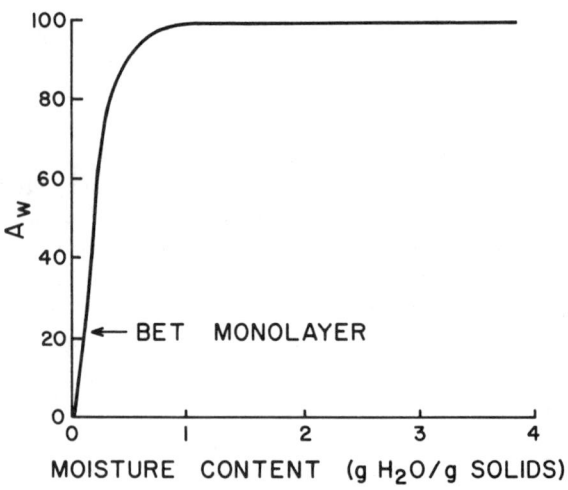

FIG. 11.1. A TYPICAL SORPTION ISOTHERM

1938). Above this point water acts to dissolve solutes, mobilizes them, and allows them to react within the aqueous environment. The BET monolayer thus describes the moisture content below which most reactions cease (Labuza 1970). Below this point (see Fig. 11.1) water is thought to be nonliquid and to be tightly bound or absorbed to specific sites on various constituents in foods. Additional discussions of the states of water can be found in the literature on water activity and sorption (Bone 1969; Karel 1973; Labuza 1968; Rockland 1969).

The state of water has a significant effect on the loss of nutrients. As noted, above the monolayer value, the water acts as a solvent for reactants and catalysts. Water may also be a product of some reversible reactions and, therefore, could slow the forward rate of reaction (Eichner and Karel 1972). Some of the reactants, such as water-soluble vitamins are present in low concentrations. As the moisture content and water activity decrease from the natural value, as would happen in drying, several important effects occur. The aqueous solutions become more concentrated. Some components of the food may form supersaturated solutions and ultimately precipitate. According to Equation 2, as the concentration in solution increases, the reaction rate increases. This leads to increased rates of losses of the nutrient as the moisture content or water activity is lowered. Some metal catalysts may lose part of the water of hydration. This can increase their catalytic effect toward unsaturated lipids. Diffusion of components in the aqueous phase may become more difficult since the observed viscosity increases. This would also decrease reac-

EFFECTS OF MOISTURE REMOVAL ON NUTRIENTS

tion rates. Finally, when all liquid water has been removed, reactions which proceed in water cease. This usually occurs at the BET monolayer.

It is important to distinguish between reaction of water-soluble nutrients and reactions of oil-soluble nutrients when discussing the effects of water on chemical reactions. Some of the water-soluble vitamins are highly soluble. It is unlikely that these highly soluble vitamins become supersaturated until the moisture content is greatly reduced. For example, ascorbic acid which is easily oxidized to dehydroascorbic acid is present at approximately 0.05 wt% in orange juice. Since the solubility of ascorbic acid is approximately 25 wt%, ascorbic acid will not become supersaturated until the moisture content has decreased from the natural >50% db to 1.3% db (data from Watt and Merrill 1963). Until this low moisture content is reached ascorbic acid becomes more concentrated as water is removed. This increase in concentration causes an increased reaction rate as shown in Equation 1. However, although the reaction rate is greater, the percentage lost during processing is independent of the concentration as shown by Equation 4. This conclusion holds when the nutrient degrades by a first order chemical reaction. Nonenzymatic unimolecular reactions are likely to be first order. Bimolecular reactions are not likely to be first order. The Maillard reaction between reducing sugars and amines which make lysine unavailable and the degradation of unsaturated lipids are not first order kinetics and the loss of these nutrients during processing should be dependent on the concentration. Unfortunately, kinetic data are not available to predict the effect of concentration on the loss of most water-soluble nutrients.

Some "water-soluble" nutrients such as riboflavin are not very soluble in water. These compounds would form saturated and supersaturated solutions during drying and concentration. Should these nutrients actually precipitate, the losses would be reduced.

The concentration of oil-soluble nutrients such as the essential fatty acids and vitamins A, D, E, and K, is extremely low in the aqueous phase of foods. Since a large part of these nutrients is found in the dispersed phase, their concentration does not change as water is removed. Other phenomena which occur during water removal are more important. Water is the solvent for heavy metals which catalyze the free radical oxidation of some unsaturated nutrients. As the moisture is decreased, catalyst mobility is decreased. At very low moisture contents in dried food, diffusion of catalysts decreases (Duckworth 1962); however, they are no longer hydrated and their effectiveness may increase (Labuza 1971). Finally, water may act as a free radical quencher to reduce the rate of reaction (Labuza 1971). These effects are not the result of direct water-nutrient interaction and remain complicated in terms of reaction rate predictions.

The overall effects of water on chemical and enzymatic reactions are shown in Fig. 11.2. The reaction rate for destruction of water-

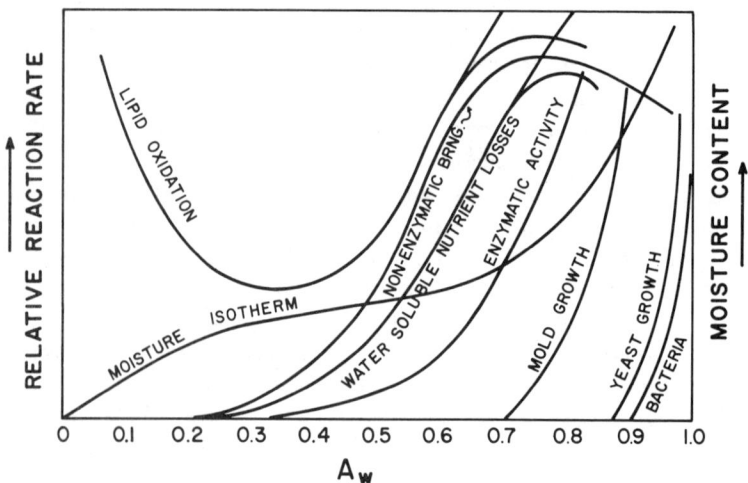

FIG. 11.2. GENERAL SCHEME OF REACTION RATES AS A FUNCTION OF WATER ACTIVITY

soluble nutrients may go through a maximum at a water activity of approximately 0.7. The maximum is due to dilution of the reactant concentration and possible product (water) inhibition. Below a water activity of 0.7, the reaction rate decreases because solutes which are reactants either precipitate out or the viscosity of the aqueous phase becomes high enough to slow diffusion. As seen, these reactions all cease at the BET monolayer which is the point where an aqueous phase ceases to exist.

For the lipid-soluble nutrients there is a minimum in the reaction rate at a_w = 0.3–0.4. This minimum is caused by the balance between catalyst hydration, mobility of catalysts, hydration of intermediates in the sequence, and free radical quenching. Considering the general picture of reactions during concentration and dehydration, the conditions to be avoided are high temperature at intermediate moisture contents.

This brief theoretical discussion of some of the factors which contribute to nutrient retention during drying and concentration provides some insight into the mechanisms of changes. These mechanisms are not completely understood at this time and the relative importance of one mechanism versus another cannot be predicted. Further research is needed to establish the important mechanisms which lead to significant loss of nutrients. Once the mechanisms are understood, process development work can proceed on a rational basis.

CONCENTRATION

The production of concentrated fruit juices, purées, jams, soups, condensed milk, and dried dairy product ingredients are among the products which are concentrated as part of production. Evaporation of water is by far the most commonly-used method of reducing moisture content. Membrane processes and freeze concentration are relatively new processes which are likely to find increasing use in the future. The obvious advantage of concentrated products is the reduction of weight and volume by processing. In addition, if the product is to be dried after concentration, the total cost of processing will be substantially lower when part of the moisture has been removed by concentration before drying. Recent work has shown that concentration before drying can lead to better volatile flavor retention during the drying operation (Thijssen 1971). Few concentrated products are produced without some additional processing.

Evaporation

Evaporation is by far the most common method of concentrating liquid food products. This process can be viewed as the simple boiling off of water at temperatures which vary depending on the product and process. Because water requires approximately 1000 BTU per lb for vaporization from the liquid, heat must be supplied to the liquid during the evaporation. The equipment used to transfer heat into the food is varied and has undergone considerable technical development with the objective of producing products with minimal amounts of change due to processing. Single evaporators are rarely used in industry for economic reasons. The important aspects of evaporators and processing schemes will be considered here with the purpose of understanding the cause of the nutritional changes occurring during evaporation. Additional processing technology can be found in several other texts (Armerding 1966; Brennan et al. 1969; Charm 1971; Perry et al. 1964).

It is useful to consider a simple batch evaporation in a steam jacketed kettle to illustrate the principles of evaporation. The steam-jacketed kettle is a simple piece of equipment which finds uses in the production of candy, jams and jelly, and some condensed milk products but is not widely used for other products. Once the original liquid has been filled into the kettle, the steam is fed into the jacket surrounding the bottom. The steam transfers heat into the liquid and boiling begins. If the kettle is open to the atmosphere then the boiling temperature at the beginning of evaporation is near 212°F. After boiling begins heat is continuously fed to the kettle to supply the heat required to evaporate water. As the water boils off and the solution becomes more concentrated several important changes occur which effect the ease of evaporation and, therefore, the nutrient retention. First, reactive solutes are more concentrated and the rate of chemical deterioration can increase. Secondly, the boiling point of the liquid rises slightly as predicted by Raoult's Law. The boiling point of a 60 wt% solution of sucrose is 3°C above the boiling point of water at atmospheric pressure. Third, the viscosity of the solution can increase dramatically. The viscosity of a 50 wt% solution of grape juice is 40 cp, 40 times more viscous than water (Saravacos 1970). The viscosity is important because it affects the ease of heat transfer to the boiling liquid. As the viscosity increases, it is more difficult to heat the liquid. This difficulty results in a nonuniform temperature distribution in the liquid food. Hot spots and burning on the wall of the kettle can result. This has a large effect on the nutrient retention resulting in large decreases in quality.

The steam-jacketed kettle can be used to perform an evaporation under vacuum. When a liquid is boiled under a partial vacuum, the boiling point is lowered considerably. Water boils at 150°F when the vacuum is 22.3 in. Hg (3.7 psia). This has the effect of lowering the temperature of processing but does not effect the rate of evaporation greatly. Lower temperatures in evaporation result in higher nutrient retentions. Other reasons for performing the evaporation under a partial vacuum will be considered later.

The steam-jacketed kettle is the simplest piece of equipment to be used to perform an evaporation. Other types of evaporators are considered by Brennan et al. (1969) and Perry et al. (1964). For the purpose of this discussion, it is useful to classify all evaporators into two classes: evaporators with a considerable amount of liquid hold-up and evaporators with minimal hold-up. The hold-up of an evaporator refers to the amount of liquid in the piece of equipment. If the hold-up of an evaporator is large, relative to the net liquid flow rate through the evaporator, the boiling liquid is in contact with the hot heat transfer surface of the equipment for a long period of time. A schematic diagram of an evaporator with a large liquid hold-up is shown in Fig. 11.3. In this example, the feed liquid enters the evaporator and mixes with the concentrated liquid. The short tubes with

EFFECTS OF MOISTURE REMOVAL ON NUTRIENTS 297

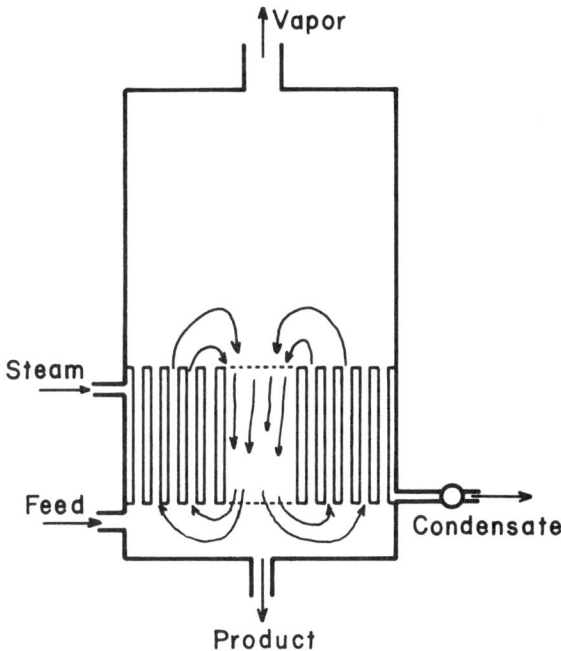

FIG. 11.3. RECIRCULATING EVAPORATOR

steam on the outside heat the liquid and force it upward. Natural convection currents cause a circulatory flow of liquid as shown by the arrows in Fig. 11.3. The amount of liquid in the evaporator is large because of the low heat transfer rates. This results in a long residence time for the liquid at the elevated temperatures of evaporation. Evaporators with long residence times are the short tube evaporators, the WURLING evaporator (Carlson et al. 1967), and evaporators with recirculating fluid. The ideal type of evaporator for maximum nutrient retention has a minimum retention time of the liquid in the equipment. Such evaporators, like the plate, rising film and falling film evaporators, pass the liquid through the equipment only once with little or no recirculation. The design of these evaporators ensures that heat transfer is efficient and the evaporation rate is high. High evaporation rates are required for single-pass processing. Evaporation with single-pass equipment can result in a reduction by a factor of ten in the time of processing at elevated temperatures (Moore and Hesler 1963).

Evaporations are usually not performed in a single evaporator. The economics of steam usage requires that the steam produced in one evaporator be used in another evaporator to boil off more water from the product of the first evaporator. The second evaporator is operated at a lower pressure which results in boiling at a lower temperature. This usage of the steam from one evaporator to heat the liquid in another evaporator is called "multieffect evaporation." Chemical processing generally uses large numbers of effects. However, the food industry, because of its concern for the quality of the final concentrate, uses no more than 3 or 4 effects. Wiegrand (1971) reports on a triple effect evaporation of milk with boiling temperatures of 70°, 60° and 42°C in the 3 effects. A larger number of effects in this case would require a longer residence time in the equipment and a higher temperature of evaporation.

Evaporation provides a rapid, cheap method of producing concentrated products. Reverse osmosis and freeze concentration are more expensive (Bomben *et al.* 1973). The temperature is high but the time of processing can be very short. The short time of processing can result in products with close to 100% nutrient retention.

Freeze Concentration

Although useful reports on the nutrient retention of freeze-concentrated products have not appeared in the literature, it is appropriate to consider this process here as a competitor of evaporation. The process involves the freezing of liquids with carefully controlled conditions to produce large ice crystals and the separation of the ice from the remaining concentrate. Thijssen (1970) has reviewed the equipment available. This process is performed at low temperatures below the freezing point of feed liquid. As a low temperature process, it is expected that the nutrient retention of freeze-concentrated products would be close to 100%. The only loss would be in any solute loss that remained with the ice or fluid adhering to the ice.

Membrane Processes

The membrane processes of reverse osmosis and ultrafiltration are finding increasing uses for final product production and ingredient manufacture. Reverse osmosis is a concentration process with the objective of removing only the water (Leightell 1972). Ultrafiltration is a concentration and purification process. Applications of ultrafiltration have been reviewed by Porter and Michaels (1970). Both processes pass the liquid to be concentrated (and purified) through equipment holding a membrane. The membrane allows the selective passage of water and perhaps other compounds. In reverse osmosis, compounds which are soluble in the membrane can be lost. Most nutrients would not be soluble in a reverse osmosis membrane. In most applications of ultrafiltration, all low molecular weight material is allowed to pass through the membrane.

DEHYDRATION

A large number of techniques are available for the production of dehydrated foods. Only the concepts of drying and a limited number of the processes used can be discussed here. Further discussions are available in the literature (Van Arsdel 1963: Van Arsdel and Copley 1964; Charm 1971; Holdsworth 1971; Masters 1972; Williams-Gardner 1971).

Before considering the individual properties of different drying procedures, a short discussion of the concepts of dehydration is required. In most types of drying, heat is supplied to the food and moisture in the vapor state is removed. The methods of supplying heat and transporting the moisture and the product are the basic variations between the different techniques of drying. As heat is supplied to a food, either the temperature of the food is raised or water is evaporated. During the initial stages of most drying operations the rates of heat transfer to the food and moisture transfer from the food are balanced and the temperature of the drying food remains at the wet-bulb temperature of the air. The wet-bulb temperature of the air does not vary over a wide range in relation to the normally measured or dry-bulb temperatures for fairly dry air as shown in Fig. 11.4. Air at 80°C and 0.01 lb H_2O per lb dry air is only at 3% RH. As seen in Fig. 11.4, the temperature of the food

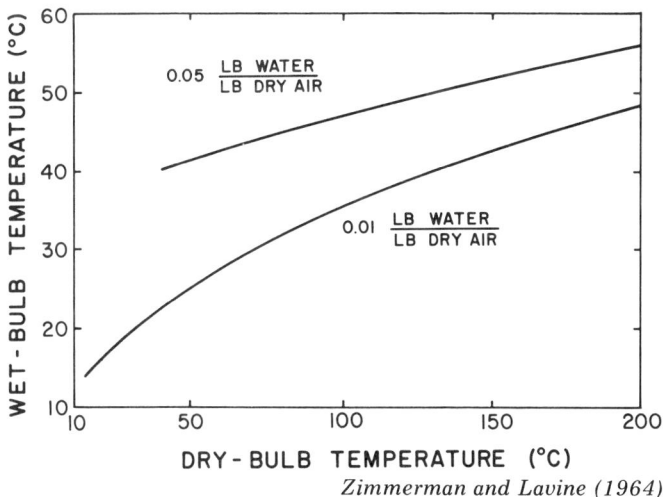

Zimmerman and Lavine (1964)
FIG. 11.4. WET-BULB AND DRY-BULB PROPERTIES OF AIR

during the first period of drying is relatively independent of air dry-bulb temperature and is much lower than the air temperature. During this first period, which is usually called the constant rate period, the moisture content of the piece of food decreases uniformly to a level which is still above the maximum in the chemical reaction rate curve (Fig. 11.2). The absolute moisture is still high enough that it does not affect the rate. The rate of drying remains at this constant and high level until the surface of the food begins to lose water at a high water activity. It would be desirable to increase the length of time a food dries in the constant rate period if nutrient retention is critically important. During this period the temperature is low. However, the constant rate period is also the time when the greatest amount of shrinkage and other undesirable changes occur and for these quality reasons this period should be short for most products.

Once the constant rate period is over, the drying rate decreases. This occurs because the surface of the piece is not completely saturated with water. The interface between water and air recedes into the piece and moisture content and water activity vary with the location in the piece. The temperature of the piece rises from the wet-bulb temperature and ultimately approaches the dry-bulb temperature of the air. It is during this time that the conditions of moderate moisture content and high temperature are present and these conditions promote chemical reactions. Actual drier designs recognize this and the effects of chemical reactions on product organoleptic quality. Most driers which are designed to produce high quality products should also produce products with reasonably high nutrient retention.

The conditions during all drying operations are changing. The time, temperature, and moisture content of the food during drying have the greatest effects on the rate of reactions. The presence of dissolved oxygen and sulfur compounds which can have an effect on nutrient retention is controlled by the operations preceding drying. It would be useful to have a sufficiently accurate description (model) of the phenomena which occur during drying and the influence of the dynamic conditions on the rates of nutrient degradation. Unfortunately, these descriptions do not exist at this time and further research is needed in this area to be able to predict and explain the effects of processing variables on the nutritional quality of dehydrated foods (Labuza 1972).

Sun Drying

Sun drying is still of importance throughout the world. Fruits, fish, meat, and grain are spread out in the sun. The radiant energy of the sun provides the heat to evaporate the water. Drying proceeds well in warm and dry weather. At night and during the rainy seasons drying will not take place. The temperatures of the food during sun drying are usually 5°-15°C above ambient temperature. The time of drying is 3-4 days and longer depending on the product and conditions.

Tunnel Drying

Tunnel driers of various types are an extremely important class of driers. Fruits and vegetables can be dried by this method. There are several types of tunnel driers and these will be discussed later. All types follow the same basic operations. The food is spread onto trays or a conveyor and passed into a high velocity air stream. The basic discussion of drying presented earlier can be applied to tunnel drying. The rate of drying is related to the air velocity, the loading of the product, the wet-bulb and dry-bulb temperatures of the air, and the thickness and other properties of the food. During the constant rate period, the properties of the air are most important in determining the rate of drying. The properties of the food materials are more important during the other periods of drying. Tunnel dryers are usually designed to take advantage of these characteristics.

Tunnel driers are classified based on the food moving mechanisms and the direction of air flow. Trays or conveyors are used to move the food through the tunnel. The directions of air flow are either parallel to the food movement, countercurrent to the food movement, and through or across the food bed. If air is introduced parallel to the food movement, the initial conditions for drying are optimal but as the air picks up moisture from the drying food, the ability of the air for further drying is decreased. If counterflow drying is used exclusively, the conditions for drying are optimal near the product end of the drier but not at the feed end. Usually, these 2 methods of air flow are combined into a 2-stage drier. Parallel flow is used in the first stage and countercurrent flow in the second stage of the drier. This provides the best compromise for efficient drying and good product quality. The temperatures of the air used depend on the product and are in the range of 70° to 90°C in the first stage and 55° to 70°C in the second stage which results in a drying time of 8-16 hr. In many cases, the product is removed from the tunnel before drying is completed, and the final moisture content is lowered in bins operated at 40°-55°C for 7 hr or more (Van Arsdel and Copley 1964).

Flow of air through a bed of food provides better contact between the heated air and the product. Higher drying rates are obtained and the times of drying are decreased to a few hours. The temperatures of drying are in the range of two-stage drying. Fluid-bed drying is conceptually similar to flow-through driers. In the fluid bed, high velocity air is forced upward through the food and suspends the particles in warm air. Drying is usually faster. Fluid-bed drying, foam-mat drying, and drying of puffed foods have been reviewed by Holdsworth (1971).

Spray Drying

Spray drying is an important process for milk and dairy products, coffee, eggs, and juices. In spray drying, a liquid feed is sprayed into a stream of hot air. The small size of the drops, which average approximately 100 μ in diameter, results in a very large surface which dries quickly. Although the air dry-bulb temperature is approximately 200°C, the air wet-bulb temperature rarely exceeds 55°C and the time in the drier is very short. Drying takes place over the first few seconds and the dried particles are removed from the drier usually within 30 sec. The temperatures of the particles during drying can range from the wet-bulb temperature of the inlet air to above 100°C as they exit in the dry state. Particles are usually at a temperature of 60°–80°C when removed from the drier. Although these temperatures are high in comparison to tunnel drying operations, the moisture near the end is usually near the BET monolayer and the time of exposure is very short. Thus, deterioration during processing is minimal. It is extremely difficult to predict the time, temperature, and moisture content history of a particle and to use this to predict the nutrient retention during spray drying.

Masters (1972) has discussed many modifications of spray driers which are likely to result in better nutrient retentions. Cooling the walls of the drier and injecting dehumidified cool air into the drier to lower the temperature of the product offers advantages for certain products. Multiple zones of air temperatures will allow the engineer to obtain optimum drying and a product of maximum nutrient retention. The combination of spray drying with other methods of drying, such as flow-through bed drying, offers new opportunities for the future (Meade 1973).

Drum Drying

Viscous slurries such as mashed potatoes and sweet potatoes are dried by drum driers. The slurry is fed into the trough between two steel drums heated from the inside with steam. The drums are rotating and spread a thin film of the slurry over the surface. As the drums turn, the product dries. Drum temperatures are in the range of 120°–170°C and drying time is 20 sec to 3 min. Although drum drying is one of the cheapest methods of drying (Greig 1971), the product is in contact with the hot drums and ultimately leaves the drums at a high temperature. This should result in deterioration which is greater than that resulting from spray drying or tunnel drying. Usually, drum-dried products cannot be dried by spray driers and tunnel driers.

Freeze Drying

Freeze drying should result in the highest nutritional quality of all the drying processes (Calloway 1962). The food which has been previously frozen is placed into a chamber which is evacuated. When the pressure has been reduced to 1 torr, heat is supplied to the frozen food. The ice does not melt because of the low pressure and the rapid removal of heat by the water vapor escaping from the food. The heating plate temperature varies during processing from above $100°C$ to about $55°C$. The heating plate temperature is highest when the food is at the lowest temperature. As drying proceeds, the temperatures of the food and the plate approach each other at an intermediate temperature. The warmest parts of the food, the outside surfaces, have already achieved a low moisture content. Therefore, the problem of nutritional deterioration is minimized. The time, temperature, and moisture content relationships during freeze drying have received considerable attention and are reviewed by King (1971). Because freeze drying results in minimal damage, these relationships are of little interest here.

Other Drying Processes

New drying processes and important modifications of current techniques are continually being developed. One interesting new technique is the use of a solvent to extract all or part of the moisture in foods. The use of a concentrated sugar solution to extract water from fruits has been called osmotic dehydration. Peaches produced by the method are soaked in $75°$ Brix for 4–6 hr to remove part of the water and then dried further (Farkas and Lazar 1969; Anon. 1973). Fish protein concentrate is produced by the extraction of water with isopropyl alcohol (Finch 1970). Hieu and Schwartzberg (1973) have proposed the use of various water-soluble solvents for the dehydration of shrimp. Most of these processes do not decrease the moisture content to a suitable level for storage and require further processing to remove the solvent and the remaining water.

Many other drying processes are used in the food industry. Many of these are reviewed by Williams-Gardner (1971) and others. Foam spray drying (Brennan and Priestly 1972), foam mat drying (Hertzendorf and Moshy 1970), pneumatic drying, and other processes could be considered but are not used to any degree by the food industry because of cost and their special nature. Many new drying processes are likely to be developed in the future. If nutrient retention data is lacking, as it usually is, the only possible statement is that nutrient retention is likely to be higher in a high quality product than in a low quality product.

NUTRIENT LOSSES

Although there has been some new work since the publication of the first edition (Harris and von Loesecke 1960), little of it can be used to predict the effects of processing variables on nutrient retention. Work which has been performed has been concerned with the retention of nutrients in an established product and process and not with the effects of a wide range of conditions available for the process. It would be useful to examine the approach required to provide this information. This approach integrates the kinetics of deterioration with the time-temperature-moisture content history of the sample. This has been applied successfully to packaging studies by Mizrahi et al. (1970A) and Quast and Karel (1972). First, the reaction rate is determined as a function of temperature and moisture content using constant conditions. Some ingenuity will have to be used to overcome some of the problems of holding foods at high temperatures or high humidities for extended periods of time. Once those data have been collected, a suitable model of drying behavior which is sensitive to the variables of air wet- and dry-bulb temperature, air velocity, food properties, and drier characteristics can be integrated with the deterioration data. The drying model must be suitably refined to be able to account for the variation in moisture content and temperature throughout a single piece of food. The model required for the nutrient retention of concentrated products is simpler because the liquid is well mixed and the moisture and temperature gradients are small.

Hendel et al. (1955) and Kluge and Heiss (1967) applied data on the browning reaction obtained during storage to the drying of potato to predict the extent of browning after drying. The reaction rate as a function of moisture content and temperature is shown in Fig. 11.5. The reaction rate of browning shows the typical response to temperature and moisture content (Labuza 1970; Loncin et al. 1968; Mizrahi et al. 1970B). There is a peak in the reaction rate at a moisture content near $a_w = 0.7$ and the rate increases as temperature increases. The rate is zero order which means the rate of brown pig-

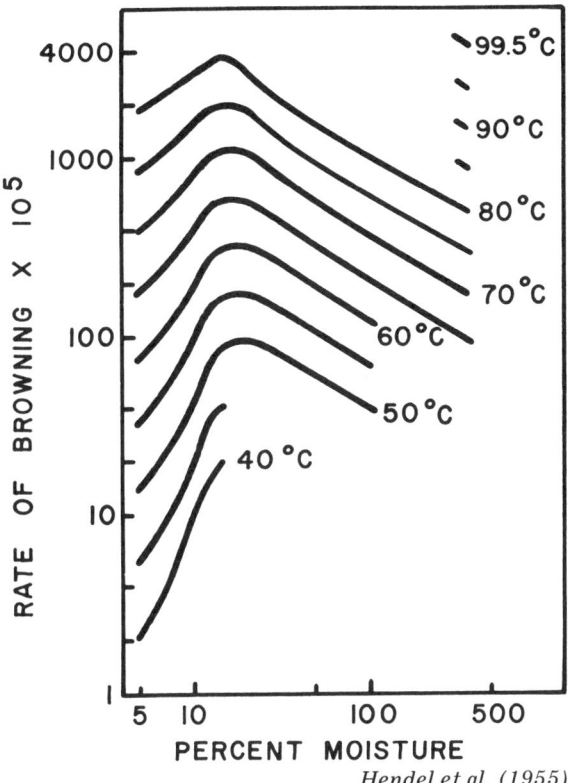

Hendel et al. (1955)

FIG. 11.5. BROWNING RATE AS A FUNCTION OF MOISTURE CONTENT AND TEMPERATURE

ment production is linear with time. This indicates that pigment may be formed without significant decreases in the concentrations of reducing sugar and lysine in the potato. Hendel et al. (1955) used the average moisture content and temperature measured at a point to represent the conditions in a piece of potato. Because the reaction rate does not vary linearly with respect to moisture content, and the center temperature of a piece underestimates the temperature elsewhere in the piece, the predicted values for the extent of browning deviate from the measured values. The moisture distribution and temperature gradients depend on the air, food, and drier properties. If these properties had been measured and a suitable model of moisture and temperature distributions had been used, the calculated results would have been in better agreement with the experimental results. This approach can be applied to predict nutrient retention. The results of this work would contribute to knowledge in the areas of drying behavior and chemical deteriorations during drying. Similar work on browning of applesauce was done by Escher and Neukom (1970, 1971).

As will be shown, the losses of nutritive values during most drying and concentration processes are small in relation to losses during cooking. Many factors preceding the drying or concentration operation have an influence on the nutrient retention measured. For example, the presence of sulfur dioxide is known to protect ascorbic acid but decreases thiamin retention. The presence of copper or iron affects ascorbic acid retention. Heavy metals are likely to catalyze the oxidation of carotenes. Each of these factors will be considered in the appropriate sections.

Proteins

Damage to proteins in the processing of foods has recently been reviewed by Bender (1972). He suggests that many diets are limited by the sulfur-containing amino acids and that reactions of lysine which is in surplus is not critical. However, the reaction of lysine, more particularly the ϵ-amino group, with reducing sugars is one of the most studied reactions in foods and is heat-sensitive. This provides a ready index for the extent of other deteriorations. Woodham (1973), in addition, has reviewed the area of vegetable protein concentrates and notes that destruction of naturally present antinutritional factors is just as important.

The reaction of lysine and reducing sugars is also important because it is involved in the reaction mechanism for one scheme of nonenzymatic browning. Browning reactions can be desirable as in the case of bread crust, syrups, and coffee, but are considered undesirable in products which are not expected to have a brown color such as dried milk products.

Work in the area of protein deterioration during drying can be divided into two broad areas: work on animal feeds and protein sources at temperatures not usually encountered in drying operations and work with foods under normal processing conditions. Not very many studies exist which relate the deterioration taking place to changes in processing conditions over a wide range of variables and which can be used as a model.

Studies with soy product drying must consider the inactivation of naturally occurring toxicants such as the trypsin inhibitor. It would be useful to design a process which accomplishes this destruction with minimal heat damage to other proteins. Drying after inactivation of the trypsin inhibitor is required because of the usual addition of water to the dry meal to increase heat transfer and effect the rate of inhibitor destruction. The data of Taira et al. (1966) for lysine loss indicate that the reaction rate constant assuming a first order reaction is 0.166 hr^{-1} with an activation energy of 30 Kcal per gm-mole in the wet state. The destruction of trypsin inhibitor occurs at a rate approximately 100 times faster with an activation energy of 18.5 Kcal per gm-mole (Hackler et al. 1965). Because lysine is more sensitive to the ultimate temperature reached (higher activation energy), trypsin inhibitor destruction should be performed at the lowest temperature possible. Hackler et al. (1965) present data to substantiate this and this is shown in Table 11.1. The PER

TABLE 11.1

EFFECT OF DRYING ON SOY MEAL[1]

	Trypsin Inhibitor Retained (%)	PER
Spray drying with air inlet at (°C)		
166	10	2.22
182	8	2.10
227	4	1.99
277	5	1.63
316	3	0.16
Drum drying		
Air drum, 150°C	5	2.19
Vacuum drum, 29 in. Hg, 108°C	10	2.22
Freeze drying, 1000 µHg	10	2.14

[1] From Hackler et al. (1965); initial value 14% inhibitor retention after cooking.

(Protein Efficiency Ratio) measured reflects the destruction of sulfur-containing amino acids which are limiting but it can be used as an index for lysine loss in dried soy milk. During spray-drying experiments, trypsin inhibitor destruction increases as the air inlet temperature increases. The PER shows only small changes until the temperature of 277°C and decreases dramatically as the temperature is increased. In this case, the air inlet temperature of 277°C is optimum when considering the balance between trypsin inhibitor destruction and PER decrease. It is difficult to explain the quantitative change in PER without further engineering data on the spray-drying technique used. It appears that when the air inlet temperature is increased to 277°C, the drops are drying very fast. These drying conditions lead to an increase in droplet temperature before the inner portions of the drop is dry. These are the conditions which favor reactions of the water-soluble nutrients.

The data of Hackler et al. (1965) can also be used to compare spray drying to drum drying and freeze drying. Drum drying in air with a drum temperature of 150°C results in a significant reduction in trypsin inhibitor without affecting the PER. Drying in this case occurs at a lower temperature during the intermediate water activity range than in the high temperature spray drying runs. The time of drying would be longer. A high temperature-short time (HTST) drying process favors the lowering of PER over trypsin inhibitor destruction. Drum drying in a vacuum of 29 Hg and freeze drying are low temperature processes which do not show marked changes in PER or trypsin inhibitor.

Livingston et al. (1971) dried alfalfa in a rotary drier at temperatures above the normal range for drying foods. These results are shown in Fig. 11.6 and are useful to establish the upper limits of loss

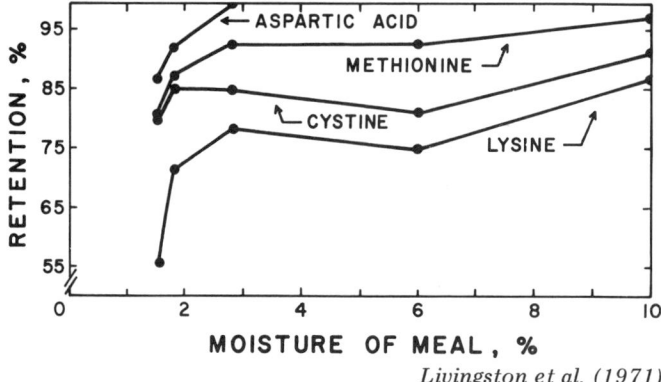

Livingston et al. (1971)

FIG. 11.6. RETENTION OF AMINO ACIDS DURING DRYING OF ALFALFA

in this product. Again, sufficient data are not available for a complete quantitative description of the losses of amino acids. However, a simplified analysis is possible. Air dry-bulb temperatures to the drier ranged from 650° to 950°C and outlet temperatures were 100° to 170°C. These conditions are typical of high efficiency drying without concern for product quality. Estimated air wet-bulb temperatures are in the range from 70° to 120°C. Drying begins at the wet-bulb temperature and the temperature of the pieces quickly increases. As the air temperature at the inlet is increased, the temperature of the piece throughout drying is increased, the outlet temperature is increased, and the final moisture content is decreased. Because the heat treatment is more severe as the final moisture content is lowered, the destruction of amino acids is increased as the final moisture content is lowered. The data are shown in Fig. 11.6. Myklestad et al. (1972) heated fish meal prepared from herring at different temperatures, times, and moisture contents. Table 11.2

TABLE 11.2

EFFECT OF HEAT AND MOISTURE ON FISH MEAL

Conditions of Treatment			Percentage of Freeze-dried Control			
			Available	Pepsin	NPU	
Temp (°C)	Moisture (%)	Time (Min)	Lysine (%)	Digestibility (%)	Rats (%)	Chicks (%)
96	7.7	30	94	88	—	98.6
	8.8	60	96	84	—	102.0
	10.8	120	87	76	—	98.1
	36.0	60	87	71	97.7	98.6
116	6.4	120	94	78.1	95.3	96.8
	7.5	60	100	78.2	97.0	98.8
	8.4	30	96.0	80.0	97.4	99.7
132	2.5	120	97	58.4	91.8	97.1

Source: Myklestad et al. (1972).

shows the available lysine loss, protein digestibility by an *in vitro* test, and two *in vivo* tests on the processed fish meal. As noted, there seems to be substantial decreases in available lysine and chemical digestibility with pepsin when moisture content of the meal during heating was high or heating time was long. However, because of the scattered conditions used, no real pattern can be established. What is more interesting is that the *in vivo* studies show only a 4-5% nutritional loss of the meal. This makes suspect many of the considerations presented in the literature based solely on chemical tests. It also indicates that even high temperature-high moisture contents during drying do not affect protein nutritional values significantly although available lysine decreases.

Aitken et al. (1967) also showed that based on rat NPU (Net Protein Utilization) studies, there was no significant difference between freeze-dried cod meal versus that which was rapidly dried at either 110°C, 115°C, or by slow drying with salt and pressing at 27°C.

The biological quality of the protein in 12 different foods both before and after drying was measured by de Groot (1963). The effects of the drying methods were small and in most cases insignificant. The products used were meats, egg, legumes, leafy vegetables, and sweet corn. The drying methods included hot air drying, vacuum drying, spray drying, and freeze drying. The vegetables were sulfited and the conditions are in the range of commercial practice. A slightly larger decrease in NPU was found for green beans dried at 71°C for 0.5 hr compared to 60°C for 1.0 hr or 49°C for 20 hr when compared to the undehydrated sample. However, when the experiment was repeated the results were not significant and the NPU difference between the experiments was 8%, approximately 5 times the standard error. This indicates the importance of good controls and use of the same lot of sample. For the above reasons, experiments which just report the chemical protein quality or vitamin content after a certain process do not always represent the true effect of processing.

Milk is particularly susceptible to the loss of lysine by reaction with lactose. This is especially true in whey processing (Rolls and Porter 1973). Considerable work has been done in this area because of the industrial importance of dried milk. MacDonald (1966) took samples of milk from commercial spray-drying and drum-drying installations. Spray drying resulted in losses of lysine of 0–4.1%. Drum drying losses of lysine from 5 drum drying installations ranged from 3 to 16%. The larger losses due to drum drying can be explained based on the time-temperature-moisture content relationships for the two drying methods. Drum drying creates a sheet of material in contact with the hot drum during the time when the milk is in the intermediate moisture content range and has the greatest reaction rate. Spray drying has much faster drying rates which indicate that the milk passes through this maximum in the reaction rate curve (Fig. 11.2 and 11.5) while still at a low temperature. What is also interesting about the work is the variability in available lysine from different production plants although this may be due to analytical technique. Rolls and Porter (1973) report that in efficient spray drying, available lysine only decreases by 3–10% whereas with roller-dried milk the loss was 5 to 40%. Methioninc loss in severely heated samples was about 10%. The biological value (animal tests) was higher for the spray-dried product.

If milk is to be used as the sole source of food, which would only be the case for a limited number of infants, the losses in the limiting sulfur amino acids are of greater concern than lysine loss since lysine is at high concentration. Bickel and Mauron (1959) showed that severely roller-dried milk was satisfactory for infant feeding. However, when the dry milk is incorporated into a diet which makes lysine limiting, as was shown by Van den Beuel et al. (1972), the NPU for rats was decreased. They fed them gluten with diets in

EFFECTS OF MOISTURE REMOVAL ON NUTRIENTS 311

which the available lysine of the milk was 64 and 86%. This shows the problem with man versus animal studies.

Mauron (1960) reported data on the availability of lysine for spray-dried and three grades of drum-dried milk. The tests were conducted by chemical methods and rat feeding studies with a diet supplemented with methionine, the limiting amino acid in milk. The addition of methionine makes lysine the limiting amino acid. Spray-dried powder showed negligible reduction in the availability of lysine. Commercial drum-dried milk showed a reduction of available lysine of approximately 30%. A slightly scorched sample had reductions in available lysine of approximately 75%. Tryptophan and tyrosine were not affected by any of the drying methods used. Methionine was inactivated to the extent of 20% in the most severely heat processed drum-dried sample. The dried milk available to the consumer has been processed further by rewetting and drying to produce an instant powder. Posati *et al.* (1974) reported reductions in lysine, methionine, and cystine due to instantizing for commercial samples of instantized dried milk in the range of 0 to 4%. Since spray drying with instantizing is the major industrial production method today, nonfat dry milk can be expected to have amino acid retentions of over 90%.

Milk and dairy products are the major products concentrated with significant protein contents. Because of the presence of reducing sugar, the major deterioration of any consequence is the reduction in lysine availability. Mauron (1960) reports that approximately 80% of the lysine is available in evaporated milk which has been retorted at 113°C for 15 min. The lysine availability of sweetened condensed milk which is not retorted was 97%. The evaporations were performed at 50°-55°C. At this low temperature the loss of lysine is minimal. Further processing, such as retorting, results in additional losses. The temperatures which can be used for evaporation cover a wide range depending on the design of the evaporators. The process described by Wiegrand (1971) uses temperatures above those used by Mauron (1960). Nutritional data and retention times were not given by Wiegrand. It is not possible to predict the lysine availability of evaporated milk from this multieffect installation; however, Shields (1973) reports that work of Mitchell at the University of Illinois in the 1930's and 40's showed no loss of biological value or vitamins during proper evaporation or commercial drying operations.

Water-Soluble Vitamins

Water-soluble vitamins are considered separately from the oil-soluble vitamins because of the difference in deteriorative mechanisms. The most unstable water-soluble vitamin is ascorbic acid. Retention of ascorbic acid is very sensitive to the presence and type of heavy metals, such as copper and iron, light, and dissolved oxygen. Because of the high sensitivity to variables which are not well controlled, the losses of ascorbic acid vary widely. In the early work reviewed in Harris and von Loesecke (1960), the losses of ascorbic acid on drying ranged from 10 to 50%. Washing, blanching, and other pretreatments covered elsewhere account for part of these losses. Although sulfite protects ascorbic acid, it reacts with and reduces the availability of thiamin. Since dried products which are sulfited are not major sources of thiamin, the overall nutritional effect is positive since sulfite prevents browning reactions and thus protein loss.

Considerable data have been collected on the loss of ascorbic acid in dried foods. These data can be used to explain the losses during drying. The loss of ascorbic acid is very sensitive to water activity. The reaction rate constant varies over three orders of magnitude for the whole water activity range. This relationship is shown in Fig. 11.7 for several works. At low water activity, ascorbic acid is relatively stable. At high water activity it is rapidly destroyed. The differences in the reaction rate constant between different foods and model systems is considerable; however, at intermediate a_w the half life is less than two months even in foods. It is interesting to note that the systems that contain high soluble sugar or glycol concentrations, such as the orange juice crystals used by Karel and Nickerson (1964), the glycerol model system of Lee and Labuza (1975) and sucrose solutions of Kyzlink and Curda (1970) have higher destruction rate constants than cereals or cereal-soy mixes such as used by Vojnovich and Pfeifer (1970) or the dehydrated cabbage of Gooding (1962). The difference may be explained by water sorption phenomena. Sorption of water with high sugar foods is likely to contain more liquid water at lower water activities since the BET monolayer is very low (below 0.1 a_w). If more liquid water is present, the volume available for reaction is greater and the aqueous phase has a lower viscosity. In studies with a model system, Lee and Labuza (1975) have shown that by using the sorption hysteresis phenomena and NMR (Nuclear Magnetic Resonance) data, the viscosity of the aqueous environment is indeed one of the most important factors in controlling the destruction of vitamin C. The higher the viscosity, the lower is the rate. If this is the mechanism responsible for the difference, the addition of soluble proteins or some other viscosity agent before drying high sugar products (Mizrahi et al. 1967; Brennan et al. 1971) may have a nutritional as well as a process aid justification. The additive should, however, not be a glycol, such as glycerol, which has a high water retention.

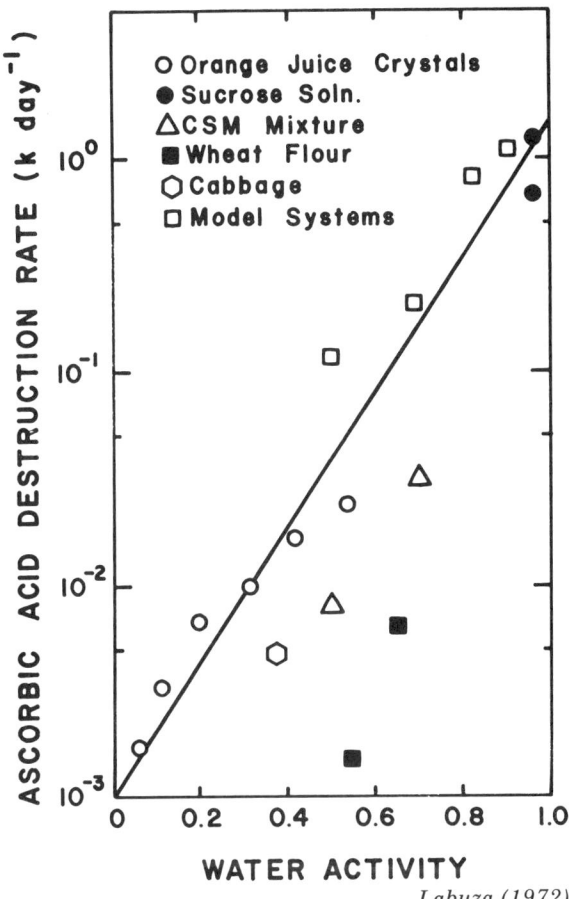

FIG. 11.7. ASCORBIC ACID DESTRUCTION RATE AS A FUNCTION OF WATER ACTIVITY

The reaction is also very sensitive to the temperature of processing. The activation energy depends on the product and the moisture content. Data for other products has been reported by Labuza (1972). In general, the activation energy is greatest at high water activities. The activation energy may change from 10 to 30 Kcal per gm-mole over the water activity range although it is usually about 20 Kcal per gm-mole. This high sensitivity to temperature at high water activity suggests that ascorbic acid retention is very dependent on the wet-bulb temperature at the beginning of drying and the temperature of evaporation.

The data on the retention of ascorbic acid after drying are so highly variable that few studies can be discussed. The results pre-

TABLE 11.3

DRYING EFFECT ON ASCORBIC ACID LOSS IN VEGETABLES

Tray[1]	Air Temperature, Dry-Bulb (°F)		Drying Time		Loss of Ascorbic Acid (%)
			First Stage (hr)	Second Stage (hr)	
Metal grid	200 ⟶ 170[2]	120 ⟶ 140[3]	0.63	2.47	16
Wood slat	200 ⟶ 170[2]	120 ⟶ 140[3]	0.92	4.08	24
Metal grid	200 ⟶ 130[2]	120 ⟶ 140[3]	1.05	2.28	26
Wood slat	200 ⟶ 130[2]	120 ⟶ 140[3]	2.25	3.67	37

[1] Tray loading, 1 lb/sq ft.
[2] Simulated first stage parallel-flow tunnel drier. Wet-bulb temperature, 120°F.
[3] Simulated second stage counterflow tunnel drier. Wet-bulb temperature, 85°F. Change from first stage to second stage conditions at circa 40% moisture. Product moisture content, 5.2–6.0%.

sented in Harris and von Loesecke (1960) provide insight into the best drying technique for high retention of ascorbic acid in vegetables. These data are presented in Table 11.3. A flow-through drier was used to simulate two-stage drying. The inlet and outlet dry-bulb temperatures and wet-bulb temperatures of the two stages were held constant. The type of tray holding the product and the outlet temperature from the first stage are variable. The ascorbic acid lost is indirectly proportional to the time in the first stage. The temperature of the potatoes in the first stage is near 50°C. The time and drying rate in the first stage is dependent on the difference between air wet-bulb and dry-bulb temperatures. The experimental samples having the least loss had the greatest difference in temperature. The effect of the metal grid and wood slat on ascorbic acid loss is due to the influence of the heat transfer rate through the tray with metal being faster and thus allowing faster drying. These results show that it is important to control drying rate particularly during the first stage of drying.

Fruits are major sources of ascorbic acid. The drying of fruits, which are high in sugar, is a difficult problem. The temperature of the product must be kept low. However, the losses in drying are usually a small part of the total loss occurring. Escher and Neukom (1970) showed that for apple flakes there was 8% loss of vitamin C in slicing, 62% loss in blanching, 10% loss during purée preparation and only 5% loss in drum drying. Many new drying methods have been reported for the drying of fruits and purees but ascorbic acid retention data are not usually included probably because, as shown, the other steps are the major problems. Low temperature drying processes, such as the vacuum drying method of Kaufman et al.

(1955), report no loss of ascorbic acid for tomato concentrates. The ultimate temperatures of 65° and 89°C were reached when the product was dried and ascorbic acid is most stable. More attention should be given to rapid blanching processes.

The evaporation of fruit juices can result in ascorbic acid losses if not performed properly. Usually, the pressed juice is deaerated and evaporated at low temperatures. Henshall (1973) reports that concentrated products after freezing have ascorbic acid retentions of 92-97%. The same range should apply to freeze-concentration processes.

Of all the B vitamins, the one most studied and probably most sensitive to temperature is thiamin, vitamin B-1. Farrer (1955) has analyzed the data available up to that time. Labuza (1972) has corrected the activation energies reported by Farrer. The activation energies for the destruction of thiamin are approximately 20 Kcal per gm-mole. Rice et al. (1944) report data which can be used to calculate the rate of thiamin degradation. At 63°C, the highest temperature reported, 20 hr are required for a 50% loss of thiamin in dried pork. For samples held at 49°C at moisture contents of 0, 2, 4, 6, and 9%, the losses of thiamin were 9, 40, 80, 90, and 89% respectively. These losses are on the order for losses of ascorbic acid. However, the activation energies for thiamin loss are slightly less than those for ascorbic acid loss at high moisture contents and the losses of thiamin in drying are less than the losses of ascorbic acid.

Most data in the literature report low loss levels of thiamin. Many of the reports are summarized in Table 11.4. Hein and Hutchings

TABLE 11.4

THIAMIN LOSS IN DRYING

Product	Conditions	Loss (%)	Reference
Freeze-dried pork	?	30	Karmas et al. (1962)
Freeze-dried chicken	?	5-6	Rowe et al. (1963)
Freeze-dried pork	-40°C	5	Thomas and Calloway (1961)
Freeze-dried chicken	1000 μHg	5	Thomas and Calloway (1961)
Freeze-dried beef		5	Thomas and Calloway (1961)
Vegetables			
Beans[1]	air dried	5	Harris and von Loesecke (1960)
Cabbage[1]	?	9	Harris and von Loesecke (1960)
Corn[1]		4	Harris and von Loesecke (1960)
Peas[1]		3	Harris and von Loesecke (1960)
Air-dried pork	?	50-70	Calloway (1962)

[1] Does not include blanching losses.

(1971) report average losses of thiamin after blanching for air-dried vegetables. These values range from a low of approximately 5% for snap beans, beets, corn, peas, and rutabagas to high values of 29% for carrots and 25% for potato. Thiamin losses are sensitive to the presence of sulfite which may account for some of the high losses found.

Data for the other water-soluble vitamins are sparse. Schroeder (1971) lists losses which vary over a wide range. Vitamin B-6 losses of 0–30% and pantothenic acid losses of 20–30% were reported for freeze-dried fish. These losses seem somewhat high when compared to other data and are suspect. Hein and Hutchings (1971) reported loss of riboflavin, niacin, and pantothenic acid for 9 vegetables; in only 2 cases did the losses of any of these vitamins exceed 10% when the blanched product is considered as 100%. Including blanching losses, the average losses for the 3 vitamins were approximately 10%. Rowe et al. (1963) report losses of riboflavin in freeze-dried chicken of 4–8%. In general, the losses of thiamin and other water-soluble vitamins, excluding ascorbic acid, are less than 10% in conventional drying. However, Miller et al. (1973) reported on vitamin losses in drum drying of bean powders (double drum 260°F for 30 sec) and found about a 20% loss for thiamin, pyridoxine, niacin, and folacin. This suggests that the other B vitamins have a stability similar to that of thiamin at least during drying. It is interesting to note that when the beans were acid treated to pH 3.5, niacin loss was only 1%, thiamin loss increased to 35%, folacin loss to 60%, and pyridoxine remained at 20%.

Mossman et al. (1973) reported on the loss of thiamin during the hot air toasting of wheat for rolled wheat flakes. Various initial moisture contents and toasting times from 10 to 40 sec were used with an air temperature of 620°F. Because of the low air humidity, the product dried as it was toasted. The results are shown in Fig. 11.8. It is obvious that less thiamin is lost at higher initial moisture contents. This is opposite to the effect expected from chemical kinetics because the rate of thiamin loss should increase with moisture content. However, as the moisture content of the feed increases, two changes occur which modify the prediction based on chemical kinetic considerations. First, more drying occurs at the lower wet-bulb temperature. Second, as more moisture is evaporated into the air, the air dry-bulb temperature decreases further and the product temperature during and after drying is lower. This is confirmed by the exit temperature shown in Fig. 11.8. In fact, higher moisture contents of the feed in these experiments result in a lower temperature toasting or drying operation. The lower temperatures of processing result in better thiamin retentions. The

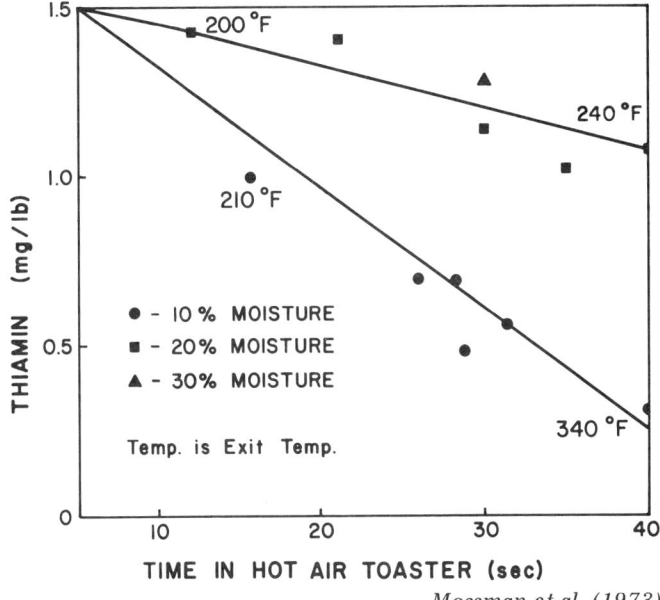

FIG. 11.8. EFFECT OF TOASTING MOISTURE AND TIME ON THIAMIN LOSS IN WHEAT

same concepts and reasoning apply to protein quality losses during toasting. However, the changes in quality as measured biologically are nonlinear with respect to chemical changes. This accounts for the fact that only the sample toasted at the most detrimental conditions, 40 sec and 10% initial moisture content, showed a significant drop in PER from 1.25 for the original to 0.88 for the toasted product.

Loss of water-soluble vitamins during concentration processes has not received much attention. Most of the B vitamins in milk are not affected by evaporation (Harris and von Loesecke 1960). Thiamin is the single exception and these losses range from 14 to 27%. Glover (1971) reports the losses of ascorbic acid, pantothenic acid, nicotinic acid, riboflavin, biotin, B-12, B-6, thiamin, and folic acid after the ultrafiltration and reverse osmosis processing of skim milk, whole milk, and whey. Ultrafiltration was performed with a membrane with a nominal molecular weight cutoff of 12,000. Ascorbic acid retention was only 13%, probably due to oxidation. Folic acid and B-12, which are associated with proteins, were retained 100%. Oil-soluble vitamins in whole milk were not measured but losses would be small. All other vitamins were lost to a significant extent. The losses appear to vary with the amount of concentration. The average of all vitamins lost, excluding ascorbic acid, are shown in Table 11.5. The results indicate that the low molecular weight vitamins pass through the membrane in proportion to the amount of water recorded. Reverse osmosis which is performed at moderate temperatures appears to be a good process for vitamin retention.

TABLE 11.5

RETENTION OF VITAMINS IN ULTRAFILTRATION

Product	Solution Retained (%)	Average Water-Soluble Vitamins Retained[1] (%)
Whole milk	50	63
Skim milk	64	77
Whey	34	39

[1] Pantothenic acid, nicotinic acid, riboflavin, biotin, thiamin, and B-6 (Glover 1971).

Fat-Soluble Vitamins

The fat-soluble vitamins have been separated from the water-soluble vitamins because of a difference between the deteriorative mechanisms. The fat-soluble vitamins would be expected to degrade by a free-radical oxidation mechanism. The free-radical oxidation of lipids has been reviewed recently by Labuza (1971). The reaction is characterized by a low activation energy (10–15 Kcal per mole) and a long induction time.

Della Monica and McDowell (1965) dried carrots by the procedures shown in Table 11.6. They measured total β-carotene and the *trans*-isomer which is the form with 100% provitamin A activity. The losses of total β-carotene are higher than would be expected if the loss rate was due solely to a free-radical oxidation mechanism. Thus, some direct thermal reaction must be taking place as evidenced by the loss in freeze drying. Sweeney and Marsh (1971) report a loss of 13% for freeze-dried carrots. Foda et al. (1970) report losses of β-carotene of approximately 4% for freeze-dried orange juice of various concentrations.

Reports of other oil-soluble vitamin retentions are limited to milk. There is little or no loss of vitamin A or D during the spray drying, drum drying, or evaporation of milk (Hartman and Dryden 1965; Harris and von Loesecke 1960).

There are also no reports in the literature of vitamin E loss during

EFFECTS OF MOISTURE REMOVAL ON NUTRIENTS

TABLE 11.6

RETENTION OF CAROTENE IN DRYING CARROTS TO 3% H_2O

Drying Process	Temperature, Time, Conditions	Total β-carotene (%)	Trans β-carotene (%)
Tray air drying[1]	200°F, 2 hr 150°F, 6 hr	74	60
Explosion puff[1]	200°F, 2 hr in tray; exploded at 35 lb/in^2; finished at 150°F; total time 5.5 hr.	81	60
Freeze dried[1]	160°F, 1000 μHg, 4-5 hr	85	80
Freeze dried[2]	?	90	80

[1] Della Monica and McDowell (1965).
[2] Sweeney and Marsh (1971).

drying. Considering that it is also involved in the lipid oxidation scheme it would be expected to be low. In oilseed processing, for example, where very high temperatures are used, it is assumed that only 5% of the vitamin is lost.

Conclusion

Drying processes appear to offer good nutrient retentions with the exceptions of ascorbic acid and β-carotene losses. Of the concentration procedures, ultrafiltration shows the highest losses. Protein quality deteriorations due to drying or evaporation are minimal in products currently in production (Mauron 1960; de Groot 1963). The losses of water-soluble vitamins other than ascorbic acid during drying average approximately 5% (Hein and Hutchings 1971). Oil-soluble vitamins in milk are not lost in either drying or evaporation (Hartman and Dryden 1965; Harris and von Loesecke 1960). Losses of β-carotene in dried carrots are significant (Della Monica and McDowell 1965), but small in freeze-dried orange juice (Foda et al. 1970). Ultrafiltration results in highly significant losses of most water-soluble vitamins (Glover 1971). Ascorbic acid losses are high in drying but concentration offers minimal losses (Henshall 1973). No experimental reports of nutrient retentions after freeze concentration could be found but this process should offer excellent nutrient retention.

The real problem with the current literature is that little work has been done on the kinetics of nutrient losses at constant temperatures and humidities. With this knowledge and the time-temperature-moisture content distribution in the product during drying, process optimization procedures could be calculated where necessary.

BIBLIOGRAPHY

AITKEN, A., JASON, A. C., DILLEY, J., and PAYNE, P. R. 1967. Effects of drying, salting and high temperatures on the nutritive value of dried cod. Fish News Intern. Sept., 42-43.

ANON. 1973. Superior, natural fruit ingredient produced by new drying process. Food Prod. Develop. 7, 40.

ARMERDING, G. D. 1966. Evaporation methods as applied to the food industry. Advan. Food Res. 15, 303-358.

BENDER, A. E. 1966. Nutritional effects of food processing. J. Food Technol. 1, 261-289.

BENDER, A. E. 1972. Processing damage to protein food: A review. J. Food Technol. 7, 239-250.

BICKEL, H., and MAURON, J. 1959. Investigation of the lysine requirements of young infants; feeding studies with a milk product of low lysine content. Ann. Pediatrics 193, 55.

BOMBEN, J. L., BRUIN, S., and THIJSSEN, H. A. C. 1973. Aroma recovery and retention in concentration and drying of foods. Advan. Food Res. 20, 1-111.

BONE, D. 1969. Water activity: Its chemistry and applications. Food Prod. Develop. 3, 81, 84-85, 88, 90, 92, 94.

BRENNAN, J. G., BUTTERS, J. R., COWELL, N. D., and LILLY, A. E. V. 1969. Food Engineering Operations. Elsevier Publishing Co., Amsterdam.

BRENNAN, J. G., HERRERA, J., and JOWITT, R. 1971. A study of some of the factors affecting the spray drying of concentrated orange juice, on a laboratory scale. J. Food Technol. 6, 295-307.

BRENNAN, J. G., and PRIESTLY, R. J. 1972. Foam-spray drying using a centrifugal atomizer. Proc. Biochem. 7, No. 3, 25-26.

BRUNAUER, S., EMMETT, H. P., and TETLER, E. 1938. Adsorption of gases in multimolecular layers. J. Am. Chem. Soc. 60, 309.

CALLOWAY, D. H. 1962. Dehydrated foods. Nutr. Rev. 20, 257-260.

CARLSON, R. A., RANDALL, J. M., GRAHAM, R. P., and MORGAN, A. I. 1967. The rotary steam-coil vacuum evaporator. Food Technol. 21, 194-196.

CHARM, S. E. 1971. Fundamentals of Food Engineering, 2nd Edition. Avi Publishing Co., Westport, Conn.

DE GROOT, A. P. 1963. The influence of dehydration of foods on digestibility and the biological value of protein. Food Technol. 17, 339-343.

DELLA MONICA, E. S., and McDOWELL, P. E. 1965. Comparison of beta-carotene content of dried carrots prepared by three dehydration processes. Food Technol. 19, 141-143.

DUCKWORTH, R. B. 1962. Diffusion of solutes in dehydrated vegetables. In Recent Advances in Food Science, Vol. II. J. Hawthorn, and J. M. Leitch, (Editors). Butterworths, London, England.

EICHNER, K., and KAREL, M. 1972. The influence of water content and water activity on the sugar-amino browning reaction in model systems under various conditions. J. Agr. Food Chem. 20, 218-223.

ESCHER, F., and NEUKOM, H. 1970. Studies on drum-drying apple flakes. Trav. Chimie Aliment Hyg. 61, 339-348. (German)

ESCHER, F., and NEUKOM, H. 1971. Non-enzymatic browning and the optimization of the drying conditions for the drum drying of apple sauce. Lebensm. Wiss. Technol. 4, 145-151.

FARKAS, D. F., and LAZAR, M. E. 1969. Osmotic dehydration of apple

pieces: Effect of temperature and syrup concentration on rates. Food Technol. 23, 688-690.
FARRER, K. T. H. 1955. The thermal destruction of vitamin B_1 in foods. Advan. Food Res. 6, 257-311.
FINCH, R. 1970. Fish protein for human foods. CRC Crit. Rev. Food Technol. 1, 519-580.
FODA, Y. H., HAMED, M. G., and ABD-ALLAH, M. A. 1970. Preservation of orange juice and guava juices by freeze-drying. Food Technol. 24, 74-80.
GLOVER, F. A. 1971. Concentration of milk by ultrafiltration and reverse osmosis. J. Dairy Res. 38, 373-379.
GOODING, E. G. B. 1962. The storage behavior of dehydrated foods. In Recent Advances in Food Science, Vol. II. J. Hawthorn and J. M. Leitch (Editors). Butterworths, London, England.
GREIG, N. 1971. Economics of Food Processing. Avi Publishing Co., Westport, Conn.
HACKLER, L. R. et al. 1965. Effect of heat treatment on nutritive value of soy milk protein fed to weanling rats. J. Food Sci. 30, 723-728.
HARRIS, R. S., and VON LOESECKE, H. 1960. Nutritional Evaluation of Food Processing. John Wiley & Sons, New York. Reprinted in 1971 by Avi Publishing Co., Westport, Conn.
HARTMAN, A. M., and DRYDEN, L. P. 1965. Vitamins in milk and milk products. ADSA, USDA, Washington, D. C.
HEIN, R. E., and HUTCHINGS, I. J. 1971. Influence of processing on vitamin-mineral content and biological availability in processed foods. Symp. Vitamins Minerals Processed Foods, Am. Med. Assoc. Council Foods Nutr. Food Ind. Liaison Comm, New Orleans.
HENDEL, C. E., SILVEIRA, V. and HARRINGTON, W. O. 1955. Rates of non-enzymatic browning of white potato during dehydration. Food Technol. 9, 433-438.
HENSHALL, J. D. 1973. Fruit and vegetable products. Proc. Nutr. Soc. 32, 17-22.
HERTZENDORF, M. S., and MOSHY, H. J. 1970. Foam drying in the food industry. CRC Crit. Rev. Food Technol. 1, 25-70.
HIEU, T. C. and SCHWARTZBERG, H. G. 1973. Dehydration of shrimp by distillation. AIChE Symp. Ser. 69, No. 132, 70-80.
HOLDSWORTH, S. D. 1971. Dehydration of food products. J. Food Technol. 6, 331-370.
KAREL, M. 1973. Recent research and development in the field of low moisture and intermediate moisture foods. CRC Crit. Rev. Food Technol. 4, 329-373.
KAREL, M., and NICKERSON, J. T. R. 1964. Effect of relative humidity, air, and vacuum on browning of dehydrated orange juice. Food Technol. 18, 1214-1218.
KARMAS, E., THOMPSON, J. E., and PERYAM, D. B. 1962. Thiamin retention in freeze-dehydrated irradiated pork. Food Technol. 16, 107-108.
KAUFMAN, V. F., WONG, F., TAYLOR, D. H., and TALBURT, W. F. 1955. Problems in production of tomato juice powder in vacuum. Food Technol. 9, 120-123.
KING, C. J. 1970. Recent developments in food dehydration technology. Proc. 3rd Intern. Congr. Food Sci. Technol., pp. 565-574.
KING, C. J. 1971. Freeze-Drying of Foods. CRC Press, Cleveland.
KLUGE, G., and HEISS, R. 1967. Studies to improve the quality of dried foods during various conditions of freeze-drying. Verfahrens Technik. 1, 251-258. (German)
KYZLINK, V., and CURDA, D. 1970. Influence of sucrose concentration and oxygen access on the course of L-ascorbic acid oxidation in liquid medium. Z. Lebensm-Untersuch. -Forsch. 143, 263-273.

LABUZA, T. P. 1968. Sorption phenomena in foods. Food Technol. 22, 263-265, 268, 270, 272.
LABUZA, T. P. 1970. Properties of water as related to the keeping quality of foods. Proc. 3rd Intern. Congr. Food Sci. Technol., pp. 618-635.
LABUZA, T. P. 1971. Kinetics of lipid oxidation in foods. CRC Crit. Rev. Food Technol. 2, 355-405.
LABUZA, T. P. 1972. Nutrient losses during drying and storage of dehydrated foods. CRC Crit. Rev. Food Technol. 3, 217-240.
LEE, S. H., and LABUZA, T. P. 1975. Destruction of ascorbic acid as a function of water activity. J. Food Sci. 40, 370-373.
LEIGHTELL, B. 1972. Reverse osmosis in the concentration of food. Proc. Biochem. 7, 40-42.
LIVINGSTON, A. L., ALLIS, M. E., and KOBLER, G. O. 1971. Amino acid stability during alfalfa dehydration. J. Agr. Food Chem. 19, 947-950.
LONCIN, M., BIMBENET, J. J., and LENGES, J. 1968. Influence of the activity of water on spoilage of foodstuffs. J. Food Technol. 3, 131-142.
MACDONALD, F. J. 1966. Available lysine content of dried milk. Nature 209, 1134.
MASTERS, K. 1972. Spray Drying. CRC Press, Cleveland.
MAURON, J. 1960. The concept of amino acid availability and its bearing on protein evaluation. Intern. Conf. Protein Needs, Washington, D.C., Aug. 21-22.
MEADE, R. E. 1973. Combination process dries crystallizable materials. Food Technol. 27, 18, 20, 24-26.
MILLER, C. F., GUADAGNI, D. and KOW, S. 1973. Vitamin retention in bean powders: cooked canned and instant. J. Food Sci. 38, 493-495.
MIZRAHI, S., BERK, Z., and COGAN, U. 1967. Isolated soybean protein as a banana spray drying aid. Cereal Sci. Today 12, 322, 324-325.
MIZRAHI, S., LABUZA, T. P., and KAREL, M. 1970A. Computer-aided predictions of extent of browning in dehydrated cabbage. J. Food Sci. 35, 799-803.
MIZRAHI, S., LABUZA, T. P., and KAREL, M. 1970B. Feasibility of accelerated tests for browning in dehydrated cabbage. J. Food Sci. 35, 804-807.
MOORE, J. G., and HESLER, W. E. 1963. Evaporation of heat sensitive materials. Chem. Eng. Progr. 59, 87-92.
MOSSMAN, A. P., ROCKWELL, W. C., and FELLERS, D. A. 1973. Hot air toasting and rolling whole wheat. J. Food Sci. 38, 879-884.
MYKLESTAD, O., BJØRNSTAD, J., and NJAA, L. 1972. Effects of heat treatment on composition and nutritive value of herring meal. Fiskerdinetoratets Skrifter Ser. Technol. Undersøk. 5, No. 10, 1-15.
PERRY, R. H., CHILTON, C. H., and KIRKPATRICK, S. D. (Editors) 1964. Chemical Engineer's Handbook. McGraw-Hill Book Co., New York.
PORTER, M. C., and MICHAELS, A. S. 1970. Applications of membrane ultrafiltration. Proc. 3rd Intern. Congr. Food Sci. Technol., pp. 462-473.
POSATI, L. P., HOLSINGER, V. H., DEVILBISS, E. D., and PALLANSCH, M. J. 1974. Effect of instantizing an amino acid content of non-fat dry milk. J. Dairy Sci. 57, 258-260.
QUAST, D. G., and KAREL, M. 1972. Computer simulation of storage life of foods undergoing spoilage by two interacting mechanisms. J. Food Sci. 37, 679-683.
RICE, E. E. et al. 1944. Preliminary studies on stabilization of thiamin in dehydrated foods. Food Res. 9, 491-499.
ROCKLAND, L. B. 1969. Water activity and storage stability. Food Technol. 23, 1241-1246, 1248, 1251.
ROLLS, B. A., and PORTER, J. G. 1973. Some effects of processing and storage of milk and milk products. Proc. Nutr. Soc. 32, 9-15.
ROWE, D. M., MOUNTREY, G. J., and PRUDENT, I. 1963. Effect of freeze-

drying on the thiamin, riboflavin and niacin content of chicken muscle. Food Technol. *17*, 1449-1450.
SARAVACOS, G. D. 1970. Effect of temperature on viscosity of fruit juices and purées. J. Food Sci. *35*, 122-125.
SCHROEDER, H. A. 1971. Losses of vitamins and trace minerals resulting from processing and preservation of foods. Am. J. Clin. Nutr. *24*, 562-572.
SHIELDS, J. B. 1973. Personal communication. American Potato Co., Blackfoot, Idaho.
SWEENEY, J. P., and MARSH, A. C. 1971. Effect of processing on provitamin A in vegetables. J. Am. Diet. Assoc. *59*, 238-243.
TAIRA, H., TAIRA, H., and SUKARAI, Y. 1966. Studies on amino acid contents of processed soybeans. 8. Effect of heating on total lysine and available lysine in defatted soybean flour. Japan. J. Nutr. Food. *18*, 359-362.
THIJSSEN, H. A. C. 1970. Freeze concentration of food liquids. Proc. 3rd Intern. Congr. Food Sci. Technol., pp. 491-498.
THIJSSEN, H. A. C. 1971. Flavour retention in drying preconcentrated food liquids. J. Appl. Chem. Biotechnol. *21*, 372-377.
THOMAS, M., and CALLOWAY, D. 1961. Nutritional value of dehydrated food. J. Am. Diet. Assoc. *39*, 105-116.
VAN ARSDEL, W. B. 1963. Food Dehydration, Vol. I. Principles. Avi Publishing Co., Westport, Conn.
VAN ARSDEL, W. B., and COPLEY, M. J. 1964. Food Dehydration, Vol. II. Products and Technology. Avi Publishing Co., Westport, Conn.
VAN DEN BEUEL, A., JAMNSKENS, P., and MOL, J. 1972. Availability of lysine in skim milk powders processed under various conditions. Neth. Milk Dairy J. *26*, 19.
VOJNOVICH, C., and PFEIFER, V. F. 1970. Stability of ascorbic acid in blends with wheat flour, CSM and infant cereals. Cereal Sci. Today *15*, 317-322.
WATT, B. K., and MERRILL, A. L. 1963. Composition of Foods. Agriculture Handbook *8*. USDA.
WIEGRAND, J. 1971. Falling-film evaporators and their applications in the food industry. J. Appl. Chem. Biotechnol. *21*, 351-358.
WILLIAMS-GARDNER, A. 1971. Industrial Drying. Leonard Hill, London, England.
WOODHAM, J. A. 1973. The effects of processing on the nutritive value of vegetable-protein concentrates. Proc. Nutr. Soc. *32*, 23-29.
ZIMMERMAN, O. T., and LAVINE, I. 1964. Psychrometric Tables and Charts. Industrial Research Service, Dover, New Hampshire.

CHAPTER 12

Ivan D. Jones | Effects of Processing by Fermentation on Nutrients

Fermentation of food for preservation, enhancement of nutritive value, improvement of flavor, or preparation of exhilarating beverages has been practiced probably since prehistoric times by peoples of nearly every civilization. Many of the fermentation practices employed for these purposes are still of major importance in our modern food industries. Within the last century food preservation by thermal processing, dehydration, and freezing has advanced in the more industrially developed countries. The perfection of these alternative methods of preservation has been accompanied by a proportionately large increase in consumption of processed foods and a concomitant reduction in the consumption of fresh or unprocessed foods.

Fermentation practices in the future will serve as increasingly important means of providing food supplements and new foods as well as continuing to be highly significant for preservation purposes. To meet the demands of a rapidly growing population, there is urgent need not only for additional food but also for new sources of food and food supplements. Currently there is recognition that the solution of these problems lies, in part, increasingly in the realm of the industrial microbiologist. Such was indicated by Bunker (1964) in the statement "insofar as food shortage is likely to be in supplies of protein, lipids, and vitamins, microorganisms not only can be employed to obtain all of these but they can supply them quickly."

Attention in this chapter will be directed to the nature and extent of nutritional value changes reported to occur in the manufacture of the following fermented foods: cheeses and fermented milk preparations; pickled and salted vegetables; wines and beers; and some Oriental fermented foods. Such changes are caused by and are characteristic for the specific processes followed. These changes may be due to chemical destruction; solution; alteration in solubility; gain or loss associated with microbial growth; or loss or concentration by partition in cheese manufacture. Attention will also be drawn to industrial fermentations as sources of protein and of certain vitamins and amino acids.

FERMENTED MILK PRODUCTS

Although the milk of many animals is an important source of food for man, our major interests are with respect to cow's milk and milk products made therefrom. For the population of the United States, the importance of dairy products, including fermented dairy foods, for providing major nutrients may be indicated by the data presented in Table 12.1. Whole milk is also a source of some importance for

TABLE 12.1

CONTRIBUTION OF DAIRY PRODUCTS, EXCLUDING BUTTER, TO FOOD ENERGY AND NUTRIENT SUPPLIES (U.S. CIVILIAN POPULATION 1968)[1]

Energy and Energy Sources	%	Minerals	%	Vitamins	%
Food Energy	11.8	Calcium	78.2	Vitamin A	11.8
Protein	22.6	Phosphorus	36.7	Ascorbic Acid	4.7
Fat	13.4	Iron	2.2	Thiamin	9.9
Carbohydrate	7.2			Riboflavin	43.1
				Niacin	1.7

Source: U.S. Dept. of Agr. (1968).

[1] Values reported as percentage of total contribution of major food groups for 1968.

the fat-soluble vitamins D, E, and K; water-soluble vitamins B-6, B-12, pantothenic acid, folic acid, and biotin; and for more than 20 minerals other than those listed in Table 12.1. For extensive discussion of milk composition and factors responsible for variation in the different components of milk and milk products see Hartman and Dryden (1965, 1974) and the many reviews by Kon and coworkers summarized by Kon (1959, 1961).

Fermented milk products as a class are primarily important for their contribution of protein, fat, carbohydrate, calcium, phosphorus and vitamins A, thiamin, and riboflavin. This review will be limited largely to consideration of changes in these nutrients associated with manufacturing treatments.

Pasteurization

Pasteurization of milk is usually the first processing step to be taken in the manufacture of fermented milk products. Much research has been conducted on the influence of pasteurization upon the nutritive value of milk. Wanner (1960) reported that the findings from numerous studies on infants, school children, adults, and laboratory animals indicated "no nutritional superiority of raw milk over pasteurized milk," that "milk proteins are only slightly affected by heat of pasteurization." Smith (1970) concluded that there was not any "significant decline in digestibility or biological value as the result of pasteurization or spray drying."

Wanner (1960), Hartman and Dryden (1974), Lampert (1970), and Kon (1961) indicate no destruction of vitamin A and little or only slight destruction of niacin by normal pasteurization. These reviewers reported pasteurization loss of thiamin ranging from 3 to 20%; and that although riboflavin is thermostable, pasteurization losses of this vitamin had been reported, probably due to light destruction. The ascorbic acid content of milk may be appreciably reduced during pasteurization. Kon (1961) has placed this loss at 10%, and Hartman and Dryden (1974) at about 20%.

Cheese Making

According to Walter and Hargrove (1969) cheese "is made by coagulating or curdling milk, stirring and heating the curd, draining off the whey, and collecting or pressing the curd." In the manufacture of most cheeses there is a curing or ripening operation following collecting or pressing the curd. Factors responsible for change in nutrient levels of the product in the manufacture of cheese are: the method of curd coagulation; the extent of whey removal and method employed for such; and the ripening process.

Curdling or coagulation causes a partition of the milk nutrients, with the water and water-soluble substances forming whey; and with protein and fat, if fat is present, forming curd. The proteins are coagulated either by rennet which is added, or by lactic acid formed during microbial fermentation. The extent of separation of whey from curd is dependent upon variations in stirring, heating, and acidity of the whey during and following coagulation and also upon variation in the pressing of the curd. Whey acidity is a factor in curd and whey composition, particularly with respect to calcium and phosphorus.

Losses of milk nutrients in cheese making are greatest at the whey separation stage (Kon 1959). To illustrate the nature and magnitude of such losses Kon cited changes occurring during manufacture of Cheddar cheese. The information Kon provided has been slightly reorganized and is presented, in part, in Table 12.2. The total protein loss through partition is about 25%; the soluble proteins which remain in the whey are the albumins and globulins. As would be expected, most of the vitamin A remains with the fat in the curd. Nearly ⅔ of the calcium remains in the rennet-coagulated curd. The

EFFECTS OF PROCESSING BY FERMENTATION ON NUTRIENTS

TABLE 12.2

PARTITION OF NUTRIENTS IN MAKING OF CHEDDAR CHEESE[1]

Nutrient	In Whey	In Curd
Water	94	6
Fat	6	94
Total Solids	52	48
Casein	4	96
Soluble proteins	96	4
Lactose	94	6
Calcium	38	62
Vitamin A	6	94
Thiamin	85	15
Riboflavin	74	26
Vitamin C	84	6

Source: After Kon (1959).
[1] Quantities in flowing whey and curd at pitching, expressed as percentages of total in original milk.

riboflavin and thiamin retention of about 25% and 15% respectively, is somewhat greater than would be expected from the partition of the water. This greater retention is attributed to an association between these vitamins and the protein.

Vitamin C retention is low, indicating destruction as well as loss through partition. Kon and Thompson (1957) have reported niacin losses of about 90% at the whey separation stage.

Nutrient level changes in the cheese during curing or ripening are dependent upon the kind or type of cheese. Although more than 400 cheeses have been described, Walter and Hargrove (1969) indicate that there are only about 18 distinct types and have suggested the classification with some examples as given below. This classification is based upon a consideration of the finished product texture and of ripening or curing practices followed.

1. Very hard (grating): (A) Parmesan, Romano.
2. Hard: (A) Ripened by bacteria, without eyes: Cheddar, Colby, Granular or Stirred-curd. (B) Ripened by bacteria, with eyes: Swiss, Gruyere.
3. Semisoft: (A) Ripened principally by bacteria: Brick and Muenster. (B) Ripened by bacteria and surface microorganisms: Limburger, Trappist. (C) Ripened principally by blue mold in the interior: Roquefort, Gorgonzola, Blue, Stilton.
4. Soft: (A) Ripened: Camembert, Hand, Neufchatel (as made in France). (B) Unripened: (Acid-curd) Cottage, Baker's, Cream. (Whey cheeses) Ricotta, Mysost, Primost.

Ripening changes which may take place in cheese are: proteolysis;

lipolysis; lactose fermentation and increased acid formation in residual whey; reduction in calcium and phosphorus through loss of whey; utilization of residual vitamins by microorganisms growing in and on the cheese; and for mold-ripened cheeses in particular, synthesis of vitamins by such microorganisms. In general, such changes are more rapid and more pronounced in the softer, more moist, ripened cheeses.

The nutritive value of the milk protein is not significantly affected in the manufacture and maturing of cheese, though the protein level in different types of cheese will be dependent upon manufacture, according to Smith (1970). He further cited the excellence of cheese as an amino acid source, particularly for lysine. Kon (1959) indicated that cheese proteins rate slightly lower nutritionally than milk proteins, due to a lower content of sulfur-amino acids. Cottage cheese, in a comparison with whole egg protein, was reported to be an outstanding source of essential amino acids by Edwards and Allen (1958).

Mattick (1951) investigated the retention of calcium and phosphorus in the hard-pressed Cheddar and the unpressed Stilton cheeses during curing. Ripe Cheddar retained about 58% of the original milk calcium and about 50% of the milk phosphorus. The Stilton showed 40% retention in the new curd but in the ripened cheese the retention was only about 8% for calcium and 27% for phosphorus. Kon (1959) indicated that the calcium contained in cottage cheese is principally in the retained whey. Lactic acid cheeses such as cottage or cream retain only about 20% each of calcium and phosphorus of the milk from which they are made (Lampert 1970).

The nature and extent of changes in the vitamin content of cheese associated with ripening processes have been reported in the comprehensive review on vitamins in milk and milk products by Hartman and Dryden (1965, 1974). Extensive additional pertinent research data by many investigators are to be found in reports published since 1965.

Because of the great differences in the manufacturing practices followed in cheese making no attempt will be made to set forth here the details of research findings on the influence of ripening on vitamin changes in cheese. Rather, the brief summary by Kon (1972) will be presented as follows:

> .... the small amount of vitamin C present in the original curd is rapidly lost in the initial stages of ripening. The other vitamins, including vitamin A in cheeses containing fat, are on the whole stable during ripening and storage. Moreover, bacterial and mold activity leads to synthesis of several vitamins of the B complex. These newly-formed vitamins are, as a rule,

EFFECTS OF PROCESSING BY FERMENTATION ON NUTRIENTS 329

on the surface of the cheese and thus may contribute little nutritionally since the rind or crust is often not eaten.

Shown in Table 12.3 are compositional values of cheeses representative of distinctly different types. This Table provides specific information concerning the influence of processing treatment on product composition.

Fermented Milk Preparations

Fermented milk preparations are extensively produced in many countries, principally by dairies from cow's milk, but also by traditional methods from the milk of many animals (Pederson 1971; Borgstrom 1968). These preparations are from cream, whole milk, skim milk or whole or skim milk to which has been added either milkfat or nonfat-milk-solids.

Fermentation is controlled through inoculation either with cultures of lactic acid bacteria or of mixtures of lactic acid bacteria and yeast (Pederson 1971). Sour cream, buttermilk from churned cream, cultured buttermilk, Bulgarian buttermilk, acidophilus milk, and yoghurt are well-known examples of acid preparations resulting from inoculation by selected cultures of lactic acid bacteria. Kefir and kumiss are acidic-alcoholic products of fermentation by cultures of lactic acid bacteria and yeast.

Nutritionally, most of these fermented preparations are quite comparable to the milks from which they were prepared. Fermentation of lactose to lactic acid or alcohol causes only slight change in caloric value. The riboflavin content of some products may be increased as the result of fermentation (Kon 1972).

Buttermilk from churned cream may have a fat content of 0.2–0.5% (Pederson 1971). Cultured buttermilks are made from skimmed milk, partly skimmed milk, or skimmed milk with butterfat added (Hargrove 1970). Kosikowski (1966) cited a 1965 USDA report indicating a fat content range of from 0.6 to 1.7% in cultured buttermilks. Butterfat over the range indicated not only contributes materially to the food energy level of the product but also provides a vitamin A potency ranging from 17 to 45% of that of whole milk.

Some dairies add from 0.5 to 2.0% milk-solids-not-fat to skim milk in making cultured buttermilk (Kristofferson and Gould 1966). Such practice increases the protein, mineral, and B vitamin content of the buttermilk by from 5 to 21% of that contributed initially by the skimmed milk used.

Yoghurt made by modern industrial methods is fermented in consumer containers and possesses textural characteristics intermediate between fermented beverages and soft cheese. There is no separation

TABLE 12.3

COMPOSITION OF CHEESES OF DIFFERENT TYPES
(PER 100-GM PORTIONS)[1]

Cheese Variety	Type[2]	Water (%)	Food Energy Calories	Protein (gm)	Fat (gm)	Calcium (mg)	Phosphorus (mg)	Vitamin A (IU)	Thiamin (mg)	Riboflavin (mg)	Niacin (mg)
				On Wet Weight Basis							
Cheddar-domestic	Hard. Ripened-bacteria	37.	398	25.0	32.2	750	478	(1310)	0.03	0.46	0.1
Roquefort-type	Semi-soft. Ripened-mold	40.	368	21.5	30.5	315	184	(1240)	0.03	0.61	1.2
Camembert-domestic	Soft. Ripened-mold	52.2	299	17.5	24.7	105	339	(1010)	0.04	0.75	0.8
Cottage	Soft. Unripened	79.0	86	17.0	0.3	90	175	(10)	0.03	0.28	(0.1)
				On Dry Weight Basis							
Cheddar-domestic	Hard. Ripened-bacteria		632	39.7	51.1	1190	759	(2079)	0.05	0.73	0.16
Roquefort-type	Semi-soft. Ripened-mold		613	35.8	50.8	525	352	(2067)	0.05	1.02	2.00
Camembert-domestic	Soft. Ripened-mold		573	33.5	47.3	201	565	(1935)	0.08	1.44	1.53
Cottage	Soft. Unripened		410	81.0	1.4	429	833	(48)	0.14	1.33	(0.48)

Source: Data of Watt and Merrill (1963).

[1] Numbers in parentheses denote values imputed—usually from another form of the food or from similar food.
[2] Taken from classification of Walter and Hargrove (1969).

EFFECTS OF PROCESSING BY FERMENTATION ON NUTRIENTS

of whey and curd in this product and therefore no loss of milk components as during cheese manufacture. Yoghurt made from whole milk is very similar in chemical composition to whole milk (Watt and Merrill 1963).

The mix or basic medium used in the manufacture of yoghurt may be varied considerably depending on the type of product desired. Whole milk may be used; however, partly skimmed, skimmed, evaporated, or dried milks may also be used and the butterfat and solids content may range from 0 to 5% and 9 to 20% respectively (Hargrove 1970). Fruit purée, preserved fruit, or fruit flavoring with or without 3-5% sugar may be added (Hargrove 1970; Lampert 1970). The final acidity of high quality yoghurt is about 1.0% as lactic acid (Kosikowski 1966; Hargrove 1970); vitamin C is practically completely destroyed during yoghurt manufacture (Sulc 1960; Watt and Merrill 1963).

Yoghurt prepared by traditional methods differs decidedly from that made by modern commercial methods. Traditionally, it is common to boil whole milk in open kettles to reduce the volume and increase the solids by about ⅓. Such severe heat treatment must be decidedly destructive of heat-sensitive components. Furthermore, following fermentation this yoghurt may be separated into curd and whey by straining (Lampert 1970; Kon 1972). Such a practice would result in the loss of water-soluble nutrients, much as occurs in the draining of whey from cheese curd as discussed earlier.

The suggestion has been made that yoghurt may be a food suitable for use by lactose-intolerant individuals (Huang and Bayless 1968; Bayless and Huang 1969; Baer 1970). Such suggestion must have been made with reference to traditionally prepared yoghurt, the curd of which would be low in lactose due to loss through partition. Yoghurt prepared as outlined by modern methods would contain most of the lactose originally in the milk or milk mix.

The beverages kefir and kumiss are products of a combined lactic acid and alcohol fermentation, the final stage of which is carried out in sealed containers for the purpose of product carbonation. They originated in the countries of Eastern Europe and the Caucasus, kefir from the milk of the sheep and goat and kumiss from mare's milk. These beverages are particularly popular in Eastern Europe and Russia and are now made by modern commercial methods from cow's milk, partly skimmed or skimmed. Kefir will contain lactic acid and alcohol at levels of about 0.8 and 1.0% respectively. Kumiss may contain added sucrose at the beginning of the fermentation; the final lactic acid and alcohol content levels will range from 0.7 to 1.8% and 1.0 to 2.5% respectively. For details of preparation and characteristics of these fermented milks see Kosikowski (1966),

Borgstrom (1968), Hargrove (1970), and Pederson (1971). These beverages are nutritionally rather similar to the milks from which they were prepared (Kon 1972) with the exception that the lactose content must be somewhat lower.

FERMENTED VEGETABLE PRODUCTS

Many vegetables are preserved by brining or salting procedures which permit and favor fermentation by lactic acid-forming microorganisms. Pederson (1971) has pointed out that such preservation is "dependent upon the combined effect of salt, acid, carbon dioxide, low-oxidation-reduction potential, and other minor factors." The salt added will vary from 2.0 to 15% according to the vegetable to be preserved and the product to be formed. Acid formation will range from 0.2 to 2.5% depending upon the vegetable salted, the salting or brining treatment, and environmental factors.

Pickles

Major fermented vegetable products are pickles and sauerkraut manufactured both by modern industrial procedures and by traditional methods of ancient origin. Pickled products such as fresh-pack pickles are not fermented vegetable products and therefore will not be considered in this discussion.

Commercially manufactured sweet and sour cucumber pickles and mixed vegetable pickles are usually made from vegetables temporarily preserved by fermenting and storing in brines varying in salt content from about 7 to 15%. This product, known as salt-stock, is further processed by nearly completely desalting followed by "finishing" in suitably spiced, sweetened vinegar. The finished product is then generally subjected to a mild thermal processing or pasteurization.

The brining preservation of cucumber salt-stock as described by Fabian et al. (1932), Jones and Etchells (1943), Etchells and Jones (1943A), Pederson and Ward (1949), and others, followed by the subsequent processing for pickle production is responsible for large loss or destruction of most of the nutrients originally present. The water-soluble constituents are leached into the brine during curing and are further largely discarded in the desalting operation. The sugars are primarily fermented to carbon dioxide and acids, principally lactic acid.

There is little published literature related specifically to the nutritional effects of cucumber brining and subsequent processing in pickle manufacture. Camillo et al. (1942) in studies of nutrient change during the manufacture of a variety of pickle products re-

EFFECTS OF PROCESSING BY FERMENTATION ON NUTRIENTS 333

ported losses of the B vitamins ranging from 75 to 85% for salt-stock pickles and 33 to 87% for genuine dill pickles. Vitamin C losses were about 100%. β-Carotene content values were reported to be appreciably higher for salt-stock than for fresh cucumbers. This surprising observation was not explained. The β-carotene level of the salt-stock, however, was not high enough to be of nutritional significance.

When viewed from a practical standpoint, the significance of pickle products in the diet is probably chiefly through the contribution to variety, flavor, and zest rather than of nutrients. Listed in Table 12.4 is the composition of raw and pickled cucumbers according to Watt and Merrill (1963). Included in this Table are values for raw cauliflower, onions, and peppers. These vegetables are often components of mixed pickles and are given the same preservation and processing treatments as cucumber salt-stock.

Bell et al. (1972) have recognized the need for providing acceptable low-sodium pickle products for use in sodium-restricted diets. Cucumbers were initially preserved by conventional procedures in sodium chloride brine; were desalted to a sodium chloride content of less than 0.1%; and were then finished with a suitably spiced, acidified liquor containing potassium chloride. Sweet pickles containing 1-2% potassium chloride were rated good; dill spears containing 2 and 4% potassium chloride were rated poor and fair, respectively.

Pickled olives are made from ripe or green olives fermented in brine. Lactic acid bacteria and yeast are responsible for the fermentations. Ripe olives and Spanish-type olives are subjected to lye treatments and subsequent leaching of lye from the fruit prior to brining, in order to reduce the content of bitter glucoside present. Such treatments are accompanied by marked reduction in water-soluble components. Vaughn et al. (1943) indicated that the Spanish-type green olives contain little vitamin C or vitamin B complex and that the small amount of vitamin A reported to be present was of no "particular significance." The soluble nutrients of ripe olives are probably leached away and fermented to an extent similar to that reported for green olives.

Sauerkraut and Related Fermented Vegetables

Lactic acid fermentation of a mixture of shredded or chopped cabbage and salt forms sauerkraut. The salt is added at such a low level, about 2.25%, that the desalting of the product is not required. Accordingly, the major loss of water-soluble nutrients which accompanies the leaching operation may be avoided. The sugar present is fermented principally to lactic acid. Although cabbage is not considered to be a food of relatively high nutritive value, it has been

TABLE 12.4

COMPOSITION OF FOODS
(100-GM EDIBLE PORTION)

	% Water	Calories Food Energy	% Protein	% Fat	gm Carbohydrate Total	gm Ash	mg Calcium	mg Phosphorus	IU Vitamin A Value	mg Thiamin	mg Riboflavin	mg Niacin	mg Ascorbic Acid
Vegetables Before Brining for Use in Pickles													
Cauliflower	91.0	27	2.7	.2	5.2	.9	25	56	60	.05	.10	.7	78
Cucumbers	95.1	15	.9	.1	3.4	.5	25	27	250	.03	.04	.2	11
Onions	89.1	38	1.5	.1	8.7	.6	27	36	40	.03	.04	.2	10
Peppers													
Green	93.4	22	1.2	.2	4.8	.4	9	22	420	.08	.08	.5	128
Red	90.7	31	1.4	.3	7.1	.5	13	30	4450	(.08)	(.08)	(.5)	204
Pickles													
Cucumber Pickles													
Dill	93.3	11	.7	.2	2.2	3.6	26	21	100	Tr.	.02	Tr.	6
Sour	94.8	10	.5	.2	2.0	2.5	17	15	100	Tr.	.02	Tr.	7
Sweet	60.7	146	.7	.4	36.5	1.7	12	16	90	Tr.	.02	Tr.	6
Olives-green	78.2	116	1.4	12.7	1.3	6.4	61	17	300	—	—	—	—
ripe	80.0	129	1.1	13.8	2.6	2.5	84	16	60	Tr.	Tr.	—	—
Vegetables for Use in Sauerkraut and Sauerkraut-like Products													
Cabbage	92.4	24	1.3	.2	5.4	.7	49	29	130	.05	.05	.3	42

Kale	87.5	38	4.2	.8	6.0	1.5	179	73	8900	—	—	125
Radish, Oriental	94.1	19	.9	.1	4.2	.7	35	26	10	.03	.02	32
Turnips	91.5	30	1.0	.2	6.6	.7	39	30	Tr.	.04	.07	36
Sauerkraut (from cabbage)	92.8	18	1.0	.2	4.0	2.0	36	18	250	.03	.04	14
Vegetable Before Brining for Non-Pickle Use												
Beans												
Snap, green	90.1	32	1.9	.2	7.1	.7	56	44	600	.08	.11	19
Lima	67.5	123	8.4	.5	22.1	1.5	52	142	290	.24	.12	29
Carrots	88.2	42	1.1	.2	9.7	.8	37	36	11000	.06	.05	8
Corn, sweet	72.7	96	3.5	1.0	22.1	.7	3	111	400	.15	.12	12
Okra	88.9	36	2.4	.3	7.6	.8	92	51	520	(.17)	(.21)	31
Peas, green	78.0	84	6.3	.4	14.4	.9	26	116	640	.35	.14	27
Grapes												
American type	81.6	69	1.3	1.0	15.7	.4	16	12	100	(.05)	(.03)	4
European type	81.4	67	.6	.3	17.3	.4	12	20	(100)	.05	.03	4
Wine												
Table	85.6	85	.1	0	4.2	.2	9	10	—	Tr.	.01	—
Dessert	76.7	137	.1	0	7.7	.2	8	—	—	Tr.	.01	0
Beer	92.1	42	.3	0	3.8	.2	5	30	—	Tr.	.03	—
Milk	87.4	65	3.5	3.5	4.9	.7	118	93	140	.03	.17	1

Source: Watt and Merrill (1963).

reported by many investigators to be a very good source of vitamin C (Pederson and Albury 1969). Sauerkraut is generally recognized as a good dietary source of vitamin C, or ascorbic acid. Pederson *et al.* (1939) reported that the vitamin C content of kraut during the active fermentation period was equal to that of the original cabbage. A slow, progressive loss of vitamin C was found to occur during vat storage of kraut after the fermentation was complete. During preheating in the canning operation further oxidative destruction occurred ranging from 25 to 33%. Stability of the vitamin in the canned product was found to be relatively good by these investigators. More recently, Pederson *et al.* (1956) have demonstrated marked ascorbic acid loss in canned kraut stored at elevated temperatures. Loss of vitamin C following canning was reported by Gangopadhyay and Mukherjee (1971). The composition of sauerkraut and cabbage according to Watt and Merrill (1963) is shown in Table 12.4.

Fermented vegetable foods are extensively prepared by traditional methods. Such foods are important in the diets of many peoples, especially in Eastern Europe and in the Orient.

Orillo *et al.* (1969) reported on the preparation of sauerkraut-type vegetable blends into which boiled soybeans were incorporated at levels of 18 and 32%. Salt was added at a rate of 2%. Fermentation was similar to that of sauerkraut. The soybeans were well preserved; possessed a chewy, peanut-like character; and greatly increased the protein content of a commonly-prepared fermented food. Pederson (1971) briefly discusses the preservation of many foods by traditional fermentation methods.

Vegetables Salted or Brined for Nonpickle Use

Okra is commercially preserved by salting for use in the manufacture of certain canned products such as vegetable soup mixtures. The salted okra, which contains about 20% salt, is not desalted before use but rather is added to the product in such quantity that the need for salt for seasoning is met. Soluble nutrient losses from salted okra are at a minimum as a result of this manner of use of the preserved product.

Brining and salting has served as an alternative to canning or dehydration in the emergency preservation of corn, peas, lima beans, green beans, carrots, and other vegetables for nonpickle use. Reports of such brining and salting studies and of nutrient alteration caused by the procedures followed have been made by Blum and Fabian (1943), Etchells and Jones (1943B), Etchells *et al.* (1943), Wadsworth and Fabian (1944), Jones and Etchells (1944), and Fabian and

Wadsworth (1945). Although nutrient retention was variable, dependent upon the vegetable and the details of the preservation treatment, the following summary is indicative of the findings of the above-mentioned investigators. If a desalting operation was required, losses of ascorbic acid and of the B vitamins were nearly 100%; and of β-carotene about 30–50%. If desalting was not practiced losses of ascorbic acid were about 40–80% and of β-carotene about 10–30%. Jones and Etchells (1944) reported protein losses of 6–53% during the brining of green beans, dependent upon treatment. Quantitative carbohydrate retention studies were not reported. The results of other studies would lead to the expectation that with treatments favoring fermentation or requiring desalting the soluble carbohydrate losses would approach 100%.

FOODS PRODUCED MAINLY BY FUNGAL FERMENTATION

Throughout the Orient and parts of Africa and South America many of the traditional foods are fermentation products made from soybeans, rice, wheat, beans, and other high protein sources, including fish. Fermentation is usually brought about by an appropriate mold. End products are foods of pronounced and characteristic flavor and aroma.

Many of the fermented foods make important contribution to the diet as sources of protein, energy, and vitamins. Other fermentation products, such as soy sauce and fish paste, are important principally as condiments (Van Veen and Steinkraus 1970). The number of such traditional foods is large; certain of these are characteristic for a region or people. Others are common to a large area of the Orient but may be known under different names in different regions. Hesseltine (1965) and Pederson (1971) have listed and characterized many such foods.

Shown in Table 12.5 are the names and certain descriptions of several important fermented foods, mainly products of soybeans with or without cereals or other high protein substances. Also listed are the principal fermentative organisms and commonly-used substrates.

Tempeh

Tempeh is a mold-fermentation product of partially cooked soybeans high in protein and unsaturated oil. It is extensively served as a main dish in Indonesian meals.

Traditionally, dry soybeans are soaked overnight in cool water or for an hour or more in hot water to hydrate and to loosen the skins. Following removal of skins the beans are partially cooked by boiling or steaming for 30–120 min; cooled and inoculated with scrapings

NUTRITIONAL EVALUATION OF FOOD PROCESSING

TABLE 12.5

SOME NUTRITIOUS FERMENTED FOODS

Name	Organisms Used	Substrate	Additives	Use	Product Nature	Region	References[1]
Tempeh	*Rhizopus oligosporus*	Soybeans	None	Protein source	Solid	Indonesia and vicinity	1, 2, 3, 4, 5, 6, 34
Miso	*Aspergillus oryzae*	Soybeans, rice or barley	Salt 4–13%	Soup, Spreads	Paste	Japan, China. Some other parts of Orient	1, 4, 6, 18, 34
Natto	*Bacillus subtilis*	Soybeans	None	Protein source	Solid	Japan	1, 4, 6, 18, 36
Shoyu (soy sauce)	*Aspergillus oryzae*, *Lactobacillus*, *Hansenula* and *Saccharomyces*	Soybeans, wheat	Salt 18%	Condiment	Liquid	China, Japan, Phillipines. Other parts of Orient.	1, 4, 6, 38
Idli	*Leuconostoc mesenteroides*	Black gram, rice	None	Protein source	Thin dough	South India	1, 22
Sufu	*Actinomucor elegans*	Soybean curd (tofu)	Brine after fermentation	Protein, calcium	Soft cheese	China	1, 4, 25, 26, 28, 34
Sierra rice	*Aspergillus flavus*	Rice	None	Diet staple	Solid	Ecuador	6, 31
Fish sauce, paste	Enzymes native to fish	Small fish	Salt 12–20%	Relishes	Liquid, sauce	Southeast Asia	1, 29, 37
Kisk	Lactic acid bacteria	Wheat parboiled	Yoghurt	Diet staple	Solid	Near East	1, 30
Ontjom	*Rhizopus oligosporus* or *Neurospora sitophiia*	Peanut press cake	Carbohydrate such as tapioca, potato	Protein source	Solid White Red	South India	1, 4, 35

Source: From Hesseltine (1965), modified.

[1] References:
1) Hesseltine (1965); 2) Hesseltine et al. (1963); 3) Steinkraus et al. (1960); 4) Hesseltine and Wang (1972); 5) Platt (1964); 6) Pederson (1971); 18) Sano (1961); 22) Khandwala et al. (1962); 25) Wang and Hesseltine (1970); 26) Chang and Murray (1949); 28) Hyashi (1957B); 29) Tan et al. (1967); 30) Van Veen et al. (1969); 31) Van Veen et al. (1968B); 34) Hesseltine (1967); 35) Van Veen et al. (1968A); 36) Sakurai and Nakano (1961); 37) Van Veen (1965); 38) Smith (1949).

from previously made tempeh; and permitted to ferment for 24-48 hr at room temperature. Tempeh is usually consumed the day it is made but may be preserved by drying.

In experimental studies modeled rather closely after the traditional methods, Steinkraus et al. (1965) reported that with whole beans total solids losses varied from about 22 to 27% depending upon variety and method of dehulling. These losses were distributed as follows: 8–12% for dehulling, 9–12% for soaking and cooking, and about 2–4% during fermentation. Nitrogen losses, according to Smith et al. (1964), during dehulling, soaking and cooking for two varieties were 3.9 and 8.0% respectively; during fermentation, losses were 0.8 and 1.7% respectively. Mechanically dehulled soybean grits are suitable for use in the manufacture of tempeh. With grits, the soaking time may be greatly reduced but the total solids losses during soaking and cooking range from 33 to 38% (Smith et al. 1964).

During fermentation, crude protein content changed only slightly but solubilization occurred to the extent of about 50% (Murata et al. 1967; Van Buren et al. 1972). Similarly, only slight change in fat content was reported but lipid hydrolysis was about 35% (Wagenknecht et al. 1961; Van Buren et al. 1972).

Smith et al. (1964) reported that values for the amino acids in tempeh, fermented up to 30 hr, were not significantly different than those for cooked soybeans. Stillings and Hackler (1965) observed that, in general, the amino acid content of tempeh remained unchanged or declined slightly during a 72-hr fermentation period and that such losses were attributed in part to the utilization by the fermentative mold. These investigators reported, however, that the tryptophan values increased significantly during the 0–24-hr fermentation period but declined thereafter.

Steinkraus et al. (1961) reported losses of 4 and 10% for methionine and lysine respectively for a 36-hr fermentation period; and of 11 and 24% for methionine and lysine respectively for a fermentation period of 60 hr. Thiamin content was reduced by 50% or more, but riboflavin content was more than doubled and niacin and B-12 contents were increased many fold during tempeh manufacture (Steinkraus et al. 1961; Roelofsen and Talens 1964; Murata et al. 1967).

Miso

Miso is a very commonly-used food of Japan, China, and other parts of the Orient. The composition of miso will vary greatly, dependent upon the proportionate amounts of salt, soybeans, and rice or barley used. The preparation of the soybeans is similar to that in the making of tempeh; therefore, solids and nitrogen losses associated with dehulling, soaking, and cooking of soybeans should be comparable with such losses in tempeh manufacture. Chemically, the composition of miso was reported as follows: moisture, 48-52%; protein, 8.3-23.5%; crude oil, 1.6-10.5%; sugar, 3.2-20.4%; starch and dextrin, 3.2-10.0%; and sodium chloride, 4.0-12.8% (Shibasaki and Hesseltine 1962). The fermentation period will vary from 1 month or less to more than 1 yr, dependent upon the salt content, prevailing temperature, and the ratio of soybeans to rice or barley (Shibasaki and Hesseltine 1962; Hesseltine and Wang 1972). Shibasaki and Hesseltine (1962) state that, in general, miso is not a good vitamin source but cite studies by Mogi et al. (1952) of the fermentation of miso with the riboflavin-producing strain of A. oryzae for vitamin enrichment.

Natto

Natto is another extensively-used soybean food of Japan. The treatments followed in the preparation of soybeans for fermentation for natto production are similar to those used in the manufacture of tempeh and will account for similar solids losses. Natto is a product of fermentation by *B. subtilis* rather than *R. oligosporus* which is used for tempeh production. According to Hayashi (1957A), there is a synthesis of thiamin and riboflavin to the extent that the thiamin content of natto is equivalent to that of the soybeans before cooking and the riboflavin content considerably exceeds that present in the original soybeans. Sano (1961) reported that the thiamin and riboflavin content of natto was about 3-fold that of the original soybeans and the B-12 content was increased nearly 5-fold.

Shoyu

Soy sauce, known as shoyu in Japan and chiang-yu in China, is the most important fermented food product of these countries (Hesseltine and Wang 1972). In Japan, shoyu is made from a mixture of soybeans and wheat, according to a ratio of 45:55. In China, chiang-yu may be derived from soybeans only or from a mixture of soybeans and grain, in which soybeans represent the larger portion. In the manufacturing process the soybeans are soaked, autoclaved at 10-13 lb steam pressure for 1 hr, cooled, and mixed with roasted or steamed wheat or wheat bran and with salt brine. Fermentation is carried out for 3-12 months, at which time the liquid is pressed from the cake and pasteurized (Smith 1949; Hesseltine 1965). Solubilization of the proteins, lipids, and starch is nearly complete as the result of the fermentations. For a discussion of the composition of soy sauce see Smith (1949), Yokotsuka (1960).

Idli

In the preparation of idli, the dehulled black gram is soaked in water for 6 hr; is drained and ground to a paste with previously parboiled or soaked rice. The proportions of black gram and rice may be 1:1 or may vary considerably, depending upon the flavor desired and the economics of preparation. Water is added to the paste in amounts of 1.5-2.2 parts per part of dry ingredients. The fermentation period is 15 hr or more (Steinkraus et al. 1967).

The soaking of the ingredients favors the loss of soluble nutrients, for example the B vitamins in particular. Data pertaining to such loss is exceedingly limited. With respect to changes in vitamin content during the fermentation process, the results are somewhat contradictory. Rajalakshmi and Vanaja (1967) reported a 2-3-fold increase in thiamin and riboflavin content of the fermented over the unfermented idli; Steinkraus et al. (1967) observed a small but significant loss of riboflavin during fermentation; and Ananthachar and Desikachar (1962) concluded that the thiamin content of idli was not altered by fermentation. Rao (1961) reported that the methionine content of idli was 20% greater than that of the unfermented preparation.

Sufu

As indicated in Table 12.5, sufu is the mold-fermented product of soybean curd or tofu. Following a fermentation of from 3-7 days the sufu is aged and stored in brine to which may have been added wine, fermented rice mash, or hot pepper for flavoring purposes. During aging, the protein and lipids of the soybean curd are digested by the enzymes of the mold mycelia.

In the preparation of tofu, soybeans are ground or milled with water to form soybean milk which is strained to remove the insoluble solids; and a curd is formed by coagulation of the heated "milk" with a solution of magnesium or calcium salt. The curd is then formed into a block of soft cheese in a mold lined with cloth to serve as a filter. The weight of water used in making the milk is about ten times that of the dry soybeans. Tofu is a bland, fiber-free, high protein product having a composition approximating 80% water, 10% protein, and 4% lipids (Wang and Hesseltine 1970). The curd represents, by rough estimate, a recovery of 40% of the protein and 33% of the lipids of the original quantity of soybeans taken. The fraction termed as a carbohydrate, representing about 30% of the dry soybean, is not a constituent of importance in the curd and is essentially discarded.

If tofu is formed by addition of calcium salt, the calcium content of tofu is about 170 mg per 100 gm (Chang and Murray 1949; Miller *et al.* 1952), which is somewhat higher than that of cow's milk. Chang and Murray (1949) reported that when soybean curd was separated from the milk, 50% of the thiamin and 25% of the nicotinic acid were retained. The riboflavin content of the curd was similar to that of the beans from which it was made. Miller *et al.* (1952) indicated a retention of thiamin, riboflavin, and niacin of approximately 20% in the preparation of soybean curd.

Other Fermented Foods

In the manufacture of low salt fish paste, Tan *et al.* (1967) reported small increases in riboflavin content during fermentation. Fish paste is a good source of calcium.

Van Veen *et al.* (1969), in studies of the preparation of fermented milk-wheat combinations (kisk), reported slight increases in riboflavin content following fermentation. The product analyzed undoubtedly had not been subjected to the riboflavin-destructive exposure which would occur in the sun-drying operation traditional in regions where kisk is an important item in the diet.

During the laboratory preparation of Sierra rice by fermentation with *Aspergillus flavus* and with *Bacillus subtilis*, riboflavin content was increased by 2-fold and 5-fold, respectively (Van Veen *et al.* (1968B).

Some Advantages of Food Fermentation

The significance of food fermentation as it relates to the high protein products which have been listed in Table 12.5 may be considered from many viewpoints. Evidence has been given concerning the preservative or spoilage-limiting effect and of the enhancement of the product through vitamin synthesis. Hesseltine (1965) cites the probable importance of the presence of desirable enzymes of microbial origin. In tempeh, for example, the strongly proteolytic enzymes of the *Rhizopus* bring about rapid hydrolysis of protein to amino acids; in miso, the starch of the rice as well as the protein of the soybeans are hydrolyzed by the enzymes of the *Aspergillus oryzae*. Such changes undoubtedly improve the digestibility of the foods and reduce cooking time requirements. Wang and Hesseltine (1970) refer to the production of an antibacterial compound by *R. oligosporus* and other strains of *Rhizopus*, which has been demonstrated to not only minimize infection from certain microorganisms but to exert a growth-stimulating effect on animals fed an inadequate diet. They consider that the true nutritional value of tempeh and certain other fermented products is explained, in part, by the fact that antibiotics are formed by the fermentative molds.

Hesseltine (1967) and Hesseltine et al. (1966) have reported studies relative to aflatoxin formation by *Aspergillus oryzae* and by *Aspergillus flavus*. These investigators indicated that there was no evidence of the formation of the toxin by the 53 strains of *A. oryzae* used commercially. Furthermore, no toxin was found in products tested which had been fermented by *A. oryzae*. These products had been prepared using pure tane koji, a starter made with pure cultures of *A. oryzae*. The investigators stated that poor koji starters obtained from some Oriental sources could be expected to be contaminated with *A. flavus*.

Liener (1972) discussed the nutritive value of tempeh, natto, miso, and soybean curd and compared the quality of these foods with unfermented soybeans and soybean products. The nutritional quality and wholesomeness of a number of fermented foods have been evaluated by Van Veen and Steinkraus (1970) and by Steinkraus and Van Veen (1971).

FERMENTATION OF ALCOHOLIC BEVERAGES

Alcoholic beverages of great variety have been made by the people of many lands since prehistoric times. Many of these beverages are prized primarily because of their euphoric and intoxicating nature. Wines and beers, however, have for centuries been important in the diet of man; currently, they are made industrially under carefully controlled conditions and also by traditional methods.

Wine

In the strictest sense, wine is the naturally fermented juice of fresh grapes. For the purpose of evaluating wine-making practices, attention will be directed to the industrial preparation of grape wine. Steps taken during wine making which distinctly alter one or more of the nutrients present in the fruit are: grape crushing, sulfiting of must or juice, and fermentation of juice.

The composition of grapes and grape musts varies greatly with respect to sugar and vitamins, according to a survey of the literature by Amerine et al. (1972). Such variation is not surprising considering the large number of grape varieties and the wide range of climate and soils of grape production areas. Ascorbic acid was reported to be present in fresh grapes, varying from 1 to 18 mg per 100 gm but is usually found at levels below 8 mg. This vitamin is rapidly destroyed following grape crushing. Thiamin ranged from 0.1 to 1.2 mg per liter in musts but is lost to a marked extent as the result of sulfiting. Riboflavin was found in musts in amounts up to 1.5 mg per liter but is subject to loss by sulfiting and during fining. Such loss may approximate 50%. The nicotinic acid content of musts ranged from 0.3 to 8.8 mg per liter. The concentration of pyridoxine, pantothenic acid, biotin, inositol, p-aminobenzoic acid, choline, and folic acid in musts were also reported.

The vitamin content values shown above indicate the grape to be a relatively poor source of vitamins. In the processing of grapes into wine, changes in vitamin content have been recorded. Peynaud and Lafourcade (1957) as reported by Amerine et al. (1972) have indicated the magnitude of changes of nine B vitamins in the preparation of white and red wines from grape musts. Observations pertaining to thiamin, nicotinic acid, and riboflavin will be discussed.

For thiamin the loss was 97% or greater and for niacin losses approximating 50% were found. The riboflavin content of the musts in this study was very low. When musts were made into white and red wines the average riboflavin content was increased about 50 and 800% respectively. Even with such a large increase, the maximum riboflavin content of red wine was of the order of 0.025 mg per 100 ml, hardly at a level to make this beverage an important source of this vitamin.

The sugar content of wines is very greatly reduced during fermentation. Ethanol and carbon dioxide are principal end products. Depending upon the type of wine produced, sugar may be added in varying amounts following the fermentation.

Without doubt the pleasing flavor and delightful aroma of grape wine are ample justification for the high position which this beverage has attained in the diet. The composition of two types of grapes and of a dessert wine and a table wine as shown by Watt and Merrill (1963) is indicated in Table 12.4.

Wine-like beverages from the juices of many fruits, berries, and certain vegetables are made by industrial and traditional methods. Honey and the fruits and saps of palms, agaves, and cacti are also used in the preparation of fermented drinks, primarily according to traditional practices. Many of these wine-like beverages are briefly discussed and characterized by Pederson (1971).

Beer

Beer is the beverage formed by the alcoholic fermentation of wort, the water extract of malted cereal. Malt adjuncts and flavoring substances may be added to the malted cereal before wort formation. Malt adjuncts are sources of additional fermentable carbohydrates.

In the industrial brewing of beer, malt adjuncts may consist of grits, meal, or flakes of corn or rice; or possibly starch, sugar, or syrups. Hops may be added for flavoring and to aid in beverage clarification. The major brewing operations are malting or sprouting of barley; kilning to dry the malt and produce desired flavor and color characteristics; mashing, or preparation of the wort; separation of the wort and sparging of the grain residues; wort boiling to sterilize the wort and to extract hop flavor; fermentation; maturing; and packaging (Prescott and Dunn 1959; Tenney 1954).

The starches of the malt and adjuncts are converted to dextrins and soluble carbohydrates by the amylolytic enzymes of the malt during mashing. Most of the sugar in the wort is fermented to alcohol by yeast; some sugar is utilized in the production of yeast; and the dextrins and a small quantity of sugar remain in the beer.

A portion of the protein of the malt is hydrolyzed to peptones, peptides, and amino acids during mashing by the proteolytic enzymes of the malt. The amino acids serve as nutrients for the yeast. Soluble proteins are precipitated during wort boiling. Beer contains small quantities of peptones, peptides, and amino acids.

Changes in vitamin content take place during brewing, as indicated according to Scriban (1969). Thiamin from the barley is quite stable during malting and mashing, but is chiefly utilized by the yeast during fermentation and little remains in the finished beer. Riboflavin increases during malting; it is synthesized and excreted into the wort during fermentation by the yeast in such amounts as to be nutritionally valuable. Nicotinic acid, pyridoxine, biotin, and pantothenic acid increase during malting. These vitamins undergo little change during mashing and wort boiling; and are reduced to some extent during fermentation but remain in the beer in considerable amounts. Ascorbic acid is formed during malting but is destroyed during the kilning operation. Scriban (1969) suggests the enrichment of beer with vitamins by conservation of vitamins previously formed or by supplementation.

Nutritive by-products of the brewing industry are the extracted brewers' grains and surplus yeast. Both of these products are rich sources of vitamins of the B complex and are good protein sources (Tenney 1954). The brewers' grain is used as an animal feed; the yeast as human food and animal feed.

Beer will vary in composition depending upon brewing practices. Shown in Table 12.4 is the composition of beer as listed by Watt and Merrill (1963).

Beer-like beverages are products of nearly every country and are made from basic ingredients typical of the area. Microorganisms for enzyme production and beverage fermentation vary greatly depending upon the ingredients and area of production.

Sake is a very popular Japanese beverage and is brewed on a large commercial scale. Rice is the starch source and the starter, or koji, used for saccharification is prepared from steamed rice inoculated with the mold *Aspergillus oryzae*. Sake contains about 12-15% alcohol, 3% solids, and 0.3% lactic acid (Pederson 1971). Ohwaki and Lewis (1970) report alcohol contents of 14-20%.

Kaffir beer and related beer-like beverages prepared by traditional methods are time-honored examples of products prepared from cereals enhanced in vitamin content by fermentation. Platt (1964) refers to this native method of food fermentation, widely practiced in Africa, as "biological ennoblement" that is "improvement of the nutritive value of foods and dietary regimens by biological agencies."

In the preparation of Kaffir or maize beer as reported by Platt and Webb (1946) and Platt (1964), a malt is formed from sprouted Kaffir corn, maize, or millet possessing diastatic activity arising primarily from the growth of *Aspergillus oryzae* or *Mucor rouxii* on the grain during malting. A mash of ground malt and macerated plantains, cassava, or unmalted grain is brewed for from 1-5 days. Saccharification and alcoholic and lactic acid fermentation take place, arising from naturally-occurring inoculation of the grain during malting. The vessels used for malting are the source of the inoculum. End products of the fermentation are soluble carbohydrates, proteins, alcohol, lactic acid, and B vitamins. The beer is coarsely strained, has the consistency of a thin gruel, and will contain most of the constituents of the materials used as malt and as starch sources. Average daily consumption of the beer is about 5 pt per person. Platt (1964) indicated that the intake of thiamin, riboflavin, and nicotinic acid was increased by about 3, 105, and 74% respectively through the utilization of 40% of the maize of the daily diet for the preparation of 5 pt of beer. Such diversion of maize in the manufacture of the beer represents a caloric intake reduction of less than 3%.

Discussion of the preparation of nutritive beers by traditional methods in various parts of the world and of the characteristics of these beverages are to be found in reports by Pederson (1971), Miracle (1965), and others.

NUTRIENT SYNTHESIS

Microorganisms may serve as sources of nutrients and of foods and food supplements. Only brief mention will be made here relative to the status of and the prospects for the fermentation industries in nutrient synthesis. More detailed discussion of these subjects is to be found in references which will be cited.

Microbial Food

Yeast has been extensively used as a protein and vitamin supplement and for flavor enhancement. Yeast for this purpose has been obtained as a by-product of the brewing of beer or by fermentation in the utilization of wastes from the dairy, sugar refining, fruit and vegetable processing, and the pulp and paper industries (Prescott and Dunn 1959; Nickerson and Brown 1965; Peppler 1967; Frazier 1967; Bhattacharjee 1970; Marth 1970).

Currently, there is rapid expansion in research on the production of microbial food and single-cell protein (Casida 1968; Bhattacharjee 1970; Peppler 1968; Litchfield 1967; Bunker 1963, 1964, 1968; Gray 1959, 1970). Such food is the product of fermentations by algae, bacteria, fungi, and yeasts grown on a wide range of substrates and synthetic media including hydrocarbons, cellulose, pentosans, and solutions of inorganic salts.

Fats

Fats or lipids are synthesized in considerable quantity by certain algae, yeasts, yeast-like fungi, and molds (Bunker 1963; Nickerson and Brown 1965; Frazier 1967). Recovery of lipids is by solvent extraction. To date, lipid production from microorganisms has been employed only at times of emergency such as during a war.

Vitamins

Yeasts are the recognized source of many of the B-complex vitamins, either as the result of synthesis or of absorption from the medium. The addition of certain precursors to the medium is reported to favor synthesis of thiamin and pantothenic acid (Frazier 1967).

Riboflavin is a by-product of the butanol-acetone fermentation of molasses, grain, or whey by *Clostridium acetobutylicum* and related species. Riboflavin may also be produced by direct fermentation by a number of fungi, especially *Eremothecium ashbyii* and *Ashbya gossypii* (Frazier 1967; Goodwin 1963; Hanson 1967; Casida 1968). Demain (1972) has comprehensively reviewed riboflavin production by bacteria, yeasts, and molds. According to Perlman (1970) "chemically synthesized riboflavin has completely displaced the fermentation product on the U.S. market."

Vitamin B-12 may be synthesized by many different fermentation processes. It is found in activated sludge in considerable amounts. It may be recovered as a by-product of the production of the antibiotics streptomycin and Aureomycin. Widely-used commercial synthetic processes are based on fermentation by strains of *Propionibacterium* species (Perlman 1967; Frazier 1967).

β-Carotene is produced by various species of molds, yeasts, and algae. However, according to Ciegler (1965), the most intensive efforts to develop an industrial fermentation for β-carotene production have been with members of the *Choanephoraceae* family in the order of the *Mucorales*. Very high yields of fermentation β-carotene have been reported by Ciegler (1965) but were considered to be relatively low for good economic competition with chemically synthesized vitamin A (Casida 1968).

Ergosterol or provitamin D is produced industrially by synthesis by special strains of the yeast *Saccharomyces carlsburgenesis*. Conversion of ergosterol to vitamin D is accomplished by the ultraviolet irradiation of ergosterol-containing yeast cells or of the purified compound (Frazier 1967). There is not industrial production of the fat-soluble vitamins A, D, E, or K (Frazier 1967).

Organic Acids

Industrially produced lactic acid is primarily the product of homofermentative lactic acid bacteria or related acid formers. These may be *Lactobacillus delbrueckii*, *L. plantarum*, *L. bulgaricus*, *L. brevis*, and others. The raw materials may be industrial wastes such as those from sugar refining; the pulp and paper, and the dairy industries; or agricultural wastes (Prescott and Dunn 1959; Frazier 1967).

Citric acid is a product of mold fermentation principally by *Aspergillus niger*. Molasses is the preferred carbohydrate source (Prescott and Dunn 1959; Frazier 1967).

Amino Acids

Interest in large-scale production of certain amino acids arises from the potential use of these substances as food and feed supplements. Lysine, threonine, methionine, and tryptophan are the essential amino acids generally recognized as those most often limiting in plant protein. Glutamic acid has been required industrially in increasingly greater quantities in recent years for use in preparation of monosodium glutamate, a flavor-enhancing agent.

The amino acids may be made quite economically as racemic mixtures of DL-isomers by chemical synthesis (Casida 1968). It is only the L-isomers of the amino acids, however, which are biologically active and only L-glutamic acid which possesses value as a flavor-enhancer. Fermentation processes are the sources of pure L-isomers of the amino acids.

Lysine, glutamic acid, threonine, tryptophan, isoleucine, aspartic

EFFECTS OF PROCESSING BY FERMENTATION ON NUTRIENTS 349

acid, phenylalanine, tyrosine, and proline were products of commercial processes in 1970 (Perlman 1970). Production of amino acids by fermentation is briefly discussed by Kinoshita (1963), Frazier (1967), Dulaney (1967), and Casida (1968).

Methionine is extensively used in the United States, primarily in the supplementation of chick feed. For this purpose the chemically synthesized DL-methionine is used (Dulaney 1967). Production of methionine by fermentation processes based on the use of precursors has been reported by Dulaney (1967).

When the need for the commercial production of methionine and other amino acids by fermentation is established, suitable processes probably can be developed (Frazier 1967; Dulaney 1967; Casida 1968). A great need for lysine, threonine, tryptophan, and methionine has been recognized by Jansen and Howe (1964) who propose the use of these nutrients as supplements to avoid or overcome protein malnutrition in regions of the world in which the major part of the calorie intake is furnished by cereals, corn, or rice.

SUMMARY

Fermentation by selected acid-forming bacteria, yeasts, and molds is the basis for the preservation of a wide variety of foods by time-honored traditional methods and by refined, controlled industrial procedures. The addition of sodium chloride in varying amounts and at different stages of the preservation process is essential with certain fermentation methods.

Losses of nutrients such as sugars and water-soluble minerals and vitamins from the foods to be preserved may occur as the result of leaching of water-soluble constituents; destruction through exposure to light, heat, and oxygen; or by utilization by microorganisms associated with the fermentation process. Although such losses are in part unavoidable, their magnitude can be effectively controlled by adherence to good processing practices and by the employment of scientifically developed preservation procedures.

The fermentation process not only exerts preservative effects but is responsible for the development of changes in the texture, flavor, and aroma of food materials to make the fermented product more appealing, digestible, and nutritious. Enhancement of the vitamin content of the food through increase in the amounts of B complex vitamins, especially riboflavin, is characteristic of many of the fermentation methods of food preservation. Certain fungal fermentations may develop growth-stimulating antibiotic potency.

Nutrients commercially synthesized by industrial microbial processes include high quality protein, lipids, β-carotene, riboflavin, vitamin B-12, ethanol, lactic acid, citric acid, and at least nine amino acids.

BIBLIOGRAPHY

ANANTHACHAR, T. K., and DESIKACHAR, H. S. R. 1962. Effect of fermentation on the nutritive value of idli. J. Sci. Ind. Res. (India) *21C*, 191-192.
AMERINE, M. A., BERG, M. S., and CRUESS, W. V. 1972. The Technology of Wine Making, 3rd Edition. Avi Publishing Co., Westport, Conn.
BAER, D. 1970. Lactose deficiency and yoghurt. Social Biol. *17*, 143.
BAYLESS, T. M., and HUANG, S. 1969. Inadequate intestinal digestion of lactose. Am. J. Clin. Nutr. *22*, 250-255.
BELL, T. A., ETCHELLS, J. L., KELLING, R. E., and HONTZ, L. H. 1972. Low-sodium pickle products for modified diets. J. Am. Dietet. Assoc. *60*, 213-217.
BHATTACHARJEE, J. K. 1970. Microorganisms as potential sources of food. *In* Advances in Applied Microbiology, Vol. 13. Academic Press, New York.
BLUM, H. B., and FABIAN, F. W. 1943. The influence of salting on vitamins A and C in vegetables. Fruit Prod. J. *22*, 273-275, 283.
BORGSTROM, G. 1968. Principles of Food Science, Vol II. Macmillan Co., New York.
BUNKER, H. J. 1963. Microbial food. *In* Biochemistry of Industrial Organisms. C. Rainbow, and A. H. Rose (Editors). Academic Press, New York.
BUNKER, H. J. 1964. Microbial food. *In* Global Impacts of Applied Microbiology. M. P. Starr (Editor). John Wiley & Sons, New York.
BUNKER, H. J. 1968. Sources of single-cell protein; perspective and prospect. *In* Single-Cell Protein. R. I. Mateles, and S. R. Tannenbaum (Editors). MIT Press, Cambridge, Mass.
CAMILLO, L. J., HOPPERT, C. A., and FABIAN, F. W. 1942. An analytical study of cucumbers and cucumber pickles. Food Res. *7*, 339-352.
CASIDA, L. E., JR. 1968. Industrial Microbiology. John Wiley & Sons, New York.
CHANG, I. C., and MURRAY, H. C. 1949. Biological value of the proteins and the mineral, vitamin and amino acid content of soymilk and curd. Cereal Chem. *26*, 297-306.
CIEGLER, A. 1965. Microbial carotenogenesis. *In* Advances in Applied Microbiology, Vol. 7. W. W. Umbreit (Editor). Academic Press, New York.
CRUESS, W. V. 1931. Pickling green olives. Univ. Calif. Coll. Agr. Bull. *498*.
DEMAIN, A. L. 1972. Riboflavin oversynthesis. *In* Annual Review of Microbiology. C. E. Clifton, S. Raffel, and M. P. Starr (Editors). Annual Reviews, Inc., Palo Alto, Calif.
DULANEY, E. L. 1967. Microbial production of amino acids. *In* Microbial Technology. H. J. Peppler (Editor). Reinhold Publishing Corp., New York.
EDWARDS, C. H., and ALLEN, C. H. 1958. Cystine, tyrosine and essential amino acid content of selected foods of plant and animal origin. J. Agr. Food Chem. *6*, 219-223.
ETCHELLS, J. L., and JONES, I. D. 1943A. Bacteriological changes in cucumber fermentation. Food Ind. *15*, No. 2, 54-56.
ETCHELLS, J. L., and JONES, I. D. 1943B. Commercial brine preservation of vegetables. Fruit Prod. J. *22*, 242-246, 251, 253.
ETCHELLS, J. L., JONES, I. D., and HOFFMAN, M. H. 1943. Brine preservation of vegetables. Proc. Inst. Food Technologists, 176-182.
FABIAN, F. W., BRYAN, C. S., and ETCHELLS, J. L. 1932. Experimental work on cucumber fermentation. Mich. Agr. Expt. Sta. Tech. Bull. *126*.
FABIAN, F. W., and WADSWORTH, C. K. 1945. Salting beets, carrots, corn, green beans and spinach. Fruit Prod. J. *24*, 231-237, 249.

FRAZIER, W. C. 1967. Food Microbiology. McGraw-Hill Book Co., New York.
GANGOPADHYAY, H., and MUKHERJEE, S. 1971. Effect of different salt concentrations on the microflora and physico-chemical changes in sauerkraut fermentation. J. Food Sci. Technol. (Mysore) 8, 127-131.
GOODWIN, T. W. 1963. Vitamins. In Biochemistry of Industrial Microorganisms. C. Rainbow, and A. H. Rose (Editors). Academic Press, New York.
GRAY, W. D. 1959. The Relation of Fungi to Human Affairs. Henry Holt & Co., New York.
GRAY, W. D. 1970. The Use of Fungi as Food and In Food Processing. CRC Press, Cleveland.
HANSON, A. M. 1967. Microbial production of pigments and vitamins. In Microbial Technology. H. J. Peppler (Editor). Reinhold Publishing Corp., New York.
HARGROVE, R. E. 1970. Fermentation products from skim milk. In Byproducts from Milk, 2nd Edition. B. H. Webb, and E. O. Whittier (Editors). Avi Publishing Co., Westport, Conn.
HARTMAN, A. M., and DRYDEN, L. P. 1965. Vitamins in milk and milk products. J. Dairy Sci. Assoc., Am. Dairy Sci. Assoc., Champaign, Ill.
HARTMAN, A. M., and DRYDEN, L. P. 1974. Vitamins in milk and milk products. In Fundamentals of Dairy Chemistry, 2nd Edition. B. H. Webb, A. H. Johnson, and J. A. Alford (Editors). Avi Publishing Co., Westport, Conn.
HYASHI, S. 1957A. Manufacture of natto. Soybean Dig. 17, No. 5, 30-31.
HAYASHI, S. 1957B. Tofu takes large volume of soybeans. Soybean Dig. 17, No. 10, 26-27.
HESSELTINE, C. W. 1965. A millenium of fungi, food and fermentation. Mycologia 57, 149-197.
HESSELTINE, C. W. 1967. Fermented products—miso, sufu and tempeh. Proc. Intern. Conf. Soybean Protein Foods. USDA, Peoria, Ill. USDA Agr. Res. Serv. 71-35.
HESSELTINE, C. W., SHOTWELL, O. L., ELLIS, J. J., and STUBBLEFIELD, P. D. 1966. Aflatoxin formation by Aspergillus flavus. Bacteriol. Rev. 30, 795-805.
HESSELTINE, C. W., and WANG, H. L. 1972. Fermented soybean food products. In Soybeans: Chemistry and Technology, Vol. 1. A K. Smith, and S. J. Circle (Editors). Avi Publishing Co., Westport, Conn.
HUANG, S. S., and BAYLESS, T. M. 1968. Milk and lactose intolerance in healthy Orientals. Science 160, 83-84.
JANSEN, G. R., and HOWE, E. E. 1964. World problems in protein nutrition. Am. J. Clin. Nutr. 15, 262-274.
JONES, I. D., and ETCHELLS, J. L. 1943. Physical and chemical changes in cucumber fermentation. Food Ind. 15, No. 1, 62-64.
JONES, I. D., and ETCHELLS, J. L. 1944. Nutritive value of brined and fermented vegetables. Am. J. Public Health 34, 711-718.
KHANDWALA, P. K., AMBEGAOKAR, S. D., PATEL, S. M., and RAO, M. V. RADHAKRISHNA. 1962. Studies in fermented foods. Part 1. Nutritive value of idli. J. Sci. Ind. Res. (India) 21C, 275-278.
KINOSHITA, S. 1963. Amino acids. In Biochemistry of Industrial Microorganisms. C. Rainbow, and A. H. Rose (Editors). Academic Press, New York.
KON, S. K. 1959. Milk and Milk Products in Human Nutrition. Food Agr. Organ. United Nations, Rome. FAO Nutr. Studies 17.
KON, S. K. 1961. Nutritional effects on milk of chemical additives and processing. Federation Proc. 20, No. 1, Part III, 209-216.
KON, S. K. 1972. Milk and Milk Products in Human Nutrition. Food Agr. Organ. United Nations, Rome. FAO Nutr. Studies 27.
KON, S. K., and THOMPSON, S. Y. 1957. Measurement of vitamins in the control of milk processing. Milchwissenshaft 12, 166-172.

KOSIKOWSKI, F. 1966. Cheese and Fermented Milk Foods. Published by the author, Ithaca, N.Y.
KRISTOFFERSON, T., and GOULD, I. A. 1966. Manufacturing practices for cultured buttermilk in Ohio dairy plants. J. Diary Sci. *49*, 690-693.
LAMPERT, L. M. 1970. Modern Dairy Products. Chemical Publishing Co., New York.
LIENER, I. E. 1972. Nutritional value of food protein products. *In* Soybeans: Chemistry and Technology, Vol. 1. A. K. Smith, and S. J. Circle (Editors). Avi Publishing Co., Westport, Conn.
LITCHFIELD, J. H. 1967. Submerged culture of mushroom mycelium. *In* Microbial Technology. H. J. Peppler (Editor). Reinhold Publishing Corp., New York.
MARTH, E. H. 1970. Fermentation products from whey. *In* Byproducts from Milk, 2nd Edition. B. H. Webb, and E. O. Whittier (Editors). Avi Publishing Co., Westport, Conn.
MATTICK, E. C. V. 1951. The calcium and phosphorus of Cheddar and Stilton cheese. J. Dairy Res. *18*, 305-316.
MILLER, C. D., DENNING, H., and BAUER, A. 1952. Retention of nutrients in commercially prepared soybean curd. Food Res. *17*, 261-267.
MIRACLE, M. P. 1965. Food technology in tribal Africa. *In* Food Technology The World Over. M. S. Peterson, and D. K. Tressler (Editors). Avi Publishing Co., Westport, Conn.
MOGI, M., NAKALIMA, S., IQUCHI, N., and YOSHIDA, F. 1951, 1952. Fortified miso (fermented soybean paste). I. Trial brewing of miso with riboflavine-producing strain of *Aspergillus oryzae*. J. Ferment. Technol. *29*, 302-310 (1951): *30*, 363-369 (1952). Cited by H. Shibasaki and C. W. Hesseltine. Econ. Botan. *16*, 180-195.
MURATA, K., IKEHATA, H., and MIYAMOTO, T. 1967. Studies on the nutritional value of tempeh. J. Food Sci. *32*, 580-586.
NICKERSON, W. J., and BROWN, R. G. 1965. Uses and products of yeast and yeastlike fungi. Advances in Applied Microbiology, Vol. 7. W. W. Umbreit (Editor). Academic Press, New York.
OHWAKI, K., and LEWIS, M. J. 1970. Sake brewing. Wallerstein Lab. Commun. *31*, No. 111, 105-113.
ORILLO, C. A., SISON, E. C., LUIS, M., and PEDERSON, C. S. 1969. The fermentation of vegetable blends. Appl. Microbiol. *17*, No. 1, 10-13.
PEDERSON, C. S. 1971. Microbiology of Food Fermentations. Avi Publishing Co., Westport, Conn.
PEDERSON, C. S., and ALBURY, M. M. 1969. The sauerkraut fermentation. N.Y. State Agr. Expt. Sta. Bull. *824*.
PEDERSON, C. S., MACK, G. L., and ATHAWES, W. L. 1939. Vitamin C content of sauerkraut. Food Res. *4*, 31-45.
PEDERSON, C. S., and WARD, L. 1949. The effect of salt upon the bacteriological and chemical changes in fermenting cucumber. N.Y. State Agr. Expt. Sta. Bull. *273*.
PEDERSON, C. S., WHITCOMBE, J., and ROBINSON, W. B. 1956. The ascorbic acid content of sauerkraut. Food Technol. *10*, 365-367.
PEPPLER, H. J. 1967. Yeast technology. *In* Microbial Technology. H. J. Peppler (Editor). Reinhold Publishing Corp., New York.
PEPPLER, H. J. 1968. Industrial production of single-cell protein from carbohydrates. *In* Singe-Cell Protein. R. I. Mateles, and S. R. Tannenbaum (Editors). MIT Press, Cambridge, Mass.
PERLMAN, D. 1967. Microbial production of therapeutic compounds. *In* Microbial Technology. H. J. Peppler (Editor). Reinhold Publishing Corp., New York.
PERLMAN, D. 1970. Some prospects for the fermentation industries. Wallerstein Lab. Commun. *33*, No. 112, 165-173.
PEYNAUD, E., and LAFOURCADE, S. 1957. B vitamins in grape and wine.

Congres International Etude Scientifique Vin et Raisin, Bordeaux 1957, 65-70. Cited by Amerine, M. A., Berg, M. S., and Cruess, W. V. 1972. The Technology of Wine Making, 3rd Edition. Avi Publishing Co., Westport, Conn.
PLATT, B. S. 1964. Biological ennoblement: improvement of the nutritive value of foods and dietary regimens by biological agencies. Food Technol. *18*, No. 5, 68-76.
PLATT, B. S., and WEBB, R. A. 1946. Fermentation and human nutrition. Proc. Nutr. Soc. *4*, 132-140.
PRESCOTT, S. C., and DUNN, C. G. 1959. Industrial Microbiology, 3rd Edition. McGraw-Hill Book Co., New York.
RAJALAKSHMI, R., and VANAJA, K. 1967. Chemical and biological evaluation of the effects of fermentation on the nutritive value of foods prepared from rice and grains. Brit. J. Nutr. *21*, 467-472.
RAO, M. V. RADHAKRISHNA. 1961. Some observations on fermented foods. *In* Meeting the Protein Needs of Infants and Children. Natl. Acad. Sci.—Natl. Res. Council Publ. *843*.
ROELOFSEN, P. A., and TALENS, A. 1964. Changes in some B vitamins during molding of soybeans by *Rhizopus oryzae* in the production of tempeh kedelee. J. Food. Sci. *29*, 224-225.
SAKURAI, Y., and NAKANO, M. 1961. Production of high protein food from fermented soybean products. *In* Meeting the Protein Needs of Infants and Children. Natl. Acad. Sci.—Natl. Res. Council Publ. *843*.
SANO, T. 1961. Feeding studies with fermented soy products (natto and miso). *In* Meeting Protein Needs of Infants and Children. Natl. Acad. Sci.—Natl. Res. Council Publ. *843*.
SCRIBAN, R. 1969. Vitamins in barley, malt and beer. Brasserie *24*, No. 211, 489-503 (tr.). Abstr. Wallerstein Lab. Commun. *33*, No. 111, 135 (1970).
SHIBASAKI, K., and HESSELTINE, C. W. 1962. Miso fermentation. Econ. Botany *16*, 180-195.
SMITH, A. K. 1949. Oriental methods of using soybeans as food with special attention to fermented foods. USDA Northern Regional Res. Lab. Rept. *234*, June.
SMITH, A. K. *et al.* 1964. Tempeh: nutritive value in relation to processing. Cereal Chem. *41*, 173-181.
SMITH, J. A. B. 1970. Milk and milk products. *In* Proteins as Human Food. R. A. Lawrie (Editor). Avi Publishing Co., Westport, Conn.
STEINKRAUS, K. H. *et al.* 1960. Studies on tempeh—an Indonesian fermented soybean food. Food Res. *25*, 777-788.
STEINKRAUS, K. H., HAND, D. B., VAN BUREN, J. P., and HACKLER, L. R. 1961. Pilot plant studies on tempeh. Proc. Conf. Soybean Products for Proteins in Human Food, USDA, Peoria, Illinois. USDA Agr. Res. Serv. Rept. *71-22*.
STEINKRAUS, K. S., VAN BUREN, J. P., HACKLER, L. R., and HAND, D. B. 1965. A pilot-plant process for the production of dehydrated tempeh. Food Technol. *19*, No. 1, 63-68.
STEINKRAUS, K. H., and VAN VEEN, A. G. 1971. Biochemical, nutritional and organoleptic changes occurring during production of traditional fermented foods. *In* Global Impacts of Applied Microbiology (Third International Conference). Y. M. Freitas, and F. Fernandez (Editors). IBP—UNESCO Symposium.
STEINKRAUS, K. H., VAN VEEN, A. G., and THIEBEAU, D. B. 1967. Studies on idli—an Indian fermented black gram-rice food. Food Technol. *21*, 110-113.
STILLINGS, B. R., and HACKLER, L. R. 1965. Amino acid studies on the effect of fermentation time and heat-processing of tempeh. J. Food Sci. *30*, 1043-1048.
SULC, J. 1960. The production of yoghurt (using) vitamin C as enrichment.

Prumysl Potravin *11*, 33; Vyziva Lidu *15*, 6-7. (Czech) Cited by M. E. Gregory, J. Dairy Res. *29*, 220-232. 1962.

TAN, T. H., VAN VEEN, A. G., GRAHAM, D. C. W., and STEINKRAUS, K. H. 1967. The manufacture of low salt fish paste. Philippine Agriculturist *51*, No. 7, 626-636.

TENNEY, R. I. 1954. The brewing industry. In Industrial Fermentations, L. A. Underkofler and R. J. Hickey (Editors). Chemical Publishing Co., New York.

U.S. DEPT. OF AGR. 1965. USDA Fluid Milk and Cream Report, May 1965. Cited by F. Kosikowski. Cheese and Fermented Milk Foods, Published by the author, Ithaca, N.Y., 1966.

U.S. DEPT. OF AGR. 1968. National Food Situation, Nov. USDA *NFS-126*.

VAN BUREN, J. P., HACKLER, L. P., and STEINKRAUS, K. H. 1972. Solubilization of soybean tempeh constituents during fermentation. Cereal Chem. *49*, 208-211.

VAN VEEN, A. G. 1965. Fermented and dried seafood products in Southeast Asia. In Fish as Food, Vol. III. Georg Borgstrom (Editor). Academic Press, New York.

VAN VEEN, A. G., GRAHAM, D. C. W., and STEINKRAUS, K. H. 1968A. Fermented peanut press cake. Cereal Sci. Today *13*, No. 3, 96-98.

VAN VEEN, A. G., GRAHAM, D. C. W., and STEINKRAUS, K. H. 1968B. Fermented rice, a food from Ecuador. Archivos Latinoamericanos Nutr. *18*, No. 4, 363-373.

VAN VEEN, A. G., GRAHAM, D. C. W., and STEINKRAUS, K. H. 1969. Fermented milk-wheat combinations. Trop. Geograph. Med. *21*, 47-52.

VAN VEEN, A. G., and STEINKRAUS, K. H. 1970. Nutritive value and wholesomeness of fermented foods. J. Agr. Food Chem. *18*, 576-578.

VAUGHN, R. H., DOUGLAS, H. C., and GILLILAND, J. R. 1943. Production of Spanish-type Green Olives. Univ. Calif. Coll. Agr. Bull. *678*.

WADSWORTH, C. K., and FABIAN, F. W. 1944. Salting early June and Alaskan peas. Fruit Prod. J. *23*, 298-301, 316.

WAGENKNECHT, A. C., MATTICK, L. R., LEWIN, L. M., and HAND, D. B. 1961. Changes in soybean lipids during tempeh fermentation. J. Food Sci. *26*, 373-376.

WALTER, H. E., and HARGROVE, R. C. 1969. Cheese varieties and descriptions. USDA Agr. Handbook *54* (Revised).

WALTER, H. E., and McDONOUGH, F. E. 1970. Cheese and cheese products. In Byproducts from Milk, 2nd Edition. B. H. Webb, and E. O. Whittier (Editors). Avi Publishing Co., Westport, Conn.

WANG, H. L., and HESSELTINE, C. W. 1970. Sufu and lao-chao. J. Agr. Food Chem. *18*, 572-575.

WANNER, R. L. 1960. Effects of commercial processing of milk and milk products on their nutrient content. In Nutritional Evaluation of Food Processing. R. S. Harris, and H. Von Loesecke (Editors). John Wiley & Sons, New York. Reprinted in 1971 by Avi Publishing Co., Westport, Conn.

WATT, B. K., and MERRILL, A. L. 1963. Composition of Foods, Raw, Processed, Prepared. USDA Agr Handbook *8*.

YOKOTSUKA, T. 1960. Aroma and flavor of Japanese soy sauce. In Advances in Food Research, Vol. 10. C. O. Chichester, E. M. Mrak, and G. F. Stewart (Editors). Academic Press, New York.

CHAPTER 13

Effects of Processing by Additives on Nutrients

Henryk Daun

PART 1
Effects of Salting, Curing, and Smoking on Nutrients of Flesh Foods

Curing has been historically defined as an act of preserving meat and fish by salting, drying, and other related methods. Currently it is considered as a treatment of these flesh foods with a mixture containing sodium chloride, sodium nitrite, and/or sodium nitrate, as well as other components.

The function of salting, curing, and smoking has been changed with the development of newer preservation techniques, especially canning and freezing. They are now used mainly for the purpose of improving the organoleptic properties of food product. Accordingly, drastic chemical and physical treatments employed previously in these processes have been changed into milder ones, having less influence on the loss of nutritive quality.

It should be stressed, however, that the preservative aspect of salting, curing, and smoking remains important in many cases. In some countries heavy salting and smoking is still being used for the preservation of flesh foods.

SALTING

Sodium chloride is an indispensable component of food. At lower concentrations it contributes significantly to the flavor. At higher concentrations it exhibits an important bacteriostatic action. Salt is easily available and not expensive. For all these reasons, salting is commonly used in meat, fish, and poultry processing.

The most often-used concentrations correspond to the organoleptic requirements. The effect of these low concentrations of salt on the nutritive value of flesh food has not been studied extensively. Nutritional evaluation of processed foods containing salt shows indirectly, however, that the influence of low concentrations of salt is most probably not significant.

In some countries salting is still being used for the purpose of preservation. High concentrations of sodium chloride may alter many factors contributing to the nutritive value of various foods. Detailed discussion of different aspects of food salting is beyond the scope of this chapter and may be found in reviews (Joslyn and Timmons 1967; Dempster 1973). Only the nutritional effects of salting will be considered here.

Effect of Salting on Losses and Nutritive Value of Edible Fish Proteins

Salting is especially important in the fish industry. There are many methods of fish salting, depending on species, climate, and local traditions (van Veen 1953, 1965; Cutting 1961; Voskresensky 1965; Ingram and Kitchell 1967; Sikorski 1971; Olley 1972; Mendelsohn 1974). In some cases, significant losses occur during the operations preceding the salting (transport from the sea, trimming, etc.). The importance of these factors, which determine the nutritive value in its broadest sense, should be evaluated by a technological study and not only by the analysis of the products alone (Cutting 1961).

In general, fish is salted by mixing with dry salt (dry salting) or by soaking in a salt solution (brine salting). In both cases, the concentration of salt is much higher on the outer surface than inside the fish. As a result, a mass transfer takes place in both directions. With the liquids which exude from the fish flesh, some water-soluble proteins are lost. The nature of the interaction between salt and fish flesh is complicated. It was observed that light salting, in which the flesh takes up from 0.5 to 1% of salt significantly decreases the formation of drip after the fish have been frozen and thawed (Tarr 1962; Miyauchi 1965). On the other hand, with the initial low salt concentration in the flesh, cell proteins are being dissolved. The migration of proteins into the brine is significant during the first hours after the salting when the level of salt does not exceed 7%. Reay (1936) observed protein losses from 1% in dry salted herring to 5% in herring salted in a brine.

After salt concentration in the flesh is increased, cell proteins are being salted-out and their solubility decreases. The use of 10–25% salt, based on the dry weight of the fish, results in complete conversion of salt-soluble proteins to a salt-nonextractable form (Leonova 1970).

It is well known that fish proteins are denatured by salt (Duerr and Dyer 1952; Linko and Nikkilä 1961). Very little information, however, is available concerning how this denaturation may affect the digestability of the protein. According to Cutting (1961) and Munro and Morrison (1965), it is not associated with any loss of

EFFECTS OF PROCESSING BY ADDITIVES ON NUTRIENTS 357

nutritive value. The essential amino acids are not affected by salting (Adrian 1957; Chari and Venkataraman 1957). NPV (Net Protein Value) value of 69% has been reported by Orraca-Tetteh (1961). Ito (1962, 1970) maintains, however, that digestibility of fish protein with pepsin and pancreatin decreases significantly with increased salt concentration. Proximate composition of salted and/or smoked fish is presented in Table 13.1 according to Watt and Merrill (1963).

TABLE 13.1

PROXIMATE COMPOSITION OF SALTED AND/OR SMOKED FISH

Fish	Water (%)	Protein (%)	Fat (%)	Carbohydrate (%)	Energy (Cal/100Gm)	Ash (%)
Eel, smoked	50.2	18.6	27.8	0	330	2.4
Haddock, smoked	72.6	23.2	0.4	0	103	3.1
Halibut, smoked	49.4	20.8	15.0	0	224	15.0
Herring						
Bismark type, pickled	59.4	20.4	15.1	0	223	4.0
Salted or brined	53.8	19.0	15.2	0	218	12.0
Smoked, bloaters	64.0	19.6	12.4	0	196	3.2
Smoked, hard	34.6	36.9	15.8	0	300	13.2^1
Smoked, kippered	61.0	22.2	12.9	0	211	4.0^2
Mackerel						
Salted	43.0	18.5	25.1	0	305	13.0
Smoked	59.4	23.8	13.0	0	219	2.0
Salmon, smoked	58.9	21.6	9.3	0	176	9.4^3
Whitefish, lake, smoked	68.2	20.9	7.3	0	155	3.7^4

Source: Watt and Merrill (1963).
[1] Sodium 6,231 mg/100 gm; potassium, 157 mg/100 gm.
[2] Calcium 66 mg/100 gm; phosphorus 254 mg/100 gm; iron 1.4 mg/100 gm.
[3] Calcium 14 mg/100 gm; phosphorus 245 mg/100 gm.
[4] Calcium 22 mg/100 gm; phosphorus 274 mg/100 gm.

Composition may vary due to seasonal and other causes of the variability of the raw material (Burgess and Shewan 1970).

Fish sauces and fish pastes constitute products with high salt concentration. They are being prepared according to local traditions and represent an important source of protein in Southeast Asia (Amano 1961; van Veen 1953, 1965; Olley 1972). The essential amino acids of 19 Thai fish products were compared with values reported for teleost fish muscle. Most of the products showed fairly good essential amino acid patterns. Two products were very low in arginine (Sorasuchart 1972).

In general, the findings of Sorasuchart (1972) are in agreement with the data published by Cabat and Standal (1965). The statement made by Proctor and Lahiry (1956), that the methods of processing and preservation of fish do not affect the amino acid composition of the product, does not hold for fermented products.

Effect of Salting on Vitamin and Mineral Content of Fish

A low content of salt, from 0.5 to 1%, in fish flesh largely eliminates exudation of free liquid after the fish have been frozen and thawed. Most probably it reduces the loss of water-soluble vitamins and minerals at the same time. The data on the losses of these materials with liquids which exude from salted fish have not been found in available literature. It is obvious, however, that such losses take place and are somehow related to water losses occurring especially during the first days of fish salting. Weight losses of baltic herring salted for 48 hr (1:5 ratio of salt) at 10°C reached 17% (Sikorski 1971).

Previous reviews concerning the vitamin content in processed fish, including some salted products, have been published by Lovern (1944), Cutting (1961), and Tarr (1960, 1962).

It is generally assumed that salting does not have an appreciable effect on vitamins. It must be stressed, however, that this conclusion is made on the basis of vitamin assays on salted fish and comparative studies on identical raw material are in most instances not available. Vitamin content of salted and/or smoked fish is presented in Table 13.5.

The contents of the B-vitamins (thiamin, riboflavin, niacin, vitamin B-12, vitamin B-6, pantothenic acid, and biotin) were determined in 19 samples of Thai fish products by Sorasuchart (1971). The results were, in general, of the order reported for salted fish on a worldwide basis. B-vitamins varied fairly extensively in different products with thiamin being low in all products.

Salt exhibits prooxidant activity in fish flesh (Castell *et al.* 1965; Tarr 1969) and may increase losses of some vitamins and undesirable oxidative changes of fat.

No correlation has been found between salt content of the salted fish and the death rate from gastric cancer in Japan (Matsuhisa 1965).

CURING

Curing may be considered as an advanced form of salting where not only salt but nitrite and/or nitrate, sugar, and other ingredients are added to the raw material. The modern objective of curing is rather the production of a thermally stable meat pigment and formation of characteristic meat flavor than meat preservation. The preservative

EFFECTS OF PROCESSING BY ADDITIVES ON NUTRIENTS 359

aspect of curing remains, however, an important factor (Pivnick et al. 1967; Leistner et al. 1973). Detailed consideration of many aspects of the curing process is beyond the scope of this chapter and may be found in reviews (Lawrie 1966; Brissey and Goeser 1967; Bard and Townsend 1971; Townsend and Bard 1971. This chapter deals only with nutritional aspects of the curing process.

Effect of Curing on Protein Quality and Loss in Meat

There are numerous curing methods depending on the kind of raw material, available equipment, and tradition (Gibbons 1953; Lawrie 1966; Bard and Townsend 1971; Dempster 1973). Usually, meats are subjected to the curing; in some countries, however, fish is treated similarly (Tanikawa 1963, 1965).

Generally speaking the following ways are used to incorporate curing agents into the meat: 1. Dry rubbing (dry curing). 2. Immersion (brine curing, wet curing). 3. Arterial pumping. 4. Needle injection (stitch pumping). 5. Modification and combination of above mentioned methods.

As a result of the curing operations, liquids exude from the meat and some water-soluble proteins are lost. Losses of ham protein in the pickle solution are reported to be relatively small, ranging from 0.35 to 1.2% of total protein (Hoagland et al. 1947). Quick chilling appears to be associated with an increase of the fluid loss during curing. The lowest protein losses occur when a concentration of 26% salt is used in the brine (Daunoraviciute et al. 1966). In a newer method (Michels 1971), meat protein which normally diffuses into brine is retained when the brine is stirred and curing time is decreased. It is known that curing agents interact with meat proteins. The nature of these interactions is complicated and not fully understood. The denaturating effect of sodium chloride was observed with a brine concentration higher than 5% (Motoc and Banu 1967). Hydrolytic decomposition of proteins and some accumulation of free amino acids takes place in cured pork (Bolshakov et al. 1965). Several gases are formed when nitrite interacts with meat components under curing conditions (Walters and Casselden 1972; Woolford et al. 1972). Nitrites probably interact with SH-groups in meat (Mirna and Hofmann 1969).

It is usually assumed that the influence of curing agents on the digestability of cured meat proteins and the availability of essential amino acids is very small (Dunker et al. 1953; Schweigert and Lushbough 1960; Burger and Walters 1973). These conclusions, however, are made on the basis of protein assays of cured meat products and comparative studies on identical raw material are in most instances not available. Schweigert and Payne (1956) published information on nutritive value of several processed meats including amino acid composition. The nutrient content of the processed meat products is similar per unit of protein to that of unprocessed cuts of meat (Table 13.2). The same conclusions may be reached on the basis

TABLE 13.2

SUMMARY OF THE AMINO ACID COMPOSITION OF PORK, BEEF, LAMB, AND PROCESSED MEAT SAMPLES
(AMINO ACIDS AS PERCENTAGE IN THE PROTEIN)

Amino Acid	Pork	Lamb	Beef	Processed Meats
Leucine	7.53	7.42	8.40	7.36
Valine	4.97	5.00	5.71	5.24
Isoleucine	4.89	4.78	5.07	4.94
Methionine	2.50	2.32	2.32	2.21
Threonine	5.12	4.88	4.04	3.92
Phenylaline	4.14	3.94	4.02	3.95
Arginine	6.35	6.86	6.56	6.57
Histidine	3.23	2.68	2.94	2.83
Lysine	7.77	7.65	8.37	7.38
Tryptophan	1.35	1.32	1.10	1.04
Glutamic acid	14.51	14.35	14.35	12.90
Aspartic acid	8.92	8.46	8.75	9.10
Proline	4.60	4.80	5.40	5.23
Tyrosine	3.02	3.21	3.24	2.87
Glycine	6.10	6.74	7.11	7.98
Serine	3.97	3.93	3.77	4.18
Gystine	1.31	1.34	1.35	1.47
Alanine	6.30	6.30	6.40	6.40
Total nitrogen accounted for, %	85.4	85.6	87.8	85.7

Source: Schweigert and Payne (1956).

of newer data (see Table 13.3). Some authors (Belenkij 1965) maintain, however, that curing significantly influences the nutritive value of proteins.

Effect of Curing on Vitamin and Mineral Content of Meat

The data on losses of water-soluble vitamins and minerals with liquids which exude from cured meat have not been found in available literature. It is logical to assume, however, that such losses take place and are proportional to the water losses occurring during first days of curing.

Older works concerning the losses of vitamins as influenced by curing or by curing and smoking have been reviewed by Schweigert

EFFECTS OF PROCESSING BY ADDITIVES ON NUTRIENTS

TABLE 13.3

PROXIMATE COMPOSITION OF CURED AND/OR SMOKED MEATS

Meat	Water (%)	Protein (%)	Fat (%)	Carbohydrate (%)	Energy (Cal/100 Gm)	Ash (%)
Bacon, raw	19.3	8.4	69.3	1.0	665	2.0
Bacon, Canadian	61.7	20.0	14.4	0.3	216	3.6
Bologna	56.2	12.1	27.5	1.1	304	3.1
Boston butt, cured, raw	55.7	17.2	24.1	0	291	3.0
Country-style sausage	49.9	15.1	31.1	0	345	3.9
Corned beef, cooked	43.9	22.9	30.4	0	372	2.9
Frankfurters	55.6	12.5	27.6	1.8	309	2.5
Ham						
Canned	65.0	18.3	12.3	0.9	193	3.5
Country style	42.0	16.9	35.0	0.3	389	5.4
Cured, raw	56.5	17.5	23.0	0	282	3.0
Luncheon meat (pork)	54.9	15.0	24.9	1.3	294	3.9
Mortadella	48.9	20.4	25.0	0.6	315	5.1
Picnic, cured, raw	56.7	16.8	23.6	0	285	2.9
Polish-style sausage	53.7	15.7	25.8	1.2	304	3.6
Pork sausage, raw	38.1	9.4	50.8	tr	498	1.7
Salami, dry	29.8	23.8	38.1	1.2	450	7.1
Salt pork	8.0	3.9	85.0	0	783	3.5
Tongue, canned or pickled	56.6	19.3	20.3	0.3	267	3.5
Tongue, smoked	48.9	17.2	28.8	—	328	—
Vienna sausage	63.0	14.0	19.8	0.3	240	2.9

Source: Watt and Merrill (1963).

and Lushbough (1960) (see Table 13.4). The results given by Fields and Dunker (1952) and Dunker et al. (1953) were similar to the observations of Hoagland et al. (1947), but comparisons with uncured controls are not available and percentages of retained vitamins cannot be considered.

The vitamin and mineral content of processed meats is similar per unit of protein to that of unprocessed cuts of meat (see Tables 13.6 and 13.7) (Schweigert and Payne 1956; Watt and Merrill 1963; Kiernat et al. 1964). Losses of specific nutrients due to the curing appear to be very small. Thiamin is the only vitamin consistently lost in processes involving heat treatment (Beuk et al. 1950; Cremer and Büttner 1961). Thiamin losses may be increased considerably when cured meat is canned and pasteurized (Stoytchev et al. 1969).

Curing and Nitrosamines

Nitrosamines constitute a group of chemical compounds which have been shown to be carcinogenic (Magee and Barnes 1956; Druckrey and Schmähl 1962; Heath and Magee 1962; Ender et al. 1964;

Druckrey et al. 1967, 1969; Magee and Barnes 1967; Terracini et al. 1967; Lijinsky and Espstein 1970; Greenblatt and Lijinsky 1972). Nitrosamines exhibit different carcinogenic intensity and organ specificity. Some of them produce tumors in rats with daily doses of 0.005–1.0 mg per kg of body weight (Druckrey et al. 1967, 1969). It should be emphasized that even one single treatment with nitrosamines of 30 mg per kg of body weight was reported to induce cancer (Druckrey et al. 1969).

A high incidence of esophageal cancer among people in certain areas of Africa is being linked with local food containing dimethylnitrosamine (McGlashan et al. 1968; DuPlessis et al. 1969).

Nitrosamines have been detected in cured and smoked flesh food (Ender and Ceh 1968; Howard et al. 1970; Möhler and Mayerhofer 1969; Mirna 1970; Fiddler et al. 1972A, B, 1973; Wolff and Wasserman 1972; Sen et al. 1969B, 1973; Fazio et al. 1973A).

The concentrations found are far below levels known to be carcinogenic for experimental animals. However, the action of very small doses over a long period of time has not been established and needs further studies. Moreover, carcinogenic action may be exhibited by two or more carcinogens in subthreshold concentrations by summation of their action (Nakahara and Fukuoka 1960).

There is a concern that nitrosamines may be generated while curing is conducted (Ender and Ceh 1971; Eisenbrand 1973; Pfeil and Liepe 1973). Smoking may also cause the formation of nitrosamines since nitrogen oxide has been detected in wood smoke and secondary amines are present in flesh food.

Nitrosamines may be generated in the human body when precursors, i.e., amines and nitrites or nitrates, contact each other in the stomach (Sander 1967; Sander et al. 1968; Sander and Seif 1969; Sen et al. 1969A; Greenblatt et al. 1971; Hawksworth and Hill 1971; Lijinsky and Greenblatt 1972).

Mirvish et al. (1972) made a suggestion that ascorbic acid may be used to block the formation of N-nitrosamines from sodium nitrite and amines. It was confirmed in other reports that adequate levels of ascorbic acid or sodium ascorbate reduce or even eliminate the formation of nitrosamines (Greenblatt 1973; Ivankovic et al. 1973; Kamm et al. 1973; Mirvish et al. 1972, 1973; Newmark et al. 1974). Effective level of sodium ascorbate must be higher than presently allowed, i.e., 500 ppm.

Vitamin A may somehow be involved in reducing the formation of nitrosamines by interaction with nitrites and nitrates (Roberts and Sell 1963). Carotene and vitamin A are oxidized by nitrogen oxide. Nitrites are more active than nitrates in vitamin A destruction (Pugh

EFFECTS OF PROCESSING BY ADDITIVES ON NUTRIENTS   363

and Garner 1963). Vitamin A content in the liver was significantly reduced when animals were fed nitrate-containing feed (O'Dell et al. 1960; Bruggeman and Tiews 1964; Mitchell et al. 1965; Wood et al. 1967; Hoar et al. 1968). The direct influence of vitamin A on nitrosamine formation *in vivo* would be an interesting subject for further studies. On the basis of available information, it seems more appropriate to decrease the limit of the allowed level of nitrites from 200 ppm to 10-20 ppm rather than to leave it unchanged as suggested by the Committee on Nitrate Accumulation (1972).

SMOKING

Traditional smoking is a process in which characteristic properties of smoked products are formed by the combined action of heat, smoke components, and flow of gases. Smoke-curing has undergone significant changes during the last half century. What was once a practice based on trial and error and tradition has become a rational process with strictly controlled parameters such as smoke quality and quantity, heat transfer, level of humidity, and circulation of gases.

The progress has been made possible by the design and construction of modern smoking kilns with sophisticated recording and regulating instruments. The development of new methods of making smoke, such as the friction-type generation or the fluidized-bed technique, the introduction of "liquid smoke" concentrates, and the application of the electrostatic principle for accelerated smoke deposition, are further indications of the advancement in the field of smoke-curing.

New smoking methods are milder and produce a better, more uniform, hygienic, and wholesome product of higher nutritive value. It should be remembered, however, that in many countries, heavy smoking is still used primarily for flesh food preservation.

More detailed information concerning smoke-curing process may be found in reviews (Kurko 1960; Draudt 1963; Cutting 1965; Tilgner 1967, 1970; Gorbatov et al. 1971; Sikorski 1971; Hofmann 1972A, B; Reuter 1972).

Influence of Smoke Components on Nutritive Value of Smoked Flesh Foods

According to available literature, there are about 300 individual chemical compounds which have been identified in curing smoke (Hofmann 1972B). It is, however, generally assumed that the real number of smoke components is much higher.

The following groups of chemical compounds are considered to be present in wood smoke: carbonyls (aldehydes and ketones), organic acids, phenols, organic bases, alcohols, hydrocarbons (including polycyclic aromatic), and gases such as carbon dioxide, carbon monoxide, oxygen, and nitrogen.

On the basis of their influence on the nutritive value of smoked products, the smoke components may be divided into the following four groups:

1. Substances protecting losses of the nutritive value of smoked products by counteracting against undesirable chemical and biological changes (for example, antioxidants and bactericides).

2. Components exhibiting no action from the nutritive value point of view.

3. Compounds interacting with food components and decreasing the nutritive value of smoked products.

4. Toxic components.

Antioxidative Action of Smoking

The nutritive value of flesh food may be affected by oxidation. The undesirable reactions include the oxidative changes of fats, decrease of the biological value of proteins, and destruction of some vitamins. Smoked products are known to have an increased resistance to oxidative changes. Some recent results and literature review have been published by Daun (1969). The method of smoke generation may influence its antioxidant activity (Tilgner and Daun 1970). It was demonstrated that the particle smoke phase exhibits much higher antioxidant properties than the vapor phase (Tilgner et al. 1965), and that the method of separation of smoke components significantly influences their activity (Daun and Grabowska 1967). It is generally accepted that phenolic substances are mainly responsible for the antioxidative properties of curing-smoke (Kurko 1959; Tilgner et al. 1967). The antioxidant activity of curing-smoke components is an important property protecting losses of nutritive value of smoked products due to their oxidation.

Bactericidal Action of Smoking

Bactericidal properties of curing smoke are widely utilized in order to protect the nutritional quality of smoked products. It is especially important in the parts of the world where possibilities of canning and freezing are limited. The bactericidal action of smoking is a result of the combined influence of heating, drying, and chemical components of the smoke. Smoke components, however, constitute the major bactericidal factor in the overall process (Shewan 1949;

EFFECTS OF PROCESSING BY ADDITIVES ON NUTRIENTS 365

Kochanowski 1962). A detailed discussion of this subject has been published by Kurko (1960). The bactericidal action of smoking is a significant factor in the protection of the nutritive value of smoked products against biological destruction.

Effect of Smoking on Protein Quality and Loss in Flesh Foods

Usually, smoking follows salting or curing. It means that even before the beginning of smoking some proteins are lost and others are changed in their nutritive value, as discussed in the sections dealing with salting and curing. Smoking, however, produces further losses and changes caused by the action of heat, flow of gases, and interaction of smoke components and proteins.

In spite of the fact that heat treatment and drying are integral parts of the smoking process, their influence on the nutritive value of flesh foods will not be considered in detail here, because they are discussed in other chapters of this book. These processes are known to reduce the biological value of meat and fish proteins sometimes exceeding 50% (Platt 1961).

Under the influence of heat, proteins of flesh foods are denaturated. The process of denaturation already starts at 40°C; the major part of it taking part from 65° to 68°C. At 70°C meat becomes grey (if previously not cured) due to the denaturation of myoglobin and hemoglobin. Sufficient heating time is required in order to soften collagen in the connective tissue. In fish products, adequate texture is achieved already during the hot smoking procedure. The connective tissue of meat softens after longer heating and usually smoking is followed by additional heat treatment. Changes in the nutritive value of flesh foods proteins during smoking as influenced by heat are comparable to the changes caused by heat in other processes.

Classical smoking involves flow of gases and that, in turn, causes drying of smoked products. The chief alteration here is loss of water and an increase of protein and fat content per unit weight of flesh food as purchased. Weight losses may range from 3 to 30% (Shewan 1949). Any changes of the nutritive value that would take place because of regular dehydration may be expected to happen under smoking conditions. Additionally, losses of water cause an increase in the concentration of salt, other curing agents, and smoke components. This phenomenon is rather specific for smoking and most probably produces some additional changes in the nutritive value of smoked products.

Many technological aspects of heating and drying during the smoking process have been reviewed (Kurko 1960; Sikorski 1971).

The nutritive value of flesh food proteins may be changed by interaction with smoke components. Krylova and Bazarova (1960) observed increased chemical activity of functional groups as proteins were denaturated during smoking. Krylova et al. (1962) demonstrated that phenols and polyphenols react with sulfhydryl groups whereas carbonyls react with amino groups. The solubility properties of proteins of chicken and pork muscles were significantly decreased by smoking (Kihara 1962). Yuditskaya (1962) evaluated penetration and interaction of smoke components with fish flesh. One group of smoke components formed polymers with some of the substances in fish tissues being retained on the surface of the skin whereas the second group penetrated deeper into the flesh. Dvorak and Vognarova (1965A, B) and Inagami and Horii (1966) investigated the effects of smoking upon the biological qualities of meat and meat products. They observed 12% reduction in available lysine. The biological availability of the amino acids was detected by their digestability. Losses caused by long-term smoking were attributed to the effect of the formaldehyde present in the smoke. These losses increased logarithmically with the content of formaldehyde. Munro and Morrison (1965) reported, however, that smoking of cod had no effect on the availability of the amino acids and Nicora et al. (1966) maintained that smoked meat showed almost complete digestibility of all of the proteins. Some hydrolysis of the protein took place during storage. Kurko (1967) demonstrated that phenolic components of smoke interact with amino acids. Kako (1968) showed that proteins are being changed during smoking. It was established that carbonyls and reductone-like components of smoke also interact with amino acids (Kurko and Schmidt 1969). Ziemba (1969) showed that the color of the smoked products is formed at least partially by the interaction of carbonylic smoke components and proteins similar to the Maillard reaction. Randall and Bratzler (1970A, B, C) reported changes in the various protein properties of pork muscle after smoking. Untreated, heated, and heated and smoked samples were evaluated. The myofibrillar protein nitrogen fraction increased in the heated samples and decreased in the heated and smoked samples. The stroma fraction increased considerably in the heated and smoked samples. These studies indicated that smoke definitely caused changes in the solubility of protein, pH, and free sulfhydryl groups. These changes most probably may be explained by reactions of smoke components with functional groups of meat proteins. Acid phosphatase activity was also affected by smoking. Chen and Issenberg (1972) reported a loss of 44% of the available lysine in an uncured lean beef strip exposed to wood smoke for 10 hr. Heating in air caused a 15% loss

EFFECTS OF PROCESSING BY ADDITIVES ON NUTRIENTS 367

under the same conditions (65°C). Beef homogenate, treated with acidic, phenolic, or neutral fractions of smoke condensate, lost 14, 38, and 45% of the available lysine, respectively.

Effect of Smoking on Vitamin and Mineral Content of Flesh Foods

Data published before 1960 concerning the losses of vitamins as influenced by salting, curing, and smoking were reviewed for meat by Schweigert and Lushbough (1960) and for fish by Tarr (1960). These data are compiled for meat in Table 13.4 and for fish in Table 13.5.

TABLE 13.4

EFFECT OF CURING AND/OR SMOKING ON VITAMIN LOSSES OF MEAT

| Product | Treatment | Percentage Loss of | | | Reference |
		Thiamin	Riboflavin	Niacin	
Ham	curing and smoking	20	3	0	Schweigert et al. (1944)
Ham	curing	1-5	1-5	1-5	Rice et al. (1947)
Ham	curing and smoking	15-20	Very small	Very small	Rice et al. (1947)
Ham	artery pumped	15	—	—	Hoagland et al. (1947)
Ham	dry rubbed	14	—	—	Hoagland et al. (1947)
Ham	brine immersed	26	—	—	Hoagland et al. (1947)
Bacon	brine immersed and smoked	26	11	19	Jackson et al. (1945)
Sides and backs	dry rubbed and smoked	16	43	4	Jackson et al. (1945)

The results reported by Taarland et al. (1958) have been included in Table 13.5.

As it may be observed from these Tables, losses of riboflavin and niacin due to smoking were very small (Schweigert et al. 1944; Rice et al. 1947). Smoking after curing resulted in 15-20% higher losses of thiamin as compared to curing alone (Rice et al. 1947). These losses may be attributed to the heating effect. Dry-curing followed by smoking was less destructive to the thiamin (retained 84%) than wet curing (retained 74%) under the same conditions (Jackson et al. 1945). Niacin behaved similarly (retention 96 and 81%, respectively). The effect on riboflavin was reversed, however; dry-curing and subsequent smoking was characterized by 57% retention whereas

TABLE 13.5

VITAMIN CONTENT OF SALTED AND/OR SMOKED FISH

Product	Treatment	Vitamin A (Mg/100 Gm)	Thiamin (Mg/100 Gm)	Riboflavin (Mg/100 Gm)	Niacin (Mg/100 Gm)	Pantothenic Acid (Mg/100 Gm)	Vitamin B-12 (µg/100 Gm)	References
Atlantic mackrel	Smoked	0.07	—	—	—			Willstaedt and Jensen (1937)
Atlantic herring	Smoked	0.05	—	—	—			Willstaedt and Jensen (1937)
Atlantic cod	Smoked	—	0.023–0.080 (0.056)[1]	0.024–0.040 (0.034)[1]	2.1–3.5 (2.93)[1]			Hoogland (1953)
Haddock	Smoked	—	—	0.055–0.103	—			Hoar and Barberie (1945)
Haddock	Smoked	—	—	0.017–0.037 (0.024)[1]	3.9–5.2 (4.56)[1]			Hoogland (1953)
Atlantic herring	Salted	—	—	0.19	—			Hoar and Barberie (1945)
Atlantic cod	Dry-salted	—	—	0.037				Hoar and Barberie (1945)
Atlantic halibut	Salted (8 days)	—	—	0.14				Hoar and Barberie (1945)
Greenland halibut	Smoked	—	—	0.17 (0.67)[2]	1.5 (5.9)[2]	0.72 (2.84)[2]	0.6 (2.4)[2]	Taarland et al. (1958)
Salmon	Smoked	—	0.11 (0.30)[2]	0.19 (0.52)[2]	5.0 (14.0)[2]	0.71 (1.92)[2]	7.0 (19.0)[2]	Taarland et al. (1958)
Fat herring	Cold smoked	—	—	0.28	4.2	0.88	15.0	Taarland et al. (1958)
Fat herring	Hot smoked	—	—	0.26	4.7	0.99	14.0	Taarland et al. (1958)
Autumn mackerel	Smoked	—	—	0.37 (0.96)[2]	6.6 (17.0)[2]	0.52 (1.34)[2]	12.0 (30.8)[2]	Taarland et al. (1958)

[1] Mean value.
[2] On a dry weight basis.

EFFECTS OF PROCESSING BY ADDITIVES ON NUTRIENTS 369

TABLE 13.6

VITAMIN CONTENT OF CURED AND/OR SMOKED MEATS

Meat	Thiamin (Mg/100 Gm)	Riboflavin (Mg/100 Gm)	Niacin (Mg/100 Gm)	B-6 (Mg/100 Gm)	Pantothenic Acid (Mg/100 Gm)	Biotin (µg/100 Gm)	Cholin (µg/100 Gm)	B-12 (µg/100 Gm)	Folic Acid (Mg/100 Gm)
Bacon, raw	0.36	0.11	1.8	0.30	0.44	7.6	80	—	—
Canadian	0.83	0.22	4.7	—	—	—	80	—	—
Bologna	0.16	0.22	2.6	0.12	0.43	—	60	—	—
Boston butt, cured, raw	0.71	0.19	4.0	—	—	—	—	—	—
Country-style sausage	0.22	0.19	3.1	—	—	—	—	—	—
Corned beef, cooked	0.02	0.18	1.5	0.15	—	—	—	—	—
Frankfurters	0.16	0.20	2.7	0.13	0.43	—	57	—	—
Ham Canned	0.53	0.19	3.8	—	—	—	—	—	—
Cured, raw	0.72	0.19	4.1	0.39	0.64	5.0	122	1.0	10.6
Luncheon meat (pork)	0.31	0.21	3.0	—	—	—	—	—	—
Picnic, cured, raw	0.69	0.19	3.9	—	—	—	—	—	—
Polish-style sausage	0.34	0.19	3.1	—	—	—	—	—	—
Pork sausage, raw	0.43	0.17	2.3	—	—	—	48	—	11.5
Salami, dry	0.37	0.25	5.3	0.24	—	—	—	—	—
Tongue Smoked	0.04	0.21	3.0	—	0.60	—	—	—	—
Vienna sausage	0.08	0.13	2.6	—	—	—	—	—	—

Source: Watt and Merrill (1963).

89% riboflavin was retained when wet-curing was applied. Beuk et al. (1950) studied the retention of thiamin, riboflavin, and niacin in 15 processed meats and sausages. Only thiamin was consistently lost. The losses ranged from 2 to 25%. Newer data on vitamin and mineral content of cured and/or smoked meats are presented in Tables 13.6 and 13.7 according to Watt and Merrill (1963).

TABLE 13.7

MINERAL COMPOSITION OF CURED AND/OR SMOKED MEATS

Meat	Calcium (Mg/100 Gm)	Phosphorus (Mg/100 Gm)	Iron (Mg/100 Gm)	Sodium (Mg/100 Gm)	Potassium (Mg/100 Gm)
Bacon, raw	13	108	1.2	680	130
Canadian	12	180	3.0	1891	392
Bologna	7	128	1.8	1300	230
Boston butt, cured, raw	10	152	2.6	—	—
Country-style sausage	9	168	2.3	—	—
Corned beef, cooked	9	93	2.9	1740	150
Frankfurters	7	133	1.9	1100	220
Ham					
Canned	11	156	2.7	1100	340
Cured, raw	10	162	2.6	—	—
Luncheon meat (pork)	0	108	2.2	1234	222
Mortadella	12	238	3.1	—	—
Picnic, cured, raw	10	150	2.5	—	—
Polish-style sausage	9	176	2.4	—	—
Pork sausage, raw	5	92	1.4	740	140
Salami, dry	14	283	3.6	—	—
Salt pork	tr	tr	6	1212	42
Vienna sausage	8	153	2.1	—	—

Source: Watt and Merrill (1963).

Comparative data, however, based on raw and processed samples from the same raw material are very rarely available. The edible muscle portion of fish has little vitamin A, and it is probably not destroyed to any significant extent by smoking (Tarr 1960). Similarly, smoking exhibits no appreciable effect on riboflavin, niacin (Tarr 1960), pantothenic acid, and vitamin B-12 (Hoogland 1953). Smoking of haddock, cod, and herring (cold and hot smoked) did not cause any appreciable changes in thiamin, riboflavin, niacin, panthotenic acid, and vitamin B-12 content. Loss of all these vitamins except B-12 has been observed in smoked salmon (Taarland et al. 1958). No data were found on losses of fat-soluble vitamins accompanying fat exudation during smoking.

Smoking and Polycyclic Aromatic Hydrocarbons

Some of the compounds belonging to the group of polycyclic aromatic hydrocarbons exhibit carcinogenic properties. This activity is linked with a specific structure of their molecules (Herndon 1974). One of the best known and most active member of this group is benzo/a/pyrene. The effective dose of this compound is in the range of micrograms (Schoental 1964).

Sulman and Sulman (1946) reported that a tar from smoking kilns shows carcinogenic activity. Benzo/a/pyrene was identified later in this material (Berankova and Sula 1953). According to recent reviews, over 25 polycyclic aromatic hydrocarbons have been identified in wood smoke (Tilgner and Daun 1969; Hofmann 1972B; Lenges 1972). Besides, there are about 40 of these substances present in wood smoke but not yet identified (Sikorski 1965; Daun 1968).

Polycyclic aromatic hydrocarbons are deposited and absorbed by the smoked product during smoking. Over 16 of these compounds have been isolated and identified from smoked products by many authors. Detailed information may be found in reviews (Grimmer and Hildebrandt 1967; Toth 1969; Tilgner and Daun 1969; Hofmann 1972B; Lenges 1972). The development of new accurate methods (Howard *et al.* 1966A, B; Fazio *et al.* 1973B; O'Hara *et al.* 1974) make comparisons of results obtained in different laboratories possible.

Benzo/a/pyrene is the most often searched-for carcinogen. The amount of this substance in smoked fish varies from 1.7 to 53.0 µg per kg (Voitelovich *et al.* 1957). The majority of newer publications report a rather low content (of the order of 1 ppm) of benzo/a/pyrene for slightly smoked flesh foods (Malanoski *et al.* 1968; Toth 1969, 1971; Filipovic and Toth 1971; Masuda and Kuratsune 1971; Toth and Blaas 1972A, C; Wierzchowski and Gajewska 1972). However, even small amounts of polycyclic aromatic hydrocarbons present in smoked flesh food have to be considered as a potential hazard to human health. A high incidence of all neoplasms and especially gastro-intestinal cancer was observed among fishermen eating considerable quantities of smoked fish (Voitelovich *et al.* 1957; Bailey and Dungal 1958; Dungal 1961). A similar phenomenon has been reported among the workers engaged in smoking of fish and meat (Kaufmann *et al.* 1959).

Truhaut (1957) suggested prohibiting the sale of all products in which benzo/a/pyrene can be detected. Tilgner and Miler (1963) appealed for "legislative rules which should forbid the designing and building of new smoke houses equipped with smoke generators of the smoldering type, exploitation of such smoke generators, and distribution of goods smoked with smoldering type smoke" and to "establish a fixed date after which only smoke generators are permitted which produce curing smoke free from carcinogens." Some other authors propose to institute limits on concentrations of benzo/a/pyrene in smoked foods.

The amount of polycyclic aromatic hydrocarbons in smoked products is affected by many factors including the method of smoke generation, temperature of combustion and air supply, length of smoke ducts, and density and temperature of the smoke. Polycyclic aromatic hydrocarbons in the smoke are in a relatively constant ratio to the benzo/a/pyrene (Toth 1971). This finding seems to support a hypothesis of Tilgner and Daun (1969) that these compounds are formed in the smoke from thermally generated methylene radicals. Therefore, it should be possible to eliminate polycyclic aromatic hydrocarbons from the smoke by applying sufficiently low temperature of the thermal wood destruction and oxidation.

A polynuclear-free smoke generated under controlled conditions (Miler 1962; Tilgner and Miler 1963; Tilgner and Daun 1964, 1965) and "steam smoke" produced at 280°-380°C containing only traces of benzo/a/pyrene (Reuter and Heinz 1969) provide experimental evidence to back up the above discussed hypothesis. However, Dikun et al. (1967) claim that benzo/a/pyrene may originate also at low temperatures (300°C).

Many ways are suggested to reduce the amount of polycyclic aromatic hydrocarbons present in the smoke. The application of an electrostatic filter reduces the amount of benzo/a/pyrene in the smoke (Rusz et al. 1969, 1971). Even a simple filtration through cotton wool eliminates more than 90% of polycyclic aromatic hydrocarbons (Toth and Blaas 1972C). The use of cellulosic casings reduces the amount of benzo/a/pyrene that migrates into the products (Simon et al. 1969; Rhee and Bratzler 1970). The amount of polycyclic aromatic hydrocarbons is reduced significantly in the process of "Liquid Smoke" preparation (Toth and Blaas 1972B).

Polycyclic aromatic hydrocarbons in the smoke and smoked food products are only one possible source of carcinogens. For example, benzo/a/pyrene is present in a very low amount in smoked food when compared with barbecued meat (Lijinsky and Shubik 1965; Fritz 1973). Thermal decomposition of all organic substances, especially with an access of air, may lead to the formation of polycyclic aromatic hydrocarbons (Tilgner and Daun 1964, 1965). Many further studies are necessary to clarify the significance of a large variety of carcinogenic substances that might be absorbed by man with foods or as pollutants of the environment.

BIBLIOGRAPHY

ADRIAN, J. 1957. Composition and nutritive value of fish preserved under different conditions: salted and dried African samples, commercial fish meal and nuoc-mam. Ann. Nutr. Aliment. *11*, 27.
AMANO, K. 1961. The influence of fermentation on the nutritive value of fish with special reference to fermented fish products in Southeast Asia. *In* Fish in Nutrition. E. Heen and R. Kreuser (Editors). Fishing News, London.
BAILEY, E., and DUNGAL, N. 1958. Polycyclic hydrocarbons in Icelandic smoked food. Brit. J. Cancer *12*, 348.
BARD, J., and TOWNSEND, W. E. 1971. Meat curing. *In* The Science of Meat and Meat Products, 2nd Edition. J. F. Price, and B. S. Schweigert (Editors). W. H. Freeman & Co., San Francisco.
BELENKIJ, N. G. 1965. Biological evaluation of principal methods of heat preservation. 11th European Meeting of the Meat Res. Workers, Belgrad, Yugoslavia. (Russian)
BERANKOVA, Z., and SULA, J. 1953. Isolation and identification of 3,4 benzpyrene in the wood tar of a smokehouse. Casopis Lekaru Ceskych *92*, 195. (Czech.)
BEUK, J. F., FRIED, J. F., and RICE, E. E. 1950. Nutritive values of sausage and other table-ready meats as affected by processing. Food Res. *15*, 302-307.
BOLSHAKOV, A., KORNIENKO, A., FOMIN, A., and SHABANOVA, V. 1965. Change in the content of amino acids in pickled pork during storage. Myasn. Industr. USSR *36*, No. 4, 47-48. Chem. Abstr. *64*, 1262d.
BRISSEY, G. E., and GOESER, P. A. 1967. Aging, curing and smoking of meats. *In* Fundamentals of Food Processing Operations. Ingredients, Methods, and Packaging. J. L. Heid, and M. A. Joslyn (Editors). Avi Publishing Co., Westport, Conn.
BRUGGEMANN, J., and TIEWS, J. 1964. Effect of NO_2^- ions and different environmental temperatures on carotene metabolism of chickens. Intern. Z. Vitaminforsch. *34*, 233-240.
BURGER, I. H., and WALTERS, C. L. 1973. The effect of processing on the nutritive value of flesh foods. Proc. Nutr. Soc. *32*, 1-8.
BURGESS, G. H. O., and SHEWAN, J. M. 1970. Intrinsic and extrinsic factors affecting the quality of fish. *In* Proteins as Human Food. Proceedings of the Sixteenth Easter School in Agricultural Science, University of Nottingham. R. A. Lawrie (Editor). Avi Publishing Co., Westport, Conn.
CABAT, F. S., and STANDAL, B. R. 1965. The composition of essential and certain nonessential amino acids in selected Hawaii fish. J. Food Sci. *30*, 172-177.
CASTELL, C. H., MACLEAN, J., and MOORE, B. 1965. Rancidity in lean fish muscle. IV. Effect of NaCl and other salts. J. Fisheries Res. Board Can. *22*, 929-944.
CHARI, S. T., and VENKATARAMAN, R. 1957. Semidrying of prawns and its effects on the amino acid composition. Indian J. Med. Res. *45*, 81.
CHEN, L., and ISSENBERG, P. 1972. Interactions of some wood smoke components with amino groups in proteins. J. Agr. Food Chem. *20*, 1113-1115.
COMMITTEE ON NITRATE ACCUMULATION. 1972. Accumulation of nitrate. Natl. Acad. Sci.—Natl. Res. Council.
CREMER, H. D., and BÜTTNER, W. 1961. Influence of industrial processing and storage on nutritional value of meats, wheat products and legumes. 5th

Intern. Congr. Nutr., Washington, D.C., 1960. Federation Proc. 20, Suppl. 7, 237-246.
CUTTING, C. L. 1961. Influence of drying, salting, and smoking on the nutritive value of fish. Conf. Fish Nutr. Working Papers, Washington, D.C. United Nations Publication Service, New York.
CUTTING, C. L. 1965. Smoking. In Fish as Food, Vol. III, Part 1. G. Borgstrom (Editor). Academic Press, New York.
DAUN, H. 1968. Research on polycyclic aromatic hydrocarbons. (Unpublished data.)
DAUN, H. 1969. Antioxidative properties of selected smoked meat products. 15th European Meeting Meat Res. Workers, Helsinki, Finland, 274-281.
DAUN, H., and GRABOWSKA, J. 1967. The antioxidant activity of smoke components as influenced by the method of separation. Roczniki Technol. Chem. Zywnosci 14, 109.
DAUNORAVICIUTE, R., RAKAUSKAS, A., and VENSKEVICIUS, J. 1966. Salt curing of pork. Mokslas ir Tech. No. 1, 27. Chem. Abstr. 64, 16530a.
DEMPSTER, J. F. 1973. Curing meat products. Process Biochem. 8, No. 3, 25-27, 30.
DIKUN, P. P. et al. 1967. Generation of 3,4 benzpyrene as a result of wood pyrolysis at 300°-400°C. Vopr. Onkol. 13, 80. (Russian)
DRAUDT, H. N. 1963. The meat smoking process: A review. Food Technol. 17, 1557-1598.
DRUCKREY, H., PREUSSMANN, R., and IVANKOVIC, S. 1969. N-nitroso compounds in organotropic and transplacental carcinogenesis. Ann. N.Y. Acad. Sci. 163, 676-696.
DRUCKREY, H., PREUSSMANN, R., IVANKOVIC, S., and SCHMÄHL, D. 1967. Organotropic carcinogenic effects of 65 different N-nitroso-compounds on BD-rats. Z. Krebs Forsch. 69, 103-201. (German)
DRUCKREY, H., and SCHMÄHL, D. 1962. Quantitative analysis of experimental carcinogenesis. Naturwissenschaften 49, 217-228. (German)
DUERR, J. D., and DYER, W. J. 1952. Proteins in fish muscle. IV. Denaturation by salt. J. Fisheries Res. Board Can. 8, 325-331.
DUNGAL, N. 1961. The special problem of stomach cancer in Iceland. J. Am. Med. Assoc. 178, 789.
DUNKER, C. F., BERMAN, M., SNIDER, G. G., and TUBIASH, H. S. 1953. Quality and nutritive properties of different types of commercially cured hams. III. Vitamin content, biological value of the protein and bacteriology. Food Technol. 7, 288-291.
DuPLESSIS, L. S., NUNN, J. R., and ROACH, W. A. 1969. Carcinogen in a Traskeian Bantu food additive. Nature 222, 1198-1199.
DVORAK, Z., and VOGNAROVA, I. 1965A. Effect of the Maillard reaction and smoking upon the biological qualities of meat and meat products. Prumysl Potravin 16, No. 4, 172-176.
DVORAK, Z., and VOGNAROVA, I. 1965B. Available lysine in meat and meat products. J. Sci. Food Agr. 16, 305-312.
EISENBRAND, G. 1973. What might be the results of forbidding or reducing the addition of nitrate and nitrite curing salt to meat products? Medical point of view. Fleishwirtschaft 53, 352-355.
ENDER, F. et al. 1964. Isolation and identification of hepatotoxic factor in herring meal produced from sodium nitrite preserved herring. Naturwissenschaften 51, 637.
ENDER, F., and CEH, L. 1968. Occurrence of nitrosamines in foodstuffs for human and animal consumption. Food Cosmet. Toxicol. 6, 569-571.
ENDER, F., and CEH, L. 1971. Conditions and chemical reaction mechanism by which nitrosamines may be formed in biological products with reference to their possible occurrence in food products. Z. Lebensm.-Untersuch-Forsch. 145, 133-142.

FAZIO, T., WHITE, R. H., DUSOLD, L. R., and HOWARD, J. W. 1973A. Nitroso pyrrolidine in cooked bacon. J. Assoc. Offic. Anal. Chem. 56, 919.
FAZIO, T., WHITE, R. H., and HOWARD, J. W. 1973B. Collaborative study of the multicomponent method for polycyclic aromatic hydrocarbons in foods. J. Assoc. Offic. Anal. Chem. 56, No. 1, 68-70.
FIDDLER, W. et al. 1972A. Effect of sodium nitrite concentration of N-nitrosodimethylamine formation in frankfurters. J. Food Sci. 37, 668-670.
FIDDLER, W., PENSABENE, J. W., DOERR, R. C., and WASSERMAN, A. E. 1972B. Formation of N-nitrosodimethylamine from naturally occurring quarternary ammonium compounds and tertiary amines. Nature 236, No. 5345, 307.
FIDDLER, W., PIOTROWSKI, E. G., PENSABENE, J. W., and WASSERMAN, A. E. 1973. Studies on nitrosamine formation in foods. 33rd Ann. Meeting Inst. Food Technologists, Miami Beach, Florida.
FIELDS, M. D., and DUNKER, C. F. 1952. Quality and nutritive properties of different types of commercially cured hams. 1. Curing methods and chemical composition. Food Technol. 6, 329.
FILIPOVIC, J., and TOTH, L. 1971. Polycyclic hydrocarbons in Yugoslav smoked meat products. Fleischwirtschaft 51, 1323-1325.
FRITZ, W. 1973. On generation of carcinogenic hydrocarbons by the thermal treatment of foods. Duetsche Lebensm.-Rundschau 69, 119-122. (German)
GIBBONS, N. E. 1953. Wiltshire Bacon. In Advances in Food Research, Vol. 4. E. M. Mrak and G. F. Stewart (Editors). Academic Press, New York.
GORBATOV, V. M. et al. 1971. Liquid smokes for use in cured meats. Food Technol. 25, No. 1, 71-77.
GREENBLATT, M. 1973. Ascorbic acid blocking of aminopyrine nitrosation in NZO/BI mice. J. Natl. Cancer Inst. 50, 1055.
GREENBLATT, M., and LJINSKY, W. 1972. Nitrosamine studies: Neoplasms of liver and genital mesothelium in nitrosophyrrolidine treated MRC rats. J. Natl. Cancer Inst. 48, 1687.
GREENBLATT, M., MIRVISH, S., and SO, B. T. 1971. Nitrosamine studies: Induction of lung adenomas by concurrent administration of sodium nitrite and secondary amines in Swiss mice. J. Natl. Cancer Inst. 46, 1029-1034.
GRIMMER, G., and HILDEBRANDT, A. 1967. Hydrocarbons in the environment of man. V. The content of polycyclic hydrocarbons in meat and smoked food products. Z. Krebsforsch. 69, 223-229.
HAWKSWORTH, G., and HILL, M. J. 1971. The formation of nitrosamines by human intestinal bacteria. Biochem. J. 122, 28.
HEATH, D. F., and MAGEE, P. N. 1962. Toxic properties of dialkylnitrosamines and some related compound. Brit. J. Ind. Med. 19, 276-282.
HERNDON, W. C. 1974. Theory of carcinogenic activity of aromatic hydrocarbons. Trans. N.Y. Acad. Sci., Ser. II, 36, No. 2, 200-217.
HOAGLAND, R. et al. 1947. Composition and nutritive value of hams as affected by method of curing. Food Technol. 1, 540-552.
HOAR, W. S., and BARBERIE, M. 1945. Distribution of riboflavin in fresh and processed fish. Can. J. Res. 23E, 8-18.
HOAR, D. W., EMBRY, L. B., and EMERICK, R. J. 1968. Nitrate and vitamin A interrelationships in sheep. J. Animal Sci. 27, 1727-1733.
HOFMANN, K. 1972A. The influence of method of preservation on the quality of meat and meat products. Fleischwirtschaft 52, 1403-1404, 1407-1408, 1411-1412, 1415.
HOFMANN, K. 1972B. The preservation of meats and meat products: its influence on the nutritive and gustatory values and its possible effects on human health. Ber. Landwirtschaftswiss. 50, 700.
HOOGLAND, P. L. 1953. The B-vitamins of cod and haddock. Fisheries Res. Board Can., Atlantic Coast Sta. Progr. Rept. 55, 11.
HOWARD, J. W., FAZIO, T., and WATTS, J. O. 1970. Extraction and gas

chromatographic determination of N-nitrosodimethylamine in smoked fish: Application to smoked nitrite-treated chub. J. Assoc. Offic. Anal. Chem. *53*, 269-274.
HOWARD, J. W., TEAGUE, R. T., JR., WHITE, R. H., and FRY, B. E., JR. 1966A. Extraction and estimation of polycyclic aromatic hydrocarbons in smoked foods. I. General method. J. Assoc. Offic. Anal. Chem. *49*, 595-611.
HOWARD, J. W., WHITE, R. H., FRY, B. E., JR., and TURICCHI, E. W. 1966B. Extraction and estimation of polycyclic aromatic hydrocarbons in smoked foods. II. Benzo (a) pyrene. J. Assoc. Offic. Anal. Chem. *49*, 611-617.
INAGAMI, K., and HORII, M. 1966. Change of available lysine in food protein by heating and smoking. Kyushu Daigaku Nogakjby Gakugei Zasshi *22*, No. 2, 191-198. Chem. Abstr. *65*, 11233a.
INGRAM, J., and KITCHELL, A. G. 1967. Salt as a preservative for foods. J. Food Technol. *2*, 1-15.
ITO, K. 1962. Changes of digestibility of fish protein by different methods of cooking. IV. Effect of salt concentration on digestibility of salted fish. Kaseigaku Zasshl *13*, 229-232.
ITO, K. 1970. On the change in digestibility of fish protein by different cooking methods. Part 5. Shizen Kagaku *21*, No. 1, 91-95.
IVANKOVIC, S., PREUSSMANN, R., SCHMÄHL, D., and ZELLER, J. 1973. Prevention of nitrosamide-induced hydrocephali by ascorbic acid after prenatal administration of ethylurea and nitrite to rats. Z. Krebsforsch. *79*, 145-147.
JACKSON, S. H., CROOK, A., and DRAKE, T. G. H. 1945. The retention of thiamine, riboflavin, and niacin in cooking pork and in processing bacon. J. Nutr. *29*, 391-403.
JOSLYN, M. A., and TIMMONS, A. 1967. Salt—use in food processing. *In* Fundamentals of Food Processing Operations. Ingredients, Methods, and Packaging. J. L. Heid, and M. A. Joslyn (Editors). Avi Publishing Co., Westport, Conn.
KAKO, Y. 1968. Studies on muscle proteins. II. Changes of beef-, pork-, and chicken-proteins during the meat products manufacturing processes. Mem. Fac. Agr. Kogashima Univ. *6*, 175.
KAMM, J. J., DASHMAN, T., CONNEY, A. H., and BURNS, J. J. 1973. Protective effect of ascorbic acid on hepatoxicity caused by sodium nitrite plus aminopyrine. Proc. Natl. Acad. Sci. U.S. *70*, 747.
KAUFMANN, B. D., MIRNOVA, A. I., and SHABAD, L. M. 1959. An attempt to evaluate frequency of malignant tumors of workers in some factories of food industry. Vopr. Onkol. *5*, 314. (Russian)
KIERNAT, B. H., JOHNSON, J. A., SIEDLER, A. J. 1964. A summary of the nutrient content of meat. AMIF Bull. *57*.
KIHARA, S. 1962. Changes in chicken plasma during preservation and processing. III. Absorption of smoke ingredients and the changes of plasma proteins and free amino acids during smoking. Tokyo Nogyo Daigaku Nogaku Shuho *8*, 4-8. Chem. Abstr. *61*, 88204.
KOCHANOWSKI, J. 1962. Bacteriostatic properties of liquid smoke. Tehnol. Mesa, Spec. Edition, 29-33,
KRYLOVA, N. N., and BAZAROVA, K. I. 1960. Denaturation of meat proteins in preparation of smoked sausage. Chem. Abstr. *56*, 3862e.
KRYLOVA, N. N., BAZAROVA, K. I., and KUZNETSOVA, V. V. 1962. Interaction of smoke components with meat constituents. 7th European Meeting Meat Res. Workers.
KURKO, V. I. 1959. Antioxidative properties of curing smoke components. Miasnaya Ind. *30*, No. 3, 19.

KURKO, V. I. 1960. Physical-Chemical and Chemical Basis of Smoking. Pischchepromizdat, Moscow. (Russian)
KURKO, V. I. 1967. Some chemical aspects of aroma development of smoked food products. 13th European Meeting Meat Res. Workers.
KURKO, V. I., and SCHMIDT, T. A. 1969. Interaction of carbonyl and reductone-like components of smoke with amino acids. 15th European Meeting Meat Res. Workers. Helsinki, Finland.
LAWRIE, R. A. 1966. Meat Science. Pergamon Press, London, England.
LEISTNER, L., HECHELMANN, H., and UCHIDA, K. 1973. What might be the result of forbidding or reducing the addition of nitrate and nitrite curing salt to meat products? Bacteriological point of view. Fleischwirtschaft 53, 371-378.
LENGES, Y. 1972. Smoking of meat products. Rev. Ferment. Ind. Aliment. 27, No. 2, 53-60.
LEONOVA, A. P. 1970. Change in muscle tissue proteins of fish during salting. Tr. Baltlisk. Nauch.-Issled. Inst. Morsk. Rybn. Khoz. Okeanogr. 4, 416-430. Chem. Abstr. 75, 10864d.
LIJINSKY, W., and ESPSTEIN, S. S. 1970. Nitrosamines as environmental carcinogens. Nature 225, 21-23.
LIJINSKY, W., and GREENBLATT, M. 1972. Carcinogen dimethylnitrosamine produced in vivo from nitrite and aminopyrine. Nature New Biol. 236, 177.
LIJINSKY, W., and SHUBIK, P. 1965. Polynuclear hydrocarbon carcinogens in cooked meat and smoked food. Ind. Med. Surgery 34, 152.
LINKO, R. R., and NIKKILÄ, O. E. 1961. Inhibition of the denaturation of salt of myosin in Baltic herring. J. Food Sci. 26, 606-610.
LOVERN, J. A. 1944. The effect of preservation process on the vitamin content of fish. Proc. Nutr. Soc. 2, 100.
MAGEE, P. N., and BARNES, J. M. 1956. Production of malignant primary hepatic tumors in rat by feeding dimethylnitrosamine. Brit. J. Cancer 10, 114.
MAGEE, P. N., and BARNES, J. M. 1967. Carcinogenic nitroso compounds. Advan. Cancer Res. 10, 163.
MALANOSKI, A. J. et al. 1968. Survey of polycyclic aromatic hydrocarbons in smoked foods. J. Assoc. Offic. Anal. Chem. 51, 114-121.
MASUDA, Y., and KURATSUNE, M. 1971. Polycyclic aromatic hydrocarbons in smoked fish. Katsuobushi, Gann 62, No. 1, 27-30.
MATSUHISA, T. 1965. Fluorine and other related components of Japanese foods. V. Fluorine and salt contents of miso and other foods rich in salt, and their geographical correlation with mortality from gastric cancer. Eiyo To Shokuryo 18, No. 4, 253-257. Chem. Abstr. 65, 12769d.
McGLASHAN, N. D., WALTERS, C. L., and McLEAN, A. E. M. 1968. Nitrosamine in African alcoholic spirits and oesophageal cancer. Lancet 2, 1017.
MENDELSOHN, J. M. 1974. Rapid techniques for salt-curing fish. A review. J. Food Sci. 39, 125.
MICHELS, P. W. 1971. A new method of rationalizing the manufacture of cooked hams. Fleischwirtschaft 51, 335-343.
MILER, K. 1962. Possibilities of curing smoke generation free of 3,4 benzopyrene and 1,2,5,6-di benzathracene. Ph.D. Thesis, Politechnika Gdanska.
MIRNA, E. 1970. The reaction of nitrite in meat products and its distribution in the various fractions. 16th European Meeting of Meat Res. Workers, Varna, Bulgaria.
MIRNA, E., and HOFMANN, K. 1969. The behavior of nitrite in meat products. Fleischwirtschaft 49, 1361-1363, 1366.
MIRVISH, S. S., CANDESA, A., WALLCAVE, L., and SHUBICK, P. 1973. Effect of sodium ascorbate on lung adenoma induction by amines plus nitrite. Proc. Am. Assoc. Cancer Res. 89, 1147.

MIRVISH, S. S., WALLCAVE, L., EAGEN, M., and SHUBICK, P. 1972. Ascorbate-nitrite reaction: possible means of blocking the formation of carcinogenic N-nitroso compounds. Science *177*, 65.
MITCHELL, G. E., JR., LITTLE, C. O., and GREATHOUSE, T. R. 1965. Influence of nitrate and nitrite on carotene disappearance from the rat intestine. Life Sci. *4*, 385-390.
MIYAUCHI, D. T. 1965. Drip formation in fish. I. A review of factors affecting drip. U.S. Fish Wildlife Serv., Bur. Com. Fisheries, Fishery Ind. Res. *2*, No. 2, 13.
MÖHLER, K., and MAYERHOFER, O. L. 1969. The influence of different factors on the formations of nitrosamines in meat products. 15th European Meeting Meat Res. Workers, Helsinki, Finland.
MOTOC, D., and BANU, C. 1967. The biochemistry of meat curing. Fleischwirtschaft *47*, 257-601, 263. Chem. Abstr. *67*, 20713w.
MUNRO, I. C., and MORRISON, A. B. 1965. Effects of salting and smoking on protein quality of cod. J. Fisheries Res. Board Can. *22*, 13-16.
NAKAHARA, W., and FUKUOKA, F. 1960. Summation of carcinogenic action of chemically different carcinogens. Naturwissenschaften *47*, 44-45. (German)
NEWMARK, H. L. et al. 1974. Stability of ascorbate in bacon. Food Technol. *28*, No. 5, 28-31, 60.
NICORA, L. M., LAMBERT, R., and WIAUX, A. 1966. Chemical alterations in smoke-cured meats during storage. Ind. Aliment. Agr. (Paris) *83*, 1623-1628.
O'DELL, B. L. et al. 1960. Effects of nitrite containing rations in producing vitamin A and vitamin E deficiencies in rats. J. Animal Sci. *19*, 1280.
O'HARA, J. R., CHIN, M. S., DAINIUS, B., and KILBUCK, J. H. 1974. Determination of benzo(a)pyrene in smoke condensates by high pressure rapid liquid-liquid chromatograph. J. Food Sci. *39*, 38-41.
OLLEY, J. 1972. Unconventional sources of fish protein. CSIRO Food Res. Quart. *32*, No. 2, 27-32.
ORRACA-TETTEH, R. 1961. Problems and some solutions: Ghana. Proc. Nutr. Soc. *20*, 109-112.
PFEIL, E., and LIEPE, H. U. 1973. Nitrosamine problem. Fleischwirtschaft *53*, 387-388. (German)
PIVNICK, H. et al. 1967. Effect of sodium nitrite and temperature on toxicogenesis by *Clostridium botulinum* in perishable cooked meats vacuum-packed in air-impermeable plastic pouches. Food Technol. *21*, 204-206.
PLATT, B. S. 1961. Problems and some solutions: Introduction. Proc. Nutr. Soc. *20*, 93-95.
PROCTOR, B. E., and LAHIRY, N. J. 1956. Evaluation of amino acids in fish processed by various methods. Food Res. *21*, 91-92.
PUGH, D. L., and GARNER, G. B. 1963. Reaction of carotene with nitrite solutions. J. Agr. Food Chem. *11*, 528-529.
RANDALL, C. J., and BRATZLER, L. J. 1970A. Effect of smoking process on solubility and electrophoretic behavior of meat proteins. J. Food Sci. *35*, 245-247.
RANDALL, C. J., and BRATZLER, L. J. 1970B. Changes in various protein properties of pork muscle during the smoking process. J. Food Sci. *35*, 248-249.
RANDALL, C. J., and BRATZLER, L. J. 1970C. Effect of smoke upon acid phosphatase activity of smoked meat. J. Food Sci. *35*, 250.
REAY, G. A. 1936. The salt curing of herring. J. Soc. Chem. Ind. (London) *55*, 309T-315T.
REUTER, H. 1972. Smoking meat products. Fleishwirtschaft *52*, 1116-1118, 1121.
REUTER, H., and HEINZ, G. 1969. Hot smoking of meat products using vapour smoke. Fleischwirtschaft *49*, 169-172.

RHEE, K., and BRATZLER, L. J. 1970. Benzo(a)pyrene in smoked meat products. J. Food Sci. *35*, 146-149.
RICE, E. E., BEUK, J. F., and FRIED, J. F. 1947. Effect of commercial curing, smoking, storage, and cooking operations upon vitamin content of pork hams. Food Res. *12*, 239-246.
ROBERTS, W. K., and SELL, J. L. 1963. Vitamin A destruction by nitrite *in vitro* and *in vivo*. J. Animal Sci. *22*, 1081-1885.
RUSZ, J., HUJNAKOVA, D., KOPALOVA, M. 1969. Improved technology for smoking food products. Prumysl Potravin *20*, No. 3, 84-86. Chem. Abstr. *70*, 95507c.
RUSZ, J., KOPALOVA, M., and PELIKANOVA, K. 1971. The influence of some factors on the contents of 3,4 benzpyrene in smoked pork back fat. Prumysl Potravin *22*, 106. (Czech.)
SANDER, J. 1967. Can nitrite in human food be a cause of a cancer via nitrosamine generation? Arch. Hyg. Bakteriol. *151*, 22-28. (German)
SANDER, J., SCHWENSBERGER, F., and MENZ, H. P. 1968. Studies on the formation of carcinogenic nitrosamines in the stomach. Hoppe-Seylers Z. Physiol. Chem. *349*, 1691-1697.
SANDER, J., and SEIF, F. 1969. Bacterial reduction of nitrate in the human stomach as a cause for nitrosamine formation. Arzneimittel-Forsch. *19*, 1091-1093.
SCHOENTAL, R. 1964. Carcinogenesis. *In* Polycyclic Hydrocarbons. E. Clar (Editor). Academic Press, New York.
SCHWEIGERT, B. S., and LUSHBOUGH, C. H. 1960. Effects of processing on meat products. *In* Nutritional Evaluation of Food Processing. R. S. Harris, and H. von Loesecke (Editors). John Wiley & Sons, New York. Reprinted in 1971 by Avi Publishing Co., Westport, Conn.
SCHWEIGERT, B. S., McINTIRE, J. M., and ELVEHJEM, C. A. 1944. The retention of vitamins in pork hams during curing. J. Nutr. *27*, 419-424.
SCHWEIGERT, B. S., and PAYNE, B. Y. 1956. A summary of the nutrient content of meat. Am. Meat Inst. Found., Bull. *30*.
SEN, N. P., DONALDSON, B., IYENGAR, J. R., and PANALAKS, T. 1973. Nitrosopyrrilidine and dimethylnitrosamine in bacon. Nature *241*, No. 5390, 973-974.
SEN, N. P., SMITH, D. C., and SCHWINGHAMER, L. 1969A. Formation of *N*-nitrosamines from secondary amines and nitrite in human and animal gastric juice. Food Cosmet. Toxicol. *7*, 301-307.
SEN, N. P., SMITH, D. C., SCHWINGHAMER, L., and MARLEAU, J. J. 1969B. Diethylnitrosamine and other *N*-nitrosamines in foods. J. Assoc. Offic. Anal. Chem. *52*, 47-52.
SHEWAN, J. M. 1949. The biological stability of smoked and salted fish. Chem. Ind. (London) *49*, 501-505.
SIKORSKI, Z. E. 1965. Chemical changes of curing smoke in corona discharge field. Zeszyty Nauk. Politech. Gdansk. *69*, Chemia No. 10. (Polish)
SIKORSKI, Z. E. 1971. Technology of marine origin food. Wydawnictwa Naukowo-techniczne, Warszawa. (Polish)
SIMON, S. *et al.* 1969. Effect of cellulose casing on absorption of polycyclic hydrocarbons in wood smoke by absorbents. J. Agr. Food Chem. *17*, 1128-1134.
SORASUCHART, T. 1971. The nutritive value of Thai fish products. I. The vitamin content. Rept. Technological Res. Concerning Norwegian Fish Ind. *5*, No. 7.
SORASUCHART, T. 1972. The nutritive value of Thai fish products. II. Amino acid composition. Rept. Technological Res. Concerning Norwegian Fish Ind. *5*, No. 9.
STOYTCHEV, M., KOMINKOV, L., and KRUSTEVA, Y. 1969. Changes in some B-complex vitamins contents during production of pasteurized canned

ham, shoulder, and loin. 15th European Meeting Meat Res. Workers, Helsinki, Finland.
SULMAN, S., and SULMAN, F. 1946. The carcinogenicity of wood soot from the chimney of a smoked-sausage factory. Cancer Res. 6, 366.
TAARLAND, T., MATHIESEN, E., OVSTHUS, O., and BRAEKKAN, O. R. 1958. Nutritional values and vitamins of Norwegian fish and fish products. Tidsskr. Hermetikind 44, 405.
TANIKAWA, E. 1963. Fish, sausage and ham industry in Japan. In Advances in Food Research, Vol. 12. C. O. Chichester, E. M. Mrak, and G. F. Stewart (Editors). Academic Press, New York.
TANIKAWA, E. 1965. Marine products in Japan. Lab. Marine Food Technol., Fac. Fisheries, Hokkaido Univ., Hakodate, Japan.
TARR, H. L. A. 1960. Effects of processing on fish products. In Nutritional Evaluation of Food Processing. R. S. Harris, and H. von Loesecke (Editors). John Wiley & Sons, New York. Reprinted in 1971 by Avi Publishing Co., Westport, Conn.
TARR, H. L. A. 1962. Changes in nutritive value through handling and processing procedures. In Fish as Food, Vol 2. G. Borgstrom (Editor). Academic Press, New York and London.
TARR, H. L. A. 1969. Nutritional value of fish muscle and problems associated with its preservation. Can. Inst. Food Technol. J. 2, 42-45.
TERRACINI, B., MAGEE, P. N., and BARNES, J. M. 1967. Hepatic pathology in rats on low dietary levels of dimethylnitrosamine. Brit. J. Cancer 21, 559-565.
TILGNER, D. J. 1967. The effects of smoke-curing and active substances in curing smoke. Fleischwirtschaft 47, 373-374. (German)
TILGNER, D. J. 1970. Principles of modern smoke-curing. Fleischwirtschaft 50, 650-652. (German)
TILGNER, D. J., and DAUN, H. 1964. The influence of cellulose and lignin on the presence of carcinogenic components by the two stage generation of curing aerosol. Przemysl Spozywczy 18, 303.
TILGNER, D. J., and DAUN, H. 1965. The influence of cellulose and lignin upon the presence of carcinogens in curing smoke. Roczniki PZH 16, 45.
TILGNER, D. J., and DAUN, H. 1969. Polycyclic aromatic hydrocarbons (polynuclears) in smoked foods. Residue Rev. 27, 19-41.
TILGNER, D. J., and DAUN, H. 1970. Antioxidative and sensory properties of curing smokes obtained by three basic smoke generation methods. Lebensm.-Wiss. Technol. 3, No. 5, 77-82.
TILGNER, D. J., DAUN, H., and RUDNICKI, A. 1967. Antioxidant activity of phenolic compounds present in curing smoke. Rocz. Inst. Przem. Miensnego 4, No. 1, 15.
TILGNER, D. J., and MILER, K. 1963. Possibilities of elimination of carcinogenic compounds from curing smoke. Przemysl Spozywczy 17, 85.
TILGNER, D. J., SIKORSKI, Z. E., DAUN, H., RUDNICKI, A. 1965. Antioxidative activity of the dispersed and dispersing phase of curing smoke. XX-Lecie Politech. Gdansk. Sesja Nauk. CH 45, 344.
TOTH, L. 1969. Carcinogenic substances in smoked meat products. Fleischwirtschaft 43, 1611-1614.
TOTH, L. 1971. Polycyclic hydrocarbons in smoked ham and belly fat. Fleischwirtschaft 51, 1069-1070.
TOTH, L., and BLAAS, W. 1972A. The effect of smoking technology on the content of carcinogenic hydrocarbons in smoked meat products. I. The effect of various smoking methods. Fleischwirtschaft 52, 1121-1124.
TOTH, L., and BLAAS, W. 1972B. 3,4-Benzopyrene content of various smoke preparations. Fleischwirtschaft 52, 1171.
TOTH, L., and BLAAS, W. 1972C. Effect of smoking technology on the content of carcinogenic hydrocarbons in smoked meat products. II. Effect of the

temperature at which the wood smoulders and of cooling, washing, and filtration of the smoke. Fleischwirtschaft *52*, 1413-1422.

TOWNSEND, W. E., and BARD, J. 1971. Factors influencing quality of cured meats. *In* The Science of Meat and Meat Products, 2nd Edition. J. F. Price, and B. S. Schweigert (Editors). Freeman & Co., San Francisco.

TRUHART, R. 1957. Evolution of the ideas about toxicity of foreign substances intentionally or accidentally incorporated into food—Comments upon the recommendations of the previous international meetings. 3rd Intern. Symp. Foreign Substances Food, Como, Italy. (French)

VAN VEEN, A. G. 1953. Fish preservation in Southeast Asia. *In* Advances in Food Research, Vol. 4. E. M. Mrak, and G. F. Stewart (Editors). Academic Press, New York.

VAN VEEN, A. G. 1965. Fermented and dried seafood products in Southeast Asia. *In* Fish and Food, Vol. 3. G. Borgstrom (Editor). Academic Press, New York.

VOITELOVICH, F. A., DIKUN, P. P., and SHOBOD, L. M. 1957. Comparative study of the frequency of malignant tumors in Tukunsk region of Latvian SSR. Vopr. Onkolog. *3*, 351. (Russian)

VOSKRESENSKY, N. 1965. Salting of herring. *In* Fish as Food, Vol. 3. G. Borgstrom (Editor). Academic Press, New York.

WALTERS, C. L., and CASSELDEN, R. J. 1972. The gaseous products of nitrite incubation with skeletal muscle. Z. Lebensm.-Untersuch-Forsch. *150*, 335.

WATT, B. K., and MERRILL, A. L. 1963. Composition of Foods; Raw, Processed, Prepared. USDA Agr. Handbook *8*, Revised.

WIERZCHOWSKI, J., and GAJEWSKA, R. 1972. Determination of 3,4-benzypyrene in smoked fish. Bromatol. Chem. Toksykol. *5*, 481-486.

WILLSTAEDT, H., and JENSEN, H. B. 1937. Svensk Kem. Tidskr. *49*, 260. (Cited by Tarr 1960).

WOLFF, I. A., and WASSERMAN, A. E. 1972. Nitrates, nitrites, and nitrosamines. Science *177*, 15-19.

WOOD, R. D., CHANEY, C. H., WADDILL, D. G., and GARRISON, G. W. 1967. Effect of adding nitrate or nitrite to drinking water on the utilization of carotene by growing swine. J. Animal Sci. *26*, 510-513.

WOOLFORD, G. *et al.* 1972. Gaseous products of the interaction of sodium nitrite with porcine skeletal muscle. Biochem. J. *130*, 82.

YUDITSKAYA, A. I. 1962. Histochemical investigation of fish tissues. Tr. Vses. Nauchn.-Issled. Inst. Morsk. Rybn. Koz. Okeanogr. *45*, 56-59. Chem. Abstr. *59*, 15862g.

ZIEMBA, Z. 1969. Effect of wood smoke components on the surface color of smoke-cured meat products. Rocz. Technol. Chem. Zywn. *15*, 153-169. Chem. Abstr. *72*, 11461p.

R. Thiessen, Jr.

PART 2
Effects of High-Sugar Processing on Nutrients

One of the less thought-of processing procedures for foods is that involving the use of high sugar concentrations. As a matter of fact, high-sugar processing methodology has been so taken for granted that little, if any, data appear in the literature relative to the nutrient content of such sugar-containing products nor data relative to the effect of processing or storage on the nutrient contents.

The principal members of the high-sugar class of products are jams, jellies, preserves, honey, molasses, hard candy, jelly beans, marshmallows, candied fruit, marmalades, etc. Generally speaking, these are usually considered to be fun foods, or edulcorates for improving the organoleptic acceptance of a meal or part of a meal. In this usage the daily consumption is usually small. Furthermore, the levels of vitamins and minerals in this type of product are not usually significant from a nutritional standpoint.

It is probably because of this low usage level and low level of nutrients that little attention has been paid to the effect of processing on the nutrients in this type of food.

Table 13.8 offers examples of the levels of nutrients in this type of food.

TABLE 13.8

LEVELS OF NUTRIENTS IN SOME HIGH-SUGAR FOOD PRODUCTS

Nutrient	Honey	Molasses (Medium)	Jams	Jellies	Marmalade
		Per 100 Gm			
Vitamin A (IU)	—	—	10	10	—
Vitamin C (mg)	—	—	2	4	6
Thiamin (mg)	—	—	0.01	0.01	0.02
Riboflavin (mg)	0.04	0.12	0.03	0.03	0.02
Niacin (mg)	0.3	1.2	0.2	0.2	0.1
Iron (mg)	0.5	6.0	1.0	1.5	0.6
Calcium (mg)	5	290	20	21	35
Vitamin B-6 (mg)	0.02	0.2	0.02	—	0.016
Pantothenic acid (mg)	0.2	0.35	—	—	—

Source: *Composition of Foods*, USDA Agr. Handbook *8*, except vitamin B-6 and pantothenic acid from USDA Home Econ. Res. Rept *36*.

Winifred M. Cort

PART 3
Effects of Treatment with Chemical Additives

Food additives[1] are substances other than the basic foodstuffs which are put into foods during production, processing, and packaging. Additives are used in commercial processing not only to reduce microbial hazards but also to reduce chemical and physical spoilage and to aid processing. Processing additives may be used as anticaking agents, chemical preservatives, emulsifiers, and stabilizers; to modify or improve flavor, texture, or color; and they also may include nutrients added to enhance nutritional value. The number of food additives greatly exceeds the number of nutrients. Since the function of food additives is so varied, the total effect on nutrients often is not known. If one makes a list of all the food additives (for example, the list in Title 21, 121.101 U.S. Code of Federal Regulations, published in 1973, CFR) and a list of all the nutrients, it becomes evident that potential interactions have hardly been explored.

The effects of food additives will be divided into two sections, namely, additives which decrease the nutrient content of foods and additives which are beneficial to nutrients. Unfortunately, as can be seen, at times this type of division will be ambiguous since some additives will be deleterious to one nutrient and beneficial to another.

ADDITIVES DETRIMENTAL TO NUTRIENTS

Sulfites

The deleterious effect of sodium bisulfite and sulfur dioxide on thiamin has been known for many years. In fact, sodium bisulfite is not allowed in meats, or in foods recognized as a source of thiamin in the CFR 121.101. Sulfites cause a nucleophilic displacement on the methylene bridge of thiamin and it is cleaved into free thiazole and pyrimidylmethane-sulfuric acid. The reaction occurs faster at a neutral pH. One might expect sulfites to react with some of the other less stable vitamins. In model systems, De Ritter (1969) reports that sulfur dioxide and acid will cleave folic acid to pteridine and

---

[1] This is a deviation of broad definition used by Food Protection Committee, National Academy of Sciences—National Research Council (see Anon. 1973).

p-aminobenzoyl glutamic moieties and the latter undergoes further destruction of the free amino group. Other than the beneficial effect of sulfur dioxide on ascorbic acid, which will be discussed later, reactions with other vitamins in food systems have not been explored.

Alkalies

Thiamin also is destroyed in alkaline foods such as chocolate cake containing sodium bicarbonate and carbonate and lime-treated corn used for tortillas. In fact, aluminum sodium sulfate and the di-, and tri-sodium potassium phosphates also aid in the destruction of thiamin. Dwivedi and Arnold (1973) have identified a number of breakdown products from alkaline oxidation of thiamin. These include thiamin disulfide which has thiamin activity, and thiochrome and thiothiazalone which do not have any thiamin bioactivity. In addition, thiamin is cleaved to the pyrimidine and thiazole moieties and the thiazole may degrade further to a number of compounds including 3-mercaptopropanal, hydrogen sulfide, 2-methylfuran, 2-methylthiophene, 2-methyl-4, 5-dihydrothiophene, and 2-methylthio-5-methylfuran. The thiophenes are associated with the odor from the decomposition of thiamin.

In tortilla studies, Massieu et al. (1949) also noted 30% loss of tryptophan, threonine, histidine, and arginine. Many of the basic amino acids are prepared as the hydrochloride because the free bases are alkali-labile. The addition of alkali to lysine hydrochloride forms the free base which decomposes and produces an odor. We would expect lysine to decompose in chocolate cake and tortillas. Amino acids also will racemize in alkaline solution and thus reduce their biological activity.

Although it has not been demonstrated in alkaline foods, from information in the pharmaceutical literature, one would expect losses of pantothenic acid, vitamin K-1, cysteine, cystine, and vitamin D by slow degradation; riboflavin by conversion to lumiflavin; and the essential fatty acids by isomerization. In a food publication, Sistrunk and Cash (1970) showed a loss of ascorbic acid in squash as the pH was adjusted from 5 to 7.5.

Acids

Citric, phosphoric, lactic, malic, tartaric, fumaric, adipic, acetic, hydrochloric, and sulfuric acids have been added to foods. In liquid or semiliquid foods adjusted to pH 4 or below, vitamin A will deesterify and isomerize to the less bioactive cis forms. Folic acid, pantothenic acid, and threonine would be expected to decompose under acidic conditions. Moore and Folkers (1968) showed that dilute acids cleaved vitamin B-12 to the inactive corrin nucleus. Proteins with acid isoelectric points can be denatured and thus are not available for enzymatic hydrolysis and subsequent utilization. Miller et al. (1973) studied vitamin loss in bean products and slurries adjusted with hydrochloric acid to pH 3.5. They showed a loss of folacin although there was no loss of pyridoxine, and thiamin or decrease in Protein Efficiency Ratio (PER).

Nitrite

The recent recognition that nitrite can react with susceptible amines to form carcinogenic nitrosamines, and that ascorbic acid will block this reaction, has resulted in renewed interest in nitrite reactions. Ascorbic acid reacts directly with nitrogen dioxide to form nitric oxide gas and dehydroascorbic acid. Recently, Newmark et al. (1974), at Hoffmann-LaRoche laboratories, reported losses of 30% ascorbic acid in bacon after processing and an additional 30% loss after 6 months' storage and frying.

The effect of curing and smoking was covered in Part 1 of this chapter and the reader should refer to this section for further information.

Copper and Iron

Copper is generally not added but it is present in many foods. Only within the broad definition of a food additive can the presence of copper be considered. On the other hand, iron is added to many foods. Because their interactions are somewhat similar and more detrimental to other nutrients than generally recognized, these reactions will be discussed together.

In Chap. 18, Part 2, the author briefly mentions destruction of ascorbic acid by copper. The ascorbic acid first is converted to dehydroascorbic acid which has antiscorbutic activity but is unstable and degrades to diketogulonic acid which is inactive. Simultaneously the Cu^{2+} is converted by ascorbic acid to the lower oxidation state(s) and Fe^{3+} to Fe^{2+} as shown by Laurence and Ellis (1972).

Fe^{3+} and Cu^{2+} oxidation and destruction of organic free radicals have been reviewed by Jukes (1974). Recently the interactions of Fe^{3+} and Cu^{2+} with all the phenolic antioxidants were shown (Cort 1974; Cort et al. 1974).

Furthermore, Cu^{2+} reacts with tocopherol to form the p-tocopherolquinone which has no vitamin E or antioxidant activity. Reduction of Fe^{3+} to Fe^{2+} and the subsequent measurement of Fe^{2+} has been the basis of the colorimetric assay for tocopherol for many years. The tocopherol will go to the inactive p-tocopherolquinone under most conditions. However, if EDTA is present, the dimeric keto-ether is formed (Cort et al. 1974B).

Less than 3% loss has been found when 95% tocopherol is kept in thin layers for 4 yr and alcoholic solutions have been stable for 1 yr when metals are not present. If Cu^{2+} or Fe^{3+} are added to the latter, 40–50% of the tocopherol is decomposed within 15 days. However, the lower oxidation states, Fe^{2+} and Cu^{1+} and ground state copper, do not react with tocopherol.

Destruction of the tocopherols and phenolic antioxidants would lead to the loss of natural and added vitamin A. Furthermore, the Fe^{3+}- and Cu^{2+}-caused breakdown of tocopherols in vegetable oils, especially gamma-tocopherol which is a better antioxidant than the alpha homolog, decreases the oxidative stability of vegetable oils. The loss of antioxidants leads to oxidation of fatty acids in the naturally-occurring glycerides. The unsaturated fatty acids oxidize to peroxides which decompose to acids, aldehydes, ketones, alkanes, etc., and represent a loss of essential fatty acid. Also, oxidized fats have been shown to react with essential amino acids and make them nonavailable. Aylward and Haisman (1969) have shown that oxidized oils will oxidize vitamin A and β-carotene.

Thiamin is also destroyed by Cu^{2+} as shown by Dwivedi and Arnold (1973). Borenstein (1971) in a review article has reported that copper catalyzes folic acid destruction. Methionine and linolenic acid, as well as a number of other substrates, have been degraded to ethylene by ascorbic acid and copper as shown by Lieberman and Kunishi (1967).

Other Detrimental Interactions

Feller and Macek (1955) have shown that thiazole produced by decomposed thiamin promotes the breakdown of vitamin B-12. This is a problem in pharmaceuticals; and, in fact, most of our information (De Ritter 1969) on vitamin interactions is found in pharmaceutical research and publications. Fortunately in most foods, the thiamin level is sufficiently low that this is not a major pathway for vitamin B-12 decomposition. Gambier and Rahn (1957) have shown that riboflavin aids the oxidation of thiamin to thiochrome in concentrated solutions.

Niacinamide and ascorbic acid react to form a yellow nicotinamide-ascorbic acid complex listed as a source of niacin and vitamin C in CFR 121.1095. However, in some food systems, the bright yellow color is not acceptable.

Amen (1973) reports that phosphates, phytates, and oxalates in spinach complex iron and make it unavailable. He also claims that calcium acts as a zinc antagonist by competing for absorption sites when phytates or phosphates are present. Furthermore, low protein diets result in decreased apoferritin production and ultimately cause iron deficiency. Carbonyls present in flavor adjuncts can react with pyridoxamine to form inactive Schiff-bases and pyridoxal could react with amines similarly. Schroeder (1971) reports losses of natural pyridoxal and pyridoxamine as well as pantothenic acid, biotin, folacin, choline, and inositol in food processing, and some of the B-6 losses may be due to interactions.

Morrison and McLaughlin (1972) reviewed the availability of amino acids in foods. Casein and glucose combination when heated together markedly reduced the biological value of the casein; this could be corrected by the addition of supplemental lysine. Propanal formed from oxidized lipids inactivated lysine. Bressani et al. (1972) reported that amino acids react with proteins to form enzyme-resistant bonds, and that proteins heated with carbohydrates resulted in loss not only of lysine, but also methionine, arginine, histidine, and tryptophan. Hydrogen peroxide used in milk for cheese manufacture causes a 30% loss in bioavailable methionine. Cuq et al. (1973) demonstrated oxidation of methionine to the sulfoxide in casein decreases the enzymatic digestion of casein which further explains the previous workers' results.

Studies on vitamin A fortification of rice in our laboratory revealed a vitamin A and talc interaction which turned green and destroyed the vitamin A. As a result, in rice premix containing talc, the vitamin A had to be added separately and coated.

ADDITIVES BENEFICIAL TO NUTRIENTS

Sulfites

As stated in the introduction, compounds which are detrimental to some nutrients can be beneficial to others. Sulfites react in solution to scavenge oxygen. Ascorbic acid reacts by the same mechanism and decomposes in the process. Thus, sodium metabisulfite, sodium sulfite, and cysteine increase the stability of ascorbic acid in solution. Very recently Bolin and Stafford (1974) revealed sulfur dioxide protection of ascorbic acid and β-carotene. In addition, sodium metabisulfite is used to extract and stabilize cobalamines present in nature.

Ascorbic acid

Ascorbic acid converts Fe^{3+} to Fe^{2+} and Cu^{2+} to lower oxidation state(s) and as a result protects against Fe^{3+}- and Cu^{2+}-promoted reactions (Cort et al. 1975). It will, therefore, protect essential fatty acids, essential amino acids, vitamin A, vitamin E, thiamin, folic acid, and make iron more available. In addition to its protective effect, Schudel et al. (1972) have shown that it will convert the keto-ether dimer of tocopherol to "bi-α-tocopheryl" and tocopheroxide to α-tocopherol.

Antioxidants

Vitamin A with 5 conjugated double bonds, β-carotene with 11, and apocarotenal with 9 are susceptible to oxidation. Apparently the double-bonds in the β-ionone ring oxidize first to the epoxides in the carotenoids and probably in vitamin A, and the epoxides have reduced vitamin A activity. Classically BHT, BHA, and α-tocopherol are used to stabilize food-grade vitamin A. In general, the antioxidants are added to the vitamin A palmitate which is homogenized into gelatin and made into beadlets. The gelatin acts as an impermeable barrier to oxygen. Unstabilized vitamin A in thin films will lose potency in one day which fully explains the necessity of antioxidants. Vitamin D-2 and D-3 also contain oxidizable double bonds and should be protected similarly.

Water-Solubilizers and Mechanisms to Solubilize

The vitamin manufacturers have made a number of water-soluble, stabilized market forms of the water-insoluble vitamins. Homogenization into thick matrixes such as acacia, dextrins, and gelatin results in reduction of the particle size of the vitamin A to 1-3 μ. The emulsions are made into beadlets or used as is. In addition to vitamin A and antioxidants, these emulsions also contain preservatives such as the parabens or sodium benzoate and sorbic acid.

More recently acacia emulsions have been spray dried in order to have a particle size below 100 μ to facilitate adding to flour because it will not rebolt out. Spray-dried vitamin A is useful also in premixes made with other fine particulate vitamins. The dried forms contain other additives such as sucrose and lactose to increase stability and sometimes coconut oil to increase flavor stability. Often "slip" agents such as silicic acid are added to increase flow and prevent caking of dried products.

Water-dispersible solutions of water-insoluble vitamins are made in Polysorb 80. Other additives, such as ethanol, and propylene glycol are added to make the Polysorb soluble in cold water. Antifoam agents sometimes are added. These dispersible solutions also are stabilized with antioxidants and are used to fortify foods. A description of the commercial vitamin A application forms available and related stability were published recently (Bauernfeind and Cort 1973).

Tocopheryl acetate, on the other hand, does not require additives to prevent oxidation since it is an extremely stable, water-insoluble, viscous liquid. In order to make it into dry products, it has been homogenized to 1-3 μ particle size into gelatin, or dextrins and made into beadlets or spray dried for water-soluble applications. Free tocopherol is not as stable as the acetate. On standing in air it will form tocored which is the quinone on the 5- and 6-position. However, as has been shown recently, this never exceeds 0.7% tocored and is not a major breakdown product. As previously described, it is susceptible to Cu^{2+}- and Fe^{3+}-caused breakdown and protected by ascorbic acid and chelating agents.

In skimmed milk, the fat-soluble vitamins are removed with the cream. Stabilized water-soluble vitamin A is added to nonfat milk products but not tocopherol. Experimentally, water-soluble forms of vitamin E have been added to skim milk, nonfat dry skim milk, filled milk, and imitation milk at 1.5, 15 and 150 IU per qt with 100% stability for 4 weeks in the liquid products and 1 yr in the dry products.

Special Additives for B-Vitamins

Since thiamin breakdown products cause off-odors and riboflavin and niacin can be bitter, special coatings have been applied. These usually are mixtures of mono- and diglycerides and, in fact, are water-insoluble. The vitamins are usually 33% concentration. They are useful in dry products; although in some foods, Borenstein (1968) reports that the coated thiamin still causes off-flavor. Coated B-vitamins are used in chewable vitamin tablets.

Other Additives

Ascorbic acid has been lightly coated with ethocel (97% ascorbic acid) and fat-coated (30% ascorbic) to increase stability since at 3% moisture and above, ascorbic acid becomes tan. The speculation has been that free radicals on the 2- and 3-positions are formed. Levandoski et al. (1971) isolated monodehydroascorbic acid-ascorbic acid complex which is yellow and may also be involved in color formation by ascorbic acid.

Cysteine has also been proposed as a protectant to thiamin, although levels which protect have caused an off-odor from cysteine in the Hoffmann-LaRoche laboratories.

In pharmaceuticals, iron salts and chelating agents (disodium EDTA or citrates) are used to stabilize vitamin B-12 in solution (Newmark 1958; Federal Register 1962). Natural materials such as liver contain sufficient iron to stabilize vitamin B-12 as shown by

Shenoy and Ramasarma (1955). The stability of B-12 in foods during processing will depend on the content of iron and chelates.

BIBLIOGRAPHY

ANON. 1973. The use of chemicals in food production, processing, storage and distribution. Nutr. Rev. *31*, 191-198.
AMEN, R. J. 1973. Trace minerals as nutrients. Food Prod. Develop. 7, No. 8, 74-78.
AMES, S. R. 1972. Tocopherols occurrence in foods. In The Vitamins, Vol. 5. W. H. Sebrell, Jr., and R. S. Harris (Editors). Academic Press, New York.
AYLWARD, F., and HAISMAN, D. R. 1969. Oxidation systems. Advan. Food Res. *17*, 1-76.
BAUERNFEIND, J. C. 1970. Vitamin fortification and nutrification foods. 3rd Intern. Congr. Food Sci. Technol., Washington, D.C.
BAUERNFEIND, J. C., and CORT, W. M. 1973. Nutrification of foods with added vitamin A. Critical Rev. Food Technol. (CRC) *4*, No. 3, 337-375
BAUERNFEIND, J. C., and CORT, W. M. 1974. Tocopherols. In Encyclopedia of Food Technology. A. H. Johnson and M. S. Peterson (Editors). Avi Publishing Co., Westport, Conn.
BAUERNFEIND, J. C., and PINKERT, D. M. 1970. Food processing with added ascorbic acid. Advan. Food Res. *18*, 219.
BIELSKI, B. H. J., COMSTOCK, D. A., and BOWEN, R. A. 1971. Ascorbic acid free radicals. 1. Pulse radiolysis study of optical absorption and kinetic properties. J. Am. Chem. Soc. *93*, 5624-5629.
BOLIN, H. R., and STAFFORD, A. E. 1974. Effect of processing and storage on provitamin A and vitamin C in apricots. Food Sci. *37*, 1034-1036.
BORENSTEIN, B. 1968. Vitamins and amino acids. In Handbook of Food Additives. T. E. Furia (Editor). CRC Press, Cleveland.
BORENSTEIN, B. 1969. Vitamin A for flour and corn meal. Northwest Miller *276*, 18.
BORENSTEIN, B. 1971. Rationale and technology of food fortification. Critical Rev. Food Technol. (CRC) *2*, No. 2, 171-186.
BORENSTEIN, B., and CORT, W. M. 1970. Vitamin A for flour. French Pat. 2,011,564, Mar. 6.
BRESSANI, R., ELIAS, L. G., and GOMEZ BRENES, R. A. 1972. Improvement of protein quality by amino acid and protein supplementation. In International Encyclopedia of Food and Nutrition, Vol. 11. E. J. Bigwood (Editor). Pergamon Press, Elmsford, N.Y., and London, England.
BUNNELL, R. H. *et al.* 1968. Vitamin E stability in milk products. (Unpublished data.)
BUTTERFIELD, S., and CALLOWAY, D. H. 1972. Folacin in wheat and selected foods. J. Am. Diet. Assoc. *60*, 310-314.
CORT, W. M. 1973. Vitamin A and talc in rice. (Unpublished data.)
CORT, W. M. 1974. Antioxidant activity of tocopherols, ascorbyl palmitate and ascorbic acid and their mode of action. J. Am. Oil Chemists' Soc. *51*, 321-325.
CORT, W. M. *et al.* 1974. Antioxidant Activity and Stability of a Number of Chromans. Presented at 65th Ann. Spring Meeting, Am. Oil Chem., Mexico, Apr. 27. Abstr. J. Am. Oil Chemists' Soc. *51*, 279A.
CORT, W. M. *et al.* 1975. Antioxidant activity and stability of 6-hydroxy-2,5,7,8-tetramethylchroman-2-carboxylic acid. J. Am. Oil Chemists' Soc. *52*, 174-178.
CUQ, J. L., PROVANSAL, M., GUILLEUX, F., and CHEFTEL, C. 1973. Oxidation of methionine residues of casein by hydrogen peroxide. J. Food Sci. *38*, 11-13.

DE RITTER, E. 1969. Vitamin liquid formulation. Presented at Meeting of Soc. of Pharmacists, St. Louis. (Unpublished data.)
DWIVEDI, B. K., and ARNOLD, R. G. 1973. Chemistry of thiamin degradation in food products and model systems: A review. J. Agr. Food Chem. *21*, 54-60.
FEDERAL REGISTER. 1962. Disodium EDTA for use with iron salts to stabilize vitamin B_{12}. Federal Register 27:883, Jan. 31.
FELLER, B. A., and MACEK, T. J. 1955. Effect of thiamin hydrochloride on the stability of solutions of crystalline vitamin B_{12}. J. Am. Pharm. Assoc. *XLIV*, No. 11, 662-665.
GAMBIER, A. S., and RAHN, E. P. G. 1957. The combination of B-complex vitamins and ascorbic acid in aqueous solutions. J. Am. Pharm. Assoc., Sci. Edition *46*, 134-140.
JUKES, A. E. 1974. The organic chemistry of copper. *In* Advances in Organometallic Chemistry, Vol. 12. F. G. A. Stone, and R. West (Editors). Academic Press, New York.
LAURENCE, G. S., and ELLIS, K. J. 1972. The detection of a complex intermediate in the oxidation of ascorbic acid by ferric ion. J. Chem. Soc. 1667-1670, Dalton Trans.
LEVANDOSKI, N. G., BAKER, E. M., and CANHAM, J. E. 1971. A monodehydro form of ascorbic acid in the autoxidation of ascorbic acid to dehydroascorbic acid. Biochemistry *3*, 1465-1469.
LIEBERMAN, M., and KUNISHI, A. T. 1967. Propanal may be a precursor of ethylene in metabolism. Science *158*, 938.
MARKAKIS, P., and EMBS, R. J. 1966. Effect of sulfite and ascorbic acid on mushroom phenol oxidase. J. Food Sci. *31*, 807-811.
MASSIEU, G. H., GUZMAN, J., and CRAVIOTA, R. O., and CALVO, J. 1949. Determination of some essential amino acids in several uncooked and cooked Mexican foodstuffs. J. Nutr. *38*, 293-304.
MAURON, J. 1974. Influence of industrial and household handling of food protein quality. *In* International Encyclopedia of Food and Nutrition, Vol. II. E. J. Bigwood (Editor). Pergamon Press, Elmsford, N.Y., and London, England.
MILLER, C. F., GUADAGNI, D. G., and KON, S. 1973. Vitamin retention in bean products: cooked, canned and acid. J. Food Sci. *38*, 493-495.
MOORE, H. W., and FOLKERS, K. 1968. Vitamin B_{12}. II. Chemistry. *In* The Vitamins, Vol. 2. W. H. Sebrell, Jr., and R. S. Harris (Editors). Academic Press, New York.
MORRISON, A. B., and McLAUGHLIN, J. M. 1972. Availability of amino acids in foods. *In* International Encyclopedia of Food and Nutrition, Vol. II. E. J. Bigwood (Editor). Pergamon Press, Elmsford, N.Y., and London, England.
NEWMARK, H. L. 1958. Stable vitamin B_{12} containing solution. U.S. Pat. 2,823,167, Febr. 11.
NEWMARK, H. L. *et al.* 1974. Stability of ascorbate in bacon. Food Technol. *28*, 28-31, 60.
O'DELL, B. L. 1969. Effect of dietary components upon zinc availability. Am. J. Clin. Nutr. *22*, 1315-1322.
OLLIVER, M. 1967. Ascorbic acid occurrence in food. *In* The Vitamins, Vol. 1. W. H. Sebrell, Jr., and R. S. Harris, (Editors). Academic Press, New York.
RUBIN, S. H., and CORT, W. M. 1968. Aspects of vitamin and mineral enrichment. *In* Protein-Rich Cereal Foods for World Needs. M. Miller (Editor). American Association of Cereal Chemists, St. Paul, Minn.
SCHROEDER, H. A. 1971. Losses of vitamins and trace minerals resulting from processing and preservation of foods. Am. J. Clin. Nutr. *24*, 562-573.
SCHUDEL, P., MAYER, H., and ISLER, O. 1972. Tocopherol chemistry. *In*

The Vitamins, Vol. 5. W. H. Sebrell, Jr., and R. S. Harris (Editors). Academic Press, New York.

SHENOY, K. G., and RAMASARMA, G. B. 1955. Iron as a stabilizer of vitamin B_{12} activity in liver extracts. Arch. Biochem. Biophys. 55, 293-295.

SISTRUNK, W. A., and CASH, J. N. 1970. Ascorbic acid and color changes in summer squash as influenced by blanch, pH, and other treatments. J. Food Sci. 35, 645-648.

SOCIETE CIVILE DE RECHERCHES SCIENTIFIQUES ET INDUSTRIELLES. 1964. Stable aqueous polyvitamin solutions. French Pat. 1,372,408, Aug. 10.

WADDELL, J. 1974. Bioavailability of iron sources. Food Prod. Develop. 8, No. 1, 80-86.

CHAPTER 14

Edward S. Josephson
Miriam H. Thomas
William K. Calhoun

Effects of Treatment of Foods with Ionizing Radiation

The effects of ionizing radiation on the nutritional value of foods are not markedly different in degree from those of other methods of preservation. Protection of nutrients is improved by holding the food at low temperature during irradiation (Thomas and Josephson 1970) and by reducing or excluding free oxygen from the radiation milieu (Metlitskii et al. 1968; Kharlamov and Shubnyakova 1964; Southern and Rhodes 1967).

Because of the protective qualities inherent in foods, the sensitivity to radiation of nutritional components in food is less than that of the same constituents irradiated in pure form or in artificial solutions and mixtures (Bregyadze and Bokeriya 1971; Metlitskii et al. 1968). For this reason, we have focused our attention on the effects of ionizing radiation on nutrients in the foods themselves. The tabular data selected for this chapter are those where most or all of the essential parameters are defined for processing foods under conditions envisaged for commercial production for the consuming public.

TYPES OF RADIATIONS

The three basic types of ionizing radiation used for food processing are electrons,[1] X-rays[2] produced by electrons in an X-ray target, and gamma rays[3] from ^{60}Co and ^{137}Cs. All three types cause ionization in the food by either the primary electrons or by the secondary electrons resulting from gamma or X-ray interactions in the food with little rise in temperature and little total chemical change. The ionized and activated molecules form unstable secondary products, notably free radicals and peroxides.

---

[1] The negatively charged particle which is a common constituent of all atoms and having a mass (m) = 9×10^{-28} gm. Electron volt (eV) is the amount of energy gained by an electron when accelerated by a potential of 1 volt. One electron volt = 1.6×10^{12} erg. 1 million eV = 1 MeV.

[2] Electromagnetic radiation of short wavelength produced when a beam of fast electrons in a high vacuum bombards a metallic target.

[3] Electromagnetic radiation of very short wavelength emitted by the nuclei of radioactive substances during decay. Gamma rays are similar to X-rays of short wavelength.

COMPOSITE DIETS

The great body of evidence accumulated during the past two decades, with few exceptions, indicates that irradiated foods are as nutritious as thermally processed foods (Raica et al. 1972). Raica and Howie (1966) reported no changes in digestibility of proteins, fats, and carbohydrates exposed to absorbed doses of 5.6 Mrad.[4] Read et al. (1961) found no differences in metabolizable energy and available protein, fat, and carbohydrate between radappertized (5.6 Mrad) and nonirradiated control composite diets fed to rats for 4 generations. Kennedy (1965) observed little change in nutritive value of animal feeds (protein concentrate) using doses of 0.5 and 1.0 Mrad and no nutritional changes with frozen eggs irradiated at 0.5 and 5.0 Mrad. Ley (1972) found no adverse effect on the nutritive value of animal feeds irradiated at 0.1 and 0.5 Mrad. He concluded that radiation at 2.0 Mrad is superior to heat processing with respect to retention of protein quality. Ley (1972) described excellent results with radappertized feed for germ-free rat and mouse colonies maintained for 5 yr. Kraybill (1960) reported that the biological value of proteins and metabolizable energy value of diets were not altered by radappertization[5] (5.6 Mrad).

Metlitskii et al. (1968) reported that a 3-Mrad dose had no effect on assimilability, energy balance, or biological value of food products. Coates et al. (1963) found only a slight nutritional loss in diets sterilized at 5-Mrad doses and fed to germ-free chicks and no nutritional loss at doses ranging from 2.0 to 3.0 Mrad. Schoen and Hiller (1971) concluded that use of sterilizing gamma radiation doses improved the digestibility of crude fat and protein in their rodent diet whereas steam sterilization decreased their digestibility to less than those of unsterilized control feed.

In 1969 the Health Authorities of the United Kingdom and the Netherlands approved radappertized foods for hospital patients in reversed barrier isolation (Anon. 1973). No apparent nutritional problems have been reported as a consequence, even though patients subsist entirely on these foods for several months. Radappertization

---

[4] Unit of absorbed ionizing radiation dose. One rad = 100 ergs of energy absorbed per gram of matter. 1000 rad = 1 Kilorad (Krad). 1 million rad = 1 Megarad (Mrad).

[5] Exposure to ionizing radiation of food in hermetically sealed packaging at doses necessary to kill all organisms of food spoilage or public health significance. Doses used are greater than 1 Mrad. Analagous to thermal sterilization (canning).

EFFECTS OF TREATMENT WITH IONIZING RADIATION 395

and radicidation[6] have been used in preference to thermal sterilization to sustain germ-free and specific pathogen-free rats, mice, pigs, and chickens (Sato 1970; Schoen and Hiller 1971; Udes et al. 1971; Ley et al. 1969; Coates et al. 1963).

PROTEINS AND AMINO ACIDS

Ley et al. (1969) reported on work of B. O. Eggum who radappertized a sample of rat diet in which the protein was supplied in the form of soya, meat-and-bone, and fish meals at doses up to 7.0 Mrad. The data indicate no significant effect of irradiation on protein quality and are summarized in Tables 14.1 and 14.2.

TABLE 14.1

EFFECT OF IRRADIATION ON THE PROTEIN QUALITY OF RAT DIET

Dose (Mrad)	True Digestibility	Biological Value	Net Protein Utilization
0	85.6	80.5	68.9
0.5	83.6	75.8	63.5
1.0	86.5	81.7	70.6
2.5	87.0	78.1	68.0
3.5	84.8	77.3	65.4
7.0	85.3	76.4	65.2

Source: Ley et al. (1969).

Based upon the amino acid analyses shown in Table 14.3, the protein in beef radappertized (4.7-7.1 Mrad) under conditions proposed for commercial processing showed no significant loss in nutritional quality when compared with beef preserved by freezing ($-18°C$) or heat sterilization (F_o = 5.8).

Brooke et al. (1966) concluded that: storage in ice or at $-21°C$ had no effect on amino acids in haddock fillets analyzed in the raw state and after steaming when the fillets were air-packaged, irradiated at 0.25 Mrad, and stored 30 days at $0.6°C$; or vacuum-packed, irradiated at 0.15 Mrad, and stored 30 days at $0.6°C$. These data along with similar data for irradiated clams (Brooke et al. 1964) which are summarized in Table 14.4 indicate that radicidation and raduriza-

---

[6] Exposure of food to ionizing radiation at doses necessary to kill all nonspore-forming pathogens. Doses used are generally below 1 Mrad. Analagous to pasteurization.

TABLE 14.2

EFFECT OF A RADIATION DOSE OF 7.0 MRAD ON THE AMINO ACID COMPOSITION OF THE PROTEIN IN RAT DIET

Amino Acid	Unirradiated Diet (Gm/16 Gm N)	Irradiated Diet (Gm/16 Gm N)
Asparagine	8.85	8.38
Threonine	3.80	3.73
Serine	4.17	4.16
Glutamic acid	15.70	15.61
Glycine	5.82	5.79
Alanine	5.61	5.54
Valine	4.78	4.68
Isoleucine	3.99	3.99
Leucine	7.44	7.47
Tyrosine	3.28	3.38
Phenylalanine	4.12	4.28
Lysine	5.72	5.82
Histidine	2.29	2.37
Arginine	6.04	6.05
Methionine	2.33	2.11
Cystine	1.34	1.44
Tryptophan	1.16	1.32

Source: Ley et al. (1969).

tion[7] do not significantly alter the nutritional quality of the protein in haddock fillets and clams.

Notwithstanding the rather extensive literature on the effects of ionizing radiation to preserve fruits, there are very few reports dealing with the effects of radiation on the nutritive aspects of their amino acids and proteins. Loaharanu (1971) stated that no significant differences were detected in terms of nutritive values between mangoes, papayas, rumbatans, and longans gamma irradiated for insect disinfestation and their nonirradiated controls. Clarke and Fernandes (1961) subjected pears to doses of gamma radiation ranging from 100 to 200 Krad and observed enhanced protein synthesis during storage for 100 days at 2.8°C. Funes (1970) found no destructive effect on the amino acids of apple and grape juices exposed to 0.8 Mrad of gamma radiation from ^{60}Co. Obara et al. (1958) observed no significant changes in amino acids in orange juice exposed to 1 Mrad. Bregvadze and Bokeriya (1971) observed no differences between the amino acid composition of semisweet wines ex-

---

[7] Exposure of food to ionizing radiation to reduce populations of organisms in order to delay onset of spoilage. Doses used are generally below 1 Mrad. Analagous to pasteurization.

TABLE 14.3

EFFECTS OF DIFFERENT PROCESSING METHODS UPON THE AMINO ACID AND VITAMIN CONTENT OF ENZYME-INACTIVATED BEEF

	Treatment and Length of Storage						
	Frozen Control		Thermally Sterilized (F_o = 5.8)		^{60}Co (4.7–7.1 Mrad)[1]		Electron (10 MeV) (4.7–7.1 Mrad)[1]
Nutrient	0	15 (Months)	0	15 (Months)	0	15 (Months)	0	15 (Months)
Amino acids (wt %)								
Aspartic acid	2.12	1.99	2.22	2.11	2.27	2.07	2.15	2.24
Threonine	0.91	1.02	0.95	0.99	1.02	1.00	0.91	0.98
Serine	0.78	0.92	0.81	0.91	0.84	0.88	0.80	0.87
Glutamic acid	3.84	3.63	3.94	3.60	4.10	3.70	3.89	3.91
Proline	1.06	1.17	1.08	1.11	1.12	1.06	1.07	1.05
Glycine	1.39	1.44	1.41	1.40	1.45	1.45	1.28	1.39
Alanine	1.43	1.44	1.47	1.42	1.50	1.46	1.40	1.48
Valine	1.00	1.06	1.07	1.06	1.09	1.09	1.04	1.08
Isoleucine	0.95	0.97	1.02	1.00	1.03	1.04	0.98	1.03
Leucine	1.78	1.84	1.87	1.84	1.95	1.91	1.76	1.94
Tyrosine	0.77	0.84	0.82	0.85	0.85	0.86	0.80	0.88
Phenylalanine	0.89	0.96	0.93	0.96	0.94	0.95	0.89	1.00
Lysine	2.02	2.05	2.09	2.09	2.12	2.01	2.09	2.06
Histidine	0.76	0.80	0.83	0.78	0.81	0.78	0.75	0.83
Arginine	1.43	1.57	1.33	1.52	1.61	1.39	1.55	1.62
Cystine[2]	—[3]	0.16	0.36	0.16	0.24	0.24	0.24	0.32
Methionine	0.53	0.54	0.57	0.54	0.62	0.54	0.56	0.55
Tryptophan	0.33	0.26	0.28	0.26	0.30	0.25	0.27	0.26
Total amino acids	21.99	22.66	23.05	22.60	23.86	22.68	22.43	23.49
Vitamins (%)								
Thiamin (mg)	0.05	0.056	0.02	0.017	0.02	0.015	0.03	0.019
Riboflavin (mg)	0.51	0.099	0.44	0.120	0.49	0.120	0.49	0.085
Niacin (mg)	4.74	4.75	4.92	4.75	4.97	4.75	5.00	6.19
Pyridoxine (mg)	0.49	0.099	0.41	0.065	0.44	0.030	0.41	0.060

Source: Office of the Surgeon General, Department of the Army, Contract No. DADA 17-71-C-1030. Industrial Bio-Test, Inc., Contractor.

[1] Packaged beef air-evacuated to internal pressure (IP) of approximately 100 mm Hg. IP at start and after irradiation and thawing was approximately 250 and 350 mm Hg, respectively. Temperature of product was −40 to −5°C during irradiation.
[2] Not including cysteic acid.
[3] Chromatographic peak not resolved from methionine.

TABLE 14.4

EFFECT OF IRRADIATION ON THE TOTAL AMINO ACID
(% OF PROTEIN) CONTENT OF SEA FOODS

Amino Acid	Clams[1]			Haddock[2]		
	0[3]	Krad 450 AP[4]	Krad 350 VP[5]	0	Krad 250 AP	Krad 150 VP
Tryptophan	1.10	1.24	1.15	1.27	1.18	1.18
Lysine	6.89	6.69	7.35	10.78	9.40	9.82
Histidine	1.31	1.74	1.35	2.26	1.75	2.22
Threonine	3.49	4.05	4.15	4.02	3.72	4.54
Valine	3.89	4.12	3.99	4.50	4.70	4.89
Methionine	2.18	2.30	2.12	3.00	3.11	3.31
Isoleucine	3.75	4.00	3.68	4.64	4.76	5.35
Leucine	6.27	6.50	5.89	5.32	7.54	8.47
Phenylalanine	2.88	3.43	2.68	3.32	3.40	4.15
$1/2$ Cystine	1.09	1.02	1.05	0.99	0.83	1.13
Ammonia	1.42	1.78	2.01	1.51	1.43	1.31
Arginine	6.24	6.79	6.93	6.66	6.07	5.13
Aspartic acid	7.46	7.60	7.75	9.30	9.78	11.05
Serine	3.47	4.08	3.81	3.91	3.79	4.97
Glutamic acid	11.35	12.11	12.41	13.33	11.12	15.75
Proline	2.85	3.24	3.14	2.97	3.14	3.57
Glycine	6.45	6.85	7.01	3.91	4.22	4.55
Alanine	7.62	7.76	7.97	5.41	5.73	6.08
Tyrosine	2.88	3.11	2.51	2.83	3.17	3.76
Protein (%)	10.75	8.97	10.05	19.00	17.49	17.63
Moisture (%)	85.50	88.28	86.30	79.56	81.14	79.10

Source: Derived from Brooke et al. (1964, 1966).
[1] Stored 30 days at 0°C postirradiation.
[2] Stored 30 days in ice.
[3] Fresh.
[4] Air packed.
[5] Vacuum packed.

posed to 1 Mrad to stabilize them microbiologically and nonirradiated controls.

Vakil et al. (1973) using rats found no significant changes in protein, fat, and mineral content of wheat, gamma irradiated at 20 and 200 Krad for insect disinfestation. Total amino acid profiles and available lysine content of irradiated wheat revealed no change, but an overall increase of about 8% in free amino acid levels was observed on irradiation up to 1 Mrad. They concluded that the changes in physicochemical properties of the starch and protein in wheat were of no major significance nutritionally. These findings are in agreement with those of Pape (1973), Doguchi (1969), Nair and Brownell (1965), and Metlitskii et al. (1968). Similar results have been reported by Leonova and Sosedov (1972) for rice and buckwheat, by

Metta and Johnson (1959) for wheat and corn, and by Revetti (1973) for maize and kidney beans.

Johnson (1961) concluded that radappertization even at doses up to 10 Mrad caused no serious damage to the protein of corn, wheat, beef, beans, peas, and milk although there was a slight depression of biological value of milk (8%) and pea protein at 6 Mrad. The destruction of lysine and arginine in peas and beans was not sufficient for these two amino acids to become limiting. At 3 Mrad doses there was no loss in metabolizable energy values of fat, carbohydrate, or protein (Johnson 1964).

Metlitskii et al. (1968) reviewed the literature on potatoes irradiated to inhibit sprouting during storage and reported that nitrogenous substances changed little. Jaarma and Henricson (1964) reported that the nutritional adequacy of irradiated (14-15 Krad for pigs and 200 Krad for rats) and control potatoes was equal. Varela and Urbano (1971) found that the irradiation of potatoes (8 Krad) and storage for 190 days had no significant effect on the digestibility or biological value of the potato protein. Lang and Bässler (1966A) reported no differences in protein efficiency in irradiated potatoes (10 Krad) fed to rats for 1 yr when compared with nonirradiated controls. Fujimaki et al. (1968) found no differences in the amino acid composition of proteins between gamma-irradiated (7-30 Krad) and nonirradiated potato tubers.

No data pertaining to the nutritional quality of proteins in onions irradiated (5-15 Krad) for sprout inhibition during storage have been noted.

Several instances of improvement in protein level and nutritional quality have been reported in the edible portion of wheat, rice, lupine, and maize plants grown from seeds of mutants induced by low dose ionizing radiation (Anon. 1972; Swaminathan et al. 1969; Dumanović and Denić 1969; Zaben'kova et al. 1972).

In summary, the evidence indicates that by carefully controlling all the conditions of food processing at the radiation doses being considered for commercial production, there should be no significant impairment in the nutritional quality of the protein constituents.

LIPIDS

The effects of ionizing radiations upon lipids are not unlike changes due to heat and oxidative processes. Several reviews and articles dealing with irradiation of lipids have appeared: Mitchell (1957), Partmann (1962), Chipault (1962), Merritt (1966), Nawar (1972). Irradiation results in a large number of compounds. The main reactions involve oxidation, polymerization, decarboxylation, and dehydration. The effects of irradiation upon lipids depend heavily upon the fatty acid composition and, as expected, unsaturated fatty acids are more easily oxidized than are the saturates. Chemical changes are minimized by irradiating the products at low temperature and in the absence of light and oxygen. Gel'fand (1970) found that steaks packaged in polyethylene-foil-cellophane to exclude light and irradiated to 0.8 Mrad resulted in fewer lipid oxidation products than similar samples packaged in transparent polyethylene-cellophane. The difference increased further upon storage for up to 6 months. Rao and Novak (1973) found that ^{60}Co irradiation of a chicken-based wet pet food product to 4.5 Mrad did not change the relative composition of the total lipid extract or of the triglyceride fraction compared to the nonirradiated control.

Few studies of the nutritional consequences of eating irradiated fats have been reported. Plough et al. (1957) fed human subjects pork irradiated to 2.79 Mrad and stored for 1 yr at room temperature. Identical apparent digestibility values for irradiated and unirradiated fat were obtained.

Schreiber and Nasset (1959) investigated the digestion by dogs of lard irradiated to 5.58 Mrad by an electron source. Irradiation slowed the rate of absorption of fat due to delayed emptying of the stomach contents. However, overall digestibility was unaffected, indicating that lipolysis and absorption of end-products were not seriously disturbed by feeding irradiated lard.

Moore (1961) fed rats corn oil irradiated to either 2.79 or 5.58 Mrad. Irradiation did not adversely affect the digestibility of the lipid.

Read et al. (1961) studied the availability of fat derived from major food components and irradiated by spent fuel rods to 5.58 Mrad. In rats, the availability of irradiated fat was 95.8% compared to 94.8% for the unirradiated control.

Lang and Bässler (1966B) compared the digestibility in rats of soybean oil electron irradiated to 100 Mrad in air at room temperature with an unirradiated control. Digestibility of the irradiated oil was only slightly diminished compared to the control. This radiation dose is 20–30 times the level required for sterilization of foods. These authors found that soybean oil irradiated to 2.5 Mrad was not adversely affected.

Based upon results obtained when lipids are irradiated under conditions anticipated for use in commercial food processing, which is not anticipated to exceed 7 Mrad, it may be concluded that irradiation does not result in significant loss of nutritional value.

EFFECTS OF TREATMENT WITH IONIZING RADIATION

CARBOHYDRATES

The main effects of radiation upon carbohydrates are those of hydrolysis and oxidative degradation. Polysaccharides are depolymerized and cellulose is made more susceptible to enzymatic hydrolysis. Pectin substances lose jelling powers. In short, complex carbohydrates are converted into simpler compounds by radiation energy.

Although irradiation may cause changes in the physical and chemical properties of high carbohydrate foods such as grains and vegetables, these have not been shown to be of any nutritional significance. Read et al. (1961) determined with rats that the availability of the carbohydrates of 8 foods was unaffected by a sterilizing dose of 5.58 Mrad.

Lang and Bässler (1966A) fed rats 72% of the diet as potatoes irradiated to either 10 or 100 Krad. No differences in utilization of starch calories were found between the irradiated and control products.

Using an intermediate dose of radiation, Saint-Lèbe et al. (1973) irradiated dry maize starch with ^{60}Co to either 300 or 600 Krad. They fed the starch both raw and cooked as 62% of the diet to rats for 1 yr. No significant differences were found in growth or reproduction between groups receiving either irradiated or nonirradiated starch. Along with other investigators, these authors found aldehydes, sugar acids, and hydrogen peroxide formed by irradiation, albeit in small quantities.

In summary, it does not appear that irradiation of food carbohydrates to levels anywhere near those contemplated for use in food processing for insect disinfestation (grains, papayas), mold inhibition (strawberries, oranges), retardation of ripening (bananas) or for sprout inhibition of tubers damages their nutritional value.

VITAMINS

The stability of micronutrients in food is influenced by many factors. When a vitamin is bombarded by ionizing radiation, destruction may occur, the kind and degree depending upon the sensitivity of the vitamin itself, the amount of energy to which it is exposed, and the nature and physical state of the medium in which it is present.

Proctor and Goldblith (1949) reported niacin to be the most resistant of the water-soluble vitamins to irradiation damage, followed by riboflavin, and that ascorbic acid was the most sensitive. Intermediate in sensitivity to irradiation were vitamin B-12 and p-amino benzoic acid. Pyridoxine is also radiosensitive and thiamin in some media may be even more sensitive to damage than ascorbic acid. Changes occurring during the irradiation of vitamin solutions are considerably more drastic than those occurring in natural foods. In concentrated solutions, less destruction takes place than in dilute solutions. The extent of destruction is usually a function of radiation dose and temperature of the medium during irradiation.

Sensitivity to irradiation of vitamins in food can be minimized by keeping the product frozen during the irradiation treatment. Studies have been made comparing the retention of thiamin, riboflavin, niacin, and pyridoxine in irradiated and thermally-processed beef (Table 14.3) and pork (Table 14.5). Results show that irradia-

TABLE 14.5

EFFECT OF PROCESSING ON THE VITAMIN CONTENT
OF SHELF-STABLE CANNED PORK LOIN

Vitamin	Treatment	Mg/100 Gm[1]	Retention (%)
Thiamin	Control	3.69 ± 0.22^2	
	4.5 Mrad @ $-80°C \pm 5°$	3.14 ± 0.25	85
	Thermally processed	0.76 ± 0.08	20
Riboflavin	Control	1.02 ± 0.28	
	4.5 Mrad @ $-80°C \pm 5°$	0.79 ± 0.06	78
	Thermally processed	0.82 ± 0.02	81
Niacin	Control	20.3 ± 5.1	
	4.5 Mrad @ $-80°C \pm 5°$	15.9 ± 2.6	78
	Thermally processed	13.2 ± 1.8	65
Pyridoxine	Control	0.76 ± 0.05	
	4.5 Mrad @ $-80°C \pm 5°$	0.75 ± 0.07	98
	Thermally processed	0.63 ± 0.07	84

Source: Thomas and Josephson (1970).
[1] Moisture, fat, salt-free basis.
[2] Mean ± S.D., three samples per treatment.

tion at low temperature is no more destructive to these vitamins than is heat processing. Similar data have been obtained for ham (Thomas and Josephson 1970).

Recently, the use of diets sterilized by ^{60}Co radiation at 5 Mrad and 6 ± 0.5 Mrad has proven to be advantageous for germ-free animal feeding studies. The effect of this treatment on various vitamins in several diets is indicated in Table 14.6.

Radurization with less than 300 Krad has been effective in delaying the softening of mature fruit, inhibiting sprouting in potatoes during storage, disinfesting wheat and wheat products, and extending the shelf-life of fresh unfrozen fish and poultry products. Analyses of

TABLE 14.6

THE EFFECT OF GAMMA IRRADIATION (^{60}Co) ON
THE VITAMIN RETENTION (%) OF VARIOUS DIETS

	Diet					
	Chick			Guinea Pig	Cat	Mouse
Dose in Mrad	2	3	5^1 5^2	2.5	2.5	6.0 ±0.5
Thiamin	104	104	72 66	90	56	87
Riboflavin	119	105	100 108	94	98	109
Niacin	100	100	103 97	102	94	103
Pyridoxine	91	91	73 78	122	55	79
Vitamin A	88	79	72 65	94	7	99
Carotene	75	69	50 61	95	36	107
Vitamin E	32	31	49 90	76	121	78
Pantothenic acid	95	92	110 100	79	88	—
Folacin	—	—	118 102	92	78	—
Vitamin B-12	82	91	100 100	112	102	—
Biotin	79	79	100 100	107	116	—

Source: Calculated from Ley (1972), Thomas (1972), and Coates et al. (1963).
[1] Air packed.
[2] Vacuum packed.

radurized oysters (0.2-Mrad dose) for thiamin, riboflavin, niacin, pantothenic acid, biotin, pyridoxine, folic acid, and vitamin B-12 indicated that losses were extensive only for thiamin and pyridoxine (Liuzzo et al. 1966). Similar changes took place in air-packed clams irradiated at 450 Krad at 0°C or vacuum-packed clams irradiated at 350 Krad (Brooke et al. 1964) after 30 days' storage in ice. Haddock fillets (Brooke et al. 1966) show the same pattern of vitamin retention after irradiation at these doses and storage conditions (Table 14.7). Kennedy and Ley (1971) have obtained comparable results with fillets of cod.

Approximately 80–90% retention of thiamin, riboflavin, and niacin was obtained in irradiated wheat at either 20 or 200 Krad (Vakil et al. 1973) (Table 14.8). Gamma irradiation of bleached, enriched, hard wheat flour in the range of 30–50 Krad had no detrimental effect on the thiamin, riboflavin, niacin, or pyridoxine content (Heiligman et al. 1973). Furthermore, the nutritive quality of bread made from this flour was not affected.

Vitamin B-12, p-amino benzoic acid, pantothenic acid, and folacin are all radiosensitive in aqueous solution. Fortunately, such is not the case in food. Considerable reduction in the radiosensitivity of vitamin B-12 was obtained in raw whole milk. Irradiation of ground

404 NUTRITIONAL EVALUATION OF FOOD PROCESSING

TABLE 14.7

EFFECT OF IRRADIATION ON VITAMIN RETENTION (%) OF SEA FOODS

Vitamin	Clams[1] Krad 450 AP[3]	Clams[1] Krad 350 VP[4]	Haddock[2] Krad 250 AP	Haddock[2] Krad 150 VP
Thiamin	80	67	37	78
Riboflavin	99	111	105	100
Niacin	84	97	106	100
Pyridoxine	63	93	125	115
Pantothenic acid	115	115	164	178
Vitamin B-12	92	91	110	90

Source: Calculated from Brooke et al. (1964, 1966).
[1] Stored 30 days at 0°C postirradiation.
[2] Stored 30 days in ice.
[3] Air packed.
[4] Vacuum packed.

TABLE 14.8

EFFECT OF IRRADIATION ON VITAMIN RETENTION (%) OF WHEAT AND WHEAT PRODUCTS

Krad	Thiamin	Riboflavin	Niacin	Pyridoxine
20[1]	88	91	88	—
200[1]	88	87	91	—
30–50[2]	100	100	89	100
30–50[3]	100	100	117	100

Source: Calculated from Vakil et al. (1973) and Heiligman et al. (1973).
[1] Wheat.
[2] Flour.
[3] Bread from irradiated flour.

pork with gamma rays from spent fuel rods produced less than 10% destruction of pantothenic acid and no destruction of folacin with doses up to 5.58 Mrad (Sheffner and Spector 1957). Moreover, irradiated diets fed to chicks had no significant decrease in folacin activity (Richardson 1955).

Ascorbic acid, like thiamin, is very radiosensitive, less so in foods than as the pure compound. The irradiation doses (5–15 Krad) approved for commercial processing of white potatoes to prevent sprouting during storage results in minimal losses of ascorbic acid. At 10 Krad, some researchers have reported 15% loss, while others have reported no loss in ascorbic acid (McKinney 1971). Doses up

EFFECTS OF TREATMENT WITH IONIZING RADIATION 405

to 12 Krad had little effect on the ascorbic acid content of onions (McKinney 1971). Fruits have shown loss of ascorbic acid even with low-dose irradiation (Dennison and Ahmed 1971-1972; Dollar et al. 1970). Table 14.9 shows that the vitamin C retention in oranges, tangerines,

TABLE 14.9

EFFECT OF RADIOPASTEURIZATION ON ASCORBIC ACID RETENTION (%) IN FRUIT

Product	Dose (Krad)	Retention (%)
Oranges, Temple	100	97
	200	72
Tangerines	40	104
	80	94
	160	94
Tomatoes	100	86
	200	86
	300	91
Papayas	125	110

Source: Calculated from Dennison and Ahmed (1971-1972) and Wenkam and Moy (1968).

tomatoes, and papayas varies from 100% to 72% with radiation doses from 40 to 300 Krad. Wenkam and Moy (1968) report no significant difference in ascorbic acid or carotene content between nonirradiated mangoes and papayas and those irradiated at 25 Krad. Ascorbic acid retention in vegetables radappertized at ambient temperature are presented in Table 14.10.

There is considerably less information about the effects of irradiation on fat-soluble vitamins than for water-soluble vitamins. We have found no reports on food which would, if irradiated in the frozen state, make a considerable contribution of fat-soluble vitamins to the dietary (e.g., liver). Goldblith and Proctor (1949) reported that carotene was radiosensitive to cathode rays. Studies by Kung et al. (1953) showed that irradiation of whole milk with 440 Krad resulted in the destruction of 40% of the carotenoids, 70% of the vitamin A, and 60% of the tocopherols. Other studies with dairy products indicate from 31 to 68% losses of vitamin A. Carotene destruction during radiation treatment can be minimized by the addition of ascorbic acid. Kuzin and Abdurakhmanov (1970) have reported that doses of from 20 to 40 Krad cause an "increase" in the carotene content of carrots, and that doses of from 10 to 80 Krad are "ineffective" or

TABLE 14.10

ASCORBIC ACID AND CAROTENE RETENTION
IN IRRADIATED VEGETABLES

Product	Dose	Percentage Retention	
		Ascorbic Acid	Carotene
Carrots	Krad		
	10		102
	20		115
	30		113
	40		109
	80		99
	Mrad		
Beans, green	4.8	73	169
Carrots	4.8	78	87
Corn	4.8	71	56

Source: Calculated from Thomas and Calloway (1961) and Kuzin and Abdurakhmanov (1970).

cause a decrease in carotene content (Table 14.10). The increase is attributed to the formation of acetyl-CoA, the first compound required for carotene synthesis, and to the decomposition of the starch producing sufficient glucose to synthesize one molecule of carotene.

The sensitivity of vitamin A is influenced by the media in which it is exposed. It is more stable to irradiation effects in margarine than in butter. Sheffner and Spector (1957) suggest that the vitamin A esters used to supplement margarine are more resistant to irradiation than the natural vitamin A in butter.

Vitamin D is obtained by the ultraviolet irradiation of ergosterol, yet irradiation of ergosterol with gamma photons results in only traces of vitamin D. Also, biological evidence indicates that the vitamin D activity for chicks is decreased by gamma irradiation of the total diet with 2.79 Mrad at ambient temperature (Sheffner and Spector 1957).

Vitamin E is another radiosensitive fat-soluble vitamin. Kung et al. (1953) estimated radiation destruction in whole milk to be 61% at a dose of 400 Krad. Diehl (1969) presented evidence that a vitamin E-destroying factor is present in irradiated oil which is favored by the presence of oxygen but not completely inhibited by its absence. Irradiation with 100 Krad caused 17% loss of vitamin E in rolled oats and in hazelnuts.

Investigations by Richardson et al. (1956) showed no loss of vitamin K content in alfalfa leaf meal and fresh spinach irradiated to a dose of 2.79 Mrad at ambient temperature. No destruction of K-1,

K-3, or K-5 in irradiated semisynthetic diets was apparent when they were fed to chickens.

The possibility that irradiation will partially supplant other methods of processing for the preservation of food requires that the nutritional consequences be known. The data presented here demonstrate that the situation with the vitamins is not entirely clear-cut and testify to the extreme variability that can be expected within and among different classes of foods. This is largely due to the lack of specificity in the description in the literature of the radiation conditions which make it impossible to make absolute comparisons. However, where it has been possible to compare the vitamin content of foods processed by irradiation technology as anticipated for commercial use with foods processed by conventional means, irradiation was no more destructive to vitamins than other food preservation methods.

CONCLUSION

When food is preserved by ionizing radiation under processing conditions proposed for commercial production, nutrient destruction is no greater than that which occurs when food is preserved by more conventional means. This conclusion is based upon data derived from *in vivo* and *in vitro* studies and chemical and physical analyses.

The data presented were selected from the vast literature on the basis of relevancy to the nutrition of the consumer. For this reason, less attention was devoted to the effects upon nutrient constituents irradiated in pure form or in artificial solutions and mixtures although there could be application for intravenous use if radappertized. Wherever possible the data presented were those where the processing parameters were known and were directly related to future application on a commercial scale.

There are several references which carry a wealth of information on all aspects of processing of foods by ionizing radiation including the effects on nutrients. Among these are the proceedings of symposia on food irradiation published by the International Atomic Energy Agency (1966, 1973) and Hearings on Food Irradiation conducted by the Joint Committee on Atomic Energy, Congress of the United States (1956, 1960, 1962, 1963, 1965, 1966, 1968).

In the 1965 Hearing, when the results of over 10 yr of research sponsored by the U.S. Army were completed, the Army Surgeon General concluded that "foods irradiated up to absorbed doses of 5.6 megarads with a cobalt 60 source of gamma radiation or with electrons with energies up to 10 million electron volts have been found to be wholesome; i.e., safe, and nutritionally adequate."

The data supporting this statement are summarized by Reber et al. (1966).

BIBLIOGRAPHY

ANON. 1965. Statement on wholesomeness of irradiated foods by the Surgeon General, Department of the Army. In Radiation Processing of Foods: Hearings before the Subcommittee on Research, Development, and Radiation of the Joint Committee on Atomic Energy, Congress of the United States, June 9 and 10, 1965. U.S. Govt. Printing Office, Washington, D.C.

ANON. 1972. Induced Mutations and Plant Improvement. Proc. Study Group Meeting, Buenos Aires, Nov. 16-20, 1970. International Atomic Energy Agency, Vienna.

ANON. 1973. Progress and future tasks in food irradiation. Intern. Atomic Energy Agency Bull. 15.

BREGVADZE, U. D., and BOKERIYA, N. M. 1971. Effect of γ-irradiation on amino acid composition of wines. Tr. Gruz. Nauch-Issled. Inst. Pishch Prom. 4, 90-96. (Russian)

BROOKE, R. O., RAVESI, E. M., GADBOIS, D. F., and STEINBERG, M. A. 1964. Preservation of fresh unfrozen fishery products by low-level radiation. III. The effects of radiation pasteurization on amino acids and vitamins in clams. Food Technol. 18, 1060-1064.

BROOKE, R. O., RAVESI, E. M., GADBOIS, D. F., and STEINBERG, M. A. 1966. Preservation of fresh unfrozen fishery products by low-level radiation. V. The effects of radiation pasteurization on amino acids and vitamins in haddock fillets. Food Technol. 20, 1479-1482.

CHIPAULT, J. R. 1962. High-energy irradiation. Lipids and their oxidation. In Symposium on Foods. H. W. Schultz (Editor). Avi Publishing Co., Westport, Conn.

CLARKE, I. D., and FERNANDES, S. J. G. 1961. Effect of γ-radiation on the protein content of apples and pears. Intern. J. Appl. Radiation Isotopes 11, 186-189.

COATES, M. E. et al. 1963. A comparison of the growth of chicks in the Gustafson germ-free apparatus and in a conventional environment, with and without dietary supplements of penicillin. Brit. J. Nutr. 17, 141-150.

DENNISON, R. A., and AHMED, E. M. 1971-1972. Effects of low level irradiation on the preservation of fruit: a 7-year summary. Isotopes Radiation Technol. 9, 194-200.

DIEHL, J. F. 1969. Combined effects of irradiation, storage and cooking on the vitamin E and B levels of foods. Presented at 33rd Annual Meeting Am. Inst. Nutr., Atlantic City, N.J., Apr. 14. Federation Proc. 28, 305. (Abstr.).

DOGUCHI, M. 1969. Effects of gamma radiation on wheat gluten. Agr. Biol. Chem. 33, 1769-1774.

DOLLAR, M. et al. 1970. Physiological, chemical, and physical changes during ripening of papaya. Ripening parameter NVO-374-17, 1, 85-100.

DUMANOVIĆ, J., and DENIC, M. 1969. Variation and heritability of lysine content in maize. Proc. Panel, Röståivga, Sweden, 1968. In New Approaches to Breeding for Improved Plant Protein. International Atomic Energy Agency, Vienna.

FUJIMAKI, M., MAKOTO, T., and MATSUMOTO, T. 1968. Effect of gamma-irradiation on the amino acids of potatoes. Agr. Biol. Chem. 32, 1228-1231.

FUNES, E. 1970. Free amino acids in fruit juices. Effects of gamma radiation. Ann. Inst. Nacl. Invest. Agron. 19, 63-82. (Spanish)

GEL'FAND, S. Y. 1970. Effect of packaging materials on the changes of the intramuscular lipids of culinary irradiated meat products. Proc. Sci.-Technol. Conf. Util. Ionizing Radiation Natl. Economy. Issue 3, 90-95. Prioks Book Printing Office, Tula, USSR. (Russian)

GOLDBLITH, S. A., and PROCTOR, B. E. 1949. Effect of high-voltage X-rays and cathode rays on vitamins (riboflavin and carotene). Nucleonics 5, 50-58.
HEILIGMAN, F. et al. 1973. Irradiation disinfestation of flour. II. Storage studies of irradiated flour. Presented at 33rd Ann. Meeting, Inst. Food Technologists.
JAARMA, M., and HENRICSON, B. H. 1964. On the wholesomeness of gamma-irradiated potatoes. Acta Vet. Scand. 5, 238.
JOHNSON, B. C. 1961. Summary and evaluation of findings to date from our participation in the food irradiation contract program. In Proc. 7th Contractors Meeting, QMC, Radiation Preservation Foods Project, 1961. QMFCI Repts. 14-61, OTS, AD 265492. U.S. Dept. Com., Washington, D.C.
JOHNSON, B. C. 1964. On the nutritive value of the major nutrients and appraisal of the toxicity of irradiated foods. Final report for period May 1, 1954-February 29, 1964. Contract No. DA-49-007-MD-544. Defense Documentation Center, Alexandria, Virginia.
KENNEDY, T. S. 1965. Studies on the nutritional value of foods treated with γ-radiation. II. Effects on the protein in some animal feeds, egg and wheat. J. Sci. Food Agr. 16, 433-437.
KENNEDY, T. S., and LEY, F. J. 1971. Studies on the combined effect of gamma radiation and cooking on the nutritive value of fish. J. Sci. Food Agr. 22, 146-148.
KHARLAMOV, V. T., and SHUBNYAKOVA, L. P. 1964. Radioactive decomposition of methionine by gamma-radiation. ZH. Prikl. Khim (Leningrad) 37, 1714-1718. (Russian)
KRAYBILL, H. F. 1960. The wholesomeness of irradiated foods. In National Food Irradiation Research Program. Hearings before the Joint Committee on Atomic Energy, Congress of the United States, Jan. 14-15, 1960. Part I, pp. 308-314. U.S. Govt. Printing Office, Washington, D.C.
KUNG, H., GADEN, E. L., and KING, C. G. 1953. Vitamins and enzymes in milk. Effect of gamma radiation on activity. J. Agr. Food Chem. 1, 142-144.
KUZIN, A. M., and ABDURAKHMANOV, A. 1970. Postradiation intensification of carotenogenesis in carrots. Proc. Sci.-Technol. Conf. Util. Ionizing Radiation Natl. Economy, Issue 3, 109-117. Prioks Book Printing Office, Tula, USSR. (Russian)
LANG, K., and BÄSSLER, K. H. 1966A. Nutritional value of irradiated potatoes. In Food Irradiation, Proc. Symp. Karlsruhe, June 6-10, 1966. International Atomic Energy Agency, Vienna.
LANG, K., and BÄSSLER, K. H. 1966B. Biological effects of irradiated fats. In Food Irradiation, Proc. Symp. Karlsruhe, June 6-10, 1966. International Atomic Energy Agency, Vienna.
LEONOVA, T. A., and SOSEDOV, N. I. 1972. Action of γ-irradiation on the amino acid composition of rice and buckwheat grains. Radiobiologiya 12, 629. (Russian)
LEY, F. J. 1972. The use of irradiation for the treatment of various animal feed products. Food Irradiation Inform. 1, 8-22.
LEY, F. J., BLEBY, J., COATES, M. E., and PATERSON, J. S. 1969. Sterilization of laboratory animal diets using gamma radiation. Lab. Animals 3, 221-254.
LIUZZO, J. A., BARONE, W. B., and NOVAK, A. F. 1966. Stability of B-vitamins in gulf oysters preserved by gamma radiation. Presented at 50th Meeting Federation Am. Soc. Exptl. Biol., Atlantic City, April 11-16; Federation Proc. 25, 722 (Abstr.).
LOAHARANU, P. 1971. Recent research on the influence of irradiation of certain tropical fruits in Thailand. In Disinfestation of Fruit by Irradiation. (Proc. Panel on Use of Irradiation to Solve Problems in the Intern. Fruit Trade, Honolulu, 1970). International Atomic Energy Agency, Vienna.

410 NUTRITIONAL EVALUATION OF FOOD PROCESSING

McKINNEY, F. E. 1971. Wholesomeness of irradiated food, especially potatoes, wheat, and onions. Isotopes Radiation Technol. *9*, 188-193.

MERRITT, C., JR. 1966. Chemical changes induced by irradiation in meats and meat components. *In* Food Irradiation, Proc. Symp. Karlsruhe, June 6-10, 1966. International Atomic Energy Agency, Vienna.

METLITSKII, L. V., ROGACHEV, V. N., and KRUSHCHEV, V. G. 1968. Radiation processing of food products. Status of the food irradiation program. *In* Hearings before the Subcommittee on Research, Development, and Radiation of the Joint Committee on Atomic Energy, Congress of the United States, July 18 and 30, 1968. U.S. Govt. Printing Office, Washington, D.C.

METTA, V. C., and JOHNSON, B. C. 1959. Biological value of gamma-irradiated corn protein and wheat gluten. J. Agr. Food Chem. *7*, 131-133.

MITCHELL, J. H., JR. 1957. Action of ionizing radiations on fats, oils, and related compounds. *In* Radiation Preservation of Food. United States Army Quartermaster Corps. U.S. Govt. Printing Office, Washington, D.C.

MOORE, R. O. 1961. The influence of irradiated foods on the enzyme systems concerned with digestion. Final Rept.: Dept. Agr. Biochem., Ohio State Univ. Res. Found., Columbus. Contract *DA-49-007-MD-787*. Defense Documentation Center, Alexandria, Virginia.

NAIR, K. K., and BROWNELL, L. E. 1965. The clearance of gamma-irradiated wheat and the international importance of this act. *In* Radiation Preservation of Foods (Proc. Intern. Conf., Boston, 1964). Natl. Acad. Sci.—Natl. Res. Council Publ. *1273*.

NAWAR, W. W. 1972. Radiolytic changes in fats. Radiation Res. Rev. *3*, 327-334.

OBARA, T., SHIMOTSUURA, A., SHIMAZU, F., and WATANABE, W. 1958. Preservation of vegetable foods irradiated with radioactive rays. I. Influence of γ-ray irradiation upon components contained in citrus juice. Radioisotopes *7*, 127-132.

PAPE, G. 1973. Some observations regarding irradiated wheat. *In* Radiation Preservation of Food (Proc. Symp. Bombay, 1972). International Atomic Energy Agency, Vienna.

PARTMANN, W. 1962. The effect of ionizing radiation on lipids. *In* Report of the Brussels meeting on the Wholesomeness of Irradiated Foods (with exclusive reference to the evaluation of nutritional adequacy and safety for consumption). FAO, United Nations, Rome.

PLOUGH, I. C. *et al.* 1957. An evaluation in human beings of the acceptability, digestibility, and toxicity of pork sterilized by gamma radiation and stored at room temperature. U.S. Army Medical Nutrition Laboratory, Denver, Rept. *204*.

PROCTOR, B. E., and GOLDBLITH, S. A. 1949. Effect of soft X-rays on vitamins (niacin, riboflavin, and ascorbic acid). Nucleonics *5*, 56-62.

RAICA, N., JR., and HOWIE, D. L. 1966. Review of the United States Army wholesomeness of irradiated food program (1955-1966). *In* Food Irradiation, Proc. Symp. Karlsruhe, June 6-10, 1966. International Atomic Energy Agency, Vienna.

RAICA, N., JR., SCOTT, J., and NIELSEN, W. 1972. The nutritional quality of irradiated foods. Radiation Res. Rev. *3*, 447-457.

RAO, M. R. R., and NOVAK, A. F. 1973. Fatty acids in an irradiated chicken product. *In* Radiation Preservation of Food (Proc. Symp. Bombay, 1972). International Atomic Energy Agency, Vienna.

READ, M. S. *et al.* 1961. Successive generation rat feeding studies with a composite diet of gamma-irradiated foods. Toxicol. Appl. Pharmacol. *3*, 153-173.

REBER, E. F., RAHEJA, K., and DAVIS, D. 1966. Wholesomeness of irradiated foods. An annotated bibliography. Federation Proc. *25*, 1530-1579.

REVETTI, L. M. 1973. Preservation of maize (*Zea mais L.*) and kidney beans

(*Phaseolus vulgaris* L.) by gamma irradiation. *In* Radiation Preservation of Food (Proc. Symp. Bombay, 1972). International Atomic Energy Agency, Vienna. (Spanish)
RICHARDSON, L. R. 1955. A long range investigation of the nutritional properties of irradiated food. Progress Report III, 1 Sept. 1954-1 July 1955, Texas Agr. Expt. Sta., College Park. Contract *DA-49-007-MD-582*. Defense Documentation Center, Alexandria, Virginia.
RICHARDSON, L. R., WOODWORTH, P., and COLEMAN, S. 1956. Effect of ionizing radiations on vitamin K. Federation Proc. *15*, 924-926.
SAINT-LÈBE, L., BERGER, G., MUCCHIELLI, A., and COQUET, B. 1973. Toxicological evaluation of the starch of irradiated maize: An account of work in progress. *In* Radiation Preservation of Food (Proc. Symp. Bombay, 1972). International Atomic Energy Agency, Vienna. (French)
SATO, T. 1970. Sterilization of diet for laboratory animals. Shokuhin Kogyo *13*. No. 24, 89-92. (Japanese)
SCHOEN, A., and HILLER, H. H. 1971. Digestibility of rat and mice diets sterilized by gamma irradiation or steam in conventional rats. Z. Tierphysiol. Tierernaehr. Futtermittelk. *27*, 338-343. (German)
SCHREIBER, M., and NASSET, E. S. 1959. Digestion of irradiated fat *in vivo*. J. Appl. Physiol. *14*, 639-642.
SHEFFNER, A. L., and SPECTOR, H. 1957. Action of ionizing radiations on vitamins, sterols, hormones and other physiologically active compounds. *In* Radiation Preservation of Food. United States Army Quartermaster Corps. U.S. Govt. Printing Office, Washington, D.C.
SOUTHERN, E. M., and RHODES, D. N. 1967. Radiation chemistry of polyamino acids in aqueous solutions. *In* Radiation Preservation of Foods. Am. Chem. Soc. Advan. Chem. Ser. *65*, 58-77.
SWAMINATHAN, M. S., AUSTIN, A., KAUL, A. K., and NAIK, M. S. 1969. Genetic and agronomic enrichment of the quantity and quality of proteins in cereals and pulses. *In* Proc. Panel, Röstånga, Sweden, June 17-21, 1968, New Approaches to Breeding for Improved Plant Protein. International Atomic Energy Agency, Vienna.
THOMAS, M. H. 1972. Unpublished data. U.S. Army Natick Laboratories, Natick, Mass.
THOMAS, M. H., and CALLOWAY, D. H. 1961. Nutritional value of dehydrated foods. J. Am. Dietet. Assoc. *39*, 105-116.
THOMAS, M. H., and JOSEPHSON, E. S. 1970. Radiation preservation of foods and its effect on nutrients. Sci. Teacher *37*, 59-63.
UDES, H., HILLER, H. H., and JUHR, N. C. 1971. Quantitative and qualitative changes in a rat and mouse diet by means of various sterilization procedures. Z. Versuchstierk *13*, 160-166. (German)
VAKIL, U. K. *et al.* 1973. Nutritional and wholesomeness studies with irradiated foods: India's Program. *In* Radiation Preservation of Food (Proc. Symp. Bombay, 1972). International Atomic Energy Agency, Vienna.
VARELA, G., and URBANO, G. 1971. Effects of irradiation on the nutritive quality of proteins in potatoes. Atti Simp. Intern. Agro-Chim. *8*, 563-568. (Spanish)
WENKAM, N. S., and MOY, A. P. 1968. Nutritional composition of irradiated fruit. I. Mango and papaya. Annual Report, 1 June 1967-31 May 1968, Hawaii Univ. Coll. Tropical Agr. AEC Rept. *UH-235-P-5-4*, 126-135.
ZABEN'KOVA, K. I., VOLODIN, V. G., and AVRAMENKO, B. I. 1972. Effect of radiomutants on the protein level and amino acid composition of spring wheat. Sel'skokhoz. Biol. *7*, 544-547. (Russian)

CHAPTER 15

Marcus Karel
and
Norman D. Heidelbaugh

Effects of Packaging on Nutrients

NATURE OF PACKAGING MATERIALS

The protection offered by a package is determined by the nature of the packaging materials and by the type of package construction.

Glass

Glass containers have been used for many centuries and still are among the important packaging media. Physically, glass is a supercooled liquid of very high viscosity. Chemically, it is a mixture of inorganic oxides of varying composition. Most container glass is of the soda-lime-silica type, with lesser amounts of other ingredients. The important packaging properties of glass, such as moldability, inertness, transparency and strength, can be modified by relatively minor changes in the glass composition. Color, for instance, may be controlled by inclusion of small amounts of oxides of various metals such as chromium, cobalt, and iron, and semiopacity imparted by addition of fluorine compounds.

The major packaging design problems involving glass arise out of its mechanical properties. Glass is brittle, and its tensile strength depends drastically on the condition of the surface. Commercial glass containers have only a small fraction of the potentially possible tensile strength because the surface conditions cannot be maintained at the perfect level required for maximum strength. As a consequence the containers must be made of fairly thick glass, and therefore have a substantial weight. Recent developments in glass packaging have included an attempt by the manufacturers to improve the mechanical properties and to reduce weight. Shatter-resistant bottles are said to be under development and one product already is on the market.

The properties of the containers depend also on type of construction. The types of containers commonly used include bottles, jars, jugs, and tumblers. A variety of closures is available, and the recent trend toward maximum convenience for the consumer has resulted in substantial efforts toward development of easy-to-open closures. Convenience as well as aesthetic considerations are often paramount in package design, and require substantial ingenuity of the design engineer.

Glass containers utilizing proper type of closure can be used for practically all types of packaging applications including heat processing, gas packaging, and other types of preservation requiring hermetically-closed containers. For more detailed discussion of glass in packaging the reader is referred to the following: (Heiss 1970; Griffin and Sacharow 1972; and Paine 1967).

Metals

Rigid Metal Containers.—The most common container for packaging of foods other than beer and carbonated beverages is the so-called tin can. It is traditionally used for heat-sterilized products, and is made of tinplate, which consists of a base sheet of steel with a coating of tin applied to it by "hot dipping," or by the electrolytic process. The amount of tin coating depends on the process and the type of can. A typical tinplate may be 0.01-in. thick, with the tin coating contributing only about 0.00005-in. to this thickness. To make the can more suitable for specific packaging applications, enamels and linings may be applied to the tin or to the tinless steel sheets. The composition of these linings varies with intended use. Plastics, shellacs, resins, glass, and inorganic oxides are among materials used for this purpose.

In recent years, very substantial changes have occurred in the can industry, and a great variety of cans different from the can made of the standard tinplate has been developed. In particular, these developments have been aimed at the beer and carbonated beverage canning applications, but may have additional applications in other foods. Tin-free steel utilizing direct coating with chromium-containing coatings, or with very thin thermoplastic coatings such as nylon, has been applied. The other innovation is the utilization of cans with cemented or welded rather than soldered side seams. This development is, in part, connected with the utilization of tin-free steel. However, in recent years there has been increasing concern with the potential contamination of foods by lead from the soldered side seams.

Cans made of aluminum have also made a very substantial impact on the beverage market, but have not been used extensively for heat-sterilized food products. It is known however that successful sterilization, and a satisfactory shelf-life can be attained with properly coated all-aluminum cans.

In terms of rigid metal construction the variety of container types is very large and they cannot be all discussed here. The reader is referred to Paine (1967) and to Brody (1971).

Metal Foils.—The most important food packaging foils consist of essentially pure aluminum. They are used in thicknesses ranging from 0.00025 to 0.006 in. The thickness of uncoated aluminum foil determines largely its protective properties. Foils of low thickness have microscopic discontinuities (pinholes) that allow limited diffusion of gases and vapors. According to Cooke (1955) foil 0.0015-in. thick or heavier is conceded to have essentially zero water-vapor permeability, indicating a complete absence of pinholes. The properties of thinner foils can be improved by combination with one or more plastics in the form of coatings or laminations.

The use of foils made of metals other than aluminum is very limited. Lead, tin, and zinc foils are mainly of historical interest, and the recently developed steel foils have yet to find major packaging uses.

Paper

A very considerable proportion of packaged foods is stored and distributed in packages made of paper or paper-based materials. It seems probable that, because of its low cost, ready availability, and great versatility, paper will retain its predominant packaging position for some time to come.

Paper consists primarily of cellulose fibers which are obtained from wood or other cellulose-containing materials by one of the several pulping processes. The discussion of paper manufacture is beyond the scope of this chapter, and the interested reader is referred to standard works on paper, such as Casey (1952).

The packaging properties of paper vary considerably, depending on the manufacturing processes and on additional treatments to which the finished paper sheet may be subjected. Its strength and mechanical properties depend on mechanical treatment of the fibers and on the inclusion of fillers and binding materials. The physicochemical properties, such as permeability to liquids, vapors, and gases, can be modified by impregnating, coating, and laminating. The materials used for this purpose include waxes, plastics, resins, gums, adhesives, asphalt, and other substances.

The quality of materials made by the above converting processes varies considerably. Some laminated, waxed, and coated papers deserve the designation of protective materials. Certain grades of converted papers, however, offer little more than protection from light and mechanical damage.

Papers may be used as flexible packaging materials or as materials for construction of rigid paper containers. The flexible packaging applications include the use of papers for wrapping materials, as well as the manufacture of bags, envelopes, liners, and overwraps. Some of the more important types of paper used for these purposes include kraft paper, greaseproof papers, glassines, and waxed papers, as well as papers prepared by conversion of these basic types.

Rigid paper containers include a great variety of types, varying in materials and methods of construction. They may include cartons, boxes, fiber cans, drums, liquid-tight cups, tetrahedral packs, and many others. They may be constructed from paperboard, laminated papers, corrugated board, and the various types of specially treated boards and papers. Frequently the paper containers have liners or overwraps, made usually of protective grades of paper, plastics, or metal foil. The packaging applications of paper are more adequately discussed in several publications on packaging, including Griffin and Sacharow (1972) and the Modern Packaging Encyclopedia, (1972).

Plastics

Plastics are organic polymers of varying structure, chemical composition, and physical properties. A detailed description of the chemical nature and physicochemical properties of plastics is beyond the scope of this chapter, and the reader is referred to modern texts on high polymers, such as Billmeyer (1971). A brief review of the types of plastics used in food packaging is given below. Some of the properties of importance to the maintenance of nutritional value are also discussed in later sections of this chapter.

Cellophanes.—Cellophanes are frequently classified with the plastics, and this usage will be followed herein, in spite of the fact that the major constituent of cellophanes—cellulose—is a natural rather than synthetic polymer. In addition to cellulose, all cellophanes contain a plasticizer such as glycerol or ethylene glycol. (Plasticizers are materials, usually solvents of low volatility, added to plastics to reduce the attractive forces between the polymer chains. This results in plastics of better flexibility.) Plasticized cellulose—called plain cellophane—offers no protection to the diffusion of water vapor. It is usually coated with protective agents, such as nitrocellulose, waxes, resins, and synthetic polymers. The packaging properties of cellophanes are determined primarily by the nature of these coatings.

Cellulosics.—Cellophane provides the base material for the production of cellulose acetate, ethyl cellulose, and cellulose nitrate. The cellulosics have properties similar to those of cellophanes, but have fewer applications in the packaging of foods.

Polyolefins.—Polyethylene is one of the most important packaging materials of the present time. It is a hydrocarbon polymer, with the nominal formula

$$-(CH_2-CH_2)-$$

but the commercial products are produced with a variable amount of branching within this nominally linear polymer. High density polyethylene has the least branching and as a result the greatest thermal stability and the lowest permeability. High density polyethylene is used in film form as well as for production of rigid plastic containers, including milk bottles. The low density polyethylene has the advantage of maximum flexibility and low cost and is widely used for packaging of foods in bags, or as an overwrap. Polyethylene is also widely used in laminations, where it provides the inner layer requiring good heat sealability. Polypropylene is closely related to polyethylene and its properties and uses are similar.

Vinyl Derivatives.—Vinyl derivatives have the general formula

$$-(CH_2-CXY)_n-$$

where X and Y are either hydrogen atoms or other substituents such as chlorine, benzene, methyl, and hydroxyl. The properties of vinyl polymers are dependent on the nature of substituents, on molecular weight, on the spatial arrangement of groups within the chains, and especially on orientation and crystallinity.

The following generalizations about the effects of polymer structure on functional properties are of particular importance in packaging:

(1) Strength, resistance to high temperatures, resistance to action of solvents, and resistance to diffusion increase with increasing degree of crystallization.
(2) Presence of polar groups decreases resistance to diffusion of polar molecules.
(3) Linearity of chains and orientation of crystallites improve strength and impermeability.
(4) Addition of plasticizers increases permeability and solubility.

The important food-packaging materials based on vinyl polymers, and their chemical formulae, are listed below:

Saran, copolymer of vinyl chloride

$$-(CH_2-CHCl)_n-$$

and vinylidene chloride

$$-(CH_2-CCl_2)_n-$$

Polyvinyl alcohol

$$-(CH_2-CHOH)_n-$$

Polyvinyl acetate

$$-(CH_2-CHOCOCH_3)_n-$$

Polystyrene

$$-(CH_2-CH-\bigcirc-)_n-$$

Polyvinyl chloride

$$-(CHCl-CH_2)_n-$$

Polyesters.—The polyester of most importance in food packaging is polyethylene terephthalate, a condensation product of ethylene glycol and terephthalic acid, better known in the United States under the trade name Mylar. Mylar is a crystalline linear polymer of excellent strength and inertness. Mylar has been widely used in lamination, in particular as the outer, abrasion-resistant layer of laminations to be used for food pouches requiring good protective properties.

Pliofilm.—Rubber hydrochloride, or pliofilm, is used for the packaging of certain types of foods. Its properties are determined by the type and amount of plasticizer, which is always added to films made from this plastic.

Polyfluorocarbons.—Several polymers with the backbone composed of carbon and fluorine, or of hydrogen and chlorine atoms in addition to carbon and fluorine, form the basis for several packaging films with remarkable properties. These materials are very costly but have unique properties. Teflon, and FEP which are the equivalents of polyethylenes with hydrogen replaced by fluorine, have remarkable inertness, and high permeability to oxygen. In contrast trifluorochloroethylene has an extremely low permeability to all gases, and is in particular very effective as a water-vapor barrier. Polyvinyl fluoride is intermediate in permeability, but has an excellent resistance to sunlight.

Polyamides.—A number of polyamide films, which are condensation polymers of diamines with diacids, are available on the market and have found extensive applications. Nylons as these polymers are usually called in the trade, are inert, heat-resistant, and have excellent mechanical properties. They are usually coated, or used in combination with other materials to produce packaging materials of good inertness as well as low permeability.

Water-Soluble and Edible Films.—There has been a recent increase in interest in edible, and other water-soluble films. In part, this interest stems from ecological considerations, and is aimed at minimizing solid wastes. The other driving force for this type of package development is the increased trend toward convenience foods. Water-soluble films include polyvinyl alcohol, some cellulose derivatives, as well as other polysaccharides including those derived from the linear starch component—amylose. Some proteins, in particular collagen, are also suitable materials (Kroger and Igoe 1971; Morgan 1971).

Other Films.—Recent developments in packaging films include the utilization of polymeric films containing ionized groups contributing to film strength through ionin cross-links. These films, known as "ionomers," are particularly resistant to effects of low temperatures. The other major area of packaging material research has been the work on developing heat-resistant materials suitable for production of sterilizable packages (Griffin and Sacharow 1972). Polymer *alloys* are another likely future development.

Wood

Containers made of wood are used extensively for shipment and storage of food packages. Their use as immediate packages for food products is of much less importance.

Shipping containers are manufactured from different types of wood and vary widely in their construction. The types of construction commonly encountered in food packages include cases, boxes, crates, and barrels.

Combinations

For many packaging applications it is necessary to combine two or more different materials in order to obtain a package with satisfactory properties. Some of the combinations used in food packaging have been mentioned in the preceding sections. The following list is intended to show the great variety of materials that can be used for this purpose.

(A) Plastic combinations containing two or more plastic films laminated with various types of synthetic and natural adhesives.
(B) Plastic, silicone, and wax coatings on packages made of metals, paper, wood, and other materials.
(C) Rigid can made of paper, with metal bottoms and tops.
(D) Plastic films with thin metallic coating.
(E) Metal foils with thin plastic coatings.
(F) Barrier materials constructed of as many as 6-7 materials including paper, textiles, metals, plastics, waxes, and resins.

EFFECTS OF PACKAGING ON NUTRIENTS 419

EFFECTS OF PACKAGING ON FACTORS INFLUENCING THE NUTRIENT CONTENT OF FOODS

The package affects the nutritive value of foods by controlling the degree to which factors connected with processing, storage, and handling can act on components of foods. The processing and storage factors amenable to control by packaging include light, oxygen concentration, moisture concentration, heat transfer, contamination, and attack by biological agents. In addition, some factors such as interaction between food and packages arise from the use of packaging itself. A summary of some of the environmental factors controllable by packaging, and of the pertinent packaging properties is shown in Table 15.1.

TABLE 15.1

PACKAGE-ENVIRONMENT INTERACTIONS

Environmental Factors	Pertinent Package Properties
Mechanical shocks	Strength
Pressure of oxygen, water vapor, etc.	Permeability
Light intensity	Light transmission
Temperature	Thermal conductivity / Porosity / Reflectivity
Biological agents	Penetrability

This section is devoted to a review of the nature of nutritionally important factors that can be controlled by packaging, and of the packaging properties important in their control.

Light

Many of the deteriorative changes in the nutritional quality of foods are initiated, or accelerated, by light. The catalytic effects of light are most pronounced for light of the highest quantum energy, that is light in the lower wavelengths of the visible spectrum and in the ultraviolet spectrum. Specific reactions involved in deterioration of foods, however, may have specific wavelength optima, and, in particular, the presence of sensitizers may shift the effective spectrum very substantially. In particular, several types of compounds present in food may act as sensitizers, including riboflavin, β-carotene, vitamin A, and peroxidized fatty acids.

Some examples of nutritionally significant light-catalyzed reactions in foods include the following.

Oxidation of Fats and Oils.—The catalytic effect of light on the free radical reactions involved in fat oxidation is well established. A recent study by Radtke et al. (1970) has focused on establishing the quantitative relations between intensity and wavelength of the light and rates of oxidation. They found that irradiation with light at 380 nm accelerates oxidation of refined soya oil by a factor of 750 times compared with oxidation in the dark, but light in green-yellow regions (577 nm) had also a significant accelerating effect (10–15 times faster than in case of dark oxidation). Following the induction period, long wavelengths of light became quite a significant factor in accelerating oxidation. Effects of light on oxidation rate of potato chips were also established by Quast and Karel (1972A), who found in particular that the antioxidant effect of water at 40% RH was largely negated by exposure of the chips to light. According to Wildbrett (1967) the effective range for free radical formation in milkfat lies in wavelengths of 295–350 nm, but phospholipids are much more sensitive to light and even light with wavelength as high as 700 nm (red light) may be effective. Oxidation of fats and oils is not only effective in lowering the nutritional value of the fat, possibly producing toxic compounds from the fats and oils, but is of extreme importance in the destruction of fat-soluble vitamins, in particular vitamins E and A.

Destruction of Riboflavin and of Other Water-Soluble Vitamins.
Riboflavin is known to be destroyed by light and to act as a sensitizer of destruction of other vitamins. In particular, the work of Reusser (1967, 1969, 1970) has shown that some otherwise light-stable compounds, such as folic acid, biotin, and antibiotic preparations are sensitized by the presence of riboflavin. Ascorbic acid is known to be quite sensitive to light (Sacharow 1969; Kiermeier and Waiblinger 1969), and in particular it is known to be able to interact with other food components during light exposure (Reusser 1967, 1970).

Changes in Proteins and Amino Acids.—Several amino acids and amino acid residues are known to have substantial light sensitivity. Histidine, tryptophan, tyrosine, phenylalanine, and the sulfur-containing amino acids are known to be readily attacked by light, and the effects of light can be potentiated by the presence of various activators, including riboflavin and peroxidized fatty acids (Smith and Hanawalt 1969; Karel 1973A).

The above examples suffice to point out that light can play a role in deterioration of nutrients. The protection offered by the package falls into two categories:
(1) Direct protection by absorption or reflection of all or of part of the incident light.
(2) Indirect protection by preventing access of a component necessary for the light-catalyzed reaction. Protection against the access of atmospheric oxygen is most important in this respect and will be discussed in later sections.

The degree of direct protection afforded by the package depends on the light transmission characteristics of the packaging materials. The total amount of light absorbed by the food is given by the following formula (Wildbrett 1967):

$$I_{abs} = I_o \cdot Tr_p \cdot \frac{1 - R_f}{1 - R_f \cdot R_p} \tag{1}$$

where

I_{abs} = intensity of light absorbed by the food
I_o = intensity of incident light
Tr_p = fractional transmission by the packaging material
R_p = fraction reflected by the packaging material
R_f = fraction reflected by the food

The fraction of the incident light transmitted by any given material may be considered to follow the Beer Lambert Law:

$$I = I_o \, e^{-kx} \tag{2}$$

where

I = intensity of light transmitted by the packaging material
k = a characteristic constant (absorbance) for the packaging material
x = thickness of the packaging material

The absorbance k, varies not only with the nature of the material but also with the wavelength. The transmission of light through a given material will therefore give a characteristic spectrum of transmitted light dependent on incident light and the properties of the package. Figure 15.1 shows light-transmission curves for several food packaging materials. It is readily apparent that several of the plastics, while transmitting similar amounts of light in the visible range, give varying degrees of protection against the damaging ultraviolet wavelengths. The protection against UV light by different packaging materials can

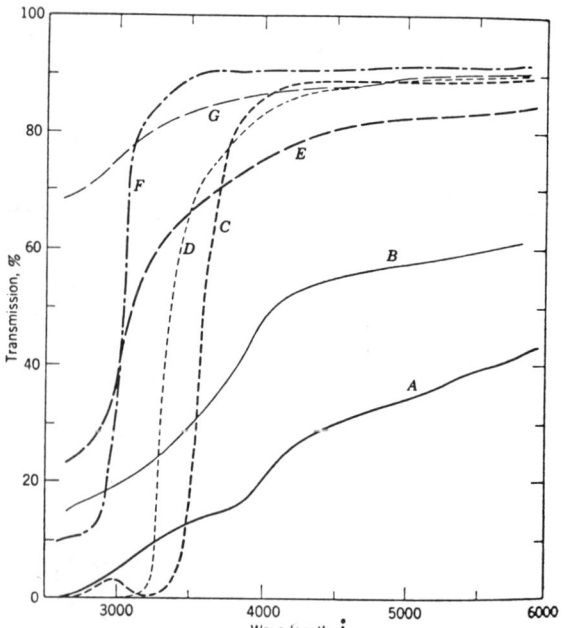

FIG. 15.1. LIGHT TRANSMISSION THROUGH VARIOUS FLEXIBLE PACKAGING MATERIALS

A—Low-pressure polyethylene, 0.0035-in. thick. B—Translucent wax paper, 0.0035-in. thick. C—Saran, 0.0011-in. thick. D—Mylar, 0.0014-in. thick. E—Pliofilm, 0.0013-in. thick. F—Cellulose acetate, 0.001-in. thick. G—Conventional polyethylene, 0.0015-in. thick.

often be characterized by a "cut-off" wavelength, below which transmission of light becomes negligible, and these are sometimes listed in compilations of package properties.

The protective characteristics of a packaging material can be improved by special treatments. Glasses are frequently modified by inclusion of color-producing agents or by application of coatings. Figure 15.2 shows the differences in light transmission between transparent and colored bottle glass. The light transmission curves of a window glass of comparable thickness are included for comparison.

Modification of plastic materials may be achieved by incorporation of dyes or by application of coatings. Figure 15.3 shows the effect of such treatments on cellophanes. Specific light screening agents continue to be developed. Geigy Chemical Corporation, for instance, has developed a UV absorber (Tinuvin® P) which is designed for

EFFECTS OF PACKAGING ON NUTRIENTS   423

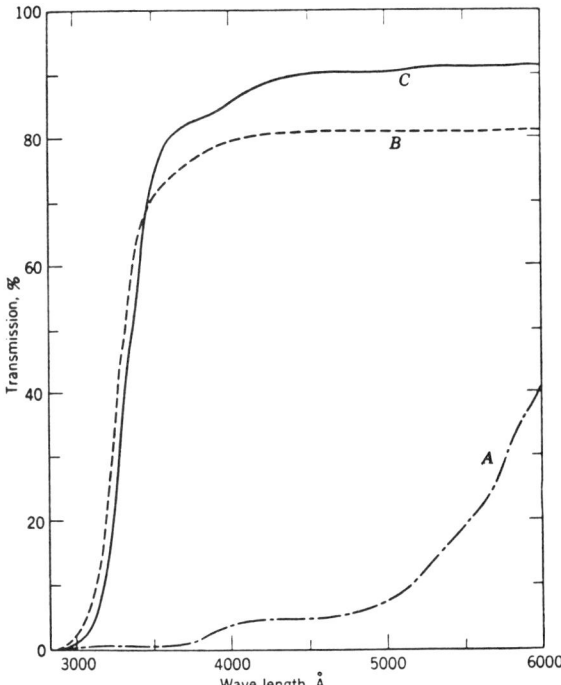

FIG. 15.2. TRANSMISSION OF LIGHT THROUGH THREE
TYPES OF GLASS

A—Amber bottle glass, 0.118-in. thick. B—Transparent milk bottle glass, 0.118-in. thick. C—Window glass, 0.119-in. thick.

incorporation in polyvinylchloride bottles. A UV-absorbing compound for cellophanes used in cheese packaging was tested by Ellickson and Hasenzahl (1958). These examples illustrate the wide range of light transmission characteristics that can be achieved in packages made of the same type of base material. The effectiveness of light screening treatment has been demonstrated by many authors. Wildbrett (1967), for instance, cites data showing that ascorbic acid losses in milk stored in uncolored glass are 14 times greater than those in brown glass, and that milk in blue paper cartons loses 5 times more Vitamin C than in red paper. Similar data are cited by Sacharow (1969): after 2 hr exposure milk in white polyethylene film lost 93% of Vitamin C, but in milk in white polyethylene overwrapped by black polyethylene loss was only 16%. Riboflavin losses were also retarded by the polyethylene containing black pigments.

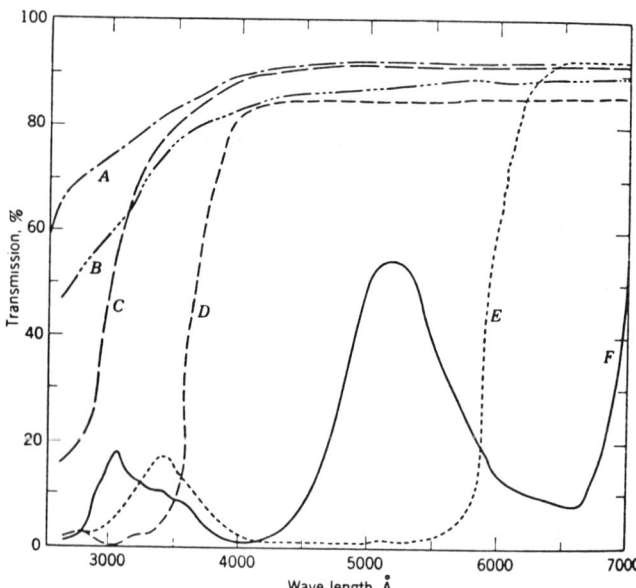

FIG. 15.3. TRANSMISSION OF LIGHT THROUGH VARIOUS TYPES OF CELLOPHANE

A—Plain cellophane, 0.0009-in. thick. B—Saran-coated cellophane, 0.0015-in. thick. C—Nitrocellulose-coated cellophane (conventional moistureproof cellophane), 0.001-in. thick. D—Moistureproof cellophane coated with a special ultraviolet absorbing composition, 0.001-in. thick. E—Dyed cellophane (red), 0.001-in. thick. F—Dyed cellophane (dark green), 0.001-in. thick.

Temperature

The effect of temperature on nutrients is reviewed elsewhere in this book. This section is devoted solely to the related problem of temperature control by the application of packaging methods.

The application of packaging engineering can have a considerable effect on retention of nutrients by affecting the rate of heat transfer to and from the packaged food products. The heat transfer may take place by conduction, convection, and radiation; and is accordingly affected by thermal conductivity, porosity, and reflectivity of packaging materials, which collectively determine the insulating value of the package.

The temperature rise in refrigerated food products, exposed for a short period of time to elevated temperatures, can be greatly retarded by the use of insulating containers. Ratzlaff (1955) studied the temperature changes in pasteurized milk during bottling, cooling, storage, and distribution. He found that the temperature of milk depended on the type of the container and on the type of outer packing case. Precooled (40°-43°F) milk, exposed to ambient temperatures of 77°-79°F, reached the temperature of 70°F in 6 hr when packaged in glass bottles in wire or wood packing cases, and in 10.5 hr when packaged in waxed paper cartons in wire packing cases. Milk in waxed paper cartons in wood packing cases had a temperature of less than 70°F after 12 hr of storage.

Insulating containers that slow down heat transfer to food products having temperatures below the ambient condition also retard the cooling of warm food products. In the study mentioned above, the time necessary to cool milk from 60° to 50°F varied from 2.5 to 17 hr depending on the type of packaging.

The insulating properties of packages are of even greater importance in the handling and distribution of frozen foods. Thawing rates of frozen foods depend on the type of container. Single containers of food defrost more rapidly than containers packaged in shipping cases. Other packaging factors of importance are size of package and location in transporting vehicles.

The package has also an important effect on the rate of freezing. The rate of freezing of foods is recognized as a factor in the preservation of quality of foods, including their nutritional value. Dunker and Hankins (1953) compared the freezing rates of unwrapped meat with those of meats wrapped in different packaging materials. The meat was frozen in an air-blast freezer at 2°F. The materials studied included butcher paper, cellophane, aluminum foil, white parchment paper, Cry-O-Rap, and polyethylene. The effect of plastic wraps on freezing rates of foods may be much less important in liquid-immersion freezing. Lentz and van den Berg (1951), for instance, found that packaging of liquid-immersion-frozen poultry in Cryovac bags did not significantly affect freezing rates.

Large variations have also been observed with rigid containers for frozen foods (Joslyn and Hohl 1948). A recent review of temperature-controlled shipping containers was presented by Bond (1973). He notes that two elements are necessary to provide simple systems for packaging control of temperature: good insulator as the packaging material and an energy source for temperature adjustment and control. Among the available insulating materials the synthetic foams of polystyrene and of polyurethane have excellent insulation properties with thermal conductivities in the range of 0.11 to 0.24 (BTU) (in.) (ft^{-2}) (hr^{-1}) (°F^{-1}). Bond (1973) discussed the potential for eutectic temperature control units. In these units substances undergoing phase transition are included within the container to provide controlled energy adsorption and release. The case of utilizing ice packs and of salt-ice mixtures is, of course, well established but other eutectic substances might be utilized, including organic compounds such as glycols, hydrocarbons, and waxes. Bond's article provides excellent illustrations of available eutectic substances and of applications. According to Bond (1973), for instance, a commercial unit utilizing a 3 × 6 × 12-in. module, and a commercially available eutectic compound can be maintained for 120 hr at 40°F while the ambient temperature is at 0°F, or for 101 hr while the ambient temperature is at 70°F, or for 49 hr at ambient temperatures of 110°F.

The potential for utilization of eutectic coolants and "warmers" other than ice is only beginning to be realized. An interesting problem in insulative packaging is the control of temperature changes in food packages placed in refrigerated display cabinets. In this case, most of the cooling takes place by conduction and convection, while simultaneously there is a heat input by radiation from fluorescent lamps used for lighting. Under these conditions the combination of high reflectivity and high conductivity of aluminum foil offers advantages, and it appears that these advantages can be significant for products such as ice cream in refrigerated display cabinets. A review of thermal properties of aluminum foil of importance in packaging was given by McKee and Strohm (1961).

Oxygen

Atmospheric oxygen is capable of affecting the nutritive quality of foods. In general, its effects are detrimental and it is desirable to maintain certain types of foods at a low oxygen tension, or at least to prevent a continuous supply of oxygen into the package. The reactions due to atmospheric oxygen include the oxidation of fats and oils, deterioration of the biological value of proteins, and the destruction of some vitamins.

The dependence of fat oxidation on oxygen tension is well known and has recently been discussed by Labuza (1971). The effect of oxygen pressure is complex as shown in Fig. 15.4, which summarizes the effects of extent of oxidation, of relative humidity, and of oxygen pressure on the oxidation of potato chips (Karel 1973B). In the case of impermeable containers the most important factor is the total amount of oxygen present in the container. If this amount is kept low, then the oxidation cannot be very severe, since no matter what the rate of oxidation, its total extent cannot exceed the limited amount of oxygen which is available. In permeable containers, however, resupply of oxygen through the package can occur and the package permeability may be the controlling factor.

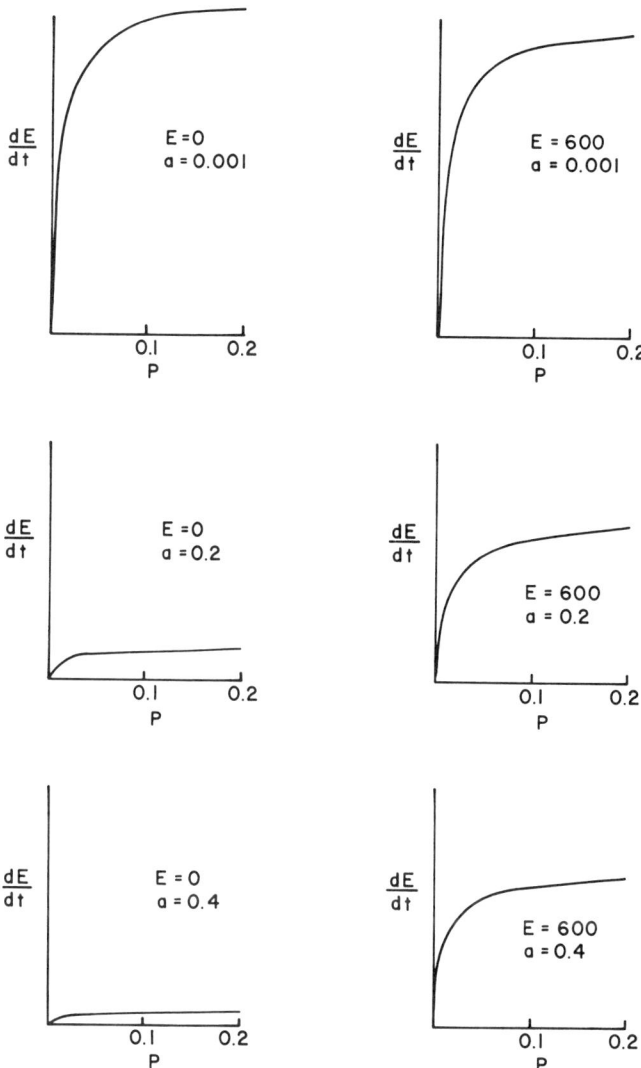

FIG. 15.4. OXIDATION OF POTATO CHIPS AS A FUNCTION OF OXYGEN PRESSURE, WATER ACTIVITY, AND EXTENT OF OXIDATION

dE/dt = Rate of oxidation in microliter per gram per hour.
E = Extent of oxidation, microliters of oxygen per gram of product.
a = Water activity.
P = Partial pressure of oxygen (atm).

The effect of oxygen on loss of protein quality is less well known than its effect on lipid deterioration. Hodson and Krueger (1947) compared the essential amino acid content of freshly prepared dry skim milk with that of dry skim milk stored for 4 yr in 2 different types of containers. They found that milk stored in containers that excluded air compared favorably with the freshly prepared milk in palatability, appearance, and amino acid content. Milk stored in containers allowing some gas interchange, however, not only lost its acceptability but also showed significant losses of arginine, histidine, lysine, and methionine.

More recent work has established that a number of amino acids, and amino acid residues of proteins are highly susceptible to oxidation, and that this oxidation may be catalyzed by peroxidizing fatty acids, or by light (Chio and Tappel 1969; Roy and Karel 1973; Karel, 1973A; Tannenbaum et al. 1969; Tomita et al. 1969).

The oxidation effect on vitamins is also well known, and ascorbic acid, vitamin E, and vitamin A are particularly susceptible to oxygen.

Of interest is also the storage behavior of prepackaged fruits and vegetables. When an adequate supply of atmospheric oxygen is available, the normal respiratory processes result in the breakdown of

TABLE 15.2

PERMEABILITY TO OXYGEN OF VARIOUS PACKAGING MATERIALS AT 25–30°C

Material	Permeability Range $(Cc) \cdot (Mil) \cdot (Day^{-1}) \cdot (M^{-2}) \cdot (Atm^{-1})$
Conventional polyethylene	6,000–15,000
High density polyethylene	1,500–3,000
Pliofilm	200–5,000[1]
Saran	10–350
Plain cellophane	20–5,000[2]
Polyvinyl fluoride	25–100
Mylar	50–100
(Poly)-Trifluorochloroethylene	50–1,000[1]
Cellulose acetate	1,000–3,000[2]
Silicon rubber	Over 250,000
Coated and waxed papers	100–15,000[3]
Cellulose acetate	2,000–5,000
Foil laminations	0[4]
Plastic laminations	10–400[5]

[1] Depends on type and amount of plasticizer.
[2] Depends on humidity.
[3] Depends on type, quality, and amount of coating, humidity, conditioning, and other factors.
[4] For undamaged samples.
[5] Primarily for cellophane-polyethylene and Mylar-polyethylene laminations.

sugars to CO_2 and water. At low oxygen tensions, however, anaerobic respiration results in the formation of ethyl alcohol. The detrimental effect of this fermentation on acceptability of fresh fruits and vegetables is well known, but its nutritional significance is obscure.

The availability of oxygen to packaged foods is dependent on the nature of the container. Adequately sealed cans and glass containers effectively prevent the interchange of gases between the food product and atmosphere. In flexible packaging, however, the diffusion of gases depends not only on the effectiveness of closure but also on the permeability of the packaging material. The diffusion may take place through pores and discontinuities in the materials, or through intermolecular spaces of very small dimensions. In the first case, sometimes described as simple diffusion, the amount of gas diffusion depends only on the total area of pores and on molecular weight of the gas. Of much more importance in packaging is the second case, that of activated diffusion, in which the permeability is dependent primarily on the physicochemical structure of the barrier. It is this mechanism which accounts for the large differences between the permeabilities of different packaging materials. The ranges of oxygen permeabilities of the various materials used in flexible packaging are shown in Table 15.2.

Moisture

Packaging is the decisive factor in controlling moisture changes in stored foods. The effect of moisture changes on the nutritional value of foods is a problem of greater complexity than the effect of oxygen, however, and the extent to which moistureproof packaging contributes to the preservation of foods depends on specific food type and on conditions of storage.

In dehydrated foods many of the deteriorative changes can be prevented by elimination of moisture infiltration. Nonenzymatic browning, which is frequently correlated with decrease in nutritional value, is effectively inhibited by maintenance of low moisture content. A similar situation prevails with respect to other food deterioration reactions (Karel 1973B). Typical dependence patterns of food deterioration reactions on water activity are shown in Fig. 15.5. Typical values of water vapor permeability of food packaging materials are shown in Table 15.3.

INTERACTION BETWEEN PRODUCT AND PACKAGE

Packages are made from many components having varying chemical and physical characteristics. Some of these components may react

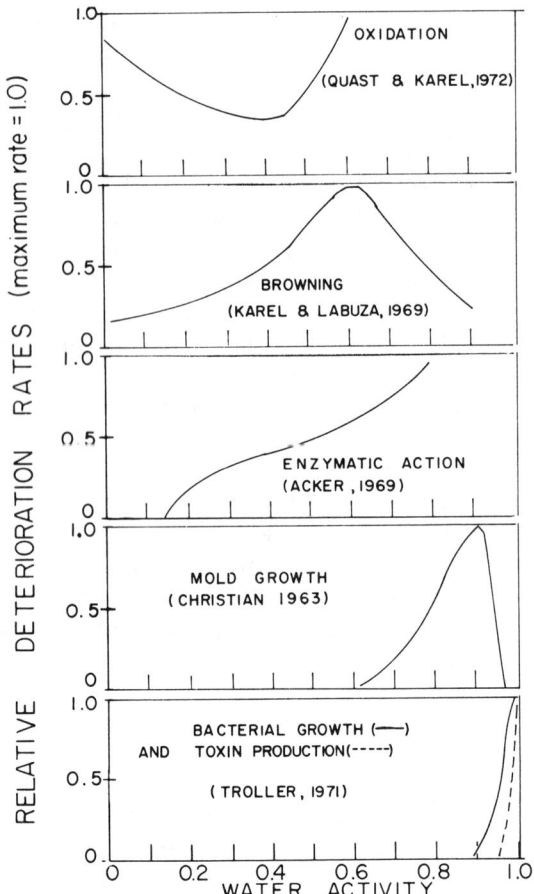

FIG. 15.5. DEPENDENCE OF RELATIVE RATES OF FOOD DETERIORATION ON WATER ACTIVITY

Bacterial growth (after Troller 1971): *S. aureus.* Toxin production (after Troller 1971): Enterotoxin B. Mold growth (after Christian 1963): *Xeromyces bisporus.* Enzymatic activity (after Acker 1969): oat lipase acting on monoolein in ground oats. Brown (after Karel and Labuza 1969): nonenzymatic browning of pork bites. Oxidation (after Quast and Karel 1972): lipid oxidation in potato chips.

with the package contents. The toxicological consequences, and potential hazards of such interactions, have received a great deal of attention. The 1958 Food Additive Amendment has specifically included in its definition of food additives substances which become a component of food or affect characteristics of food as a consequence of packaging. The passage of this amendment has stimulated

TABLE 15.3

PERMEABILITY OF VARIOUS PACKAGING MATERIALS TO WATER
VAPOR AT 100°F AND 95% VERSUS 0% RELATIVE HUMIDITY

Material	Permeability Range (Gm) (Mil) $(24\ \text{Hr})^{-1}$ $(100\ \text{Sq In.})^{-1}$
Plain Cellophane	20–100
Nitrocellulose-coated cellophane	0.2–2.0
Saran-coated cellophane	0.1–0.5
Polyethylene, conventional	0.8–1.5
Polyethylene, low pressure	0.3–0.5
Saran	0.1–0.5
Vinyl-chloride-based films	0.5–8.0
Aluminum foil, 0.00035 in. thick	0.1–1.0
Aluminum foil, 0.0014 in. thick	Less than 0.1
Plastic paper foil laminations	Less than 0.1
Waxed papers	0.2–15.0
Coated papers	0.2–5.0
Mylar	0.8–1.5
(Poly)-trifluorochloroethylene	0.01–0.1
Silicone rubber	>200
Polypropylene	0.2–0.4

research into food-package interactions. The problem is a complex and difficult one. The determination of transfer of individual substances from package to food is difficult and regulations have been based on the less difficult determination of extraction into food-simulating solvents including: (a) water, (b) 3% aqueous NaCl, (c) 3% aqueous $NaHCO_3$, (d) 3% acetic acid, (e) 3% lactic acid, (f) 20% aqueous sucrose, (g) lard or vegetable oil (Griffin and Sacharow 1972). Other countries use different food-simulating solutions (Institute of Packaging 1971).

The problem of extraction into fatty substances is particularly difficult, because some components of packaging materials are difficult to differentiate from normal components of fatty foods (Karel and Wogan 1963). Accordingly, fat-simulating solvents, such as heptane, are often used to test the inertness of packaging materials. An even more imposing problem is the correlation of type and amount of extractive with physiological effects in animals (Frawley 1967).

Some trends in the field of toxicological implications of food packaging are:

(1) New techniques have been developed for the determination of trace quantities of specific components of packaging materials and enamels (Ramsey 1967; Figge 1972; Garlock and Paynter 1965).

(2) Specific substances of especially high potential danger have

been tested utilizing extremely sensitive methods. Polychlorinated biphenyls (PCB), for instance, fall in this category (Garty 1972).

(3) Recent concern about potential toxicity of various heavy metals has resulted in ongoing research on transfer of contaminants from various metallic containers.

Apart from safety considerations, the problem of acceleration of nutritive deterioration by package components has also received some recognition and can be solved by the application of proper packaging methods.

In metal containers, application of inert coatings, enamels, and lacquers guards against reaction between food and package. Where, as in the case of canned foods of high acidity, some corrosion does occur, its effect on retention of nutrients appears to be beneficial rather than detrimental (Clifcorn 1948). This is probably caused by the reducing action of tin.

Glass has an excellent inertness, and when inert liners are used for closures, there is no interaction between food and package. Most of the flexible packages also possess adequate inertness. In some cases, however, flexible materials may contain excessive concentrations of metallic contaminants. Thompson and Kocher (1950) found that some cartons used for packaging of biscuits contained up to 600 ppm of copper and 735 ppm of iron. They concluded that metal contamination was an important factor in oxidative deterioration of fat in the biscuits.

A related problem concerns the penetration of the package by free oil from the packaged food products. Free oil is defined as the fat available for staining of the package. Penetration of the packaging materials by free oil sets up a focal point for rancidity development that can affect the nutritional value of the packaged food. The importance of greaseproof materials in controlling fat and oil penetration is self-apparent.

The effects of specific processing methods on the total package-food system must be considered. In the case of heat-processing and radiation-processing, FDA approval was sought and obtained by U.S. Army Natick Laboratory for specific packages to be used with specified foods in these processes. Similar considerations would, of course, apply in other food processing operations involving the package-food combination.

EFFECTS OF PACKAGING ON PRESERVATION OF NUTRIENTS IN SPECIFIC FOOD PRODUCTS

The preceding sections discuss general mechanisms by which packaging can modify food storage conditions. All of these mechanisms can affect the preservation of nutrients. Their predominating individual and collective effects, as well as their nutritional significance to the consumer, can vary with different food products and with different systems of food distribution.

Based on knowledge of stability of individual nutrients, certain broad generalizations can be hypothesized regarding the effects of the package on retention of nutrients in foods. Where such speculation seems justified, these generalizations have been offered in the following discussions of specific food types. The present status of exact knowledge based upon published experimental evidence is, however, rather limited in scope concerning food types and limited in depth within food types.

This lack of information is evidence that there is a need to arouse interest in the collection of experimental data concerning the effects of packaging on retention of nutrients in foods. This need for more exact nutrient information arises from the (1) need for nutritionists to better understand the nutrient content of foods as they are consumed, (2) need to know effects on nutrients in transporting foods to areas of food shortage, (3) need of manufacturers to know the stability of nutrients which they have claimed on their labels in order to comply with labelling laws, and (4) need for consumers to be better informed regarding what to expect in regard to nutrition from the foods they purchase.

The functions of food packages are determined primarily by such factors as: the nature of the food, local customs, the food distribution system, merchandising, economics, sanitation, and legal constraints. In any given society, the frequency of food package changes tends to accelerate with increasing affluence. These factors result in a wide variety of packaging being used for comparable products in different locales. In general, it is fair to say that the problem of nutrient preservation, per se, is not characteristically the prime determinant in selecting the food package.

Bakery Products

Bread.—Bread offers a good example of the acceleration of change in food packaging with increasing affluence. In the United States prior to 1910, most bread was sold unwrapped. In the decade 1910-1920, wax paper began to be used, followed in 1920 by cellophane. By 1930, coated cellophanes dominated the bread market. Following World War II, a rapid succession of varieties of polyethylene and polypropylene films were utilized. Today, about 80% of all bread in the United States is packaged in low density polyethylene bags.

It is unlikely that the caloric value of bread or the biological value of bread protein is affected by the package. Under relatively ideal storage conditions, the gluten quality of wheat was shown by Kozlova and Nekrasov (1956) to deteriorate with time. Low quality grain retained its quality for 4 yr; average quality grain retained its quality for 5-6 yr; and high quality grain retained its quality for 10 yr.

Of the vitamins added to bread, thiamin and niacin are conceded to have good stability under normal distribution conditions. The retention of riboflavin, however, is a controversial problem. Loy et al. (1951) studied the retention of riboflavin in enriched bread and rolls. The samples were exposed to artificial light and to sunlight passing through clear glass windows. The bread was exposed unwrapped and in waxed paper wraps, the rolls unwrapped and wrapped in clear cellophane. Considerable losses were found in unwrapped bread and in unwrapped and wrapped rolls exposed to sunlight. Exposure to artificial light also resulted in losses, but they were less severe than in the case of sunlight. The authors concluded that waxed paper wraps offered good protection but that the cellophane had little protective value. Similar results were obtained by Kanninen and Hardy (1955) who studied the effects of diffuse daylight and of fluorescent light, and concluded that waxed paper was considerably more effective than cellophane in protecting riboflavin in enriched white bread. Morgareidge (1956) studied the retention of riboflavin in enriched bread packaged in unprinted cellophane, printed cellophane, and in waxed paper. Exposure to light simulated commercial conditions. No difference in riboflavin retention was found between bread packaged in cellophanes and that packaged in waxed paper. Morgareidge (1956) suggested that conditions of experiments of Loy et al. (1951) were more severe than those encountered in practice, and this could explain the discrepancies in the results of the two studies. Similar conclusions were reached by the American Institute of Baking (Anon. 1955) which expressed the opinion that, under conditions of practice, riboflavin is retained in enriched bread packaged in either of the two materials. Birdsall and Teply (1957) compared the retention of riboflavin in partially baked rolls wrapped in aluminum foil, waxed paper, and cellophane. The authors described their experimental lighting conditions as comparable to those "in a rather brightly illuminated food market." Their results indicated that aluminum foil offered considerably better protection than waxed paper, which, in turn, was more protective than cellophane. Stephens and Chastain (1959) studied partially baked rolls under simulated grocery conditions, and found little riboflavin loss in cellophane-packaged rolls.

It appears from evidence available to date that the optical properties of packaging material are of minor importance in preservation of the riboflavin content of bakery products with a well-developed crust. In partially baked rolls, and similar bakery goods, however, the retention of riboflavin appears to be determined in part by the light-transmission characteristics of the package.

Biscuits, Bakery Goods, and Cake Mixes.—The main purpose of protective packaging of biscuits is the prevention of undesirable changes in flavor, color, and appearance. Similar considerations prevail in packaging of cake mixes. These commodities are packaged extensively in moistureproof materials of various nature. From the nutritional point of view, prevention of oxidative changes is probably most important. Oxygen impermeability and inertness are the packaging properties retarding oxidation. Since most moistureproof materials have these two properties, it is probable that the prevention of organoleptic changes is accompanied by preservation of nutritive value.

Dairy Products

Fluid Milk.—In the United States, glass bottles were used for virtually all fresh milk sold in 1950. Today, 90% of U.S. milk is sold in plastic-coated paperboard. This trend away from glass has not occurred in many affluent countries.

The light-transmission characteristics of the milk package are important. Light can cause a variety of undesirable effects in fluid milk including the production of "oxidized" off-flavors and the degradation of ascorbic acid, riboflavin, and methionine (Patton 1954). Concomitant with the shift from glass bottles to paperboard cartons for milk distribution in the United States, there has been a shift of milk distribution through supermarkets. In the supermarket, the milk container is often exposed to high intensity lighting.

In a study of the effect of light intensities and wavelengths typical of supermarket display cabinets, Sattar and de Man (1973) concluded that the paperboard carton and returnable plastic jug were both inadequate for protection of milk from loss of flavor and nutritional value. In this study, fresh milk at 4°C was exposed to 100 or 200 foot candles of light for various lengths of time. The milk was packaged in 1 of 4 types of containers: (1) standard paperboard carton, (2) commercial plastic jug, (3) clear polyethylene, and (4) black-pigmented opaque polyethylene. The carton, jug, and clear package had significant deterioration of flavor in 12 hr compared to the opaque pouch. At 100 foot candles, the percentage losses of ascorbic acid after 24 hr were 23.6, 86.2 and 90.9% for the carton, jug, and clear package, respectively, compared to 13.0% loss in the opaque pouch. Under the same conditions, the riboflavin losses were 7.1, 11.1, and 22.3% for the carton, jug, and clear package, respectively, compared to essentially no loss in the opaque pouch. These workers noted that some supermarkets have light intensities of 400-500 foot candles at which even more rapid deteriorations could be expected. The retention of vitamins as a function of various milk containers to light has been studied by several other investigators.

Losses of riboflavin and ascorbic acid were correlated with light transmission through containers, with opaque materials giving better protection. Dunkley et al. (1962) studied milk from three retail markets and found great variation in light-caused damaged between different samples, but flavor protection and protection of riboflavin and of ascorbic acid by opaque containers was evident. Sacharow (1969) also reported significant protection of vitamins in milk by opaque containers. Smith and MacLeod (1955) exposed milk in bottles to fluorescent and incandescent light and found that the exposure resulted in rapid destruction of ascorbic acid. Cox et al. (1957) studied the effect of exposure of pasteurized whole and skim milk on retention of vitamin A. The milk was packaged in amber glass bottles, plain glass bottles, and waxed paperboard cartons, and was exposed to diffused daylight, and to short-time irradiation with direct sunlight. Care was taken to avoid temperature changes. It was found in this study that amber glass bottles and paperboard cartons afforded good protection, but plain glass bottles did not prevent considerable losses of the vitamin. Waxed paperboard cartons were slightly more protective than amber glass. It is interesting to note in this connection that amber glass has been cited as giving better protection against light-activated off-flavors than waxed paperboard cartons (Pont 1956).

Karel and Cornell (1956) studied the effect of overwraps on the light-activated off-flavors in milk packaged in glass bottles. They found that overwraps screening out portions of visible spectrum as well as ultraviolet light were more effective than overwraps screening out only ultraviolet light. The above-mentioned study of Cox suggests that similar effects might be expected in the protection of vitamins.

According to Kon and Thompson (1953), up to 50% of riboflavin in milk may be lost in distribution during the summer months, and up to 20% during the rest of the year. Even greater losses may be expected in milk stored for prolonged periods of time. The need for and effectiveness of various opaque packages for the protection of nutrient and organoleptic quality of fresh milk has been a recurring subject of older as well as more recent literature (Kiermeier 1969; Kiermeier and Waiblinger 1969; Nordlund et al. 1970). It is evident that in societies in which brightly lighted supermarkets have been introduced, but clear glass milk bottles remain in use, significant nutrient and quality losses in fresh milk can be anticipated, not only in riboflavin content but also in ascorbic acid and possibly other nutrients.

Butter.—Data on the effect of packaging on nutritional value of butter is scarce. It appears, however, that packaging of butter affects primarily its stability with nutritional effects being of minor importance.

It might be expected that protection against light, oxygen, and metal contamination would be of considerable importance. Some reports confirm these theoretical assumptions (Barnicoat 1947; Mummery et al. 1950). Other published studies, however, indicate that the effect of inertness and impermeability of the package on nutritional value of butter may be of only secondary importance. MacDowell and MacDowell (1953) found that metal contamination did not affect the vitamin A content of stored butter. Melnick et al. (1953) found that margarine stored at room and refrigerated temperatures became unsalable because of appearance defects before losing a significant amount of vitamin A. Downey and Murphy (1968) have reported the light-barrier properties of various butter wrapping materials. Gilchrist et al. (1968) confirmed that prevention of exposure to light is the only important packaging factor in prevention of surface oxidation of butter.

Cheeses.—In protective packaging of cheeses, consideration is given primarily to the preservation of consumer acceptance. The need for protective packages such as laminated foils, low-permeability plastics, and similar materials is based on prevention of moisture loss, off-flavor development, and appearance defects. It is probable that nutritive losses directly ascribable to packaging materials are less common. Several recent studies have, in fact, considered the effect of packaging on oxidative rancidity and other quality defects in cheese, but they do not give data on losses of nutrients (Kristoffersen et al. 1964; Kiermeier and Wolfseder 1972; Hartmann 1972).

Dehydrated Foods

Nonenzymatic browning and oxidation are, in general, the two predominating mechanisms causing loss of nutritional value in stored dehydrated foods. The rates of both of these reactions can be influenced by packaging.

Nonenzymatic browning is independent of the oxygen content of the package, but it is influenced by moisture content of the product. In general, below water activity of 0.4, the rate of browning is slow because of lack of sufficient solvent water. As water activity increases above 0.4, the rate of browning increases to a maximum at about water activity = 0.65. Above this water activity, the rate of browning begins to decrease because reactants are diluted, and because water being a product of the browning reaction tends to diminish the reaction rate by the law of mass action (Eichner and Karel 1972).

The rate of oxidation can be retarded by protection of the food from atmospheric oxygen, light, and other pro-oxidants. Oxidation is affected by moisture content in at least three important ways: an antioxidant effect due to hydration of metal catalysts, which decreases their catalytic action; an antioxidant effect due to bonding of hydroperoxides, which reduces their reactivity; a pro-oxidant effect due to increasing the mobility of reactants and catalysts.

As water activity increases, the two antioxidant effects of water are generally seen before the pro-oxidant effect becomes significant. With continued increase in water activity, the pro-oxidant effect tends to predominate. This explains the existence of a critical water activity up to which the continued increase of water activity is increasingly antioxidant, but above which any further rise in water activity is increasingly pro-oxidant. Knowledge of a system's critical water is, therefore, essential for determining the predominant effect of moisture content on rates of oxidation (Heidelbaugh et al. 1971).

The critical water activity and the critical water content depend strongly on the composition of a food. The literature contains apparently conflicting reports regarding the effects of water content on the oxidation stability of foods. Some studies report that addition of water retards oxidation and others report the opposite. A critical review shows that reports indicating that water retarded oxidation were dealing either with systems having low water-binding capability (e.g., adding water to bulk oil) or with relatively low water activities. On the other hand, reports showing a pro-oxidative effect of increasing water dealt generally with high water activities.

The packaging properties which foster conditions for minimization of oxidation and browning and thereby serve to preserve nutritional value coincide with those properties which are optimal for preservation of consumer-acceptance characteristics. This fact has stimulated study of application of protective food packaging, which includes the following methods (Strashun and Talburt 1954; Copley et al. 1956; Bishov et al. 1971): (1) use of chemically-inert packaging materials,

(2) use of materials of low permeability to moisture and gases, (3) packaging under vacuum or under inert gases, (4) use of in-package desiccation, and (5) use of in-package oxygen scavenger.

Dehydrated foods can be grouped into two categories based on their susceptibility to the above two chief mechanisms of nutrient deterioration (oxidation and browning). These food categories are: (1) dehydrated fruits and vegetables in which lowering of nutritive value results primarily from oxidation of vitamins or vitamin precursors (ascorbic acid, carotene), and (2) dehydrated high-protein foods in which changes in the protein and lipid fractions, as well as vitamin destruction, can be of nutritional significance, and such degradations result from oxidation or nonenzymatic browning. These factors are the subject of a comprehensive review by Labuza (1972).

Dehydrated Fruits and Vegetables.—Among the earliest investigations on dehydrated foods was a series of studies by the Continental Can Co. Research Staff (1944–1945). Included in the study were cans sealed in air, cans sealed in inert gases, and cartons with laminated inner bags which were heat-sealed in air. In these studies, it was found that: (1) Riboflavin and thiamin retention was not affected by any of the packaging variables. (2) Ascorbic acid in apple nuggets, cabbage, and rutabagas was retained better in gas-packed than in air-packed cans and cartons. In sweet and white potatoes there were differences between cans and cartons, as well as between air-packed and gas-packed cans. In tomato flakes and cranberries, the retention of ascorbic acid was independent of the method of packaging. (3) Carotene retention was not affected by any of the packaging variables in tomato flakes and onions, and only by package atmosphere in sweet potatoes and carrots.

Heiss (1956) presented data showing that retention of ascorbic acid and carotene is affected by moisture changes as well as by package atmosphere. He emphasized the need for moistureproof packaging of dehydrated foods. De Lange (1953) found that ascorbic acid was retained better in cans than in polyethylene and pliofilm bags. Of the two plastics, polyethylene was slightly more protective than pliofilm.

Dehydrated tomato juice, carrots, cabbage, onions, beets, white potatoes, and sweet potatoes have been studied during storage. Ascorbic acid retention in these products was found to depend markedly on package atmosphere and on moisture content. Excellent retention of ascorbic acid was achieved by combination of nitrogen packaging with in-package desiccation. The use of in-package desiccation without nitrogen packing was, in general, more effective than nitrogen without in-package desiccation (Legault et al. 1954; Wong et al. 1956).

Protection against moisture is the most important function of the container and is essential to retention of ascorbic acid in dehydrated fruits and vegetables. Protection from oxygen is usually less important for protection against loss of ascorbic acid than protection from water (Karel and Nickerson 1964; Goldblith and Tannenbaum 1966). In completely dry fruits and vegetables containing lipids or other readily oxidizable components (potatoes, carrots) the package should protect against oxygen as well, and packing in nitrogen is desirable. Carotene protection and flavor protection required exclusion of light as well (Bolin et al. 1964).

Dehydrated High-Protein Foods.—Hodson and Krueger (1947) compared the essential amino acid content of freshly prepared dry skim milk with that of dry skim milk stored for 4 yr in 2 different types of containers. They found that milk stored in containers that excluded air compared favorably with the freshly prepared milk in palatability, appearance, and amino acid content. Milk stored in containers allowing some gas interchange, however, not only lost its acceptability but also showed significant losses of arginine, histidine, lysine, and methionine. The oxidation effects on vitamin content of foods have been the subject of many investigations. Results of Olsen et al. (1948) indicate that moistureproofness of the package may have an importance in thiamine retention.

Almquist (1956) showed that deteriorative changes in protein and fat fractions of fish meal, including the lowering of protein digestibility and of fat extractability, could be prevented by storage of fish meal in glass tubes evacuated to 10 nm pressure.

Whitmore et al. (1948) studied the storage behavior of dehydrated meat and found that packaging had no effect on development of changes in the peroxide number and in organoleptic characteristics. Work on freeze-dried meats, however, shows that packaging may have an important effect on nutritive value. Tappel (1956) found that the most important oxidative changes in freeze-dried meats involved the protein and the fat fractions, and that they could be prevented by packaging in inert atmospheres or at high vacuum. Similar work was reported by Bengtsson and Bengtsson (1968). Nonoxidative changes in freeze-dried meats were studied by Regier and Tappel (1956) who found that browning reactions were of most importance. They showed that browning was independent of oxygen tension, but could be retarded by low moisture content. Similar conclusions with respect to browning of dehydrated eggs were reached by Olcott (1952).

Fats and Oils

Oxidative changes in fats and oils can result in destruction of fat-soluble vitamins and essential fatty acids, and in production of potentially toxic substances and off-flavors.

In a study of the long-term storage of commercially prepared soybean and cottonseed oils, it was found that bottled oils had a better stability compared to oils in screw-cap metal cans (Evans et al. 1973). All these studies were performed in the dark. Any level of oxygen contamination in the headspace was found, in these studies, to be detrimental to flavor. The best storage method for oils is to exclude all oxygen from the package.

The following list summarizes some of the packaging techniques available which prevent or retard degradation of fats and oils in storage:

(1) The inclusion of antioxidants on the internal surface of the package.
(2) Use of opaque containers and of colored wrapping materials and colored glasses.
(3) Use of containers preventing access to the food of chemical pro-oxidants.
(4) Use of methods limiting the access of atmospheric oxygen.
(5) Use of greaseproof packages to prevent penetration by free oil.
(6) Use of oxygen scavengers in the package (Bishov et al. 1971).

Fresh Fruits and Vegetables

The primary aim of packaging of fresh fruits and vegetables is the protection of consumer acceptance characteristics. The choice of packaging materials has usually been based, therefore, on factors such as protection from contamination, mechanical damage, protection from wilting through water loss, and simultaneous assurance of sufficient ventilation to avoid anaerobic respiration. The optimal conditions of storage for a great variety of fruits and vegetables have been studied and the scientific bases for optimizing these conditions developed. Review articles by Smith (1963) and the handbook by Wright et al. (1963) are particularly useful in this respect. The shelf-life of produce is controlled in particular by temperature, relative humidity, and by oxygen and carbon dioxide concentration in the atmosphere surrounding the fruit. Because of the dependence on concentration of the last-mentioned gases, permeability of the packaging materials chosen is of particular importance.

A number of studies have demonstrated that optimal gas compositions can be achieved by placing fruits and vegetables in packages with controlled permeability characteristics.

It is desirable to slow respiration rates of fruits and vegetables by lowering oxygen tension, but the O_2 level must also be maintained above the critical minimum content at which plant tissues ferment. For fruits which have respiration rates depending only on O_2 concentration and which absorb O_2 and evolve carbon dioxide in equimolar concentration, it is possible to use very simple graphical methods to predict the steady-state concentrations of these gases (Karel and Go 1964; Jurin and Karel 1963). The steady-state concentrations are attained rapidly for bananas stored in polyethylene; and similar results have been attained with apples and other fruit. Recently, Henig and Gilbert (1972) applied similar analyses to packaging of bananas and tomatoes.

The steady-state concentration of oxygen can be predetermined by choosing packaging materials with a given permeability. It is theoretically possible to control both O_2 and CO_2 levels, but in practice only a few combinations of specific permeabilities to O_2 and to CO_2 are available if the choice is limited to single films or to their laminations. In an attempt to overcome this limitation studies were conducted in which apples were stored in a container made of an impermeable material having "windows" made of two different materials with different permeabilities to O_2 and to CO_2 (Veeraju and Karel 1966). By varying the areas of each of the three materials, which, of course, represent mass transfer resistances arranged in parallel, we were able to control independently both the O_2 and the CO_2 levels in the package. Similar approaches to control of atmosphere inside produce packages were undertaken by French workers (Marcellin 1971).

The ability to calculate package properties needed for packaging fruits and vegetables may allow, at least on a limited scale, the use of packages to control atmosphere instead of storage facilities requiring continuous control of gas concentration. The value of the approach has been attested by recent industrial interest in semipermeable packaging for control of fruit respiration (Badran et al. 1969; Rosenfield 1971).

With respect to retention of specific nutrients, the influence of packaging is not as clear as its role in maintaining organoleptic characteristics. Mapson (1952) states that wilting and mechanical damage are also the important factors affecting nutritive quality. Salunkhe and Wu (1973) observed that oxygen concentration in storage, which is controllable by packaging, affects β-carotene content and sugar content of tomatoes. Smith (1963) reports that oxygen and carbon dioxide concentrations affected ascorbic acid content of lemons, with high CO_2 concentrations seriously reducing the vitamin C level in the stored fruit. Similar results were apparently obtained with other fruits. From the point of view of ascorbic acid retention, therefore, reduction of oxygen tension without concomitant CO_2 buildup appears desirable. Choice of packaging materials has been shown to affect the percentage of weight of lettuce remaining edible after storage, and thus the availability of all of its nutrients (Cook et al. 1958). Wilting of lettuce has also been suggested as an index of loss in initial provitamin A content (Harris and Mosher 1941). It appears, therefore, that protection of organoleptic properties is also usually effective in protecting nutritive value, but there still is a lack of information on experimental correlation of packaging properties and retention of nutrients in fresh fruits and vegetables.

Heat-Processed Foods

After a review of the literature, Clifcorn (1948) reported that of the vitamins studied in heat-processed foods only ascorbic acid was affected by type of container. Retention of ascorbic acid in canned foods of low pH was slightly better in plain tin cans than in enameled tin cans, and considerably better than in glass containers. The retention was also improved by inclusion of tin strips in the glass containers. Clifcorn concluded that the differences in retention were due to consumption of residual oxygen by reactions between tin and food acids, and that the differences would decrease with increasing pH or decreasing residual oxygen content.

Mohr (1950) cites additional evidence indicating that type of container is important in retention of ascorbic acid. His conclusions were based in part on studies of German workers, including Wendland (1948), who found that retention of ascorbic acid was lower in cans made of sparingly tinned steel plate and of phosphated black plate than in some types of lacquered black plate cans.

Kuprianoff and Gutschmidt (1951) investigated the effect of type of tinplate on ascorbic acid retention and found that the effect differed with different types of canned foods. Gutschmidt (1952) reached similar conclusions. His data indicated that ascorbic acid content of beans and spinach was preserved better in black plate than in white plate cans. Glass containers were less effective than cans.

Pederson and Robinson (1952) in work on sauerkraut found that ascorbic acid was retained better in cans than in glass jars. Buffa et al. (1955) studied ascorbic acid content of citrus and tomato juices as a function of type of container, and found that best retention could be obtained with certain types of cans made of unlacquered tinplate.

Guerrant and O'Hara (1953) studied vitamin retention in tin- and glass-packed peas and lima beans, and found no significant differences in the type of container. Vitamins studied in this investigation included ascorbic acid, thiamin, riboflavin, niacin, and carotene. Sedky et al. (1952) determined ascorbic acid retention in sauerkraut packaged in laminated pliofilm bags and found that retention depended on amount of oxygen present in the bags. Hu et al. (1955) studied retention of ascorbic acid which was added to applesauce processed in Mylar and Trithene bags and found no differences in retention between the two plastics.

The effects of the package on changes in heat-processed foods were studied by Heidelbaugh and Karel (1970), by comparing the changes in cranberry sauce (pH 2.7), vegetables and bacon (pH 5.2), and pork (pH 6.4) when stored in (1) metal cans, (2) aluminum laminate film (0.5-mil polyester/0.005-in. aluminum/2-mil polyethylene), (3) 2 layers of 2.5-mil polyester film, or (4) 1 layer of 2.5-mil polyester film. The values of oxygen permeance (ml O_2 per 24 hr per sq in.) were: (1) aluminum film less than 0.001 gm; (2) double polyester, 0.648 gm; and (3) single polyester, 0.324 gm.

All the packaged foods in these studies were heat processed to prevent any change due to microbial action. The packages were filled under vacuum to minimize headspace gases, and the packages had essentially the same dimensions. Therefore, the only significant

TABLE 15.4

SUMMARY OF TEST RESULTS ON CRANBERRY SAUCE (pH 2.7)

Test	Prior to Storage	After Storage for 5 Weeks at 37°C at 80% RH in Packages Made of:			
		No. 2 Lined Tin Can	Aluminum Laminate	Polyester Film (2 Layers)	Polyester Film (1 Layer)
pH	2.7	2.7	2.7	2.7	2.7
Moisture (%)	56.1	55.9	56.0	54.5	53.5
Acceptability Hedonic scale (mean)	7.3	6.7	7.2	6.3	5.3
Range of scores	5-9	3-8	6-8	3-8	3-7
Browning, OD at 400 nm	0.154	0.233	0.234	0.338	0.367
Ascorbic acid, mg per 100 gm	89.20	73.40	71.72	0.40	0.00

Note:
Acceptability on a 9 point hedonic scale with 5 being null point.
Results on cranberry sauce in original commercial pack held for 5 weeks at room temperature were similar to values of test on cranberry sauce prior to process.

EFFECTS OF PACKAGING ON NUTRIENTS

TABLE 15.5

SUMMARY OF TEST RESULTS ON VEGETABLES AND BACON (pH 5.2)

Test	Prior to Process	Prior to Storage	After Storage for 5 Weeks at 37°C at 80% RH in Packages Made of:			
			Aluminum Metal Can	Aluminum Laminate	Polyester Film (2 Layers)	Polyester Film (1 Layer)
pH	5.2	5.2	4.8	4.8	4.7	4.7
Moisture (%)	84.2	84.5	84.8	84.1	84.5	83.1
Acceptability Hedonic scale (mean)	5.2	5.6	6.2	6.0	2.9	2.8
Range of scores	3.7	2-8	4-9	4-8	1-7	1-7
TBA (OD)	0.205	0.219	0.428	0.381	0.401	0.452
Peroxide No.	1.51	2.04	1.21	1.16	3.60	14.53
Ascorbic acid mg per 100 gm	3.7	3.0	0.6	0.6	0.0	0.0

Note:
Acceptability on a 9 point hedonic scale with 5 being null point.
TBA, optical density (OD at 530 nm).
Peroxide number, milliequivalents peroxide per kilogram of fat.
Test results on vegetables and bacon in original commercial pack held for 5 weeks at room temperature were similar to values of test on vegetables and bacon prior to process.

differences between the packages were the moisture and oxygen permeances. The results of these studies are summarized in Tables 15.4, 15.5, and 15.6. The chemical and organoleptic tests indicated that oxidative reactions were the chief pathway for degradations in the foods. As oxygen permeability of the package increased, the ascorbic acid

TABLE 15.6

SUMMARY OF TEST RESULTS ON PORK (pH 6.4)

Test	Prior to Process	Prior to Storage	After Storage for 5 Weeks at 37°C at 80% RH in Packages Made of:			
			Aluminum Metal Can	Aluminum Laminate	Polyester Film (2 Layers)	Polyester Film (1 Layer)
pH	6.4	6.4	6.3	6.3	6.3	6.3
Moisture (%)	80.3	79.3	80.7	79.3	79.3	78.4
Acceptability Hedonic scale (mean)	4.5	4.0	6.5	5.8	3.0	2.5
Range of scores	1-7	1-7	2-9	2-8	1-7	1-6
TBA (OD)	0.205	0.213	0.258	0.258	0.294	0.70
Peroxide No.	2.65	3.54	5.45	4.74	15.40	12.22
Thiamin, μg per 100 gm	360.8	223.8	66.0	54.4	62.1	40.0

Note:
Acceptability on a 9 point hedonic scale with 5 being null point.
TBA, optical density (OD at 530 nm).
Peroxide number, milliequivalents peroxide per kilogram of fat.
Test results on pork in original commercial pack held for 5 weeks at room temperature were similar to values of test on pork prior to process.

content of the food decreased and peroxide values and thiobarbituric acid values both increased.

Ascorbic acid, as the nutrient which has been reported as most often significantly affected by package differences during storage of properly heat-processed foods contained in impermeable packages, is strongly dependent upon the degree of permeability of those packages.

The availabiltiy of iron in foods canned in glass jars and cans of different composition was studied by Theriault and Fellers (1942). The foods included peaches, asparagus, sweet corn, green snap beans, spinach, lima beans, and red sea-perch fillets. The results indicated that there was no change in total or available iron in foods packed in glass jars. Iron content of foods packaged in cans depended on pH and on presence of enamels. Foods of low pH gained significant amounts of available iron when packaged in plain tin cans. Alkaline foods lost some available iron. The greatest losses occurred in the fish fillets.

Frozen Foods

The emphasis in protection of frozen foods has traditionally been placed on prevention of desiccation and oxidation. Among the requirements for all frozen food packages are: inertness, impermeability to liquids, resistance to mechanical damage at low temperature, and impermeability to gases and vapors (Karel 1956). In addition, special requirements may arise from specialized methods of distribution or display (Anon. 1972). The extent to which protective packaging affects nutritive value of frozen foods depends on time and temperature of storage and on the nature of the food. The effect of packaging on temperature changes in frozen foods has been discussed previously. Some of the other packaging effects are mentioned below.

Frozen foods are packaged in a variety of containers made from paper, plastics, glass, and metals (Shelor and Woodroof 1954; Nickerson and Karel 1964; Anon. 1972). Some properties of these packages may have an important effect on retention of nutrients. In this respect some data have been published in the literature about fruits, vegetables, and flesh foods.

Packaging of Frozen Fruits and Vegetables.—Ascorbic acid in frozen fruits and vegetables is highly susceptible to oxygen and hence its retention would be expected to depend on the oxygen permeability of containers. This contention is supported by the findings of researchers at the USDA Western Regional Laboratory who have pioneered in evaluation of storage stability of frozen foods (Van Arsdel *et al.* 1969). Specifically, ascorbic acid retention in frozen strawberries was found to be much greater in tin cans than in composite cans (Guadagni and Nimmo 1957). In frozen cauliflower and beans ascorbic acid was found to be retained better in aluminum foil or in plastic than in paper wrapping (Offergaard and Nordness 1958). Frozen orange juice stored in cans was found to lose no ascorbic acid in 6 months' storage at 0°F, but a 10% loss occurred when the juice was stored in polyethylene freezer cups. Addition of 3% metaphosphoric acid prevented the loss in the polyethylene cups (Miller and Tingleff 1960). Kefford et al. (1959) also found that presence of oxygen in frozen juice allowed slow oxidation of ascorbic acid.

The moistureproofness of the package appears to have a lesser effect on retention of ascorbic acid, although it is of primary importance in protection of organoleptic quality. Volz et al. (1949) and Gutschmidt and Wolodkewitsch (1950) found no correlation between retention of ascorbic acid and water-vapor permeability of containers used in storage of selected frozen fruits and vegetables.

Frozen Flesh Foods.—Protective packaging of frozen meats, fish, and poultry is designed to prevent desiccation and oxidative changes. The choice of packaging materials is based, therefore, on their water-vapor permeability, oxygen permeability, and light-transmission characteristics. These properties may also be effective in preventing loss of nutritional value.

Of particular interest in this respect appears to be the prevention of oxidation, since oxidative changes may lower nutritive value by destroying vitamins and by deteriorating the protein and fat fractions. Of these effects only fat oxidation has been studied extensively. Packaging properties have been found to affect it considerably. Palmer et al. (1953) found a correlation between the moistureproofness of packages and retardation of fat oxidation in frozen pork. Similar results were obtained in studies on ground beef (Shelor and Woodroof 1954), frozen cooked poultry (Hanson et al. 1950) and frozen salmon steaks (Heerdt and Stansby 1955). Additional work on frozen poultry (Klose et al. 1959; Hanson et al. 1959) confirmed the importance of prevention of oxidation through protective packaging. Whether desiccation of the surfaces of flesh foods during frozen storage has nutritional consequences, beyond reduction of palatability, is not entirely clear. Kaess and his co-workers have studied extensively the condition known as freezer-burn (Kaess and Weidemann 1971) and found it related to package protection. Nikkila and Linko (1956) state that desiccation of the surface of fish flesh may lead to denaturation of myosin.

Thaw Indicators.—Retention of nutrients in frozen foods depends foremost on control of temperature. As mentioned previously, packaging can have only a limited role in maintaining a low temperature in frozen foods stored for prolonged periods of time and external cooling is usually required. Another aspect of packaging interaction with temperature control is the incorporation in the package of "thaw indicators." While the idea of indicators of defrosting is old, there have been in this area significant new technical developments. A thorough survey of available indicators was made recently by Schoen and Byrne (1972) and Hayakawa (1972). There is at the moment, however, no general agreement on the feasibility and desirability of widespread inclusion of thaw indicators in—or on—commercial packages of frozen foods (Semling 1967).

QUANTITATIVE ANALYSIS OF NUTRIENT RETENTION DURING PROCESSING AND STORAGE OF PACKAGED FOODS

General Approach

The quality and wholesomeness of foods as they reach the consumer depend not only on the composition of the raw materials but also on various changes occurring during processing, storage, and distribution. The natural environment of the atmosphere is deleterious to long-time storage, since many of its components promote deteriorative reactions of foods. The processing measures themselves often have an undesirable effect on nutrient retention, which is, however, justified by the need to accomplish the primary objective of the process which is the elimination of danger of microbial spoilage.

The deterioration during storage may often be minimized by protective packaging. The extent to which a package renders protection to the product is determined in the first place by its ability to act as a barrier between the external environment and the internal environment with which the food is in contact.

Given the great complexity of factors involved in the package-product interactions, the choice of proper packaging methods for a given product, and prediction of effectiveness of a given packaging system have been determined either by extended storage tests, or by judicious guessing. This approach, however, has substantial economic penalties, and severely limits freedom of choice resulting usually in over-packaging. This is not only wasteful, but is a severe impediment to development of entirely new packaging concepts, of which those aiming at reduction of environmental pollution through disposable and biodegradable packaging are particularly timely.

Recently it has become possible to apply a scientific analysis of mechanisms of food deterioration and in particular of mechanisms of nutrient degradation to develop techniques by which laboratory tests on food systems and on packaging materials may be used to predict shelf-life of a particular food-package combination; or, conversely, by which protective packaging requirements for a given food may be calculated if the needed shelf-life is known. These techniques are based on combination of kinetic parameters of food deterioration reactions with mass transfer properties of the packaging materials, and result in analytical or numerical solutions useful for prediction of the desired quantities either with or without computer assistance.

The work on prediction of storage life as a function of efficiency of packaging measures and of environmental storage conditions was initiated on a large scale following World War II. Initial work was stymied by the complexity of the factors involved in this problem. The general conclusion was often voiced that the complexity of these factors precluded laboratory prediction of shelf-life. Preliminary evaluation of packaging problems on the basis of laboratory tests, however, can greatly simplify the overall task of eliminating obvious failures, and pinpointing likely successes. Such an analysis requires that answers be obtained to the following questions:

(1) What are the optimal conditions for storage of the particular food product in terms of the major significant environmental factors?
(2) What are the expected external environmental conditions to which the package is likely to be exposed?
(3) What barrier properties will be required in order to maintain the internal environment at the desired optimal conditions?

In a series of studies reviewed recently by Karel (1973C), an approach was taken based on the following assumptions:

(A) Properties of food which determine quality (F) depend on the initial condition of the food (F_0) and on reactions which change these properties with time (t); these reactions, in turn, depend on internal environment of the package (I).

It is assumed that deteriorative mechanisms limiting shelf-life and their dependence on environmental parameters (oxygen pressure, water activity, temperature) can be described by a mathematical (though not necessarily analytical) function. As an example, consider the case shown in Equation 3, which describes a deterioration dependent on the three above-mentioned parameters.

$$\frac{dD}{dt} = f(RH, P_{0_2}, T) \quad (3)$$

where

D = deteriorative index
t = time

RH = equilibrium relative humidity
P_{O_2} = oxygen pressure
T = temperature

(B) Maximum acceptable deterioration level D_{MAX} can be determined by correlating objective tests of deterioration with organoleptic or toxicological parameters.

(C) The internal environment (I) depends on the condition of the food (F), on package properties (B) and on external environment (EO). It is assumed that changes in environmental parameters can be related to food and package properties. For instance,

$$a = f(a_0, t, RH, k_1 \ldots k_n, T \ldots) \tag{4}$$

where

a = water activity in the food at any time
a_0 = initial a
$k_1 \ldots k_n$ = constants characterizing sorptive and diffusional properties of food and of package

(D) Barrier properties of the package can, in turn, be related to internal and external environments (I and EO).

(E) The various equations can be combined and solved with or without the aid of a digital computer. The solutions predict storage life, or required package properties for a given storage.

Moisture-Sensitive Foods

Many of the nutrient deterioration reactions depend on moisture content. Two types of dependence can be distinguished:

(1) Those reactions for which a definite critical moisture content can be established, below which the rate of spoilage is insignificant. Among nutrient deterioration mechanisms following this pattern are nutrient losses due to enzyme reactions (Acker 1970) and those due to growth of molds and of bacteria (Scott 1957; Troller 1973).

(2) Reactions which proceed at all moisture contents, but whose rate depends strongly on the moisture content. Many of the nutrient deterioration mechanisms follow this pattern including nonenzymatic browning which impairs availability of lysine and other amino acids, and may have even wider detrimental nutritional effects (Patton 1955; Osner and Johnson 1968). Nonenzymatic browning depends on water content in a complex manner with a maximum reaction rate occurring at intermediate moisture contents (Eichner and Karel 1972; Karel and Nickerson 1964; Loncin et al. 1968). The rate of loss of ascorbic acid in dehydrated food also varies strongly with moisture (Karel and Nickerson 1964; Krebes and Behun 1963).

The prediction of moisture protection required has been achieved for the first case (that of a single critical moisture content) by a number of workers. Oswin (1954), Paine (1963), and Heiss (1956), for instance, have analyzed this case on the basis of simplified assumptions including a linear moisture isotherm and a package permeability which is independent of relative humidity. More recently, significant progress has been made in developing more realistic approximations of shelf-life for products having a single critical moisture content, and even, more importantly, in analyzing more complicated types of dependence on moisture content. In a study conducted in connection with space-food requirements, fairly accurate storage life predictions were attained for several sensitive rations (Karel and Labuza 1969).

The study then concentrated on the detailed analysis of one system, freeze-dried cabbage. The mathematical model was based on a combination of kinetic data for the browning reaction, the sorption properties of the cabbage, and the permeability characteristics of the packages. Thus, the following functions had to be developed:

(A) A function relating extent of browning to time of storage and to moisture content.
(B) A function relating moisture content within the food to partial pressure of water (sorption isotherm).
(C) Function(s) relating change of moisture content in the samples to properties of the package, the food, and the environment.

It was then determined that simple functions relating to the above parameters could be used to give excellent predictions of browning in packaged cabbage stored at $37°C$ (Mizrahi et al. 1970A). In a subsequent study, it was demonstrated that tests at elevated temperatures and moisture contents could be used to obtain the data necessary for the prediction of storage life of dehydrated cabbage, reducing the time required to evaluate package requirements from 1 yr to 2 weeks (Mizrahi et al. 1970B).

The above studies were made on packaging materials with constant permeabilities. Following this, packaging materials whose water-vapor permeability depended on relative humidity were incorporated into the analysis. Under these conditions the rate of water transfer changes continuously during storage, not only because the driving force changes as the food absorbs water, but also because the resistance of the container alters with changes in the relative humidity in equilibrium with the food inside the package. Incorporating these complications, excellent correlations were obtained between prediction and experiment, using dehydrated cabbage stored in polyamide (nylon) pouches (Fig. 15.6). Good results were obtained also in

452 NUTRITIONAL EVALUATION OF FOOD PROCESSING

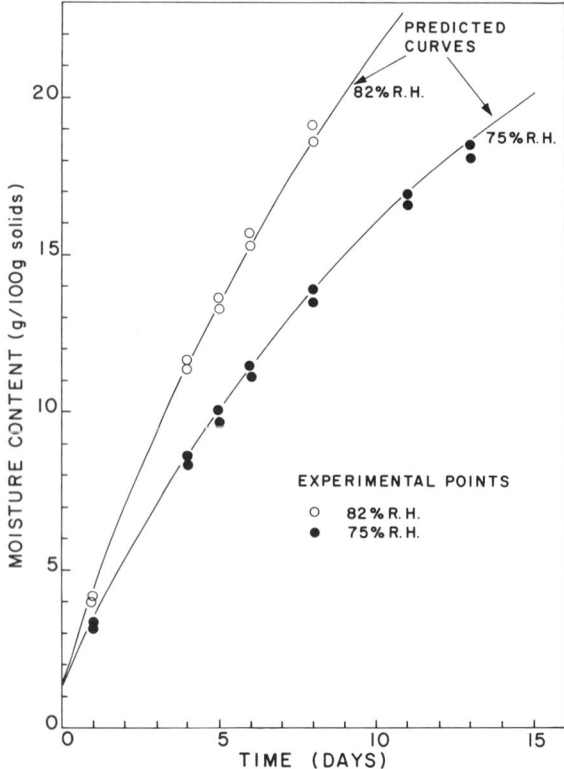

FIG. 15.6. PREDICTED AND ACTUAL INCREASE IN MOISTURE CONTENT OF CABBAGE PACKAGED IN POLYAMIDE POUCHES

applying this analysis to hydrophilic films coated with polyethylene (Karel et al. 1971).

Accelerated Testing

As mentioned previously, accelerated tests are necessary in order to allow the development of valid storage life predictions within a reasonable time span. The work of Mizrahi et al. (1970B) has already been alluded to, in which acceleration was achieved by raising temperature and by increasing moisture content. Other workers have made significant progress in utilizing acceleration due to increased temperature. Most base their calculations on the assumption that the Arrhenius equation is applicable.

Charm et al. (1972) have used the Arrhenius equation and organoleptic data to estimate refrigeration shelf-life of fish. Hu (1972) has used an oxidizable chemical system inside a package, and assumed validity of Arrhenius equation in shelf-life estimation of products packaged in flexible pouches. Wanninger (1972) used the dependence of ascorbic acid decomposition on moisture and temperature in a simplified model predicting stability of this vitamin. Kwolek and Bookwalter (1971) attempted to describe the storage stability dependence on temperature in terms of several simple models.

There have also been some good studies on accelerated tests based on temperature exaggeration in the area of stability of pharmaceuticals, particularly by Kay and Simon (1971), Greiff and Greiff (1972), and Zoglio et al. (1968)—all of whom have developed techniques for nonisothermal stability studies, in which temperature is changed during the accelerated test in some predicted manner (often linearly with time) and the kinetic constants are extracted from the test results with the aid of suitable mathematical techniques.

With respect to nutrients, Labuza (1972) reviewed the situation in analysis of losses in dehydrated foods, and points out scarcity of data. Even in situations in which the water content remains constant and the only factor is changing temperature, the development of nutrient retention prediction has only recently begun to make progress (Teixera et al. 1969).

Oxygen-Sensitive Foods

Another packaging problem is the control of oxygen transfer for the protection of food containing unsaturated fatty acids. These foods undergo oxidative reactions the rate of which is dependent on the oxygen concentration in the package. As mentioned previously, among the oxygen-sensitive nutrients are several vitamins, all of the essential fatty acids, and many amino acid residues.

Past studies on measurement of food storage life for foods deteriorating through oxidation have been limited to packaging the foods in systems allowing large differences in oxygen and water transport through the package. The foods were stored at different external conditions and tested periodically either for chemical deterioration, for organoleptic acceptability, or for nutritional value.

Recent research improved this trial-and-error method significantly through the use of mathematical models and of computers. In studying oxidation in potato chips, these researchers were able to formulate quite accurate predictions of storage life and of required packaging protection (Quast et al. 1972; Quast and Karel, 1972A, B, 1973).

It is their belief that laboratory tests combined with mathematical modeling will allow scientists in the future to greatly improve the calculation or required packaging protection for nutrients in packaged foods.

EFFECTS OF PACKAGING ON DELIVERY OF NUTRIENTS TO THE CONSUMER

The 20th Century has been characterized by the application of technological innovations which have revolutionized industrial societies. Modern food packaging is one of these technological innovations. Man's newly gained ability to store and distribute a wide variety of food in mass quantities has enabled such major phenomena as: the development of great urban centers, the distribution of food to areas of mass food shortage, the transport of food to support large scale military operations, and the ability of man to enjoy dinner on the moon. Today's social complexity, which is largely taken for granted, would be impossible utilizing food packaging as practiced 50 yr ago.

Recently, there has been an upsurge in romanticizing the "good old days" along with "old fashioned" or "natural foods." In fact, however, the average individual in those "good old days" was significantly more poorly nourished compared to average individuals in societies applying modern food technology. Reasoned analyses of the total food system impact of modern food packaging reveal that man is better nourished whenever this technology is properly exploited. The romanticized "good old days," when examined unemotionally, reveal life in former times to have been harsh, sordid, and violent compared to modern times.

There are many instances of deterioration of individual nutrients in foods packaged and processed under modern methods. Typical instances of losses due to package types have been identified in this chapter. The reader must guard against oversimplified conclusions based upon such data. To properly evaluate the significance of these nutrient losses, we must critically examine the entire food system. Individual events within segments of the food system do not necessarily portray the total reality. Specifically, the significance of nutrient losses reported in this chapter must consider that in many instances in which nutrients degraded there would have been a total loss of the food due to its becoming inedible if modern packaging techniques had not been applied. Claims that modern food packages somehow reduce the "natural" benefits of foods cannot be substantiated when they are placed in the context of the total food system in a modern society. Such fears are promulgated by individuals who have not performed an analysis of the entire food system. It should be noted that the relationship between loss of nutrients and type of food package was scientifically investigated long before the recent outcries for "health foods" and "natural foods." Scientists can be credited with the insight to examine foods for nutritional factors of packaging long before such concern became popularized. These same scientists, however, can be criticized for not relating to the consumer the total system's impact of their findings. Fortunately, recent publications have begun to emphasize these types of factors (Birch et al. 1972; Margolis 1973).

The kinetics and mechanisms of nutritional degradations in foods as a function of the package need to be better understood. Such data are needed so that systems for preservation of nutrients can be optimized, and so that practicing nutritionists and dietitians can implement proper countermeasures. Nobile and Woodhill (1973) have shown that selected consumer groups in affluent societies can suffer subclinical nutritional deficiencies. In the problems of distributing food to areas of shortages, there is a need for better data regarding methods to preserve nutrients and deliver them to the ultimate consumer who often lives in a food distribution system quite foreign to that in which the food was produced. Specialized food delivery systems can be developed by exploiting the nutrient and quality preservation as well as the convenience and organoleptic enhancing capability of modern food packages. Examples of such applications in clinical situations have been discussed by Heidelbaugh et al. (1973). In these regards, it should be recalled that packaging is more than preservation of nutrients—it involves enhancement of food acceptance (hence nutritional status) by improved organoleptics, convenience, and economics.

The experience of the U.S. military regarding adequate use of food packaging has demonstrated shelf-life up to seven or more years accompanied by reliable preservation of nutrients (Cecil and Woodroof 1962). Perhaps the most dramatic application of specialized food packaging in recent times has been its application in space food systems. In these systems, packaging has been optimized to preserve and deliver nutrients (Smith et al. 1971; Huber et al. 1973). For example, in the Skylab program, the packaging is configured such that known quantities (within a tolerance of ± 2%) of protein, calcium, magnesium, sodium, and phosphorus are delivered to and consumed by the astronauts for periods up to 84 days (Heidelbaugh et al. 1973). These exacting nutritional standards have been achieved even though weightlessness imposed severe design constraints on the package (Rambaut et al. 1972). The success of the space food program in delivery of nutrients to its consumers is an example of what can be achieved in the delivery of highly acceptable and nutritious food under extremely constraining conditions. Astute application of the techniques of modern packaging could make comparable contributions in the solution to problems in all sectors of societies.

BIBLIOGRAPHY

ACKER, L. 1969. Enzymatic changes in low-moisture foods. Z. Ernährungswiss. Suppl 8, 45. (German)
ACKER, L. 1970. Water content and enzyme activity. Food Technol. 23, 1257.
ALMQUIST, H. J. 1956. Changes in fat extractability and protein digestibility in fish meal during storage. Agr. Food Chem. 4, 638.
ANON. 1955. Am. Inst. Baking Spec. Bull. 82, June.
ANON. 1972. Recommendation for the Processing and Handling of Frozen Foods. Institut International du Froid, 75 Paris 17^e, France.
BADRAN, A. M., WOODRUFF, R. E., and WILSON, L. G. 1969. Method of packaging ripening perishable plant foods, to prolong storage life. U.S. Pat. 3,450,544, June 17.
BARNICOAT, C. R. 1947. Experiments on the packing and storage of butter. VI. The effect on keeping quality of butter on exposure to light during manufacture. New Zealand J. Sci. Technol. 29A, 185.
BENGTSSON, O., and BENGTSSON, N. E. 1968. Freeze-drying of raw beef. III. Influence of some storage variables. J. Sci. Food Agr. 19, 486.
BILLMEYER, F. W., JR. 1971. Textbook of Polymer Science, 2nd Edition. John Wiley & Sons, New York.
BIRCH, G., GREEN, L. F., and PLASKETT, L. J. 1972. Health and Food. John Wiley & Sons, New York; Toronto, Canada.
BIRDSALL, J. J., and TEPLY, L. J. 1957. Effects of packaging on partially baked rolls: observations on the retention of riboflavin, moisture and flavor. Food Technol. 11, 608.
BISHOV, S. J. et al. 1971. Quality and stability of some freeze-dried foods in "zero" oxygen headspace. J. Food Sci. 36, 532.
BOLIN, H. R., NURY, F. S., and BLOCH, F. 1964. Effect of light on processed dried fruits. Food Technol. 18, 151.
BOND, T. 1973. Temperature-controlled shippers. Mod. Packaging 46, No. 7, 47.
BRODY, A. 1971. Food canning in rigid and flexible packages. CRC Critical Rev. Food Technol. 2, No. 2, 187.
BUFFA, A., CAPUANO, G., and AMBANELLI, G. 1955. Technological observations on the behavior of preserved fruit juices in relation to the nature of the containers. Ind. Conserve (Parma) 30, 20.
CASEY, J. P. 1952. Pulp and Paper—Chemistry and Chemical Technology, Vol. 1 and 2. Interscience Publishers, New York.
CECIL, S. R., and WOODROOF, J. G. 1962. Long Term Storage of Military Rations. Quartermaster Research and Engineering Command, Quartermaster Food and Container Institute for the Armed Forces, Chicago.
CHARM, S. E., LEARSON, R. J., RONSIVALLI, L. R., and SCHWARTZ, M. 1972. Organoleptic technique predicts refrigeration shelf life of fish. Food Technol. 26, No. 7, 65.
CHIO, K. S., and TAPPEL, A. L. 1969. Inactivation of ribonuclease and other enzymes by peroxidizing lipids and by malonaldehyde. Biochemistry 8, 2827.
CHRISTIAN, J. H. B. 1963. Water activity and the growth of microorganisms. In Recent Advances in Food Science, Vol. 3. J. M. Leitch and D. N. Rhodes (Editors). Butterworths, London, England.
CLIFCORN, L. E. 1948. Factors influencing vitamin content of canned foods. Advan. Food Res. 1, 42.
CONTINENTAL CAN CO. (Research Staff). 1944-1945. New facts about packaging and storing dehydrated food. Food Ind. 16, 171, 267, 366, 458, 635, 702, 815, 903, 991; 17, 147.
COOK, H. T., PARSONS, C. S., and McCOLLOCH, L. P. 1958. Methods to extend storage of fresh vegetable aboard ships of the U.S. Navy. Food Technol. 12, 548.

COOKE, N. A. 1955. Know your packaging materials. Am. Management Assoc. Packaging Ser. *46*, No. 3.
COPLEY, M. J., KAUFMAN, V. F., and RASMUSSEN, C. L. 1956. Recent developments in fruit and vegetable powder technology. Food Technol. *10*, 589.
COX, D. H., COULTER, S. T., and LUNDBERG, W. O. 1957. Effect of NDGA and other factors on stability of added vitamin A in dry and fluid milks. J. Dairy Sci. *40*, 564.
DE LANGE, P. 1953. Nutritive value of milk in storage. Voeding *14*, 16.
DOWNEY, W. K., and MURPHY, M. F. 1968. Light-barrier properties of various butter wrapping materials. J. Soc. Dairy Technol. *21*, No. 2, 104.
DUNKER, C. F., and HANKINS, O. G. 1953. Rates of freezing and thawing meats. Food Technol. *7*, 505.
DUNKLEY, W. L., FRANKLIN, J. D., and PANGBORN, R. M. 1962. Effects of fluorescent light on flavor, ascorbic acid and riboflavin in milk. Food Technol. *16*, No. 9, 112.
EICHNER, K., and KAREL, M. 1972. The influence of water content and water activity on the sugar-amino browning in model systems under various conditions. J. Agr. Food Chem. *20*, 218.
ELLICKSON, B. E., and HASENZAHL, V. 1958. Use of a light screening agent for retarding oxidation of process cheese. Food Technol. *12*, 577.
EVANS, C. D., LIST, G. R., MOSER, H. A., and COWAN, J. C. 1973. Long term storage of soybean and cottonseed salad oils. J. Am. Oil Chemists' Soc. *50*, 218.
FIGGE, K. 1972. Migration of additives from plastic films into edible oils and fat simulants. Food Cosmet. Toxicol. *10*, 815.
FRAWLEY, J. P. 1967. Scientific evidence and common sense as a basis for food packaging regulations. Food Cosmet. Toxicol. *5*, 293.
GARLOCK, E. A., and PAYNTER, O. E. 1965. Determination of extractable substances in food packaging materials. Residue Rev. *10*, 40.
GARTY, O. M. 1972. Improved test for PCB's. Mod. Packaging *45*, No. 8, 46.
GILCHRIST, M. R., VIJAY, I. K., and HUMBERT, E. S. 1968. Factors affecting suborganoletpic levels of light-induced oxidation. Can. Inst. Food Technol. J. *1*, No. 4, 133.
GOLDBLITH, S. A., and TANNENBAUM, S. R. 1966. The nutritional aspects of the freeze drying of foods. Proc. 7th International Congress of Nutrition, Vol. 4. Vieweg & Son, Hamburg, Germany.
GREIFF, D., and GREIFF, C. 1972. Linear nonisothermal, single step, stability studies of dried preparations of influenza virus. Cryobiology *9*, 34.
GRIFFIN, R. C., and SACHAROW, S. 1972. Principles of Package Development. Avi Publishing Co., Westport, Conn.
GUADAGNI, D. G., and NIMMO, C. C. 1957. The time-temperature tolerance of frozen foods. III. Effectiveness of vacuum, oxygen removal, and mild heat in controlling browning in frozen peaches. Food Technol. *11*, 43.
GUERRANT, N. B., and O'HARA, M. B. 1953. Vitamin retention in peas and lima beans after blanching, freezing, processing in tin and in glass, after storage, and after cooking. Food Technol. *7*, 473.
GUTSCHMIDT, J. 1952. The influence of the can material on the ascorbic acid content of fruits and vegetables. Ind. Obst- /Gemüseverwert. *37*, 325.
GUTSCHMIDT, J., and WOLODKEWITSCH, N. 1950. Effect of drying on frozen fruits and vegetables. Kaeltetechnik *2*, 49.
HANSON, H. L., FLETCHER, L. R., and LINEWEAVER, H. 1959. Time-temperature tolerance of frozen foods. XVII. Frozen fried chicken. Food Technol. *13*, 221.
HANSON, H. L., WINEGARDEN, H. M., HORTON, M. B., and LINEWEAVER, H. 1950. Preparation and storage of frozen cooked poultry and vegetables. Food Technol. *4*, 430.
HARRIS, R. S., and MOSHER, L. M. 1941. Effect of reduced evaporation on

the provitamin A content of lettuce in refrigerated storage. Food Res. 6, 387.
HARTMANN, G. 1972. Behavior of hard cheese and sliced cheese portions in flexible packages. Verpackungs Rundschau. Reprint from issue 6, 1972. (German)
HAYAKAWA, K. 1972. An evaluation of frozen food indicators now on the market. Quick Frozen Foods 35, No. 4, 72.
HEERDT, M., and STANSBY, M. E. 1955. Freezing and cold storage of Pacific Northwest fish and shellfish. Com. Fisheries Rev. 17, No. 3, 13.
HEIDELBAUGH, N. D., and KAREL, M. 1970. Changes in pouched heat-processed foods. Mod. Packaging 43, No. 11, 80.
HEIDELBAUGH, N. D., YEH, C. P., and KAREL, M. 1971. Effects of model system composition on autoxidation of methyl linoleate. Agr. Food Chem. 19, No. 1, 140.
HEIDELBAUGH, N. D. et al. 1973. Clinical nutrition applications of space food technology. J. Am. Dietet. Assoc. 62, No. 4, 383.
HEISS, R. 1956. Packaging of Moisture-Sensitive Products. Springer-Verlag, New York; Berlin, Germany. (German)
HEISS, R. 1970. Principles of Food Packaging. An International Guide. P. Keppler Verlag. KG, Heusenstamm, Germany.
HENIG, Y., and GILBERT, S. G. 1972. Analysis of the variables affecting the internal atmosphere in packaged products. 32nd Ann. Meeting, Inst. Food Technologists, Minneapolis, May 21–25, 1972. Food Technol. Abstr. 207.
HODSON, A. Z., and KRUEGER, G. M. 1947. Changes in the essential amino acid content of the proteins of dry skim milk on prolonged storage. Arch. Biochem. 12, 51.
HU, K. H. 1972. Time-temperature indicating systems "write" status of product shelf life. Food Technol. 26, No. 8, 56.
HU, K. H., NELSON, A. I., LEGAULT, R. R., and STEINBERG, M. P. 1955. Feasibility of using plastic film packages for heat processed foods. Food Technol. 9, 236.
HUBER, C. S., HEIDELBAUGH, N. D., RAPP, R. M., and SMITH, M. C. 1973. Nutrition systems for pressure suits. Aerospace Med. 44, No. 8, 905.
INSTITUTE OF PACKAGING. 1971. Food packaging and health: migration and legislation. Conf. Proc. Inst. Packaging, London, England.
JOSLYN, M. A., and HOHL, L. A. 1948. The commercial freezing of fruit products. Calif. Agr. Expt. Sta. Bull. 703.
JURIN, V., and KAREL, M. 1963. Studies on control of respiration of McIntosh apples by packaging methods. Food Technol. 17, No. 6, 104.
KAESS, G., and WEIDEMANN, J. F. 1971. On the formation of freezer burn in liver tissue protected with plastic film. J. Food Sci. 36, 1135.
KANNINEN, W. H., and HARDY, W. L. 1955. The retention of riboflavin in commercially wrapped enriched breads. Proc. Inst. Food Technologists 15th Ann. Meeting, Columbus, Ohio, June 12–16.
KAREL, M. 1956. Protective packaging of frozen foods. Quick Frozen Foods 19, No. 1, 201.
KAREL, M. 1973A. Protein-lipid interactions. J. Food Sci. 38, 756.
KAREL, M. 1973B. Recent research and development in the field of low-moisture and intermediate-moisture foods. CRC Critical Rev. Food Technol. 3, 329.
KAREL, M. 1973C. Quantitative analysis of food packaging and storage problems. AIChE Symp. Ser. 69, No. 132, 107.
KAREL, M., and CORNELL, A. 1956. The effect of sunlight on milk packaged in glass bottles. MIT Dept. Food Technol. (Unpublished rept.)
KAREL, M., and GO, J. 1964. Control of respiration of bananas by packaging methods. Mod. Packaging 37, No. 6, 123.
KAREL, M., and LABUZA, T. P. 1969. Optimization of protective packaging

of space foods. MIT Rept. Contract Res. Project *F41-609-68-C-0015* with Aerospace Med. Div., U.S. Air Force.
KAREL, M., MIZRAHI, S., and LABUZA, T. P. 1971. Computer prediction of food storage. Mod. Packaging *44*, No. 8, 54.
KAREL, M., and NICKERSON, J. T. R. 1964. Effects of relative humidity, air, and vacuum on browning of dehydrated orange juice. Food Technol. *18*, 104.
KAREL, M., and WOGAN, G. 1963. Migration of substances from flexible containers for heat-processed foods. Rept. Contract Res. U.S. Army Natick Laboratories. Contract *DA 19-129-OM-2080*. MIT, Cambridge, Mass.
KAY, A. I., and SIMON, T. H. 1971. Use of analog computer to simulate and interpret data obtained from linear nonisothermal stability studies. J. Pharm. Sci. *60*, 205.
KEFFORD, J. F., McKENZIE, H. A., and THOMPSON, P. C. O. 1959. Effects of oxygen on quality and ascorbic acid retention in canned and frozen orange juices. J. Sci. Food Agr. *1*, 51.
KIERMEIER, F. 1969. On the use of plastics in the dairy industry. Deut. Molkerei-Z. *90*, 2197. (German)
KIERMEIER, F., and WAIBLINGER, W. 1969. Influence of light, especially from fluorescent lamps, on the content of vitamins C and B_2 in polyethylene-packed milk. Z. Lebensm. Untersuch. Forsch. *141*, 320. (German)
KIERMEIER, F., and WOLFSEDER, H. 1972. On the behavior of cheese in the package. Z. Lebensm. Untersuch. Forsch. *149*, 218. (German)
KLOSE, A. A., POOL, M. F., CAMPBELL, A. A., and HANSON, H. L. 1959. Time-temperature tolerance of frozen foods. XIX. Ready-to-cook cut-up chicken. Food Technol. *13*, 479.
KON, S. K., and THOMPSON, S. Y. 1953. The effect of light on the riboflavin in milk. Proc. 13th Intern. Dairy Congr. *2*, 363.
KOZLOVA, L. I., and NEKRASOV, B. P. 1956. Changes in wheat quality during prolonged storage. Chem. Abstr. *55*, 6715b.
KREBES, T., and BEHUN, M. 1963. Quality and manufacture of freeze dried fruit nectars. Proc. 11th Intern. Congr. Refrig., Vol. 11, 1597.
KRISTOFFERSEN, T., STUSSI, D. B., and GOULD, I. A. 1964. Consumer packaged cheese. I. Flavor stability. J. Dairy Sci. *47*, 496.
KROGER, M., and IGOE, R. S. 1971. Edible containers. Food Prod. Develop. *5*, No. 7, 74.
KUPRIANOFF, J., and GUTSCHMIDT, J. 1951. Suitability of white- and black-plated cans for the conservation of fruit and vegetables and meat. Ind. Obst- /Gemüseverwert. *36*, 131. (German)
KWOLEK, W. F., and BOOKWALTER, G. N. 1971. Predicting storage stability from time-temperature data. Food Technol. *22*, 263.
LABUZA, T. P. 1971. Kinetics of lipid oxidation in foods. CRC Critical Rev. Food Technol. *2*, 355.
LABUZA, T. P. 1972. Nutrient losses during drying and storage of dehydrated foods. CRC Critical Rev. Food Technol. *3*, No. 2, 217.
LEGAULT, R. R., HENDEL, C. E., and TALBURT, W. F. 1954. Retention of quality in dehydrated vegetables through in-package-desiccation. Food Technol. *8*, 143.
LENTZ, C. P., and VAN DEN BERG, L. 1957. Liquid immersion freezing of poultry. Food Technol. *11*, 247.
LONCIN, M., BIMBENET, J. J., and LENGES, J. 1968. Influence of activity of water on spoilage of foodstuffs. Food Technol. *3*, 131.
LOY, H. W., JR., HAGGERTY, J. F., and COMBS, E. L. 1951. Light destruction of riboflavin in bakery products. Food Res. *16*, 360.
MacDOWELL, A. K. R., and MacDOWELL, F. H. 1953. The vitamin A potency of New Zealand butter. J. Dairy Res. *20*, 76.
MAPSON, L. W. 1952. Factors in distribution affecting the quality and nutri-

tive value of foodstuffs—loss of nutrients during the transport and distribution of fruits and vegetables. Chem. Ind. (London), 25.
MARCELLIN, P. 1971. Preservation of fruits in an atmosphere controlled by means of plastic packages. Chim. Ind. Genie Chim., *104*, 2141. (French)
MARGOLIS, S. 1973. Health Foods Facts and Fakes. Walker & Company, New York.
McKEE, A. B., and STROHM, D. B. 1961. Aluminum foil's thermal properties offer packaging advantages. Package Eng. *6*, No. 10, 76.
MELNICK, D., LUFKMANN, F. H., and VAHLTEICH, H. W. 1953. Retention of preformed vitamin A and carotene in margarine based upon physicochemical assays. Food Res. *18*, 504.
MILLER, E. V., and TINGLEFF, A. J. 1960. Retention of ascorbic acid in frozen orange juice. Proc. Penn. Acad. Sci. *34*, 17.
MIZRAHI, S., LABUZA, T. P., and KAREL, M. 1970A. Computer-aided predictions of browning in dehydrated cabbage. J. Food Sci. *35*, 799.
MIZRAHI, S., LABUZA, T. P., and KAREL, M. 1970B. Feasibility of accelerated tests for browning in dehydrated cabbage. J. Food Sci. *35*, 804.
MODERN PACKAGING ENCYCLOPEDIA. 1972. McGraw Hill Book Co., New York.
MOHR, W. 1950. The preservation of vitamins in canned foods. Ind. Obst-/Gemüseverwert. *35*, 107.
MORGAREIDGE, K. 1956. The effect of light on vitamin retention in enriched white bread. Cereal Chem. *33*, 213.
MORGAN, B. H. 1971. Edible packaging update. Food Prod. Develop. *5*, No. 4, 75.
MUMMERY, W. R., MacDOWELL, F. H., and MacDOWELL, A. K. R. 1950. Quality of vegetable parchment wrappers in relation to the quality of butter. New Zealand J. Sci. *32A*, No. 3, 1.
NICKERSON, J. T. R., and KAREL, M. 1964. Preservation of food by freezing. In Food Processing Operations, Vol. III. M. A. Joslyn, and J. L. Heid (Editors). Avi Publishing Co., Westport, Conn.
NIKKILA, D. E., and LINKO, R. R. 1956. Freezing, packaging and frozen storage of fish. Food Res. *21*, 42.
NOBILE, S., and WOODHILL, J. M. 1973. A survey of the vitamin content of some 2000 foods as they are consumed by selected groups of the Australian population. Food Technol. Australia *22*, No. 2, 2.
NORDLUND, J., JUNKKARIEN, L., and KREULA, M. 1970. The effect of the packaging method on the quality of beverage milk. XVIII Intern. Dairy Congr. *1E*, 167.
OFFERGAARD, E., and NORDNESS, T. 1958. Retention of ascorbic acid in frozen vegetables. Experiments with different scalding periods and wrapping materials. Nutr. Abstr. *28*, 1093, No. 5123.
OLCOTT, H. S. 1952. Surv. Progr. Military Subsistence Probl. Quartermaster Food Container Inst., Ser. *I*, No. 1, 51.
OLSEN, A. L., WEYBREW, J. A., and CONRAD, R. M. 1948. Thiamine stability in spray-dried whole egg. Food Res. *13*, 184.
OSNER, R. C., and JOHNSON, R. M. 1968. Nutritional changes in proteins during heat processing. Food Technol. *3*, 81.
OSWIN, C. R. 1954. Protective Wrappings. Cam Publishers, London, England.
PAINE, F. A. 1963. Fundamentals of Packaging. Blackie & Sons, London, England.
PAINE, F. A. 1967. Packaging Materials and Containers. Blackie & Sons, London, England.
PALMER, A. Z., BRADY, D. E., NAUMANN, H. D., and TUCKER, L. N. 1953. Deterioration in frozen pork as related to fat composition and storage treatments. Food Technol. *7*, 90.
PATTON, S. 1954. The mechanism of sunlight-flavor formation in milk with special reference to methionine and riboflavin. J. Dairy Sci. *37*, 446.

PATTON, S. 1955. Browning and associated changes in milk and its products: A review. J. Dairy Sci. *38*, 457.
PEDERSON, C. S., and ROBINSON, W. B. 1952. The quality of sauerkraut preserved in tin and in glass. Food Tech. 7, 46.
PONT, E. G. 1956. Milk—protection from sunlight. Food Manuf. *31*, No. 12, 523.
QUAST, D. G., and KAREL, M. 1972A. Effects of environmental factors on the oxidation of potato chips. J. Food Sci. *37*, 584.
QUAST, D. G., and KAREL, M. 1972B. Computer simulation of storage life of foods undergoing spoilage by two interacting mechanisms. J. Food Sci. *37*, 679.
QUAST, D. G., and KAREL, M. 1973. Simulating shelf life. Mod. Packaging *46*, No. 3, 50.
QUAST, D. G., KAREL, M., and RAND, W. M. 1972. Development of a mathematical model for oxidation of potato chips as a function of oxygen pressure, extent of oxidation and equilibrium relative humidity. J. Food Sci. *37*, 673.
RADTKE, R., SMITHS, P., and HEISS, R. 1970. The influence of light of different intensity and wavelength regions on the oxidative deterioration of edible oils. II. Results and discussion. Fette, Seifen, Anstrichmittel *72*, 497.
RAMBAUT, P. C. et al. 1972. Some flow properties of foods in null gravity. Food Technol. *26*, No. 1, 58.
RANSEY, L. L. 1967. Analytical methods and the preclearance food protection amendments to the FD&C Act. JAOAC *50*, 1014.
RATZLAFF, A. 1955. Observations on temperature changes in pasteurized milk during bottling, storage and distribution. J. Milk Food Technol. *18*, 195.
REGIER, L. W., and TAPPEL, A. L. 1956. Freeze dried meat. III. Nonoxidative deterioration of freeze dried meat. Food Res. *21*, 630.
REUSSER, P. 1967. On the effect of riboflavin photolysis on biotin, and its prevention by sodium thiosulfate in the biotin determination in multiple vitamin preparations. Z. Anal. Chem. *231*, 272. (German)
REUSSER, P. 1969. On the photochemical inactivation of chlortetracycline by riboflavin and its prevention by ascorbic acid and sodium thiosulfate. Z. Anal. Chem. *248*, 317. (German)
REUSSER, P. 1970. Study on the photochemical inactivation of folic acid in the presence of riboflavin and its inhibition by ascorbic acid. J. Intern. Vitaminolog. *39*, 64. (French)
ROSENFIELD, D. 1971. Process for lengthening the post-harvest time of certain perishable fruits and vegetables. Can. Pat. 868 298, Apr. 13.
ROY, R. B., and KAREL, M. 1973. Reaction products of histidine with peroxidizing methyl linoleate. J. Food Sci. *38*, 869.
SACHAROW, S. 1969. Light as a factor in food deterioration. Food Prod. Develop. *3*, No. 5, 67.
SALUNKHE, D. K., and WU, M. T. 1973. Effects of low oxygen atmosphere storage on ripening and associated biochemical changes of tomato fruits. J. Am. Soc. Hort. Sci. *98*, No. 1, 12.
SATTAR, A., and DEMAN, J. M. 1973. Effect of packaging material on light induced quality deterioration of milk. Can. Inst. Food Sci. Technol. J. *6*, No. 3, 170.
SCHOEN, H. M., and BYRNE, C. H. 1972. Defrost indicators. Food Technol. *26*, No. 10, 46.
SCOTT, W. J. 1957. Water relations of food spoilage microorganisms. Advan. Food Res. *7*, 84.
SEDKY, A., STEIN, J. A., and WECKEL, K. G. 1952. Factors affecting color and flavor of sauerkraut packaged in Pliofilm bags. Food Technol. *6*, 377.
SEMLING, H. V., JR. 1967. Thaw indicators safeguard or nuisance? Food Process. Marketing *28*, No. 2, 16.

SHELOR, E., and WOODROOF, J. G. 1954. Frozen food containers. Food Technol. *8*, 490.
SMITH, A. C., and MacLEOD, P. 1955. The effect of artificial light on milk in cold storage. J. Dairy Sci. *38*, 870.
SMITH, H. W. 1963. The use of carbon dioxide in the transport and storage of fruits and vegetables. Advan. Food Res. *12*, 96.
SMITH, K. C., and HANAWALT, P. C. 1969. Molecular Photobiology. Academic Press, New York.
SMITH, M. C., HUBER, C. S., and HEIDELBAUGH, N. D. 1971. Apollo 14 food system. Aerospace Med. *42*, No. 11, 1185.
STEPHENS, L. C., and CHASTAIN, M. F. 1959. Light destruction of riboflavin in partially baked rolls. Food Technol. *13*, 527.
STRASHUN, S. I., and TALBURT, W. F. 1954. Stabilized orange juice powder. Food Technol. *8*, 40.
TANNENBAUM, S. R., BARTH, H., and LEROUX, J. P. 1969. Loss of methionine in casein during storage with autoxidizing methyl linoleate. J. Agr. Food Chem. *17*, 1353.
TAPPEL, A. L. 1956. Freeze dried meat. II. The mechanism of oxidative deterioration of dried beef. Food Res. *21*, 195.
TEIXERA, A. A., DIXON, J. R., ZAHRADNIK, J. W., and ZINSMEISTER, G. E. 1969. Computer optimization of nutrient retention in the thermal processing of conduction heated foods. J. Food Sci. *33*, 845.
THERIAULT, F. R., and FELLERS, C. R. 1942. Effect of freezing and canning in glass and in tin on available iron content of foods. Food Res. *7*, 503.
THOMPSON, J. B., and KOCHER, R. B. 1950. Trace metals in food packaging. Mod. Packaging *23*, No. 7, 119.
TOMITA, M., MASACHIKE, I., and UKITA, T. 1969. Sensitized photooxidation of histidine and its derivatives. Products and mechanism of the reaction. Biochemistry *8*, 5149.
TROLLER, J. 1973. The water relations of food-borne bacterial pathogens. A review. J. Milk Food Technol. *36*, No. 5, 276.
TROLLER, J. A. 1972. Effect of water activity on enterotoxin A production and growth of *Staphylococcus aureus*. Appl. Microbiol. *24*, 440.
VAN ARSDEL, W. B., COPLEY, M. J., and OLSON, R. L. 1969. Quality and stability of frozen foods. Time-temperature tolerance and its significance. Wiley-Interscience, New York.
VEERAJU, P., and KAREL, M. 1966. Control of atmosphere inside a fruit container. Mod. Packaging *40*, No. 2, 168.
VOLZ, F. E., GORTNER, W. A., and DELWICHE, C. V. 1949. The effect of desiccation on frozen vegetables. Food Technol. *3*, 307.
WANNINGER, L. A. 1972. Mathematical model predicts stability of ascorbic acid in food products. Food Technol. *26*, No. 6, 42.
WENDLAND, G. 1948. The influence of cans on the vitamin C content of fruits and vegetables. Z. Lebensm. Untersuch. Forsch. *88*, 618.
WHITMORE, R. A., SELIGSON, D., KRAYBILL, H. R., and WEBB, B. H. 1948. Packaging dehydrated meat. Food Res. *13*, 19.
WILDBRETT, G. 1967. Plastics in the milk industry—A critical study. III. Behavior of packaged dairy products and its dependence on permeability characteristics of plastic films. Fette, Seifen, Anstrichmittel *69*, 781. (German)
WONG, F. F., DIETRICH, W. C., HARRIS, J. G., and LINDQUIST, F. E. 1956. Effect of temperature and moisture on storage stability of vacuum dried tomato juice powder. Food Technol. *10*, 97.
WRIGHT, R. C., ROSE, D. H., and WHITEMAN, T. M. 1963. The commercial storage of fruits, vegetables, and florist and nursery stocks. USDA Agr. Handbook *66*.
ZOGLIO, M. A. *et al.* 1968. Linear nonisothermal stability studies. J. Pharm. Sci. *57*, 2080.

SECTION IV

Effects of Preparation and Service of Food on Nutrients

CHAPTER 16

Paul A. Lachance

Effects of Food Preparation Procedures on Nutrient Retention with Emphasis upon Food Service Practices

In the first edition, Fenton (1960) and Harris (1960) divided and discussed this topic as it pertained to foods of animal origin and foods of plant origin, respectively. Because the amount of new information is meager and more and more foods utilized in food service are convenience products which include both animal and plant matter, the subdivision is not being emphasized by separate subchapters in this edition. The important original data has been synopized and retained. New, although limited, information has been added and specific examples given in order to make interpretations more meaningful.

It is deplorable that the study of nutrient retention as affected by food service practices has received so little attention. Since systematic funding of research on the nutritive value of foods has never existed, it is understandable that this research would have a low priority relative to other food and nutrition programs. The fact that the 1973-74 FDA nutrition labeling regulations do not as yet apply to food service situations has removed a potentially significant stimulus to such research. However, if adopted, other federal proposals, such as nutrient standard menu planning in the school food service programs administered by the USDA Food and Nutrition Service, may serve to prod interest in nutrient composition and the effects of preparation practices on nutrient retention. In any event, it is very evident that food service as an industry is growing very rapidly. The importance of research on the retention of nutritive value in food service is of significant concern because the responsibility for food as the input to nutrition, and therefore, resultant health is not only that of the individual and the family but must include the management of food service operations.

THE AWAY-FROM-HOME EATING MARKET

The business of feeding the American public away from home is one that has been increasing since World War II. It is also increasing in other industrialized nations. Some of the reasons for this rapid increase are: (1) more women are working outside the home, with as many as 44% holding full-time occupations; (2) the American public with its higher disposable incomes and paid vacations is travel-conscious. The travel activity ranges from back packing and ownership of recreational vehicles to patronizing chains of motels and international air travel.

Burke (1952) estimated that in 1948 the retail value of civilian food eaten outside the home was 14.5 billion dollars or 29.6% of all food consumed exclusive of the Armed Services. According to McIntosh (1974) the revenue, including government food services is approaching 50 billion dollars or 41.7% of all food dollars expended with 80% being food service industry sales and 11% being government food services. The remainder includes hospital, institutional, and school food services. Table 16.1 provides an overview of the

TABLE 16.1

PUBLIC EATING ESTABLISHMENTS AND INSTITUTIONS WITH FOOD SERVICE: PERCENTAGE DISTRIBUTION OF FOODS RECEIVED, BY KIND OF BUSINESS, UNITED STATES (1969)

Kind of Business	Quantity of Food (%)	Cost Value of Food (%)
Separate eating places	57.7	59.0
Separate drinking places	4.0	4.6
Drug or proprietary stores	1.3	1.3
Retail stores	4.9	4.1
Hotels, motels, or tourist courts	5.0	5.9
Recreation or amusement places	3.0	3.3
Civic, social, or fraternal associations	0.5	0.6
Factories, plants, or mills	3.4	3.2
Other public eating places	2.2	2.3
Total public eating establishments	82.0	84.3
Hospitals	5.9	5.6
Sanatoria, convalescent or rest homes	1.3	1.0
Homes for children, handicapped, or mentally ill	2.1	1.5
Colleges, universities, professional or normal schools	6.8	6.0
Other institutions	1.9	1.6
Total institutions	18.0	15.7
Grand Total	100.0	100.0

Source: U.S. Dept. of Agr. (1971).

EFFECTS OF FOOD SERVICE PRACTICES ON NUTRIENTS 465

distribution of the away-from-home business. Food service comprises the fourth largest industry in the United States for sales volume, and is first among the service industries in the number of persons served by some 600,000 food service establishments.

In the spring 1965 USDA study of the food and nutrient intake of individuals in the United States (U.S. Dept. of Agr. 1972), 38% of all individuals reported having food or beverage away from home (not taken from home) on at least 1 daily occasion (18% at lunch time and 10% between 4:00-8:00 p.m.); 12% ate or drank 3 or more times per day away from home. Away-from-home eating or drinking or both accounted for 13% of the protein in the 1-day diets of all persons in the survey. It should be evident that away-from-home food consumption can have a significant bearing on the nutritive quality of the dietary as a whole.

A survey conducted by the USDA in cooperation with the food service industry in 1969 (U.S. Dept. of Agr. 1971) reveals that even though about 4000 foods are used in food service, as few as 40 account for the majority of the foods served. Table 16.2 lists the food categories and items with the highest percentage of usage. To date, the selection of foods for the study of the effects of preparation have had no evident pattern. This suggests that any future systematic research program should concentrate on those foods having the highest utilization in the food service marketplace.

Table 16.2 also provides a relative index of the ratio of foods from animal origin (41%) as compared to foods from plant origin (40%), with fats and oils (5%), sugars and sweets (7%), and soft drink and coffee (7%) being the remainder. Those foods which are best described (Lachance 1974) as partitioned—relatively pure chemical food ingredients such as sugar and oils—make up 12% of the total. At present, there is no reliable measure of the quantity of foods which could be labeled as fabricated foods, meat extenders, or formulated meal replacement products; but these products are now definitely established in the marketplace and all indications are that their lower cost and convenience will result in their increased use.

Fabricated foods such as soybean-based imitation meat products are used in Seventh Day Adventist hospitals but probably in few other large-scale food service operations. Meat extenders are used in increasing quantities, particularly in the involuntary food service operations of schools and institutions (Lachance 1972). Formulated meal replacement products are used in hospital situations and range from infant formulas to ready-to-eat formulas for adults based on whole protein, or amino acids, plus added vitamins, minerals, and energy sources. No information is available on the effect of prepara-

TABLE 16.2

PUBLIC EATING ESTABLISHMENTS AND INSTITUTIONS WITH FOOD SERVICE: PERCENTAGE DISTRIBUTION OF SELECTED FOOD ITEM/QUANTITIES RECEIVED, UNITED STATES (1969)

Food Product	Distribution (%)	Food Product	Distribution (%)
Dairy products and ices	18.8	Fish and shellfish	2.7
Fresh whole milk	52.2	Shrimp	17.7
Ice cream, bulk and		Haddock	10.0
packaged	8.6	Clams	9.6
Cheese	6.5	Lobster	8.6
Sweet cream	4.4	Cod	7.9
Chocolate milk	3.4	Sugar and sweets	6.7
Ice milk	3.3	Beverage fountain syrup	54.7
Fats and oils	4.8	Sugar, white	20.8
Solid shortening, deep		Fountain toppings	8.0
frying	24.7	Candy	5.9
Butter	12.6	Vegetables	18.1
Margarine	11.2	Potatoes, white, fresh	39.6
Mayonnaise, regular	11.2	Potatoes, white, frozen	15.2
Flour and cereal	3.1	Lettuce	10.8
Flour	43.9	Onions	6.8
Prepared flour mixes	17.9	Tomatoes, fresh	6.2
Dry pastas	9.2	Cabbage	5.3
Rice	7.9	Beans, green snap, string	3.2
Bakery products	9.6	Tomatoes, canned	3.0
Hamburger buns	28.0	Fruits	3.6
Bread, white buttermilk,		Melons	16.4
potato	17.9	Apples	13.2
Pie	9.6	Oranges	8.5
Hot dog rolls	7.1	Lemons	8.4
Beef	8.6	Peaches	8.1
Patties, including		Juices, ades, and drinks	2.1
hamburger	28.0	Orange juice, single-	
Steak	22.5	strength	20.6
Roast	12.6	Tomato juice, single-	
Ground, regular	8.6	strength	18.4
Ground, lean	7.4	Orange juice, concentrated	13.4
Other meats	4.3	Beverages	7.0
Ham	19.9	Cola	43.6
Frankfurters	14.2	Soft drinks, fruit type	19.0
Bacon	10.6	Coffee, bean or ground	16.1
Poultry and eggs	5.2	Prepared foods and mixtures	0.9
Broilers and fryers	43.8	Nuts, condiments and	
Eggs, fresh, in shell	36.4	leavenings	3.2
Turkey	8.5	Tomato catsup	29.8
Soups, sauces and gravies	1.0	Pickles	27.3
Soups	79.2	Mustard	7.6
Sauces and gravies	16.1	Salt	7.4
Soup bases and Bouillon	4.6	Nuts and peanut butter	6.1

Source: U.S. Dept. of Agr. (1971).

EFFECTS OF FOOD SERVICE PRACTICES ON NUTRIENTS

tion practices on these products. In general, label claims are met by the addition of nutrients sufficient to meet FDA allowances during processing. These products are primarily susceptible to deterioration with long-term storage, particularly at temperatures above ambient (Borenstein 1968); but since they are centrally processed, dating for shelf stability is automatic although not always "open," i.e., intelligible, to the end user.

NUTRIENT STANDARD MENU PLANNING

The computer has accelerated and sophisticated management practices in general, and an increasing number of food service operations have not only adopted computer programming for cost accounting and inventory control but also for computer assisted menu planning (CAMP). One of the programmable constraints in the system is nutrient level. To date, the nutrient file has been USDA Agriculture Handbook 8 (Watt and Merrill 1963) nutritive value food composition data. The greatest use of CAMP has been in hospital food service (Balintfy 1973); however, a modification of the concept emphasizing the nutritive profile of meals has been evolved (Jansen and Harper 1974) and is termed "Nutrient Standard Menu Planning." The USDA is pursuing such a concept for the "National Child Nutrition Programs" (school lunch, breakfast, supplemental feeding), and the "Nutrition Programs for the Elderly," administered by the U.S. Dept. of Health, Education and Welfare, require that meals assure $^1/_3$ RDA. Menus specifying food combinations approximating this goal are provided the institution preparing food, in lieu of nutrient tabulations because of the lack of available data. With the advent of nutrition labeling, the U.S. RDA as a nutrient standard has been adopted for product development by the food industry and labeling use, as well as consumer use. The USDA has evolved a totalizer of the U.S. RDA nutrients called the NUTRIMETER for the consumer to add the data from labels and thus determine the contribution of his diet to the daily U.S. RDA standards (Hill 1975). In any event, it is important to recognize that the nutritive data on the label does correct for or otherwise consider processing losses for those nutrients listed, but the values do not correct for or otherwise consider subsequent nutrient losses during preparation and prior to actual consumption. This fact emphasizes the interest nutritionists and food service personnel should have in better understanding nutrient retention associated with large-scale and home preparation practices.

COMPARISON OF LARGE-SCALE AND HOME FOOD PREPARATION

Amount of Food Served

There is no clear-cut division between amounts of food served in differentiating between large-scale and home cooking. In general, amounts to serve 25 or more are considered large-scale. However, most recipes in food service cookbooks provide 100 portions or servings; and those in home cookbooks, 4–6 portions or servings.

The amount of food handled and/or cooked in one lot in food service depends on many factors, including the number and size of portions to be served, recipe used, available equipment, pan size, the load capacity of given equipment (e.g., oven), the number of pieces of similar equipment (e.g., the number of ovens), temperature and rate of cooking, and the product itself.

In reality, there are only a few research studies that have specifically compared the effects of large-scale and home preparation cooking practices for given foods and none have done so for more than one nutrient. This chapter emphasizes food service practices but the information is often derived from laboratory-scale analyses which are as applicable to a discussion of the effects of home practices. Much of the information in this chapter is not repeated in the following chapter on the effects of home practices and the reader is urged to also read the applicable segments of both chapters, if a more comprehensive viewpoint is desired.

Length of Time and Temperature of Cooking

Thermal processing in the food processing industry has become a very sophisticated science with the end point being safety based upon the thermal death time of specific organisms. Some research has attempted to correlate the time and temperature parameters with nutrient retention (Hayakawa 1969). In contrast, the food service industry has situations where either older equipment may require larger food loads than home food preparation, to have longer cooking time, or have forced air ovens and/or steamers and/or microwave equipment which capitalize on high temperature, short time techniques. This may be further complicated in food service operations by the products being frozen or partially tempered at the onset of the cooking process. Not only does the research literature contain few reports considering these variables but some reports cannot be fully considered simply because conditions are not sufficiently described and/or the techniques used are of questionable accuracy. What is really needed is a study that would consider the most frequently used foods, the most frequently used equipment, and time and temperature variables.

EFFECTS OF FOOD SERVICE PRACTICES ON NUTRIENTS 469

Length of Time and Temperature of Holding

The issue of nutrient losses during storage prior to food preparation are discussed in other chapters of this book. Losses of nutrients during "storage," i.e., holding after cooking, is another problem. Warming ovens, steam tables, and insulated heated transport equipment are often used in food service to provide convenience to the consumer or to expedite the delivery of hot meals to scattered individuals such as in a hospital. Again the variables are time and temperature vis-a-vis the product and the equipment. A few studies of this problem have been reported and are discussed in a later section of this chapter. The point should be made that no research has been systematically carried out in this area.

Variable Types of Food Service Systems

Food service has evolved from the extension of a home kitchen to feed more people to manufacturing kitchens which prepared preplated, prefrozen entrées or meals to be delivered miles away at a hotel or in an airplane over the ocean. The point is that the large manufacturing kitchens are really small food processing operations with a capacity to freeze food—in an entrée or meal form—that is in a more convenient form to meet the consumer and/or situation demand.

No research of significance has been performed on nutrient retention of food and meals as affected by this type of manufacturing kitchen—food service production. Calculations (Miskimin et al. 1974) of the nutritive value of frozen school lunch meals prepared at the industrial level reveal data which coincide with prior food processing experience at the manufacturing level, and therefore, the general principles discussed in other chapters of this book are relevant in terms of guidelines. Precise information is otherwise lacking. Without a doubt, the most ill-defined food service nutrient retention information concerns the effects of holding times which can range from long-term (weeks, months) interim frozen storage to long-term (hours) cafeteria steam table holding of bulk pan foods, and to analogous periods of holding preplated trays in heated insulated mobile carts, used in catering and hospital food service.

The following broad categorization of food service systems is provided as an overview of the diverse approaches being used, and therefore the complexity of the problems of nutrient retention:

(1) Cold and/or hot sandwich commissary (school district or vendor).

(2) Hot, recently prepared (cooked or reheated), preplated meal system delivered in some form of insulated carrier.

(3) Previously frozen entrée and/or preplated meal system heated and served just prior to use.

(4) Hot, recently prepared or reconditioned (previously frozen) bulk pan tray system of food delivered in some form of insulated carrier and then assembled into meals at the serving site.

(5) Flexible packaged precooked ready-to-eat or dried ready-to-rehydrate entrée and/or meal system for camping and specialty feeding applications such as military bivouac needs.

A broad listing of the types of food service equipment involved which should be considered in the conduct of systematic nutrient retention studies are:

(1) Cooking equipment: kettles, ovens, steamers, fryers.

(2) Holding equipment: (a) Cold: refrigerators, freezers. (b) Hot: insulated cabinets, heated cabinets.

(3) Conditioning equipment: convection ovens, steamers.

(4) Serving equipment: insulated containers, steam tables, infrared lamps.

LOSSES PREPARATORY TO COOKING

In spite of the fact that nutrient losses associated with the effects of preparatory practices are to be expected because humans otherwise rarely consume peelings and trimmings, etc., the practice is real and occurs at the manufacturing kitchen as well as the home level. This topic is discussed at length as it occurs at the commercial level in Chap. 7.

For foods of animal origin the losses of major interest are those from fat trim and thaw juices. For foods of plant origin the losses of major interest are those from trimming, washing and soaking, and chopping.

FOODS OF ANIMAL ORIGIN

Trimming

The only report on trimming found was on fat trim. The present tendency is for trimming to be done at the packing plant rather than in the kitchen. Prefabricated meats, often referred to as oven-prepared or portion-ready meats, are finding increased acceptance in institutions, as are also frozen meats. The Armed Services, which use large quantities of 4-way boneless beef, purchase meats according to specifications.

In the trade, specifications for trimming of prefabricated meat have not yet been standardized. Certain guides such as "outside fat shall not be over $1/2$ inch" and "length of oven-prepared ribs of beef shall

not be over 7 inches from the inside of vertebrae to rib end" are being used (Kotschevar et al. 1953).
Many institutions are buying meat to their own specifications. The USDA offers the service of aiding private institutions in drawing up specifications. They also offer a follow-up by official graders to check on whether specifications are being met as orders are filled.
A study of fat trim from oven roasts in 4-way-boneless beef cut to Army specifications was reported by Toepfer et al. (1955). Specifications for the roasts limited external fat to $3/4$-in. in thickness.

TABLE 16.3

RETENTION OF FOOD ENERGY, PROTEIN, FAT, AND ASH IN BONELESS BEEF AFTER THAWING AND COOKING

Type of Beef and Type of Cooking	Weight[1]		Retention[2]			
	Raw Frozen (Kg)	Drained Cooked (Kg)	Food Energy (Cal)	Protein (%)	Fat (%)	Ash (%)
Oven roasts						
Blade roll	1.670	1.148	85.4	90.4	83.4	111.7
Inside of round	7.385	4.568	82.9	91.9	77.4	86.6
Knuckle of round	4.086	2.463	88.1	92.5	83.2	72.3
Loin strip	4.652	3.132	85.0	91.3	82.9	84.2
Sirloin butt	6.162	3.874	81.5	87.1	79.5	94.3
Spencer roll	4.306	2.794	76.7	95.0	72.8	74.2
Tenderloin	2.411	1.574	72.5	95.1	65.8	97.9
Pot roasts						
Clod	7.156	5.132	92.3	95.9	91.1	107.6
Chuck roll	5.086	3.098	79.8	92.5	73.3	76.1
Chuck tender	1.074	0.631	91.6	101.4	82.4	88.1
Outside of round	5.777	3.716	76.8	92.8	69.4	80.4
Rump butt	2.367	1.520	82.8	86.1	81.5	64.2
Griddle-broiled steaks						
Blade roll	1.538	1.060	99.0	87.2	103.8	77.5
Inside of round	7.386	4.710	89.9	81.4	90.9	77.9
Knuckle of round	4.147	2.558	89.1	89.4	87.7	82.2
Loin strip	5.074	3.495	90.8	88.4	91.1	68.9
Spencer roll	4.735	3.381	88.2	85.9	88.6	95.2
Tenderloin	2.825	1.952	85.6	97.9	82.2	69.1
Swiss steaks						
Clod	7.202	5.842	93.7	87.4	89.7	71.0
Chuck roll	4.776	3.064	71.8	80.3	63.7	64.2
Chuck tender	1.144	0.747	92.2	91.4	86.5	52.2
Outside of round	5.474	4.104	87.9	93.4	80.6	45.0
Rump butt	2.188	1.747	90.1	94.3	85.4	79.8

Source: Toepfer et al. (1955).
[1] Weight average values. The inside of round represents more than four times that of blade roll.
[2] Calculated from data given in Table "Average Composition and Energy Value per 100 Gm of Cuts and Forms of 4-Way Boneless Beef, Raw and Cooked and Drippings" and Table "Average Weights of Raw and Cooked Items...." (Toepfer et al. 1955).

Seven cuts, which were assigned to oven roasting (Table 16.3) were prepared for serving without removal of this fat, another 5 cuts were trimmed of fat to about 1/4 in. Blade roll and knuckle of round were not trimmed since the layer of fat did not exceed 1/4 in.

Trim amounted to 6% of the average weight of the cuts. Roasts lost only 12% of the protein, but 24.4% fat to trim.

Fat trim was mostly fat, from 82.6 to 87.5%, with only small amounts of protein, 2.1 to 4.5% (Table 16.4).

TABLE 16.4

COMPOSITION OF FAT TRIM[1]

Type of Oven Roast	Water (%)	Fat Trim Protein (%)	Fat (%)
Inside of round	12.2	4.5	82.6
Loin strip	9.5	3.0	86.6
Sirloin butt	10.8	3.2	84.0
Spencer roll	8.3	2.1	87.6
Tenderloin	10.8	4.0	85.6

Source: Toepfer et al. (1955).

[1] An average of 6% of the original weight of the raw beef was lost to the fat trim.

Thaw Juices

Increasing amounts of frozen meats are being used in institutions. However, only 3 studies were found on thaw juice; 2 studies were on boneless beef cut to Army specifications and the third was on liver but included some work on muscle meats.

Weight Loss.—Toepfer et al. (1955) reported 4.2% average loss to thaw juice. Ground beef lost the least; diced stew meat with its relatively large surface lost the most, 5%.

Kotschevar et al. (1955) reported greater drip loss from liver slices, 12–15%, which had a greater cut-surface area per volume than from liver pieces, 8–11%. A thawing drip range of from 0.9 to 5.7% was reported from 5 cuts of beef, 3 of pork, and 2 of lamb. In addition to the area of cut surface per volume, Kotschevar reported that the ratios of bone and fat to lean meat were governing factors.

The U.S. Fish and Wildlife Service (U.S. Dept. of Interior 1955) reported thaw juice loss of 4.5–15.2% from frozen fish. Composition of the thaw juice was not reported.

In most instances, trimming beef roasts to 1/4 in. of fat resulted in a slightly higher loss to thaw juice (Toepfer et al. 1955).

EFFECTS OF FOOD SERVICE PRACTICES ON NUTRIENTS

Protein Loss.—Protein loss from beef to thaw juices (Toepfer et al. 1955) was small and followed a similar pattern to weight loss: from 1.4% in ground meat for meat loaf to 3.1% in diced meat for stew (Table 16.5).

TABLE 16.5

SUMMARY OF RETENTION OF PROTEIN AND FAT IN THAW JUICES, DRIPPINGS, AND COOKED MEAT FROM VARIOUS CUTS AND FORMS OF BEEF

Cut or Form of Beef	Protein			Fat[1]	
	Thaw Juices (%)	Drippings (%)	Cooked Beef or Recipe (%)	Drippings (%)	Cooked Beef or Recipe (%)
Beef cuts					
Oven roasts	2.6	1.8	95.6	20.6	79.4
Pot roasts	2.5	7.6	89.9	31.1	68.9
Griddle-broiled steaks	3.0	0.4	96.6	4.2	95.8
Swiss steaks	1.6	14.7	83.7	39.2	60.8
Diced beef					
Stew	3.1	—	96.9	—	100.0
Ground beef					
Hamburger	2.4	0.4	97.2	18.2	81.8
Meat loaf	1.4	0	98.6	12.6	87.4

Source: Toepfer et al. (1955).
[1] No fat was reported in thaw juice.

From 8.8 to 12.1% of the thaw juice, depending upon cut, consisted of protein (Table 16.6).

Fat Loss.—No fat was found in the thaw juice (Toepfer et al. 1955).

Vitamin Loss.—A study on the stripped-out longissimus dorsi muscle of a beef carcass showed the following losses of B vitamins to the thaw juice: thiamin, 12.2%; riboflavin, 10.3%; niacin, 14.5%; pyridoxine, 33.3%; and folic acid, 8.1% (Pearson et al. 1951).

Kotschevar et al. (1955) found that, in general, thawing drip from frozen liver had gram for gram about as much or slightly more thiamin, slightly less riboflavin, and about as much or slightly more niacin than remained in the thawed liver. He stated that percentages of drip obtained from liver, steaks, and some roasts were high enough to indicate significant loss during thawing, although not from certain other muscle meats.

Mineral Loss.—No reports were found on losses to thaw juice, but losses of relatively soluble minerals seem likely.

Use of Thaw Juice.—No reports of the effect of adding thaw juice to the cooking pan on nutritive value of meat and pan drippings were

TABLE 16.6

THAW JUICE LOSSES FROM BEEF

Type of Beef	Weight of Frozen Beef (%)	Thaw Juice Water in Thaw Juice (%)	Protein in Thaw Juice (%)
Blade roll			
Untrimmed	6.7	88.2	10.2
Inside of round			
Untrimmed	5.8	87.4	10.6
Trimmed	6.6	88.2	9.8
Knuckle of round			
Untrimmed	7.9	89.4	8.8
Loin strip			
Untrimmed	7.4^1	86.1^1	12.1^1
Trimmed	7.4^1	86.4^1	11.4^1
Sirloin butt			
Untrimmed	6.3^1	87.8^1	9.9^1
Trimmed	7.0^1	85.5^1	11.3^1
Spencer roll			
Untrimmed	2.7	87.0	11.3
Trimmed	5.5	87.4	11.0
Tenderloin			
Untrimmed	4.7	89.0	9.1
Trimmed	4.8	89.0	9.0
Average for above cuts			
Untrimmed	6.0		
Trimmed	6.6		

Source: Toepfer et al. (1955).
[1] Single item.

found. Apparently, it is common practice in large-scale preparation, and also in some homes, to discard thaw juices. Studies are needed to show whether these juices should be added to meat, to gravy, or be discarded. However, it is very likely that emphasis on sanitation overrides any consideration for the nutritive value of thaw juices.

Cooked Frozen Versus Thawed.—Studies are needed comparing nutritive value of a meal cooked large-scale from the frozen state with that cooked after thawing. Such comparisons on home-scale amounts show little, if any, difference in total retention of thiamin (Westerman et al. 1949) or riboflavin (Westerman et al. 1949; Fenton et al. 1956).

FOODS OF PLANT ORIGIN

Trimming

The losses that occur in the preparation of vegetables vary with the variety of food, condition and freshness of the food, the season of the year, the cultural habits of the consumer, etc. When root vegetables are peeled or scraped; when the wilted, discolored, or fragmented leaves and tough stems of green vegetables are removed; and when the skins of fruits are discarded, a considerable amount of plant tissue may be lost.

McCance et al. (1936) have published figures for waste expressed as percentage of the purchased weight of vegetables used for boiling. These losses vary between 21 and 61% in leafy vegetables, with a mean value of 25%. In root vegetables, the range given is 14 to 26% for peelings, and only 4 to 5% for scrapings. The losses from fresh peas and broad beans are 63 and 71%, respectively.

Andross (1946) presented data on waste resulting from the preparation of vegetables in Scotland as follows: broad beans, 72%; broccoli, 58%; brussels sprouts, 45%; cabbage, 40%; cauliflower, 58%; cress, 45%; French beans, 4%; kale, 45%; lettuce, 25%; parsley, 64%; green peas, 64%; spinach, 64%; beetroot, 50%; carrot, 50%; celery, 29%; leek, 30%; parsnip, 35%; potato, 40%; and turnip, 37%.

Watt and Merrill (1950) list the following refuse losses from fruits and vegetables in the United States: apples, 12%; apricots, 6%; asparagus, 25%; avocado, 25%; banana, 33%; lima bean, 60%; snap bean, 10%; beet, 47%; beet greens, 25%; broccoli, 39%; brussels sprouts, 23%; cabbage, 27%; cantaloupe, 53%; carrot, 37%; cauliflower, 55%; celery, 37%; chard, 14%; cherry, 6%; collard, 55%; sweet corn, 62%; cress, 37%; cucumber, 30%; currant, 3%; date, 13%; egg plant, 13%; endive, 48%; grapefruit, 34%; grape, 22%; guava, 13%; honeydew melon, 37%; kale, 36%; kohlrabi, 46%; lemon, 38%; lettuce, 31%; mango, 34%; mustard greens, 27%; okra, 12%; olive, 16%; onion, 6% or 59%; orange, 28%; papaya, 32%; parsnip, 22%; peach, 12%; peanut, 28%; pear, 17%; pea, 55%; pepper, 16%; pineapple, 47%; plum, 15%; potato, 16%; prune, 15%; pumpkin, 31%; radish, 51%; rhubarb, 32%; spinach, 18%; squash (summer), 3%; squash (winter), 26%; strawberry, 4%; sweet potato, 14%; tangerine, 29%; tomato, 12%; turnip, 34%; watermelon, 54%. It will be noted that these estimates are, in general, much more conservative than those of Andross (1946).

When foods of plant origin are trimmed, the nutrient losses generally exceed the weight losses, because these nutrients are usually found in higher concentrations in the outer leaves of vegetables and in the outer layers of seeds, tubers, roots, and fruits. Ascorbic acid is present in highest concentrations just beneath the corklike peel layer in the potato; because of this, the peeling loss may range between 12 and 35% (Wager et al. 1945; Wertz and Weir 1946; Streightoff et al. 1946A). The peel of carrots is expecially rich in thiamin, niacin, and riboflavin (Streightoff et al. 1946B). The outer leaves of lettuce and spinach are rich in the B vitamins and in ascorbic acid; on the other hand, the outer leaves of cabbage are similar in composition to the inner leaves (Wood et al. 1946). The loss of protein due to trimming is usually the same as the weight loss (Andross 1946). The outer green leaves are often richer in mineral salts (McCance et al. 1936) than the inner leaves.

The leaves of many vegetables contain from 2 to over 6 times as much ascorbic acid as the stems; the leaves of spinach may contain 20 times as much as the stems. Leaf blades contain 5 to 20 times more carotene than the petioles, and up to 99% of the total carotene is in the leaf. Leaves also contain more minerals than do the stems; for instance, iron is 2 to 4 times more concentrated in the leaves than in the stems (Sheets et al. 1941). Broccoli buds are richer in ascorbic acid than the stalks (Fenton 1941). There is 1.5-3 times as much iron, 1.5 times as much ascorbic acid, and 21 times as much carotene in the green leaves of cabbage as in the bleached leaves. Lettuce and asparagus follow the same pattern (Sheets et al. 1941).

Washing and Soaking

Preliminary washing and soaking of vegetables and tubers before cooking permit the extraction of water-soluble constituents, but these losses are generally insignificant.

Peeled potatoes stored under refrigeration for 20 hr, lost little ascorbic acid (Wertz and Weir 1946). Potatoes soaked 2 hr in water lost 11.9% thiamin, and sweet potatoes soaked 5 hr in water lost 21.1% thiamin. Peeled potatoes soaked for 28 hr in water lost 14% ascorbic acid, 8% thiamin, 6% riboflavin, and 14% niacin (Gleim et al. 1946A). Peeled new and old potatoes lost 5 and 10% ascorbic acid, respectively, while standing in water (Branion et al. 1947). Peeled apples contained 50-75% as much ascorbic acid as an equal weight of unpeeled apples (Todhunter 1936). In another study, the loss in peeled apples was only 8-17% (Curran et al. 1937).

The loss of nutrients during the washing of processed cereals may be considerable. For instance, as much as 60% niacin and 60% thiamin was lost when raw-milled rice was washed (Swaminathan 1941, 1942).

Chopping

The losses in ascorbic acid when cabbage and other salad greens are diced and minced can be considerable (Pyke 1944; Van Duyne et al. 1944). Munsell et al. (1949) reported losses of 19% of the reduced ascorbic acid content when cabbage was minced or shredded, but the losses in total ascorbic acid were only 3 and 6%, respectively. No further losses in total or reduced ascorbic acid occurred when shredded cabbage was held for 3 or for 22 hr, or in minced cabbage when held for 3 hr. However, when the minced cabbage was held for 22 hr, it showed a further loss of 8% in total, and 9% in reduced ascorbic acid. Coleslaw treated with French dressing and held for 22 hr showed still greater losses. Clayton and Goos (1947) observed a 53% loss in 2 hr when vinegar was used on coleslaw.

Walker and Arvidsson (1952) reported 40% loss in minced cabbage. Wood et al. (1946) reported losses of 9-15% in ascorbic acid, 0-5% in thiamin, and 0-3% in riboflavin in diced cabbage, with no appreciable further loss when held at room temperature for 2 hr. Wilcox and Neilson (1947) found a 10% loss in diced cabbage, and a further loss of 4% on holding 90 min. Millross et al. (1973) reported a vitamin C loss of 35-37% when spring cabbage was shredded and washed in a period of 25 min.

Large losses in ascorbic acid in mashed potatoes have been reported: 80% by Wertz and Weir (1946); and 30% by Hellstrom (1952-1953). Bring et al. (1963) reported a 36% (23% dry weight basis) loss for mashed potatoes. Bring and Raab (1964) later reported losses as high as 53% (40% dry weight basis). The potatoes used were from two different potato processing plants which may have handled the product differently. Since retention variations in mashed potatoes were negligible over 6 months of sampling in spite of the decreasing vitamin C content of the raw potatoes with storage (29.3 mg per 100 gm to 10.6 mg per 100 gm), these data are probably realistic.

Workers have reported that cucumbers lost 22% ascorbic acid during slicing, 33-35% during standing for 1 hr, and 41-49% during standing for 3 hr; and grated radish lost 27% thiamin in 24 hr at 10°C; while grated sweet potatoes lost 21% thiamin. During 24 hr of refrigeration, cantaloupe slices lost 35% ascorbic acid without syrup, and 25% with syrup (Wolfe et al. 1949). Tomato pulp lost little ascorbic acid during 4 hr of storage, but in 24 hr the loss was 40% (Hellstrom 1952-1953). It has been reported that orange juice held at 9°C lost 17% thiamin in 24 hr. Sliced bananas exposed to air at 25°C lost 12% ascorbic acid in 20 min, 22% in 80 min, 50% in 5 hr, then remained the same for 22 hr (Harris and Poland 1939). Quartered apples both peeled and unpeeled, lost 20% ascorbic acid in 1-2 hr, and 33% in 3 hr, at 25°C (Burrell and Ebright 1940).

Cucumber salads lost 22% ascorbic acid during preparation, and a further 8% and 11% during standing for 1 and 3 hr, respectively. The losses in tomato salads were 8% in 3 hr.

LOSSES DURING COOKING

Cooking Food of Animal Origin

Fenton (1960) reported that no studies could be found in the open literature on the effect on the nutritive value of the large-scale preparation of fish, cheese, milk, and eggs. With the exception of one abstract (Tooley 1972) on the effect of frying on the availability of fish lysine (approximately 17% loss of available lysine during deep frying in fresh oil, compared with a loss of 25% when using thermally abused oil) no other reports through mid 1974 were found.

Milk and cheese are foods which are often eaten cold and, therefore, an important consideration would be nutrient retention during commercial processing and storage. This is discussed elsewhere in this volume.

Meats and Poultry.—*Broiling.*—In the Toepfer et al. (1955) study, steaks broiled on a griddle without added fat retained most of their protein, 96.6%, and fat, 95.8% (Table 16.5). The finding on protein was similar to that in oven roasts. The retention of fat during broiling was higher, no doubt because of the shorter cooking time. According to Noble (1964), Tucker et al. (1946) reported thiamin retentions of 77 and 70%, respectively, and a riboflavin retention of 92% in beef loin steaks broiled to interior temperatures of 58° and 70°C, while Causey et al. (1950A) reported retentions of slightly over 80% for both thiamin and riboflavin in frozen ground beef which had been broiled to 74°C. McIntire et al. (1943) found retentions of 70 and 79% of thiamin and riboflavin, respectively, in broiled sirloin of lamb, and Causey et al. (1950B) in the neighborhood of 85 to 55%, respectively, in frozen ground lamb broiled to 85°C.

Noble (1964) demonstrated that pan and oven broiled cuts were not significantly different in percentages of thiamin or riboflavin retained. Comparisons based on averages for individual cuts within each type of meat showed that thiamin retention was significantly higher in broiled beef loin steaks and 1-serving patties (78%) than in rib steaks and 3-serving patties (average 62%); likewise, in lamb chops and 1-serving lamb patties (average 66%) than in 3-serving lamb patties (55%); and in Canadian-style bacon (79%) than in ham slices (65%). Riboflavin retentions overlapped considerably, with that for broiled rib and loin steaks (90%) not significantly different from that for either 1- or 3-serving patties, and that for Canadian-style bacon (82%) not significantly different from that for ham slices. Riboflavin retention in lamb chops (96%) was significantly higher, however, than in lamb patties (average 75%).

Comparisons based on averages for each of the three types of meat

TABLE 16.7

MEAN THIAMIN AND RIBOFLAVIN IN RAW AND BROILED MEAT AND PERCENTAGE RETENTION AFTER BROILING

Cut	Thiamin			Riboflavin		
	Raw Meat (Mg/100 Gm)	Cooked Meat (Mg/100 Gm)	Retention (%)[1]	Raw Meat (Mg/100 Gm)	Cooked Meat (Mg/100 Gm)	Retention (%)[1]
Beef						
Ground beef, 3-portion patty	0.066 ± 0.006[2]	0.066 ± 0.006	65	0.120 ± 0.006	0.192 ± 0.010	100
Rib steak	0.072 ± 0.001	0.062 ± 0.001	59	0.136 ± 0.003	0.184 ± 0.006	91
Loin steak	0.062 ± 0.002	0.066 ± 0.002	78	0.146 ± 0.002	0.180 ± 0.005	89
Ground beef, 1-portion patty	0.092 ± 0.013	0.099 ± 0.008	78	0.152 ± 0.011	0.181 ± 0.012	84
Lamb						
Ground lamb, 3-portion patty	0.072 ± 0.007	0.076 ± 0.008	55	0.145 ± 0.019	0.205 ± 0.013	73
1-portion patty	0.087 ± 0.002	0.111 ± 0.006	67	0.147 ± 0.009	0.204 ± 0.008	78
Chops	0.111 ± 0.008	0.104 ± 0.005	66	0.185 ± 0.006	0.257 ± 0.009	96
Pork						
Ham slices	1.775 ± 0.092	1.922 ± 0.167	65	0.157 ± 0.006	0.208 ± 0.013	80
Canadian-style bacon	1.550 ± 0.091	1.960 ± 0.195	79	0.208 ± 0.008	0.271 ± 0.016	82

Source: Noble (1964).

[1] Any two means not side-scored by the same line may be considered significantly different at the 5% level.
[2] Standard error.

showed that thiamin retentions in beef and pork were not significantly different (average 70%) but were higher than in lamb (62%), while riboflavin retentions in beef (92%) were significantly higher than in pork and lamb which were not significantly different (average 82%). The data are presented in Table 16.7 and although retail cuts were used, it is presented here because of its uniqueness to the field.

In a sophisticated study, Meyer et al. (1963) studied the effect of broiling in an electric oven (5 min per side at a distance of 3 in. from the broiler unit) on the content of niacin, thiamin, and riboflavin of pork gluteus medius (loin) muscle. Pale, soft, and watery (PSW) and dark, firm, and dry (DFD) muscles were subjectively classified, and the classification objectively substantiated with Hunter color values and expressible juice ratios. PSW pork had about twice as much niacin ($P < 0.005$) as DFD, both fresh and cooked, (see Table 16.8).

TABLE 16.8

VITAMIN CONTENT OF 4 PALE, SOFT, WATERY AND 4 DARK, FIRM DRY GLUTEUS MEDIUS MUSCLES

Vitamin[1]		Pale, Soft Watery		% Loss in Cooking[2]	Dark, Firm Dry		% Loss in Cooking[2]
		Fresh	Cooked		Fresh	Cooked	
Niacin	$\bar{x}$	87.2	74.7	41.2	32.4	35.2	20.7
	$s_{\bar{x}}$	2.9	8.2	6.2	2.2	8.9	7.3
Riboflavin	$\bar{x}$	2.27	3.14	8.5	2.38	2.50	22.8
	$s_{\bar{x}}$	0.06	0.44	7.1	0.06	0.21	3.9
Thiamin	$\bar{x}$	13.0	13.8	29.72	14.2	16.5	16.6
	$s_{\bar{x}}$	0.50	0.40	5.4	1.4	2.1	2.0

Source: Meyer et al. (1963).

[1] Expressed as µg/gm fresh tissue.

[2] Vitamin loss calculated to dry weight: $\dfrac{\mu g/gm\ raw - \mu g/gm\ cooked \times \%\ cooked\ wt}{\mu g/gm\ raw}$

Fresh DFD muscle had slightly higher riboflavin and thiamin contents, and in the cooked form contained a higher thiamin level. PSW muscle showed greater exudate formation, more expressible juice, and consequently significantly higher cooking weight loss, (30–45% versus 25–32.5%). This higher cooking loss contributed to a significantly higher nutrient loss on a fresh-weight basis. When equal amounts of the vitamins were present in the fresh form, the PSW muscle lost a greater quantity of vitamins during cooking, either by destruction or due to "drip." The PSW muscle had a higher niacin level in the cooked form, even though these muscles had a considerably higher cooking loss. It is postulated that the niacin differences reflect a difference in the pyridine nucleotide coenzyme content of the PSW and DFD muscles.

Frying.—Kotschevar et al. (1955), who cooked liver slices in margarine melted in a pan over a low gas flame, reported a loss in fresh liver of 15% of the thiamin, 6% of the riboflavin, and 13% of the niacin. Amounts of liver cooked in one pan were not given. He found greater cooking losses of thiamin from frozen liver slices but an increase in riboflavin. Results with niacin were variable, ranging from a cooking loss of 15% to a gain of 30%.

Braising.—Toepfer et al. (1955) reported (Table 16.3) that retentions in Swiss steaks were very similar to those in pot roasts and showed the effect of addition of water on losses. Swiss steak retained 83.7% of the protein and 60.8% of the fat; pot roasts retained 89.9% of the protein and 68.9% of the fat. On the other hand, over 95% of the protein was retained during methods of cooking in which no water was added.

Schlosser et al. (1957) reported that braised turkey meat contained as much lysine as did that steamed under 5- or 10-lb pressure. The only meaningful studies of vitamin retentions in braised meats is that of Noble (1965, 1970) who studied several cuts of beef, veal, and pork (Table 16.9).

For the beef cuts, thiamin retention was significantly higher in the round pot roasts and braised steaks (40% of the amount present in the raw samples) than in the braised chuck and short ribs (average of 24%). It was intermediate in the flank steaks, which retained neither a significantly lower percentage than the round cuts on the one hand nor a significantly higher percentage than the chuck and short ribs on the other. Riboflavin retention showed a pattern different from that of thiamin. It was highest in chuck, flank steaks, and round roasts, which were not significantly different and averaged 73%, and lowest in short ribs and round steaks, which were not significantly different and averaged 62%.

For the veal cuts, thiamin retention in the round steaks (48%) was significantly higher than in chops (38%), but riboflavin retention was not significantly different and averaged 74%.

For the pork cuts, thiamin retention was significantly different among the different cuts, tenderloin retaining the highest (57%), spareribs the lowest (26%), and chops the intermediate percentage (44%). Riboflavin retention was also highest in tenderloin but lowest in chops and intermediate in spareribs, the retention in which was not significantly different from that in either tenderloin or chops.

Thiamin and riboflavin retentions were very similar from one type of animal to another when all cuts were considered. Thus, the mean

TABLE 16.9

MEAN THIAMIN AND RIBOFLAVIN IN RAW AND BRAISED MEAT AND RETENTION AFTER BRAISING

Cut	Thiamin			Riboflavin		
	Raw Meat (Mg/100 Gm)	Cooked Meat (Mg/100 Gm Cooked Wt)	Retention[1] (%)[2]	Raw Meat (Mg/100 Gm)	Cooked Meat (Mg/100 Gm Cooked Wt)	Retention[1] (%)[2]
Beef						
Short ribs	0.070 ± 0.003[3]	0.040 ± 0.003	25	0.077 ± 0.002	0.106 ± 0.006	58
Chuck	0.076 ± 0.005	0.033 ± 0.004	23	0.204 ± 0.015	0.234 ± 0.015	74
Flank steak	0.056 ± 0.001	0.032 ± 0.005	30	0.082 ± 0.003	0.112 ± 0.010	72
Round (roast)	0.073 ± 0.020	0.062 ± 0.005	40	0.208 ± 0.021	0.250 ± 0.023	73
Round (steak)	0.081 ± 0.003	0.058 ± 0.002	40	0.229 ± 0.009	0.269 ± 0.004	65
Veal						
Chops	0.148 ± 0.008	0.105 ± 0.005	38	0.217 ± 0.012	0.289 ± 0.019	73
Round steak	0.125 ± 0.004	0.119 ± 0.015	48	0.200 ± 0.005	0.303 ± 0.016	76
Pork						
Chops	1.990 ± 0.012	1.600 ± 0.168	44	0.155 ± 0.019	0.178 ± 0.021	64
Spareribs	0.680 ± 0.016	0.355 ± 0.004	26	0.262 ± 0.014	0.372 ± 0.003	72
Tenderloin	2.728 ± 0.202	2.569 ± 0.035	57	0.302 ± 0.015	0.425 ± 0.016	83

Source: Nobel (1965).

[1] Any two means not side-scored by the same line may be considered significantly different at the 5% level. The interrupted line beside the riboflavin retention figures is interpreted as follows: the retentions for short ribs and round steaks are not significantly different from each other.
[2] Calculated on the basis of the entire sample, before and after cooking.

thiamin value for all beef cuts (32%) was significantly lower than those for either veal or pork, but the values for the last two and all the riboflavin means were not significantly different (averages of 42 and 72%, respectively).

The cooking liquids from the beef cuts contained approximately 25% thiamin; those from the veal chops and round, 17 and 33% respectively; and those from the pork tenderloin, spareribs, and chops 5, 1, and 13%, respectively, of the thiamin originally present in the raw samples. Liquids from the beef and pork cuts also contained approximately 20% and those from the veal chops and round 17 and 24%, respectively, of the riboflavin originally present.

The thiamin and riboflavin contents and retentions of braised or simmered sweetbreads, beef kidney, lamb heart, and pork heart are given in Table 16.10. For the meats as a group, braising as compared to simmering was found to cause a significantly higher retention of thiamin (46% versus 39%), but not riboflavin (retention 70% for both cooking methods).

The kind of meat made a significant difference in the average amount of thiamin and riboflavin retained in the combined braised and simmered samples. Thus, sweetbreads retained the largest percentage of thiamin (60%—the average retention in braised and simmered) and beef, veal, and pork heart the lowest (average, 29%). Veal, beef, lamb, and pork hearts, on the other hand, retained the highest percentage of riboflavin (average, 75%) and beef kidney the lowest (55%).

The cooking liquids contained from 12 to 25% (average, 19%) of the thiamin and from 13 to 22% (average, 16%) of the riboflavin originally present in the raw sample.

The loss in weight of the various meats during braising ranged from 42 to 50% (average, 46%) and during simmering from 34 to 47% (average, 40%).

Roasting.—In a study of boneless beef cut to Army specifications, Toepfer et al. (1955) showed the effect of several factors on losses of calories, protein, fat, and ash during oven roasting and pot roasting.

Weights of the cuts varied (Table 16.3) and, as a result, probably also cooking time. Protein suffered little loss in oven roasting: 95.6% was retained in the drained roast. Since 2.6% of the protein had been lost to thaw juice, and 1.8% to pan drippings, all of the protein was accounted for.

The effect of the addition of water was evident in the smaller retention of protein, 89.9%, in the pot roast and the larger loss to the drippings, 7.6%. The retention of fat was also slightly lower in the drained pot roast.

TABLE 16.10

MEAN THIAMIN AND RIBOFLAVIN CONTENT OF VARIETY MEATS AND RETENTION AFTER COOKING

Meat	No. of Purchase Lots	Mean Raw Content (Mg/100 Gm)	Braised Content		Simmered Content		Retention		Dissolved in Cooking Liquid	
			Mean, on Cooked Weight Basis (Mg/100 Gm)	Mean, on Raw Weight Basis (Mg/100 Gm)	Mean, on Cooked Weight Basis (Mg/100 Gm)	Mean, on Raw Weight Basis (Mg/100 Gm)	Braised (%)	Simmered (%)	Braised (%)	Simmered (%)
Thiamin										
Sweetbreads	9	0.081 ± 0.008[1]	0.087 ± 0.006	0.053	0.063 ± 0.003	0.045	66	55	16	35
Beef kidney	4	0.790 ± 0.053	0.673 ± 0.050	0.355	0.554 ± 0.043	0.290	45	37	25	53
Lamb heart	5	0.983 ± 0.084	0.931 ± 0.066	0.501	0.773 ± 0.084	0.483	51	49	23	31
Pork heart	5	0.999 ± 0.083	0.636 ± 0.046	0.340	0.474 ± 0.031	0.297	34	30	12	23
Riboflavin										
Sweetbreads	9	0.244 ± 0.021	0.268 ± 0.047	0.149	0.266 ± 0.037	0.168	61	69	13	27
Beef kidney	4	3.677 ± 0.413	4.577 ± 0.730	2.380	3.262 ± 0.600	1.705	65	46	22	37
Lamb heart	5	1.111 ± 0.020	1.621 ± 0.048	0.852	1.294 ± 0.155	0.805	77	72	15	31
Pork heart	5	1.314 ± 0.056	1.899 ± 0.116	0.944	1.504 ± 0.063	0.907	72	69	13	30

Source: Nobel (1970).
[1] Standard error.

Cover et al. (1949) reported the effect of high oven temperature on retention of B vitamins after large-scale roasting of beef and pork. One-muscle roasts were used: longissimus dorsi (eye of rib), semimembranosus (inside of top round), and biceps femoris (outside of bottom round) of both beef and pork. These isolated muscles vary both in shape (Tucker et al. 1952) and weight (Wellington 1954).

Beef roasted at 150°C (302°F) to an internal temperature of 80°C (176°F) retained 61% of the thiamin, that roasted at 205°C (450°F) and cooked to an internal temperature of 98°C (209°F) retained only 47%. Pork roasted at 150°C (302°F) to an internal temperature of 84°C (183°F) retained 64% of the thiamin; that roasted at 205°C (459°F) to an internal temperature of 98°C (209°F) retained 54% (Table 16.11).

The lower temperatures resulted in practically complete retention of the riboflavin, niacin, and pantothenic acid in the drained meat plus the drippings; about 75% of each vitamin was in the drained meat. At the higher temperature, pantothenic acid retention in the meat was 13% less than at the lower temperature, but the retention of the heat-stable vitamins, riboflavin and niacin, was only slightly less. Pan drippings at the higher temperature were unusable.

Statistical analysis showed that beef roasts cooked at 150°C (302°F) to an internal temperature of 80°C retained significantly more thiamin, pantothenic acid, and riboflavin than did those roasted at 204°C (450°F) to a temperature of 98°C. In the pork, however, thiamin was the only vitamin retained in statistically significantly greater quantities at the lower temperature.

Roasting times required by the above muscles at the two temperatures were not reported. However, the destructive effect of increased cooking on thiamin is well known (Farrer 1955).

In 1960, Noble and Gomez clarified the work of Cover et al. (1949) by roasting beef to a constant internal temperature. Standing rib pairs of choice beef were roasted at 149° and 177°C, and rolled and standing rib pairs were roasted to the same interior temperature at 177°C to learn if the longer heating period required by the first treatment would affect thiamin retention. The other cuts roasted were: top round, rump, tenderloin, and beef loaf, the meat for which came from the chuck arm. All roasts were heated to an interior temperature of 71°C; the loaves were heated to 75°C (Table 16.12).

Longer heating did not affect the thiamin retention in the cooked meat. Roasted beef loaf retained an average of 70% of the amount of thiamin in the raw ingredients. This retention was significantly higher than that of cooked top round and rib roasts, which averaged 54%. It was not significantly different, however, from the retention

TABLE 16.11

VITAMIN RETENTION IN BEEF AND PORK AFTER LARGE-SCALE ROASTING

Oven Temperature, Type and Internal Temperature of Meat	Thiamin			Pantothenic Acid			Niacin			Riboflavin		
	Meat Only (%/Pan)	Drippings (%/Pan)	Total (%/Pan)	Meat Only (%/Pan)	Drippings (%/Pan)	Total (%/Pan)	Meat Only (%/Pan)	Drippings (%/Pan)	Total (%/Pan)	Meat Only (%/Pan)	Drippings (%/Pan)	Total (%/Pan)
Low oven temperature (150°C)												
Beef to 80°C[1]	61	6	67	73	20	93	76	16	92	75	16	91
Pork to 84°C	64	17	81	65	25	90	69	26	95	73	19	92
High oven temperature (205°C)												
Beef to 98°C[1]	47	2	—	60	2	—	71	2	2	68	2	—
Pork to 98°C	54	2	—	63	2	—	67	2	2	69	2	—

Source: Cover et al. (1949).

[1] Isolated muscles: longissimus dorsi, semimembranosus, biceps femoris of beef and pork, and semitendinosus of beef only.
[2] Drippings from roasts cooked at high oven temperature were charred and unusable.

TABLE 16.12

MEAN THIAMIN AND RIBOFLAVIN IN RAW AND ROASTED BEEF AND RETENTION AFTER ROASTING

Cut of Beef	Mean Thiamin			Mean Riboflavin		
	Raw Roast (Mg/100 Gm)	Cooked Roast (Mg/100 Gm Cooked Wt)	Retention[1] (%[2])	Raw Roast (Mg/100 Gm)	Cooked Roast (Mg/100 Gm Cooked Wt)	Retention[1] (%[2])
Top round	0.102 ± 0.01[3]	0.084 ± 0.02	54	0.182 ± 0.01	0.227 ± 0.05	81
Ribs, 6–12	0.058 ± 0.01	0.044 ± 0.01	54	0.126 ± 0.02	0.171 ± 0.04	88
Rump	0.098 ± 0.01	0.084 ± 0.01	61	0.193 ± 0.08	0.236 ± 0.07	86
Tenderloin	0.091 ± 0.02	0.082 ± 0.02	63	0.197 ± 0.03	0.230 ± 0.02	84
Beef loaf prepared from ground chuck arm[4]	0.162 ± 0.02	0.184 ± 0.01	70	0.213 ± 0.02	0.284 ± 0.05	80

Source: Nobel and Gomez (1960).

[1] Any two means not side-scored by the same line may be considered significantly different at the 5% level, except for the riboflavin value for top round. This mean is not significantly different from those for tenderloin and beef loaf. Shortest significant ranges at the 5% level of significance are: for thiamin, 11 for 5 means, 10 for 2 means; for riboflavin, 6 for 5 means, 5 for 2 means.
[2] Calculated on entire roast basis.
[3] Standard deviation.
[4] Mean thiamin and riboflavin content of beef from which loaves were prepared: 0.186 and 0.195 mg per 100 gm raw weight, respectively.

in rump and tenderloin roasts, both of which showed intermediate values.

Roasted beef loaf and top round retained the lowest proportion, about 80%, of the riboflavin originally present in the raw samples, while rib and rump roasts retained the highest proportion, an average of 87%. Tenderloin roasts were intermediate in riboflavin retention and not significantly different in this respect from any of the other roasts.

The raw meat findings are in agreement with the values for raw meat as reported by others (Watt and Merrill 1950; Schweigert and Payne 1956; Cover et al. 1949). The cooked values are in agreement with the values reported by Watt and Merrill (1950) and Leverton and Odell (1958). In an earlier publication, Noble and Gomez (1958) provided results for roast lamb and the results are given in Table 16.13. The results were subsequently verified by Noble and Gomez (1960) in a study comparing the electronic cooking of lamb and bacon with conventional roasting. This latter paper is discussed under Electronic Cooking.

Mahon et al. (1956) roasted approximately 40-lb lots of smoked ham consisting of 4 hams at 300°F to an internal temperature of 170°F. The mean cooking times of the 40-lb lots was 4 and 9 min. Before roasting, the lean ham had a mean thiamin content of 0.527 mg per 100 gm on the moist, fat, salt basis, and 2.14 mg per 100 gm on the moisture-, fat-, chloride-free basis. After roasting, the mean thiamin content was 0.476 mg per 100 gm and on the moisture-, fat-, chloride-free basis, it was 1.71 mg per 100 gm (Leeking et al. 1956).

Broasting.—This is a relatively new food service process involving pressure frying which requires specialized equipment. The system because of pressure is more rapid than regular deep fat frying and results in less absorption of fat into the product. There is no known data on nutrient retention advantages and disadvantages.

Pressure Cooking.—Only one report was found on nutrient losses in large-scale pressure cooking of foods of animal origin. This was by Schlosser et al. (1957) and was on one nutrient, lysine, in older turkeys. These workers found that cooking cut-up older, fresh-killed tom turkeys by steaming at 5- or 15-lb pressure in a self-contained high-compression steamer or covered with aluminum foil and braised in an oven at 325°F made no appreciable difference in the lysine content of the cooked meat. Percentage retention of lysine during cooking was not obtained. Steaming under pressure required ⅕ the time of braising.

Infrared Cooking.—No reports were found on losses of nutrients during infrared cooking of foods of animal origin.

TABLE 16.13

MEAN THIAMIN AND RIBOFLAVIN IN RAW AND ROASTED LAMB AND RETENTION AFTER ROASTING

Cut	Mean Thiamin			Mean Riboflavin		
	Raw Roast Mg/100 Gm[1]	Cooked Roast Mg/100 Gm[2]	Retention %[3]	Raw Roast Mg/100 Gm[1]	Cooked Roast Mg/100 Gm[2]	Retention %[3]
Leg	0.176 ± 0.054[4]	0.194 ± 0.09	64 ± 8	—	—	—
Shoulder	0.088 ± 0.02	0.078 ± 0.02	63 ± 10	0.164 ± 0.07	0.207 ± 0.07	93 ± 11
Rack	0.110 ± 0.06	0.082 ± 0.04	62 ± 6	0.205 ± 0.04	0.219 ± 0.04	85 ± 9
Loin	0.169 ± 0.04	0.138 ± 0.02	66 ± 5	0.333 ± 0.06	0.242 ± 0.05	88 ± 12

Source: Nobel and Gomez (1958).

[1] Raw weight.
[2] Cooked weight.
[3] Calculated on entire roast basis.
[4] Standard deviation.

TABLE 16.14

EFFECT OF ELECTRONIC AND CONVENTIONAL COOKING ON RETENTION OF VITAMINS AND LYSINE IN GROUND MEAT PATTIES

Ground Meat Patties	Method of Cooking	Thiamin (%)	Vitamin Content[1] (Mg/100 Gm)	Riboflavin (%)	Niacin (%)	Amino Acid Lysine (%)	Laboratory
Beef	Electronic range	100		100	—	—	Proctor and Goldblith (1948)
	Fried	96		88	—	—	Thomas et al. (1949)
	Electronic range	77		99	89	—	
	Grilled	55		105	91	—	
Beef, frozen	Electronic range	84		69	—	90	Causey et al. (1950B)
	Fried	84		87	—	89	
	Electric oven	89		84	—	89	
Pork	Electronic range	91		87	81	—	Thomas et al. (1949)
	Grilled	79		102	84	—	
Pork, frozen	Electronic range	94		92	99	—	Causey et al. (1950A)
	Fried	86		84	104	—	
	Electric oven	89		81	99	—	
Lamb, frozen	Electronic range	82		45	—	85	Causey et al. (1950C)
	Fried	88		60	—	92	
	Electric oven	94		61	—	85	
Pork, frozen	Electronic range	70	2.80	—	—	—	Lim et al. (1959)
	Electric oven	—	2.74	—	—	—	
	Electric skillet	—	2.66	—	—	—	
Pork, frozen, irradiated, 3×10^6 rep	Electronic range	77	1.74	—	—	—	
	Electric oven	—	1.36	—	—	—	
	Electric skillet	—	1.40	—	—	—	
Pork, frozen, irradiated, 6×10^6 rep	Electronic range	81	1.13	—	—	—	
	Electric oven	—	0.79	—	—	—	
	Electric skillet	—	0.70	—	—	—	

Source: Fenton (1960).
[1] Dry, fat-free basis.

Electronic Cooking.—Wing and Alexander (1972) point out that microwave heating, accomplished by high-energy, electro-magnetic radiations, is an extremely efficient process that permits virtually no heat loss. It has been estimated that the average coupling efficiency of microwave radiations into most food products is about 80%; in conventional heating, this efficiency is only a few percent. Microwave heating is also extremely rapid and lends itself to use in continuous flow systems. In view of these factors, the application of microwave radiations in the food industry is growing.

Fenton (1960) reviewed the early work and compiled Tables 16.14 and 16.15. One notes the relatively good correlation in retention

TABLE 16.15

EFFECT OF ELECTRONIC AND CONVENTIONAL COOKING ON RETENTION OF VITAMINS IN ROAST MEAT

Meat	Method of Cooking	Retention of Vitamins			Laboratory
		Thiamin (%)	Riboflavin (%)	Niacin (%)	
Beef	Electronic range	63	84	63	Thomas *et al.* (1949)
	Electric oven	75	90	75	
	Electronic range	97	113	—	Campbell *et al.*
	Electric oven	98	105	—	(1958)
Pork	Electronic range	95	81	—	Campbell *et al.*
	Electric oven	100	87	—	(1958)

Source: Fenton (1960).

between electronic cooking and conventional cooking. Causey and Fenton (1951A) published data on the effect of electronic heating of frozen precooked dishes on thiamin retention as compared to electric oven results and reported no essential difference.

Thomas *et al.* (1949) reported higher thiamin retention, but similar riboflavin retention, when beef patties were cooked by microwaves than when grilled by a conventional method. Beef roasts retained slightly more riboflavin and thiamin after conventional than after microwave cooking. Kylen *et al.* (1964) reported that beef roasts retained less thiamin when cooked by microwave than by conventional methods, but that pork roasts and beef and ham loaves retained similar amounts by either method. Noble and Gomez (1962) found no significant difference in thiamin and riboflavin retention in lamb roasts cooked conventionally and by microwaves. Goldblith *et al.*

(1968) showed that microwave radiation *per se* had no destructive effect on thiamin and that loss of the vitamin was the result of heat only. Van Zante and Johnson (1970) obtained slightly higher thiamin and riboflavin retentions in conventionally than in electronically heated aqueous solutions buffered to simulate the pH of pork, but the losses by either heating method were small and of little practical importance.

In 1972, Bowers and Fryer studied turkey muscle (pectoralis) for thiamin and riboflavin, after cooking or heating in a microwave or a gas oven. The type of oven did not significantly affect thiamin content, and neither oven type or treatment (reheated after one day of refrigerated or frozen storage), significantly affected riboflavin values calculated on a net loss or as percentage retention (range of 86 to 90% retention). In fact, variation was greater among birds than between ovens or among treatments. Wing and Alexander (1972) processed chicken breasts for 1.5 min at 2,450 MHz which retained more vitamin B-6 than did chicken roasted conventionally. Conventional cooking resulted in a larger drip volume and vitamin B-6 loss but microwave heating resulted in greater product weight loss and less moisture retention (Table 16.16).

TABLE 16.16

VITAMIN B-6 CONTENT OF RAW AND COOKED CHICKEN BREAST
(10 SAMPLES) AND DRIP SAMPLES (4 SAMPLES)[1]

	Raw	Microwave[2]	Conventional[3]
Breast: mcg/gm mean[4]	20.2 ± 0.8	18.5 ± 0.9 (91%)[5]	16.7 ± 0.6 (83%)
Drip: mcg/gm	na	5.4 ± 0.8 (1.5%)	15.6 ± 1.5 (5.4%)

Source: After Wing and Alexander (1972).
[1] All samples freeze-dried before analysis; data based on dried tissue. (Expressed in terms of mcg/100 gm of moist chicken meat, average values were: raw, 525; microwave, 638; and conventional, 543.)
[2] Heated for 1.5 min.
[3] Heated for 45 min.
[4] The three means are significantly different from each other by Duncan's new multiple range test (P<0.05). The minimum significant difference between means is 0.4 mcg/gm for breast and 0.6 mcg/gm for drip.
[5] Retention calculated as a percentage of raw content.

It should be evident that the differences for at least three vitamins (thiamin, riboflavin, and pyridoxine) between the two methods of cooking are not particularly marked. This supports the rationale for urging that food service, such as in hospitals, use preplated frozen meals reheated in a microwave oven in lieu of holding hot, freshly prepared foods. Kahn and Livingston (1970) reported better thiamin retention for frozen meals (beef, chicken, and shrimp dishes) than in freshly prepared food both held at 82°C.

As to the effects of electronic cooking on amino acids, Causey *et al.* (1950B,C) reported no statistically significant differences in lysine retention in frozen ground beef patties cooked electronically, 90%, or conventional, 89%, or in frozen lamb patties electronically, 85%, or conventionally, 92 and 85% (Table 16.14). Campbell *et al.* (1958) found losses of 5 of the essential amino acids during electronic and conventional cooking of beef to be about the same, 15%.

Cooking and the Protein Quality of Meat.—Siedler (1961) reviewed the effects of standard cooking and processing methods on the nutritional value of meat protein, and on the basis of the reports of Lushbough *et al.* (1957), Schweigert and Guthneck (1954), and Heller *et al.* (1961) concluded that little of the essential amino acids tryptophan, methionine, and lysine is destroyed or lost by heat treatment using standard cooking or processing procedures, and that their biological availability is not impaired. In view of this and the price of protein food of animal origin, it would appear wise to de-emphasize meat as a source of vitamins.

Drippings During Cooking.—*Protein.*—Toepfer *et al.* (1955) showed the effect of the addition of water on loss of protein to pan drippings. Beef cooked without added water (oven roasts, griddle-broiled steaks and hamburgers, and meat loaf) lost only from 0.4 to 1.8% of protein to the pan drippings, while that cooked with added water (pot roasts and Swiss steak), lost 7.6 to 14.7% (Table 16.3). All of the beef was apparently cooked at low or moderate temperatures. The authors concluded that trimming of fat did not influence the protein loss to pan drippings.

Vitamins.—Cover *et al.* (1949) showed the damaging effect of high oven temperatures, 205°C, and cooking to a high internal temperature, 98°C; the pan drippings of both pork and beef were so burned that they were not usable. On the other hand, pan drippings of beef roasted at 150°C to an internal temperature of 80°C contained 6% of the thiamin, 20% of the pantothenic acid, and 16% of the niacin and riboflavin; those of pork roasted at the same temperature but to a higher internal temperature, 84°C, contained a higher percentage of the vitamins: 17% of the thiamin, 25% of the pantothenic acid, 26% of the niacin, and 19% of the riboflavin (Table 16.11).

Thomas *et al.* (1949) reported appreciable amounts of thiamin, 19%, riboflavin, 17%, and niacin, 20%, in the juice from beef patties cooked in an electronic oven. Causey *et al.* (1950B) found from 1 to 4% of the thiamin and from 3 to 12% of the riboflavin in the cooking drip of frozen beef patties.

Wing and Alexander (1972) recovered 5.4% vitamin B-6 in the drip from roasting chicken breast in conventional ovens for 45 min, as compared to a 1.5% loss when the breasts were cooked in an electronic oven for 1.5 min.

These and other studies show that appreciable B-6 vitamins, in addition to moisture and fat, may be lost to pan drippings. Pan drippings are sometimes used in soups and casserole dishes to give flavor, and are an integral part of meat dishes such as Swiss steak and stews. The practice of making gravies from drippings and serving gravies with meals is much less common, and the various dry gravy mixes marketed for their convenience in preparation have not been analyzed for vitamin content.

Carving Juice.—In the carving of beef, pork, lamb, and veal roasts, and of roasted poultry, appreciable cutting juice may be lost if carving is done while the roasts are very hot. No studies on nutritive loss to carving juices were found.

Holding on Steam Table.—Westerman (1948) found that 91% of the thiamin in sliced roast pork was retained during holding of the slices in an aluminum pan over rapidly boiling water for 30 min.

Erikson and Boyden (1947B) concluded that the institution practice of reheating turkey and holding it hot on the steam table had little or no destructive action on thiamin or riboflavin content.

Kahn and Livingston (1970) reported that the losses of thiamin in 4 common dishes (beef stew, chicken a la king, shrimp Newburg, and peas in cream sauce) freshly prepared and held hot at 180°F (82.2°C) for 1, 2 or 3 hr, were greater than those when the same foods were prepared, frozen, stored at −10°F (−23.3°C) and reheated in a microwave or an infrared oven to 194°F (90°C). When reheated by immersion in boiling water, thiamin retention was also greater than in the fresh food held hot, with the exception of peas in cream sauce, where the retention was the same for the frozen-reheated product and the fresh product held hot for 1 hr (Table 16.17).

Based on the averages of the 4 products studied, a difference of as much as 0.26 µg of thiamin per gram of food could occur between fresh food held hot for 3 hr and the microwave-heated frozen food. In an institution where 2 hot meals per day are served, assuming a total intake of 20 oz of entrées and vegetables per day, the thiamin difference is equivalent to as much as 18.4% of the daily recommended allowance for certain age groups.

Lachance et al. (1973) studied the thiamin retention of commercial chicken pot pies which were frozen, and either reheated in a convection oven, infrared oven, or conventional electric oven to an internal temperature of 180°F. The effect of reheating the product covered with aluminum foil and their holding uncovered product on a steam

EFFECTS OF FOOD SERVICE PRACTICES ON NUTRIENTS

TABLE 16.17

THIAMIN RETENTION IN FRESH AND FROZEN LABORATORY-MADE PRODUCTS AND IN COMMERCIALLY-PREPARED PRECOOKED FROZEN PRODUCTS SUBJECTED TO VARIOUS HEATING TREATMENT

	Thiamin Retention (%)[1]				
Sample/Treatment Laboratory Samples	Beef Stew	Chicken a la King	Shrimp Newburg	Peas in Cream Sauce	Avg of All Products
Freshly prepared	100	100	100	100	100.0
After $-10°$F storage	96	96	96	96.5	96.1
Frozen, microwave heated	95	94	92.5	93	93.5
Frozen, infrared heated	91	90	88	92	90.4
Frozen, immersion heated	85	87	86	87	86.0
Fresh, held at 180°F 1 hr	73.5	75	76	87	78.2
Fresh, held at 180°F 2 hr	68	70	73	83	73.9
Fresh, held at 180°F 3 hr	63	63	66	76	67.4

Source: Kahn and Livingston (1970).

[1] Percentage retention of thiamin in relation to freshly prepared products, except for commercial samples, where percentage retention in relation to frozen products is shown.

table at 180°F for 30 min were examined. The results are given in Table 16.18.

Whereas the product retained thiamin rather well during reheating except during the longer time required in an electric oven, subsequent losses occurred during holding.

Holding Roast Beef.—Boyle and Funk (1972) studied the thiamin content of beef roasts held over dry heat for 90 min before and after slicing, or refrigerated for 24 hr, sliced and reheated and compared with that of roasts sliced and served immediately. The U.S. Choice loin roasts had been cooked to an internal temperature of 54°C in a

TABLE 16.18

EFFECT OF RECONDITIONING UNIT AND HOLDING TO MAINTAIN 180°F ON PERCENTAGE RETENTION OF THIAMIN IN CHICKEN POT PIES

Oven	Covered (%)	Holding (Steam) (%)
Control, 77 ± 7 µg/100 gm	100	100
Convection, (30 min)	99	82
Infrared, (70 min)	97	79
Electric, (100 min)	68	77

Source: Lachance et al. (1973).

149°C oven following the procedure outlined by Boyle and Funk (1972). The 6 roasts for each of the 4 treatments stood undisturbed for 30 min following removal from the oven; during this time, the internal temperature rose to 60°C. During holding, maximum internal temperatures recorded for the roasts held unsliced and sliced were 58° and 59°C, respectively, while reheated slices reached a maximum temperature of 60°C during the reheating period, which averaged 67.2 min.

Average thiamin values of 0.202 ± 0.002, 0.195 ± 0.010, 0.173 ± 0.001, and 0.201 ± 0.001 mg per 100 gm dry weight basis, were calculated for roasts (a) held unsliced, (b) held sliced, (c) refrigerated, sliced, and reheated, and (d) sliced and served immediately. Thus, 79.2, 76.5, 67.8, and 78.8%, respectively, of the thiamin content was retained during cooking and the subsequent treatments.

According to Fenton (1960), Westerman (1948) reported that holding cooked pork roasts in the refrigerator (40°F) overnight resulted in a thiamin retention of 92.8%. Some of this pork was sliced and reheated in a 350°F oven in a pan with a small amount of water for 35 min and 93.1% of the thiamin was retained.

State of the Art.—Fenton (1960) found 12 reports applicable to the topic of nutrient retention in the large-scale cooking of foods of animal origin. This review has included about a dozen more and some of these were not strictly large-scale preparation experiments; however, they are included because the experiments were well controlled and studied quantities or cuts (leg of lamb) which are different in the institutional milieu only in the sense that a number of them may be cooked at the same time.

It is a meager group of studies and reflects poorly on the programming of research funding in this field.

The findings to date should be viewed as indicative rather than conclusive, although certain trends are obvious. The list does not include veal, fish, eggs, cheese, or milk. Many more studies on each group are needed—well-controlled studies which report details of initial quality and treatments so that results from several laboratories may be compared. In spite of the many interrelationships, information leading to better retention of nutrients can be obtained. Studies of the effect of large-scale cooking on each of the limiting nutrients in the dietary such as vitamin B-6, folacin, and trace minerals are essentially nil.

Cooking Food of Plant Origin

Boiling.—*Proportion of Water to Solid Material.*—Great variations occur in the proportion of water to solid material commonly used;

EFFECTS OF FOOD SERVICE PRACTICES ON NUTRIENTS

the ratio of water to solids usually varies from 1:2 to 2:1. Krehl and Winters (1950) measured the effect of cooking approximately 1-lb amounts of 12 vegetables to the same state of doneness in varying amounts of water on losses of both minerals and vitamins. In the pressure saucepan, ½ cup of water was added; in "waterless" cooking none was added. These methods were compared with boiling in water to cover and in ½ cup of added water. In all cases the pan was kept covered. Results of ascorbic acid and carotene losses are given in Table 16.19.

TABLE 16.19

EFFECT OF COOKING VEGETABLES IN VARYING AMOUNTS OF WATER ON LOSSES OF ASCORBIC ACID AND CAROTENE

	Percentage of Ascorbic Acid and Carotene Retained							
	Pressure Cooked		Water to Cover		½ Cup Water		Waterless	
Vegetable	Ascorbic Acid	Carotene	Ascorbic Acid	Carotene	Ascorbic Acid	Carotene	Ascorbic Acid	Carotene
Asparagus	67.6	78.5	45.2	64.6	66.4	92.3	69.4	101.5
Beets	93.8	81.4	74.0	72.4	87.3	82.8	81.1	96.2
Broccoli	68.0	88.6	50.6	76.0	68.7	84.3	70.2	97.7
Cabbage	75.5	96.8	44.3	73.3	57.4	89.7	68.4	95.6
Carrots	79.1	88.4	63.1	84.5	75.1	86.3	72.5	98.9
Cauliflower	75.5	89.8	47.3	80.7	54.0	83.7	70.7	97.4
Corn	74.9	88.2	60.2	86.4	65.1	87.3	69.6	93.1
Green beans	76.1	94.4	58.5	85.6	64.0	90.3	74.8	96.3
Peas	73.7	89.7	51.3	83.2	70.0	89.4	78.8	91.2
Potatoes	57.3	86.3	41.0	78.9	48.4	80.5	79.4	85.8
Squash	65.3	92.3	50.5	82.4	66.5	84.2	74.8	91.9
Spinach	61.7	74.8	49.1	80.7	51.7	87.2	70.0	91.3

Source: Krehl and Winters (1950).

These results show clearly that the greatest retention of both vitamins is obtained when vegetables are cooked without added water, and that least retention is associated with cooking in the largest amount of water, i.e., water to cover. Losses when a pressure saucepan or a small amount of water were used were intermediate. Other studies show that cooking by steaming above water gives retentions similar to those for cooking in a pressure saucepan or in a small amount of water. Krehl and Winters (1950) also obtained data on the minerals: calcium, iron, and phosphorus; and the vitamins: thiamin, riboflavin, and niacin. Losses of these nutrients were in the same direction as those of ascorbic acid and carotene so far as they were affected by the amount of water added. Also, the percentage

TABLE 16.20

ASCORBIC ACID IN RAW VEGETABLES AND PERCENTAGE[1] RETENTION AFTER COOKING

Vegetable	In raw vegetable		Retention in cooked vegetable[2]			Retention in cooking water		
	Mean (Mg/100 Gm)	Standard Error (Mg/100 Gm)	Pressure Saucepan (%)	"Waterless" Saucepan (%)	Boiling Water (%)	Pressure Saucepan (%)	"Waterless" Saucepan (%)	Boiling Water (%)
Broccoli	131.7	13.27	85	51	45	5	3	52
Brussels sprouts	97.6	2.84	78	63	60	4	none	21
Cabbage	49.5	2.65	74	56	41	6	4	38
Cauliflower	60.0	3.03	88	69	53	7	none	23
Rutabagas	30.6	0.84	85	75	46	15	2	33
Turnips	22.5	1.36	69	76	43	16	1	27

Source: Gordon and Noble (1964).

[1] Each percentage is the average of 4 separate cookings.
[2] Within each vegetable, any two means not underscored by the same line are significantly different. Shortest significance range, 5% level of significance, equals 8.8 for 18 means, 7.3 for 2 means.

EFFECTS OF FOOD SERVICE PRACTICES ON NUTRIENTS 499

of other nutrients lost was usually more than that for carotene and less than that for ascorbic acid. Thus, retention of ascorbic acid is therefore used as a measure of the relative desirability of a cooking procedure for foods but it is sad that so much work has been done on vitamin C alone.

In waterless cooking, retentions of vitamin C in the 12 vegetables ranged from about 70% to about 80%. The results of Gordon and Noble (1964) conducted on 4 serving portions did not show as good a retention for vitamin C as in the more exhaustive Krehl and Winters study; however, the percentage retention of ascorbic acid during cooking in the pressure saucepan was greater than during cooking by either the "waterless" technique or "boiling water to cover" technique for 6 out of 7 vegetables tested. The waterless method was superior to boiling for 5 out of 7 vegetables (Table 16.20). These data and those from other studies demonstrate that it is possible to retain to a high degree the original nutritional value of a vegetable in cooking. On the other hand, an amount of water enough to cover usually resulted in a retention of $\frac{1}{2}$ or less of the original vitamin. In any case, since the lost nutrients are for the most part in the cooking water, utilizing the cooking liquids is definitely worth the effort, but it is a disappearing practice.

Cooking Without Paring or in Large Pieces.—As might be expected there is evidence that solution losses in cooking are affected by the amount of surface, especially cut surface, exposed to the water. The skins of potatoes are such an effective barrier that boiled or steamed whole potatoes show little loss of vitamin C. When peeled, losses may amount to about 13% (Van Duyne et al. 1945). "Frenched" green beans (cut in lengthwise strips) retain only 28% of the original vitamin C as compared with 54% retention in those cooked whole, and 52% retention in those cut in 1-in. lengths (Noble and Worthington 1948).

Vitamin Retention During Boiling.—Harris (1960) prepared a summary of the many studies on the effects of large-scale boiling upon nutrient content. Table 16.21 is an abbreviated version of the original table. More recent studies of the effect of boiling consider ascorbic acid retention only and were not added to the Table.

Vitamin Retention During Pressure Cooking and Steam Cooking.—Invariably, the loss of nutrients is considerably less during pressure cooking or steam cooking of vegetables as compared to boiling at atmospheric pressures. This is readily evident from Tables 16.22 and 16.23 on vitamin retention during pressure cooking and steaming respectively. What is startling is that no new studies of any consequence have been published in the open literature since the 1950's.

TABLE 16.21

VITAMIN RETENTION DURING BOILING, LARGE-SCALE COOKING

	Cooking Time (Min)	Carotene (%)	Thiamin (%)	Riboflavin (%)	Niacin (%)	Ascorbic Acid (%)
Cereals						
Corn (canned)	30		78	95		32
Rice, white	20		46	52	59	
Rice, white enriched	20		46	62	59	
Vegetables						
General		100	70-75			25-60
General			46-100	50-97	36-98	8-100
Student meals		90	90	100	87	60
Earth vegetables						
Carrot		96	50	66	53	27
Carrot	30	96			99	
Carrot (dehyd)	25-45	81-99	52-75	55-89	58	
Carrot (canned)	30		95			38
Carrot	45		48	29	55	77
Carrot		96	50	66	53	27
Carrot	23		100			
Onion	15		89			
Parsnips	30					30
Parsnips	15					91
Parsnips	10		97			
Parsnips						64
Potato	10		87-92			
Potato	20					69
Potato	15-20		79			
Potato (dehyd, sulfited)	35-45				56-97	56-84
Potato	36		80			
Potato	done		83	97	83	89
Potato	26					48
Potato	15		89-92			
Potato			70-80	85	75	25-60
Potato	25		96	97	100	98
Pumpkin	15		100			
Swede	45					54
Sweet potato			92	100	100	
Sweet potato		90				100
Squash	10		89			
Turnip	15-20		60			
Taro	35		89			
Fruit vegetables						
Tomatoes (canned)	40	89-110			94-106	88-102
Tomatoes	30	95				
Tomatoes (canned)	30		71-94			99
Herbage vegetables						
Asparagus		97	71	78		81
Asparagus (canned)	30		87	99		93

EFFECTS OF FOOD SERVICE PRACTICES ON NUTRIENTS 501

TABLE 16.21 (*Continued*)

	Cooking Time (Min)	Carotene (%)	Thiamin (%)	Riboflavin (%)	Niacin (%)	Ascorbic Acid (%)
Beans (snap)	45		74			91
Beans (canned)	30		92	100		53
Beans	11		55-85	56-79		44-55
Beans	18-140		59-77	48-83		8-98
Beans			30-59			
Beans (frozen)	25-45		45-85	64-96	68	26-34
Beans	40-85		41-100			
Beans	17-19					62-66
Beans				57-71		25-84
Broccoli	5-6					53-82
Broccoli (frozen)	7-7					80
Broccoli (frozen)	8-20		44-61	48-71		46-61
Broccoli	13		80	95		74
Broccoli	30	100	95	73	88	67
Brussels sprouts	2					77
Brussels sprouts	5-20					39-62
Brussels sprouts	15					53
Brussels sprouts	30-40					44
Cabbage (dehyd)						67
Cabbage			50	59		50
Cabbage	120	84-87	33-43	39-50		18-30
Cabbage	4-8		64-85	66-90		57-74
Cabbage	20		73			
Cabbage (red)	60		18			5
Cabbage	13		46	58		51
Cabbage	12-20					30
Cabbage	18					38
Cabbage	30-40					60
Cabbage	7					57
Cauliflower	30-40					73
Cauliflower	3					81
Cauliflower	15		52			
Cauliflower	12		87			
Chard, Swiss	10					42
Corn, cob			80	97	87	63
Kale	6		65			50
Kale	25					23
Kale	180-240					3-53
Kale	30		100			
Lima beans (frozen)	14-81		56-82	59-89		52-76
Lima beans (frozen)	25-50		37-65	67	78-55	23-73
Lima beans	17					62
Lima beans (canned)	30		88	86-95		45
Okra	20					55-75
Peas	2					70
Peas	9		93	89		63
Peas	9		97	100		86

TABLE 16.21 (Continued)

	Cooking Time (Min)	Carotene (%)	Thiamin (%)	Riboflavin (%)	Niacin (%)	Ascorbic Acid (%)
Peas	30	90	94	93	97	86
Peas						40-80
Peas	8	96	80	69	87	50
Peas	12		91			
Peas (canned)	65		61-66	66	63	35
Peas	8		80	65		45-60
Peas (frozen)	20-27		50-75	62	59	32
Peas	35-135				45	
Peas	15-129		66-94	69-98		38-92
Peas	20		70			
Spinach	5			65		23
Spinach	30-40					38
Spinach	7		78			
Spinach	3					70
Spinach (frozen)	23-45		36-78	78		20-35
Spinach	6	97	50	52		24
Spinach	5		100			
Spinach	10-15	85-88	13-34	19-24	17-42	4-14
Spinach	7		50	53		33
Spinach (canned)	30		92-99	95		75-78
Soya sprouts	13-21		63-72			27-38
Turnips						40
Turnip greens						16-53
Legumes						
Legumes					8	
Beans	35-135				45	
Beans	225					84-95

Source: Harris (1960).

Meanwhile, the use of steamers has augmented in food service applications whereas the use of pressure cookers in the home has probably diminished. Less cooking is being done from fresh foods at both the home and institutional level. More frozen raw foods and precooked frozen foods are now in use. The food service steamer is used as much, if not more, for conditioning convenience foods as it is for cooking in the classical sense. Since nutrient retention is generally better in frozen products and the duration of cooking is less and the amount of water necessary used in steaming less, all indications are that nutrient retention is probably improved, but there are no systematic studies to confirm or deny such a hypothesis as it applies to the foods, equipment, and practices in use today.

Vitamin Losses into Cooking Water.—Even the very unstable vitamin, ascorbic acid, can be found in cooking water. Although only

EFFECTS OF FOOD SERVICE PRACTICES ON NUTRIENTS

TABLE 16.22

VITAMIN RETENTION DURING PRESSURE COOKING

	Cooking Time (Min)	Pressure (Lb)	Thiamin (%)	Riboflavin (%)	Ascorbic Acid (%)
Earth vegetables					
Carrots	10	7	75		49
Potato			96		89
Potato	20	6	95		100
Fruit vegetables					
Chili		10			71-93
Bitter gourds		10			71-93
Herbage vegetables					
Broccoli (frozen)	6	15			72
Broccoli	5-7	5-15	76-95	84-100	77-83
Broccoli (frozen)	5-6	5-15			75-81
Brussels sprouts	0.33	15			97
Cabbage	10	15			60
Cauliflower	0.5	15			92
Lima beans	10				64
Okra (115°C)	2				82
Peas	1	15			88
Peas	9-10	5-15	98	86-92	68-72
Squash	45	6	52		42
Spinach	0.75	15			80
Turnips	30	6	55		100

Source: Harris (1960).

three vitamins have ever been repeatedly studied, it appears probable that other vitamins could also be expected in cooking water. Table 16.24 is a condensation of the information tabulated by Harris (1960). Other investigations show that time of cooking is evidently a critical variable; for example, broccoli boiled for 2, 5½, and 11 min lost 25%, 32%, and 33%, respectively, of the ascorbic acid content into the cooking water (Barnes et al. 1943B). These losses can be reduced considerably by limiting the volume of the cooking water. For instance, Barnes et al. (1943B) observed ascorbic acid retentions of 82%, 57%, and 53% when broccoli was cooked in 100 cc, 500 cc, and 1000 cc of water, respectively. McIntosh et al. (1942) observed retentions of 84%, 63%, and 65% ascorbic acid when cauliflower was cooked in 120 cc, 250 cc, and 480 cc of water. When 400 gm of cabbage was cooked in 200 cc and 800 cc water, the ascorbic acid retention was 74% and 47% respectively (Van Duyne et al. 1948). Gilpin et al. (1959) reported similar ascorbic acid changes in broccoli. Leaching into the cooking liquid rather than destruction by heat is therefore to be expected.

TABLE 16.23

VITAMIN RETENTION DURING STEAM COOKING

	Cooking Time (Min)	Carotene (%)	Thiamin (%)	Riboflavin (%)	Niacin (%)	Ascorbic Acid (%)
Cereals						
Corn			85	100	100	64
Rice	32		95			
Earth vegetables						
Carrots						70-83
Carrots	20					86
Carrots		93	82	92	84	62
Carrots (dehyd)	15	91	49-57	63-67	21-57	
Parsnips	25					86
Potato	60		86			
Potato	53		84	72	78	88
Potato	60					46
Potato	50					95
Sweet potato						98
Sweet potato	25-64		70-93	71-100		87-100
Herbage vegetables						
Asparagus						83
Beans, snap						73
Beans, snap	36		81-95	78-93		37-46
Broccoli	16					80
Broccoli	8-9		56-83	70-86		59-87
Brussels sprouts	7					89-94
Brussels sprouts	20					60-64
Brussels sprouts						73
Cabbage	30					52
Cabbage			89	95		68
Cabbage	20		88	100		67
Cabbage	9		85	82		84
Cauliflower	10					71-83
Cauliflower						76
Kale	30-50					40-67
Lima beans						65
Peas	12					68
Peas						74
Peas (frozen)			89	91		53
Spinach			90	85		76
Spinach (frozen)	10					71-73
Spinach	90	98	61	74	72	14
Spinach			82	78		30
Spinach	5-6		79-81	80-89		50-67
Legumes and oilseeds						
Beans			79	83		28
Soybeans						81

Source: Harris (1960).

EFFECTS OF FOOD SERVICE PRACTICES ON NUTRIENTS   505

TABLE 16.24

VITAMIN LOSSES INTO COOKING WATER

	Cooking Time (Min)	Thiamin (%)	Riboflavin (%)	Niacin (%)	Ascorbic Acid (%)
Cereals					
Corn		15	12	14	30
Corn (canned)	30	22	27		
Earth vegetables					
Carrots (dehyd)	25-45	52-93	55-89	42-58	
Carrots (canned)	30	19			12
Potato				2-19	
Potato					48-82
Potato		2-45	2-48		8-38
Potato	25				5
Sweet potato					15-19
Sweet potato				3-4	
Herbage vegetables					
Asparagus (canned)	30	22	22		23
Bean, snap				8-41	
Bean, snap	30				7
Bean, (canned)	30	31	30		23
Broccoli				6-13	
Broccoli	5-8				13-15
Broccoli	5-13	17-22	11-14		10-12
Broccoli	9-16	40-46	41-54		16-35
Brussels sprouts					30-64
Brussels sprouts					49-53
Cabbage	20-120	46-72	53-78		26-33
Cabbage		6-40	6-41		3-35
Cabbage	4-8	24-35	20-53		14-38
Cabbage	8				37
Cauliflower					57-65
Cauliflower				14-18	
Lima beans				5-13	
Lima beans (canned)	30	22	19		16
Okra	20				13
Peas			1	15-33	
Peas		8-62	1-69		4-44
Peas	9				9-14
Spinach	5	14-43	10-39		18-43
Spinach		5-26	7-47		7-30
Spinach (canned)	30	26-32	23		21-28

Source: Harris (1960).

Mineral Losses with Boiling and Steaming.—McCance et al. (1936) investigated the effect of food preparation upon calcium content. Baking, frying, roasting, and steaming had no important effect upon calcium content. During boiling, measurable amounts of calcium are extracted from vegetables. When scarlet runner beans were boiled for 40 min and carrots for 120 min, between 12 and 20% of the calcium passed into the cooking water. At the same time, over 60% of the chlorine and potassium were extracted from the beans. The addition of alkali to the water had no effect.

Krehl and Winters (1950) reported that over 20% of the calcium was extracted from cabbage during boiling, and only 9% during pressure cooking. The average results with 11 vegetables showed nearly

TABLE 16.25

STEAM VERSUS BOILING[1]

Vegetable	Cooking Method	Loss of Dry Matter (%)	Loss of Protein (%)	Loss of Calcium (%)	Loss of Magnesium (%)	Loss of Phosphorus (%)	Loss of Iron (%)
Asparagus	Boiled	14.0	20.0	16.5	8.8	25.8	34.4
	Steamed	7.9	13.3	15.3	1.4	10.4	20.0
Beans, string	Boiled	24.6	29.1	29.3	31.4	27.6	38.1
	Steamed	14.2	16.6	16.3	21.4	18.8	24.5
Beetgreens	Boiled	29.7	22.2	15.9	41.6	44.9	43.1
	Steamed	15.7	6.9	3.8	14.1	14.0	24.5
Cabbage	Boiled	60.7	61.5	72.3	76.1	59.9	66.6
	Steamed	26.4	31.5	40.2	43.4	22.0	34.6
Cauliflower	Boiled	37.6	44.4	24.6	25.0	49.8	36.2
	Steamed	2.1	7.6	3.1	1.7	19.2	8.3
Celery	Boiled	45.4	52.6	36.1	57.1	48.7	—
	Steamed	22.3	22.3	11.6	32.4	15.7	—
Celery cabbage	Boiled	63.2	67.1	49.7	61.6	66.1	67.6
	Steamed	38.3	33.5	16.3	32.6	30.2	44.1
Spinach	Boiled	33.9	29.0	5.5	59.1	48.8	57.1
	Steamed	8.4	5.6	0.0	17.8	10.2	25.7
Beets	Boiled	30.9	22.0	18.7	30.9	33.6	—
	Steamed	21.5	5.4	1.5	29.4	20.1	—
Carrots	Boiled	20.1	26.4	8.9	22.8	19.0	34.1
	Steamed	5.1	14.5	5.1	5.6	1.1	20.7
Kohlrabi	Boiled	33.6	23.2	27.8	40.4	27.7	51.7
	Steamed	7.6	1.0	1.0	14.3	7.7	21.3
Onions	Boiled	21.3	50.2	15.6	27.8	40.2	36.1
	Steamed	11.0	30.7	7.1	15.7	31.5	15.9
Parsnips	Boiled	21.9	13.3	11.4	46.8	23.7	27.6
	Steamed	4.6	20.0	4.2	8.2	5.7	8.1
Potatoes	Boiled	9.4	—	16.8	18.8	18.3	—
	Steamed	4.0	—	9.6	14.0	11.7	—
Sweet potatoes	Boiled	29.0	71.5	38.3	45.3	44.4	31.5
	Steamed	21.1	15.0	22.1	31.5	24.5	25.1
Rutabagas	Boiled	45.8	48.6	37.1	42.7	57.2	50.0
	Steamed	13.2	15.7	13.4	3.4	24.6	14.3
Average for all vegetables	Boiled	39.4	43.0	31.9	44.7	46.4	48.0
	Steamed	14.0	16.0	10.7	18.6	16.7	21.3

Source: *Cooking for Profit*, Aug. 1965, p. 15.

[1] This chart shows the dramatic savings of nutrients when steam is used in preference to a stock pot. There is a close relationship between nutrients, color, flavor, and texture.

EFFECTS OF FOOD SERVICE PRACTICES ON NUTRIENTS 507

25% of the calcium was leached when the vegetable was covered with water during cooking, and less was lost when less water was added. Table 16.25 provides a comparison of boiling and steaming losses for calcium, magnesium, phosphorus, and iron from several vegetables (Anon. 1956). Some of the cooking losses can be associated with the dry matter and protein losses but in several instances mineral losses, especially for magnesium and iron, are substantial. The very beneficial effects of steaming as compared to boiling on mineral retention are self evident.

Frying.—Okra fried with fat for 15 min retained 55% of its original ascorbic acid content (Walker and Arvidsson 1952). Levy (1937) noted only 20–45% retention of ascorbic acid when potatoes were fried, and 55–80% retention when baked. Fenton (1940) and Richardson et al. (1937) observed ascorbic acid retentions of 67% in fried, and 60% in baked potatoes. Domah et al. (1974) studied the effect of frying potatoes at 140°C for 10, 20 or 30 min and at 180°C for 5 min. The results are given in Table 16.26. The retention of

TABLE 16.26

INFLUENCE OF FRYING ON STABILITY OF AA AND DAA IN POTATOES
(RESULTS ARE AVERAGE VALUES OF 8 EXPERIMENTS)

Sample	Dry Matter	DAA Mg/100 Gm Dry Matter	AA Mg/100 Gm Dry Matter	Total Content of Vitamin C Mg/100 Gm Dry Matter
Raw peeled potatoes before frying	26.20	7.4	44.6	52.0
Fried potatoes (140°C/10 min)	83.01	29.7	20.6	50.3
Fried potatoes (140°C/20 min)	84.00	33.7	7.3	41.0
Fried potatoes (140°C/30 min)	88.00	42.7	0.0	42.7
Fried potatoes (180°C/5 min)	89.10	42.8	0.0	42.8

Source: Domah et al. (1974).

total vitamin C was good and, in fact, better than that in boiled potatoes (see Table 16.27). However, ascorbic acid is oxidized to dehydroascorbic acid (DAA) more rapidly with frying but the hydrolysis of DAA is slowed by the dehydration of the product during frying and therefore DAA accumulates in the fried potato. During boiling, DAA is hydrolyzed to 2,3 diketogluconic acid.

Absolutely no retention data exists on the several other foods, in

508 NUTRITIONAL EVALUATION OF FOOD PROCESSING

TABLE 16.27

INFLUENCE OF SODIUM CHLORIDE ON STABILITY OF AA AND DAA OF POTATOES[1] (RESULTS ARE AVERAGE VALUES OF 4 EXPERIMENTS)

Sample	DAA Mg/100 Gm Dry Matter	AA Mg/100 Gm Dry Matter	Total Content of Vitamin C Mg/100 Gm Dry Matter
Raw peeled potatoes	7.4	43.1	50.5
Cooked peeled potatoes in water	9.0	17.1	26.1
Infusion	4.9	5.4	10.3
Cooked peeled potatoes in 1% NaCl	9.1	13.1	22.2
Infusion	5.7	5.0	10.7
Cooked peeled potatoes in 5% NaCl	7.1	11.2	18.3
Infusion	7.6	4.1	11.7
Cooked peeled potatoes in 10% NaCl	5.8	8.9	14.7
Infusion	7.0	6.6	13.6

Source: Domah et al. (1974).
[1] Cooking time: 25 min.

particular fast food such as chicken, fish, onion, etc., which are commonly deep fat fried on a large scale. The effect of frying several vegetables on a small scale is briefly discussed in Chap. 17.

Vitamin Losses in Baking Vegetables.—Baking destroys significant amounts of unstable nutrients in some foods. Spiers et al. (1945) reported that baked potatoes retained 76% carotene and 96% ascorbic acid, while parallel samples of boiled potatoes retained 90% carotene and 110% ascorbic acid. Similarly, Pearson and Luecke (1945) compared baked and boiled sweet potatoes, and reported respective retentions of 76% and 92% thiamin, 89% and 103% riboflavin, 86% and 101% niacin, and 77% and 100% pantothenic acid. Kahn and Halliday (1944) compared baked (in skin), pared-baked, and pared-cut baked, and French-fried potatoes, and observed ascorbic acid retentions of 80%, 80%, 42%, and 77%, respectively. After standing approximately 1 hr, the retentions were 41%, 52%, 11%, and 71%, respectively. Page and Hanning (1963) studied 58 samples of boiled and baked potatoes for B-6 and niacin retention. The results are given in Table 16.28. Baking losses (9% for B-6 and 4% for niacin) were less than for boiling (20 and 18% respectively). The difference was essentially found in the cooking liquid.

Microwave Cooking.—No significant differences in either vitamin retention or palatability were found by Stevens and Fenton (1951)

EFFECTS OF FOOD SERVICE PRACTICES ON NUTRIENTS 509

TABLE 16.28

NIACIN AND VITAMIN B-6 RETENTION IN BOILED AND BAKED POTATOES

Location and Variety	No. of Samples	Mean Retention in Boiled Potatoes (%)	Mean Loss in Cooking Liquid (%)	Mean Retention in Baked Potatoes (%)
		Niacin		
Wisconsin				
Cobbler	7	80.7 ± 2.73[1]	17.1 ± 1.47[1]	91.2 ± 6.25[1]
Triumph	14	82.4 ± 4.45	15.5 ± 2.32	99.1 ± 4.60
Minnesota, Cobbler	12	82.6 ± 6.76	16.1 ± 6.11	93.6 ± 3.05
Kentucky, Cobbler	9	82.6 ± 3.14	17.6 ± 2.04	91.1 ± 4.66
Indiana, Chippewa	10	79.0 ± 2.36	173. ± 1.15	94.1 ± 4.93
Colorado, McClure	3	84.8 ± 5.01	16.8 ± 1.91	105.4 ± 5.36
Idaho, Russet Burbank	3	84.7 ± 2.92	16.2 ± 3.57	90.6 ± 0.85
Overall mean		81.9	16.6	95.8
		Vitamin B-6		
Wisconsin				
Cobbler	7	81.0 ± 3.03	15.9 ± 1.32	90.9 ± 9.17
Triumph	14	78.5 ± 6.24	14.5 ± 2.14	92.8 ± 9.44
Minnesota, Cobbler	12	81.8 ± 8.03	15.8 ± 5.67	93.2 ± 8.52
Kentucky, Cobbler	9	78.2 ± 10.70	15.2 ± 1.84	88.6 ± 12.10
Indiana, Chippewa	10	79.8 ± 8.10	14.0 ± 1.11	88.8 ± 5.32
Colorado, McClure	3	79.0 ± 4.31	15.4 ± 1.35	91.8 ± 8.06
Idaho, Russet Burbank	3	84.7 ± 8.26	18.4 ± 5.40	91.3 ± 2.85
Overall mean		80.0	15.2	91.2

Source: Page and Hanning (1963).
[1] Standard deviation.

when frozen peas were cooked in a small amount of water in a stewpan or in a package in an electronic oven. Campbell et al. (1958) concluded that the retention of ascorbic acid was practically equivalent after conventional and microwave when solidly frozen broccoli and peas were heated for the length of time required for defrosting and cooking to pleasing tenderness. Higher retentions for ascorbic acid were obtained by Gordon and Noble (1959A) in fresh cabbage, broccoli, and cauliflower cooked in an electronic range designed for home use than in those prepared by pressure cooking or boiling. The retention is given in Table 16.29. The Gordon and Noble (1959B) work used considerably larger amounts of water and shorter cooking times for boiling and less water and shorter times for microwave cooking than were used by Kylen et al. (1961) who studied 7 fresh and 3 frozen (4 and 8 months) vegetables and found no statistically significant differences in the amounts of ascorbic acid retained

TABLE 16.29

SEPARATION OF MEAN PERCENTAGE RETENTION OF ASCORBIC ACID INTO HOMOGENEOUS GROUPS ACCORDING TO DUNCAN'S MULTIPLE RANGE TEST (9)

Vegetable	Method of Cooking	Mean Retention[1] (%)
Cauliflower	Electronic range	90
Broccoli	Electronic range	87
Cauliflower	Pressure saucepan	82
Broccoli	Pressure saucepan	81
Cabbage	Electronic range	80
Cauliflower	Boiling water	73
Cabbage	Pressure saucepan	70
Broccoli	Boiling water	45
Cabbage	Boiling water	38

Source: Gordon and Noble (1959A).

[1] Any two means not side-scored by the same line may be considered significantly different. Shortest significant ranges, 5% level of significance = 5.8 for 9 means, 5.0 for 2 means. Each percentage is the average of 4 determinations.

(Tables 16.30 and 16.31). The issue of the effect of the quantity of water used was further investigated by Eheart and Gott (1964).

Ascorbic acid retentions in vegetables cooked with and without water in the microwave oven and by the conventional method are reported in Table 16.32. Frozen peas, frozen broccoli, and fresh potatoes showed retentions which were not significantly different when the two cooking methods were compared. Only in frozen spinach was a significant difference found in percentage retention of ascorbic acid. Microwave-cooked spinach retained an average of 64.7% of the vitamin, while conventionally-cooked spinach contained only 49.3%. The difference was significant at the 5% level.

Eheart and Gott (1964) also reported no change in the carotene content of frozen peas cooked by microwave or boiling. In 1965 Eheart and Gott studied reduced ascorbic acid retention in green beans and broccoli cooked by stir-fry, microwave, and boiling at two different water-food ratios. Table 16.33 provides the data. In the case of green beans there was no difference between the three methods except when minimal water was used in boiling which gave the best retention. However, broccoli showed as good a retention with stir-frying as with the minimal water cooking—both superior to microwave or boiling with eight times as much water.

In general, it appears that microwave cooking provides for ascorbic acid retention as well as does minimum water cooking. Until data on

TABLE 16.30
MOISTURE AND ASCORBIC ACID CONTENT OF RAW AND COOKED FRESH VEGETABLES

Vegetable	Raw		Cooked				Mean Retention (Raw Weight Basis)	
	Moisture[1] (%)	Ascorbic Acid[1] (Mg/Gm)	Cooking Method	Moisture[1] (%)	Ascorbic Acid[1]		Vegetable (%)	Cooking Water (%)
					Vegetable (Mg/Gm)	Cooking Water (Mg/Ml)		
Broccoli	89.1 ± 0.08	1.46 ± 0.042	Conventional[2]	87.7 ± 0.67	1.23 ± 0.033	0.54 ± 0.065	83	10
			Microwave[2]	86.3 ± 0.28	1.33 ± 0.052	0.55 ± 0.022	79	11
Cabbage	92.9 ± 0.12	0.47 ± 0.023	Conventional	92.3 ± 0.14	0.37 ± 0.025	0.20 ± 0.014	69	14
			Microwave	91.4 ± 0.27	0.41 ± 0.024	0.17 ± 0.009	72	11[3]
Cauliflower	91.0 ± 0.89	0.82 ± 0.056	Conventional	89.4 ± 1.05	0.78 ± 0.062	0.28 ± 0.031	92	7
			Microwave	89.0 ± 1.12	0.76 ± 0.045	0.14 ± 0.022	87	6
Peas	80.3 ± 0.59	0.38 ± 0.018	Conventional	79.2 ± 0.59	0.29 ± 0.011	0.38 ± 0.020	73	23
			Microwave	77.4 ± 0.64	0.30 ± 0.017	0.41 ± 0.027	74	15[3]
			Microwave	78.2 ± 0.48	0.30 ± 0.009	0.38 ± 0.021	74	17[2]
Green beans	91.5 ± 0.40	0.18 ± 0.009	Conventional	91.5 ± 0.95	0.15 ± 1.010	0.06 ± 0.010	74	9
			Microwave	90.4 ± 1.12	0.16 ± 0.012	0.07 ± 0.006	78	7[4]
Soybeans	68.7 ± 1.33	0.24 ± 0.009	Conventional	65.7 ± 1.04	0.20 ± 0.004	0.25 ± 0.071	79	11
			Microwave	66.0 ± 1.92	0.20 ± 0.013	0.30 ± 0.023	76	10[5]
Spinach	92.8 ± 0.16	0.52 ± 0.025	Conventional	88.9 ± 0.33	0.52 ± 0.032	0.44 ± 0.160	61	5
			Microwave	89.4 ± 0.27	0.42 ± 0.015	0.31 ± 0.027	56	9

Source: Kylen et al. (1961).

[1] Mean values of four replications and standard deviations of the means.
[2] Cooking procedure modified by adding vegetable to boiling instead of cold water.
[3] Indicates that the mean percentage retention after microwave cooking is significantly lower at the 1% level.
[4] Indicates that the mean percentage retention after microwave cooking is significantly lower at the 5% level.
[5] Means for these samples include five instead of four values.

TABLE 16.31

MOISTURE AND ASCORBIC ACID CONTENT OF RAW AND COOKED FROZEN VEGETABLES

Vegetable	Fresh (Raw)		Length of Freezer Storage (Months)	Cooking Method	Frozen				Mean Retention (Fresh Raw Weight Basis)	
	Moisture (%)	Ascorbic Acid (Mg/Gm)			Moisture[1] (%)	Ascorbic Acid[1]				
						Vegetable (Mg/Gm)	Cooking Water (Mg/Ml)		Vegetable (%)	Cooking Water (%)
Broccoli	89.4 ± 0.32	1.29 ± 0.062	4	Raw[2]	91.6 ± 0.20	0.85 ± 0.022	—		70	—
			4	Conventional	89.7 ± 0.25	0.66 ± 0.024	0.30 ± 0.025		48	12
			4	Microwave	89.2 ± 0.20	0.74 ± 0.026	0.41 ± 0.045		52	9
			8	Raw[2]	91.3 ± 0.34	0.70 ± 0.024	—		60	—
			8	Conventional[2]	89.2 ± 0.31	0.62 ± 0.017	0.38 ± 0.007		47	11
			8	Microwave[2]	89.5 ± 0.07	0.63 ± 0.012	0.39 ± 0.015		48	7[3]
Green beans	95.5 ± 0.15	0.16 ± 0.009	4	Raw	92.6 ± 0.14	0.10 ± 0.011	—		62	—
			4	Conventional	90.5 ± 0.05	0.08 ± 0.011	0.05 ± 0.011		38	8
			4	Microwave	90.9 ± 0.31	0.08 ± 0.012	0.05 ± 0.010		38	8
			8	Raw	93.7 ± 0.25	0.09 ± 0.011	—		53	—
			8	Conventional	91.0 ± 0.33	0.06 ± 0.009	0.04 ± 0.005		30	5
			8	Microwave	91.8 ± 0.30	0.06 ± 0.007	0.03 ± 0.007		28	6
Spinach	92.8 ± 0.016	0.52 ± 0.025	4	Raw	92.1 ± 0.25	0.29 ± 0.018	—		40	—
			4	Conventional	88.5 ± 0.74	0.21 ± 0.011	0.14 ± 0.015		22	10
			4	Microwave	88.8 ± 0.29	0.25 ± 0.020	0.20 ± 0.018		26	5[3]
			8	Raw	91.9 ± 0.19	0.23 ± 0.027	—		36	—
			8	Conventional	89.7 ± 0.22	0.16 ± 0.016	0.09 ± 0.015		20	8
			8	Microwave	88.5 ± 0.33	0.20 ± 0.023	0.13 ± 0.041		23	5[4]

Source: Kylen et al. (1961).

[1] Mean values for four replications and standard deviations of the means.
[2] Means for these samples include three instead of four values.
[3] Indicates that the mean percentage retention was significantly lower at the 5% level.
[4] Indicates that the mean percentage retention was significantly lower at the 1% level.

EFFECTS OF FOOD SERVICE PRACTICES ON NUTRIENTS 513

TABLE 16.32

RETENTION OF ASCORBIC ACID IN VEGETABLES COOKED BY CONVENTIONAL AND MICROWAVE METHODS

Vegetable	Conventional Cooking		Microwave Cooking				F Values	Least Significant Difference
			With Water		Without Water			
	Mean[1] (%)	Range (%)	Mean[1] (%)	Range (%)	Mean[1] (%)	Range (%)		
Broccoli	74.8	65.6 - 87.1	76.4	67.8 - 95.3	82.2	71.3 - 101.3	1.47	
Peas	68.4	57.4 - 81.0	62.3	45.6 - 71.2	65.0	52.8 - 80.7	0.86	
Potatoes	79.9	63.9 - 94.5	76.5	53.4 - 102.7	74.4	63.1 - 87.9	0.24	
Spinach	49.3	38.3 - 64.0	64.7	42.9 - 87.4	67.2	49.0 - 88.5	4.32[2]	13.7

Source: Eheart and Gott (1964).
[1] Eight replications, except 6 for potatoes.
[2] Significant at the 5% level.

other water-soluble nutrients are obtained, the convenience of microwave seems to be advantageous both from the viewpoint of time and nutrient retention. Food service systems based on satellite microwave conditioning of frozen foods, as in hospitals, may be superior to holding conventionally-heated foods, but systematic studies are certainly needed to verify this apparent likelihood.

Holding Refrigerated Before Reheating.—There is little information on this practice which relates to both the issue of holding leftovers

TABLE 16.33

ASCORBIC ACID RETENTION OF BROCCOLI AND GREEN BEANS COOKED BY FOUR METHODS

Vegetable and Cooking Method	Amount		Cooking Time (Min)	Ascorbic Acid Retention (%)
	Water (Ml)	Vegetable (Gm)		
Broccoli				
Raw				100
Stir-fry	240	360	10[1]	76.6
Microwave	240	300	11	56.8
Uncovered[2]	1200	300	15	44.8
Covered[3]	150	300	20	74.2
Green Beans				
Raw				100
Stir-fry	180	360	15[4]	57.5
Microwave	240	300	10	58.9
Uncovered[2]	1200	300	20	59.6
Covered[3]	150	300	20	76.0

Source: Eheart and Gott (1965).
[1] 1 min at 350°F plus 9 min at 250°F.
[2] 4 : 1 H_2O to vegetable.
[3] 0.5 : 1 H_2O to vegetable.
[4] 5 min at 350°F plus 10 min at 250°F.

and the preparation of cooked foods which are preplated but not frozen because only 4 to 24 hr will elapse before delivery to a satellite location and reconditioning in a convection oven. Loss of thiamin by meat held on the steam table may not be great according to Rice and Beuk (1945) since they found that the rate of loss at temperatures below 170°F decreased for 16 to 24 hr, after which it remained constant. Vail and Westerman (1946) noted that pork roasts sliced and held over steam for 30 min retained 91% of the thiamin of the cooked meat, that roasts held overnight kept 93%, and when the latter were sliced and reheated they retained 90%. Erikson and Boyden (1947A) reheated roast turkey meat, both white and dark (which had been held in the refrigerator for 24 hr) for 20 or 50 min, and found practically no loss of thiamin beyond that produced by the roasting. The same was true of riboflavin although both release of bound riboflavin and its partial destruction may have occurred. Holding under refrigeration and reheating of cooked meat does not appear to be harsh on either thiamin or riboflavin. This, however, should not remove the concern over losses with these practices as they may apply to food of plant origin. The data of Branion et al. (1947) relevant to ascorbic acid in potatoes may be indicative of the concern needed about such practices as they may affect other labile nutrients which may already be limiting in the dietary, such as folacin and vitamin B-6. Table 16.34 summarizes the Branion et al. data and demonstrates that number of servings involved is not as important as the practice itself since there

TABLE 16.34

CHANGES IN ASCORBIC ACID CONTENT OF BOILED OLD[1] POTATOES LEFT STANDING IN REFRIGERATOR

Condition When Assayed	100 Servings		4 Servings	
	Ascorbic Acid[2] (Mg/100 Gm)	Loss (%)	Ascorbic Acid[2] (Mg/100 Gm)	Loss (%)
Potatoes as purchased	11.3		11.9	
Potatoes, edible portion	11.5		12.6	
Potatoes boiled	4.9		8.5	
After standing in refrigerator 1 day	2.8	43	2.5	71
After standing in refrigerator 2 days	0.8	84	2.1	75
After standing in refrigerator 3 days	0.1	98	0.4	95
After standing in refrigerator 4 days	0.0	100	0.1	99
After standing in refrigerator 5 days	0.0	100	0.0	100

Source: Branion et al. (1947).

[1] February, 1944.
[2] Average of three determinations.

is no ascorbic acid remaining in boiled potatoes after standing in the refrigerator three days.

Holding Before Freezing (Institutional Meals).—Some school and hospital food service systems, as well as commercial operations, prepare and cook foods and within a short period of time (up to 2 hr) freeze the food, either individually packaged or assembled into preplated meal trays. There is no data on the changes that may occur in this interim period. Millross et al. (1973B) in their treatise on the "Cook-Freeze System for School Meals" do mention this problem, and show the effect of delays between cooking and blast freezing on the vitamin C content of spring cabbage. Table 16.35 demonstrates

TABLE 16.35

VITAMIN C CONTENT OF COOKED SPRING CABBAGE

Time Between End of Cooking and Entry into Freezer (Min)	Vitamin C Content at End of Cooking Process (Mg/100 Gm)	Vitamin C Content Prior to Freezing (Mg/100 Gm)	Loss (%)
18	50.5	46.0	8.9
25	41.0	29.8	27.3
30	48.5	40.0	21.7
55	55.4	33.7	39.2

Source: Millross et al. (1973).

that the longer the delay, the greater the loss, at least for vitamin C. With the cook-freeze system gaining use in the United States and abroad, it becomes imperative that nutrient changes in this interim be systematically researched.

Steam Table.—It is common practice, especially in cafeteria-type restaurants, to hold cooked foods on heated surfaces and steam tables for as long as 1, 2, and even 3 hr until served. Under these conditions the food slowly loses nutrients, primarily through oxidative changes. Thomas (1942) studied 9 vegetables for ascorbic acid by home and institutional cooking methods and measured retention after holding 1 hr over steam (Table 16.36). Whereas the differences in retention between cooking small and large quantities were lower when institutional methods were used, they were not dramatic for 5 of the vegetables; however, holding 1 hr was substantial for 6 of 7 of the vegetables tested.

Branion et al. (1947) reported extensive data on ascorbic acid losses from old and new potatoes cooked in 100-serving lots by

TABLE 16.36

PERCENTAGE RETENTION OF ASCORBIC ACID IN VEGETABLES COOKED BY THE HOME AND INSTITUTIONAL METHODS

Vegetable	Home Cooked (%)	Institutional Cooked	
		Tested Immediately (%)	Tested After Holding Over Steam 1 Hr (%)
Carrots	56	48	26
Potatoes	42	21	14
Green beans	38	34	—
Spinach	24	21	12
Broccoli	27	15	9
Brussels sprouts	31	23	11
Cabbage	34	32	0
Cauliflower	30	25	20
Turnips	33	21	

Source: Thomas (1942).

several methods and held on a steam table for 15 to 120 min. Table 16.37 is a condensation of this data as it applies to what the authors described as old potatoes meaning not freshly harvested potatoes, which are of course very seasonal. The results with new potatoes reflect the substantially higher initial ascorbic acid content expected and, therefore, better retentions but the rate of losses with holding is definitely similar.

The data in Table 16.38 were compiled by Harris (1960) and clearly indicate that the nutrient content of cooked foods after standing on steam tables, or similar equipment, may be greatly reduced.

Campbell et al. (1958), in a study of microwave and conventional cooking, studied reduced ascorbic acid retention in frozen broccoli and green peas after conventional cooking by simulated steam table holding for 1 hr at 185°F. The retention for broccoli only dropped from 69.5 to 63.5% but the percentage retention for green peas dropped from 82.5 to 58.2%. Since the retention by microwave cooking can be superior or equal to that of conventional cooking, the advantages of microwave cooking just prior to serving would appear superior to holding conventionally-cooked foods hot.

Kahn and Livingston (1970) measured thiamin retention in four recipes (beef stew, chicken a la king, shrimp Newburg, and peas in cream sauce); the results are given in Table 16.17. Average retention after 1 hr at 180°F was 78.2%; 2 hr, 73.9%; and 3 hr, 67.4%. Considering the practices of institutional and commercial cafeterias one would expect that much more sophisticated data would be available.

TABLE 16.37

PERCENTAGE LOSS IN ASCORBIC ACID IN POTATOES BY VARIOUS METHODS IN QUANTITIES OF 100 SERVINGS[1]

Method	With Jackets Baked	Without Jackets Boiling	Without Jackets Steaming	Mashing	After Boiling Oven Browned	Creaming	Boiled and Diced Pan Fried	After Blanching French Fried
Immediately after	20	47	20	77	46	67	80	62
After holding 15 min	27	58	38	88	58	65	81	64
After holding 30 min	44	62	48	98	59	58	100	60
After holding 45 min	42	80	36	100	56	73	95	48
After holding 60 min	40	71	49	96	65	72	100	79
After holding 90 min	60	86	38	100	74	60	97	61
After holding 120 min	51	87	50	98	91	63	91	58

Source: After Branion et al. (1947).
[1] Data based on an average of 3–4 determinations.

TABLE 16.38

EFFECT OF STEAM TABLE STANDING ON VITAMIN CONTENT OF VEGETABLES (% LOSS)

Vegetables	Time, Temp	Carotene (%)	Ascorbic Acid (%)	Thiamin (%)	Riboflavin (%)	Niacin (%)
Beans, snap	3 hr., hot		52	30		
Cabbage	18 min, hot		22			
Cabbage	45 min, hot		36			
Cabbage	105 min, hot.		80			
Cabbage	0 min, hot		0			
Cabbage	15 min, hot		25			
Cabbage	30 min, hot		40			
Cabbage	45 min, hot		50			
Cabbage	60 min, hot		60			
Cabbage	75 min, hot		70			
Cabbage	90 min, hot		75			
Cabbage	boiled, held hot	15	11	6	8	
Cabbage	cooked 7 min		56	63	62	
Cabbage	15 min, steam		25			
Cabbage	30 min, steam		43			
Cabbage	120 min, steam		50	59	58	
Cabbage	cooked 7 min		29	74	77	
Cabbage	15 min, hot		51			
Cabbage	120 min, hot		61	72	65	
Cabbage, red	3 hr, hot		0	4		
Carrots	3 hr, hot		39	64		
Cauliflower	3 hr, hot		0	57		
Kale	18 min, hot		20			
Kale	30 min, hot		38			
Kale	60 min, hot		50			
Peas	3 hr, hot		94	66		
Potatoes	3 hr, hot			24		
Spinach	3 hr, hot		16	45		
Squash	3 hr, hot		37			
Turnip	3 hr, hot			11		

Source: Harris (1960).

Holding in Insulated Containers.—Thermal containers which hold several hot meals for hours are being used with increasing frequency in small school food service and "Meal on Wheels" programs. Meals are prepared in a central kitchen and dispatched to several different sites in insulated containers.

Only one very limited study (Wagner 1971) was found and the results for total vitamin C loss in several vegetables are reproduced in Table 16.39. The effect is dramatic loss and should prompt a thorough study of a profile of nutrients.

TABLE 16.39

LOSSES REFERRING TO THE VITAMIN C CONTENT OF VEGETABLES IN COOKED CONDITION

Food	Thermophor[1] (3 Hr)	Multi-Serv[2]		
		After 24 Hr	After 48 Hr	After 72 Hr
Potatoes	78.4	44.3	52.2	53.9
Brussels sprouts	88.1	23.7	29.2	40.5
Cauliflower	72.8	23.5	26.2	30.2
Red cabbage	82.1	12.2	22.1	34.6
White cabbage	89.1	39.3	49.3	61.7
Savoy cabbage	76.0	46.5	53.5	63.3
Paprika, green	72.8	32.4	45.4	54.2
Spinach	(1 hr) 53.8	22.8	41.2	54.4
	(2 hr) 68.1			
	(3 hr) 75.3			
Green cabbage	(1 hr) 61.0	17.8	35.4	53.1
	(2 hr) 71.9			
	(3 hr) 80.0			

Source: Wagner (1971).
[1] Held hot in insulated container.
[2] Held refrigerated and heated in convection oven 320°F 15 min.

SUMMARY

Standards of living are being constantly raised in this country. One of the most important of these is adequate nutrition. There must be constant regard for methods of preparation which will ensure the public that it is getting the nutrients away from home that it expects to get. This is especially true of food served in institutions in which people may eat all meals on the premises, such as schools, colleges, hospitals, and homes for the aged.

With increased knowledge of essential nutrients and their importance in maintaining optimum health, with innovations in equipment (e.g., broasting), and changes in form of the products received in the kitchen, constant checking of nutrient retention is needed as well as checks for convenience, cost, flavor, appearance, and reasonable uniformity. The latter items are more tangible and to date have received more attention, but are no more important than nutrient content. Fortunately, incidations are that the best treatments for the accomplishment of convenience, flavor, appearance, and reasonable uniformity are also best for the retention of nutrients.

It may seem strange to use as a summary to this chapter, the 1954 work of Patton and Green (1954) on vitamin C losses in cabbage cookery (Table 16.40), but it is unique because it represents the only study that considered the effects of a cross-section of food service practice including various cooking methods. As such it represents a model for the research which must be undertaken in this field. It is absolutely embarrassing to acknowledge the expanding volume of the food service business in developed countries and its associated sophistication in management and technology, on the one hand, and to also admit to the paucity of information on the effects

TABLE 16.40

LOSS OF VITAMIN C IN CABBAGE COOKERY

Method	Preparation	Vitamin C Loss (%)
Boiled	Shredded	78
Boiled	Coarsely shredded	74
Boiled	Cabbage wedges	65
Boiled	Cabbage 4 serving lots	62
Boiled	Cabbage 100 serving lots	72
Boiled	Cabbage army mess lots	88
Low pressure steamer	Cabbage 100 serving lots	70
Institutional steamer	Cabbage 100 serving lots	58
Pressure cooker	Cabbage 100 serving lots	47-77
Fried	Cabbage 100 serving lots	51.6
Electronic	Cabbage 100 serving lots	41

Loss in Cooking

Adding cabbage to boiling water slowly resulted in less loss. Large pieces lose less than small pieces in the ratio of 58-24%.

	(% Loss)
After 30 min cooking	46
After 1 hr cooking	56
After 2 hr cooking	67

Addition of salt as affecting loss is controversial.
Addition of soda in small amounts seems to have no effect.

Loss in Holding Cooked

	(% Loss)
Refrigerator equals	25
Steam table equals	64
	(at end of 1 hr) 72
	(at end of 2 hr) 92

Source: Patton and Green (1954).

Note: Cabbage boiled in open saucepan (etc.) showed large losses in vitamin C. Usually this loss is associated with the high concentration in the liquid drained from the cabbage. The higher retention of vitamin C was found in cooking with a small amount of water in a tightly covered pan.

of food service practices on the nutritive value of food—the input to nutrition.

BIBLIOGRAPHY

ANDROSS, M. 1946. Losses of nutrients in the preparation of foodstuffs. Proc. Nutr. Soc. 4, 155.
ANON. 1956. Toxicity of heated and aerated oils. Nutr. Rev. 14, 28.
AUGHEY, E., and DANIEL, E. P. 1940. Effect of cooking upon the thiamine content of foods. J. Nutr. 19, 285.
BALINTFY, J. L. 1973. Mathematical modeling and human nutrition. Science 181, 581.
BARNES, B., TRESSLER, D. K., and FENTON, F. 1943A. Effect of different cooking methods on the vitamin C content of quick-frozen broccoli. Food Res. 8, 13.
BARNES, B., TRESSLER, D. K., and FENTON, F. 1943B. Thiamine content of fresh and frozen peas and corn before and after cooking. Food Res. 8, 420.
BORENSTEIN, B. 1968. Vitamins and amino acids. In Handbook of Food Additives. Chemical Ruber Co., Cleveland.
BOWERS, JANE A., and BRYER, BETH A. 1972. Thiamine and riboflavin in cooked and frozen reheated turkey. J. Am. Dietet. Assoc. 60, 399.
BOYLE, MARY ANN, and FUNK, KAYE. 1972. Thiamine in roast beef held by three methods. J. Am. Dietet. Assoc. 60, 398.
BRANION, H. D., ROBERTS, J. S., CAMERON, C. R., and McCREADY, A. M. 1947. The loss of ascorbic acid in the preparation of old and freshly harvested potatoes. J. Am. Dietet. Assoc. 23, 414.
BRANION, D. H., ROBERTS, J. S., CAMERON, C. R., and McCREADY, A. M. 1948. Ascorbic acid content of cabbage. J. Am. Dietet. Assoc. 24, 101.
BRESSANI, R., MARCUCCI, E., ROBLES, C. E., and SCRIMSHAW, N. S. 1954. Nutritive value of Central American beans. Food Res. 19, 263.
BRIANT, A. M., MACKENZIE, V. E., and FENTON, F. 1946A. Vitamin retention in frozen peas and frozen green beans in quantity food service. J. Am. Dietet. Assoc. 22, 507.
BRIANT, A. M., MACKENZIE, V. E., and FENTON, F. 1946B. Vitamin content of frozen peas, green beans and lima beans, and market-fresh yams prepared in a navy mess hall. J. Am. Dietet. Assoc. 22, 605.
BRING, SHIRLEY V., GRASSL, CAROL, HOFSTRAND, JOYCE T., and WILLARD, M. J. 1963. Total ascorbic acid in potatoes. J. Am. Dietet. Assoc. 42, 320.
BRING, SHIRLEY V., and RAAB, P. FRANCES. 1964. Total ascorbic acid in potatoes. J. Am. Dietet. Assoc. 45, 149.
BROWN, E. J., and FENTON, F. 1942. Losses of vitamin C during cooking of parsnips. Food Res. 7, 218.
BURK, M. C. 1952. Distribution of the food supply of the United States. Agr. Econ. Res. 4, 83.
BURRELL, R. C., and EBRIGHT, V. R. 1940. The vitamin C content of fruits and vegetables. J. Chem. Educ. 17, 180.
CAMPBELL, CAROL, TUNG YU LIN, L., and PROCTOR, B. E. 1958. Microwave vs conventional cooking. J. Am. Dietet. Assoc. 34, 365.
CAUSEY, K., and FENTON, F. 1950. Effects of four cooking pressures on commercially frozen broccoli. J. Home Econ. 42, 649.
CAUSEY, K., and FENTON, F. 1951A. Effect of reheating on palatability, nutritive value, and bacterial count of frozen cooked foods. I. Vegetables. J. Am. Dietet. Assoc. 27, 390.

CAUSEY, K., and FENTON, F. 1951B. Effect of reheating on palatability, nutritive value, and bacterial count of frozen cooked foods. II. Meat dishes. J. Am. Dietet. Assoc. *27*, 491.
CAUSEY, K. et al. 1950A. Effect of thawing and cooking methods on palatability and nutritive value of frozen ground meat. I. Food Res. *15*, 237.
CAUSEY, K., HAUSRATH, M. E., RAMSTAD, P. E., and FENTON, F. 1950B. Effect of thawing and cooking methods on palatability and nutritive value of frozen ground meat. II. Food Res. *15*, 249.
CAUSEY, K., HAUSRATH, M. E., RAMSTAD, P. E., and FENTON, F. 1950C. Effect of thawing and cooking methods on palatability and nutritive value of frozen ground meat. III. Food Res. *15*, 256.
CLAYTON, M. M., and GOOS, C. 1947. Effect of French dressings, vinegars and acetic acid on rate of loss of vitamin C in raw cabbage. Food Res. *12*, 27.
CONNOLLY, F., HILTZ, M. C., and ROBINSON, A. D. 1947. Thiamine in Manitoba vegetables. Can. J. Res. *25*, 43.
COVER, S., DILSAVER, E. M., HAYS, R. M., and SMITH, W. H. 1949. Retention of B vitamins after large-scale cooking of meat. II. Roasting by two methods. J. Am. Dietet. Assoc. *25*, 949.
COVER, S., and SMITH, W. H. 1948. Effect of thiamine retention of adding a carbohydrate vegetable to beef stew. Food Res. *13*, 475.
CURRAN, K. M., TRESSLER, D. K., and KING, C. G. 1937. Losses of vitamin C during cooking of northern Spy apples. Food Res. *2*, 549.
CUTLAR, K., JONES, J. B., HARRIS, K. W., and FENTON, F. 1944. Ascorbic acid, thiamine, and riboflavin retention in fresh spinach in institution food service. J. Am. Dietet. Assoc. *20*, 757.
DOMAH, AABMUD A. M. B., DAVIDEK, J., and VELISEK, J. 1974. Changes of L-ascorbic and L-dehydroascorbic acids during cooking and frying of potatoes. Z. Lebensm. Unters. -Forsch. *154*, 272.
EHEART, MARY S., and GOTT, CLAIRE. 1964. Coventional and microwave cooking of vegetables. J. Am. Dietet. Assoc. *44*, 116.
EHEART, MARY S., and GOTT, CLAIRE. 1965. Chlorophyl, ascorbic acid and pH changes in green vegetables cooked by stir-fry, microwave, and conventional methods and a comparison of chlorophyll methods. Food Technol. *19*, 185.
ERIKSON, S. E., and BOYDEN, R. E. 1947A. Effect of different methods of cooking on the thiamine, riboflavin, and niacin content of pork. Kentucky Agr. Expt. Sta. Bull. *503*.
ERIKSON, S. E., and BOYDEN, R. E. 1947B. Thiamine and riboflavin content of turkey cooked by institution methods. Kentucky Agr. Expt. Sta. Bull. *504*.
FARRER, K. T. H. 1955. Thermal destruction of vitamin B_1 in foods. In Advances in Food Research, Vol. 6. E. M. Mrak and G. F. Stewart (Editors). Academic Press, New York.
FENTON, F. 1940. Vitamin C retention as a criterion of quality and nutritive value in vegetables. J. Am. Dietet. Assoc. *16*, 524.
FENTON, F. 1941. Nutritive value of quick frozen foods. Refrig. Eng. *42*, 140.
FENTON, F. 1945. Retention of vitamin values in large-scale food service. Proc. Inst. Food Technologists, *35*.
FENTON, F. 1960. Effects of large-scale preparation on nutrients of foods of animal origin. In Nutritional Evaluation of Food Processing. R. S. Harris and H. Von Loesecke (Editors). Avi Publishing Co., Westport, Conn.
FENTON, F., and TRESSLER, D. K. 1938. Losses of vitamin C during the cooking of certain vegetables. J. Home Econ. *30*, 717.
FENTON, F., TRESSLER, D. K., CAMP, S. C., and KING, C. G. 1937. Losses of vitamin C during the cooking of Swiss chard. J. Nutr. *14*, 631.
FENTON, F., TRESSLER, D. K., CAMP, S. C., and KING, C. G. 1938. Losses

of vitamin C during commercial freezing, defrosting, and cooking of frosted peas. Food Res. *3*, 403.
FENTON, F. *et al.* 1945. Effect of quantity preparation procedures on vitamin retention: canned peas. J. Am. Dietet. Assoc. *21*, 700.
FENTON, F. *et al.* 1946. Retention of vitamins and palatability in large-scale food service of sulfited, dehydrated cabbage. Food Res. *11*, 475.
FENTON *et al.* 1956. Study of three cuts of lower and higher grade beef, unfrozen and frozen, using two methods of thawing and two methods of braising. Cornell Univ. Agr. Expt. Sta. Mem. *341*.
FLYNN, L. M., and HOGAN, A. G. (Editors). 1946. Nutrients consumed by army students at the University of Missouri. J. Am. Dietet. Assoc. *22*, 8.
GILPIN, G. L., SWEENEY, J. P., CHAPMAN, V. J., and EISEN, J. N. 1959. Effect of cooking methods on broccoli. J. Am. Dietet. Assoc. *35*, 359.
GLEIM, E., TRESSLER, D. K., and FENTON, F. 1944. Ascorbic acid, thiamine, riboflavin, and carotene contents of asparagus and spinach in the fresh, stored, and frozen states both before and after cooking. Food Res. *9*, 471.
GLEIM, E. *et al.* 1946A. Ascorbic acid, thiamine, riboflavin, and niacin content of potatoes in large-scale food service. Food Res. *11*, 461.
GLEIM, E. *et al.* 1946B. Retention of carotene, thiamine, riboflavin, and niacin during large-scale food service of dehydrated carrots. Food Res. *11*, 61.
GLEIM, E. *et al.* 1946C. Effects of quantity preparation procedures on vitamin retention: canned tomatoes. J. Am. Dietet. Assoc. *22*, 29.
GLEIM, E. *et al.* 1947. Palatability and vitamin retention in large-scale food preparation of dehydrated sulphited potato strips. J. Am. Dietet. Assoc. *23*, 322.
GOLDBLITH, S. A., TANNENBAUM, S. R., and WANG, D. I. D. 1968. Thermal and 2450 MHz microwave energy effect on the destruction of thiamine. Food Technol. *22*, 1266.
GORDON, JOAN, and NOBLE, ISABEL. 1959A. Comparison of electronic vs conventional cooking of vegetables. J. Am. Dietet. Assoc. *35*, 241.
GORDON, JOAN, and NOBLE, ISABEL. 1959B. Effect of cooking method on vegetables. J. Am. Dietet. Assoc. *35*, 578.
GORDON, JOAN, and NOBLE, ISABEL. 1964. "Waterless" vs boiling water cooking of vegetables. J. Am. Dietet. Assoc. *44*, 378.
HALLIDAY, E. G., and NOBLE, L. T. 1936. Recent research in foods. J. Home Econ. *28*, 15.
HARRIS, P. L., and POLAND, G. L. 1939. Variations in ascorbic acid content of bananas. Food Res. *4*, 317.
HARRIS, R. S. 1960. Effects of large-scale preparation on nutrients of foods of plant origin. *In* Nutritional Evaluation of Food Processing. R. S. Harris and H. von Loesecke (Editors). Avi Publishing Co., Westport, Conn.
HAYAKAWA, K. 1969. New parameters for calculating mass average sterilizing value to estimate nutrients in thermally conductive food. Can. Inst. Food Technol. J. *2*, 165.
HELLER, B. S. *et al.* 1961. Utilization of amino acids from foods by the rat. V. Effects of heat treatment on lysine in meat. J. Nutr. *73*, 113.
HELLER, C. A., McCAY, C. M., and LYON, C. B. 1943. Losses of vitamins in large-scale cooking. J. Nutr. *26*, 377.
HELLSTROM, V. 1952-1953. The durability of vitamin C in raw vegetables after mashing. Z. Vitamin-, Hormon-, Fermentforsch. *5*, 98.
HILL, MARY. 1975. The NUTRIMETER—A device for totaling the number of calories and the percentages of the U.S. RDA from a day's food. USDA Agr. Res. Serv. Govt. Printing Office, Superintendent of Documents, Washington, D.C.
HINMAN, W. F., BRUSH, M. K., and HALLIDAY, E. G. 1944. The nutritive

value of canned foods: VI. Effect of large-scale preparation for serving on the ascorbic acid, thiamine, and riboflavin content of commercially-canned vegetables. J. Am. Dietet. Assoc. 20, 752.

HOLLINGER, M. E. 1944. Ascorbic acid value of the sweet potato as affected by variety, storage, and cooking. Food Res., 9, 76.

HOLLINGER, M. E., and COLVIN, D. 1945. Ascorbic acid content of okra as affected by maturity, storage, and cooking. Food Res. 10, 255.

IRESON, M. G., and EHEART, M. S. 1944. Ascorbic acid losses in cooked vegetables: Cooked uncovered in a large amount of water and covered in a small amount of water. J. Home Econ. 36, 160.

JANSEN, G. R., and HARPER, J. M. 1974. Nutritional aspects of nutrient standard menus. Food Technol. 28, 38.

JOHNSTON, C. H., SCHAUER, L., RAPOPORT, S., and DEUEL, H. J., JR. 1943. The effect of cooking with and without sodium bicarbonate on the thiamine, riboflavin, and ascorbic acid content of peas. J. Nutr. 26, 227.

JONES, J. B. et al. 1944. Ascorbic acid, thiamine, and riboflavin retention in quick-frozen broccoli in institution food service. J. Am. Dietet. Assoc. 20, 369.

KAHN, LESLIE N., and LIVINGSTON, G. E. 1970. Effect of heating methods on thiamine retention in fresh or frozen prepared foods. J. Food Sci. 35, 349.

KAHN, R. M., and HALLIDAY, E. G. 1944. Ascorbic acid content of white potatoes as affected by cooking and standing on steam table. J. Am. Dietet. Assoc. 20, 220.

KIK, M. C. 1945. Effect of milling, processing, washing, cooking and storage on thiamine, riboflavin and niacin in rice. Arkansas Agr. Expt. Sta. Bull. 458.

KOTSCHEVAR, L. H., MOSSO, A., and TUGWELL, T. 1953. Utility and economy factors in using prefabricated meats. J. Am. Dietet. Assoc. 29, 878.

KOTSCHEVAR, L. H., MOSSO, A., and TUGWELL, T. 1955. B-vitamin retention in frozen meat. J. Am. Dietet. Assoc. 31, 589.

KREHL, W. A., and WINTERS, R. W. 1950. Effect of cooking methods on retention of vitamins and minerals in vegetables. J. Am. Dietet. Assoc. 26, 966.

KYLEN, ANNE M. et al. 1961. Microwave cooking of vegetables. J. Am. Dietet. Assoc. 39, 321.

KYLEN, ANNE M. et al. 1964. Microwave and conventional cooking of meat. J. Am. Dietet. Assoc. 45, 139.

LACHANCE, P. A. 1972. Update: Meat extenders and analogues in child feeding programs. Proc. Meat Ind. Res. Conf. Ann. conf. sponsored by Am. Meat Sci. Assoc. in cooperation with Am. Meat Inst. Found., Chicago.

LACHANCE, P. A. 1974. Sorting out the confusion in the nutrition revolution era. Food Prod. Develop. 8, 63.

LACHANCE, P. A., RANADIVE, A. S., and MATAS, J. 1973. Symposium: Effects of processing, storage, and handling on nutrient retention in foods. Food Technol. 27, 36-38.

LAMPITT, L. H., CLAYSON, D. H. F., and BARNES, E. M. 1944. The destruction of ascorbic acid during the cooking of green vegetables. Observations on the mechanisms involved. J. Soc. Chem. Ind. (London) 63, 193.

LANE, R. L., JOHNSON, E., and WILLIAMS, R. R. 1942. Studies of the average American diet. I. Thiamine content. J. Nutr. 23, 613.

LEE, F. A., and WHITCOMBE, J. 1945. Blanching of vegetables for freezing, effect of different types of potable water on nutrients of peas and snap beans. Food Res. 10, 465.

LEEKING, P. et al. 1956. The quality of smoked hams as affected by adding antibiotic and fat to the diet and phosphate cure. III. Moisture, fat, chloride, and thiamine content. Food Technol. 10, 274-276.

LEVERTON, R. M., and ODELL, G. V. 1958. The nutritive value of cooked meat. Oklahoma Agr. Exp. Sta. Misc. Publ. *MP-49*.

LIM, E., YEN, J., and FENTON, F. 1959. Effect of irradiation on quality of ground pork and of cooking conventionally and electronically. Food Res. *24*, 645.

LUNDE, G., KRINGSTAD, H., and OLSEN, A. 1940. Retention of vitamin B_1 with the cooking and preservation of vegetables. Angew. Chem. *53*, 123. (German)

LUSHBOUGH, C. H., PORTER, T., and SCHWEIGERT, B. S. 1957. Utilization of amino acids from foods by the rat. IV. Tryptophan. J. Nutr. *62*, 513.

MACK, G. L., TAPLEY, W. T., and KING, C. G. 1939. Vitamin C in vegetables. X. Snap beans. Food Res. *4*, 309.

MAHON, P. et al. 1956. The quality of smoked ham as affected by adding antibiotic and fat to the diet and phosphate to the cure. Food Technol. *10*, 265.

MAYFIELD, H. L., and RICHARDSON, J. E. 1940. Ascorbic acid content of parsnips. Food Res. *5*, 361.

McCANCE, R. A., WIDDOWSON, E. M., and SHACKLETON, L. 1936. The nutritive value of fruits, vegetables and nuts. Med. Res. Council Spec. Rept. *213*.

McINTIRE, J. M., SCHWEIGERT, B. S., HENDERSON, L. M., and ELVEHJEM, C. A. 1943. The retention of vitamins in meat during cooking. J. Nutr. *25*, 143.

McINTOSH, J. A., TRESSLER, D. K., and FENTON, F. 1940. The effect of different cooking methods on the vitamin C content of quick-frozen vegetables. J. Home Econ. *32*, 692.

McINTOSH, J. A., TRESSLER, D. K., FENTON, F. 1942. Ascorbic acid content of five quick frozen vegetables. J. Home Econ. *34*, 314.

McINTOSH, R. 1974. General principles of food service. *In* Food Service Science. Laura Lee W. Smith and L. J. Minor (Editors). Avi Publishing Co., Westport, Conn.

MEYER, J. A., BRISKEY, E. J., HOEKSTRA, W. G., and WECKEL, K. G. 1963. Niacin, thiamine, and riboflavin in fresh and cooked pale, soft, watery versus dark, firm, dry pork muscle. Food Technol. *17*, 485.

MILLER, C. D., BAUER, A., and DENNING, H. 1952. Taro as a source of thiamine, riboflavin, and niacin. J. Am. Dietet. Assoc. *28*, 435.

MILLROSS, JANICE, SPEHT, A., HOLDSWORTH, KATHLEEN, and GLEW, G. 1973. The utilization of the cook-freeze catering system for school meals. The University of Leeds. W. S. Maney and Son, Leeds, England.

MISKIMIN, DOROTHY, BOWERS, J., and LACHANCE, P. A. 1974. Nutrification of frozen preplated school lunches is needed. Food Technol. *28*, 52.

MULAY, IL, DHOPESHWARKAR, G. A., and MAGAR, N. G. 1952. Losses of nutrients during cooking of vegetables. Indian J. Med. Res. *40*, 443.

MUNSELL, H. E. et al. 1949. Effect of large-scale methods of preparation on the vitamin content of food. III. Cabbage. J. Am. Dietet. Assoc. *25*, 420.

MYERS, PATRICIA W., and ROEHM, GLADYS HARTLEY. 1963. Ascorbic acid in dehydrated potatoes. J. Am. Dietet. Assoc. *42*, 325.

NAGEL, A. H., and HARRIS, R. S. 1943. Effect of restaurant cooking and service on vitamin content of foods. J. Am. Dietet. Assoc. *19*, 23.

NOBLE, ISABEL. 1964. Thiamine and riboflavin retention in broiled meat. J. Am. Dietet. Assoc. *45*, 447.

NOBLE, ISABEL. 1965. Thiamine and riboflavin retention in braised meat. J. Am. Dietet. Assoc. *47*, 205.

NOBLE, ISABEL. 1970. Thiamine and riboflavin retention in cooked variety meats. J. Am. Dietet. Assoc. *56*, 225.

NOBLE, ISABEL, and GOMEZ, LUCILLE. 1958. Thiamine and riboflavin in roast lamb. J. Am. Dietet. Assoc. *34*, 157.

NOBLE, ISABEL, and GOMEZ, LUCILLE. 1960. Thiamine and riboflavin in roast beef. J. Am. Dietet. Assoc. 36, 46.

NOBLE, ISABEL, and GOMEZ, LUCILLE. 1962. Vitamin retention in meat cooked electronically. J. Am. Dietet. Assoc. 41, 217.

NOBLE, I., and WORTHINGTON, J. 1948. Ascorbic acid retention in cooked vegetables. J. Home Econ. 40, 129.

OLLIVER, M. 1936. The ascorbic acid content of fruits and vegetables with special reference to the effect of cooking and canning. J. Soc. Chem. Ind. 55, 153T.

OLLIVER, M. 1943. Ascorbic acid values of fruits and vegetables for dietary surveys. Chem. Ind. (London) 62, 146.

OSER, B. L., MELNICK, D., and OSER, M. 1943. Influence of cooking procedure upon retention of vitamins and minerals in vegetables. Food Res. 8, 115.

PAGE, EDNA, and HANNING, FLORA M. 1963. Vitamin B_6 and niacin in potatoes. J. Am. Dietet. Assoc. 42, 42.

PATTON, MARY B., and GREEN, MARY E. 1954. The effect of cooking method on loss of vitamin C. Ohio Agr. Expt. Sta. Res. Bull. 742.

PEARSON, A. M. et al. 1951. Vitamin losses in drip obtained upon defrosting frozen meat. Food Res. 16, 87.

PEARSON, P. B., and LUECKE, R. W. 1945. The B vitamin content of raw and cooked sweet potatoes. Food Res. 10, 325.

POTGIETER, M., and GREENWOOD, M. L. 1950. Influence of cooking method on ascorbic acid and thiamine contents of four varieties of kale (Brassica oleracea u. acephala) Food Res. 15, 223.

PROCTOR, B. E., and GOLDBLITH, S. A. 1948. Radar energy for rapid food cooking and blanching, and its effect on vitamin content. Food Technol. 2, 95.

PYKE, M. 1944. Food supplied for collective feeding. Proc. Nutr. Soc. (Engl. and Scot.) 1, 92.

RICE, E. E., and BEUK, J. F. 1945. Reaction rates for decomposition of thiamine in pork at various temperatures. Food Res. 10, 99.

RICHARSON, J. E., DAVIS, R., and MAYFIELD, H. L. 1937. Vitamin C content of potatoes prepared for table use by various methods of cooking. Food Res. 2, 85.

RUSSELL, W. C., TAYLOR, M. W., and BUEK, J. F. 1943. The nicotinic acid content of common fruits and vegetables as prepared for human consumption. J. Nutr. 25, 275.

SCHLOSSER, G. E., SEAQUIST, R., REA, E. W., and DAWSON, E. H. 1957. Food yields and losses in pressure cooking. J. Am. Dietet. Assoc. 33, 1154.

SCHWEIGHERT, B. S., and GUTHNECK, B. T. 1954. Utilization of amino acids from foods by the rat. III. Methionine. J. Nutr. 54, 333.

SCHWEIGERT, B. S., and PAYNE, B. J. 1956. A summary of the nutrient content of meat. Am. Meat Inst. Found. Bull. 30.

SHEETS, O., LEONARD, O. A., and GEIGER, M. 1941. Distribution of minerals and vitamins in different parts of leafy vegetables. Food Res. 6, 553.

SIEDLER, A. J. 1961. Effect of standard cooking and processing methods on the nutritional value of meat protein. Am. Meat Inst. Found. Bull. 51.

SPIERS, M. et al. 1945. The effects of fertilizer treatments, curing storage, and cooking on the carotene and ascorbic acid content of sweet potatoes. Southern Coop. Ser. Bull. 3.

STEVENS, H. P., and FENTON, F. 1951. Dielectric versus stewpan cookery. J. Am. Dietet. Assoc. 27, 32.

STREIGHTOFF, F. et al. 1946A. Effect of large-scale methods of preparation on vitamin content of food. I. Potatoes. J. Am. Dietet. Assoc. 22, 117.

STREIGHTOFF, F. et al. 1946B. Effect of large-scale methods of preparation on the vitamin content of food. II. Carrots. J. Am. Dietet. Assoc. 22, 511.

STREIGHTOFF, F. et al. 1949A. Effect of large-scale methods of preparation on the vitamin content of food. IV. Corn. J. Am. Dietet. Assoc. 25, 687.

STREIGHTOFF, F. et al. 1949B. Effect of large-scale methods of preparation on the vitamin content of food. V. Spinach. J. Am. Dietet. Assoc. 25, 770.

SUTHERLAND, C. K., HALLIDAY, E. G., and HINMAN, W. F. 1947. Vitamin retentions and acceptability of fresh vegetables cooked by four household methods and by an institutional method. Food Res. 12, 496.

SWAMINATHAN, M. 1941. The effect of washing and cooking on the nicotinic acid content of raw and parboiled rice. Indian J. Med. Res. 29, 83.

SWAMINATHAN, M. 1942. The effect of washing and cooking on the vitamin B_1 content of raw and parboiled rice. Indian J. Med. Res. 30, 409.

TAKAHASHI, T., and TAKADA, M. 1949. J. Japan. Soc. Food Nutr. 1, 162.

THOMAS, M. H., BRENNER, S., EATON, A., and CRAIG, V. 1949. Effect of electronic cooking on nutritive value of foods. J. Am. Dietet. Assoc. 25, 39.

THOMAS (nee Higgins), MIRIAM MASON. 1942. Retention of ascorbic acid in institution and home cooked vegetables. M. S. Thesis. Univ. Chicago.

TODHUNTER, E. N. 1936. Some factors influencing ascorbic acid content of apples. Food Res. 1, 435.

TOEPFER, E. W., PRITCHETT, C. A., and HEWSTON, E. M. 1955. Boneless beef: raw, cooked, and stewed. Results of Analysis for moisture, protein, fat, and ash. USDA Bull. 1137.

TOOLEY, P. J. 1972. The effect of deep-fat frying on the availability of fish lysine. Nutr. Soc. Proc. 31, 2A.

TREFETHEN, I., CAUSEY, K., and FENTON, F. 1951. Effect of two cooking pressures on locally grown broccoli: cooking time, palatability, ascorbic acid, thiamine, and riboflavin. Food Res. 16, 409.

TUCKER, H. Q., VOEGELI, M. J., WELLINGTON, G. H., and BRATZLER, L. J. 1952. A Cross Sectional Muscle Nomenclature of the Beef Carcass. Michigan State College Press, East Lansing, Michigan.

TUCKER, R. E., HINMAN, W. F., and HALLIDAY, E. G. 1946. The retention of thiamine and riboflavin in beef cuts during braising, frying, and broiling. J. Am. Dietet. Assoc. 22, 877.

U.S. DEPT. OF AGR. 1971. The food service industry: Type, quantity and value of foods used. USDA Marketing Econ. Div., Econ. Res. Serv. Statist. Bull. 476.

U.S. DEPT. OF AGR. 1972. Food and nutrient intake of individuals in the United States, Spring 1965. Household food consumption survey, 1965. USDA Agr. Res. Serv. Rept. 11.

U.S. DEPT. OF INTERIOR. 1955. Compilation of Laboratory Data: Yields and Losses in Preparation of Foods. U.S. Fish and Wild Life Service Mimeo.

VAIL, G. E., and WESTERMAN, B. D. 1946. B-complex vitamins and meat. I. Thiamine and riboflavin content of raw and cooked pork. Food Res. 11, 425.

VAN DUYNE, F. O., CHASE, J. T., OWEN, R. F., and FANSKA, J. R. 1948. Effects of certain home practices on riboflavin content of cabbage, peas, snapbeans, and spinach. Food Res. 13, 162.

VAN DUYNE, F. O., CHASE, J. T., and SIMPSON, J. I. 1944. Effects of various home practices on ascorbic acid content of cabbage. Food Res. 9, 164.

VAN DUYNE, F. O., CHASE, J. T., and SIMPSON, J. I. 1945. Effect of various home practices on ascorbic acid content of potatoes. Food Res. 10, 72.

VAN DUYNE, F. O., OWEN, R. F., WOLFE, J. C., and CHARLES, V. R., 1951. Effect of cooking vegetables in tightly covered and pressure saucepans. J. Am. Dietet. Assoc. 27, 1059.

VAN ZANTE, H. J., and JOHNSON, S. K. 1970. Effect of electronic cookery on thiamine and riboflavin in buffered solutions. J. Am. Dietet. Assoc. 56, 133.

WAGER, H. G. et al. 1945. The drying of potatoes. Food Manuf. 20, 289-293, 321-325, 367-371, 375.

WAGNER, K. H. 1971. On the question of vitamin preservation in food which has been treated according to the Multimet-MultiServ procedure, as compared to the preservation in orthodox thermo containers. Bull. CX 167. Crown-X, Inc., Cleveland.

WALKER, A. R. P., and ARVIDSSON, U. B. 1952. The vitamin C content of braised cabbage cooked under pressure and prepared on a very large scale. S. African J. Med. Sci. 17, 143.

WATT, B. K., and ATTAYA, M. B. 1945. Vitamin retention in quantity cooking of vegetables. J. Home Econ. 37, 340.

WATT, B. K., and MERRILL, A. L. 1963. Composition of Foods—Raw, Processed, Prepared. USDA Agr. Handbook 8. (Revised Dec. 1963)

WELLINGTON, G. H. 1954. Body composition and carcass changes of young Holstein cattle. Unpublished Ph.D. Thesis, Michigan State Univ.

WELLINGTON, M., and TRESSLER, D. K. 1938. Vitamin C content of vegetables. IX. Influence of method of cooking on vitamin C content of cabbage. Food Res. 3, 311.

WERTZ, A. W., and WEIR, C. E. 1946. Effect of institutional cooking methods on vitamin contents of foods. II. Ascorbic acid content of potatoes. Food Res. 11, 319.

WESTERMAN, B. D. 1948. The Kitchen Reporter. Kelvinator Kitchen, Detroit.

WESTERMAN, B. D., VAIL, G. E., TINKLIN, G. L., and SMITH, J. 1949. B complex vitamins in meat. II. The influence of different methods of thawing frozen steaks upon their palatability and vitamin content. Food Technol. 3, 184.

WILCOX, E. B., and NEILSON, A. M. 1947. Effect of quantity preparation on the ascorbic acid content of cabbage salad. J. Am. Dietet. Assoc. 23, 223.

WING, R. W., and ALEXANDER, J. C. 1972. Effect of microwave heating on vitamin B_6 retention in chicken. J. Am. Dietet. Assoc. 61, 661.

WOLFE, J. C., OWEN, R. F., CHARLES, V. R., and VAN DUYNE, F. O. 1949. Effect of freezing and freezer storage on the ascorbic acid content of muskmelon, grapefruit sections, and strawberry purée. Food Res. 14, 243.

WOOD, M. A. et al. 1946. Effect of large-scale food preparation on vitamin retention: Cabbage. J. Am. Dietet. Assoc. 22, 677.

WU, C. H., and FENTON, F. 1953. Effect of sprouting and cooking of soybeans on palatability, lysine, tryptophan, thiamine, and ascorbic acid. Food Res. 18, 640.

CHAPTER 17

Paul A. Lachance
and
John W. Erdman, Jr.

Effects of Home Food Preparation Practices on Nutrient Content of Foods

In the first edition, Agnes Fay Morgan (1960) stated that "it is impossible to separate most effects of commercial from home processing because the physical and chemical changes involved are identical." This chapter will avoid repeating the known information on the effects of various commercial practices because these are discussed in detail in other chapters, and the topic as it pertains to food preparation procedures and food service in particular, has been discussed in the previous chapter.

The goal of this chapter is to discuss household-type consumer practices which have a bearing on the nutritive value of food. In the process, we hope to impress the reader with the paucity of the available knowledge and to stimulate systematic research into the subject.

CHANGING FOOD HABITS

In developed countries, economics and desire for the hardware of living are having a strong influence on food habits and, therefore, food preparation practices. In the United States (Lachance 1973), 30-50% of families have one or more members who fairly regularly skip breakfast; and ¾ of the families do not eat breakfast as a family unit. The first food of the day is often at coffee break time. While school children have lunch at school, the blue collar father is having a bag lunch and the white collar father is tied up in a fairly heavy business lunch. Very often Mom, whether at home or at the office, has no lunch because it is one of her ways of dieting. The children, invariably, eat after school. The main meal may be small and the degree of convenience involved depends upon how much time is available for its preparation. If mother works, the amount of time available is minimal. There are all sorts of snacking during TV time and/or a substantial snack before bed. The sacred family meal is dinner and, theoretically, it is eaten seven days a week. In reality, it can be as seldom as three days a week or less. Alternatively, a family may order food from short order establishments such as a pizzeria or go to a drive-in restaurant as a family but order different food combinations. On the average over 30% of meals and 40% of the family food dollar is spent on food outside the home. Otherwise, the various family members may individually decide their evening meal—which may range from breakfast cereal to a gourmet meal to a series of snacks.

Shopping for the family's food supply occurs on the average of once per week; however, this often is supplemented by one or more express line shopping trips. Food shopping practices vary considerably and could have effects on the nutritive quality of the food purchased but have not been studied. These practices include (1) time delays en route from the supermarket and before food can be sorted out and properly stored; (2) storage in freezers of differing types (hanging box freezer, across the top freezer and separate upright or chest freezer resulting in differences in recommended storage time for frozen foods, ranging from 1 week to 3 months); (3) dry warm storage as compared to dry cool storage; (4) the careless or careful rotation of stored supplies, etc. These problems may be a matter of education but there is little information on which to base sound recommendations.

FOOD OF ANIMAL ORIGIN

Storage

The effects of packaging and storage on foods of animal origin are discussed by Karel and Heidelbaugh in Chap. 15. The effects of various cooking procedures are discussed in the previous chapter. The following additional information is provided as it pertains to home practices and exemplifies the need for further research.

The rate of freezing in a home freezer was found by Lee *et al.* (1954) to have little effect on the B vitamin content of pork chops and various cuts of beef (Lee *et al.* 1950).

Refrigerator storage (40°F) of pork loins and ground pork for 2 weeks or less produced less than 10% loss of thiamin, riboflavin, and pantothenic acid (Rice *et al.* 1946, 1948).

Ground pork after 6-18 weeks of frozen storage was thawed (1) during cooking, (2) at room temperature to an internal temperature of 98.6°F, or (3) in running cold water, and then cooked in a variety of ways (Causey *et al.* 1950A). No effect could be seen due to method of thawing or of cooking upon the thiamin, riboflavin, or niacin content of the meat.

Storage of canned pork at 45°, 70°, and 98°F for 1 yr was found by Feaster *et al.* (1946) to result in no significant losses of riboflavin, niacin, or pantothenic acid. Thiamin, however, disappeared in storage at 98°F with a loss of 80-88%, at 70°F, 21-41%, and at 45°F, 0-11%. The loss of thiamin was dependent only on time and temperature of storage, not upon the size of can or on the final thiamin content.

Morgan et al. (1949) determined the thiamin, riboflavin, and niacin content of various chicken tissues before and after cooking and after frozen storage ($-9°F$) for 4, 8, or 12 months. Retention of thiamin in storage was good, 75-100%, except in 1 lot of small broilers. Loss of riboflavin was generally small up to 8 months of storage, as was that of niacin. After 12 months, niacin loss in leg muscle and heart, and riboflavin loss in nearly all the tissues, were significant.

Millares and Fellers (1949) reported retention of thiamin, in light and dark chicken meat, after storage for 8 months at $22°F$ of 88 and 58%, respectively, as well as 97 and 89% of riboflavin, and full retention of niacin.

Frozen ground beef patties thawed in the original wrappings at room temperature or cooked unthawed by 4 different cooking methods were found to retain 84% of the thiamin, 79% of the riboflavin, and 89% of the lysine of the original meat (Causey et al. 1950B). Much the same results were found with ground lamb patties (Causey et al. 1950C). Thiamin retention in the cooked meat was 86%, riboflavin 55%, and lysine 92%.

Another study of the effects of thawing (Westerman et al. 1949) involved the use of round steaks thawed before cooking in 4 different ways: in the refrigerator, at $163°F$, at room temperature, and in running tap water. All were cooked by braising and compared with similar steaks analyzed in the raw state. Thawing in water reduced the acceptability of the meat. Retentions of thiamin, riboflavin, and niacin were not significantly altered by method of thawing but pantothenic acid was better retained after thawing in the refrigerator and at room temperature than by the other two methods. When the vitamins in the drip were added to those in the cooked meat, it was seen that riboflavin and niacin total retentions were not affected by thawing method but that thiamin and pantothenic acid were unfavorably affected by thawing at $163°F$ or in running water. This is not surprising since these two vitamins are more labile and soluble than the others. Loss of panthothenic acid in drip from frozen beef thawed in 14-15 hr at $79°F$ was also found by Pearson et al. (1951) to be much larger (33%) than losses of the other B vitamins which ranged from 8% of the folic acid to 14.5% of the niacin.

On the basis of biological assays, Deuel and Greenberg (1953) reported 75% retention of vitamin A in margarines held 2 yr at $14°F$, 52-60 weeks at $41°F$, 17 weeks at $64°F$, and 15 weeks at $82°F$. Retention of vitamin A by butters was slightly less. Preformed vitamin A and carotene determined by physiochemical tests were stable in margarine to the extent of 97-98% during shelf-life at $45°F$.

A study of the changes in vitamin content of shell eggs during cold storage from 3 to 12 months has shown significant losses in niacin, vitamin B-6, riboflavin, folic acid, and vitamin B-12. Table 17.1 is a compilation of these results (Evans et al. 1951A,B, 1952A,B, 1953A,B, 1955). Apparently, shell eggs lose no biotin or choline but up to 47% of vitamin B-6 (pyridoxine), 6% of pantothenic acid, 27% of folic acid, 23% of vitamin B-12, and 14–18% of riboflavin and niacin. A composite loss of all the vitamins of about 6% occurred after 3 months, 11% after 6–7 months, and 19% after 12 months.

TABLE 17.1

LOSS OF VITAMINS IN SHELL EGGS IN COLD STORAGE AT $0°C$

Vitamin	Fresh Eggs	Stored Eggs					
		3 Months	Loss (%)	6–7 Months	Loss (%)	12 Months	Loss (%)
Niacin, mg/gm	0.66	0.60	9	0.54	18	—	—
Choline, mg/gm	14.9	14.4	0	15.4	0	14.9	0
Vitamin B-6, $\mu g/gm$	2.52	2.06	18	1.78	29	1.34	47
Riboflavin, $\mu g/gm$	3.49	3.32	5	2.93	16	3.07	14
Pantothenic acid, $\mu g/gm$	12.5	11.7	6	11.7	6	11.8	6
Folic acid, $m\mu g/gm$	94	93	0	80	16	74	27
Biotin, $m\mu g/gm$	225	244	0	220	0	228	0
Vitamin B-12, $m\mu g/gm$	6.54	6.07	7	6.17	5	5.03	23

Source: Evans et al. (1951A,B, 1952A,B, 1953A,B, 1955).

An increasing number of frozen precooked foods, both entrées and dinners, are on the market. However, little has been published on the comparative nutritive value of products as freshly prepared, cooked frozen, and reheated. West et al. (1959) prepared and studied the thiamin retention of three different foods: ham loaf, Italian rice, and chicken. The results are given in Table 17.2 and demonstrate remarkably good retentions.

Fortified soy protein products at the 30% level have been permitted in the Type A school lunch since February 1971 (USDA, Food and Nutrition Service Notice 219). In preparing beef loaves extended with 30% hydrated soy, 78 gm dry textured soy and 117 gm of water are used per pound (454 gm) of meat. Nielsen and Carlin (1974) studied thiamin retention in frozen raw or precooked, all-beef loaves and precooked beef-soy loaves, with and without added tripolyphosphate (0.3%). The thiamin retention results are given in Table 17.3. Although the loaves frozen raw were in the oven ($350°F$) a shorter time (127 min) than the precooked, reheated loaves (184 min), thiamin destruction was almost identical.

EFFECTS OF HOME PREPARATION PRACTICES ON NUTRIENTS

TABLE 17.2

THIAMIN CONTENTS OF HAM LOAF, ITALIAN RICE, AND CHICKEN AFTER DIFFERENT TREATMENTS

Treatment	No. of Replications	Thiamin Means and Standard Deviations of the Means as Determined (Mcg/Gm)	Means Retentions[1] and Standard Deviations of the Means (%)
Ham Loaf			
Raw	4	6.34 ± 0.775	100.0
Freshly prepared	4	6.00 ± 0.703	71.8 ± 0.48[2]
Cooked, frozen 2 months, reheated	4	6.08 ± 0.764	72.5 ± 1.57[2]
Raw, frozen 2 months, cooked	4	6.10 ± 0.623	73.3 ± 2.94[2]
Cooked, frozen 4 months, reheated	4	6.03 ± 0.749	71.9 ± 1.46[2]
Raw, frozen 4 months, cooked	4	6.01 ± 0.640	71.5 ± 1.78[2]
Italian Rice			
Freshly prepared, completely cooked	3	0.85 ± 0.042	100.0
Freshly prepared, partially cooked	3	0.88 ± 0.036	100.0
Completely cooked, frozen 2 months, reheated	3	0.73 ± 0.022	96.6 ± 2.24
Partially cooked, frozen 2 months, reheated	3	0.77 ± 0.029	98.0 ± 2.71
Completely cooked, frozen 4 months, reheated	3	0.68 ± 0.034	90.0 ± 1.82[3]
Partially cooked, frozen 4 months, reheated	3	0.71 ± 0.037	90.8 ± 1.89[3]
Chicken			
Raw	4	0.24 ± 0.054	
Freshly cooked	4	0.21 ± 0.038	
Cooked, frozen 2 months, reheated	4	0.19 ± 0.018	
Raw, frozen 2 months, cooked	4	0.18 ± 0.023	
Cooked, frozen 4 months, reheated	4	0.13 ± 0.014	
Raw, frozen 4 months, cooked	4	0.14 ± 0.011	

Source: West et al. (1959).
[1] Calculated on the basis of the original raw weights and thiamin content in the case of ham loaves and in the case of Italian rice on the basis of weights and contents of the freshly cooked samples. After freezer storage completely cooked samples of Italian rice were compared with freshly prepared completely cooked samples and partially cooked samples, with freshly prepared partially cooked samples.
[2] Mean is significantly lower at the 1% level than the mean of the raw sample.
[3] Mean is significantly lower at the 5% level than the mean of the freshly prepared sample.

TABLE 17.3

AVERAGE THIAMIN CONTENT AND THIAMIN RETENTION IN FRESH RAW, FROZEN RAW, PRECOOKED, AND REHEATED BEEF AND BEEF-SOY LOAVES

Sample	All-Beef Loaves	Beef-Soy Loaves	Beef-Soy + TP[1] Loaves
Thiamin content[2]	(Mg/100 Gm)	(Mg/100 Gm)	(Mg/100 Gm)
Fresh raw[3]	0.41	0.50	0.50
Frozen raw, cooked to 165°F[4]	0.38	—	—
Precooked to 165°F, frozen and reheated to 130°F[4]	0.37	0.29	0.29
Thiamin retention	(%)	(%)	(%)
Frozen raw, cooked to 165°F	93	—	—
Precooked to 165°F, frozen and reheated to 130°F	90	58	58

Source: Nielsen and Carlin (1974).

[1] Tripolyphosphate.
[2] Dry weight basis.
[3] Average of 6 determinations.
[4] Average of 18 determinations for loaves stored 6 months (3 loaves × 6 determinations per loaf).

After precooking and reheating, thiamin retention was approximately 90% for all-beef loaves and 58% for beef-soy loaves. One explanation for the lower retention of thiamin in the beef-soy loaves may be that the synthetic thiamin added to fortify SPC is more heat-labile than naturally occurring thiamin. Somewhat comparable results were reported by investigators of the thermolabile properties of thiamin who found that thiamin in natural food products was more resistant to heat than pure thiamin in aqueous solution (Greenwood *et al.* 1943; Melnick *et al.* 1941).

Venison is a delicacy enjoyed by some homes and so information on its thiamin content and the effect of cooking is discussed here. Guild and Raines (1972) undertook the determination of thiamin in a limited number of samples of venison which indicated that the average content in raw loin and rib chops and in sirloin tip was approximately 0.27 mg per 100 gm. The average contribution from 100 gm braised loin or rib chop and roasted round was about 0.21 mg. Retention of thiamin during braising was about ⅔ of the original amount for loin chops, but only a little over ½ for a thinner rib chop. Weight loss during braising was ⅕-¼ of the raw weight.

Venison appears to contribute about 1½ times as much thiamin as lamb, twice as much as veal, 4 times as much as beef, and from ¼-⅓ as much as pork.

Vitamin losses with cooking foods of animal origin has been dis-

EFFECTS OF HOME PREPARATION PRACTICES ON NUTRIENTS 535

cussed extensively in the previous chapter. Because of the nature of meat cuts, the differences between home and food service practices are not sufficient to make a differentiation. A general synopsis follows:

Thiamin is unstable in all heating processes. The method of cooking fresh muscle meat may gravely affect its loss. The processes which appear to be favorable to retention of thiamin in the order of retention are: broiling, 60-86%; frying, 50-89%, roasting, 40-70%, broiling and braising, 26-50%, canning, 23-44%.

Retention of riboflavin is generally fair, from 60 to 100%, regardless of method of cooking.

From $1/3 - 1/2$ of the niacin of muscle meat is lost in braising and stewing, but this is nearly all recoverable in the broth. From 3 to 27% is lost in broiling, frying, and oven roasting; most of the loss, therefore, is through leaching and dripping, and only an insignificant amount appears to be destroyed.

Cover et al. (1947) reported about 75% retention of the pantothenic acid in stewed beef and lamb. Browning increased the loss by 10%, and increased temperature, as in pressure cooking, increased the loss in beef but not in lamb. The amount of water used also affected the extraction so that $1/2$ or more of the vitamin was found in the broth when a large amount of water was used, but only about 40% when a small amount was used. The known solubility and heat instability of pantothenic acid make these results predictable. Cheldelin and Williams (1943) noted losses of 12-37% of the pantothenic acid in the cooking of meats, and Meyer et al. (1947) noted 7-25% losses in fried steaks and liver.

Using more effective extraction methods but a limited number of samples, Schweigert and Guthneck (1953) concluded that 50-85% of the pantothenic acid of pork, beef, and lamb muscle cuts was retained, the higher values being found in the broiled and the lower in the roasted samples. These values are similar to those reported earlier with the less effective extraction procedures.

Meyer et al. (1969) studied the retention of pantothenic acid in beef roasts. Data on the retention of pantothenic acid after oven-roasting and oven-braising are shown in Table 17.4. Retention in the oven-roasted loin averaged 89%; a mean of 19% was transferred to the drip. Cover et al. (1944) obtained retentions of 91 and 75%, respectively, in beef ribs roasted at 300°F and recovered less than 5% in the drip.

In the braised roasts, pantothenic acid retention averaged 56% in the meat and 44% in the drip. More than twice as much pantothenic acid was extracted from the meat by braising as by roasting. The

TABLE 17.4

RETENTION OF PANTOTHENIC ACID IN BEEF COOKED BY TWO METHODS

6 Samples	Retained In Meat	Recovered In Drip	Total Retention
		(Mean ± Standard Error)	
Oven roasted loin	89 ± 3.8	19 ± 1.7	108 ± 3.6
Oven braised round	56 ± 2.0	44 ± 4.3	99 ± 2.6

Source: After Meyer et al. (1969).

amount of pantothenic acid retained in the braised round agrees with the average retentions of 40 and 53% reported by Meyer et al. (1947) for braised heel-of-round and chuck beef roasts, respectively. Recovery in the drip, however, was considerably higher in the present study than the average retentions of 23 and 28% reported earlier. The mean total retention with oven-braising also was higher than the 78% retention in beef stew reported by Cover et al. (1947).

There was no evidence of heat destruction of pantothenic acid with either method of roasting. The high percentage of pantothenic acid transferred to the drip by braising is consistent with the generally recognized water-soluble nature of the vitamin.

Folic acid was found by Cheldelin et al. (1943) to be by far the most seriously affected of the B vitamins by cooking. The losses varied from 46% in halibut to 95% in fried pork chops. The authors speculate on the possibility that some of the vitamin may have been "bound" during the cooking and so have become unavailable to the microorganism used for assay. Schweigert et al. (1946) confirmed these findings, reporting from 8 to 67% retention of folic acid in cooked meats after duplicate assays with two microorganisms. Hurdle et al. (1968) reported no loss in the cooking of lamb liver, boiled or fried, and the white meat of fried chicken; however, egg yolk showed a 70% loss in folate when boiled and a 29% loss when fried. The research area is plagued by the methodology involved. Standardized and systematic studies are evidently needed.

A surprising loss of pyridoxine (vitamin B-6) occurred in all the cooked meats examined by McIntire et al. (1944) by the yeast-growth method. Roasting and broiling produced about 30% retention, braising and stewing only about 18%. Less than 6% of the extracted vitamin was recovered in the broth or drippings. Earlier assays by rat growth (Henderson et al. 1941) yielded somewhat similar results. Curing or boiling pork ham caused a loss of about 60%, roasting about 50%, and frying about 30% of the vitamin B-6 activity of meat.

TABLE 17.5

VITAMIN B-6 CONTENT OF FRESH AND COOKED MEATS

Meat Tested	S. carlsbergensis Assay		
	Fresh (Mg/100 Gm)	Cooked (Mg/100 Gm)	Retention (%)
Beef, standing rib roast	0.32	0.28	56
Beef, Boston cut	0.38	0.25	42
Pork, ham, uncured	0.42	0.40	57
Lamb, leg	0.26	0.16	43
Veal, leg	0.33	0.29	67

Source: Lushbough et al. (1959).

Several fresh, cooked, and processed meats were assayed for B-6 content using the *S. carlsbergensis* microbiological yeast assay by Lushbough et al. (1959). The results for beef, pork, lamb, and veal are given in Table 17.5. B-6 retention in cooked meats averaged 54%. Utilizing the same method, Meyer et al. (1969) studied B-6 retention in 2-lb beef loin and round roasts. The results are given in Table 17.6.

Retention in the loin averaged 72%, and 16% was recovered in the drip. Retention in the round averaged 49% and transference to the drip 34%. Approximately twice as much vitamin B-6 was transferred to the drip by braising as by roasting. Because vitamin B-6 is water-soluble, it seems reasonable that more of the vitamin would be recovered in the drip from meat cooked by moist- than by dry-heat methods.

Mean retentions of vitamin B-6 in the roasted and braised beef in this investigation were considerably higher than in earlier studies with beef and other meats. Reasons for the differences are not evident.

Nearly all of the choline was retained in the meat examined by McIntire et al. (1944). About 77% of the biotin was retained in the cooked meats examined by Schweigert et al. (1943). No other data were found in the published literature.

TABLE 17.6

RETENTION OF VITAMIN B-6 IN BEEF COOKED BY TWO METHODS

6 Samples	Retained In Meat	Recovered In Drip	Total Retention
	(Mean ± Standard Error)		
Oven roasted loin	72 ± 3.6	16 ± 1.9	88 ± 4.8
Oven braised round	49 ± 1.3	34 ± 1.7	83 ± 2.9

Source: After Meyer et al. 1969.

FOODS OF PLANT ORIGIN

General Preparation

To the best of our knowledge, the losses due to trimming, washing, soaking, chopping, and mincing as prepared for consumption in the home are similar in nature and extent to those that occur in large-scale food preparation. This subject was discussed at length in the previous chapter. In brief, it is better to delay the preparation of foods until a few minutes before they are to be cooked and served. Protracted soaking should be avoided. Frozen vegetables should not be thawed or washed before cooking; instead, they should be placed directly into a minimum quantity of rapidly boiling water. In this respect, boil-in-the-bag procedures offer superior conditions because losses into cooking water are completely avoided and nutrients are theoretically held by the "butter" or cream sauces usually associated with food packaged in this manner.

Salads should be prepared just before they are to be served in order to minimize losses in nutrients, especially ascorbic acid.

Boiling

The literature might be labelled extensive because of the number of reports published; however, considering the array of available foods and the approximately 50 nutrients one should examine, the existing information is sparse. Taken as a whole, the literature supports the following theorems: (1) When home-scale portions of vegetable foods

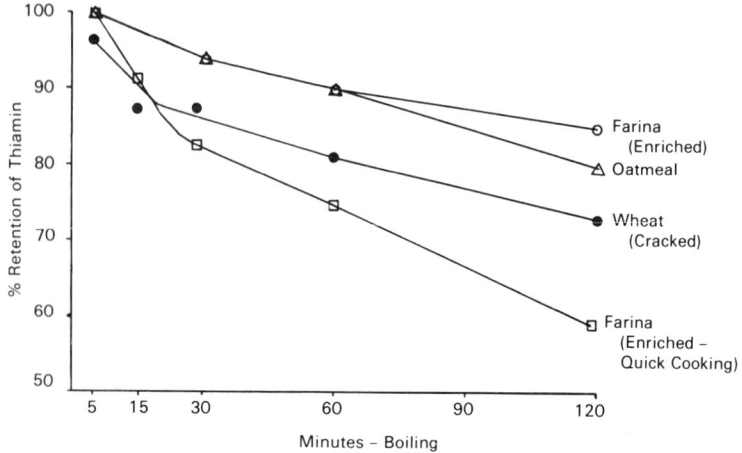

Adapted from Lincoln et al. (1944)

FIG. 17.1. PERCENTAGE OF RETENTION OF THIAMIN DURING SMALL-SCALE BOILING ON CEREAL PRODUCTS

EFFECTS OF HOME PREPARATION PRACTICES ON NUTRIENTS

TABLE 17.7

PERCENTAGE RETENTION OF VITAMINS DURING SMALL-SCALE PREPARATION (BOILING) OF FOODS OF PLANT ORIGIN

	Cooking Time (Min)	Thiamin (%)	Riboflavin (%)	Niacin (%)	Ascorbic Acid (%)
Earth vegetables					
Beets	40		100	39	
Carrots	5-20	100	100	75	
Carrots					70
Carrots	11	100			
Garlic					60
Onion					64
Onion					24
Potatoes		75			68
Potatoes	30				53-66
Potatoes	60				40-50
Potatoes	90				17
Potatoes	30		91	69	
Potatoes	20		57	73	
Potatoes (salted water)					81
Potatoes, in skins					100
Potatoes					77
Potatoes, pared		67			
Rutabagas	26-36				58-60
Squash, summer	20-30				50
Sweet potatoes			87	95	
Sweet potatoes		80			
Sweet potatoes	20				69
Sweet potatoes, whole	35				76
Sweet potatoes	12-25				131
Turnips			90	60-100	59
Herbage vegetables					
Bean, snap	45				60-70
Bean, snap			68-77		
Bean, snap (frozen)	15	78			72
Bean, snap (canned)	10	71			27
Bean, snap			73-88		
Bean, snap	5-20	120	75	70	70
Beet tops	10		86	96	
Broccoli (frozen)	5-20		55	47	100
Broccoli (frozen)	11				55
Broccoli (frozen)	2				64
Cabbage	17-19				27-35
Cabbage					21
Cabbage	60				80
Cabbage		47-74			
Cabbage	30				70-78
Cabbage	60				53-58
Cabbage	90				13
Cabbage (20 gm peanut oil at 80°)					76
Cauliflower (4-min fry)	12				15
Celery					50

540 NUTRITIONAL EVALUATION OF FOOD PROCESSING

TABLE 17.7 (Continued)

	Cooking Time (Min)	Thiamin (%)	Riboflavin (%)	Niacin (%)	Ascorbic Acid (%)
Chard, Swiss	10				42-46
Chard, Swiss	14				25-29
Collards	5				60
Collards	30				21
Lima beans	30				82
Lima beans	180				59
Peas			64-68		
Peas			62-75		
Peas		64-84			
Peas		92	70	55	100
Rhubarb	10				84-87
Spinach			55-70		
Spinach			36-58		
Spinach		70			
Spinach	7				58
Spinach		100	80	60	72
Turnip greens	slow				63-76
Turnip greens	rapid				77-84
Legumes and oilseeds					
Soybeans					52
Soybeans			68-77		
Fruits					
Apple	20-30			75	
Apple					50
Blackberry					40
Black currant					68
Cherry					50
Gooseberry					50
Green plantain					32
Pear					60
Plum					60
Tomato					76

Source: Adapted from Harris and Levenberg (1960).

are boiled, nutrient losses vary according to (a) the type of food; (b) the stability of the nutrient; (c) the amount of cooking water (Table 16.19, Chap. 16); (d) the time of cooking; (Fig. 17.1, and Table 17.7); and (e) the type of equipment. (2) Vegetables with large surface-to-weight ratios are especially sensitive. Slicing and quartering vegetables increases vitamin losses (Table 17.8). (3) Nutrient losses increase as the ratio of cooking water to food increases (Table 17.9 and Fig. 17.2). As expected ascorbic acid is the most unstable nutrient but folic acid and vitamin B-6 may be even more so, for example, in the case of reheating frozen vegetables (Tepley and Derse 1958). Under certain conditions nutrients are sensitive to other prac-

TABLE 17.8

EFFECT OF SLICING ON THE PERCENTAGE RETENTION OF VITAMINS DURING SMALL-SCALE BOILING OF FOODS OF PLANT ORIGIN

	Water/Food Ratio	Cooking Time (Min)	Thiamin (%)	Riboflavin (%)	Niacin (%)	Ascorbic Acid (%)
Potatoes, peeled, halved						87
Potatoes, peeled, whole						87
Potatoes, whole		15-20				75
Potatoes, peeled		15-20				65
Potatoes, in skins	1/1	40	83		100	93
Potatoes, pared	1/1	40	84		100	94
Potatoes, quartered	1/1	23	100		100	82
Bean, snap (Frenched)		18				28
Bean, snap (whole)		31				54
Cabbage, soup						50
Cabbage, stewed						15

Source: Adapted from Harris and Levenberg (1960).

tices, for example, the addition of bicarbonate affects thiamin (Table 17.10).

Glass, stainless steel, aluminum, enamel, and similar equipment have no measurable effect upon the nutrient content of cooked foods (Table 17.11) whereas copper, brass, and monel may be quite destructive.

Hewston *et al.* (1948) have reported the results of a comprehensive study of the effect of home preparation on the vitamin and mineral content of 20 common foods. Ascorbic acid was the most sensitive of all the nutrients studied. Retentions in the edible portions ranged from only 30 to 50%, with extremes as low as 18% in cabbage strips and 100% in boiled unpared sweet potatoes. Rententions were lowest when the volume of the cooking water was large, the time of cooking was long, and the size of food particle was small. Whole baked sweet potatoes retained 89% ascorbic acid while halved sweet potatoes retained only 31%.

Carotene was found to be the most stable vitamin, the retention being nearly 100% in most cases. Low retentions of 33% were noted for carotene in cooked green fall cabbage and 67% in green spring cabbage. Niacin was well retained in most products. Riboflavin was retained to the extent of 60-90% in boiled vegetables, and the retention did not appear to be influenced by the volume of cooking water.

Thiamin is both water-soluble and heat-labile. This will explain why the retention of thiamin in vegetables during cooking was as low as 35%; usually the retention was about 60%. The volume of cooking water has an important effect upon thiamin retention.

Because of water solubility, the retention of minerals in cooked

TABLE 17.9

EFFECT OF WATER/FOOD RATIO ON THE PERCENTAGE RETENTION OF VITAMINS DURING SMALL-SCALE BOILING OF FOODS OF PLANT ORIGIN

	Water/Food Ratio	Cooking Time (Min)	Thiamin (%)	Riboflavin (%)	Niacin (%)	Ascorbic Acid (%)
Carrots	3.5/2	30			71	
Carrots	1/30	30			99	
Potatoes	1/1	30	70	55	74	88
Potatoes	1/7.5	25	96	97	100	98
Bean, snap	2/1	15-30				48-50
Bean, snap	1/2	15				74
Broccoli (frozen)	1/1	5.5				82
Broccoli (frozen)	5/1	5.5				57
Broccoli	1/4	30	56	63	36	47
Broccoli	1/9	30	95	73	88	67
Cabbage	1/2	7				78
Cabbage	2/1	7				60
Cabbage	4.2/1	6-8				60
Cabbage	5/1	10				68
Cabbage	5/1	8				46
Cabbage	4/1	12.5				70
Cabbage	4/1	5.5				51
Cabbage	2/1	8.5				50
Cabbage	2.5/4	15				82
Cabbage	3/1	15				24
Kohlrabi	1/1	2				45
Peas (300 gm/150 cc)	1/2	15				74
Peas (300 gm/600 cc)	2/1	15-30				48-50
Peas	4/5	24	54	69	64	68
Peas	1/30	20	94	93	97	86
Peas	1/3	8				71-76
Peas	1/1					54-57
Spinach	2/4	7				47
Spinach	2/1	7				36
Spinach	1/1	2				63
Spinach	2/1	2				25
Spinach	2/1	8				24
Spinach	1/5	8				62
Soybeans	1/2	12				71
Soybeans	2/1	12-24				52-45

Source: Adapted from Harris and Levenberg (1960).

vegetables ranged between 50 and 100%. The more water used in cooking, the greater was the mineral loss.

Tepley and Derse (1958) conducted an extensive study of the retention of nutrients in 20 frozen vegetables during cooking to "optimum flavor doneness" for home serving. There was little nutrient loss by leaching into the liquor except for sodium (10-25%) in most products, and 15-25% of all vitamins except beta-carotene in turnip greens. Very little destruction of thiamin, riboflavin, and niacin occurred. In some instances, the destruction of beta-carotene, folic acid, pantothenic acid, and vitamin B-6 was as high as 50%. In most cases the destruction of ascorbic acid was less than 10%, but in cut

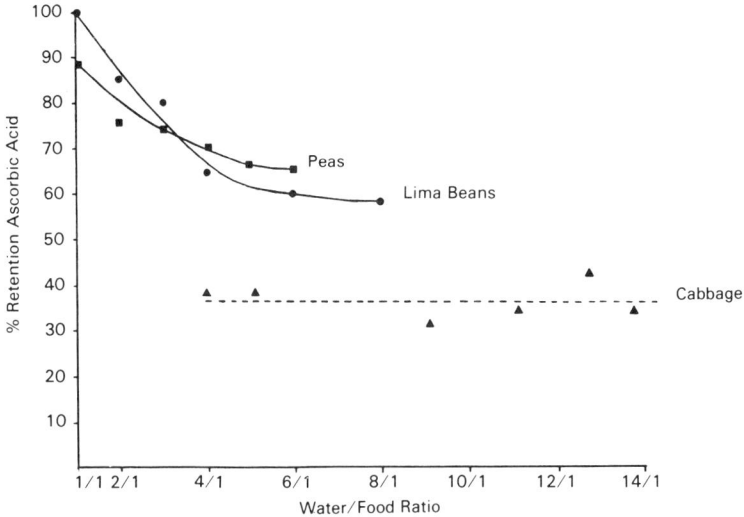

FIG. 17.2. PERCENTAGE OF RETENTION OF ASCORBIC ACID DURING SMALL-SCALE COOKING AS A RESULT OF INCREASING WATER TO FOOD RATIO

SOURCE: Peas (cooked to doneness), McIntosh et al. (1942). Lima beans (cooked 17 min), McIntosh et al. (1942). Cabbage (cooked 30 min), Olliver (1941).

TABLE 17.10

EFFECT OF COOKING WATER ADDITIVES ON PERCENTAGE RETENTION OF VITAMINS DURING SMALL-SCALE BOILING OF FOODS OF PLANT ORIGIN

	Water/Food Ratio	Cooking Time (Min)	Thiamin (%)	Riboflavin (%)	Niacin (%)	Ascorbic Acid (%)
Apricots (with sugar)		55	82	92		
Apricots (with NaHCO$_3$)		35	78	92		
Bean (4 gm NaHCO)			80			
Bean (30 ml NH$_3$)			90			
Bean, navy			100			
Bean, navy (with soda)			100			
Bean, snap			68			
Bean, snap (with soda)			41			
Peas (simmer)			80			
Peas (simmer and soda)			67			
Peas (no bicarbonate)			65			83
Peas (bicarbonate)			43			80
Peas (no bicarbonate Pyrex)			87			
Peas (bicarbonate, Pyrex)			86			
Peas (no bicarbonate, aluminum)			93			
Peas (bicarbonate, aluminum)			92			

Source: Adapted from Harris and Levenberg (1960).

TABLE 17.11

EFFECT OF KETTLE TYPE AND COVER ON PERCENTAGE RETENTION OF VITAMINS DURING SMALL-SCALE BOILING OF FOODS OF PLANT ORIGIN

	Water/Food Ratio	Cooking Time (Min)	Thiamin (%)	Riboflavin (%)	Niacin (%)	Ascorbic Acid (%)
Parsnips (enamel)		8				84
Parsnips (Pyrex)		8				81
Parsnips (stainless)		8				66
Parsnips (aluminum)		8				71
Bean, snap (old-fashioned, water, salt, and fat meat)		120				17
Bean, snap (open kettle)						62-74
Bean, snap (tight kettle)						66-68
Bean, snap (tight covered)	1/6	30				59
Bean, snap (open kettle)	4/1	30				58
Bean (Frenched and tight kettle)						50
Bean (snapped and tight kettle)						62
Bean (Frenched and open kettle)						28
Bean (snapped and open kettle)						45
Cabbage (tight cover)		18				57
Cabbage, shredded, (open kettle)		11				55
Cabbage (sections, old-fashioned with fat)		15				30
Peas (stewpan)			97	100		86
Peas (dielectric)		3¾	95	98		83

Source: Adapted from Harris and Levenberg (1960).

snap beans, collards, corn, and chopped spinach the loss ranged between 27 and 48%.

The results of Noble (1967) of the effects of cooking time on the ascorbic acid content of eight vegetables confirmed prior studies. Average retentions in the tissues cooked for the longer (50 min) and shortest periods (5 min) in boiling water, and 5 min as compared to 1 min in the pressure saucepan were respectively 35-44% and 66-78% of the original ascorbic acid.

Pressure Cooking

McIntosh et al. (1940) compared the retentions of ascorbic acid when five vegetables were cooked by boiling, by steaming, and by pressure to the same degree of doneness. The retentions by the three methods of cooking, respectively, were: Brussels sprouts (77, 91, and 97%), cauliflower (81, 77, and 92%), lima beans (62, not reported, and 75%), peas (70, 68, and 98%), and spinach (70, 72, and 80%). Thus, best results were obtained by pressure cooking.

EFFECTS OF HOME PREPARATION PRACTICES ON NUTRIENTS 545

Some data on the retentions of ascorbic acid when various foods were pressure cooked are presented in Table 17.12. Other studies on ascorbic acid retention include that of Noble (1967) and Sweeney *et al.* (1960). Little data relating to retentions of other nutrients are available. The reader should also consult the analogous section in the previous chapter.

TABLE 17.12

PERCENTAGE RETENTION OF ASCORBIC ACID DURING
SMALL-SCALE PREPARATION (PRESSURE COOKING)
OF FOODS OF PLANT ORIGIN

	Ascorbic Acid (%)
Earth vegetables	
Potato, peeled, halved (15-lb pressure)	85
Potato, pared, cut (15-lb pressure)	89
Potato, pared, cut, water soak	86
Sweet potato, 117°C	83
Fruit	
Chili (10-lb pressure)	71-93
Herbage vegetables	
Asparagus	82
Beans, snap	73
Broccoli	79
Broccoli (frozen)	72
Brussels sprouts	73
Cabbage	66
Cauliflower	76
Spinach	30
Legumes	
Lima beans	65
Peas	74
Soybeans	81

Source: Harris and Levenberg (1960).

Steam Cooking

The remarks in the previous chapter relating to steam cooking also apply here. Lower losses of nutrients occur when vegetables, especially green leafy vegetables, are steam-cooked rather than boiled (Table 17.13). However, the retention of thiamin, niacin, and folic acid in steam-cooked vegetables is remarkably low.

Although only ascorbic acid is considered, the work of Gordon and Noble (1959) best illustrates in a comparative manner the effects of boiling versus pressure-cooking versus steaming in a tightly covered saucepan. The results as given are in Table 17.14. For the vegetables

TABLE 17.13

PERCENTAGE RETENTION OF VITAMINS DURING SMALL-SCALE PREPARATION (STEAMING) OF FOODS OF PLANT ORIGIN

	Cooking Time (Min)	Carotene (%)	Thiamin (%)	Riboflavin (%)	Niacin (%)	Ascorbic Acid (%)	Pantothenic Acid (%)	Biotin (%)	Folic Acid (%)	Inositol (%)
Vegetables										
Brussels sprouts	20					60	72	100	8	65
Cabbage	30			84	81	52				
Cabbage	30									
Cauliflower	20			87	78		100	87	12	97
Cauliflower	30			83			94	72	31	54
Kale	30					67				
Lima beans	60	74		100	100	100		100	8	39
Lima beans					39					
Lima beans					41					
Lima beans (canned)			32	25	41					
Lima beans (canned)			24		25					
Okra				100	100		100	100	28	100
Peas	25				36					
Peas	25				51					
Peas					84					
Peas					24–25					
Peas					30–35					

EFFECTS OF HOME PREPARATION PRACTICES ON NUTRIENTS 547

Peas	2							
Peas (canned)	0	34	30					
Peas (canned)		35						
Peas	16							
Soybeans		22						
Spinach	3							
Spinach	10							
Spinach (canned)	30	31	73					
Spinach	10		24	94		86	16	74
Swiss chard								
Tubers and root vegetables								
Potato (unpeeled)	35				83			
Potato (peeled, new)	45				74			
Potato (new)	30				53			
Potato (old)	25				82			
Potato (old)	30				73			
Potato (old)	50				67			
Potato (unpeeled, new)	45				56			
Potato (new)	30				62			
Potato (old)	35				77			
Potato (old)					69			
Potato (peeled)	20				75–80			
Potato (peeled)					70–76			
Carrots					86			
Cereals								
Rice	25	100	100	100		100	8	39

Values 18, 37, 39, 40, 26, 26, 84, 38, 24, 44 appear in an adjacent column aligned with the Swiss chard / Tubers rows.

Source: Complied by Harris and Levenberg (1960).

TABLE 17.14

ASCORBIC ACID IN RAW VEGETABLES AND PERCENTAGE RETENTION AFTER COOKING

	In Raw Vegetable		Retention in Cooked Vegetable				Retention in Cooked Vegetable Plus Cooking Water			
	Mean (Mg/100 Gm)	Standard Error (Mg/100 Gm)	Boiling Water (%)	Pressure Saucepan (%)	Tightly Covered Saucepan (%)	Steamer (%)	Boiling Water (%)	Pressure Saucepan (%)	Tightly Covered Saucepan (%)	Steamer (%)
Asparagus	29.1	1.21	43	80	71	78	68	88	80	86
Wax beans[1]										
Series a	24.8	1.53	45	65	56	75	73	66	59	75
Series b	25.6	2.45	57	66	68	76	75	67	69	76
Series c	22.1	0.44	63	86	45	78	73	92	53	83
Broccoli	123.0	9.13	33	82	67	79	96	89	72	82
Brussels sprouts	88.4	6.76	35	84	71	86	84	88	75	88
Cauliflower	52.2	6.16	37	70	55	71	93	78	58	75
Onions	16.3	2.14	36	53	68	67	93	80	87	94
Peas[1]										
Series a	27.1	1.06	44	64	66	67	82	75	73	81
Series b	30.4	3.76	46	70	70	71	77	73	74	79
Series c	23.1	1.79	48	73	71	66	76	81	80	77
Rutabagas	36.1	9.09	36	65	59	65	75	71	66	73
Spinach	90.2	1.38	45	78	63	64	84	93	77	76
Squash	16.2	1.19	60	64	71	76	82	73	76	87
Turnips	24.7	1.12	45	63	65	61	74	71	70	69

Source: Gordon and Noble (1959).

[1] Each series came from a different year's harvest and contained four replications.

as a whole, the percentage retention in the boiling water method, 45% of the original content, was significantly smaller than in the steaming methods, 69%. Within the steaming methods, the pressure saucepan method, which uses short cooking times but high temperatures, was not significantly different from the average retention in the other two methods which use longer cooking times but lower temperatures. Within each cooking method, however, the different vegetables retained varying amounts of ascorbic acid. No analogous data for other nutrients was found in the open literature with the exception of a paper from India (Kamalanathan et al. 1974) which studied 3 vegetables, only 1 of which (*Phaselous vulgaris*) is common in Western countries. In a comparison of boiling, steaming, pressure-cooking, and panning, the retention of calcium, phosphorus, ascorbic acid, thiamin, and riboflavin were studied. Boiling was the harshest and pressure-cooking found to be the best for the array of nutrients studied.

Bicarbonate in Cooking

Sodium bicarbonate and other alkaline salts are occasionally added to the cooking water because the color of cooked green vegetables is preserved better and because the rate of cooking is increased. This practice is destructive to nutrients that are sensitive to alkali, especially thiamin and ascorbic acid (Table 17.10). Alkali should not be used in cooking vegetables.

Losses Into Cooking Water

The data referred to in the previous chapter are pertinent to cooking losses in the home also. Many studies in which small or large batches of food were used indicate that the loss of nutrients from vegetable foods during cooking is caused mostly by extraction into the cooking water rather than by destruction. Some examples are given in Table 17.15 and Fig. 17.2. For example, Munsell et al. (1949) measured the cooking water recovered from cabbage. The cooking water contained more thiamin, riboflavin, and niacin than the cabbage itself. Brush et al. (1944) demonstrated that the water phase of canned foods contain considerable amounts of vitamins extracted during canning. How much more is extracted with reheating is not clear but the practice of discarding the water phase of canned foods, and cooking or reheating water in general, is a serious loss of water-soluble nutrients. The housewife who has the time should consider this water for homemade soups. Otherwise one is tempted to encourage the use of frozen or preferably frozen boil-in-the-bag vegetables. Most shoppers are unaware that frozen vegetables are competitively priced if compared on a solids basis. Canned vegetables

550 NUTRITIONAL EVALUATION OF FOOD PROCESSING

TABLE 17.15

PERCENTAGE LOSSES OF VITAMINS DURING SMALL-SCALE PREPARATION OF FOODS OF PLANT ORIGIN (EXTRACTION INTO COOKING WATER)

	Cooking Time (Min)	Water/Food Ratio	Thiamin (%)	Riboflavin (%)	Ascorbic Acid (%)
Vegetables					
Asparagus (canned)	0		38	29	40
Beans, snap	17				26
Beans, snap (canned)			36		44
Beans, snap (canned)			33	24	36
Lima beans	17				31
Lima beans (canned)	0		32	25	41
Lima beans (canned)			24		25
Brussels sprouts	30-40				54
Brussels sprouts	20	3/1			34
Brussels sprouts		4/1			48
Brussels sprouts	30-40				54
Brussels sprouts	20	3/1			34
Brussels sprouts	20	4/1			48
Cabbage	18				56
Cabbage	40				26
Cabbage	till done		43	50	30
Cabbage	till done		72	67	33
Corn, kernel (canned)	0		33	20	52
Peas	25	3/1			51
Peas	25	12/1			84
Peas (canned)	0		34	30	37
Peas (canned)			35		39
Peas	16				40
Soybeans			22		26
Spinach	3				26
Spinach (canned)			31	24	38
Spinach (canned)	30				24
Swiss chard	10				44

Source: Adapted from Harris and Levenberg (1960).

have the advantage of requiring no refrigeration and having an extended shelf-life. The consumer rarely considers the cost of freezer operation in the home and the priorities the consumer has for the available freezer space varies considerably. Freezing bread extends shelf-life but the same space in vegetables affords nutrient advantages for a more costly type of food.

Frying

Few vegetables are fried, and there are but few reports in the literature of the effects of frying upon their nutrient content. In 1942, a study was made on the effects of frying upon 30 vegetables and losses in ascorbic acid ranging from 0.7 to 32.7% were reported. Vegetables which were boiled, then fried, lost from 22.2 to 78.8% ascorbic acid. Potatoes fried for 15 min in deep fat retained 72% ascorbic acid, according to another early study; yet potatoes fried in butter or Crisco were reported to lose no ascorbic acid (Richardson et al. 1937). Conversely, another early worker reported that potatoes fried 20 min retained only 20-45% ascorbic acid. Cauliflower fried in ghee lost 25% ascorbic acid (Rudra 1937). Cheldelin et al. (1943) reported a 74% loss of riboflavin in onions after 20 min of frying. Basu and Neogy (1948) observed the retentions of ascorbic acid in several fried foods as follows: potato, 60% (12 min); sweet potato, 19% (20 min); jack fruit, 22% (5 min); papaya, 48% (5 min); and peas, 60% (4 min).

Domah et al. (1974) studied the ascorbic acid content of potatoes after frying (see Table 16.26, Chap. 16), and reported a loss of 18% during 30 min of frying at 140°C. In comparing the reduced ascorbic acid of broccoli and green beans cooked by stir frying as compared to microwave and conventional methods, Eheart and Gott (1965) demonstrated better retention of ascorbic acid with stir frying in the case of brocolli but not green beans (Table 16.33, Chap. 16).

Frying as a means of retaining water-soluble nutrients such as ascorbic acid is often castigated and there currently is no experimental basis for such a view. However, one must recognize that the nutrient/calorie ratio of fried foods is considerably poorer than foods cooked in minimal water. If weight reduction is a concern, then fried foods are not justified.

Linoleic acid degradation occurs in oil, (safflower, cottonseed, corn) used for frying. According to Kilgore and Bailey (1970), the percentages of linoleic acid (expressed as percentages of total fatty acids) in the fresh fats were: safflower oil, 72.0; corn oil, 57.2; cottonseed oil, 55.5; and shortening, 30.2. After the fats had been used for intermittent frying periods totaling 7½ hr, during which 10 lb potatoes were fried, the percentages were: safflower oil, 69.2; corn oil, 51.7; cottonseed oil, 49.0; and shortening, 26.7. The percentage of linoleic acid in fat extracted from the 10th pound of potatoes fried was 64.8 when safflower oil was used; and 50.2, 43.3, and 26.5 for corn oil, cottonseed oil, and shortening, respectively. Similar results were reported in the work of Fleishman et al. (1963) who also studied sesame, coconut, olive, peanut, and soy oils. Possibly related to these changes is that the quality of frying oils used had an influence on the stability of ascorbic acid.

BAKING

Vegetables

The effects of baking upon the nutrient content of potatoes may be considerable. The following thiamin retention rates have been reported: 55 to 80%; 60%; 84%; and 41% (large), 65% (medium), and 78% (small) potatoes.

Baked sweet potatoes retained only 9% thiamin in an early report; whereas Pearson and Luecke (1945) found retentions to be 75% ascorbic acid, 89% thiamin, 85% niacin, and 77% pantothenic acid.

Apples retained 40% ascorbic acid when baked 60-90 min at 204°C in an open dish. The retention was 25% during baking in a covered dish and only 20% during baking in a pie (Curran et al. 1937).

Bread

A few studies have been made on the effect of baking of bread and biscuits upon the nutrient content (Table 17.16). The losses of thiamin may be considerable, especially when the pH is above 6.

TABLE 17.16

LOSSES OF VITAMINS AND LYSINE IN BAKED GOODS DURING HOME PREPARATION

Type of Baked Goods	Thiamin (%)	Riboflavin (%)	Niacin (%)	Lysine (%)
Bread, white	14-24			
Bread, white	30			
Bread, white				11-32
Bread, white (toasted)				16-40
Bread, white	24	0	0	
Bread, white	18-44			
Bread, Pakistani (leavened)	30			
Bread, white (unleavened)	11-26			
Bread, rye, Finnish sour	20			
Corn bread, pH 5.4-6.2	11-16	0-7	2-5	
Corn bread, pH 6.6	93	11	0	
Corn bread, pH 6.6	11-20	102	4	
Corn bread, 30% crust	15			
Corn muffins, 40% crust	21			
Corn muffins, commercial mix, 68%	36	26-29		
Corn muffins, lab mix		16-21	1	
Corn muffins, sour milk, whole meal	20			
Corn muffins, sour milk, bolted	17			
Corn muffins, sweet milk, whole meal	31		1	
Corn muffins, sweet milk, bolted meal	54			
Corn sticks, 68% crust	34			
Doughnuts		23	20	
Biscuits	70			
Muffins	40-65			

Source: Harris and Levenberg (1960).

Baking powders destroy thiamin. Sodium aluminum sulfate and phosphate baking powder caused a 16% loss when added at low levels (1.64 gm soda; 10.28 gm baking powder) and as much as 84% loss when included at high levels (4.5 gm soda) (Briant and Hutchins 1946; Briant and Klosterman 1950). A pH of 6 or lower is desirable for satisfactory stability of thiamin during baking.

Mild baking is less destructive than extensive baking. Zaehringer and Personius (1949) have reported that pale, medium, and dark baked bread retained 83.1, 80.4, and 73.8% thiamin, respectively. Similar data on dinner rolls were 89.5, 84.5, and 77.5%; data on clover leaf rolls were 93.0, 89.8, and 84.0%; and data on pan rolls were 92.6, 91.5, and 87.8% retention during baking.

This effect of baking appears to be caused largely by losses in the crust. From Table 17.17 it is evident that the toasting of bread causes considerable thiamin destruction.

TABLE 17.17

DESTRUCTIVE EFFECT OF TOASTING UPON THIAMIN IN BREAD

Toasted (Sec)	Unenriched White Bread (%)	Enriched White Bread (%)	100% Whole Wheat Bread (%)
0	0.0	0.0	0.0
30	9.2	5.2	4.0
40	22.0	7.0	8.2
50	19.7	13.0	12.5
60	26.7	15.0	15.3
70	31.4	17.0	21.0

Source: Adapted from Downs and Meckel (1943).

The composition of the bread formula seems important also. The retention of thiamin was 87, 84, 83, and 81% when biscuits were made with fresh milk, evaporated milk, dry milk, and water, respectively (Briant and Hutchins 1946).

The vulnerability of lysine to heat leads to serious impairment in the nutritive quality of the protein because of its relatively exposed epsilon-amino acid (Liener 1960).

The actual amount of lysine lost during baking varies considerably, but in general it is a function of such factors as time and temperature. Most destruction takes place at the crust, where the browning reaction occurs. Rosenberg and Rohdenburg (1951) and Ericson *et al.* (1961) found that the loss of lysine in bread due to baking was about 15%. Tara's *et al.* experiment (1972), utilizing a microbiological assay reported a 12.5% loss with baking. Jansen *et al.* (1964)

reported a loss of 15% lysine, added as the HCl to water bread containing 4% nonfat dry milk and baked 30 min at 450°F. It was also reported that, when the baking time was decreased to 20 min, the loss in white bread was 18% by ion-exchange chromatography. However, in experiments in which a higher (6-14) percentage of nonfat dry milk was added to bread which was baked 20 min at 450°F, the nutritive loss of lysine was estimated as 36%. They concluded that the presence of moderate amounts of nonfat dry milk increased greatly the nutritive loss of lysine during baking, presumably because of the high content of the reducing sugar D-lactose. Losses of lysine averaging 25% (22-28%) were found by Matthews *et al.* (1969) in yeast breads prepared with the *atta* of 85% extraction at 6 different levels of lysine fortification and baked 30 min in a 400°F experimental oven. On the other hand, Akino (1971) reported that the loss of amino acids added during the processing of bread and cakes is small. In bread manufacture, there is no loss during the dough making and fermentation processes, but a loss of up to 10% occurs in baking. Therefore, he recommended that 150 mg of L-lysine, or 190 mg of L-lysine-HCl be added to 100 gm of wheat flour. On the contrary, Hepburn *et al.* (1957) found that the proportions of the following amino acids, determined microbiologically, increased significantly in the conversion of flour into bread: lysine, 19%; aspartic acid, 11%; methionine, 10%; threonine, 7%; and valine, 3%. They concluded that this increased difference in the amino acid level of flour and bread resulted from inclusion of milk and yeast in the bread formula and from the baking process itself.

Thiamin is probably the most heat-sensitive of the B vitamins, especially in nonacid foods. Temperature, pH, time of heating, form of the thiamin molecule, trace metals, oxygen, and processing or storage are the most important factors contributing to the loss of thiamin in food products (Dwivedi and Arnold 1973; Labuza 1973).

Morgan and Frederick (1935) found a definitely lower level of thiamin in the crust of whole wheat bread than in the crumb; the maximum difference was 35%. A 20% loss of thiamin in bread baked at 475°F for 30 min was determined by Coppock *et al.* (1957) when working with European bread; Bottomley and Nobile (1962) reported an average loss of 29.5% of the thiamin during baking.

Borenstein (1971) and Bauernfeind and Cort (1974) have reported that, surprisingly, little "good" data have been published on the stability of not only added vitamin A, but also other vitamins in fortified food.

Retention of dry vitamin A [Type 250-SD (100 mesh), a commercial dry particulate retinyl palmitate encapsulated in gum acacia with antioxidants] in corn meal with 15% moisture content stored in a liner box at room temperature for 6 months was 86% (Bauernfeind and Cort 1974). Similar stability has been reported in wheat flour with 13.8% moisture (Anderson and Pfeifer 1970), and in wheat flour containing 13.5% moisture (Borenstein 1969). In flour of 11% moisture, typical retention was about 90% after 6 months at room temperature (Bauernfeind and Cort 1974).

Retention of vitamin A in the baking of tortillas, muffins, chapaties, and white bread was reported as good to excellent (Guttikar et al. 1965).

The tortilla in Mexico and other Latin American countries is prepared by soaking the kernels of maize in solutions prepared with lime or wood ash, then grinding the softened grain on a stone "matate," and cooking the thin, flat cakes on a hot iron plate or hot stone. Cravioto et al. (1945) have studied the composition of tortillas and the corn from which they were prepared, and the following retention values were noted: carotene, 60%; iron, 137%; phosphorus, 115%; calcium, 2010% (from lime water treatment). Massieu et al. (1949) noted retentions of 70% tryptophan, 74% threonine, 75% histidine, and 84% arginine in Mexican tortillas.

For a thorough discussion of the protein quality of maize products, (tortillas and Arepas) as used in Mexico, Central America, and South America, the reader is encouraged to consult Bressani et al. (1972).

LOSSES ON REFRIGERATOR STORAGE OF COOKED FOODS

Data presented in other sections of this book indicate that the nutrient content of vegetables, fruits, and fruit juices is reduced during storage, even in refrigerators. This is true during storage in home refrigerators also (Table 17.18 and Fig. 17.3).

Charles and van Duyne (1958) studied the effects of refrigerator storage and of reheating on the ascorbic acid content of cooked vegetables. Cooked broccoli, Brussels sprouts, shredded cabbage, sliced cabbage, cauliflower, peas, and snap beans lost significant amounts of ascorbic acid during one day's refrigeration, whereas cooked asparagus and spinach lost insignificant amounts. When these stored cooked vegetables were reheated, they all showed a further significant loss. The ascorbic acid in the cooking liquids was stable during refrigerator holding and during reheating, indicating that the vitamin in the vegetables was being destroyed. When the refrigerator storage was extended to three days, the losses were even greater.

There is a need for more extensive data. Home practices are changing considerably and leftover convenience foods on which there is no data also must be studied.

TABLE 17.18

LOSSES IN THE VITAMIN CONTENT OF FOODS DURING REFRIGERATOR AND FROZEN STORAGE

Food	Storage Temp	Storage Time	Carotene (%)	Thiamin (%)	Riboflavin (%)	Niacin (%)	Ascorbic Acid (%)	Vitamin B-6 (%)	Pantothenic Acid (%)
Asparagus	Cool storage	7 days	14	18	17		67		
Beans	Frozen	0 months		14			17		
Beans	−40°F	10 months		14					
Beans	0°F	10 months		10					
Broccoli	Refrigerator	24 hr					42		
Broccoli	Refrigerator	48 hr					19		
Broccoli	0 to −10°F	5 months					34		
Cabbage	Refrigerator	24 hr					0		
Cabbage	Refrigerator	72 hr					29		
Lima beans (shelled)	Room temperature	48 hr					24		
Lima beans	Refrigerator	48 hr					67		
Lima beans (in pods)	Room temperature	48 hr					16		
Lima beans	Refrigerator	48 hr					39		
Melon cuts	Refrigerator	24 hr					5		
Melon syrup	Refrigerator	24 hr					66		
							25		

EFFECTS OF HOME PREPARATION PRACTICES ON NUTRIENTS

Item	Condition	Duration					
Orange conc	$-8°F$	12 months					2
Orange conc	$40°F$	12 months					5
Orange juice	Refrigerator	72 hr					20
Peaches	$0°F$	10 months					9
Peas	$-17°$ to $-23°C$	1 yr				0	
Peas	Quick freeze						10–22
Snap beans	Quick freeze						4–15
Soya beans	Quick freeze						19–22
Spinach	Quick freeze						30–45
Spinach	Cold storage	7 days					35
Strawberry purée	Freezer	1 month					20
Strawberry purée	Freezer	5 months					20
Swede juice	Refrigerator	0 min	5				0
Swede juice	Refrigerator	45 min		15			10
Swede juice	Refrigerator	60 min			17		13
Swede juice	Refrigerator	120 min					19
Swede juice	Refrigerator	240 min					24
Tangerine conc	$-80°F$	12 months					6
Tangerine conc	$40°F$	12 months					10
Vegetables	Ice chest			4			38
Vegetables	Room temperature			16			70
Vegetables	Cool storage	35 days	20–50	30–65	12–17	17–41	21–29

Source: Harris and Levenberg (1960).

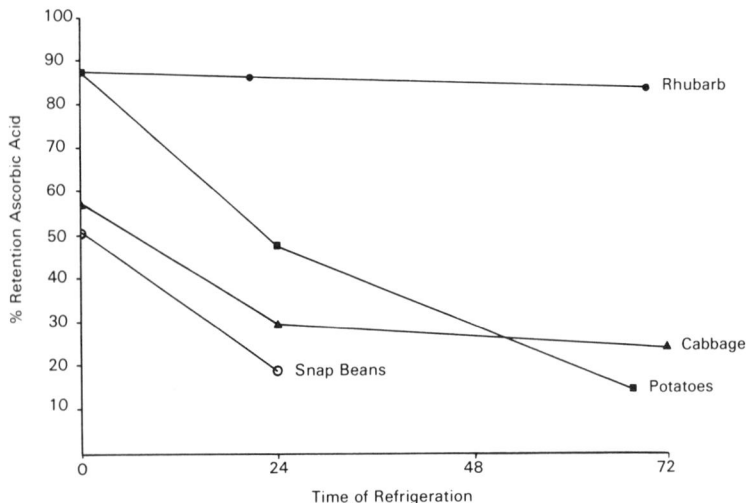

FIG. 17.3. PERCENTAGE OF RETENTION OF ASCORBIC ACID DURING REFRIGERATION STORAGE

SOURCE: Rhubarb (boiled 12 min), Van Duyne et al. (1947). Cabbage (boiled), Van Duyne et al. (1947). Potatoes (boiled), Van Duyne et al. (1945). Snap beans (boiled 17 min), Wadsworth and Wilcox (1945).

SHELF-LIFE

Retail food products have a shelf-life predetermined by the manufacturer and usually based on maintenance of acceptable organoleptic qualities. Shelf-life information is always coded and may also be a part of the "open" label in the form of an expiration date (e.g., milk) or date of manufacture (e.g., packaged cold cuts). Table 17.19 provides a listing of the typical shelf-life of an array of products. There is no published data on the changes one might expect in nutritive value for those foods having shelf-life periods exceeding a week or so. The FDA regulation on nutritional labeling requires that a product's indigenous nutrients be assured at 80% of its nutrition information label claim based upon a composite value by analysis of 12 retail units. If the product has added nutrients, then the product's nutrients must be assured at 100% of the nutrition information label claim based upon a value by analysis of 12 retail units. At this writing, there is insufficient experience with this system to determine whether it will prompt changes in commercial shelf-life expectations or adoption of "open" dating to protect, in effect, the manufacturer. Just how much of this nutrient change information will be published is unknown.

TABLE 17.19

SHELF-LIFE FOR RETAIL PRODUCTS

Product	Days	Weeks	Months
Bread, white (summer)	2-5		
Bread, white (winter)	3-7		
Bread, white (frozen)	30+		
Cake, angel		2	
Cake, cup		2	
Cake, fruit			24
Donuts	1-4		
Flour, all-purpose			15
Refrigerated dough		9-10	
Evaporated milk			12
Fluid milk	5-7		
Cottage cheese	10-15		
Creamed cheese			3
Ice cream			3
Canned apricots			36
Canned asparagus			24
Canned kidney beans			36
Canned tomatoes			30-36
Catsup			24
Canned fruit cocktail			36
Canned fruit and vegetable juices			24
Frozen lobster			3
Frozen dinners			6
Frozen foods (general)			12
Cereal, ready-to-eat			6-8
Macaroni (dry)			6-8
Spaghetti (dry)			9-12
Dehydrated gravy/sauce mixes			6-12
Sweet or dill pickles			12-15
French/Italian dressings			10-12
Pizza sauce (jar)			36
Lard			3
Vegetable oil (liquid)			4
Margarine			2-6
Canned puddings			24

Source: Anon. (1971).

FOOD COMPOSITION TABLES

Since most foods are cooked or otherwise treated before being served, some account must be taken of losses during preparation when calculating the nutritive values of dietaries. Most of the data on the composition of foods are related to raw processed and to some prepared foods. However, the data for a given food are often from different sources and therefore not necessarily comparable.

Watt and Merrill (1950) have devoted their careers to the compilation of food composition data in general and to the optimization of the quantity and quality of their data in *Composition of Foods—Raw, Processed, Prepared* (USDA Agriculature Handbook 8) which was revised in 1963. In preparation by the USDA team is an extensive update of the 1963 Revised Edition which should be available in 1975.

If data on the composition of raw foods are used in calculating the nutrient content of diets, the intake of certain nutrients may be overestimated. Clark and Fincher (1954) attempted to meet this problem by the use of factors which at least partially correct for cooking losses. The factors considered thiamin, riboflavin, niacin, and ascorbic acid only. The use of factors at this stage in our knowledge of nutrient changes as compared to the array of foods now available is questionable. The important point to remember is that cooking losses do occur and that calculated data should be interpreted accordingly. With the advent of nutritional labeling which stimulates food manufacturers to conduct nutritional composition studies, one can hope that data will also be obtained on product "as served" so that we may better understand the losses which occur through marketing channels and home use up to time of actual consumption.

CONCLUDING REMARKS

Although there are a few papers on the composition of some convenience foods, there are no publications devoted to the nutrient changes which occur with the preparation for eating of these foods. The next revision of this chapter should be able to contain such a section.

Finally, we have to become increasingly concerned not only with the nutrient value of particular foods but of food in the combinations made by consumers—by the meal and by the day—including snacks. Pennington (1976) conducted an extensive computer analysis of meal combinations of various ethnic groups in California which indicates that certain nutrients in conventional food combinations are predictive of other nutrients. These indicator nutrients are vitamin B-6, pantothenic acid, vitamin A, folacin, and the minerals, magnesium and iron. The available literature we have reviewed provides a very limited insight into these nutrients and it suggests a fertile area for research. Further investigations must be tempered with the reality that an increasing number of foods have added nutrients, the retention of which is not well understood either.

BIBLIOGRAPHY

AKINO, K. 1971. Recent development in amino acid fortification in Japan. *In* Amino Acid Fortification of Protein Foods. N. S. Scrimshaw and A. M. Altschul (Editors). MIT Press, Cambridge, Mass.

ANDERSON, R. A., and PFEIFER, V. F. 1970. Stability of vitamin A in wheat flour. Northwest Miller *277*, No. 3, 14.
ANON. 1971. Food Statiblity Survey, Vol. II. Dept. Food Sci., Rutgers Univ. Available from U.S. Govt. Printing Office, Washington, D.C.
AUGHEY, E., and DANIEL, E. P. 1940. Effect of cooking upon the thiamine content of foods. J. Nutr. *19*, 285.
BARNES, B. W., TRESSLER, D. K., and FENTON, F. 1943A. Effect of different cooking methods on the vitamin C content of quick-frozen broccoli. Food Res. *8*, 13.
BARNES, B. W., TRESSLER, D. K., and FENTON, F. 1943B. Thiamine content of fresh and frozen peas and corn before and after cooking. Food Res. *8*, 420.
BASU, N. M., and NEOGY, R. 1948. Vitamin C content of some ordinary foods before and after cooking. Indian J. Physiol. Appl. Sci. *2*, 15.
BAUERNFEIND, J. C., and CORT, W. M. 1974. Nutrification of foods with added vitamin A. Critical Rev. Food Technol. *4*, No. 3, 337.
BEDFORD, C. L., and McGREGOR, M. A. 1948. Dehydroascorbic acid in frozen and cooked frozen vegetables. Science *107*, 251.
BORENSTEIN, B. 1969. Vitamin A fortification of flour and corn meal. Northwest Miller *276*, No. 2, 18.
BORENSTEIN, B. 1971. Rationale and technology of food fortification with vitamins, minerals, and amino acids. Critical Rev. Food Technol. *2*, No. 2, 171.
BOTTOMLEY, R. A., and NOBILE, S. 1962. The thiamine content of flour and white bread in Sydney, New South Wales. J. Sci. Food Agr. *13*, 550.
BRESSANI, R., BRAHAM, J. E., and BEHAR, M. (Editors) 1972. Nutritional Improvement of Maize. INCAP Publ. L-4, Institute of Nutrition of Central America and Panama, Guatemala City, Guatemala.
BRIANT, A. M., and HUTCHINS, M. R. 1946. Influence of ingredients on thiamine retention and quality in baking powder biscuits. Cereal Chem. *23*, 512.
BRIANT, A. M., and KLOSTERMAN, A. M. 1950. Influence of ingredients on thiamine and riboflavin retention and quality of plain muffins. Trans. Am. Assoc. Cereal Chem. *8*, 69.
BRINKMAN, E. V. S., HALLIDAY, E. G., HINMAN, W. F., and HAMMER, R. J. 1942. Effect of various cooking methods upon subjective qualities and nutritive values of vegetables. Food Res. *7*, 300.
BROWN, E. J., and FENTON, F. 1942. Losses of vitamin C during cooking of parsnips. Food Res. *7*, 218.
BROWN, E. J., SCHUELE, H., and FENTON, F. 1941. Loss of vitamin C during cooking of rhubarb. Food Res. *6*, 217.
BRUSH, M. K., HINMAN, W. F., and HALLIDAY, G. 1944. The nutritive value of canned foods. V. Distribution of water soluble vitamins between solid and liquid portions of canned vegetables and fruits. J. Nutr. *28*, 131.
CAUSEY, K. *et al.* 1950A. Effect of thawing and cooking methods on palatability and nutritive value of frozen ground meat. I. Food Res. *15*, 237.
CAUSEY, K., HAUSRATH, M. E., RAMSTAD, P. E., and FENTON, F. 1950B. Effect of thawing and cooking methods on palatability and nutritive value of frozen ground meat. II. Food Res. *15*, 249.
CAUSEY, K., HAUSRATH, M. E., RAMSTAD, P. E., and FENTON, F. 1950C. Effect of thawing and cooking methods on palatability and nutritive value of frozen ground meat. III. Food Res. *15*, 256.
CHARLES, F. R., and VAN DUYNE, F. O. 1958. Effect of holding and reheating on the ascorbic acid content of cooked vegetable. J. Home Econ. *50*, 159.
CHELDELIN, V. H., and WILLIAMS, R. R. 1943. Studies of the average American diet. II. Riboflavin, nicotinic acid and pantothenic acid content. J. Nutr. *26*, 417.

CHELDELIN, V. H., WOODS, A. A., and WILLIAMS, R. J. 1943. Losses of B vitamins due to cooking of foods. J. Nutr. *26*, 477.
CHICK, H. 1940. Nutritive value of the potatoe. Chem. Ind. (London) *59*, 737.
CLARK, F., and FINCHER, L. J. 1954. Nutritive contents of homemaker's meals: four cities, winter 1948. USDA Inform. Bull. *112*.
COPPOCK, J. B. M., CARPENTER, B. R., and KNIGHT, R. A. 1957. Thiamine losses in bread baking. Chem. Ind. (London) *23*, 735.
COVER, S., DILSAVER, E. M., and HAYS, R. M. 1947. Retention of the B-vitamins in beef and lamb after stewing. 3. Pantothenic acid. J. Am. Dietet. Assoc. *23*, 693.
COVER, S., McLAREN, B. A., and PEARSON, P. B. 1944. Retention of the B-vitamins in rare and well-done beef. J. Nutr. *27*, 363.
CRANG, A., JAMES, D. P., and STURDY, M. 1948. The retention of ascorbic acid in preserved fruits. Chem. Ind. (London) *37*, 583.
CRAVIOTO, R. O. *et al.* 1945. Nutritive value of the Mexican tortilla. Science *102*, 91.
CURRAN, K. M., TRESSLER, D. K., and KING, C. G. 1957. Losses of vitamin C during cooking of Northern Spy apples. Food Res. *2*, 549.
DAWSON, E. H., KIRKPATRICK, M. E., and TOEPFER, E. W. 1952. Amount of ascorbic acid required for home freezing of Elberta peaches. Food Res. *17*, 433,
DAWSON, E. H., REYNOLDS, H., and TOEPFER, E. W. 1949. Home-canned versus home-frozen snap beans. J. Home Econ. *41*, 572.
DEUEL, H. J., JR., and GREENBERG, S. M. 1953. A comparison of the retention of vitamin A in margarines and in butter based upon bioassays. Food Res. *18*, 497.
DEUEL, H. J., JOHNSTON, C. H., SCHAUER, L., and RAPAPORT, S. 1943. The effect of sodium bicarbonate on the thiamine content of peas. Science *97*, 50.
DOMAH, AABMUD A. M. B., DAVIDEK, J. and VELISEK, J. 1974. Changes of L-ascorbic and L-dehydroascrobic acids during cooking and frying of potatoes. Z. Lebensm. Unters.-Forsch. *154*, 272.
DOWNS, D. E., and MECKEL, R. B. 1943. Thiamine losses in toasting bread. Cereal Chem. *20*, 352.
DWIVEDI, B. K., and ARNOLD, R. G. 1973. Chemistry of thiamine degradation in food products and model systems: A review. J. Agr. Food Chem. *21*, No. 1, 54.
EHEART, J. F. *et al.* 1946. Vitamin studies on lima beans. Southern Coop. Ser. Bull. *5*.
EHEART, M. S., and GOTT, C. 1965. Chlorophyll ascorbic acid and pH changes in green vegetables cooked by stir-fry, microwave, and conventional methods and a comparison of chlorophyll methods. Food Technol. *19*, No. 5, 185.
EHEART, M. S., and SHOLES, M. L. 1948. Effect of old-fashioned and modern methods of cooking on retention of nutrients in vegetables. II. Snap beans. Food Res. *13*, 227.
ERICSON, L. E., LARSSON, S., and LID, G. 1961. The loss of added lysine and threonine during the baking of wheat bread. Acta. Physiol. Scand. *53*, 85.
EVANS, R. J., BANDEMER, S. L., BAUER, D. H., and DAVIDSON, J. A. 1955. The vitamin B_{12} content of fresh and stored shell eggs. Poultry Sci. *34*, 922.
EVANS, R. J., BUTTS, H. A., and DAVIDSON, J. A. 1951A. The niacin content of fresh and stored shell eggs. Poultry Sci. *30*, 132.
EVANS, R. J., BUTTS, H. A., and DAVIDSON, J. A. 1951B. The vitamin B_6 content of fresh and stored shell eggs. Poultry Sci. *30*, 515.
EVANS, R. J., BUTTS, H. A., and DAVIDSON, J. A. 1952A. The riboflavin content of fresh and stored shell eggs. Poultry Sci. *31*, 269.
EVANS, R. J., BUTTS, H. A., and DAVIDSON, J. A. 1952B. The pantothenic acid content of fresh and stored shell eggs. Poultry Sci. *31*, 777.

EVANS, R. J., DAVISON, J. A., BAUER, D., and BUTTS, H. A. 1953A. The biotin content of fresh and stored shell eggs. Poultry Sci. 32, 680.
EVANS, R. J., DAVIDSON, J. A., BAUER, D., and BUTTS, H. A. 1953B. Folic acid in fresh and stored shell eggs. J. Agr. Food Chem. 1, 170.
EVERSON, G. J., and SMITH, A. H. 1945. Retention of thiamine, riboflavin and niacin in deep fat cooking. Science 101, 338.
FDA. 1973. Nutrition labeling and labeling of foods with information on cholesterol and fat and fatty acid composition. Federal Register 38, 6951, Mar. 14.
FEASTER, J. F., JACKSON, J. M., GREENWOOD, D. A., and KRAYBILL, H. R. 1946. Vitamin retention in processed meat. Ind. Eng. Chem. 38, 87.
FENTON, F., CAMP, S. C., and KING, C. G. 1937A. Losses of vitamin C during the cooking of Swiss chard. J. Nutr. 14, 631.
FENTON, F., TRESSLER, D. K., and KING, C. G. 1937B. Losses of vitamin C during the cooking of peas. J. Nutr. 12, 285.
FENTON, F., TRESSLER, D. K., CAMP, S. C., and KING, C. G. 1938. Losses of vitamin C during commercial freezing, defrosting, and cooking of frosted peas. Food Res. 3, 403.
FLEISCHMAN, A. I. et al. 1963. Studies on cooking fats and oils. J. Am. Dietet. Assoc. 42, 394.
FLOYD, W. W., and FRAPS, G. S. 1940. Changes in vitamin C content during boiling of turnip greens in various waters in covered and uncovered containers. Food Res. 5, 33.
GLEIM, E. G., TRESSLER, D. K., and FENTON, F. 1944. Ascorbic acid, thiamine, riboflavin, and carotene contents of asparagus and spinach in the fresh, stored, and frozen states both before and after cooking. Food Res. 9, 471.
GORDON, J., and NOBLE, I. 1959. Effect of cooking method on vegetables. J. Am. Dietet. Assoc. 35, 578.
GREENWOOD, D. A., BEADLE, B. W. and KRAYBILL, H. R. 1943. Stability of thiamine to heat. II. Effect of meat-curing ingredients in aqueous solutions and in meat. J. Biol. Chem. 149, 349.
GUILD, L., and RAINES, R. 1972. Thiamine content and retention in venison. J. Am. Dietet. Assoc. 60, 42.
GUTTIKAR, N. N. et al. 1965. Studies on processed protein foods based on blends of groundnut, bengalgram, soybean and sesame flours and fortified with minerals and vitamins. J. Nutr. Dietet. 2, 21.
HARRIS, R. S., and LEVENBERG, R. K. 1960. Effects of home preparation on nutrient content of foods of plant origin. In Nutritional Evaluation of Food Processing. R. S. Harris and H. Von Loesecke (Editors). Avi Publishing Co., Westport, Conn.
HASHMI, M. H., ULLAH, R., and AHMAD, B. 1954. Pakistan J. Sci. Res. 6, 66.
HENDERSON, L. M., WAISMAN, H. A., and ELVEHJEM, C. A. 1941. The distribution of pyridoxine (vitamin B_6) in meat and meat products. J. Nutr. 21, 589.
HEPBURN, F. N., LEWIS, E. W., JR., and ELVEHJEM, C. A. 1957. The amino acid content of wheat, flour and bread. Cereal Chem. 34, 312.
HEWSTON, E. M., DAWSON, E. H., ALEXANDER, L. M., and ORENT-KEILES, E. 1948. Vitamin and minerals content of certain foods as affected by home preparation. U.S. Dept. Agr. Misc. Publ. 628.
HOLLINGER, M. E. 1944. Ascorbic acid value of the sweet potato as affected by variety, storage, and cooking. Food Res. 9, 76.
HUGGART, R. L., HARMAN, D. A., and MOORE, E. L. 1954. Ascorbic acid retention in frozen concentrated citrus juices. J. Am. Dietet. Assoc. 30, 682.
HURDLE, A. D. F., BARTON, D., and SEARLES, I. H. 1968. A method for measuring folate in food and its application to a hospital diet. Am. J. Clin. Nutr. 21, No. 1, 1202.
IRESON, M. G., and EHEART, M. S. 1944. Ascorbic acid losses in cooked veg-

etables: Cooked uncovered in a large amount of water and covered in a small amount of water. J. Home Econ. *36*, 160.
JANSEN, G. R., EHLE, S. R., and HAUSE, N. L. 1964. Studies on the nutritive loss of supplemental lysine in baking. I. Loss in a standard white bread containing 4% nonfat dry milk. Food Technol. *18*, No. 3, 109. II. Loss in water bread and in breads supplemented with moderate amounts of nonfat dry milk. Food Technol. *18*, No. 3, 114.
KAMALANATHAN, G., GIRI, J., JAYA, T. V., and PRIYADARSANI. 1974. The effect of boiling, steaming, pressure cooking and panning on the mineral and vitamin content of three vegetables. Ind. J. Nutr. Dietet. *11*, 10.
KILGORE, L., and BAILEY, M. 1970. Degradation of linoleic acid during potato frying. J. Am. Dietet. Assoc. *56*, 130.
LABUZA, T. P. 1973. Effect of dehydration and storage. Food Technol. *27*, No. 1, 20.
LACHANCE, P. A. 1973. The vanishing American meal. Food Prod. *7*, No. 9, 36,; also Med. Dimensions *2*, No. 10, 11.
LEE, F. A., and WHITCOMBE, J. 1945. Blanching of vegetables for freezing. Effect of different types of potable water on nutrients of peas and snap beans. Food Res. *10*, 465.
LEE, F. A. et al. 1950. Effect of freezing rate on meat. Appearance, palatability, and vitamin content of beef. Food Res. *15*, 8.
LEE, F. A. et al. 1954. Effect of rate of freezing on pork quality. Appearance, palatability, and vitamin content. J. Am. Dietet. Assoc. *30*, 351.
LIENER, I. E. 1960. Effect of processing on cereal proteins, *In* Nutritional Evaluation of Food Processing. R. S. Harris, and H. Loesecke (Editors). John Wiley & Sons, New York. Reprinted in 1971 by Avi Publishing Co., Westport, Conn.
LINCOLN, H., HOVE, E. L., and HARREL, C. G. 1944. Cereal Chem. *21*, 274.
LUSHBOUGH, C. H., WEICHMAN, J. M., and SCHWEIGERT, B. S. 1959. The retention of vitamin B_6 in meat during cooking. J. Nutr. *67*, 451.
LYONS, M. E., and FELLERS, C. R. 1939. Potatoes as carriers of vitamin C. Am. Potato J. *16*, 169.
MACK, G. L., TAPLEY, W. T., and KING, C. G. 1939. Vitamin C in vegetables. X. Snap beans. Food Res. *4*, 309.
MASSIEU, G. H., GUZMAN, J., CRAVIOTO, R. O., and CALVO, J. 1949. Determination of some essential amino acids in several uncooked and cooked Mexican foodstuffs. J. Nutr. *38*, 293.
MATTHEWS, R. H., RICHARDSON, G., and LICHTENSTEIN, H. 1969. Effect of lysine fortification on quality of chapatties and yeast bread. Cereal Chem. *46*, 14.
MAYFIELD, H. L., and RICHARDSON, J. E. 1939. The vitamin content of green string beans when cooked or canned and stored. Montana Agr. Expt. Sta. Bull. *373*.
McGUCKEN, F. C., and GODDARD, V. R. 1948. Thiamine and riboflavin content of raw and cooked dried apricots. J. Am. Dietet. Assoc. *24*, 510.
McINTIRE, J. M., SCHWEIGERT, B. S., and ELVEHJEM, C. A. 1944. The choline and pyridoxine content of meats. J. Nutr. *28*, 219.
McINTOSH, J. A., TRESSLER, D. K., and FENTON, F. 1940. The effect of different cooking methods on the vitamin C content of quick frozen vegetables. J. Home Econ. *32*, 692.
McINTOSH, J. A., TRESSLER, D. K., and FENTON, F. 1942. Ascorbic acid content of five quick-frozen vegetables as affected by composition of cooking utensil and volume of cooking water. J. Home Econ. *34*, 314.
MECKEL, R. B., and ANDERSON, G. 1945. Thiamine retention and composition of U.S. Army bread. Cereal Chem. *22*, 429.
MELNICK, D., ROBINSON, W. D., and FIELD, H., JR. 1941. Fate of thiamine in the digestive secretions. J. Biol. Chem. *138*, 49.
MEYER, B. H., HINMAN, W. F., and HALLIDAY, E. G. 1947. Retention of some vitamins of the B complex in beef during cooking. Food Res. *12*, 203.

MEYER, B. H., MYSINGER, M. A., and WODARSKI, L. A. 1969. Pantothenic acid and vitamin B_6 in beef. J. Am. Dietet. Assoc. 54, 122.

MILLARES, R., and FELLERS, C. R. 1949. Vitamin and amino acid content of processed chicken meat products. Food Res. 14, 131.

MORGAN, A. F. 1960. Losses of nutrients in foods during home preparation In Nutritional Evaluation of Food Processing. R. S. Harris and H. Von Loesecke (Editors). John Wiley & Sons, New York. Reprinted in 1971 by Avi Publishing Co., Westport, Conn.

MORGAN, A. F., and FREDERICK, H. 1935. Vitamin B (B_1) in bread as affected by baking. Cereal Chem. 12, 390.

MORGAN, A. F., MacKINNEY, G., and CAILLEAU, R. 1945. Losses of ascorbic acid and four B vitamins in vegetables as a result of dehydration, storage, and cooking. Food Res. 10, 5.

MORGAN, A. F. et al 1949. Thiamine, riboflavin, and niacin content of chicken tissues, as affected by cooking and frozen storage. Food Res. 14, 439.

MULAY, I., DHOPESHWARKER, G. A., and MAGAR, N. G. 1952. Losses of nutrients during cooking of vegetables. Indian J. Med. Res. 40, 443.

MUNSELL, H. E. et al. 1949. Effect of large scale methods of preparation on the vitamin content of food. III. Cabbage. J. Am. Dietet. Assoc. 25, 420.

NIELSEN, L. M., and CARLIN, A. G. 1974. Frozen, precooked beef and beef-soy loaves. J. Am. Dietet. Assoc. 65, 35.

NOBLE, I. 1967. Ascorbic acid and color of vegetables. J. Am. Dietet. Assoc. 50, 304.

NOBLE, I., and WADDELL, E. I. 1946. Effects of different methods of cooking on the ascorbic acid content of cabbage. Food Res. 10, 246.

NOBLE, I., and WORTHINGTON, J. 1948. Ascorbic acid retention in cooked vegetables. J. Home Econ. 40, 129.

OLLIVER, M. 1943. Ascorbic acid values of fruits and vegetables for dietary surveys. Chem. Ind. 62, 146.

OSER, B. L., MELNICK, D., and OSER, M. 1943. Influence of cooking procedure upon retention of vitamins and minerals in vegetables. Food Res. 8, 115.

OSTERMAN, R. A. 1938. The ascorbic acid content of celery. J. Home Econ. 30, 715.

PACE, J. K., and WHITACRE, J. 1953A. Factors affecting retention of B vitamins in corn bread made with enriched meal. I. The relation of pH to the retention of thiamine, riboflavin, and niacin in corn bread. Food Res. 18, 231.

PACE, J. K., and WHITACRE, J. 1953B. Factors affecting retention of B vitamins in corn bread made with enriched meal. II. Relation of crust and crumb and baking utensil to retention of thiamine in corn bread. Food Res. 18, 239.

PACE, J. K., and WHITACRE, J. 1953C. Factors affecting retention of B vitamins in corn bread made with enriched meal. III. Retention of thiamine in corn muffins made with commercial mixes and in corn bread made with self rising meals. Food Res. 18, 245.

PATTON, M. B., DEVADAS, R., and WILSON, E. D. 1950. Ascorbic acid retention and palatability of cabbage as cooked in South India. J. Am. Dietet. Assoc. 26, 897.

PEARSON, A. M. et al. 1951. Vitamin losses in drip obtained upon defrosting frozen meat. Food Res. 16, 85.

PEARSON, P. B., and LUECKE, R. W. 1945. The B vitamin content of raw and cooked sweet potatoes. Food Res. 10, 325.

PENNINGTON, JEAN A. 1976. Dietary Nutrient Guide. Avi Publishing Co., Westport, Conn.

PETIT, L., GUILBOT, A., and GUILLEMET, R. 1945. Assay for the distribution of vitamin B_1 in breadmaking. Use of formic acid extraction. Bull. Soc. Chem. Biol. 27, 529. (French)

PRESSLY, H. B. 1948. Influence of cooking on ascorbic acid content of collards. Food Res. *13*, 491.

RICE, E. E., FRIED, J. F., and HESS, W. R. 1946. Storage and microbial action upon vitamins of the B complex in pork. Food Res. *11*, 305.

RICE, E. E., SQUIRES, E. M., and FRIED, J. F. 1948. Effect of storage and microbial action on vitamin content of pork. Food Res. *13*, 195.

RICHARDSON, J. E., DAVIS, R., and MAYFIELD, H. L. 1937. Vitamin C content of potatoes prepared for table use by various methods of cooking. Food Res. *2*, 85.

RICHARDSON, J. E., and MAYFIELD, H. L. 1943. Vegetable preservation handbook for wartime use. Montana Agr. Expt. Sta. War Circ. *1*.

ROHRER, V., and TREADWELL, C. R. 1944. The content and stability of ascorbic acid in orange juice under home conditions. Texas Rept. Biol. Med. *2*, 175.

ROSENBERG, H. R., and ROHDENBURG, E. L. 1951. The fortification of bread with lysine. I. The loss of lysine during baking. J. Nutr. *45*, 593.

RUDRA, M. N. 1937. Studies in vitamin C. The effect of cooking and storage on the vitamin C contents of foodstuffs. J. Indian Med. Res. *25*, 89.

SCHEUNERT, A., and KOHLEMANN, E. 1936. Concerning the vitamin C content of potato. Mitterhung: Alte Gelagerte Kartoffeln Ernte 1935. (German)

SCHWEIGERT, B. S., and GUTHNECK, B. T. 1953. Liberation and measurement of pantothenic acid in animal tissue. J. Nutr. *51*, 283.

SCHWEIGERT, B. S., NIELSEN, E., McINTIRE, J. M., and ELVEHJEM, C. A. 1943. Biotin content of meat and meat products. J. Nutr. *26*, 65.

SCHWEIGERT, B. S., POLLARD, A. E., and ELVEHJEM, C. A. 1946. The folic acid content of meats and the retention of this vitamin during cooking. Arch. Biochem. *10*, 107.

SCOULAR, F. I., and EAKLE, D. H. 1943. Loss of ascorbic acid during cooking of stored sweet potatoes. Food Res. *8*, 156.

STEVENS, H. B., and FENTON, F. 1951. Dielectric versus stewpan cookery. J. Am. Dietet. Assoc. *27*, 32.

SWEENEY, J. P., GILPIN, G. L., MARTIN, M. E. and DAWSON, E. H. 1960. Palatability and nutritive value of frozen broccoli. J. Am. Dietet. Assoc. *36*, 122.

TARA, K. A., USHA, M. S. M., and BAINS, G. S. 1972. Effects of lysine on dough and protein quality of whole wheat meal chapatis and leavened bread. J. Agr. Food Chem. *20*, No. 1, 116.

TEPLEY, L. J., and DERSE, P. H. 1958. Nutrients in cooked frozen vegetables. J. Am. Dietet. Assoc. *34*, 836.

THOMAS, K., PACE, J. K., and WHITACRE, J. 1952. Effect of enrichment on the thiamine, riboflavin and niacin of corn meal and grits as prepared for eating. Texas Agr. Expt. Sta. Bull. *753*.

TODHUNTER, E. N. and ROBBINS, R. C. 1941. Ascorbic acid (vitamin C) content of garden-type peas preserved by frozen-pack method. Wash. State Coll. Agr. Expt. Sta. Bull. *408*.

U.S. DEPT. OF AGR. 1971. Textured vegetable protein products (B-1) to be used in combination with meat for use in lunches and suppers served under child feeding program. USDA, FNS Notice *219*.

VAN DUYNE, F. O., CHASE, J. T., FANSKA, J. R., and SIMPSON, J. I. 1947. Effect of certain practices on reduced ascorbic acid content of peas, rhubarb, snap beans, soybeans, and spinach. Food Res. *12*, 439.

VAN DUYNE, F. O., CHASE, J. T., OWEN, R. F., and FANSKA, J. R. 1948. Effects of certain home practices on riboflavin content of cabbage, peas, snap beans, and spinach. Food Res. *13*, 162.

VAN DUYNE, F. O., CHASE, J. T., and SIMPSON, J. I. 1944. Effects of various home practices on ascorbic acid content of cabbage. Food Res. *9*, 164.

VAN DUYNE, F. O., CHASE, J. T., and SIMPSON, J. I. 1945. Effect of various home practices on ascorbic acid content of potatoes. Food Res. *10*, 72.
VAN DUYNE, F. O., WOLFE, J. C., and OWEN, R. F. 1950. Retention of riboflavin in vegetables preserved by freezing. Food Res. *15*, 53.
WACHHOLDER, K. 1940. Review of vitamin C content of raw and cooked vegetables and fruits. Ernahrung *5*, 79.
WADSWORTH, H. I., and WILCOX, E. B. 1945. Effect of home cooking practices on the ascorbic acid content of frozen and canned lima beans. J. Am. Dietet. Assoc. *21*, 289.
WATT, B. K., and MERRILL, A. L. 1950. Composition of Foods—Raw, Processed, Prepared. USDA Agr. Handbook 8. (Revised Dec. 1963)
WEAST, E. O., GROODY, M., and MORGAN, A. F. 1948. Utilization by dogs of the nitrogen of heated casein. Am. J. Physiol. *152*, 286.
WELLINGTON, M., and TRESSLER, D. K. 1938. Vitamin C content of vegetables. IX Influence of method of cooking on vitamin C content of cabbage. Food Res. *3*, 311.
WEST, L. C., TITUS, M. C., and VAN DUYNE, F. O. 1959. Effect of freezer storage and variations in preparation on bacterial count, palatability and thiamine content of ham loaf, Italian rice, and chicken. Food Technol. *6*, 322.
WESTERMAN, B. D., VAIL, G. E., TINKLIN, G. L., and SMITH, J. 1949. Food Technol. *3*, 184.
WILSON, E. C. G. 1943. The effect of baking powder on the vitamin B_1 content of whole meal. New Zealand J. Sci. Technol. *24*, 35.
WOLFE, R. H., OWEN, R. F., CHARLES, V. R., and VAN DUYNE, F. O. 1949. Effect of freezing and freezer storage on the ascorbic acid content of muskmellon, grapefruit sections, and strawberry purée. Food Res. *14*, 243.
WOODRUFF, R. N., and SCOULAR, F. I. 1942. Loss of vitamin C during cooking of summer squash. Food Res. 7, 267.
ZAEHRINGER, M. V., and PERSONIUS, C. J. 1949. Thiamine retention in bread and rolls baked to different degrees of browness. Cereal Chem. *26*, 384.

SECTION V
Nutrification and Nutrient Metabolism

CHAPTER 18

Nutrification of Foods

Ricardo Bressani

PART 1
Addition of Amino Acids to Foods

The efficient use of protein by man or animal requires that it contain the essential amino acids as well as nitrogen in amounts and proportions needed by the organism to meet its needs for specific as well as for general physiological functions. Most food proteins however, do not contain the essential amino acids in the amounts and proportions needed by the organism. Therefore, their utilization tends to be inefficient, particularly for protein foods derived from the vegetable kingdom. These proteins are more abundant than animal proteins and provide most of the protein consumed by the human population, particularly in developing countries. To be able to reduce the waste of nitrogen from essential amino acids not utilized because of deficiencies in the plant protein source, various approaches can be implemented either by natural ways or through a process of supplementation (enrichment). The natural way consists in consuming diets made up of various protein food sources, both from vegetable and animal sources which will complement each other. The process of enrichment or supplementation involves the addition of the essential amino acid which because of its relatively low concentration in the protein with respect to the other amino acids, limits the utilization of the protein consumed.

In the early 1900's, this process was proved to be quite efficient in increasing the utilization of food protein; however, for various reasons very little has been done in practice. The low availability of food, the inefficient and wasteful use of essential amino acid deficient protein, and the increases in world population, however, are almost demanding that this approach be implemented.

The present chapter summarizes some basic principles of amino acid supplementation, followed by experimental findings, showing the effectiveness of the approach. The final section presents practical ways to incorporate amino acids into food protein.

ESSENTIAL AMINO ACIDS

Tissue protein synthesis is a process which requires each amino acid to be present in its place at the time to form each specific protein. If a single amino acid is missing, the protein cannot be synthesized. After digestion and absorption of the amino acids at the gastrointestinal level, proteins enter the metabolic pool to be used directly for anabolism, or, through intermediary metabolic pathways to contribute to the synthesis of other amino acids. Therefore, it is not absolutely necessary for all amino acids needed for protein synthesis to be supplied by the diet. Many organisms, however, do not have the ability for rapidly synthesizing some of the amino acids in the amount needed according to the demands of anabolism. These amino acids must be supplied by food protein in adequate amounts and proportions: therefore, these have been called essential (Table 18.1) by Rose et al. (Rose 1949; Rose et al. 1948, 1949, 1950, 1951,

TABLE 18.1

ESSENTIAL AMINO ACIDS FOR VARIOUS ORGANISMS

Man (Infant)	Dog	Rat	Swine	Chick
Histidine	Arginine	Arginine	Arginine	Arginine
Isoleucine	Histidine	Histidine	Histidine	Glycine
Leucine	Isoleucine	Isoleucine	Isoleucine	Histidine
Lysine	Leucine	Leucine	Leucine	Isoleucine
Methionine	Lysine	Lysine	Lysine	Leucine
Phenylalanine	Methionine	Methionine	Methionine	Methionine
Threonine	Phenylalanine	Phenylalanine	Phenylalanine	Phenylalanine
Tryptophan	Threonine	Threonine	Threonine	Threonine
Valine	Tryptophan	Tryptophan	Tryptophan	Tryptophan
	Valine	Valine	Valine	Valine

1954). The absence from the diet of any of these essential amino acids results in slow growth, poor performance, and in the death of the organism, resulting in the need for minimum amounts or requirements. Their importance and quantitative need varies, however, with the particular living system and with the function being measured in that system such as growth, maintenance, production, and repletion of depleted tissues. Various food proteins, therefore, have a high biological value because their content of essential amino acids meet with the minimum needs of the living organisms; on the other hand, other food proteins cannot provide the organism with all essential amino acids so therefore they have a low biological value.

THE CONCEPT OF BALANCED PROTEIN

Amino Acid Balance

The concept of amino acid balance was stated for the first time in 1914 by Osborne and Mendel, who suggested that if the requirements of the individual are not satisfied, even roughly by a protein food, its intake would result in a poor performance even if it were offered at a high dietary level. Mitchell (1950) and others (Price et al. 1953; Williams et al. 1954) suggested that the high correlation observed between the proportions of the essential amino acids contained in tissue protein, and their requirements for several species of animals, could give an indication for the development of a correct amino acid pattern for the dietary protein in the food. Based on this relationship, Harper (1958) proposed the concept of amino acid balance. It means that a protein providing the essential amino acids in the amounts and proportions needed, according to the minimum requirements of the individual, can result in an optimum performance even at low levels of dietary intake. It is necessary however, that nonessential amino acids or sufficient nonprotein nitrogen be present to prevent the essential amino acids to be utilized for purposes other than tissue building functions.

Limiting Amino Acids

Many studies have been carried out to establish definite requirements for the essential amino acids for both animal and man, after the classification by Rose (1937, 1938, 1949) of the amino acid in a protein into essential and nonessential. Whatever the technique used, it was found that, with only a few exceptions, most proteins did not conform in amino acid content with the figures used as requirements. By comparing the amino acid pattern in a protein with the amino acid pattern used as reference, the concept of the first limiting amino acid was developed. It is defined as the amino acid present in the smallest amount in comparison with the reference concentration (Mitchell and Block 1946). Therefore, if a protein is characterized by having a low concentration of one or more essential amino acids, the efficiency of utilization of all other amino acids is limited by the one present in the lowest amount (Flodin 1953, 1956). An example of various selected sources of protein is shown in Table 18.2 in which the reference used is the pattern for essential amino acids found in

TABLE 18.2

ESSENTIAL AMINO ACID PATTERNS OF REFERENCE AND SELECTED PROTEINS[1]

Amino Acid	Required for Growth		Egg	Human Milk	Cow's Milk	Common Maize	*Opaque-2* Maize	Rice	Soybean	Cotton-seed
	Rat	Chick	Swine								
Arginine	0.8	2.0	0.4	1.2	0.9	0.8	0.9	2.1	1.5	1.8	3.2
Histidine	0.5	0.5	0.4	0.5	0.5	0.6	0.5	1.0	1.4	0.6	0.7
Isoleucine	1.4	1.0	1.2	1.3	1.2	1.4	1.2	1.0	1.2	1.4	1.1
Leucine	1.9	2.3	1.3	1.8	2.0	2.1	3.2	2.5	2.2	1.9	1.7
Lysine	1.6	1.8	1.7	1.5	1.4	1.7	0.72	1.3	1.0	1.6	1.2
Phenylalanine	1.3	0.8	0.8	1.2	1.0	1.0	1.1	1.3	1.3	1.2	1.5
Tyrosine	0.7	0.8	0.3	0.6	1.1	1.1	1.5	1.2	1.2	0.8	0.8
Total aromatic	2.0	1.6	1.1	1.8	2.1	2.1	2.7	2.5	2.4	2.1	2.3
Total sulfur	1.4	1.4	1.2	1.5	0.9	0.7	0.8	0.9	0.8	0.8	0.8
Threonine	1.0	1.0	1.0	1.0	1.0	1.0	1.0	1.0	1.0	1.0	1.0
Tryptophan	0.21	0.25	0.29	0.24	0.36	0.31	0.15	0.39	0.27	0.35	0.33
Valine	1.7	1.3	1.1	1.5	1.4	1.5	1.3	1.5	1.8	1.3	1.4

[1] Threonine taken as unity.

whole egg protein. Also shown are the growth requirements for the rat, the chick, and swine. Obviously, the decision whether an amino acid is or is not deficient, and its order of limitation, depends on the pattern used as reference, which for the rat, the chick, and swine, has been established experimentally. For humans, however, amino acid requirements are not well known although for purposes of comparison of food protein the amino acid pattern of the egg has been used. The amino acids in the protein sources, present in concentrations higher than the level needed in the pattern are not utilized for protein synthesis, and their utilization further decreases because of the low concentration of the most limiting amino acid.

Improvement of Amino Acid Pattern

As a basic principle for the improvement of the amino acid pattern of a protein, the first limiting amino acid should be added in such an amount that the total of this amino acid in the protein of the diet balances with the amount of the second limiting amino acid and the other components in accordance with the needs of the organism or the essential amino acid reference pattern. In doing so, both the first and the second limiting amino acids become of equal importance with respect to the quality of the protein. The addition of more supplement to the protein should include both of these amino acids in the amounts needed to maintain the proper ratio according to the requirements of reference pattern. Since there are ten essential amino acids, the additions should take them all into consideration, although this will very seldom be necessary for maximum response (Flodin 1953, 1956; Rosenberg 1959).

Although nonessential amino acids are needed mainly as nitrogen sources, there are suggestions which indicate that a good balance should exist between essential and nonessential amino acids (Rose et al. 1948; Stucki and Harper 1962; Young and Zamora 1968; Kies and Linkswiler 1965; Scrimshaw et al. 1966; Huang et al. 1966). While there are only a few studies available, examination of good quality proteins suggests that the quantity of nonessential amino acids is equal to the total quantity of the essential ones.

The importance and practical significance of the principle of amino acid supplementation have been evident in poultry and swine diets, and should be considered when human diets are to be improved in quality upon amino acid supplementation. If not properly considered, the addition of an excess of the first limiting amino acid may result in a secondary deficiency and, thus, there will be no improvement in the overall quality of the diet. The indiscriminate addition of essential amino acids may give as a result the situation described by Harper as "amino acid imbalance" (Harper 1958).

Imbalance of Amino Acids

The concepts discussed in previous sections have served as the basis for detailed studies of the interrelationship among the essential amino acids and between amino acids and other dietary factors such as vitamins, calories, and protein itself. The results from such studies lead to an appreciation of the importance of amino acid balance and imbalance in protein nutrition.

To support maximum growth, a protein must provide each essential amino acid in the quantity in which it is required by the animal (Allison 1955; Flodin 1953, 1956). Although requirements for amino acids may be affected by age (Allison 1955; Hartsook and Mitchell 1956; Henry and Kon 1957), physiological condition of the animal (Allison 1955, Allison and Fitzpatrick 1960), and various other factors (Allison 1955), it is possible in a particular situation to calculate a ratio which represents the relative needs of the individual amino acids (Block and Mitchell 1946-1947; Oser 1951). As indicated before (Table 18.2), an idea of the nutritional adequacy of a protein can be obtained from a comparison of the ratio, representing the relative proportions of the essential amino acids in the protein, with that which represents the relative essential amino acid needs of the animal. It follows that if the ratio of amino acids in a protein deviates significantly from the ratio of requirements, there is unsatisfactory growth or overall performance. This condition is classified as a deficiency of an amino acid. There are, however, other situations in which a change in the proportion of the amino acids in a diet cause a deficiency of one of them as well as additional adverse effects, which have been demonstrated many times and described by Harper as amino acid imbalance (Harper 1958, 1956). Thus, Harper defines an unbalanced protein as one in which there is a deficiency of an essential amino acid and which is, therefore, poorly utilized. On the other hand, an amino acid imbalance has been defined by Harper (1958, 1956) as the situation in which there is an additional adverse effect which accentuates that taking place from an amino acid deficiency. It should also be noted that amino acid imbalance is different from amino acid toxicity, which may occur from an excessive intake of a particular amino acid.

Examples of these various situations have been described in great detail by Harper and co-workers (Harper 1958, 1956) and others (Sauberlich 1956; Benton *et al.* 1956; Ebisuzaki *et al.* 1952; Elvehjem 1956; Grau and Kamei 1950) and attempts to elucidate the mechanism by which the situation can take place have been carried out (Kumta *et al.* 1958; Deshpande *et al.* 1958; Kumta and Harper 1960A; Sanahuja and Harper 1963; Harper *et al.* 1964). Although imbalances were first demonstrated in rats, similar results have been reported with other experimental animals (Gessert and Phillips 1956; Bressani 1963; Fisher and Shapiro 1961; Hill and Olsen 1963, 1965; Lloyd and Crampton 1961; Berry et al. 1966; Baker et al. 1969) and man (Scrimshaw et al. 1958; Salmon 1954; Kaumitz et al. 1955).

These interrelationships, demonstrated in laboratory experiments, indicate that the practical aspects of amino acid supplementation should receive close examination before programs for human populations are implemented on a large scale.

Relationship between Protein Content in the Diet and Amino Acid Needs

The amount of protein in the diet may control the amount of the first limiting amino acid that should be added (Rosenberg 1959). This relationship is more easily understandable in terms of the relation existing between available energy and protein content in the diet. Rosenberg and Baldini (1957) found in chicks that with a restricted energy content in the diet the requirements for methionine did not increase in proportion to the increase in protein. However, they found that an increase in methionine requirement was proportional to the increase in total protein as long as the diet contained sufficient energy to permit the chicks to make complete use of the balanced portion of the protein for growth and maintenance.

If energy intake was restricted, the animal used part of the protein for meeting its energy requirements. Rosenberg (1959) indicated that if the ratio of calories to protein is held constant, but there is an increase in protein in the diet, amino acid needs for best performance remain constant when expressed on the basis of a percentage of protein. If the ratio increases or decreases, amino acid needs expressed as a percentage of the protein increase or decrease as the case may be.

The literature on the subject reveals that most investigators have found an almost linear increase in lysine requirements of chicks as the protein content of the diet was raised from 5 to 30%. A similar observation for methionine and lysine was made by Almquist (1949), Twining et al. (1955), and Grau and Kamei (1950). Almquist and Merrit (1950) found a linear relationship between arginine requirement and protein levels in the range of 15 to 35% of protein with baby chicks. In the pig, Brinegar et al. (1950) observed a lysine requirement of 0.6% using a ration containing 10.6% of protein and a requirement of 1.2% with a ration containing 22% of protein. Becker et al. (1957) found increases in isoleucine needs for the weanling pig fed increased levels of dietary protein. Similar results using rats were reported by Forbes et al. (1955) with respect to isoleucine requirements, and Almquist (1952) reviewed evidence to show that as the percentage of total protein in the diet of the chick increased from 10 to 40%, the percentage of lysine and methionine required, expressed as a percentage of the protein, decreased. Bressani and Mertz (1958A) reported the lysine requirement expressed as a percentage of the diet remained essentially constant in the protein range of 16 to 40% for the weanling rat. Expressed as a percentage of the protein, lysine requirements decreased in the same range of dietary protein.

Changing protein content in the diet has been useful to measure the first limiting amino acid in a protein, and the sequence of limitation as well. Thus, Harper (1959) found that for casein the sequence was methionine, isoleucine, lysine, and tryptophan. A similar approach has been used by other investigators for other proteins such as fibrin (Kumta and Harper 1960B), soybean (Berry *et al.* 1962), and raw and roasted peanut protein (McOsker 1962).

The effect of a change in protein level in relation to amino acid requirements is a problem of practical importance for domestic animals as well as for man. In the latter case, intakes of various foods containing different concentration of protein varies. An example is shown in Fig. 18.1 for maize and beans. Since maize is lysine-deficient (Bressani *et al.* 1968; Rosenberg *et al.* 1960) and beans are

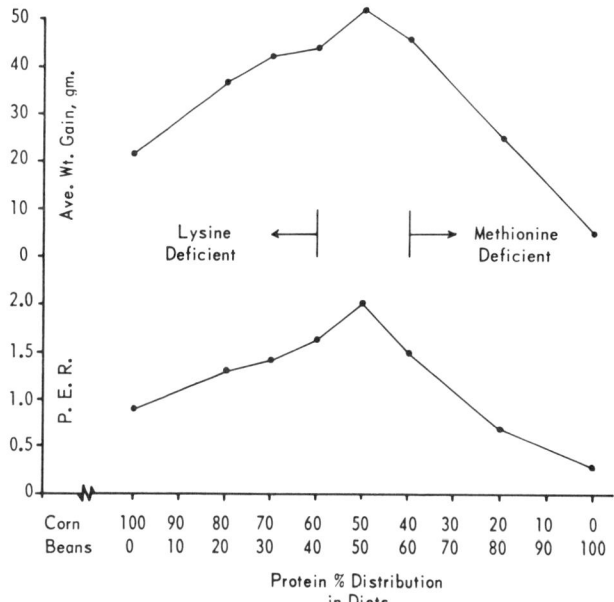

FIG. 18.1. PROTEIN QUALITY OF VARIOUS COMBINATIONS OF CORN AND BLACK BEAN PROTEINS

methionine-deficient (Bressani *et al.* 1963D; Patwardhan 1962), the limiting amino acid in the mixture will depend on the amount of protein derived from each component (Bressani *et al.* 1962A; Bressani 1969) (Table 18.3). If beans are consumed in larger

TABLE 18.3

AMINO ACID SUPPLEMENTATION OF MAIZE AND BEANS AND OPTIMUM PROTEIN QUALITY MAIZE-BEAN DIETS

Prot. Distrib.		Amino Acid Added (%)	Average Wt Gain (Gm)	PER
Maize (%)	Black Beans (%)			
100	0	None	29	1.05
100	0	Lysine, tryptophan, isoleucine	74	2.47
50	50	None	51	2.10
50	50	Methionine, tryptophan	59	2.03
50	50	Methionine, lysine	75	2.42
0	100	None	-3	-
0	100	Methionine, tryptophan, leucine	23	1.04

amounts, methionine will be the limiting amino acid, but if maize provides more protein, lysine will be the limitation in protein utilization. At the optimum complementation point both amino acids added together will improve the quality of the protein. These amino acid relationships between two foods can also be observed in Fig. 18.2 for maize-soybean protein mixtures. As the protein distribution shifts from 100 to 0 for one protein (P1) and from 0 to 100 for the second protein (P2), lysine and tryptophan which are richer in P2 will increase; whereas those amino acids found in equal concentration in both P1 and P2 do not change. Optimum performance or quality is obtained when the best amino acid balance is obtained; however, at the optimum point there is still opportunity of improvement in quality by optimizing amino acid balance. For soy-maize mixtures, the amino acids which still improve protein quality are methionine, lysine, and threonine (Bressani 1972A).

For a fixed mixture, higher intakes will also determine which amino acid limits the quality of the protein, and it will be the one required in the largest amount in relation to the need of the organism. These factors, therefore, must be taken into consideration in the practical applications of amino acid supplementation of proteins in foods.

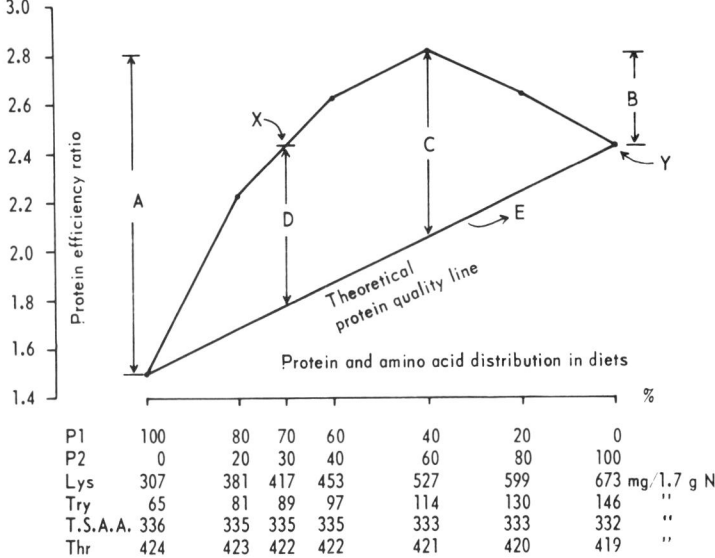

FIG. 18.2. PROTEIN QUALITY OF MIXTURES OF NORMAL MAIZE AND SOYBEAN PROTEIN

Relation between Energy Content of the Diet and Amino Acid Needs

A balanced protein can be used with the maximum efficiency only if the animal has, at the same time, ingested sufficient calories (Allison 1949, 1955). The high caloric requirement of subjects receiving amino acid diets was first indicated by Rose and co-workers (Rose, 1949; Rose et al. 1950, 1951, 1954).

During the last 15–20 yr, evidence has been obtained to show that there is a direct relationship between the energy and the protein content of the diet. This relationship has been very well demonstrated in poultry feeds (Rosenberg 1959). Thus, relationships have been shown between methionine and productive energy (Rosenberg 1959) and lysine and energy level in the diet (Rosenberg 1959). Relationships have also been demonstrated in human subjects (Calloway and Spector 1954).

Age and Amino Acid Requirements

In the context of the improvement of amino acid patterns, attention must also be given to amino acid requirements and biological value of proteins as affected by age. Mitchell (1924, 1959) and Reber et al. (1953) demonstrated that protein requirements for maximum growth decreases with increasing age and body weight. Since protein requirement is a summation of individual amino acid needs, the latter must also decrease with increasing age.

Henry and Kon (1957) found that a higher biological value was obtained with adult rats fed a 4% than an 8% level of protein, or than at either level of intake with the same rats when young. The addition of 0.4% DL-methionine increased significantly the biological value of casein only with the young rats at the 8% level of protein intake. These results suggested that the sulfur amino acid requirement of the rat decreased with age. The authors also obtained evidence to show that lysine requirements were lower for the adult rat.

Womack and Rose (1941) found that cystine could meet about $\frac{3}{4}$ of the total requirements for the sulfur-containing amino acids in the adult rat, whereas in the young rat the same authors found that only $\frac{1}{6}$ of the requirement for the sulfur-containing amino acids could be met by cystine. This was interpreted by them to indicate a decreased requirement for methionine with age, concomitant with an increased need for cystine.

Hartsook and Mitchell (1956) concluded that protein requirements as a percentage of the diet decreases in an exponential fashion as age increases. They also arrived at the conclusion that the methionine plus cystine requirement expressed as a percentage of the diet, decreases as age increases, although when expressed as a percentage of the protein requirement it increased.

Mitchell (1947) also found that diets in which lysine-deficient cereal proteins were the main nitrogen source had higher biological values for the adult rat than for the growing rat, suggesting lower needs of the amino acid for the mature rat.

Lysine requirements also decrease with age as demonstrated by Neuberger and Webster (1945) and Nasset and Ely (1952). These last workers presented evidence to show that lysine is indispensable in the nutrition of the adult rat although the requirement was small under the conditions used in the studies. The withdrawal of lysine, however, had adverse effects on nitrogen balance for the adult rat as shown by Wissler et al. (1948). Nasset et al. (1951) demonstrated that the adult rat needs threonine for maintenance of nitrogen equilibrium, although the levels needed are smaller than for the young rat.

Studies of other workers also showed lower needs by the adult rats of other essential amino acids (Nasset and Anderson 1951; Bendett et al. 1950; Womack et al. 1953).

Nakagawa and co-workers (Nakagawa et al. 1960, 1961A,B, 1962, 1965) recently reported on the amino acid requirements of children

TABLE 18.4

ESSENTIAL AMINO ACID REQUIREMENTS OF HUMANS AT VARIOUS AGES

Amino Acid	Estimated Requirements, Mg/Kg/Day				
	Infant		10-12 Yr	Men	Women
	Ref[1]	Ref[2]	Ref[3]	Ref[4]	Ref[5]
Histidine	34	28	0	0	0
Isoleucine	119	70	30	10	7.7
Leucine	229	161	45	16	12
Lysine	103	161	60	11	12
Methionine + cystine	45 + cys	58	27	16	9.5
Phenylalanine + tyrosine	90 + tyr	125	27	16	12
Threonine	87	116	35	7	5
Tryptophan	22	17	4	3.6	2.7
Valine	105	93	33	11	11

[1] Holt and Snyderman (1961).
[2] Fomon and Filer (1967).
[3] Nakagawa et al. (1965).
[4] For 70-kg men (Rose 1949).
[5] For 58-kg women (Leverton and Steel 1962).

(Table 18.4). Using nitrogen balance techniques, they were able to show that the requirements for the essential amino acids of children were in all cases higher than those reported by Rose (1949). They also reported that the values found for the children used in their studies were lower than those reported for infants by Snyderman et al. (1955, 1959).

THE ADDITION OF AMINO ACIDS

Cereal Grains

The cereal grains have been considered over the years as the most important sources of energy for man and domestic animals. As is well known, they contain a relatively low protein concentration; however, because they are consumed in large quantities, the amount and quality of their protein becomes extremely important. Recent studies by Howe et al. (1965A,B, 1967) have demonstrated the improvement in quality which can be obtained by fortifying foods with amino acids.

Wheat.—The studies performed by Osborne and Mendel (1919, 1920) demonstrated over 50 yr ago that addition of lysine to wheat protein markedly increased its protein efficiency or biological value. These findings have been confirmed with experimental animals by a number of workers (Jansen 1962; Moran 1959; Howe et al. 1965B;

Flodin 1953, 1956; Rosenberg, 1959) and in human beings by others (Bressani et al. 1960B, 1963A; Barnes et al. 1961; Hoffman and McNeil, 1949; Pereira et al. 1969).

The effect on growth rate and food consumption in experimental animals as a result of the levels of lysine supplementation has been studied in diets containing different levels of protein intake.

Hutchinson et al. (1958, 1959), for example, fed increasing amounts of added L-lysine to bread diets whose nitrogen content was 2.44% dry weight basis. With this level of protein in the diet, optimum weight gain/day and food consumed, was obtained with the addition of 0.25% L-lysine. Higher levels caused a decrease in the variables being measured. The same authors (Hutchinson et al. 1958, 1959), found that with diets containing 1.34% nitrogen, optimum response was obtained from the addition of 0.1% L-lysine. As with the higher protein-containing bread study, higher levels of lysine decreased performance.

The relationship described between the optimum amount of lysine to be added and the percentage of protein in the diet is of considerable importance in the interpretation of amino acid supplementation, particularly when other amino acids are also added together with the one that is most limiting (Moran 1959). In this respect, the effect of threonine seems to be important since the results have not always been in agreement. For example, various authors have reported that addition of lysine and threonine to wheat flour produces a much greater response than lysine addition alone. Other authors have reported the opposite (Hutchinson et al. 1958, 1959; Ericson 1960). The difference in results may be explained on the basis of protein content in the diets, since those investigators who did not find an effect of threonine used diets containing more protein than the diets used by workers who reported better results when threonine is added.

Deshpande et al. (1955) presented results which showed that with a given protein content in the diet the results obtained are influenced by the level of supplementation with lysine and threonine. It seems that the effect of threonine depends on the level of lysine added at any level of dietary protein. This again is an excellent example of the concept of limiting amino acid and amino acid balance (Harper 1958; Mitchell and Block 1946; Oser 1951; Rosenberg 1959; Elvehjem 1956).

The results obtained with experimental animals have also been found working with humans (Jansen, 1962; Barnes et al. 1961; Hoffman and McNeil 1949), and children (Pereira et al. 1969; Bressani et al. 1960B, 1963A). In this case, however, the number of studies have not been as extensive as those with lower animals.

Maize.—Maize, in the form of maize meal, lime-treated maize, or any other form, is an important staple foodstuff in Latin America, Africa, and various other countries, providing in most cases 80% of of the calories and about 70% of the protein (Flores and García 1960; Flores 1961). In the animal industry, maize occupies a position not reached by any other cereal grain or cereal grain product. Therefore, there has been much interest in improving the nutritional value of its protein by amino acid supplementation.

Among the first studies were the results of Mitchell and Smuts (1932) who obtained a definite improvement in growth when 8% maize protein diets were supplemented with 0.25% lysine. These results have been confirmed through the years by several authors (Howe et al. 1965B; Gillespie et al. 1958; Sure et al. 1953), while others (Bressani et al. 1968; Rosenberg et al. 1960; Bressani and Marenco 1963) have shown that the effect of lysine added to maize causes only a small improvement in protein quality. These results can be explained on the basis of the variation reported in the lysine content of maize varieties (Bressani and Mertz 1958A; Bressani et al. 1960A, 1962B) which led to the discovery in 1964 by Mertz et al. (1964) of a high lysine maize called *opaque-2* maize.

Some workers have reported that tryptophan rather than lysine is the first limiting amino acid in maize (Hogan et al. 1955), which can be true for some varieties with a high lysine concentration or maize products modified by some type of processing. The results on which all workers have agreed is that the addition of both lysine and tryptophan, added simultaneously, improves protein quality of maize significantly. These results have been demonstrated in experimental animals (as indicated above), adult man (Truswell and Brock 1961; Kies et al. 1965A, B), and children (Scrimshaw et al. 1958; Bressani et al. 1958, 1963A).

The improvement in quality obtained after the addition of lysine and tryptophan has been small in some studies and higher in others, upon addition of other amino acids (Benton et al. 1955; Sauberlich et al. 1953; Bressani 1963). Apparently, the amino acid which becomes limiting after lysine and tryptophan is isoleucine, as detected from animal feeding studies (Benton et al. 1955) and tests in humans (Scrimshaw et al. 1958; Bressani et al. 1958, 1963A). Most workers who reported such findings indicate that the effect of isoleucine addition is due to an excess of leucine which interferes in the absorption and utilization of isoleucine (Harper et al. 1955). It has been indicated that the high intake of leucine consumed with the protein in maize increases niacin requirements, and this amino acid could be partially responsible for pellagra.

When a response to threonine addition has been observed, it has been interpreted as effects of this amino acid to correct amino acid imbalances, caused by addition of methionine (Bressani 1963). A similar role can be ascribed to isoleucine where its addition resulted in improved performance. Similarly the addition of valine, which results in decreases in protein quality, could be counteracted by the addition of either isoleucine or threonine (Bressani 1963).

In any case, isoleucine seems to be more effective than threonine, producing more consistent results. The possible explanation for these findings, is that maize is not deficient in either isoleucine or threonine; however, some samples of maize may contain larger amounts of leucine, methionine, and valine, and they require the addition of isoleucine and threonine besides lysine and tryptophan to improve the protein quality.

Rice.—According to amino acid calculations, the first limiting amino acid in rice protein is lysine (Oser 1951; Block and Mitchell 1946-1947). However, most of the studies carried out before those of Rosenberg and Culik (1957) and Rosenberg et al. (1959) did not demonstrate any beneficial effect from the addition of lysine only, probably because the level of dietary protein obtained with rice was below 7%. Pecora and Hundley (1951) reported that the combination of lysine and threonine improved growth considerably. Rosenberg and Culik (1957) on the other hand, were able to show that 0.10% lysine added to 90% precooked rice diets produced the highest weight gains as well as the lowest feed efficiency expressed in grams of feed per gram of weight gained.

These findings proved that lysine is the first limiting amino acid in rice protein. It is likely, however, that a higher performance could have been obtained if threonine had been added.

Since these authors (Rosenberg et al. 1959; Rosenberg and Culik 1957) used precooked rice it is possible that some of the lysine in the tested rice protein was bound with carbohydrates, as it has been established that heat reduces lysine availability. If this were the case the effect of the added lysine is self-explanatory.

It has been indicated that threonine in rice protein is present in sufficient amounts, but due to its low availability to the organism it becomes limiting for growth. Harper and De Muelenaere (1963), however, have indicated that threonine is highly available from rice. Studies with humans have in some cases given inconsistent results and Hundley et al. (1957) indicated that rice protein is deficient first in total nitrogen. Other authors, however, have been able to establish that lysine is deficient in rice, for children and young adults (Parthasarathy et al. 1964). Bressani et al. (1971) found that rice protein could be improved by lysine addition, giving nitrogen retention values close to those obtained when milk was fed at isonitrogenous and isocaloric levels of intake.

Better performance was obtained when the lysine-fortified rice protein was further supplemented with methionine and threonine. Parthasarathy *et al.* (1964) found with humans aged 8-9 yr, that the addition of lysine and threonine brought about a highly significant improvement in the nitrogen retention, biological value, and NPU (Net Protein Utilization) in rice. Methionine addition further improved the three measurements of protein quality. These results corroborated the findings, in rats, reported by Sure (1955) and others (Pecora and Hundley 1951; Deshpande *et al.* 1955; Harper *et al.* 1955); who found an increase in the PER (Protein Efficiency Ratio) of rice proteins when methionine was added with lysine and threonine. It would appear, however, that the two most limiting amino acids in rice protein are lysine and threonine as indicated by Howe *et al.* (1965B) in studies carried out with weanling rats.

Oats.—Studies carried out in rats by Tang *et al.* (1958) have indicated that rolled oats are deficient in lysine, methionine, and threonine in this order. They reported that threonine addition to rolled oats supplemented with lysine and methionine resulted in an improvement in weight gain and protein efficiency ratio (PER) because this amino acid is not completely available to the rat.

Howe *et al.* (1965B) also reported an increase in weight gain and PER when rolled oats were supplemented with lysine and threonine. Other workers have reported similar findings (Bressani and Elías 1967; Mitchell and Smuts 1932; Osborne and Mendel 1920; Hischke *et al.* 1968).

Very few studies have been carried out with human subjects. In a study with children aged from 2 to 5 yr, Bressani *et al.* (1963B) reported a significant increase in nitrogen balance from the addition of threonine to rolled oats when fed at a level of 1.5 gm per kg per day. Nitrogen retention very close to that obtained from milk protein was observed only when rolled oats were supplemented with three amino acids. Leverton and Steel (1962) reported poor nitrogen retention in young women fed an oat diet, which could have been due to the overdiluted essential amino acid nitrogen to total nitrogen in the diets used by them.

Sorghum and Millet.—Pond *et al.* (1958) indicated that sorghum grains are deficient in lysine which when added increased weight gain and PER. Similar results were reported by Howe *et al.* (1965B) who found that threonine was limiting after lysine. Millet protein was reported by Mangay *et al.* (1957) and Howe *et al.* (1965B) to be improved by the addition of relatively high levels of lysine, and Howe *et al.* (1965B) further established that threonine was the second limiting amino acid.

Waggle *et al.* (1966, 1967), in studies with whole sorghum grain, found that the nutritive value of low protein sorghum grain, when compared on an equal protein basis was superior to that of a high protein sorghum grain. Amount of dibasic amino acids, especially lysine, of sulfur-containing amino acids and of threonine in high protein sorghum grain were lower than in low protein grain. Shoup *et al.* (1969) reported on the nutritional value of milled sorghum grain products, indicating that bran germ fractions were superior nutritionally to the original grain and endosperm fractions. The whole grain and endosperm were improved in protein quality as measured by growth and PER by the addition of lysine.

Rye.—Kihlberg and Ericson (1964) observed that in the growing rat, lysine was the first limiting amino acid in rye flour and threonine was the second. Sure (1954) had indicated that the addition of valine to rye would effectively improve its protein quality. However, Howe *et al.* (1965B) were not able to show such an effect. The results of these workers suggest that in rye flour protein, tryptophan is more limiting than threonine. Other workers have also reported improvement of rye protein upon supplementation with lysine and proteins containing high levels of this amino acid such as skim milk or casein (Knipfel 1969; Strømnaes and Kennedy 1957).

Other Cereal Grain Flours.—Howe *et al.* (1965B) showed barley protein to be improved by the addition of lysine; a further increase in quality resulted when threonine was also added. These results confirm the findings of other workers (Osborne and Mendel 1920). Teff, a cereal grain consumed in large quantities in Ethiopia, was extensively studied by Jansen *et al.* (1962) who reported that it had a good essential amino balance, except for being markedly deficient in lysine.

Lysine supplementation was found to raise the PER approximately to the level obtained with casein. Triticale, a hybrid cereal produced by cross breeding of wheat and rye, was studied for its protein quality by Knipfel (1969). The PER of triticale, was equal to the protein value of rye and superior to that of wheat. Amino acid composition in test diets and in blood plasma of rats fed such diets, indicated that lysine was severely limiting. Its addition caused an increase in protein quality.

Conclusion.—From the results of all the studies carried out in weanling white rats, fed the various cereal grains, it can be concluded that all have lysine as the first limiting amino acid, although its addition alone is not capable of increasing the quality to the values obtained with casein. Most studies show that for a higher improvement in quality all cereal grains require the addition, besides lysine, of the second limiting amino acid which for the most of them is threonine, with the exception of maize and rye which require tryptophan. Similar conclusions can be reached from the limited studies carried out with human subjects and Fig. 18.3 presents a summary of such re-

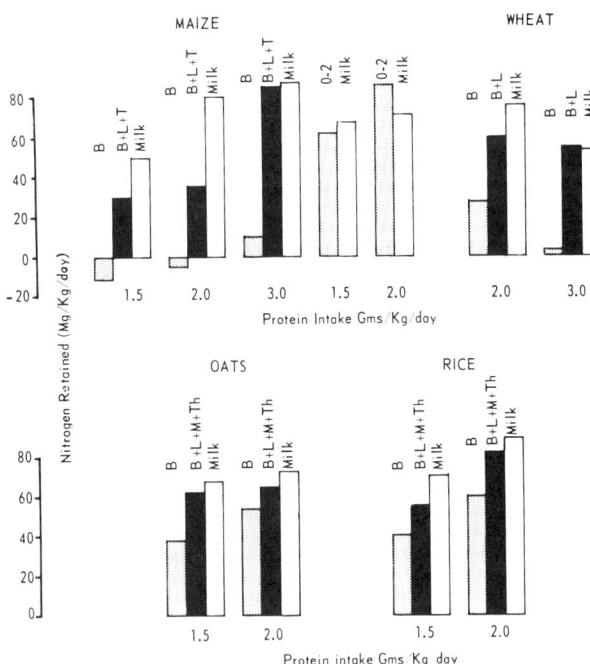

FIG. 18.3. NITROGEN RETENTION OF CHILDREN FED CEREAL PROTEINS WITH AND WITHOUT THEIR LIMITING AMINO ACIDS

sults in children. The black bars represent the nitrogen retention values when the cereal grain was fortified with the most limiting amino acid or amino acids.

The data reviewed also indicate that the level of dietary protein influences the number of amino acids to be added to show an improvement in protein quality; a fact very well demonstrated in the studies with wheat flour (Moran 1959; Bender 1958; Hutchinson et al. 1958, 1959), rice (Pecora and Hundley 1951; Deshpande et al. 1955; Hundley et al. 1957), and other cereal grains (Bressani and Elías 1967; Hogan et al. 1955; Sauberlich et al. 1953; Waggle et al. 1966). Although the rat is a rapidly growing animal with higher nutritional requirements than those of man, the available data indicate

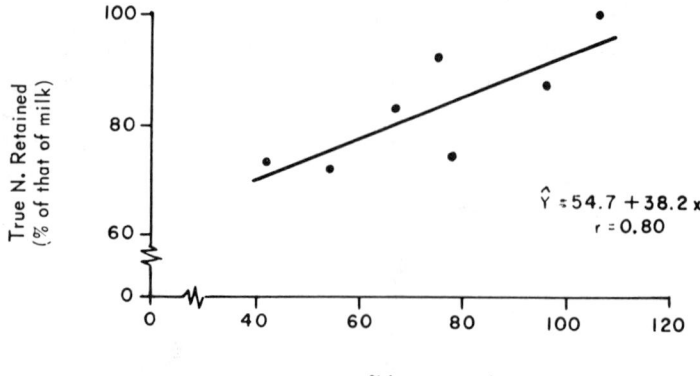

FIG. 18.4. NITROGEN BALANCE IN PRESCHOOL AGE CHILDREN

Correlation between nitrogen retention and lysine intake from different sources of protein.

that, at least for children, the cereal grains respond to the addition of the same limiting amino acids, as demonstrated by various investigators (Scrimshaw et al. 1958; Bressani et al. 1958, 1960B, 1963A,B; Barnes et al. 1961). In this respect it is of interest to study the relationship between lysine addition to maize as tested in rats and children, shown in Fig. 18.4 and 18.5.

Finally, when differences in response to amino acid supplementa-

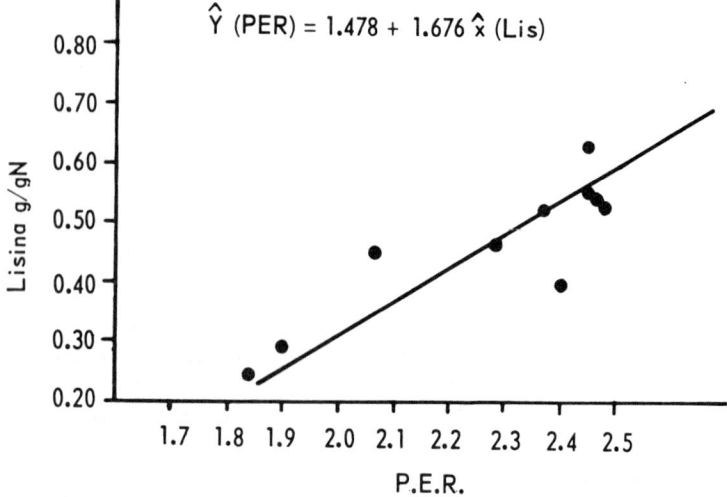

FIG. 18.5. RELATIONSHIP BETWEEN PER AND LYSINE

tion are reported it should be remembered that cereal grains are characterized by a high concentration of prolamines. The amounts in cereal grains are easily affected by fertilizer applications, variety, and environmental factors (Hogan et al. 1955; Bressani and Mertz 1958A; Bressani et al. 1960A, 1962B; Mertz et al. 1964; Sauberlich et al. 1953; Waggle et al. 1966; Waggle et al. 1967; Gunthardt and McGinnis 1957). These conditions can very well change the amino acid content of the whole kernel and, therefore, the response to amino acid supplementation.

Legume Foods

Pulses and grain legumes are very important foodstuffs in the diets of large populations in the tropics. The characteristic feature of legume seed proteins is that they are markedly deficient in methionine and tryptophan. In fact, methionine is the first limiting essential amino acid in almost all the legume grains. Due to this limitation, several investigators have demonstrated the improvement in protein quality as measured by the usual methods, from the supplementation with methionine (Bressani et al. 1963D; Patwardhan 1962; Russell et al. 1946; Schneider and Miller 1954; Jaffé 1949; Finks et al. 1922; Venkat Rao et al. 1964; Woods et al. 1943).

Russell et al. (1946) found that an addition of 0.1% of methionine improved the protein quality of *Cicer arietinum*, *Phaseolus lunatus*, *Phaseolus vulgaris*, and *Pisum sativum*. The improvement ranged from several hundred percent over the unsupplemented bean value to about 50%. Richardson (1948) reported that the addition of 0.2% methionine to *Vigna sinensis* diets doubled the growth rate of rats with 10% protein in the diet. *Phaseolus lunatus*, however, could not be improved. Similar findings were reported by Braham et al. (1965), when attempts were made to improve the quality of *Cajanus cajan*. These workers found tryptophan addition improved protein quality, further increased by the addition of methionine. This could probably explain the negative results reported by Richardson with respect to the addition of methionine to *Phaseolus lunatus* (Jaffé 1949). Sherwood et al. (1954) and Schneider and Miller (1954) also demonstrated the supplementary value of methionine on the nutritive value of *Vigna sinensis* and Alaska peas. Common Indian pulses can also be improved by methionine supplementation as indicated by other workers (Patwardhan 1962; Venkat Rao et al. 1964). Bressani et al. (1963D) found black *Phaseolus vulgaris* to be improved in protein quality upon the addition of 0.2% methionine; no increase in protein digestibility was reported by these workers, however. It should be indicated that tryptophan has been found to be a deficient amino acid in leguminous seed proteins (Patwardhan 1962; Venkat Rao et al. 1964; Braham et al. 1965; Block and Mitchell 1946-1947).

Oilseed Flours

In recent years much attention has been given to the by-product of the oil-containing seeds, as sources of protein to help meet the protein need of the human population in developing areas. Furthermore, some of these proteins are being used to extend animal protein sources. In view of this, it is essential to know of the limiting amino acids in these products in order to make a better use of these sources.

Soybeans.—The available evidence indicates that meals of soybean products, flour, or protein concentrates are improved in protein quality upon supplementation with methionine (Berry et al. 1962, 1966; Evans and McGinnis 1946; Shurpalekar et al. 1961; Smith 1961). Some studies also suggest that threonine is the second limiting amino acid. The improvement in protein quality brought about by adding methionine is so high that any other amino acid added causes only small increases. The threonine deficiency is insignificant for practical purposes if the material is to be used by itself, but may be important if used in combination with other protein sources which have similar deficiencies (Berry et al. 1962; Bressani et al. 1966).

Cottonseed.—Cottonseed protein, as flour or as protein concentrate, has been shown by many workers to be deficient in lysine (Conckerton and Frampton 1959; Martínez et al. 1961; King et al. 1962; Elías et al. 1969; Lyman 1953; Bressani et al. 1966; Phelps 1966; Howe et al. 1965A, 1967; Bressani 1965). Native cottonseed protein is already low in lysine, a deficiency which increases during processing due to reactions between lysine and carbohydrates (Martínez et al. 1961) and lysine and gossypol (Conckerton and Frampton 1959; King et al. 1962). Allison et al. (1960) reported that cottonseed flour was also limiting in methionine, which, when added together with lysine improved the nitrogen balance index of the protein as tested in dogs. These results, however, have not been confirmed by other workers; this may be due to the use of rats in one case (Elias and Bressani 1971) and of chicks in the other (Heywang and Bird 1950; Fisher 1965).

Peanut.—Peanut protein, a source of great value to the population in India, has been shown to be markedly deficient in methionine and lysine (Bressani and Elías 1968; Davidson and Boyne 1962), which must be added in the correct proportion to cause a significant improvement in protein quality. The addition of only one of the two amino acids decreases the efficiency of utilization of the protein (Davidson and Boyne 1962).

The deficiencies, particularly that of lysine, can be increased by improper processing; that is, by the use of excessive heat (Cama and Morton 1950; Buss and Goddard 1948).

Other Oilseed Proteins.—Sesame protein has been shown to have lysine as the first limiting amino acid and the quality of the protein improves upon its addition (Grau and Almquist 1944; Carter *et al.* 1961).

Sesame seed protein has been shown to be one of the highest natural sources of methionine (Almquist and Grau 1944; Krishnamurthy *et al.* 1960; Zaghi and Bressani 1969). Other oilseed proteins of minor importance have also been shown to be first limiting in lysine by McGinnis *et al.* (1948), Slinger and co-workers (1949), and others (Grau and Almquist 1945; Alexander and Hill 1952). Sunflower seed oil meal protein has been shown to be deficient, in both lyine and methionine in swine (Delie *et al.* 1963). Coconut protein, which is important in various regions of the world as a protein source, is limiting mainly in lysine (Butterworth and Fox 1963). Safflour, a potential source of protein for human food, has been shown to be deficient in methionine (Kohler 1966); as well as the protein in rubber seed (Giok *et al.* 1967).

The results have demonstrated that in this group of food products, the most limiting amino acids are methionine and lysine, the latter in most cases because of the excessive use of high temperatures during processing of the seed for oil extraction.

Animal Protein Sources

It is well recognized that animal protein sources have, in general, a higher biological value than vegetable proteins. In spite of this, proteins of animal origin, have been demonstrated to have deficiencies of the essential amino acids. Casein, a protein used commonly as a reference protein in animal assays for protein quality, has been shown to have methionine as its first limiting amino acid (Harper 1959). This deficiency, however, can only be demonstrated when casein provides less than 15% protein in the test diets. Harper (1959) also demonstrated casein to be deficient in arginine, isoleucine, tryptophan, and valine when fed at levels of dietary protein less than 8%. Very few studies are available on the limiting amino acids of meat; however, it has been shown that for this group of proteins, methionine is the first limiting amino acid (March *et al.* 1950).

Gelatin, a protein used very extensively as a dietary food was shown by various workers to be deficient in tryptophan, a fact known from its essential amino acid composition (Mitchell and Block 1946; Block and Mitchell 1946-1947). Furthermore, this protein is extremely low in methionine and has many other amino acids out of balance.

Blood meal protein has also been studied. Results indicate that this protein group is deficient in isoleucine, methionine, and arginine for the chick. These must apparently be added together, since isoleucine, which appears to be most limiting when added alone, does not materially improve nutritive quality (Fisher 1969).

Fish protein, as fish meal, very frequently used as a protein supplement in animal rations, and available in a more pure form as fish protein concentrate, has also been shown to be deficient in sulfur-containing amino acids (Smith and Scott 1965; Carpenter 1960; Stillings et al. 1969), when its processing is not carried out under well-controlled conditions.

Although under very special conditions, such as extremely low levels of dietary proteins, deficiencies have also been indicated for milk proteins, egg, lactoalbumin, and others; they are more likely to be due to the effects of the preparation steps used, and are of academic interest only (Mauron 1961; Mauron et al. 1955).

Other Protein Sources

Much attention has recently been given to single-cell proteins as sources of this nutrient for the supplementation of protein-poor diets. The available results indicate that yeast, bacteria, and algae proteins, have the sulfur-containing amino acids as the first limiting for growth (Klose and Fevold 1944; Elías and Bressani 1970). The effect of methionine addition in improving the protein quality of yeast food has been amply demonstrated by many authors (Bressani 1968).

Leaf and algae protein concentrates have also been shown to be limiting in methionine by various authors (Leiveille et al. 1962).

As pointed out before, the knowledge of the limiting amino acids in the various protein sources is of importance for situations in which they are to be combined with another protein source, with the possible partial exception of the cereal grains which are consumed in much greater amounts than other sources. Even in this situation, however, the knowledge of the limiting amino acids in these foods allows for the proper use of protein sources rich in the amino acids limiting in the cereal grain proteins.

FORTIFICATION TECHNOLOGIES

Various approaches have been proposed to add synthetic amino acids to foods. The technology involved has been reviewed by Milner (1969), Senti and Pence (1971), and Rosenfield and Berntson (1971). In this section the problem will be reviewed as it applies to industrially-produced flours, to whole grains and/or grains subjected to minimum processing, and to foods during processing.

Various procedures can be used to fortify foods with synthetic amino acids; however, the best approach depends on the form in which the food is distributed and by the habits of people consuming such foods. In developing countries, where the need for fortification seems more urgent, most cereal grains consumed are processed at home and if the food is obtained as flour it comes from small mills. Therefore, fortification would be a simple procedure for industrially-produced flour but more difficult for the home-processed food or that coming from small community mills.

Industrial Flours

Cereal grains, such as maize, wheat, sorghum and oats, are milled in most cases into a flour, a meal, or as grits. When the food is so prepared, the deficient amino acids can be added quite easily as a component of the premix containing vitamins and minerals, using standard powder chemical feeders of the milling industry. This approach will not apply for cereal grains which are not milled into flour, as for example rice, which in most countries is left as a grain without the husks and outer aleurone layers and germ. In such a situation the fortifying amino acids could also be a part of the vitamin-mineral supplement used to enrich it, as is done in Taiwan, Philippines, and other countries. In these places, one of the approaches used for vitamin fortification consists in adding the vitamin mixture in acetone solution, followed by a protein cover, such as zein, also added in an acetone solution. The grains are thus covered with a layer which protects the added nutrients from leaching out during preparation for consumption.

Whole Grains

Synthetic Kernels.—The possibility of adding the limiting amino acids as components of simulated or synthetic grains is an approach discussed by Senti and Pence (1971), which has various advantages, particularly when the cereal grains are processed as such at home. An example of this approach is the use of simulated rice kernels (Bressani 1972A). An example of the improvement in protein quality to be obtained is shown in Table 18.5.

The kernels when added to common polished rice at a level of 2% improved weight gain and PER. This approach has been also proposed for maize and for beans (Bressani 1971). One of the advantages of this approach is that the man-made kernel can be a carrier of additional amounts of protein, useful for cereal grains which are low in total protein content and may decrease the amounts of free amino acids added because the protein itself will be providing part of the total levels needed to improve the cereal grain under consideration. The

TABLE 18.5

SUPPLEMENTATION OF RICE WITH SIMULATED RICE KERNELS[1]

Addition to Rice	Protein in Diet (%)	Average Wt Gain[2] (Gm)	PER
None	7.2	45	1.94
1% simulated rice	7.3	86	3.04
2% simulated rice	7.6	108	3.32
3% simulated rice	7.7	103	3.29

[1] Contains 75.59% L-lysine HCl + 15.12% L-threonine + vitamins.
[2] Initial weight: 45 gm.

disadvantage is that the synthetic or simulated kernel might not be exactly as the natural kernel and that the man-made kernel might introduce a different taste to food.

Infusion Techniques.—Graham et al. (1968) have reported on an infusion technique for preparing high lysine level fortified wheat kernels. These were lightly scarified and the resulting product was reported to be not easily distinguishable from untreated wheat. The authors claimed such fortified kernels could be blended with untreated wheat to yield a desirable level of fortification and could be introduced into regular marketing channels to reach the consumer. Ferrel et al. (1970) reported on the storage stability of the lysine in lysine-fortified wheat.

Highly fortified kernels and blends with 0.1% fortification were stored at 90° and 100°F at moisture levels of 9, 11, and 13%. Evaluation consisted of periodic tests for added lysine, organoleptic characteristics such as odor or color changes, and biologically for lysine availability. From all these observations the authors concluded that stability of lysine remained relatively high throughout the storage period.

Similar studies have been carried out by Blessin et al. (1970) on the infusion of lysine to dent maize. The results of these studies indicated that, on the average, lysine content of whole kernels of corn was increased from 0.24% to approximately 1% during infusion with a 15% solution of lysine at 160°F for 30 min. The same authors reported that lysine fortification could be accomplished by coating dehulled maize with zein or isolated soy protein containing added lysine.

Similar studies were carried out by Gómez Brenes et al. (1974) with maize. It was found that from an aqueous 30% lysine solution at 25°C and pH 6.0, concentration in the kernel increased from 0.11 to 6.0%. For tryptophan optimum infusion conditions were obtained when using a 20% tryptophan solution in sodium hydroxide at pH 12.0 and a temperature of 40°-45°C, increasing free tryptophan levels to an average of 4.6%, or 277 times the levels present in maize. The quality improvement resulting from the addition of infused maize kernels to normal grain was tested biologically by these authors, and representative values are shown in Table 18.6.

TABLE 18.6

RESULTS OBTAINED FROM THE BIOLOGIC ASSAYS WITH GROWING RATS[1] FED MAIZE SUPPLEMENTED WITH LYSINE AND TRYPTOPHAN

Rations	Protein (Gm %)	Avg. Wt. Gain (Gm)	Avg. Food Consumed (Gm)	PER
1. Maize	8.6	30 ± 3.5	284 ± 18	1.22 ± 0.08
2. Maize + lys + maize (try)[2]	9.0	92 ± 8.0	420 ± 21	2.42 ± 0.10
3. Maize + try + maize (lys)[2]	8.9	100 ± 11.0	427 ± 27	2.59 ± 0.12
4. Maize + maize (try) + maize (lys)	8.8	91 ± 9.3	427 ± 25	2.38 ± 0.12
5. Maize + lys + try	8.9	95 ± 8.3	435 ± 23	2.45 ± 0.11
6. Casein	10.4	135 ± 6.8	473 ± 13	2.73 ± 0.07

[1] Average initial weight: 51 gm; groups of 8 rats (4 males and 4 females).
[2] Tryptophan and lysine incorporated to corn by infusion.

With popcorn, Blessin et al. (1971) found that by infusion techniques lysine could be increased from 0.32% in unprocessed kernels to 3.0% in infused grains. Cavins et al. (1972) also reported on the infusion of lysine, methionine, and tryptophan individually into sorghum kernels. They indicate that infusion is altered significantly by time, temperature, and concentration of infusion solution. They further indicated that infusion proceeds in two stages, one being concentration-dependent and the other concentration-independent, with amino acid solubility affecting the maximum level of infusion possible. Infusion techniques have also been used to increase the lysine content of rice. Ruiloba (1973) studied the best conditions for lysine and threonine infusion in polished rice. For lysine these were a 10% solution for 4 hr at 40°C with 18% initial moisture in the rice kernel at the normal pH of the solution, and for threonine a 10% solution for 4 hr at pH 6.0 at room temperature. Concentration for lysine and threonine in the infused kernels reach 1.5% and 1.2%, respectively.

For commercial application of these techniques, simple and reliable methods for controlling levels of fortification must be available. Ferrel et al. (1969), Tara and Bains (1969), and Finley et al. (1972)

have described methods for lysine and methionine level control for wheat and bulgur while Jenneskens (1969) showed the improvement in quality of lysine-fortified bulgur and the way in which it could be controlled. Gómez Brenes and Bressani (1973) described a method for the control of lysine and tryptophan levels added to maize. These techniques will become very useful when the practice of fortification with amino acids becomes a reality.

During Home Processing.—This approach is similar to a very large extent to other ways discussed previously, and it refers to the addition of the supplement whether as a powder, synthetic, or infused kernel to the food at the moment of being processed. This approach for maize being tested in Guatemala consists of adding the supplement when the cooked grain is milled before making tortillas. The added amino acids may be added alone or with protein (Molina et al. 1972). The stability has been found to be good (as shown in Table 18.7) by the recoveries of the added amino acids (Elías and Bressani 1972).

TABLE 18.7

EFFECT OF PROCESSING ON THE STABILITY OF AMINO ACID[1] ADDED TO THE DIFFERENT PREPARATIONS

Preparation	Cooking Time (Min)	Gm Lys per Gm N	Gm Try per Gm N	Recovery Lys (%)	Recovery Try (%)
Masa	0	0.198	0.053	—	—
Tortilla	3.0	0.205	0.057	—	—
Tortilla	4.5	0.201	0.054	—	—
Tortilla	6.0	0.199	0.055	—	—
Masa + 0.30% lysine + 0.10% tryptophan	0	0.370	0.080	106.3	95.2
Tortilla + 0.20% lysine + 0.10% tryptophan	3.0	0.359	0.080	103.2	94.0
Tortilla + 0.30% lysine + 0.10% tryptophan	4.5	0.311	0.099	89.4	117.8
Tortilla + 0.30% lysine + 0.10% tryptophan	6.0	0.344	0.099	98.8	117.8

[1] L-lysine HCl; DL-tryptophan.

STABILITY OF ADDED AMINO ACIDS TO INDUSTRIAL PROCESSING, STORAGE, AND HOME COOKING

As indicated in previous sections of this chapter, poor quality proteins can be improved in their quality by appropriate fortification with their limiting amino acids. Most of the studies have been carried out with the addition of amino acids to the food in the precooked state or after the food has been subjected to some type of processing. Only a few studies have been carried out, however, on the addition of amino acids before processing, for the specific purpose of learning of the stability of the added amino acids to processing conditions.

Although the information available on amino acid fortification of unprocessed foods has been of great value in providing the basis for the concept of fortification, more importance should now be given to the problem of learning the fate of added amino acids during processing, during storage of the fortified food, and on cooking before consumption.

Stability in Processing

There are numerous reports in the literature which indicate that amino acids present in the protein molecule of foods are inactivated and even destroyed by various factors during processing. These factors can be of at least two kinds: (a) those concerned with the composition of the food such as the presence or absence of reducing sugars, level of moisture, and free fatty acids; and (b) enzymatic activity of the food. Therefore, if processing conditions affect the stability of the amino acids in the protein molecule, the same would be expected to take place for the amino acids added to increase the quality of the protein in the foods.

One of the first reports to this effect was that of Rosenberg and Rohdenburg (1951) who found that the loss of L-lysine HCL in bread due to baking was about 15% as measured by microbiological assay techniques with a variation of 9.5 to 23.8%. Two forms of lysine were used to fortify the wheat flour; DL-lysine was added at a level of 0.25%. Of particular interest was the observation made by the authors that the lysine loss from the use of DL-lysine averaged 11% (a range of 2.4 to 21.8%) while that for L-lysine averaged 32% (a range of 25.0 to 36.8%). No other report using DL-lysine was found; therefore, it would appear appropriate to corroborate the above findings, even though Cremer et al. (1951), using nitrogen balance studies on human subjects fed lysine-fortified bread, reported an almost complete loss of free DL-lysine added to wheat flour before baking.

The protein quality of lysine-fortified bread, when lysine is added before and after baking, was estimated biologically by Rosenberg and Rohdenburg (1952). These authors reported increased growth responses in weanling rats when compared to the performance of rats fed unfortified bread. The experiment lasted 6 months, which suggests that if lysine-fortified bread was prepared only once for the entire study, time of storage did not reduce lysine. This, however, was not indicated by the authors. The improved protein quality of lysine-fortified bread when lysine is added before and after baking and compared to unfortified bread, has been corroborated by the results of several workers in experimental animals (Sabiston and Kennedy 1957; Culik and Rosenberg 1958; Ericson and Larson 1962). Other reports, however, show a decreased performance of rats fed lysine-fortified bread added before baking in comparison with the performance of the animals fed the lysine-fortified ingredients (Gates and Kennedy 1964; Jansen et al. 1964A,B). These authors indicated that the main factor responsible in reducing rat performance due to losses of lysine was baking time although they also indicated that nonfat dry milk increased loss probably due to the presence of lactose (Jansen et al. 1964A,B) in this supplement.

Ericson et al. (1961) reported that threonine added to bread before baking was lost in an amount that was equivalent to 20–25% as measured by microbiological assays, while lysine loss ranged from 5 to 10% using similar assaying techniques. During the fermentation of the dough, free amino acids are produced, the concentration of which decrease upon baking (Kretovich and Pnomareva 1961; Morimoto 1966; Rubenthaler et al. 1963). The decrease has been associated with the color of the crust, where larger losses take place. In some cases, a decrease in the crumb has also been reported, but to a smaller degree. The addition of lysine does not affect loaf volume, taste, or texture although browning of the crust increases (Ehle et al. 1959).

The losses of added lysine to various other foods have also been reported. Clark et al. (1959) obtained recoveries of 86.7–91.3% added lysine to baking powder biscuits. Maleki and Djazayeri (1968) reported on the effect of fortification of flour to make Arabic bread with lysine, threonine, and methionine. They reported that baking did not change the protein quality of the lysine or lysine-threonine-supplemented bread as assayed with rats. The addition of methionine was without effect before or after baking. Stillings and Hackler (1965) reported that free amino acids produced during the fermentation stages of tempeh preparation decreased after deep-fat frying, carried out for 5–7 min. On the other hand, steaming of the tempeh for 2 hr had no effect on free amino acids. During deep-fat frying, lysine and cystine were most susceptible to heat destruction. Matthews et al. (1969) reported losses of 4% of added lysine due to baking of chapatties and a 25% loss in bread.

Elías and Bressani (1972) found no destruction or biological inactivation of L-lysine HCl and L-tryptophan added to corn masa before making tortillas, a cooking operation lasting up to 5 min. The stability of the added amino acids was tested by chemical and microbiological methods as well as by biological trials with weanling rats. Strømnaes and Kennedy (1957) reported that the protein quality of rye breads made from lysine and nonfat dry milk fortified flour and from unfortified flour was lower than that of the unbaked ingredients.

Some results have been reported on the effect of cooking of aminoacid-fortified protein-rich foods. Bressani *et al.* (1964) reported that with Incaparina (INCAP formula No. 9) fortified with 0.30% L-lysine HCl subjected to cooking up to 25 min in the presence of sucrose, no losses occurred up to 15 min of cooking. From the initial value of 240 mg L-lysine per 100 gm at 15 and 20 min of cooking, 224 and 180 mg L-lysine per 100 gm were recovered respectively. Planella *et al.* (1968) reported losses in amino acid content and decreases in PER values of mixtures of maize, soybean, lysine, and methionine due to heating prior to canning of the food mixture. Gómez Brenes *et al.* (1974) reported (as shown in Table 18.8) that lysine- and

TABLE 18.8

EFFECT OF PROCESSING ON THE STABILITY OF INFUSED LYSINE AND TRYPTOPHAN IN MAIZE

	Lysine (%)		Tryptophan (%)	
Raw maize	6.00		5.00	
Toasted maize	4.64		2.91	
Percentage loss		22.7		41.8
Autoclaved maize (dry)	4.46		2.54	
Percentage loss		25.7		49.2
Autoclaved maize (moist)	4.84		2.32	
Percentage loss		18.4		53.6

tryptophan-infused maize kernels lost variable amounts of the two amino acids, when maize fortified with infused kernels was subjected to lime cooking, steam cooking, and toasting. The losses observed in some processes were explained by indicating that the infused amino acid diffused out of the kernel, since the amino acids were recovered in the cooking liquor. This is described in Table 18.9, which shows losses of 69 and 74% infused lysine and tryptophan, respectively, when maize was processed into tortilla. With respect to rice subjected to cooking processes commonly used by housewives before consumption, Bressani (1972A) showed that cooking enriched rice decreases the supplementary effect of the fortified

TABLE 18.9

LYSINE AND TRYPTOPHAN DISTRIBUTION OF
ENRICHED MAIZE WITH AMINO ACID-INFUSED
MAIZE DURING MAIZE PROCESSING INTO
TORTILLA[1]

	Lysine (Mg, %)	Tryptophan (Mg, %)
Raw mixture[2]	250.0	50.0
Cooked maize dough	83.6	12.0
Cooking water	162.4	38.0
Loss in water (%)	64.9	74.0
Tortilla	80.2	14.0
Loss from dough to tortilla (%)	4.1	—
Recovery (%)	31.0	26.0

[1] Values are on a dry weight basis.
[2] Mixture of maize and infused maize to add the equivalent of 0.25% lysine and 0.05% tryptophan.

kernel as shown in Table 18.10. It is of interest to point out that cooking improved quality when rice was unsupplemented, but the quality decreased when rice was supplemented. However, the values were at least in one case still above the unsupplemented protein quality values. Blessin *et al.* (1970) also reported losses of infused lysine in popcorn, when it was popped. Recoveries of lysine were of the order of 61%.

In summary, therefore, it appears that baking and cooking processes reduce levels of added amino acids to various food products, with the loss being detected by chemical, microbiological, and bio-

TABLE 18.10

GROWTH OF YOUNG RATS FED COMMON RICE SUPPLEMENTED
WITH SYNTHETIC KERNEL (RAW AND COOKED)

	Weight Gain[1] (Gm)	PER
Rice without supplement		
Raw	36.4 ± 3.7	1.84 ± 0.14
Cooked	41.0 ± 2.1	2.02 ± 0.05
Rice + 1% PA_1		
Raw	87.0 ± 8.4	3.03 ± 0.16
Cooked	70.9 ± 6.6	2.87 ± 0.12
Rice + 1% PA_2		
Raw	65.9 ± 7.1	2.73 ± 0.11
Cooked	40.1 ± 4.5	1.94 ± 0.13
Casein	106.8 ± 7.1	2.85 ± 0.13

[1] Group of 8 rats (4 male and 4 female); average initial weight of 46 gm.

logical assay techniques. Processing time and temperature as well as ingredient content and composition are of importance in determining the extent of the loss. It is recommended that further studies be carried out, particularly of foods cooked at home before consumption.

Stability in Storage

Very few studies are available on the effect of storage conditions on the stability of added amino acids. Hunter *et al.* (1956) found that DL-methionine added to animal feeds was stable for more than a year, even in situations with temperatures as high as 50°C. Higher temperatures decreased added methionine, but the feed itself was also deteriorated. Ferrel *et al.* (1969) reported that the stability of infused lysine in wheat kernels was relatively high even after 12 months at high storage temperatures and moisture levels as high as 13%.

Further studies are recommended on storage of foods fortified with various amino acids since no definite conclusions can be reached from the evidence available at the present time.

ECONOMIC BENEFITS FROM AMINO ACID ADDITION

The abundant evidence presented in this chapter and other reviews of the subject show that addition of the limiting amino acids to deficient proteins improve their protein quality significantly. The results have been quite conclusive with experimental animals in all cases and with human subjects in most cases.

Still, application of these findings is not at the level of the experimental results. There are, of course, various reasons which explain this lag; but today, with the world facing a serious food crisis, particularly that of protein foods, the need to make maximum use of and efficient utilization of the protein consumed suggests that it is urgent for such findings to be applied.

The purpose of this section is to show that addition of the limiting amino acids induces economic savings in protein due to the increased efficiency with which protein is utilized. For this purpose maize with and without lysine and tryptophan addition was chosen, because of the available experimental results from human subjects. The calculations made are presented in Table 18.11.

It was assumed that intake for the unsupplemented and supplemented maize is the same, 300 gm which provide a nitrogen intake of 4.80 gm. The digestibility of this protein ($N \times 6.25$) is 83%, which indicates that of the ingested nitrogen 3.44 gm are absorbed

TABLE 18.11

THE ECONOMY OF AMINO ACID SUPPLEMENTATION

	Unsupplemented Maize	Lys and Try Supplemented
Maize intake (gm)	300.0	300.0
Nitrogen intake (1.6 gm N/100 gm)	4.80	4.80
True protein digestibility (%)	82.0	82.0
Absorbed nitrogen (gm)	3.44	3.44
Fecal nitrogen (gm)	1.36	1.36
Biological value (%)	45.0	75.0
Retained nitrogen (gm)	1.55	2.58
Nitrogen loss in urine (gm)	1.89	0.86
Total nitrogen loss (gm)	3.25	2.22
Total nitrogen loss in terms of maize (gm)	203.0	139.0
Maize utilization (%)	32.3	53.7
Cost of maize lost	16.2	11.1
Cost of maize ingested (8¢/kg)	24.0	24.0

for the 2 types of maize. Fecal nitrogen is then equal to 1.36 gm. The biological value of unsupplemented maize is 45% which with the supplements of lysine and tryptophan increases to 75%. Since biological value represents the amount of protein retained from that which was absorbed, in the maize example it means a nitrogen retention of 1.55 gm from unsupplemented maize and 2.58 gm of nitrogen from supplemented maize. From these figures it was calculated a urine nitrogen loss of 1.89 gm and 0.86 gm from each material. Total nitrogen loss is equivalent to 3.25 gm and 2.22 gm, respectively. These losses are equivalent to 203 gm of unsupplemented maize and 139 gm for the supplemented. From the figures it was calculated that only 32.3% of the unsupplemented maize was used while from the supplemented maize 53.7% was used. If costs are included in these calculations, it becomes evident that supplementation represents not only better protein nutrition for the individual, but a significant economic saving as well, even if the supplemented maize has a 30-35% increased price.

BIBLIOGRAPHY

ALEXANDER, J. C., and HILL, D. C. 1952. The effect of heat on the lysine and methionine in sunflower seed oil meal. J. Nutr. 48, 149-159.
ALLISON, J. B. 1949. Biological evaluation of proteins. Advan. Protein Chem. 5, 155.
ALLISON, J. B. 1955. Biological evaluation of proteins. Phys. Rev. 35, 664-700.

ALLISON, J. B., and FITZPATRICK, W. H. 1960. Dietary proteins in health and diseases. *In* Dietary Proteins in Health Diseases. C. C. Thomas (Publisher), Springfield, Illinois.
ALLISON, J. B., WANNEMACHER, R. W., JR., and McCOY, J. R. 1960. The determination of the nutritive value of cottonseed flour. *In* Cottonseed Protein for Animal and Man. Proc. Conf. Cottonseed Protein for Animal and Man., New Orleans.
ALMQUIST, H. J. 1949. Amino acid balance at super-normal dietary levels. Proc. Soc. Exptl. Biol. Med. *72*, 179-180.
ALMQUIST, H. J. 1952. Amino acid requirements of chickens and turkeys. A review. Poultry Sci. *31*, 966.
ALMQUIST, H. J., and GRAU, C. R. 1944. Mutual supplementary effect of the proteins of soybean and sesame meals. Poultry Sci. *23*, 341-343.
ALMQUIST, H. J., and MERRIT, J. B. 1950. Protein and arginine levels in chick diets. Proc. Soc. Exptl. Biol. Med. *73*, 136.
BAKER, D. H. *et al.* 1969. Lysine imbalance of corn protein in the growing pig. J. Animal Sci. *28*, 23-26.
BARNES, L. A., KAYE, R., and VALYASEVI, A. 1961. Lysine and potassium supplementation of wheat protein. Am. J. Clin. Nutr. *9*, 331-344.
BECKER, D. E. *et al.* 1957. The isoleucine requirements of weanling swine fed two protein levels. J. Animal Sci. *16*, 26.
BENDER, A. E. 1958. Nutritive value of bread protein fortified with amino acids. Science *127*, 874-875.
BENDETT, E. P., WOOLRIDGE, R. L., STEFFEE, C. H., and FRAZIER, L. E. 1950. Studies in amino acid utilization. IV. The minimum requirements of the indispensable amino acids for maintenance of the adult well-nourished male albino rat. J. Nutr. *40*, 335-350.
BENTON, D. A., HARPER, A. E., and ELVEHJEM, C. A. 1955. Effect of isoleucine supplementation on the growth of rats fed zein or corn diets. Arch. Biochem. Biophys. *57*, 13-19.
BENTON, D. A., HARPER, A. E., SPIVEY, H. E., and ELVEHJEM, C. A. 1956. Leucine, isoleucine and valine relationships in the rat. Arch. Biochem. Biophys. *60*, 147-155.
BERRY, T. H. *et al.* 1962. The limiting amino acids in soybean protein. J. Animal Sci. *21*, 558-561.
BERRY, T. H., COMBS, G. E., WALLACE, H. D., and ROBBINS, R. C. 1966. Response of the growing pig to alterations in the amino acid pattern of isolated soybean protein. J. Animal Sci. *25*, 722-728.
BLESSIN, C. W., CAVINS, J. F., and INGLETT, G. E. 1971. Lysine infused popcorn. Cereal Chem. *48*, 373-377.
BLESSIN, C. W., INGLETT, G. E., CAVINS, J. F., and DEATHERAGE, W. L. 1970. Lysine fortification of dent corn. Cereal Sci. Today *15*, 375-394.
BLOCK, R. J., and MITCHELL, H. H. 1946-1947. The correlation of the amino acid composition of proteins with their nutritive value. Nutr. Abstr. Rev. *16*, 249-278.
BRAHAM, J. E., MADDALENO, V., BRESSANI, R., and JARQUÍN, R. 1965. Effect of cooking and supplementation with amino acids on the nutritive value of the protein of Pigeon pea (Cajanus indicus). Arch. Venezolanos Nutr. *15*, 19-32.
BRESSANI, R. 1963. Effect of amino acid imbalance on nitrogen retention. II. Interrelationships between methionine, valine, isoleucine, and threonine as supplements to corn protein for dogs. J. Nutr. *79*, 389-394.
BRESSANI, R. 1965. The use of cottonseed protein in human foods. Food Technol. *19*, 51-58.
BRESSANI, R. 1968. The use of yeast in human foods. *In* Single Cell Protein. R. I. Mateles and S. R. Tannenbaum (Editors). MIT Press, Cambridge, Mass.
BRESSANI, R. 1969. Formulation and testing of weaning and supplementary foods containing oilseed proteins. *In* Protein-Enriched Cereal Foods for

World Needs. M. Milner (Editor). American Association of Cereal Chemists, St. Paul, Minn.

BRESSANI, R. 1971. Application of food science and technology to the exploitation of new protein sources. *In* Proceedings of the Western Hemisphere Nutrition Congres III. P. L. White (Editor). American Medical Association, Chicago.

BRESSANI, R. 1972A. Complementary amino acid patterns. *In* Proceedings of the Symposium on Proteins in Processed Foods. American Medical Association, Chicago.

BRESSANI, R. 1972B. Prospects for other foods. *In* Nutritional Improvement of Maize. R. Bressani, J. E. Braham, and M. Béhar (Editors). INCAP conf. proc., Guatemala, C. A. INCAP Publ. *L-3*.

BRESSANI, R., and ELÍAS, L. G. 1967. Supplementation of oats with amino acids. Arch. Latinoamer. Nutr. *17*, 149-163.

BRESSANI, R., and ELÍAS, L. G. 1968. Processed vegetable mixtures for human consumption in developing countries. *In* Advances in Food Research, Vol. 16. C. O. Chichester, E. M. Mrak, and G. F. Stewart (Editors). Academic Press, New York.

BRESSANI, R., ELÍAS, L. G., and BRAHAM, J. E. 1966. Cottonseed protein in human foods. *In* World Protein Resources. Advan. Chem. Ser. *57*.

BRESSANI, R., ELÍAS, L. G., and BRAHAM, J. E. 1968. Supplementation of maize and tortilla with amino acids. Arch. Latinoamer. Nutr. *18*, 123-234.

BRESSANI, R., ELÍAS, L. G., JARQUÍN, R., and BRAHAM, J. E. 1964. All-vegetable protein mixtures for human feeding. XIII. Effect of cooking mixtures containing cottonseed flour on free gossypol content. Food Technol. *18*, 95-99.

BRESSANI, R., ELÍAS, L. G., SCRIMSHAW, N. S., and GUZMÁN, M. A. 1962B. Nutritive value of Central American corns. VI. Varietal and environmental influence on the nitrogen, essential amino acid and fat content of ten varieties. Cereal Chem. *39*, 59-67.

BRESSANI, R., ELÍAS, L. G., and VALIENTE, A. T. 1963D. Effect of cooking and of amino acid supplementation on the nutritive value of black beans. Brit. J. Nutr. *17*, 69-78.

BRESSANI, R., and MARENCO, E. 1963. Corn flour supplementation. The enrichment of lime-treated corn flour with proteins, lysine and tryptophan and vitamins. J. Agr. Food Chem. *11*, 517-522.

BRESSANI, R., and MERTZ, E. T. 1958A. Relationship of protein level to the minimum lysine requirement of the rat. J. Nutr. *65*, 481-492.

BRESSANI, R., and MERTZ, E. T. 1958B. Studies on corn proteins. IV. Protein and amino acid content of different corn varieties. Cereal Chem *35*, 227-235.

BRESSANI, R., SCRIMSHAW, N. S., BÉHAR, M., and VITERI, F. 1958. Supplementation of cereal protein with amino acids. II. Effect of amino acid supplementation of corn-masa at intermediate levels of protein intake on the nitrogen retention of young children. J. Nutr. *66*, 501.

BRESSANI, R., VALIENTE, A. T., and TEJADA, C. E. 1962A. All-vegetable protein mixtures for human feeding. VI. The value of combinations of lime-treated corn and cooked black beans. J. Food Sci. *27*, 394-400.

BRESSANI, R., WILSON, D. L., BÉHAR, M., and SCRIMSHAW, N. S. 1960B. Supplementation of cereal proteins with amino acids. III. Effect of amino acid supplementation of wheat flour as measured by nitrogen retention of young children. J. Nutr. *70*, 176-186.

BRESSANI, R. *et al.* 1960A. Nitrogen and essential amino acid content of different selections of maize. Arch. Venezolanos Nutr. *10*, 85-100.

BRESSANI, R. *et al.* 1963A. Supplementation of cereal proteins with amino acids. IV. Lysine supplementation of wheat flour fed to young children at different levels of protein intake in the presence and absence of other amino acids. J. Nutr. *79*, 333.

BRESSANI, R. et al. 1963B. Supplementation of cereal proteins with amino acids. V. Effect of supplementation of lime-treated corn with different levels of lysine, tryptophan, and isoleucine on the nitrogen retention of young children. J. Nutr. *80*, 80.
BRESSANI, R. et al. 1963C. Supplementation of cereal proteins with amino acids. VI. Effect of amino acid supplementation of rolled oats as measured by nitrogen retention of young children. J. Nutr. *81*, 399-404.
BRESSANI, R. et al. 1971. Effect of amino acid supplementation of white rice fed to children. Arch. Latinoamer. Nutr. *21*, 347-360.
BRINEGAR, M. J. et al. 1950. The lysine requirement for the growth of swine. J. Nutr. *42*, 129-138.
BUSS, L. W., and GODDARD, V. R. 1948. Effect of heat upon the nutritive value of peanuts. I. Protein quality. Food Res. *13*, 506-511.
BUTTERWORTH, H. M., and FOX, H. C. 1963. The effect of heat treatment on the nutritive value of coconut meal and the prediction of nutritive value by chemical methods. Brit. J. Nutr. *17*, 445-452.
CALLOWAY, D. H., and SPECTOR, H. 1954. Nitrogen balance as related to calorie and protein intake in active young men. Am. J. Clin. Nutr. *2*, 405-412.
CAMA, H. R., and MORTON, R. A. 1950. Changes occurring in the proteins as a result of processing ground nuts under selected industrial conditions. 2. Nutrition changes. Brit. J. Nutr. *4*, 297-316.
CARPENTER, K. J. 1960. The estimation of available lysine in animal protein foods. Biochemistry 77, 604-610.
CARTER, F. L., CIRINO, V. O., and ALLEN, L. E. 1961. Effect of processing on the composition of sesame seed and meal. J. Am. Oil Chemists' Soc. *38*, 148-150.
CAVINS, J. F., BLESSIN, C. W., and INGLETT, G. E. 1972. Infusion of grain sorghum with lysine, methionine and tryptophan. Cereal Chem. *49*, 605-608.
CLARK, H. E., HOWE, J. M., MERTZ, E. T., and REITZ, L. L. 1959. Lysine in baking powder biscuits. J. Am. Dietet. Assoc. *35*, 469-471.
CONCKERTON, E. J., and FRAMPTON, V. L. 1959. Reaction of gossypol with free E-amino groups of lysine in proteins. Arch. Biochem. *81*, 130-134.
CREMER, H. D., LANG, K., HUBBER, I., and CULIK, V. 1951. Improvement of the biological value of wheat protein with lysine or yeast and a comparison with oat protein. Biochem. Z. *322*, 58-67. (German)
CULIK, R., and ROSENBERG, H. R. 1958. The fortification of bread with lysine. IV. The nutritive value of lysine-supplemented bread in reproduction and lactation studies with rats. Food Technol. *12*, 169-174.
DAVIDSON, J., and BOYNE, A. W. 1962. Methionine and lysine supplementation of ground nut meal in experimental diets for laying hens. Brit. J. Nutr. *16*, 541-549.
DELIE, I., BOKOVOV, T., SRECKOVIC, A., and NIKILIC, M. 1963. Biological value of sunflower oilmeal as a protein feed for fattening pigs. Stocarstvo *17*, 464-480.
DESHPANDE, P. D., HARPER, A. E., and ELVEHJEM, C. A. 1958. Amino acid imbalance and nitrogen retention. J. Biol. Chem. *230*, 335-342.
DESHPANDE, P. D., HARPER, A. E., QUIROS-PEREZ, F., and ELVEHJEM, C. A. 1955. Further observations on the improvement of polished rice with protein and amino acid supplements. J. Nutr. *57*, 415-428.
EBISUZAKI, K., WILLIAMS, J. N., JR., and ELVEHJEM, C. A. 1952. A study of a possible mechanism of the threonine. J. Biol. Chem. *198*, 63-69.
EHLE, S. R., HAUSE, N. L., and PAUL, D. R. 1959. Taste and texture evaluation of lysine-supplemented baked products. Cereal Sci. Today *4*, 74.
ELÍAS, L. G., and BRESSANI, R. 1970. Protein quality of torula yeast and its value as complement to protein concentrates. Arch. Latinoamer. Nutr. *20*, 135-149. (Spanish)

ELÍAS, L. G., and BRESSANI, R. 1971. Amino acid and protein supplementation of defatted cottonseed flour. Arch. Latinoamer. Nutr. *21*, 149-167. (Spanish)

ELÍAS, L. G., and BRESSANI, R. 1972. Nutritional value of the protein of tortilla flour and its improvement by fortification in Central America. In Nutritional Improvement of Maize. R. Bressani, J. E. Braham, and M. Béhar (Editors). INCAP Conf. Proc., Guatemala, C.A. INCAP Publ. *L-3*.

ELÍAS, L. G., SÁNCHEZ LOARCA, S., and BRESSANI, R. 1969. Comparative study of different methods for the evaluation of the protein quality of cottonseed flour. Arch. Latinoamer. Nutr. *19*, 279-297. (Spanish)

ELVEHJEM, C. A. 1956. Amino acid imbalance. Federation Proc. *15*, 965-970.

ERICSON, L. E. 1960. Studies on the possibilities of improving Swedish wheat bread. 2. The effect of supplementation with lysine, threonine, methionine, valine and tryptophan. Acta Physiol. Scand. *48*, 295-301.

ERICSON, L. E., and LARSON, S. 1962. The loss of added lysine during the baking of a soft bread with an exceptionally high content of reducing sugars. Acta Physiol. Scand. *55*, 64-73.

ERICSON, L. E., LARSON, S., and LID, G. 1961. The loss of added lysine and threonine during the baking of wheat bread. Acta Physiol. Scand. *53*, 85-98.

EVANS, R. J., and McGINNIS, J. 1946. The influence of autoclaving soybean oilmeal on the availability of cystine and methionine for the chick. J. Nutr. *31*, 449-461.

FERREL, R. E., FELLERS, D. A., and SHEPHERD, A. D. 1969. Determination of free lysine and methionine in amino acid-fortified wheat. Cereal Chem. *46*, 614-620.

FERREL, R. E., SHEPHERD, A. D., and GUADAGNI, D. G. 1970. Storage stability of lysine in lysine-fortified wheat. Cereal Chem. *47*, 33-37.

FINKS, A. J., JONES, D. B., and JOHNS, C. V. 1922. The role of cystine in the dietary properties of the proteins of the cowpea (*Vigna sinensis*) and of the field peas (*Pisum sativum*). J. Biol. Chem. *52*, 403-410.

FINLEY, J. W., FERREL, R. E., and FELLERS, D. A. 1972. Determination of added lysine in fortified wheat and bulgur. Cereal Chem. *49*, 514-521.

FISHER, H. 1965. Unrecognized amino acid deficiencies of cottonseed protein for the chick. J. Nutr. *87*, 9-12.

FISHER, H. 1969. The amino acid deficiencies of blood meal for the chick. Poultry Sci. *47*, 1478-1481.

FISHER, H., and SHAPIRO, R. 1961. Amino acid imbalance: rations low in tryptophan, methionine or lysine and the efficiency of utilization of nitrogen in imbalance rations. J. Nutr. *75*, 395-401.

FLODIN, N. W. 1953. Amino acids and proteins. Their place in human nutrition problems. Agr. Food Chem. *1*, 222-235.

FLODIN, N. W. 1956. The philosophy of amino acid fortification of foods. Cereal Sci. Today *1*, 165.

FLORES, M. 1961. Food patterns in Central America and Panama. In Tradition Science and Practice in Dietetics. Proc. 3rd Intern. Congr. Dietet. Nm. Byles and Sons, Bradford, Yorkshire, Great Britain.

FLORES, M., and GARCÍA, B. 1960. The nutritional status of children of preschool age in the Guatemalan community of Amatitlán. I. Comparison of family and child diets. Brit. J. Nutr. *14*, 207-215.

FOMON, S. J., and FILER, L. L. 1967. Amino acid requirements of normal growth. In Amino Acid Metabolism and Genetic Variation. W. L. Nyan (Editor). McGraw-Hill Book Co., New York.

FORBES, R. M., VAUGHAN, L., and NORTON, H. W. 1955. Studies on the utilization of dietary isoleucine by the growing albino rat. I. Isoleucine requirements determined with amino acid mixtures. J. Nutr. *57*, 593-598.

GATES, J. C., and KENNEDY, B. M. 1964. Protein quality of bread and bread ingredients. J. Am. Dietet. Assoc. *44*, 374-377.

GESSERT, C. F., and PHILLIPS, P. H. 1956. Adverse effects of some amino acid supplements in low protein diets for growing dogs. J. Nutr. 58, 423-431.

GILLESPIE, G. T., FLYNN, L. M., O'DELL, B. L., and HOGAN, A. G. 1958. Nicotinic acid, lysine, tryptophan and threonine as supplements to high-protein corn. Missouri Agr. Exptl. Sta. Res. Bull. 679.

GIOK, L. T., SAMSUDIN, H., and TARWOTJO, I. 1967. Nutritional value of rubber seed protein. Am. J. Clin. Nutr. 20, 1300-1303.

GÓMEZ BRENES, R., ACEVEDO GONZÁLEZ, C. E., and BRESSANI, R. 1974. Improvement of the nutritive value of maize by means of lysine and tryptophan infusions. Arch. Latinoamer. Nutr. 24, 243-262. (Spanish)

GÓMEZ BRENES, R., and BRESSANI, R. 1973. A method for amino acid determination applicable to supplementation, plant improvement and nutritional biochemistry problems. Arch. Latinoamer. Nutr. 23, 445-464. (Spanish)

GRAHAM, R. P., MORGAN, A. I., JR., HART, M. R., and PENCE, J. W. 1968. Mechanics of fortifying cereal grains and products. Cereal Sci. Today 13, 224-253.

GRAU, C. R., and ALMQUIST, H. J. 1944. Sesame protein in chick diets. Proc. Soc. Exptl. Biol. Med. 57, 187-189.

GRAU, C. R., and ALMQUIST, H. J. 1945. Value of sunflower seed protein. Proc. Soc. Exptl. Biol. Med. 60, 373-374.

GRAU, C. R., and KAMEI, M. 1950. Amino acid imbalance and the growth requirements for lysine and methionine. J. Nutr. 41, 89-101.

GUNTHARDT, H., and McGINNIS, J. 1957. Effect of nitrogen fertilization on amino acids in whole wheat. J. Nutr. 61, 167-176.

HARPER, A. E. 1956. Amino acid imbalance, toxicities and antagonisms. Nutr. Rev. 14, 225-227.

HARPER, A. E. 1958. Balance and imbalance of amino acids. Ann. N.Y. Acad. Sci. 69, 1025-1041.

HARPER, A. E. 1959. Sequence in which amino acids of casein become limiting for the growth of the rat. J. Nutr. 67, 109-122.

HARPER, A. E., BENTON, D. A., and ELVEHJEM, C. A. 1955. L-leucine and isoleucine antagonist in the rat. Arch. Biochem. Biophys. 57, 1-12.

HARPER, A. E., and DE MUELENAERE, H. J. H. 1963. The nutritive value of cereal proteins with special reference to the availability of amino acids. Proc. 5th Intern. Congr. Biochem. 8, 82-107.

HARPER, A. E., LEUNG, P., YOSHIDA, A., and ROGERS, Q. R. 1964. Some new thoughts on amino acid imbalance. Federation Proc. 23, 1087-1092.

HARPER, A. E., WINJE, M. E., BENTON, D. A., and ELVEHJEM, C. A. 1955. Effect of amino acid supplements on growth and fat deposition in the livers of rats fed polished rice. J. Nutr. 56, 187-198.

HARTSOOK, E. W., and MITCHELL, H. H. 1956. Effect of age on the protein and methionine requirements of the rat. J. Nutr. 60, 173-195.

HENRY, K. M., and KON, S. K. 1957. Effect of level of protein intake and age of rat on the biological value of proteins. Brit. J. Nutr. 11, 305-313.

HEYWANG, B. W., and BIRD, H. R. 1950. Supplements for cottonseed meal in diets for chickens. Poultry Sci. 29, 486-495.

HILL, D. C., and OLSEN, E. M. 1963. Effect of the addition of imbalanced amino acid mixtures to a low protein diet, on weight gain, and plasma amino acids of chicks. J. Nutr. 79, 296-302.

HILL, D. C., and OLSEN, E. M. 1965. Weight gain and plasma free lysine of chicks fed an imbalanced amino acid mixture. Poultry Sci. 44, 596-601.

HISCHKE, H. H., JR., POTTER, G. C., and GRAHAM, W. R., JR. 1968. Nutritive value of oat protein. I. Variated differences as measured by amino acid analysis and rat growth response. Cereal Chem. 45, 374-378.

HOFFMAN, W. S., and McNEIL, G. C. 1949. The enhancement of the nutritive value of wheat gluten by supplementation with lysine, as determined from nitrogen balance indices in human subjects. J. Nutr. 38, 331-343.

HOGAN, A. G. et al. 1955. The percentage of protein in corn and its nutritional properties. J. Nutr. 57, 225-235.
HOLT, L. E., JR., and SNYDERMAN, S. E. 1961. The amino acid requirements of infants. J. Am. Med. Assoc. 175, 100-103.
HOWE, E. E., GILFILLAN, E. W., and MILNER, M. 1965A. Amino acid supplementation of protein concentrates as related to the world protein supply. Am. J. Clin. Nutr. 16, 321-326.
HOWE, E. E., JANSEN, G. R., and ANSON, M. L. 1967. An approach toward the solution of the world food problems with special emphasis on protein supply. Am. J. Clin. Nutr. 20, 1134-1147.
HOWE, E. E., JANSEN, G. R., and GILFILLAN, E. W. 1965B. Amino acid supplementation of cereal grains as related to the world food supply. Am. J. Clin. Nutr. 16, 315-320.
HUANG, P. C., YOUNG, V. R., CHOLAKOS, B., and SCRIMSHAW, N. S. 1966. Determination of the minimum dietary essential amino acid-to-total nitrogen ratio for beef protein fed to young men. J. Nutr. 90, 416-422.
HUNDLEY, J. M., SANDSTEAD, H. R., SAMPSON, A. G., and WHEDON, G. D. 1957. Lysine, threonine and other amino acid imbalance. Am. J. Clin. Nutr. 5, 316-326.
HUNTER, I. R., FERREL, R. E., and HOUSTON, D. F. 1956. Free amino acids of fresh and aged parboiled rice. J. Agr. Food Chem. 4, 874-875.
HUTCHINSON, J. B., MORAN, T., and PACE, J. 1958. Bread and the growth of weanling rats: the lysine-threonine balance. Nature (London), 181, 1733-1734.
HUTCHINSON, J. B., MORAN, T., and PACE, J. 1959. The nutritive value of bread protein as influenced by the level of protein intake, the level of supplementation with L-lysine and L-threonine, and the addition of egg and milk proteins. Brit. J. Nutr. 13, 151-163.
JAFFÉ, W. G. 1949. Limiting essential amino acids of some legumes. Proc. Soc. Exptl. Biol. Med. 71, 398-399.
JANSEN, G. R. 1962. Lysine in human nutrition. J. Nutr. 76, 1-35.
JANSEN, G. R. DIMAIO, L. R., and HAUSE, N. L. 1962. Cereal proteins. Amino acid composition and lysine supplementation of teff. J. Agr. Food Chem. 10, 62-64.
JANSEN, G. R., EHLE, S. R., and HAUSE, N. L. 1964A. Studies on the nutritive loss of supplemental lysine in baking. Food Technol. 18, 367-372.
JANSEN, G. R., EHLE, S. R., and HAUSE, N. L. 1964B. Studies on the nutritive loss of supplemental lysine in baking. Food Technol. 18, 372-375.
JENNESKENS, P. J. 1969. Lysine enrichment of bulgur. Cereal Sci. Today 14, 186-188.
KAUMITZ, H., SLANETZ, C. A., and JOHNSON, R. E. 1955. Dietary casein level and B-factor deficiencies produced by antagonists. Science 122, 1017-1018.
KIES, C. V., and LINKSWILER, H. M. 1965. Effect of nitrogen retention on men of altering the intake of essential amino acids with total nitrogen held constant. J. Nutr. 85, 139-144.
KIES, C., WILLIAMS, E., and FOX, H. M. 1965A. Determination of first limiting nitrogenous factor in corn protein for nitrogen retention in human adults. J. Nutr. 86, 350-356.
KIES, C., WILLIAMS, E., and FOX, H. M. 1965B. Effect of "non-specific" nitrogen intake on adequacy of cereal proteins for nitrogen retention in human adults. J. Nutr. 86, 357-361.
KIHLBERG, R., and ERICSON, L. E. 1964. Amino acid composition of rye flour and the influence of amino acid supplementation of rye flour and bread on growth, nitrogen efficiency ratio and liver fat in the growing rat. J. Nutr. 82, 385-394.
KING, W. H., KUCK, J. C., and FRAMPTON, V. L. 1962. Lysine, gossypol

and nitrogen solubility in chemically treated cottonseed meals. J. Am. Chem. Soc. *39*, 58-60.
KLOSE, A. A., and FEVOLD, H. L. 1944. Methionine deficiency in yeast protein. Proc. Soc. Exptl. Biol. Med. *56*, 98-101.
KNIPFEL, J. E. 1969. Comparative protein quality of triticale, wheat and rye. Cereal Chem. *46*, 313-317.
KOHLER, G. O. 1966. Safflour, a potential source of protein for human food. *In* World Protein Resources. R. F. Gould (Editor). Advan. Chem. Ser. Am. Chem. Soc.
KRETOVICH, V. L., and PONOMAREVA, A. N. 1961. Amino acid participation in melanoidin formation in bread making. Biokhimiya *26*, 237-242. (Russian)
KRISHNAMURTHY, K. *et al*. 1960. Studies on the nutritive value of sesame seeds and meal. Ann. Biochem. Exptl. Med. (Calcutta) *20*, 73-76.
KUMTA, U. S., and HARPER, A. E. 1960A. Amino acid balance and imbalance. III. Quantitative studies of imbalance in diets containing fibrin. J. Nutr. *70*, 141-146.
KUMTA, U. S., and HARPER, A. E. 1960B. Sequence in which indispensable amino acids become limiting for growth of rats fed diets low in fibrin. J. Nutr. *71*, 310-316.
KUMTA, U. S., HARPER, A. E., and ELVEHJEM, C. A. 1958. Amino acid imbalance and nitrogen retention in adult rats. J. Biol. Chem. *233*, 1505-1508.
LEVEIVILLE, G. A., SAUBERLICH, H. E., and SHOCKLEY, J. W. 1962. Protein value and the amino acid deficiencies of various algae for growth of rats and chicks. J. Nutr. *76*, 423-428.
LEVERTON, R. M., and STEEL, D. 1962. Nitrogen balances of young women fed the FAO reference pattern of amino acids and the oat pattern. J. Nutr. *78*, 10-14.
LLOYD, L. E., and CRAMPTON, E. W. 1961. Effect of protein level, amino acid supplementation and duration of feeding of a dry early-weaning pig ration. J. Am. Sci. *20*, 172-175.
LYMAN, C. M. 1953. Cottonseed meals by chick growth and by chemical index method. Evaluation of protein quality. J. Nutr. *49*, 679-690.
MALEKI, M., and DJAZAYERI, A. 1968. Effect of baking and amino acid supplementation of the protein quality of Arabic bread. J. Sci. Food Agr. *19*, 449-451.
MANGAY, A. S., PEARSON, W. N., and DARBY, W. J. 1957. Millet (*Setaria italica*): its amino acid niacin content and supplementary nutritive value for corn (maize). J. Nutr. *62*, 377-393.
MARCH, B. E., BIELY, J., and YOUNG, R. J. 1950. Supplementation of meat scrap with amino acids. Poultry Sci. *29*, 444-449.
MARTINEZ, W. H., FRAMPTON, V. L., and CABELL, C. A. 1961. Effects of gossypol and raffinose on lysine content and nutritive quality of proteins in meals from glandless cottonseed. J. Agr. Food Chem. *9*, 64-66.
MATTHEWS, R. H., RICHARDSON, G., and LICHTENSTEIN, H. 1969. Effect of lysine fortification on quality of chapatties and yeast bread. Cereal Chem. *46*, 14-21.
MAURON, J. 1961. The concept of amino acid availability and its bearing on protein evaluation. *In* Progress in Meeting Protein Needs of Infants and Preschool Children. Proceedings of International Congress, Washington, D.C., Natl. Res. Council Publ. *843*.
MAURON, J., MOTTA, F., BUJARD, E., and EGLI, R. H. 1955. The availability of lysine, methionine and tryptophan in condensed milk and milk powder "*in vitro*," digestion studies. Arch. Biochem. Biophys. *59*, 433-451.
McGINNIS, J., HSU, P. T., and CARVER, J. S. 1948. Nutritional deficiencies of sunflower seed oil meal for chicks. Poultry Sci. *27*, 389-393.

MCOSKER, D. E. 1962. The limiting amino acid sequence in raw and roasted peanut protein. J. Nutr. *76*, 453-459.

MERTZ, E. T., BATES, L. S., and NELSON, O. E. 1964. Mutant gene that changes protein composition and increases lysine content of maize endosperm. Science *145*, 279-280.

MILNER, M. (Editor). 1969. Protein-Enriched Cereal Foods for World Needs. American Association of Cereal Chemists, St. Paul, Minn.

MITCHELL, H. H. 1924. The nutritive value of proteins. Physiol. Rev. *4*, 424.

MITCHELL, H. H. 1947. Protein utilization by the adult rat: The lysine requirement. Arch. Biochem. *12*, 293-300.

MITCHELL, H. H. 1950. Some species and age differences in amino acid requirements. In Protein and Amino Acid Requirements of Mammals. A. A. Albanese (Editor). Academic Press, New York.

MITCHELL, H. H. 1959. Some species and age differences in amino acid requirements. In Protein and Amino Acid Nutrition. A. A. Albanese (Editor). Academic Press, New York.

MITCHELL, H. H., and BLOCK, R. J. 1946. Some relationships between the amino acid contents of proteins and their nutritive values for the rat. J. Biol. Chem. *163*, 599.

MITCHELL, H. H., and SMUTS, D. B. 1932. The amino acid deficiencies of beef, wheat, corn, oats, and soybeans for growth in the white rat. J. Biol. Chem. *95*, 263-281.

MOLINA, M. et al. 1972. The technology of maize fortification in Latin America. In Nutritional Improvement of Maize. Conf. INCAP Proc., Guatemala, C.A. R. Bressani, J. E. Braham, and M. Béhar (Editors). INCAP Publ. *L-3*, 235-255.

MORAN, T. 1959. Nutritional significance of recent work on wheat flour and bread. Nutr. Abstr. Rev. *29*, 1-16.

MORIMOTO, T. 1966. Studies on free amino acids in sponges, doughs, and baked soda crackers and bread. J. Food Sci. *31*, 736-741.

NAKAGAWA, I., TAKAHASHI, T., and SUZUKI, T. 1960. Amino acid requirements of children. J. Nutr. *71*, 176-181.

NAKAGAWA, I., TAKAHASHI, I., and SUZUKI, T. 1961A. Amino acid requirements of children: Isoleucine and leucine. J. Nutr. *73*, 186-190.

NAKAGAWA, I., TAKAHASHI, I., and SUZUKI, T. 1961B. Minimal needs of lysine and methionine based on nitrogen balance method. J. Nutr. *74*, 401-407.

NAKAGAWA, I., TAKAHASHI, T., SUZUKI, T., and KOBAYASHI, K. 1962. Amino acid requirements of children: Minimal needs of threonine, valine and phenylalanine based on nitrogen balance method. J. Nutr. *77*, 61-68.

NAKAGAWA, I., TAKAHASHI, T., SUZUKI, T., and KOBAYASHI, K. 1965. Amino acid requirements of children: Quantitative amino acid requirements of girls based on nitrogen balance method. J. Nutr. *86*, 333-336.

NASSET, E. S., and ANDERSON, J. T. 1951. Nitrogen balance index in the adult rat as affected by diets low in L- or DL-methionine. J. Nutr. *44*, 237-247.

NASSET, E. S., ANDERSON, J. T., and SILICIANO, A. M. 1951. Nitrogen balance of adult rats fed diets low in L and DL-threonine. J. Nutr. *45*, 173-182.

NASSET, E. S., and ELY, M. T. 1952. Nitrogen balance of adult rat fed diets low in L and DL-lysine or devoid of arginine. J. Nutr. *48*, 391-400.

NEUBERGER, A., and WEBSTER, T. A. 1945. The lysine requirements of the adult rat. Biochem. J. *39*, 200-202.

OSBORNE, T. B., and MENDEL, L. B. 1914. Amino acids in nutrition and growth. J. Biol. Chem. *17*, 325.

OSBORNE, T. B., and MENDEL, L. B. 1919. The nutritive value of wheat kernel and its milling products. J. Biol. Chem. *37*, 557-601.

OSBORNE, T. B., and MENDEL, L. B. 1920. Nutritive value of the proteins of the barley, oat, rye and wheat kernels. J. Biol. Chem. *41*, 275-306.

OSER, B. L. 1951. Method for integrating essential amino acid content in the nutritional evaluation of protein. J. Am. Dietet. Assoc. *27*, 246.

PARTHASARATHY, H. N. *et al.* 1964. The effect of supplementing a rice diet with lysine, methionine, and threonine on the digestibility coefficient, biological value and on the retention of nitrogen in children. Can. J. Nutr. *42*, 385-393.

PATWARDHAN, V. N. 1962. Pulses and beans in human nutrition. Am. J. Clin. Nutr. *11*, 12-30.

PECORA, L. J., and HUNDLEY, J. M. 1951. Nutritional improvement of white polished rice by the addition of lysine and threonine. J. Nutr. *44*, 101-112.

PEREIRA, S. M., ALMAS BEGUM, JESUDIAN, G., and SUNDARARAJ, R. 1969. Lysine-supplemented wheat and growth of school children. Am. J. Clin. Nutr. *22*, 606-611.

PHELPS, R. A. 1966. Cottonseed meal for poultry; from research to practical application. World's Poultry Sci. *22*, 86-112.

PLANELLA, I., OWEN, D. F., SPILLER, G., and CHICHESTER, C. O. 1968. Effect of different processes on the protein quality of cereals enriched with amino acids. Nutr. Bromatol. Toxicol. *7*, 65-72. (Spanish)

POND, W. G., HILLIER, J. C., and BENTON, D. A. 1958. The amino acid adequacy of milo (grain sorghum) for the growth of rats. J. Nutr. *65*, 493-502.

PRICE, W. A., JR., TAYLOR, M. W., and RUSSELL, W. C. 1953. The retention of essential amino acids by the growing chick. J. Nutr. *51*, 413-422.

REBER, E. F., WHITEHAIR, C. K., and McVICAR, M. 1953. The effect of level of protein fed baby pigs. J. Nutr. *50*, 451-458.

RICHARDSON, L. R. 1948. Southern peas and other legume seeds as a source of protein for the growth of rats. J. Nutr. *36*, 451-462.

ROSE, W. C. 1937. The nutritive significance of the amino acids and certain related compounds. Science *86*, 298-300.

ROSE, W. C. 1938. The nutritive significance of the amino acids. Physiol. Rev. *18*, 109-136.

ROSE, W. C. 1949. Amino acid requirements of man. Federation Proc. *8*, 546-552.

ROSE, W. C., COON, M. J., and LAMBERT, C. F. 1954. The amino acid requirements of man. VI. The role of caloric intake. J. Biol. Chem. *210*, 331-342.

ROSE, W. C., HAINES, W. J., WARNER, D. T., and JOHNSON, J. E. 1951. The amino acid requirements of man. II. The role of threonine and histidine. J. Biol. Chem. *188*, 49-58.

ROSE, W. C., JOHNSON, J. E., and HAINES, W. J. 1950. The amino acid requirements of man. I. The role of valine and methionine. J. Biol. Chem. *182*, 541-556.

ROSE, W. C., OESTERLING, M. J., and WOMACK, M. J. 1948. Comparative growth on diets containing ten and nineteen amino acids, with further observations upon the role of glutamic and aspartic acids. J. Biol. Chem. *176*, 753-762.

ROSE, W. C., SMITH, L. C., WOMACK, M., and SHANE, M. 1949. The utilization of the nitrogen of ammonium salts, urea and certain other compounds in the synthesis of non-essential amino acids *in vivo*. J. Biol. Chem. *181*, 307-316.

ROSENBERG, H. R. 1959. Amino acid supplementation of foods and feeds. *In* Protein and Amino Acid Nutrition. A. A. Albanese (Editor). Academic Press, New York.

ROSENBERG, H. R., and BALDINI, J. T. 1957. Effect of dietary protein level

on the methionine-energy relationship in broiler diets. Poultry Sci. *36*, 247-252.
ROSENBERG, H. R., and CULIK, R. 1957. The improvement of the protein quality of white rice by lysine supplementation. J. Nutr. *63*, 477-487.
ROSENBERG, H. R., CULIK, R., and ECKERT, R. E. 1959. Lysine and threonine supplementation of rice. J. Nutr. *69*, 217-228.
ROSENBERG, H. R., and ROHDENBURG, E. L. 1951. The fortification of bread with lysine. I. The loss of lysine during baking. J. Nutr. *45*, 593-598.
ROSENBERG, H. R., and ROHDENBURG, E. L. 1952. The fortification of bread with lysine. II. The nutritional value of fortified bread. Arch. Biochem. Biophys. *37*, 461-468.
ROSENBERG, H. R., ROHDENBURG, E. L., and ECKERT, R. E. 1960. Multiple amino acid supplementation of white corn meal. J. Nutr. *72*, 415-422.
ROSENFIELD, D., and BERNSTON, B. L. 1971. Economics and technology of cereal fortification. *In* Symposium: Seed Proteins. G. E. Inglett (Editor). Avi Publishing Co., Westport, Conn.
RUBENTHALER, G., POMERANZ, Y., and FINNEY, K. F. 1963. Effects of sugars and certain free amino acids on bread characteristics. Cereal Chem. *40*, 658-665.
RUILOBA, M. H. 1973. Rice supplementation with synthetic grains and by means of amino acid infusions. M.S. Thesis. Postgraduate course Food Sci. Technol. Animal Nutr., INCAP, Guatemala, C.A. (Spanish)
RUSSELL, W. C., TAYLOR, M. W., MEHRHOF, T. G., and HIRSCH, R. R. 1946. The nutritive value of the protein of varieties of legumes and the effect of methionine supplementation. J. Nutr. *32*, 313-325.
SABISTON, A. R., and KENNEDY, B. M. 1957. Effect of baking on the nutritive value of proteins in wheat bread with and without supplements of nonfat dry milk and of lysine. Cereal Chem. *34*, 94-110.
SALMON, W. D. 1954. The tryptophan requirement of the rat as affected by niacin and level of dietary nitrogen. Arch. Biochem. Biophys. *51*, 30-41.
SANAHUJA, J. C., and HARPER, A. E. 1963. Effect of dietary amino acid pattern on plasma amino acid pattern and food intake. Am. J. Physiol. *204*, 686-690.
SAUBERLICH, H. E. 1956. Amino acid imbalances as related to methionine, isoleucine, threonine and tryptophan requirements of the rat or mouse. J. Nutr. *59*, 353-370.
SAUBERLICH, H. E., CHAN, W. U., and SALMON, W. D. 1953. The amino acid and protein content of corn as related to variety and nitrogen fertilization. J. Nutr. *51*, 241-250.
SCHNEIDER, B. H., and MILLER, D. F. 1954. The biological value of *Alaska pea* protein. J. Nutr. *52*, 581-590.
SCRIMSHAW, N. S., BRESSANI, R., BÉHAR, M., and VITERI, F. 1958. Supplementation of cereal proteins with amino acids. I. Effect of amino acid supplementation of corn masa at high levels of protein intake on the nitrogen retention of young children. J. Nutr. *66*, 485.
SCRIMSHAW, N. S. *et al.* 1966. Minimum dietary essential amino acid-to-total nitrogen ratio for whole egg protein fed to young men. J. Nutr. *89*, 9-18.
SENTI, F. R., and PENCE, J. W. 1971. Technological aspects of adding amino acids to foods. *In* Proc. Intern. Conf. Amino Acid Fortification Protein Foods, Mass. Inst. Technol., Cambridge.
SHERWOOD, F. W., WELDON, V., and PETERSON, W. J. 1954. Effect of cooking and of methionine supplementation on the growth promoting property of cowpeas (*Vigna sinensis*) protein. J. Nutr. *52*, 199-208.
SHOUP, F. K., DEYOE, C. W., CAMPBELL, J., and PARRISH, D. B. 1969. Amino acid composition and nutritional value of milled sorghum grain products. Cereal Chem. *46*, 164-171.
SHURPALEKAR, S. R., CHANDRASEKHARS, M. R., SWAMINATHAN, M.,

and SUBRAHMANYAN, V. 1961. Chemical composition and nutritive value of soybeans and soybean products. Food Sci. (Mysore) 2, 52-64.
SLINGER, S. J., HILL, D. C., GARTLEY, K. M., and BRANION, H. D. 1949. Soybean oil meal and sunflower seed oil meal in rations for Broad Breasted Bronze turkeys. Poultry Sci. 28, 534-540.
SMITH, A. K. 1961. Theories on improving the nutritional value of soybean meal. In Proc. Conf. Soybean Products for Protein in Human Foods, USDA Northern Regional Res. Lab. Peoria, Illinois.
SMITH, R. E., and SCOTT, H. M. 1965. Biological evaluation of fish meal proteins as sources of amino acids for the growing chick. Poultry Sci. 44, 394-400.
SNYDERMAN, S. E. et al. 1955. The phenylalanine requirement of the normal infant. J. Nutr. 56, 253-263.
SNYDERMAN, S. E. et al. 1959. The essential amino acid requirements of infants: Valine. Am. J. Diseases Children 97, 185-191.
STILLINGS, B. R., and HACKLER, L. R. 1965. Amino acid studies on the effect of fermentation time and heat-processing of tempeh. J. Food Sci. 30, 1043.
STILLINGS, B. R., HAMMERLE, O. A., and SNYDER, D. G. 1969. Sequence of limiting amino acids in fish protein concentrate produced by isopropyl alcohol extraction of red Hake (*Vrophyeis chuss*). J. Nutr. 97, 70-78.
STRØMNAES, A. S., and KENNEDY, B. M. 1957. Effect of baking on the nutritive value of proteins in rye bread and without supplements of non-fat dry milk and of lysine. Cereal Chem. 34, 196-200.
STUCKI, W. P., and HARPER, A. E. 1962. Effect of altering the ratio of indispensable to dispensable amino acids in diets for rats. J. Nutr. 78, 278-286.
SURE, B. 1953. Protein efficiency. Improvement in whole yellow corn with lysine, tryptophan and threonine. J. Agr. Food Chem. 1, 626-629.
SURE, B. 1954. Relative nutritive value of proteins in whole wheat and whole rye and effect of amino acid supplements. J. Agr. Food Chem. 2, 1108-1110.
SURE, B. 1955. Effect of amino acids and vitamin B_{12} supplements on the biological value of proteins in rice and wheat. J. Am. Dietet. Assoc. 31, 1232-1234.
TANG, J. J. N., LANDICK, L. L., and BENTON, D. A. 1958. Studies on amino acid supplementation and amino acid availability with oats. J. Nutr. 66, 533-543.
TARA, K. A., and BAINS, G. S. 1969. Test for checking distribution of lysine impregnated grains in fortified bread. Cereal Sci. Today 14, 152-153.
TRUSWELL, A. S., and BROCK, J. F. 1961. Effects of amino acid supplements on the nutritive value of maize protein for human adults. Am. J. Clin. Nutr. 9, 715-728.
TWINING, P. F., ROMOSER, G. L., and COMBS, G. F. 1955. Studies on methionine requirements of growing chicks. Poultry Sci. 34, 1225-1226.
VENKAT RAO, S., JOSEPH, A. A., SWAMINATHAN, M., and PARPIA, H. A. B. 1964. Amino acid supplementation as a means of improving the quality and overcoming shortage of protein in developing countries. J. Nutr. Dietet. India 1, 192-200.
WAGGLE, D. H., DEYOE, C. W., and SANFORD, P. E. 1967. Relationship of protein level of sorghum grain to its nutritive value as measured by chick performance and amino acid composition. Poultry Sci. 46, 655-659.
WAGGLE, D. H., PARRISH, D. B., and DEYOE, C. W. 1966. Nutritive value of protein in high and low protein content sorghum grain as measured by rat performance and amino acid assay. J. Nutr. 88, 370-374.
WILLIAMS, H. H. et al. 1954. Estimation of growth requirements for amino acids by assay of the carcass. J. Biol. Chem. 208, 277-286.
WISSLER, R. W. et al. 1948. Studies in amino acid utilization. III. The role of

the indispensable amino acids in maintenance of the adult albino rat. J. Nutr. *36*, 245-262.
WOMACK, H., and ROSE, W. C. 1941. The practical replacement of dietary methionine by cystine for purposes of growth. J. Biol. Chem. *141*, 375-379.
WOMACK, M., HARLEN, H. A., and LIN, P. H. 1953. The influence of the non-essential amino acids on the requirement of the adult rat for isoleucine, methionine and threonine. J. Nutr. *49*, 513-526.
WOODS, E., BEESON, W. M., and BOLIN, D. W. 1943. Field peas as a source of protein for growth. J. Nutr. *26*, 327-335.
YOUNG, V. R., and ZAMORA, J. 1968. Effects of altering the proportions of essential to non-essential amino acids on growth and plasma acid levels in the rat. J. Nutr. *96*, 21-27.
ZAGHI, S. DE, and BRESSANI, R. 1969. Use of Central American food resources for the improvement of animal industry. II. Chemical composition of sesame seed and sesame meal (*Sesamum indicum*). Turrialba *19*, 34-38. (Spanish)

PART 2
Addition of Vitamins and Minerals to Foods

Benjamin Borenstein

The addition of vitamins and minerals to foods has long been viewed primarily as a public health issue and many attempts to establish policy on this subject have been made by the U.S. Food and Drug Administration, the Council on Foods and Nutrition of the American Medical Association, and the Food and Nutrition Board of the National Academy of Sciences—National Research Council.

Policy statements and proposed regulations re food fortification have shown an evolutionary change in the past decade indicating a better understanding of the diversity and changeability of the problems facing the consumer and the food industry.

To the extent that fortification is a public health issue, e.g., bread fortification, a regulated unified approach is essential even though it can slow progress, as in the argument from 1970 to 1974 on the simple issue of increasing the level of iron in Enriched White Bread from 15 mg per lb to 25 mg per lb.

The policy makers, however, must establish their domain boundaries and not prevent addition of vitamins and minerals to foods for valid reasons exclusive of public health. The Council on Foods and Nutrition (1968) and the Food and Nutrition Board published a joint policy statement which said in part:

> The Council on Foods and Nutrition and the Food and Nutrition Board endorse the addition of nutrients to foods when in keeping with all of the following circumstances:
> 1. The intake of the nutrient(s) is below the desirable level in the diets of a significant number of people.
> 2. The food(s) used to supply the nutrient(s) is likely to be consumed in quantities that will make a significant contribution to the diet of the population in need.
> 3. The addition of the nutrient(s) is not likely to create an imbalance of essential nutrients.
> 4. The nutrient(s) added is stable under proper conditions of storage and use.
> 5. The nutrient(s) is physiologically available from the food.
> 6. There is reasonable assurance against excessive intake to a level of toxicity.

The above criteria ignore the fact that a marketer of a new food should not have to prove that a nutrient is deficient in the diet of a significant number of people since his new product will partially replace existing foods in the marketplace and thus will lower the available supply of nutrients if it is not "nutritionally designed." A second point, these criteria make no provision for the restoration of micronutrients lost in food processing. In a recent review, presumably written by a nutritionist (Anon. 1973), the food technologist is criticized for not giving more attention to nutritional quality in process development and a request is made that attention be given to nutrient retention as influenced by alternative processing procedures. "It is strongly recommended that studies on nutrient retention include all of the nutrients for which that food item is considered a good or superior source." This laudable approach cannot be faulted, but it can be broadened to include restoration. Leaching of water-soluble vitamins, for example, in blanching, washing, and brine grading of vegetables may have no economically viable processing alternative and restoration can be an inexpensive way to rectify the nutritional deficits involved in many food processing operations.

For example, in the case of the nutrition quality guidelines suggested by FDA (1971) for frozen convenience dinners, the cost of adding the 5 critical micronutrients (as indicated by FDA) to 10,000 dinners would be approximately $0.70 as compared to the cost of $70.00 for a single assay for these 5 compounds to monitor their content after processing.

A third point, excluded from the 1968 joint policy statement, is fortification with anticaries agents as proposed by Harris (1970):

> It is reasonable and practical to control caries in human populations by fortifying selected foods with anticaries agents that are tasteless, colorless, nontoxic, and effective. . . . Cereals and cereal products, which are consumed daily by large population groups, and sucrose and candies, which are cariogenic, are examples of the types of foods to be considered for fortification. Fluorides, phosphates, and trace minerals are typical of the cariostatic agents that might be added.

The fact that formulated foods, new by-product ingredients, and snack foods have a significant place in the U.S. diet and present special nutrition problems has been recognized by the nutrition community and the Food and Nutrition Board issued a revised statement on the improvement of nutritive quality of foods in 1973 (Food and Nutrition Board 1973). It discusses the restoration issue as follows:

> . . . . It is, of course, desirable that foods having undergone some form of post harvest processing contain the variety and amounts of essential nutrients that were present at significant levels in the raw, preprocessed food. A significant level of a nutrient is considered to be 5 percent or more of the RDA of that nutrient in an average serving. Restoration of nutrients lost in the processing of foods has been recognized for many years as an appropriate method for improving their nutritional value. Under circumstances where restoration is undertaken through the addition of nutrients as specific chemical entities, it is important that all essential nutrients present at significant levels in the original preprocessed foods be restored to the extent that is technically feasible. The nutrient level achieved through restoration should be 1.0–1.5 times preprocessed levels. The principle of extensive restoration is of special importance when a label claim for the content of an added nutrient is made.

This 1973 policy statement further recognizes the development of new and formulated foods:

> New foods can imitate common foods in appearance, texture, flavor, and odor. Examples are foods resembling dairy products, fruit juices and meats. Other such foods do not physically resemble conventional foods. Another type of product is produced primarily to serve as a meal replacement. Because these products are replacing foods that make significant nutrient contributions, their nutritional value should at least equal the foods replaced.
>
> The composition of a new or formulated food becomes especially important when an average serving of the product it imitates or replaces contributes 5 percent or more of the recommended daily allowance of any essential nutrient or energy. . . .

The FDA issued a major policy statement on food fortification and enrichment in 1974 (FDA, 1974, General Principles Governing the Addition of Nutrients to Foods; Federal Register *39*, 116, 20900–20904, June 14). Under these proposals, if a food product does not have a Standard of Identity or a Nutritional Guideline, it can only be fortified by the principle of restoration or the principle of nutrient density. Since these principles may not become final for several years and may be greatly modified, they will not be discussed further here.

In 1973 the FDA adopted a simplified form of the Recommended Dietary Allowances of the Food and Nutrition Board (in place of the Minimum Daily Requirements established by FDA in 1941) as the official way to express nutrient potency in labeling (Table 18.12).

TABLE 18.12

U.S. RDA (RECOMMENDED DIETARY ALLOWANCES)

Vitamins and Minerals	Unit of Measurement	Infants	Children Under 4 Yr of Age	Adults and Children 4 or More Years of Age	Pregnant or Lactating Women
Vitamin A	International Units	1500	2500	5000	8000
Vitamin D	International Units	400	400	400	400
Vitamin E	International Units	5	10	30	30
Vitamin C	Milligrams	35	40	60	60
Folic acid	Milligrams	0.1	0.2	0.4	0.8
Thiamin	Milligrams	0.5	0.7	0.5	1.7
Riboflavin	Milligrams	0.6	0.8	1.7	2.0
Niacin	Milligrams	8	9	20	20
Vitamin B-6	Milligrams	0.4	0.7	2.0	2.5
Vitamin B-12	Micrograms	2	2	6	8
Biotin	Milligrams	0.05	0.15	0.30	0.30
Pantothenic acid	Milligrams	3	3	10	10
Calcium	Grams	0.6	0.8	1.0	1.3
Phosphorus	Grams	0.5	0.8	1.0	1.3
Iodine	Micrograms	45	70	150	150
Iron	Milligrams	15	10	18	18
Magnesium	Milligrams	70	200	400	450
Copper	Milligrams	0.6	1.0	2.0	2.0
Zinc	Milligrams	5	8	15	15

This is part of a major policy and rule-making procedure on nutrition labeling of foods issued in several parts by FDA (1973A,B,C). It establishes minimum information which must be supplied if any nutritional claims are made, a mandatory label format for claims, compliance procedures for claims of both fortified and unfortified foods, and reference molecular weights for the vitamins. This last point may seem obvious and unnecessary and yet, even here, there was an apparent disagreement between experts. The reference compound for vitamin B-1 in the January 1973 proposal was thiamin chloride, MW 300.8. In the August 1973 revision (FDA 1963C), the reference compound was changed to thiamin chloride hydrochloride, MW 337.3, a substantial difference to one calculating the vitamin content of foods. Table 18.13 lists the reference compounds chosen in the final version.

TABLE 18.13

VITAMIN REFERENCE FORMS

Vitamin	Name	Mol Wt
Vitamin C	L-Ascorbic acid	176.12
Folic acid	Pteroylmono-L-glutamic acid	441.41
Niacin	Nicotinic acid	123.11
Riboflavin	Riboflavin	376.37
Thiamin	Thiamin chloride hydrochloride	337.28
Vitamin B-6	Pyridoxine	169.18
Vitamin B-12	Cyanocobalamin	1355.40
Biotin	D-Biotin	244.31
Pantothenic acid	D-Pantothenic acid	219.23

Source: Federal Register, Aug. 2, 1973.

These regulations also establish Standards of Identity for vitamin and mineral supplements and this will affect food fortification practices. Supplements must contain from 50 to 150% U.S. RDA per dose and a food product which is fortified within this range must comply with these regulations on supplements unless the food has a Standard of Identity. In some cases, this will encourage processors to fortify at 45% U.S. RDA or less per serving to avoid falling into the supplement category.

Relatively little has changed with respect to fortification of standardized foods over the years. Flour, bread, corn grits, and corn meal are enriched with vitamins B-1, B-2, niacin, plus iron. The margarine standard was amended by FDA in 1973 and it is now mandatory that margarine contain not less than 15,000 units vitamin A per pound. The addition of vitamin D at 1500 units per pound is optional. Canned applesauce may contain 60 mg vitamin C per 4 oz. As of 1968, nonfat dry milk may be fortified to contain 500 units vitamin A and 100 units vitamin D per 8 fl oz reconstituted product.

Recently, foods without standards of identity have been fortified with more diverse micronutrients. Breakfast cereals have been fortified with as many as ten vitamins. Some breakfast cereals are marketed as cereal forms of vitamin supplements containing 100% MDR per serving (100% U.S. RDA of 9 mandatory vitamins per serving under 1973 Standard of Identity established for multivitamin supplements). Breakfast drinks, which are in effect replacements for orange juice, are fortified with vitamin C. Textured vegetable protein, standardized by USDA specifications (U.S. Dept. of Agr. 1971) for use with ground beef in the school lunch program, is fortified with B-1, B-2, niacin, B-6, calcium pantothenate, B-12, and iron. Instant breakfast products designed to be mixed with milk contain added vitamins and minerals and are designed to be complete meals.

All these newer applications have increased the interest in methodology of fortification and stability of added nutrients. Publications on specific applications are available. Vitamin A fortification of tea has been solved (Brooke and Cort 1972). Juice drink fortification has been reviewed (Bunnell 1968). The development of a nutritionally designed baked product with cream filling for school feeding is discussed by Cotton et al. (1971). Cereal product fortification was reviewed recently (Borenstein 1974) and a general review of fortification problems is available (Borenstein 1971).

One of the newest and most important applications, vegetable protein, has received little literature attention and will, therefore, be discussed here.

Textured soy protein has been fortified commercially both before and after granulation or "texturization" primarily due to individual preferences among the processors as to production ease and control. The two major concerns are uniformity of distribution of the added micronutrients and compliance with label claims (USDA specifications, Table 18.14). A convenient way to monitor fortification and improve physical distribution in soy fortification is to premix all the added micronutrients and thus add from 50 to 100 mg of premixed ingredients per 100 gm of soy protein instead of 0.4 mg of B-1, etc., one at a time. This premix can then be added directly to the soy flour before granulation either by batch mixing, continuous metering or with the "dough water" if the production process and equipment lends itself to this approach. Assuming this method is followed, the major concern is stability during processing. The short processing time requiring heat, pressure, and moisture to texturize does not significantly degrade the more labile vitamins—B-1, B-12, and pantothenic acid—in this system. Although it is generally advisable (Borenstein 1971) to add micronutrients after processes involving heat stress, in the case of soy fortification it does not appear necessary.

The natural micronutrient content of soy protein is high and it is

TABLE 18.14

TEXTURED VEGETABLE PROTEIN SPECIFICATIONS
(FNS 219)

	Min	Max
Protein[1] (wt %)	50.0	—
Fat (wt %)	—	30.0
Magnesium (mg/100 gm)	70.0	—
Iron (mg/100 gm)	10.0	—
Thiamin (mg/100 gm)	0.30	—
Riboflavin (mg/100 gm)	0.60	—
Niacin (mg/100 gm)	16.0	—
Vitamin B-6 (mg/100 gm)	1.4	—
Vitamin B-12 (mcg/100 gm)	5.7	—
Pantothenic acid (mg/100 gm)	2.0	—

Source: U.S. Dept. of Agr. (1971).

[1] $N \times 6.25$.

Note: All values are expressed on the dry basis and are applicable to dry or hydrated forms of the product. Moisture content of the hydrated form shall not exceed 65.0%, or be less than 60.0%.
The protein efficiency ratio, PER, of the textured vegetable protein shall be not less than 1.8 on basis of PER = 2.5 for casein. PER of a meat-textured vegetable protein combination shall be not less than 2.5.

desirable to determine the levels which can be conservatively calculated to be present and adjust the fortification addition levels accordingly. For example, soy protein products contain well over the 70 mg magnesium per 100 gm required in the USDA specification and there is no scientific reason to add any magnesium to comply with the specification. Similarly, soy contains substantial levels of iron—7-13 mg per 100 gm soy protein products—according to soy processors.

It is theoretically necessary, therefore, to add only 3 mg iron per 100 gm of soy to ensure compliance with FNS 219 (Food and Nutrition Service, USDA). In practice, since available data are probably inadequate to make a statistical assessment, it is more logical to add 5-6 mg iron per 100 gm product and be well over the label claim than to add a marginal amount and run expensive control assays.

In all fortification projects an input or overage above the label claim is essential to ensure compliance after processing and storage. Even in the case of the most stable vitamin, niacin, an overage of approximately 10% is required because of imperfect distribution in its addition to most products and inherent analytical errors in assaying the food. The less stable vitamins, as discussed in Chap. 2, require higher overages.

There is probably more published on iron with respect to fortification than the other minerals put together, but in fact, the problems of calcium fortification are probably the most difficult to solve. The major problem is the excessive weight of calcium salts required to supply a meaningful percentage of the U.S. RDA. The adult U.S. RDA for calcium is 1000 mg and more than 3 gm of the commonly-used calcium sources (calcium phosphate dibasic and calcium phosphate tribasic) are required to obtain 1000 mg of calcium. The quantities required to obtain even 10% RDA per serving can cause serious organoleptic problems—sediment, opacity, chalky and metallic flavor, sandiness. Calcium may be the most difficult organoleptic vitamin and mineral fortification problem of all at this time.

The nutritionist is particularly interested in iron fortification because of diet inadequacies and concern about bioavailability of added iron compounds. Unfortunately, almost all the bioavailability work on iron salts is in animal studies using model feeds so that little is known about the effect of processing and storage on the chemistry and, hence, bioavailability of iron added to baked goods, frozen dinners, canned foods, or other processed foods. Limited work indicates that in specific cases the iron source is of limited importance since an equilibrium Fe^{++}/Fe^{+++} ratio is reached regardless of whether ferrous sulfate, a nutritionally preferred source, or ferric orthophosphate, an inferior source, is used (Hodson 1970; Theuer et al. 1971). This is highly significant since soluble ferrous salts are organoleptically and technologically unsuitable in many foods. Ferrous sulfate is unsuitable in fluid milk (20 mg iron per quart) due to serious off-flavors (Borenstein 1971). At 10 mg iron per quart ferrous sulfate causes oxidized flavor after 2 days' storage and catalyzes the destruction of vitamins E and A (Wang and King 1973).

Ferric orthophosphate is insoluble in milk and, hence, also unsuitable. In Hodson's study, ferric orthophosphate was largely dissolved and reduced to Fe^{++} during 2-5 months' storage of canned, milk-based, liquid complete foods, but ferrous sulfate was organoleptically unsuitable. Since several months usually elapses between processing and consumption of the products he studies, he concluded that all the organoleptic benefits resulting from using ferric orthophosphate in these products could be obtained without sacrificing the nutritional benefits attributed to ferrous compounds.

In a recent human study (Cook et al. 1973) on the absorption of iron compounds from bread, sodium iron pyrophosphate was 10% and ferric orthophosphate was 33% as effective as either ferrus sulfate or hydrogen reduced iron.

The food technologist cannot ignore the potential conflict between bioavailability and technological suitability in adding iron compounds to foods.

Since both iron and copper are excellent catalysts for ascorbic acid oxidation, there are products where simultaneous fortification with iron, copper, and ascorbic acid does not appear feasible. Ferric orthophosphate has been used commercially to fortify frozen orange drink concentrate without a significant effect on the stability of ascorbic acid or flavor shelf-life of the product. In one model system study, 0.85 ppm Cu^{++} accelerated ascorbic acid degradation 500%, pH 2.9 (Timberlake 1960).

Fortification of foods requires careful attention to food regulations, rationale, and technical limitations prior to extensive project investment.

BIBLIOGRAPHY

ANON. 1973. Nutritional quality and food product development. Nutr. Rev. *31*, 226-227.
BORENSTEIN, B. 1971. Rationale and technology of food fortification with vitamins, minerals and amino acids. Critical Rev. Food Technol. *2*, 171-186.
BORENSTEIN, B. 1974. Enrichment of wheat food products. In Wheat Production and Utilization. G. E. Inglett (Editor). Avi Publishing Co., Westport, Conn.
BROOKE, C. L., and CORT, W. M. 1972. Vitamin A fortification of tea. Food Technol. *26*, 50-52.
BUNNELL, R. H. 1968. Enrichment of fruit products and fruit juices. J. Agr. Food Chem. *16*, 177-183.
COOK, J. D. et al. 1973. Absorption of fortification iron in bread. Am. J. Clin. Nutr. *25*, 861-872.
COTTON, R. H. et al. 1971. Astrofood, a fortified baked product with creamed filling. Cereal Sci. Today *16*, No. 6, 188-189.
COUNCIL ON FOODS AND NUTRITION. 1968. Improvement of nutritive quality of foods. J. Am. Med. Assoc. *205*, 160-161.
FDA. 1971. Nutritional quality guidelines for foods. Federal Register. *36* (247), 24822-24824.
FDA. 1973A. Food labeling. Federal Register. *38* (13), 2124-2164, Jan. 19.
FDA. 1973B. Food labeling. Federal Register. *38* (49), 6950-6975, Mar. 14.
FDA. 1973C. Food labeling. Federal Register. *38* (148), 20702-20750, Aug. 2.
FOOD AND NUTRITION BOARD. 1973. General policies in regard to improvement of nutritive quality of foods. Natl. Acad. Sci.—Natl. Res. Council, Washington, D. C.
HARRIS, R. S. 1970. Fortification of foods and food products with anticaries agents. J. Dental Res. *49*, 1340-1344.
HODSON, A. Z. 1970. Conversion of ferric to ferrous iron in weight control dietaries. J. Agr. Food Chem. *18*, 946-948.
THEUER, R. C. et al. 1971. Effect of processing on availability of iron salts in liquid infant formula products, experimental soy isolate formulas. J. Agr. Food Chem. *19*, 555-558.
TIMBERLAKE, C. F. 1960. Metallic components of fruit juices. III. Oxidation and stability of ascorbic acid in model systems resembling black currant juice. J. Sci. Food Agr. *11*, 258-268.
U.S. DEPT. OF AGR. 1971. Textured vegetable protein products (B-1) to be used in combination with meat for use in lunches and suppers served under child feeding programs. FNS Notice *219*, Febr. 22.
WANG, C. F., and KING. R. L. 1973. Chemical and sensory evaluation of iron-fortified milk. J. Food Sci. *38*, 938-940.

CHAPTER 19

Olaf Mickelsen
and
D. D. Makdani

Factors Affecting Nutrient Metabolism[1]

The ultimate aim of most food production and processing is to supply human beings with nutritious and tasty food throughout most of the year. Admittedly, a certain segment of the food processing industry produces food for man's pets. Even here, an unknown fraction of that food is thought to find its way to the dinner plates of some individuals.

Too frequently, the ultimate consumer is lost sight of when questions arise as to the nutritional suitability of various foods for man's use. This situation arises partly from the fact that much of the experimental work in the field of nutrition has involved studies with laboratory animals. There are many reasons why animals were used so extensively in nutritional investigations. The use of animal models may increase in the whole realm of human health-related investigations since various regulations make it more and more difficult to carry out many types of studies with human subjects. In some respects, this continuing trend toward a greater and greater reliance on animal experimentation as a substitute for human studies is unfortunate. Should that continue, we may only recognize too late, the validity of Alexander Pope's admonition, "The proper study of mankind is man."

Studies with animals are an important part of the overall evaluation of the changes that may occur in foods as a result of processing procedures. The results of animal assays provide a basis for comparing the relative nutritional value of different foods. However, the results of such animal experiments cannot be applied, in many cases, directly to human beings. This stems from the fact that there are marked differences in the quantitative nutrient requirements of man and animals. These differences arise from the rapidity with which all laboratory animals grow and the relative slowness with which an infant increases in size. For instance, most nutritional investigations have utilized the weanling laboratory rat. Such an animal, when fed a nutritionally good ration, can double its body weight in five days. On the other hand, a newborn infant which, in development, would

---

[1] Published as Journal Article 7155 from the Michigan Agricultural Experiment Station.

be comparable to the weanling rat doubles its weight only after 6-8 weeks or more. The very rapid rate of growth of the young laboratory rat is possible only when its ration contains a high concentration of the amino acids and other nutrients essential for the rapid formation of body tissue. The requirement of the rapidly growing rat for a highly concentrated source of essential nutrients in its feed means that whatever food promotes good growth in that animal should provide all the nutrients required by the human infant in more than adequate concentrations.

PROTEINS

Although the nutritional requirements of the weanling rat are high, there is a decrease in both the number and quantity of amino acids required after the animal matures. This means that if a food protein could not support growth in the weanling rat, it might maintain body weight in adult animals. An extension of that argument might be that if the protein food did not support growth in the weanling rat or maintain body weight in the adult animal, then it should not be used in the human diet. If that type of reasoning were accepted, a number of foods that have been major components of the human diet would be contra-indicated. The fallacy of that argument is well illustrated by the studies which have attempted to evaluate the biological value of the protein in commercial white wheat flour (white flour).

The incorporation of wheat flour into a ration as the only source of protein will not promote growth in the weanling rat while the adult rat loses weight when fed such a ration. This is true even when the ration is supplemented with the necessary vitamins, minerals, and essential fatty acids (Bolourchi 1963). However, when that white flour ration was supplemented with small amounts of lysine, growth of the weanling rats resulted and the PER (protein efficiency ratio) was markedly improved. On that basis, many nutrition scientists have recommended that wheat flour be supplemented with lysine to make it a nutritious food for man's use. According to such advocates, the need to supplement flour with lysine is due to the extent to which flour has been refined. The milling industry has been accused of shunting the more nutritious fraction of the wheat kernel into animal feed.

Despite the implied poor nutritional quality of wheat flour protein, it supplies an adequate amount of the essential amino acids for man's needs. This became evident when 12 college males were fed controlled diets for 70 days (Bolourchi *et al.* 1968A). During a control period of three weeks, each subject was fed a diet which provided

70 gm of protein per day from a variety of foods. The subjects were in nitrogen equilibrium throughout that period. The control period was followed by a 50-day experimental period when the diet provided 70 gm of protein with 90-95% coming from white flour; there was no animal protein in the experimental diets. As soon as the experimental period began, the subjects developed a negative nitrogen balance. That negative nitrogen balance was fairly large for the first 10 days; thereafter, it rapidly became positive and for the last 30 days of the experimental period, enough nitrogen was retained to compensate for the loss in the early phase of that period.

This study provides evidence that the results of nutritional experiments with animals cannot always be directly applied to man. Wheat flour as the primary source of protein in a diet may produce a negative nitrogen balance during the first 14 days after the initiation of the wheat diet. Furthermore, the results of that study with the college men suggest that the experimental period must be much longer than has been the case in many previous reports. When the period is sufficiently long to ensure the establishment of an equilibrium state, the conclusions may not be the same as those secured in the early phase of the study. Although this dictum appears to be self evident, too frequently, it has been honored in the breach. Many investigators have catered to the whims of their subjects and thereby lulled themselves into believing that an equilibrium state was established after 5-7 days (for review, see Vaghefi *et al.* 1974). Had that been the duration of the MSU Bread Study, the inevitable conclusion would have been that the protein in wheat would not provide an adequate amount of the essential amino acids needed for nitrogen equilibrium in normal young men.

Not only will the protein in wheat flour provide all the essential amino acids for adult males, but there are suggestions that this is also true for the growing child. The latter is based on calculations of the extent to which the amino acid requirements of a 7-yr-old child will be met when 80% of its caloric requirement is provided by wheat flour (Hegsted 1962). Hegsted's calculations are based on the report that the 7-yr-old child needs 0.71 gm of lysine per day; U.S. commercial flour in an amount to provide 80% of the calories recommended for that child would contribute 1.7 gm of lysine or 2.4 times the daily requirement. This means that the remaining 20% of the caloric intake need provide no essential amino acids. In other words, the protein in wheat makes an addition to the amino acid needs of human beings which is more than adequate on the basis of its caloric contribution.

It was calculations such as these, made independently in our laboratory, which led to the MSU Bread Study. When animal assays indicated that bread as the sole source of protein in the ration would not support the growth of weanling rats, estimates were made of the amount of lysine that an adult human subject would secure from U.S. commercial wheat flour. These calculations (Table 19.1) were

TABLE 19.1

COMPARISON OF THE EXTENT TO WHICH BREAD PROVIDES THE AMINO ACIDS REQUIRED BY ADULT HUMAN BEINGS AND GROWING RATS

Amino Acid	In Bread		For Man (Gm/Day)		For Rat (Gm/Day)	
	Gm/16Gm N^1	Gm/100Gm^2	Needed^3	Intake^4	Needed^5	Intake^6
Arginine	4.3	0.41		2.59	0.045	0.033
Histidine	2.5	0.24		1.51	0.080	0.019
Lysine	2.8	0.27	0.8	1.69	0.138	0.022
Tyrosine	3.3	0.31	1.1	1.99		0.025
Tryptophan	0.8	0.08	0.25	0.48	0.045	0.006
Phenylalanine	4.6	0.44	0.30	2.77	0.138	0.035
Cystine	3.0	0.29	0.81	1.81		0.023
Methionine	1.9	0.18	0.2	1.15	0.090	0.015
Threonine	2.4	0.23	0.5	1.45	0.080	0.018
Leucine	4.8	0.46	1.1	2.89	0.138	0.037
Valine	3.7	0.35	0.8	2.23	0.115	0.029
Glutamic Acid	29.0	2.76		17.47		0.223
Glycine	3.1	0.30		1.87		0.024
Alanine	3.0	0.29		1.81		0.023
Isoleucine	3.9	0.37	0.7	2.35	0.090	0.030

^{1} Block and Mandl (1958).
^{2} The values for grams of amino acid per 100 gm of bread were secured by multiplying the values in column 1 by the factor 9.5/16 X 6.25. The 9.5 is the value for the protein content of fresh bread (polled commercial white bread secured from a number of cities in the United States) (Block and Mandl 1958).
^{3} Taken from H. H. Williams (1959).
^{4} The calculated amino acid intakes were based on a diet consisting of 70% bread and supplying 2500 calories. The amino acid intakes as listed represent only that which is secured from bread.
^{5} Taken from Albritton (1955). These figures represent the amino acids needed for growth by a 50-gm rat.
^{6} Based on the average feed consumption of the weanling rats fed the ration containing white bread without any milk powder as the sole source of calories. Adequate amounts of minerals and vitamins were added to more than cover the animals' requirements.

made on the assumption that the flour contributed 70% of the caloric needs which were set at 2500 kcal. Under such circumstances, the flour would supply 1.69 gm of lysine per day which was twice the 0.8 gm requirement.

These calculations suggest that, for man, the so-called limiting amino acid in wheat could just as well be tryptophan as lysine. This is based on the estimate that the tryptophan in flour exceeds the requirement by a factor of 1.9 whereas the lysine exceeds man's requirement by a factor of 2.1. Very likely, these differences are insignificant when one considers the problems involved in estimating the amino acid requirements of human subjects.

The data in Table 19.1 indicate very dramatically why wheat cannot promote growth in weanling rats when it is the sole source of amino acids. The concentrations of the essential amino acids in white flour are so low that even when practically the entire ration is composed of this product, the weanling rat secures only 15% of its lysine requirement therefrom, whereas man would obtain 210% of his daily needs for the same amino acid even when the flour in his diet provided only 70% of his caloric needs. Similar computations for the other essential amino acids indicate that the contribution of white flour to the rat's amino acid needs is so small that, from purely theoretical grounds, it can be concluded that commercial white flour (about 72% extraction) is of no value for rodents. In no case does the concentration of an essential amino acid in the feed consumed by a weanling rat meet its needs (Table 19.1).

Another factor that is frequently overlooked in the nutritional evaluation of foods when human subjects are used is their innate individual characteristics. This factor becomes especially important when only a few subjects are involved in a study. Under such circumstances, the experimental findings may be biased by the choice of subjects (Ahmad et al. 1975). In other words, when the same subjects have been involved in a number of studies, their relative ranking at the termination of each study, frequently, appears to be similar for the same nutritional parameters.

The fact that human subjects eat a varied diet is frequently overlooked when the results of animal studies are evaluated. This should instill a great deal of caution into any conclusions that might be made on the basis of animal assays. For various reasons, animal studies usually involve diets in which the nutrient under study comes from only one food. Experimental protocol demands that to evaluate the biological value of the protein in wheat flour, that should be the only component of the diet which provides any protein.

Although a food may appear to be of low biological value on the basis of animal assays, that may be of only minor significance for the nutritional health of the consumer. In addition to the limitations previously mentioned about animal assays, the fact that practically no individual secures all his food from only one source means that one food may supplement the deficiency in the other. The overall effect of the combined foods in the diet may adequately meet the nutritional needs of the consumer. This means that the contribution of the food to the diet as a whole should be considered. Admittedly, that is a horrendous task since human diets vary so greatly. Under such circumstances, the better approach is to use the results of the animal assay as a guide to changes that might be produced during processing.

The concern about the nutritional changes that might be produced by processing should not be the overriding factor in determining the procedures to be used. Obviously, safety from spoilage is one of the primary considerations in evaluating food processing techniques. Another factor, and one that receives inadequate attention from many nutrition scientists, is the taste of the finished product. Unless the food appeals to the individual's taste to such extent that it is eaten in reasonable amounts, it will make very little, if any, contribution to that person's nutritional well-being.

NUTRITION LABELING

To a great extent, the taste problem appears to have been ignored in the current discussions about nutrition labeling. The statements made by most individuals who are involved in this activity are that the consumer has a right to know what is in the food she purchases. It is further implied that when such information is available, the most nutritious foods will be selected. Such reasoning ignores a number of basic facts. One of these is that the consumer, as well as most nutritionists, will not know how the food under consideration should complement the other foods included in a day's diet. For instance, which tomato juice should the consumer purchase if there are two brands, the label on one of which indicates that a serving provides 15% of the daily vitamin A requirement and 50% of the vitamin C needs, while the label on another brand lists one serving as providing, for one day, 50% of vitamin A and 15% of the vitamin C requirement? Another shortcoming of that program is that a serving for many foods is listed as 1 cup which has been defined as 240 ml (U.S. Food and Drug Admin. 1973A). To consume that amount of any one food will be possible only by those whose appetite for that food is especially hearty. That conclusion is predicated on the assumption that the taste of the highly nutritious food will appeal to the individual. This naive assumption, that the majority of consumers will consistently eat those foods that are nutritious regardless of taste, is accepted by too many armchair specialists.

The recognition that consumers are motivated to purchase foods primarily by factors such as taste and relative cost places a responsibility on everyone involved in the production and processing of food for human consumption. From an ethical, moral, as well as economic standpoint, this is a responsibility that must be assumed by everyone involved in this important activity. Although some individuals maintain that if the consumer is provided with adequate nutritional information about the foods in the marketplace, she will choose those foods that are best; by implication this suggests that the least nutritious foods will disappear from the marketplace. Whether that will happen is open to a number of questions, not the least of which is the definition of those foods that are "non-nutritious."

For various reasons, increasing efforts are being made to provide the consumer with adequate amounts of all his nutrients in one serving of a food. This activity appears to emanate from the proposal that each meal must provide all the nutrients that might be needed by the body until the next meal is eaten. Partial responsibility for that philosophy can be attributed to those nutritionists who claim that only those meals are nutritionally adequate which provide a third of the Recommended Dietary Allowance (RDA) for each nutrient. Such pronouncements ignore the fact that there is no valid evidence that each meal must provide its appropriate fraction of the daily requirement for each nutrient.

NUTRIENT ADDITION TO FOODS

Some nutrients may be added to processed foods; the product is then referred to as being restored, enriched, or fortified. The term restoration requires the addition of nutrients to "foods in order to restore those nutrients that are present naturally, but have been destroyed or lost in processing" (AMA Council on Foods and Nutrition 1973). The quantity of the nutrients that may be added under such circumstances brings the levels in the processed food up to what they had been in the fresh or raw state. Enrichment refers to "the addition of more than one nutrient in conformity with a standard developed by the government ..." (AMA Council on Foods and Nutrition 1973). Examples of the latter are enriched bread, flour and cereal products. The term fortification is restricted to "the addition of one or more nutrients that were not present or were present in small amounts ..." in the original food prior to processing (AMA Council on Foods and Nutrition 1973).

The principles that should govern the addition of nutrients to foods have been enunciated by the Council on Foods and Nutrition of the American Medical Association (1973). They are: "(1) The intake of a nutrient considered for addition to food should be judged to be below a desirable level in the diets of a significant number of people. (2) The food that is to carry the nutrient should be consumed by the segment of the population in need, and the added nutrient should make an important contribution to the diet. (3) The addition of the nutrient should not create a dietary imbalance.

(4) The nutrient added should be stable under customary conditions of storage and use. (5) The nutrient should be physiologically available from the food. (6) There should be reasonable assurance that an excessive intake to a level of toxicity will not occur. (7) The additional cost should be reasonable for the consumer."

The increasing number of foods that are being fortified with a large variety of nutrients may not necessarily enhance the consumer's health. Actually, these efforts may produce a rebound effect which might have unforeseen repercussions among both consumers and processors. Such a situation might develop when a nutrient such as iron is added to foods to overcome the reported "high" incidence of iron-deficiency anemia. The addition of this nutrient to some foods may result in the formation, during processing, of unpleasant flavors and/or odors. If no adverse reactions occur during processing, they may develop while the food is stored prior to purchase. Either one or both of these developments may result in a reduced consumption of the food, which, obviously, would not alleviate the reported iron-deficiency anemia.

The importance of a conservative approach to the fortification of foods with iron was stressed by Dr. Russell Wilder, one of the organizers of the Food and Nutrition Board and one of the first chairmen of that Board. As far back as 1952, he claimed:

> Using standards as generally accepted, anemia undoubtedly is a prevalent defect, especially among women, but, again, we find ourselves in an embarrassment. It appears to be impossible to decide what proportion of this defect—if it is a defect—is clearly nutritional in origin. Unquestionably, some anemias can be corrected by nutritional therapy. However, the very common mild hypochromic, so-called iron deficiency anemia, which represents the major part of the problem, has responded poorly, if at all, to the increased iron intake resulting from enrichment with iron of white bread. We must entertain the thought, in my opinion, that much of this anemia is in part an artifact, attributable to the use of unrealistic hemoglobin standards, and that many of these so-called anemic individuals actually suffer no ill health from their hemoglobin levels.

In the long run, the indiscriminate addition of nutrients to foods will have only minor, if any, effect in improving the health of the consumer. This is especially true of the trace minerals which have come in for a great deal of attention within the last few years. Some of this concern that our foods may be deficient in trace minerals is associated with the fact that an increasing number of these elements are being proposed as essential for the health of both animals and man (Hopkins and Mohr 1971; Nielsen 1971; Schwarz 1971). These suggestions have raised questions in the minds of some people about the extent to which the essential trace elements are absorbed by plants grown in certain parts of the country. Since a number of these minerals do not appear essential for the growth and development of plants, a few individuals have expressed concern about the adequacy of these minerals in the American diet. To assuage these fears would require a tremendous amount of sophisticated analyses and statistical data. Even such information might not allay the suspicions existing in the minds of some individuals about the quality of the food commercially available. These concerns have been accentuated by reports (Gortner 1972) that the average American diet contains as much as 70 mg EDTA (ethylenediaminetetraacetic acid). In the gastrointestinal tract, that compound may chelate with a variety of minerals, thus making them unavailable. Besides EDTA, the diverse phosphorus compounds being added to foods may pose future nutritional problems involving the trace elements unless the effects of these additives are carefully evaluated before they are used. For these reasons, both the consumer groups and some food processors have chosen to overcome these "deficiencies" in the American diet by proposing the addition of various trace minerals to foods.

Today, it is unlikely that a trace mineral deficiency will develop due to its absence from our food. This is due to the fact that many of the foods appearing on our dining tables throughout the year have come from many different regions with some from farms thousands of miles away while many have been imported from various countries throughout the world. The diverse soil conditions in the areas where the food is raised should ensure a reasonable uptake of the minerals by the plants and an overall adequacy for the consumer. Furthermore, most individuals consume a variety of foods which usually differ from day to day. This is added ensurance that, except under most unusual circumstances, the consumer who incorporates a variety of foods into his meals should be provided with an adequate intake of trace minerals.

PHYTATES

It is becoming increasingly apparent that the concentration of some minerals in foods may be of only secondary importance as far as their availability to the human subject is concerned. The mineral that has received most study from that standpoint is zinc. The availability of zinc is influenced, to a large extent, by the presence of phytates in the diet. The interrelationship between these two substances is discussed in the minerals section. The reason for mentioning it here is twofold:

(1) To stress the fact that other foods in the meal may have a

marked influence on the nutritional availability of a nutrient. This means that the listing of a mineral in a food as a percentage of the U.S. RDA may be of little significance from an overall nutritional standpoint unless there is some indication that it will be absorbed after it is consumed. Obviously, that is almost impossible to estimate since the inclusion in the meal of foods that contain large amounts of phytates may decrease the availability of the zinc present therein, as well as the zinc that may be present in foods that are free of phytates. Such might be the case when foods rich in zinc become a part of a meal which includes plant foods rich in phytates.

(2) To suggest that dietary phytates may have been responsible for at least one of the nutritional enigmas still challenging us. That involves the explanation for the increase in the heights and weights of Japanese children following World War II. Studies in that country by both Japanese and American scientists produced no hypothesis for this change other than an unknown effect of the increased animal protein content of the postwar diet (Mitchell 1962). There was no indication that the caloric content of the prewar diets limited growth or that there was any amino acid deficiency. Since the body makes no distinction between the amino acids from one source versus those from another, it is difficult to understand how the increased amount of animal protein in the postwar diet could improve the nutritional condition purely from the standpoint of protein composition. A more rational explanation involves a decrease in the phytate and an increase in the zinc content of the postwar diets. The increased animal protein content of the latter diets may have been associated with a decrease in plant proteins. If that were so, there might well have been a reduction in the phytate content of the diet, since animal products contain no phytate. Furthermore, animal proteins usually have a higher concentration of zinc than plant products. These suggestions lead to the possibility that Japanese children prior to World War II were slightly deficient in zinc and that this condition was overcome by the dietary change brought about by the termination of the war. It should be pointed out that one of the earliest symptoms of a zinc deficiency is a loss of appetite. If the zinc deficiency were only a mild one, then it might have been possible for the prewar children to grow at a slower than "normal" rate but still sufficiently fast to achieve sexual maturity.

FIBER IN THE DIET

There is increasing evidence that the nutritional spotlight will be focused on nutrients other than those listed in the U.S. RDA. The dietary substance that is currently receiving increased attention is the so-called nondigestible fiber. Until recently, that component of the diet received only cursory attention by most nutrition scientists. True, it was recognized as being important in maintaining "regularity" but since its only contribution to the diet was its presumed role as an inert filler, it was given little consideration. Now, the impetus provided initially by Burkitt's reports (Burkitt 1973; Burkitt et al. 1974; Scala 1974; Trowell 1974) has catapulted dietary fiber into the political arena. That move was suggested when a consumer group requested a federal regulation that would require the addition of fiber to the foods sold in the United States (Anon. 1974).

THE ACID-BASE POTENTIALS OF FOODS

Another area that may receive increased attention in the not-too-distant future involves the effect of foods on the acid-base balance of the body. The importance of this long-neglected field is only now becoming apparent. That the acid-base situation in the body may modify the nutritional effect of foods is suggested by changes in the blood urea level when different diets are consumed. This became evident when normal young men in an MSU Bread Study were fed a diet in which 90–95% of the protein came from wheat (Bolourchi et al. 1968B). Within 36 hr after the men started to consume the wheat diet which contained no animal protein, their blood urea levels decreased by 50%. This reduction in urea level was based on the levels during the preceding control period when the diets provided the same total protein intake except that a generous portion of it was of animal origin. Throughout the entire study, the protein intake was maintained constant to avoid the alterations in blood urea levels shown by Addis et al. (1947) to be associated with changes in dietary protein intake.

A possible explanation for the reduction in blood urea levels when the subjects consumed the wheat diet was provided by the work of Lyons et al. (1931). These English investigators observed that children with a secondary renal infection associated with scarlet fever showed a return toward normal in their blood urea levels when they were given sodium bicarbonate. The reduction in urea levels was accompanied by an almost complete remission of the symptoms referrable to the kidney infection.

Preliminary results in our laboratory suggest that sodium bicarbonate also lowers the blood urea levels in normal subjects. This occurred when the same diet was consumed on 5 consecutive days by 3 subjects. At the end of that period, 5 gm of sodium bicarbonate was taken with each meal for the next 3 days. Although the diet was kept constant throughout the 8 days both as to kinds and amounts of food, the blood urea levels decreased in all 3 subjects when the sodium bicarbonate was taken. The latter converted the acid urine excreted during the control period to a definite alkalinity. Admittedly, the physiological and medical significance of this observation has not been established. However, it should be emphasized that the level of urea in the blood is one of the major clinical criteria used in evaluating the functional activity of the kidneys.

It is difficult to predict the future significance of these findings to the food processor. He has the potential of altering, to a certain extent, the degree to which foods affect the acid-base balance in the body. This can be brought about by the relative amounts of fat and carbohydrate, by the nature and amount of protein and by the kinds of salts (e.g., the types of phosphates) that are incorporated into the processed food. That such factors may become important is suggested by the extent to which acid-producing foods are currently being used both by normal individuals and those who are restricted to certain therapeutic diets. The increasing intake of animal protein by the average individual in the United States is one reason why the average diet is likely to produce an acidic condition in the body. The latter is due to the relatively large amounts of sulfur and phosphorus compounds in animal proteins. Metabolism of those proteins means that the sulfur and phosphorus must be excreted in the urine primarily as acids. Both of these acids require strong bases for their excretion. To conserve the body's supply of cations, the kidney excretes the sulfur and phosphorus compounds in the slightly acidic form.

The most frequently-used diets that produce an acidic condition in the body are those that contain large amounts of fat. Within the past three decades those diets have been used by large numbers of individuals who desired to lose considerable amounts of body weight within a short period of time. The high fat diets were able to achieve that goal since the acidic condition they produced in the body was associated with the loss of large amounts of water. High fat diets are again being used by physicians treating patients subject to epileptic attacks. This type of therapy is being practiced since the convulsions of some patients cannot be controlled by drugs (Lasser and Brush 1973).

SUCROSE

In the past, the cost of sucrose has been one of the primary factors determining the amount of that ingredient added to foods during processing. Within the near future, it is likely that other considerations may also influence that decision. One of these is the effect that sucrose is reported to have on dental caries. Considerable evidence suggests that the oxygen bridge between glucose and fructose in sucrose is essential for the production of dextrans and levulans by *Streptococcus mutans*. This is the bacterium believed to be largely responsible for dental caries in human beings. The polysaccharides produced by *S. mutans* form the protective surface on the dental plaque essential for cariogenic activity of the bacteria (Newbrun 1967).

At present, there does not appear to be any official effort to either limit the amount of sucrose added to processed foods nor to require the declaration of the amount present in the finished product (U.S. Food and Drug Admin. 1973B). However, under the existing political climate, it is difficult to know when that may change. The refusal of FDA to lift the ban on the use of cyclamates has accentuated the search for sweetening agents that can be used in foods. Although various artificial sweeteners are being developed as substitutes for sucrose, their use in food processing is frequently limited by their decomposition at the temperatures to which foods are heated during processing.

Studies with animals suggest that hexoses are not as cariogenic as sucrose. For that reason, work is under way to evaluate the use of such compounds in various food processing procedures. Since a mixture of equal parts of glucose and fructose is practically as sweet as the same weight of sucrose, there is a potential use for such a mixture. This mixture can be used in baking cakes that ordinarily require high concentrations of sucrose in the batter. When the hexose mixture is used in place of sucrose, the resulting cake is highly acceptable to a taste panel (Thompson *et al.* 1974).

Only passing reference will be made to the proposal that the sucrose in man's diet is primarily responsible for the high cardiovascular mortality among adults in the more developed countries. That postulate, championed by Yudkin (1972), has not aroused as much concern in the food industry as the cariogenic potentials of sucrose. There is probably no more reason to become concerned about the cardiovascular repercussions of the sucrose in our foods than about the numerous other foods and nutrients that have been implicated in the development of cardiovascular diseases.

SODIUM CHLORIDE

Salt is another important ingredient in many food processing techniques that has come under public scrutiny. The primary reason for the concern expressed about the amount of that seasoning added to foods was its reported association with the development of hypertension. Both animal studies and epidemiological surveys suggested a positive correlation between the amount of salt in an individual's diet and his susceptibility to high blood pressure (Dahl 1960). One of the principal concerns in this area is related to the amount of salt present in baby foods. This was based on the report that when very young laboratory rats were fed a ration containing high levels of salt, they developed hypertension which persisted even after the salt in their ration was reduced (Dahl and Schackow 1964). Such hypertension was reportedly produced when commercial baby foods were used as the primary component in the weanling rats' ration (Dahl et al. 1963; Dahl 1968). On the basis of that report, a public clamor was raised to restrict the amount of salt in baby foods. The committee of experts convened to consider the problem, recommended that no baby food should contain more than 0.25% salt (Filer 1971). Shortly after the report was released, the baby food manufacturers voluntarily accepted the limitation on the salt concentration in their products.

MONOSODIUM GLUTAMATE

About the same time that some activists were attempting to interdict the use of salt, monosodium glutamate (MSG) as a food additive was subjected to public discussion and debate. The basis for that concern was not so much the Chinese restaurant syndrome, which initially focused attention on that compound, as its reported effects on newborn animals. These reports suggested that when large doses of MSG were injected into a day-old monkey, or fed to very young mice, a variety of nerve lesions developed (for review, see Anon. 1970A,B). Although the retinal lesions in the mice fed large amounts of MSG were ominous, their significance was abated by the recognition that such lesions could be produced only when MSG was administered in large amounts to mice prior to the time they were 10 days of age. Thereafter, a barrier appeared to develop which prevented the transfer of MSG from the blood to the brain. Since mice are born in a far less developed state than the human infant, there was and still is considerable question as to the application of these results of animal studies to human beings.

Too many protagonists condemning the use of MSG as a food additive ignored the fact that many dietary proteins contain 25% or more of this amino acid. Whether the body can distinguish between the glutamic acid derived from proteins and that from MSG is doubtful. The only possibility in this respect is that the free amino acid might be absored more rapidly than that released by the digestive processes from the proteins. It should be obvious that there may be a difference in physiological response when a large dose of a compound such as MSG is given to an animal or an individual in the fasting state (i.e., on an empty stomach) either by mouth or by injection and when the same dose of the compound is incorporated into a serving of food. The consumption of the latter, especially as part of a regular meal, will produce a much lower blood level of glutamic acid than will either of the former procedures. This difference in blood levels is due largely to the rates of stomach emptying in the two situations involving oral ingestion.

Frequently, the dire predictions of the consequences that might ensue from the use of a food additive ebb into oblivion as a new theoretical threat appears on the horizon. The major reason why the public and, for that matter, the scientific community focuses its attention on one danger for a relatively short time is the difficulty in maintaining public interest on a subject that is not of imminent importance to the average person. Another reason is that the governmental agency or the scientific laboratory initiating or involved in the uproar has secured the necessary funding to continue or to expand its proposed activities. The one thing that is often ignored is that occasionally the original observations are later shown to be artifacts or they cannot be duplicated in other animal species. The latter occurred in the case of MSG. When large doses of MSG were given to young gerbils, they did not develop any neurological lesions (Bazzano et al. 1970). That was also true when large doses of MSG were given to newborn rats (Adamo and Ratner 1970). On these bases, the significance of the original reports about the lesions produced by MSG ingestion are open to question as far as their application to human beings is concerned.

It is difficult to determine the extent to which artifacts or the restriction of an abnormality to one animal species have been involved in the results suggesting disastrous consequences from ingesting specific substances or foods. Unfortunately, many of these reports first appear in the public press. The rationale for the use of such rapid publication techniques has been that the health and welfare of the public had to be protected immediately from the impending dangers in our food, water, or air. This approach has created a tremendous amount of publicity for the individual who single-handedly has assumed the task of defending the hapless public against "malevolent" industries.

ADVENTITIOUS IODINE IN FOODS

This defense of industry in the nutritional-toxicological arena should not be misconstrued. It does not absolve the processor from the responsibility of carefully considering the nutritional and health implications of any changes that might be introduced into currently accepted techniques. Only one such illustration will be mentioned. That relates to organic iodine compounds. These are being used in ever-increasing amounts in many food industries. From many standpoints, they appear to be ideal bacteriostatic agents. The iodine from these agents used in "sterilizing" food processing equipment frequently appears in the processed foods. This results from the fact that the equipment can be used immediately after the iodine sterilization treatment. The residual solution of the organic iodine in the equipment contaminates the food processed therein. Such contamination has been considered beneficial since the iodine added thereby to the food should aid in preventing the development of enlarged thyroids, a public health problem that has plagued workers in various parts of this country.

The result of the widespread use of organic iodine compounds in maintaining sanitation in food processing plants has been a marked increase in the intake of this nutrient. Clinical examination of adolescents in parts of the United States where there was a high incidence of goiter, indicated that such children excreted in their urine amounts of iodine which exceeded their requirement (Trowbridge *et al.* 1973). This suggests that the goiters in these youths were not due to a dietary iodine deficiency and that the enlarged thyroid was probably due to some other condition.

There are a number of reports (Hemken 1970; Fisher and Carr 1974; Anon. 1973) and many unpublished findings which indicate that the iodine in our foods is far higher than it was a number of years ago. At the present time, there is nothing to suggest that these high iodine intakes, coming primarily from processed foods, are of immediate concern (Talbot *et al.* 1974). However, there is always the possibility that the use of organic iodine compounds in the food industry may come under the same intensive scrutiny as salt, MSG, cyclamates, and a host of other nutrients.

MICRONUTRIENTS

Vitamins

Vitamins are classified into two groups according to whether they are soluble in water or fat solvents. Although they are required in minute amounts, they are dietary essentials for normal health. Since each vitamin has specific metabolic functions, the factors affecting their utilization are considered separately for each vitamin.

Vitamin A.—Vitamin A is absorbed in the small intestine. In the mucosal wall of the small intestine it is esterified into vitamin A palmitate which then enters the blood stream. The absorption of vitamin A or provitamins A (carotenoids) varies depending on whether it is ingested in pure form or as it occurs in foodstuffs. The absorption of dietary carotenoids is significantly reduced when the diet is low in fat (Kramer *et al.* 1947). Also, there is great difference in the ability of different mammals to absorb dietary carotenoids. For example, man and bovine can absorb both vitamin A and carotenoids and convert carotenoids into vitamin A, but the rat and pig do not absorb significant amounts of the carotenoid pigments.

Aqueous, colloidal suspensions of vitamin A are more rapidly absorbed than are true solutions of this vitamin (Kramer *et al.* 1947). Bile salts enhance the absorption of vitamin A but are not absolutely necessary for that process (Greaves and Schmidt 1935). Because of its unsaturated character, vitamin A is readily destroyed by oxidizing agents. The presence of vitamin E protects vitamin A from oxidation and improves its utilization.

The addition of nutrients that are not present in significant amounts in a food appears possible on the basis of regulations proposed by the FDA (U.S. Food and Drug Admin. 1974). That raises a specter that may haunt some segments of the food industry at a later date. This is especially true for both vitamins A and D. These fat-soluble vitamins are retained by the body and each additional intake only adds to its stores. Should that be increased by the added intake of vitamin A from a wide variety of foods, then there is the possibility of developing a toxicity. Previously, the toxicity of these two fat-soluble vitamins has been associated primarily with the ingestion of high potency vitamin preparations (Roels and Lui 1973; Omdahl and DeLuca 1973). Arctic explorers have been most prominently associated with vitamin A toxicity resulting from the ingestion of foods. To relieve their protracted period of starvation, they ate the livers of the seals and polar bears they killed (Mickelsen *et al.* 1973).

The fortification of foods with vitamins and minerals might be undertaken, among other reasons, for humanitarian purposes. So much emphasis has been given in the public press to the poor dietary habits of the United States inhabitants that fortification of their food seemed necessary. The decision to fortify a food with either vitamin A or D or both should be considered from the standpoint of the overall intake of these nutrients. Addition of those vitamins to foods should be carefully considered since there is the possibility that a chronic vitamin A toxicity might develop among some consumers. For vitamin A, a toxicity may occur when the intake approaches 20-30 times the RDA (Roels and Lui 1973). The standard of reference for nutrition labeling purposes, is the U.S. RDA which is the maximum value listed for that nutrient recommended for a normal individual in the Recommended Daily Dietary Allowances published in 1968 (National Academy of Sciences 1968).

Toxicity of vitamin A is more likely to develop in children if vitamin fortification of foods does become widespread. The vitamin A level recommended for young children is half the maximum RDA or U.S. RDA value. Naturally, their caloric intake will be less than that of the individuals for whom the maximum RDA value was established; however, some children consume fairly large amounts of certain foods. Furthermore, the younger child may be receiving a daily supplement of a vitamin preparation which very likely contains vitamin A.

The potential toxicity of vitamin A and/or vitamin D added to foods may become critical for some individuals, especially children, if a large number of foods are fortified with these and other nutrients. This may be done providing the package is adequately labeled and the regulations pertaining to these foods are followed (Code of Federal Regulations 1974). Such foods must be labeled to prominently indicate they are "A special dietary supplement." At the present time, there appears to be no restriction to such fortification other than the requirement that the package be properly labeled.

Some food processors have initiated what may become a trend, depending on the sale of the fortified foods. One of these conventional foods which has been fortified with vitamins and iron indicates on the panel that one serving contains "100% U.S. Recommended Daily Allowance (U.S. RDA) of all mandatory vitamins and iron." That means that one serving presumably provides 5000 I.U. of vitamin A. The package states that the food is a "Multivitamin and iron supplement." As such, it would appear to be a special dietary food, but since it is located with the regular foods of the same kind, the ordinary shopper might not recognize it as containing an increased amount of vitamin A. Whether competition will force other processors to do likewise will determine the nature of the problems that the food industry may face in the not-too-distant future.

Vitamin D.—Being a fat-soluble vitamin, vitamin D is absorbed with fats in food. Therefore, any condition which interferes with the absorption of fat also interferes with the absorption of vitamin D. Bile is believed essential for the normal absorption of vitamin D (Greaves and Schmidt 1937). Dietary vitamin D can be absent without producing any deficiency provided the individual is exposed to an adequate amount of sunshine or ultraviolet light. Under such circumstances the 7-dehydrocholesterol normally present in the skin is converted to vitamin D.

The long period of quiescence in the field of vitamin D metabolism came to an end with the discovery of the first change this vitamin undergoes prior to its participation in calcium absorption. It had been known that when a rachitic animal was given a therapeutic dose of vitamin D, a number of hours elapsed before there was any improvement in the absorption of calcium from the intestine. During that period, a change in the structure of the vitamin occurs. This, according to the work of DeLuca and collaborators (1969), takes place in the liver and in the kidney. During this same time, the intestinal mucosal cells synthesize a protein which, in conjunction with the proper vitamin D metabolite, is involved in the active absorption of calcium from the intestinal lumen (Omdahl and DeLuca 1973).

As with vitamin A, there is a possibility that indiscriminate fortification of foods with vitamin D may produce toxicity symptoms especially among very young children. This appears to have occurred in England. Shortly after World War II, many infant foods in England were supplemented with vitamin D. That was associated with the appearance of a condition called idiopathic hypercalcemia (Anon. 1956). Although there was some discussion as to whether the syndrome was due to a vitamin D toxicity, Forbes (1957) claimed that overdosage with this vitamin could not be excluded as an etiologic agent.

Although a number of reports have described substances in foods that interfere with the action of vitamin D, most of that has been with animals. One investigator (Grant 1953) found a chloroform-soluble factor in green oats which, when incorporated into a ration fed to rats, decreased its vitamin D activity as estimated by bone mineralization. Unheated soybean meal and the isolated protein therefrom contain a substance which counteracts the metabolic function of vitamin D in a number of animal species (reviewed by Liener 1969). The rachitogenic activity of the soybean meal as well as that in the protein isolate is destroyed by heat. Another group of investigators (Coates et al. 1961) found that the addition of fresh or vacuum-dried pigs' liver to a ration produced rickets in chicks despite the presence of adequate amounts of vitamin D. The rachitogenic substance in pigs' liver appeared to interfere with intestinal absorption of calcium. It was water-soluble and heat-labile.

Vitamin E.—Like vitamin D, bile is essential for the optimal absorption of vitamin E (Greaves and Schmidt 1937). Vitamin E is important in maintaining the stability and integrity of the cell membrane which is probably associated with the high concentration of unsaturated fatty acids that are present in these membranes. This function of vitamin E is, very likely, due to its antioxidant properties.

The requirement of both man and animals for vitamin E depends on a number of factors. In addition to the selenium content of the diet, one of the more important factors is the amount of unsaturated fatty acids and their degree of unsaturation. Increasing the polyunsaturated fatty acids in the diet increases the vitamin E requirement or, in a vitamin E deficiency, it aggravates the deficiency symptoms (Harris and Embree 1963).

The increasing emphasis being put on the health benefits that may accrue from the replacement of some of the saturated fat in the diet with polyunsaturated fats makes this subject important to everyone involved in food processing. The drastic dietary alterations proposed for these foods make it essential to carefully evaluate their antioxidant activities. Such evaluation is important since a few studies suggest that diets containing large amounts of polyunsaturated fats may have adverse effects on the consumer. One of these reported an increased cancer mortality among such men (Dayton and Pearce 1968). A subsequent paper from the same clinic indicated that the men consuming the diet high in polyunsaturated fat over a period of years developed more bile stones than a comparable group of men eating a more normal diet (Sturdevant et al. 1973).

The increasing attention, directed to the ever-expanding older segment of our population, will only accentuate the potential importance of any proposal that diet may assist in retaining an individual's vitality. There are a number of suggestions that increased intakes of vitamin E may assist in that respect. Actually, Tappel (1973) has indicated that the aging process may be linked to lipid peroxidation. As an extension of that proposal, he suggests that the deteriorating aspects of some of these reactions "might be slowed by use of increased amounts of dietary antioxidants."

In a similar vein, the reports on the ameliorative effects of antioxidants on the hepatic injury associated with the ingestion of large amounts of ethanol or other hepatotoxic agents (DiLuzio 1973) will accentuate the pressure to increase the level of antioxidants in our food supply. The theory based on experimental results is that one of the initial hepatic reactions of these compounds "is the formation of lipoperoxides at selective subcellular sites due to an alteration in antioxidant activity of the hepatic cell" (DiLuzio 1973).

Various legumes are reported to contain substances that interfere with the action of vitamin E (for a review, see Liener 1969). The nutritional significance of these in foods that are to be used by human beings is unknown. All of the studies in this area have been limited to experimental animals.

Vitamin K.—Vitamin K requires bile and pancreatic juice for maximum absorption from the intestine. Any disorder which obstructs the delivery of bile to the small intestine reduces the absorption of vitamin K. There are certain antagonistic substance which inhibit the effectiveness of vitamin K, the best known being dicoumarol. When dicoumarol is administered orally to man or animal, it inhibits the production of prothrombin and other plasma coagulation factors, thus interrupting the reactions involved in the clotting mechanism. Another vitamin K antagonist is vitamin A. Light and coworkers (1944) reported that high doses of vitamin A in rats decrease the blood prothrombin content. This effect of high doses of vitamin A presumably is due to its interference with the absorption of vitamin K.

Thiamin.—Thiamin is readily absorbed from the small intestine. Beyond a certain point, ingested thiamin is not stored in the tissues (Mickelsen *et al.* 1947). The dietary requirement for thiamin varies with the composition of the diet. Carbohydrate is particularly important in determining the requirement while a diet providing most of its calories from fat results in a low thiamin requirement. The requirement of animals for thiamin is reduced when sulfonamides, penicillin, and similar antibiotics are added to the diet (Mickelsen 1953). An enzyme thiaminase found in certain uncooked fish and clams is capable of destroying thiamin. Thiaminases are widely found in fish, fern, bacteria, and related organisms (Mickelsen *et al.* 1973). Since these enzymes are readily destroyed by heat, they are of significance to the food processors primarily when foods containing these enzymes are being prepared for processing. Should a mixture of foods, one ingredient of which contains a high potency thiaminase, be permitted to remain at room temperature for a few hours, there may be considerable destruction of the thiamin not only in that food but in the other ingredients as well.

Another set of reactions that should be called to the attention of the food processors, especially those who produce dehydrated foods, are those involving ethylene oxide. That gas is a highly effective germicidal agent, destroying not only viable bacteria but, under proper conditions, spores as well. This procedure can be applied at the end of the line after the food has been sealed in its container thus ensuring the sterility of the product when it leaves the plant. Providing the container is made of plastic or paper, the ethylene oxide readily penetrates and permeates the contents of the package. This can be accomplished without the application of heat and after the sterilization only the minutest residue of the gas is present in the food.

The major disadvantage in using ethylene oxide for food sterilizations is that it reacts with the nutrients in some foods. It destroys practically all the thiamin present in free form such as is the case in a purified ration (Mickelsen 1957). That is also true for large amounts of riboflavin, pyridoxine, niacin, and folic acid. This means that ethylene oxide might also react with any of these nutrients that might be used in food fortification. Pantothenic acid, biotin and vitamin B-12 were not influenced by the gas. A number of factors influence the extent to which the susceptible vitamins are destroyed. One of these is the percentage moisture in the food, which, together with the reactivity of the nutrient and the temperature at which the processed food is stored influence its stability. Another is the nature of other substances that may be in contact with the vitamins. For instance, when choline chloride was present the destruction of the vitamins was much greater than in the presence of an equivalent amount of choline citrate (Bakerman *et al.* 1956). The ethylene oxide also destroys some of the amino acids present in a protein such as casein. Approximately 22% of the histidine and 17% of the methionine were destroyed when casein was sterilized with ethylene oxide (Mickelsen 1957).

Riboflavin.—Riboflavin is absorbed from the small intestine and then enters the portal vein. Relatively little concern has been elicited about the potential dangers of any toxicity that might be associated with an excessive intake of this vitamin. A large percentage of any excess riboflavin that is ingested is rapidly excreted in the urine.

The Ten State Nutrition Survey during 1968–1970 (U.S. Dept. of Health, Education and Welfare 1972A,B) was limited to the groups that were most vulnerable to nutritional deficiencies. The results thereof suggested that some degree of inadequate riboflavin intake may exist among children in states with a high percentage of low income families (U.S. Dept. of Health, Education and Welfare 1972B). In those states, the largest percentage of children with low urinary riboflavin excretions occurred among blacks, with Spanish-Americans intermediate between them and whites. With an increase in age, the percentage of individuals with "low" excretion values decreased for all races.

Riboflavin intake is frequently associated with protein intake. That is due largely to the fact that those foods rich in protein frequently contain considerable amounts of riboflavin. Since the estimated intake of protein in the United States for all age and sex groups is from 2 to 3 times the Recommended Dietary Allowance (Leveille 1975), one would anticipate little indication of a riboflavin deficiency. This appears confirmed by the clinical findings of the Ten State Nutrition

Survey. According to that report (U.S. Dept. of Health, Education and Welfare 1972A) there was "no uniform pattern of positive findings ... in the survey data. Cheilosis, cracking and dry scaling of the lips, appeared to occur somewhat more frequently in whites." Since these symptoms are usually associated with a deficiency of riboflavin and since the smallest percentage of low urinary riboflavin excretions occurred among white children, there appears little evidence to support the existence of a riboflavin deficiency among those considered to be the most susceptible to nutritional disturbances.

There are reports of a few naturally-occurring compounds that interfere with the metabolic function of riboflavin (Liener 1969). These compounds occur primarily in the fruit of a tropical plant called akee. To our knowledge, it is not a part of the food processing operations in the United States.

Riboflavin is listed among the water-soluble vitamins, but it is soluble only to the extent of 10–13 mg per 100 ml at 25°C; even at 100°C, no more than 230 mg are soluble (Wagner-Jauregg 1972). It is quite soluble in alkali; however, such solutions are unstable. It is stable in the presence of acids, air, and the common oxidizing agents. This stability is increased if the vitamin is protected from light. Exposure of milk to sunlight leads to the destruction of more than half of the riboflavin in 2 hr.

Niacin.—Niacin and nicotinamide are rapidly absorbed from the intestine.

The dietary requirement of niacin depends on the tryptophan content of the diet since tryptophan can serve as a precursor of niacin for most animals as well as man. It appears that about 50 to 100 mg of tryptophan can replace 1 mg of niacin (Hundley 1954). A high level of leucine in the diet has been reported responsible for the niacin deficiency appearing among some individuals in India. Balavady et al. (1967) suggest that leucine interferes with the conversion of tryptophan to niacin and of niacin to nicotinamide-adenine dinucleotide (NAD) and nicotinamide-adenine dinucleotide phosphate (NADP). As far as the food processors in this country are concerned, it can be stated that there is nothing to suggest that the leucine intake of any segment of the U.S. population is likely to lead to a niacin deficiency.

Niacin as such has been used in a few foods for reasons other than vitamin fortification. Apparently, one of these has been to maintain the red color in ground beef. For that purpose a mixture of nicotinic and ascorbic acids has been used. There is one report that meat thus treated produced a severe flushing reaction among some of the individuals who consumed it. Initially, the cause of the severe itching that accompanied the flushing was unknown. Since these symptoms developed in 39 out of 116 members of a sorority and 5 waiters the situation was called to the attention of the local health authorities (Press and Yeager 1962). Two subjects who were "more seriously affected" were hospitalized overnight. They, like the others, "recovered with no apparent residual effects." The acute effects, in most cases, disappeared within a few hours after the first signs became manifest. These symptoms, of which the flushing reaction is the most prominent, are associated with the ingestion of the free nicotinic acid and then only when the amounts are considerably greater than human requirements.

Of more significance is the former use of niacin to reduce the level of blood cholesterol and triglycerides. The daily ingestion of from 1.5 to 6.0 gm of niacin daily has been used therapeutically among adults who, for one reason or another, were advised by their physician to lower their serum cholesterol levels (for review, see Mickelsen 1967). A number of patients have been reported as ingesting these high doses of niacin for periods as long as 10-11 yr. Although a number of minor abnormalities, primarily apparent as slight deviations from biochemical standards, appeared in these individuals, all symptoms disappeared on termination of therapy.

Niacin is freely soluble in water. It is stabile to heat, light, air, and alkali. In fact, it is one of, if not, the most stable vitamins.

Pantothenic Acid.—Pantothenic acid is part of coenzyme A which is an important factor involved in the acylation reactions of carbohydrates, fats, and proteins. Pantothenic acid is widely distributed in natural foods. An unidentified compound from pea seedlings is reported to be an antagonist of pantothenic acid (Smashevski 1966). The effect of this antivitamin is counteracted by methionine.

Pantothenic acid is destroyed rapidly in acid or alkaline medium. It is labile to dry heat, hot acids, and hot alkalis.

Vitamin B-6.—Vitamin B-6 is involved in the metabolism of all the amino acids. That role explains why diets high in protein, especially in methionine, increase the requirement for vitamin B-6.

The utilization of vitamin B-6 is influenced by several antivitamin factors. Klosterman et al. (1967) have reported that flax seeds contain a compound which in a variety of animal species produces symptoms which can be overcome by pyridoxine. The North Dakota group of investigators isolated a compound from unheated flax seeds. That was 1-amino-D-proline which was combined via a peptide linkage with glutamic acid.

Pyridoxine is reported to be stable to heat both in acid and alkaline solutions (Keresztesy 1954). However, under some conditions, a large proportion of this vitamin is destroyed during commercial processing. One such episode which developed public health repercussions, occurred when a formula produced convulsions in some of the infants fed that product. This preparation was subjected to "prolonged heat treatment" as a means of reducing the allergenic properties of cow's milk protein (Tomarelli *et al.* 1952). As a result of such treatment, the pyridoxine content of the milk was 33–64% that in fresh milk as measured by microbiological assay (Tomarelli *et al.* 1955). Of the remaining pyridoxine, only about $1/2$ was biologically active for rats. The exact explanation for this episode does not appear to have been completely clarified. Although it has been shown that when pyridoxine is exposed to heat in the presence of a solution of amino acids, a Schiff's base-like compound is formed (Heinert and Martell 1963); the availability of the vitamin in the resulting compound when consumed by human subjects is still unresolved (Snell and Rabinowitz 1948). In the case of the infant formulas, part of the problem resulted from the fact that in addition to the increased heat treatment, there were changes in the nature of the fat and the levels of some added nutrients (Nelson 1956).

Infants fed the heat-treated formula, developed symptoms related primarily to the central nervous system (Coursin 1964). These symptoms appeared within 6 weeks to 4 months after the infant was first fed the formula. One of the more prominent symptoms and the one which was primarily responsible for focusing attention on this problem was the convulsions manifested by the affected infants. Apparently, not all infants receiving the formula showed any symptoms; the reported incidence was 3 in 1000 (Leitch and Hepburn 1961).

Vitamin B-12.—Absorption of vitamin B-12 requires a factor called the "intrinsic factor" secreted by the stomach. In pernicious anemia, the absorption of vitamin B-12 is poor primarily due to the fact that such individuals produce little, if any, intrinsic factor.

Little attention has been given to vitamin B-12 by the food processing industries. Part of this is probably due to the fact that a deficiency of this vitamin occurs only under very special conditions. One of these involves the tenacious retention of the vitamin by tapeworms that are attached to the intestinal wall near the duodenum. This tapeworm is acquired by the eating of raw fish. Any vitamin B-12 in the intestinal tract is absorbed by the tapeworm before it becomes available to the host. The result is the development of a "true" vitamin B-12 deficiency (Von Bonsdorff 1953).

The primary condition with which vitamin B-12 is associated involves patients with pernicious anemia. That disease is not a true manifestation of a dietary deficiency of the vitamin. The disease develops despite a normal intake of vitamin B-12. Those patients who have pernicious anemia are unable to synthesize an adequate amount of the intrinsic factor. The latter is essential for the abosorption of vitamin B-12 and in its absence, very little of the vitamin finds its way from the intestine to the tissues; the exception occurs when large doses of the pure vitamin are given by mouth. Under such circumstances, the vitamin B-12 crosses the intestinal wall by diffusion.

There is some confusion in the literature about the stability of vitamin B-12 to processing techniques. One report (Herbert 1973) claims that "the vitamin B-12 molecule is almost indestructible unless heated in alkali at temperatures in excess of $100°C$." Another report (Morgan 1971) is that the sterilization of milk destroyed practically all of the vitamin B-12 with roller drying producing a 20% loss and spray drying a 35% loss.

Minerals

From the nutritional and health standpoints, minerals have been subjected to diametrically opposite demands. Some minerals such as sodium have been subjected to a number of proposals that far too large amounts were introduced to the human diet especially through seasoning at the table and by food processors. Some of those who are most adamant in demanding the complete removal of salt from food processing techniques claim that the taste for this condiment is acquired. The acquisition results from the individual's exposure to foods which are seasoned with it. Their argument is that such an acquired taste is subject to "unlearning" provided the individual could get salt-free foods in adequate variety and amounts. Regardless of the validity of that argument, there are many individuals who, although restricted to low sodium diets for health reasons, find it very difficult to follow their prescribed diets. For that reason, many physicians attempt to adjust the dose of diuretic or other drugs to the lowest sodium intake that the patient can "live with." To assist these and other individuals who wish to restrict their intake of sodium, a mixture of sodium and potassium chlorides was developed which contains only $1/2$ the sodium in ordinary salt but which, for most individuals, is as salty as regular salt on an equal weight basis (Frank and Mickelsen 1969).

At the other end of the scale are the minerals, of which iron is a good example, that are proposed for inclusion in our foods at concentrations greater than those currently recommended. This would be true especially of those foods used to a large extent by individuals who are most vulnerable to a deficiency of iron. Since the most prominent manifestation of an iron deficiency is anemia, this parameter has frequently been proposed as an index of the extent of iron deficiency in a population. Although the number of such individuals in the United States is frequently listed as being very large, there may be some basis for questioning some of those values. There appears to be almost universal agreement that most of these anemic individuals occur among the younger members of low income groups and especially among those who are also members of minority groups. On that basis, a survey was made in an industrial city where low income preschool children were provided a program called "Head Start." The hemoglobin levels among these children suggested that very few if any were anemic. The hemoglobin levels for most of the children were determined both in October and the following May (Mickelsen et al. 1970).

There are three problems that have an important bearing on evaluating the incidence of anemia among various population groups. The first of these is the analytical techniques. Although the hemoglobin determination has been used very extensively and requires only simple manipulative procedures, there are a number of pitfalls that make it difficult to compare the results of one survey with those of another. The second difficulty involves the levels of hemoglobin which are used to characterize anemia. Since there are no generally accepted or universally used values for that criterion, each investigator chooses his own. As a result, different hemoglobin levels have been used to distinguish normal from anemic people (for discussion see Mickelsen et al. 1970). The level of hemoglobin chosen for that purpose becomes very important since a change in the lowest normal hemoglobin level from 10 to 9% markedly affects the percentage of subjects designated as anemic. The third problem which is frequently ignored in mass surveys to determine the incidence of iron deficiency anemia is the etiology of that condition. Many investigators accept, as a truism, the reports that iron deficiency is one of the most important nutritional problems facing the people in developed countries. On that basis, it is often assumed that all, or practically all, individuals with hemoglobin levels below the "normal limits" are iron deficient. That may be a false assumption according to the observations of Wilder (1952).

Calcium.—The nutritional concern about calcium has undergone considerable fluctuations during the past half century. For a number of years the most commonly reported deficiency in the American diet was calcium. This, together with vitamins A and C, formed the triad of the so-called nutritional problem. Very likely, this was due to the fact that the dietary intakes were compared with the RDA.

That probably these "deficiencies" were artifacts is now apparent since some of the RDAs involved in these calculated dietary deficiencies have been markedly reduced.

There is still some concern that the calcium intake may not be as high as it should be among the members of some segments of our population. This results partly from the increasing consumption of carbonated beverages which presumably are being used at the expense of milk. This use of carbonated beverages is also reported to have another adverse effect upon the calcium that may be in the individual's diet that includes large amounts of these drinks. The presence of phosphates in the carbonated beverages alters the calcium:phosphorus ratio in such a way that it interferes with calcium absorption.

The fact that carbonated beverages are used to a very large extent by teen-agers is another reason for concern. An increasing number of adolescent girls are experiencing pregnancies at a time when they would normally be needing their dietary calcium for the formation of their own skeletal systems. The concern for the pregnant adolescent girls is that when they reach adulthood they will be more susceptible to osteoporosis. The incidence of the disease osteoporosis is increasing; the increased incidence is attributed to the numbers of women who are attaining the age at which osteoporosis frequently occurs. There is still some controversy as to the etiology of osteoporosis. One of the more recent theories is that the amount of calcium in the bones of an individual at age 30–35 yr will determine whether or not osteoporosis will develop (Garn 1970). This implies that the calcium in the bones is lost at a constant rate after the period of deposition has terminated.

There are a number of factors that may become important in influencing calcium absorption. Some of these may result from the proposed changes in the dietary practices advocated for people in the United States. One of these that has been discussed is the increase in the intake of foods that are high in fiber. Such a dietary change may have two important effects insofar as calcium absorption is concerned. One of these is a decreased transit time (Holmgren and Mynors 1972). Under such circumstances, the rate of passage of nutrients through the intestinal tract may increase to the point where the more poorly absorbed substances such as calcium show a decreased absorption. Secondly, relatively large amounts of fiber have been reported to carry with it extra amounts of calcium. Whether this will become a problem in the American diet is questionable. The proposal to increase the fiber in the diet has been so recent that there has been inadequate time to assess the effect of the proposed changes on the population's nutritional status.

Of more importance is the fact that the primary source of fiber that might be incorporated into the diet would be that from bran. This would introduce an increased amount of phytate. This, in turn, might decrease calcium absorption. If that bran is incorporated into bread that has been subjected to yeast fermentation, there is a possibility that the amount of phytate will be reduced (Ter-Sarkissian et al. 1974).

Another factor that has been suggested as influencing the calcium absorption is the protein content of the diet. Work by Linkswiler and associates (1974) has indicated that as the protein content of the diet is increased, the amount of calcium retained is decreased, providing the amount of this mineral in the diet is maintained constant. The protein intakes that were studied with these normal young men were 47, 95 and 142 gm per day. None of these can be considered a very extreme intake. The primary effect of the protein in the diet appeared to be on the urinary excretion of calcium. With the higher intakes, there was a greater loss of calcium in the urine (Linkswiler et al. 1974; Anand and Linkswiler 1974; Walker and Linkswiler 1972; Johnson et al. 1970).

Phosphorus.—For many years, phosphorus was considered to be one nutrient with which there was relatively no problem. This was due largely to the fact that both plants and many animal foods contribute phosphorus to the diet. The recent concern about phosphorus in the diet has been brought about as a result of the high consumption of carbonated beverages; this has been discussed previously.

Another area that has received considerable attention insofar as phosphorus is concerned relates to dental caries. A number of reports indicated that the addition of phosphates to cariogenic diets reduced the incidence of caries in experimental animals to a very great extent. Some reports suggest that this also occurs in man. However, a 3-yr, well controlled, study indicates that the addition of phosphates to bread in the diet of children consuming large amounts of this food has no effect on dental caries experience (Mickelsen 1964; Ship and Mickelsen 1964).

CONCLUSIONS

Any food processing technique should be scrutinized not only from the standpoint of the technical contribution that it makes but also from the standpoint of the potential hazards it might present to the consumer. The concern about MSG, salt, cyclamates, and other substances are only an indication of what may become a continuing demand on the part of consumer groups.

With the increasing emphasis on energy conservation, it is very likely that we will see major changes in the nature of our dietary habits in the immediate future. This stems partly from the much greater cost in terms of energy and water to produce animal foods in comparison to plant foods used directly by man. In addition to the cost associated with the production of food there are added costs of waste disposal, storage, and processing. The marked increase in the trend to vegetarianism among certain segments of our population may be a passing fad, but it may also be an indication of a trend that may be adopted by increasing numbers of people. These changes may have dramatic health repercussions and may necessitate changes in food processing techniques.

The impact of nutrition labeling on the food processor and the consumer is open to debate. It is possible that this innovation may be accepted by the food industry and the regulatory agencies in somewhat the same manner as vitamin fortification. There were perhaps reasonable grounds for suggesting the introduction of vitamin fortification in the 1940's. Whether there is any need for such supplementation at the present time is open to question. However, it is doubtful that it will change. In the same way, nutrient labeling may come about despite the fact that there may be considerable question about its effect in improving the health and welfare of the consumer.

BIBLIOGRAPHY

ADAMO, N. J., and RATNER, A. 1970. Monosodium glutamate: lack of effects on brain and reproductive function in rats. Science *169*, 673-674.

ADDIS, T., BARRETT, E., POO, L. J., and YUEN, D. W. 1947. The relation between the serum urea concentration and the protein consumption of normal individuals. J. Clin. Invest. *26*, 869-874.

AHMAD, M., GILL, J. L., MAKDANI, D. D., and MICKELSEN, O. 1975. Minimizing bias and experimental error in human nutrition studies. (Unpublished)

ALBRITTON, E. C. 1955. Standard Values in Nutrition and Metabolism. McGregor and Werner, Dayton, Ohio.

AMA COUNCIL ON FOODS AND NUTRITION. 1973. Improvement of the nutritive quality of foods, general policies. J. Am. Med. Assoc. *225*, 1116-1118.

ANAND, C. R., and LINKSWILER, H. M. 1974. Effect of protein intake on calcium balance of young men given 500 mg calcium daily. J. Nutr. *104*, 695-700.

ANON. 1956. Idiopathic hypercalcemia. Nutr. Rev. *14*, 338-341.

ANON. 1970A. Monosodium glutamate—studies on its possible effects on the central nervous system. Nutr. Rev. 28, 124-129.
ANON. 1970B. Monosodium glutamate: blood levels and absorption after oral and intravenous administration. Nutr. Rev. 28, 158-162.
ANON. 1973. Frozen fried-chicken dinners. Consumer Reports 38, 402-405.
ANON. 1974. Scientists urge more dietary fiber. CNI Weekly Rept. 4, No. 31, 6. Community Nutrition Institute.
BAKERMAN, H. et al. 1956. Stability of certain B vitamins exposed to ethylene oxide in the presence of choline chloride. Agr. Food Chem. 4, 956-959.
BALAVADY, B., MADHAVAN, T. V., and GOPALAN, C. 1967. Production of nicotinic acid deficiency (black tongue) in pups fed diets supplemented with leucine. Gastroenterology 53, 749-753.
BAZZANO, G., D'ELIA, J. A., and OLSON, R. E. 1970. Monosodium glutamate: feeding of large amounts in man and gerbils. Science 169, 1208-1209.
BLOCK, R. J., and MANDL, R. H. 1958. Amino acid composition of bread protein. J. Am. Dietet. Assoc. 34, 724-726.
BOLOURCHI, S. 1963. A study of the biological value and composition of breads from various countries. M.S. Thesis, Michigan State University.
BOLOURCHI, S., FEURIG, J. S., and MICKELSEN, O. 1968A. Wheat flour, blood urea concentration, and urea metabolism in adult human subjects. Am. J. Clin. Nutr. 21, 836-843.
BOLOURCHI, S., FRIEDEMANN, C. M., and MICKELSEN, O. 1968B. Wheat flour as a source of protein for adult human subjects. Am. J. Clin. Nutr. 21, 827-835.
BURKITT, D. P. 1973. Some diseases characteristic of modern western civilization. Brit. Med. J. 1, 274-278.
BURKITT, D. P., WALKER, A. R. P., and PAINTER, N. S. 1974. Dietary fiber and disease, J. Am. Med. Assoc. 229, 1068-1074.
COATES, M. E., HARRISON, G. F., and HOLDSWORTH, S. E. 1961. The effect of rachitogenic factor on calcium metabolism. Brit. J. Nutr. 15, 149-155.
CODE OF FEDERAL REGULATIONS. 1974. Food and Drugs. Chap. 21, parts 10 to 129. Revised Apr. 1, 1974. Offic. Federal Register, Natl. Arch. Records Serv., Washington, D.C.
COURSIN, D. B. 1964. Vitamin B_6 metabolism in infants and children. In Vitamins and Hormones, Vol. 22. R. S. Harris, I. G. Wool, and J. A. Loraine (Editors). Academic Press, New York.
DAHL, L. K. 1960. Salt, fat and hypertension: the Japanese experience. Nutr. Rev. 18, 97-99.
DAHL, L, K. 1968. Salt in processed baby foods. Am. J. Clin. Nutr. 21, 787-792.
DAHL, L. K., HEINE, M., and TASSINARI, L. 1963. High salt content of western infants' diet: possible relationship to hypertension in the adult. Nature 198, 1204-1205.
DAHL, L. K., and SCHACKOW, E. 1964. Effect of chronic excess salt ingestion: experimental hypertension in the rat. Can. Med. Assoc. J. 90, 155-160.
DAYTON, S., and PEARCE, M. E. 1968. Controlled trial of diet high in unsaturated fat for prevention of atherosclerotic complication. Lancet 2, 1060-1062.
DELUCA, H. F. 1974. Vitamin D: the vitamin and the hormone. Federation Proc. 33, 2211-2219.
DILUZIO, N. R. 1973. Antioxidants, lipid peroxidation and chemical induced liver injury. Federation Proc. 32, 1875-1881.
FILER, L. J., JR. 1971. Salt in infant foods. Nutr. Rev. 29, 27-30.
FISHER, K. D., and CARR, C. J. 1974. Iodine in foods: chemical methodology and sources of iodine in the human diet. Federation Am. Soc. Exptl. Biol., Bethesda, Maryland.

FORBES, G. B. 1957. Overnutrition for the child: blessing or curse? Nutr. Rev. 15, 193-196.
FRANK, R. L., and MICKELSEN, O. 1969. Sodium—potassium chloride mixtures as table salt. Am. J. Clin. Nutr. 22, 464-470.
GARN, S. M. 1970. The earlier gain and the later loss of cortical bone. In The Nutritional Perspective. Thomas, Springfield, Illinois.
GORTNER, W. A. 1972. The impact of food technology on nutrient supplies. Food Technol. Australia, Oct., 504-517.
GRANT, A. B. 1953. Carotene: a rachitogenic factor in green-feeds. Nature 172, 627.
GREAVES, J. D., and SCHMIDT, C. L. A. 1935. The utilization of carotene by jaundiced and phosphorus treated vitamin A deficient rats. Am. J. Physiol. 111, 502-506.
GREAVES, J. D., and SCHMIDT, C. L. A. 1937. Relation of bile to absorption of vitamin E in rat. Proc. Soc. Exptl. Biol. Med. 37, 40-42.
HARRIS, P. L., and EMBREE, N. D. 1963. Quantitative consideration of the effect of polyunsaturated fatty acid content of the diet upon the requirements for vitamin E. Am. J. Clin. Nutr. 13, 385-392.
HEGSTED, D. M. 1962. The potential of wheat for meeting man's nutrient needs. In World's Food Supply, Report of a Conference. USDA Western Regional Research Laboratory, Albany, Calif.
HEINERT, D., and MARTELL, A. E. 1963. Pyridoxine and pyridoxal analogs. VI. Electronic absorption spectra of Schiff bases. J. Am. Chem. Soc. 85, 183-188.
HEMKEN, R. W. 1970. Iodine. J. Dairy Sci. 53, 1138-1143.
HERBERT, V. 1973. Folic acid and vitamin B_{12}. In Modern Nutrition in Health and Disease, 5th Edition. R. S. Goodhart, and M. E. Shils (Editors). Lea & Febiger, Philadelphia.
HOLMGREN, G. O. R., and MYNORS, J. M. 1972. The effect of diet on bowel transit times. S. African Med. J. 46, 918-920.
HOPKINS, L. L., JR., and MOHR, H. E. 1971. The biological essentiality of vanadium. In Newer Trace Elements in Nutrition. W. Mertz, and W. E. Cornatzer (Editors). Marcel Dekker, New York.
HUNDLEY, J. M. 1954. Niacin. II. Chemistry. In The Vitamins, Vol. 2. W. H. Sebrell, Jr., and R. S. Harris (Editors). Academic Press, New York.
JOHNSON, N. E., ALCANTARA, E. N., and LINKSWILER, H. 1970. Effect of level of protein intake on urinary and fecal calcium and calcium retention of young adult males. J. Nutr. 100, 1425-1430.
KERESZTESY, J. C. 1954. Pyridoxine and related compounds. II. Chemistry. In The Vitamins, Vol. 3. W. H. Sebrell, Jr., and R. S. Harris (Editors). Academic Press, New York.
KLOSTERMAN, H. J., LAMOUREUX, G. L., and PARSONS, J. L. 1967. Isolation, characterization, and synthesis of linatine. A vitamin B_6 antagonist from flaxseed (Linum usitatissimum). Biochemistry 6, 170-176.
KRAMER, B., SOBEL, A. E., and GOTTFRIED, S. P. 1947. Serum levels of vitamin A in children. Am. J. Diseases Children 73, 543-553.
LASSER, J. L., and BRUSH, M. K. 1973. An improved ketogenic diet for treatment of epilepsy. J. Am. Dietet. Assoc. 62, 281-285.
LEITCH, I., and HEPBURN, A. 1961. Pyridoxine: metabolism and requirement. Nutr. Abstr. Rev. 31, 389-401.
LEVEILLE, G. A. 1975. Issues in human nutrition and their probable impact on foods of animal origin. J. Animal Sci. 41, 723-731.
LIENER, I. E. 1969. Miscellaneous toxic factors. In Toxic Constituents of Plant Foodstuffs. I. E. Liener (Editor). Academic Press, New York.
LIGHT, R. F. ALSCHER, R. P., and FREY, C. N. 1944. Vitamin A toxicity and hypoprothrombinemia. Science 100, 225-226.

LINKSWILER, H. M., JOYCE, C. L., and ANAND, C. R. 1974. Calcium retention of young adult males as affected by level of protein and of calcium intake. N.Y. Acad. Sci. Trans. *36*, 333-340.
LYONS, D. M., DUNLOP, D. M., and STEWART, C. P. 1931. The alkaline treatment of chronic nephritis. Lancet *221*, 1009-1013.
MICKELSEN, O. 1953. Nutritional aspects of antibiotics. J. Am. Dietet. Assoc. *29*, 221-229.
MICKELSEN, O. 1957. Chemicals and food processing. Nutr. Rev. *15*, 129-131.
MICKELSEN, O. 1964. Specific nutrients and dental caries. Food Nutr. News *36*, No. 3, 1, National Live Stock and Meat Board.
MICKELSEN, O. 1967. Present knowledge of niacin. *In* Present Knowledge in Nutrition. The Nutrition Foundation, New York.
MICKELSEN, O., CASTER, W. O., and KEYS, A. 1947. A statistical evaluation of the thiamin and pyramin excretions of normal young men on controlled intakes of thiamin. J. Biol. Chem. *168*, 415-431.
MICKELSEN, O., SIMS, L. S., BOGER, R. P., and EARHART, E. 1970. The prevalence of anemia in head start children. Mich. Med. *69*, 569-575.
MICKELSEN, O., YANG, M. G., and GOODHART, R. S. 1973. Naturally occurring toxic foods. *In* Modern Nutrition in Health and Disease, 5th Edition. R. S. Goodhart, and M. E. Shils (Editors). Lea & Febiger, Philadelphia.
MITCHELL, H. S. 1962. Nutrition in relation to stature. J. Am. Dietet. Assoc. *40*, 521-524.
MORGAN, A. F. 1971. Effects of home preparation on nutrient content of foods of animal origin. *In* Nutritional Evaluation of Food Processing. R. S. Harris, and H. Von Loesecke (Editors). Avi Publishing Co., Westport, Conn.
NATIONAL ACADEMY OF SCIENCES. 1968. Recommended Dietary Allowances. Publ. *1694*, Washington, D.C.
NELSON, E. M. 1956. Association of vitamin B_6 deficiency with convulsions in infants. Public Health Rept. *71*, 445-448.
NEWBURN, E. 1967. Sucrose, the arch criminal of dental caries. Odontol. *18*, 373-386.
NIELSEN, F. H. 1971. Studies on the essentiality of nickel. *In* Newer Trace Elements in Nutrition. W. Mertz, and W. E. Cornatzer (Editors). Marcel Dekker, New York.
OMDAHL, J. L., and DELUCA, H. F. 1973. Vitamin D. *In* Modern Nutrition in Health and Disease, 5th Edition. R. S. Goodhart, and M. E. Shils (Editors). Lea & Febiger, Philadelphia.
PRESS, E., and YEAGER, L. 1962. Food "poisoning" due to sodium nicotinate—report of an outbreak and review of literature. Am. J. Public Health *52*, 1720-1728.
ROELS, O. A., and LUI, N. S. T. 1973. Vitamin A and carotene. *In* Modern Nutrition in Health and Disease, 5th Edition. R. S. Goodhart, and M. E. Shils (Editors). Lea & Febiger, Philadelphia.
SCALA, J. 1974. Fiber, the forgotten nutrient. Food Technol. *28*, 34-36.
SCHWARZ, K. 1971. Tin as an essential growth factor for rats. *In* Newer Trace Elements in Nutrition. W. Mertz, and W. E. Cornatzer (Editors). Marcel Dekker, New York.
SHIP, I. I., and MICKELSEN, O. 1964. The effects of calcium acid phosphate on dental caries in children: a controlled clinical trial. J. Dental Res. *43*, 1144-1155.
SMASHEVSKI, N. D. 1966. A natural antivitamin of pantothenic acid. Chem. Abstr. *65*, 2677e.
SNELL, E. E., and RABINOWITZ, J. C. 1948. The microbiological activity of pyridoxylamino acids. J. Am. Chem. Soc. *70*, 3432-3434.

STURDEVANT, R. A. L., PEARCE, M. L., and DAYTON, S. 1973. Increased prevalence of cholelithiasis in men ingesting a serum-cholesterol-diet. New Engl. J. Med. *288*, 24-27.

TALBOT, J. M., FISHER, K. D., and CARR, C. J. 1974. A review of the significance of untoward reactions to iodine in foods. Federation Am. Soc. Exptl. Biol., Bethesda, Maryland.

TAPPEL, A. L. 1973. Lipid peroxidation damage to cell components. Federation Proc. *32*, 1870-1874.

TER-SARKISSIAN, N. *et al.* 1974. High phytic acid in Iranian breads. J. Am. Dietet. Assoc. *65*, 651-653.

THOMPSON, C. M., FUNK, K., SCHEMMEL, R., and MICKELSEN, O. 1974. Dental caries: possible sugar substitutes for sucrose. Ecol. Food Nutr. *3*, 231-236.

TOMARELLI, R. M., LINDEN, E., and BERNHART, F. W. 1952. Nutritional quality of milk thermally modified to reduce allergic reaction. Pediatrics *9*, 89-93.

TOMARELLI, R. M., SPENCE, E. R., and BERNHART, F. W. 1955. Biological availability of vitamin B-6 of heated milk. Agr. Food Chem. *3*, 338-341.

TROWBRIDGE, F. L., MATOVINOVIC, J., and NICHAMAN, M. Z. 1973. Goiter prevalence and iodine status of children. Federation Proc. *32*, 954.

TROWELL, H. 1974. Dietary fibre, coronary heart disease and diabetes mellitus. Plant Foods Man *1*, 11-16.

U.S. FOOD AND DRUG ADMIN. 1973A. Food labeling. Federal Register *38*, No. 49, Part II, Title 21, Mar. 14.

U.S. FOOD AND DRUG ADMIN. 1973B. Food labeling. Federal Register *38*, No. 13, Part III, Title 21, Jan. 19.

U.S. FOOD AND DRUG ADMIN. 1974. Federal Register *39*, No. 116, Part I, June 14.

U.S. DEPT. OF HEALTH, EDUCATION AND WELFARE. 1972A. Ten-State Nutrition Survey. III. Clinical. DHEW Publ. *(HSM) 72-8131*. Center for Disease Control, Atlanta, Georgia.

U.S. DEPT. OF HEALTH, EDUCATION AND WELFARE. 1972B. Ten-State Nutrition Survey. IV. Biochemical. DHEW Publ. *(HSM) 72-8132*. Center for Disease Control, Atlanta, Georgia.

VAGHEFI, S. B., MAKDANI, D. D., and MICKELSEN, O. 1974. Lysine supplementation of wheat proteins. A review. Am. J. Clin. Nutr. *27*, 1231-1246.

VON BONSDORFF, V. B. 1953. Diphyllobothrium latum and perniziöse anämie. Acta Haematol. *10*, 129-143.

WAGNER-JAUREGG, T. 1972. Riboflavin. II. Chemistry. *In* The Vitamins, Vol. 5. W. H. Sebrell, Jr., and R. S. Harris (Editors). Academic Press, New York.

WALKER, R. M., and LINKSWILER, H. M. 1972. Calcium retention in the adult human male as affected by protein intake. J. Nutr. *102*, 1297-1302.

WILDER, R. M. 1952. Nutritional health of adults. Proc. Natl. Food Nutr. Inst. USDA Agr. Handbook *56*.

WILLIAMS, H. H. 1959. Protein and amino acid requirements in man. N.Y. State J. Med. *59*, 4008.

YUDKIN, J. 1972. Sweet and Dangerous. Peter H. Wyden, New York.

Index

Acid-base potentials, 631-632
  blood urea level, 631
  high fat diets, 632
Acidophilus milk, 329
Additives, acids, effect on, proteins, 384
    vitamins, 384
  alkalies, destruction of thiamin, 384
    effect on amino acids, 384
  antioxidants, 388
  ascorbic acid protection, 387-388
  cooking, 541-543, 553
  copper, 385-386
    destruction of, ascorbic acid, 385
      thiamin, 386
    reaction with tocopherol, 385
  detrimental interactions, 386-387
    calcium, 387
    phosphates, 386
    vitamins, 386
  enrichment, 627
  food, 383
  food-package interaction, 430-431
  fortification, 627
  health, 628
  iron, 385-386
    reaction with tocopherol, 386
  principles, 627-628
  protection of B-vitamins, 389
  restoration, 627
  sodium bisulfite, effect on thiamin, 383
  sodium nitrite, 385
  solubility mechanisms, 388-389
  stabilization of ascorbic acid, 387
  sulfur dioxide, effect on, ascorbic acid, 383
    thiamin, 383
  trace minerals, 628-629
Alcoholic beverages, fermentation, 343-346
Algae, as food, 347
Amino acid, balance, 570-572, 576, 580
  imbalance, 572-574
  pattern, 570-572
  requirements, 577-579
  supplementation, 568, 578, 580-587
    animal protein, 589-590
    cereal grains, 578-587
    economic benefits, 599-600
    legume foods, 587
    oilseed flours, 588-589
    stability, in processing, 595-598
      in storage, 599
    toxicity, 573
Amino acids, essential, 569-570
  fortification technologies, 590-594
  genetics, effect on, 37
  in, cheese, 328
    idli, 341
    miso, 342
    shoyu, 340
    tempeh, 332, 339, 342
  ionizing radiation, 395-399
  limiting, 570-576, 584-590
  location, 44
  synthesis by industrial fermentations, 348-349
Animal processing, grading and composition, 127-134
  beef, 126
  eggs, 130-131
  factors, 127
  lamb, 130
  milk, 131
  nutrients, fish, 129
    meats, 129
    poultry, 131
  official standards, 127
  pork, 130
  poultry, 130
  refrigeration, 134
    "cooler shrink," 134
    vitamin retention, 135-137
      eggs, 136-137
      fish, 135-136
      milk, 136
      pork, 134-135
  veal, 130
  vitamin content, 131-134
    eggs, 133
    fish, 133
    meats, 132-133
    milk, 133
    poultry, 133
  low-temperature storage, 125

salable product yield, 126-127
  beef, 126
  composition factors, 127
  lamb, 126
  pork, 126
  poultry, 126
  transport
Animal protein deficiencies, 589-590
Antioxidants, 388
Apples, ascorbic acid, 38, 49
  quartered, 477
Arsenic, livestock products, 72
Ascorbic acid, fruits and vegetables, 98-99, 103-104
  genetic effects on, acerola, 36
    cabbage, 35
    snap beans, 35-36
    tomato, 35
  heat processing effects, 214
  in, cheese, 327-328
    fermented milk products, 325
    milk, 325-326, 335
    pickled and salted vegetables, 333-335, 337
    sauerkraut, 334-336
    wine, 343
    yoghurt, 332
  light, effect on, 38-39
    apples, 38
    tomato, 38
    turnip greens, 38
  location, 43
  loss in milk pasteurization, 326
  maturity, 50
  nitrogen fertilization, 46-47
  oxidative destruction, 106
  season, 41-42
  size, 49-50
    apples, 49
    peaches, 49
    tomatoes, 49-50
  temperature, 41

Bacteria, on food, 347
Baked goods, packaging, nutrient retention, 435
Baking, nutrient stability, 240-242
  lysine, 242
  minerals, 240-241
  vitamins, 241-242
    thiamin, 241

Bananas, sliced, ascorbic acid, 477
Barley, amino acids, 154-155
  in foods produced by fungal fermentations, 338-339
  milling, 154
  products, 154-155
  protein, 154-155
  supplementation, 584
Beans, 335-338, 341
  amino acid, 44
  ascorbic acid, 35-36, 41
  composition, 177-180
    amino acids, 177-178
    carbohydrates, 177, 180
    fatty acids, 177, 180
    minerals, 177, 179
    proximate analyses, 177
    vitamins, 177, 179
  food uses, 181
  genetic manipulation, 28-29
  nutritional properties, 180-181
    heat, effects, 181
    oligosaccharides, 180-181
  processing, 181-183
    and cooking, 182
    products, 182
    protein isolates, 183
  structure, 176
  thiamin, 43-44
  varieties, 176
Beef, age, 64-66
  antemortem stress, 69
  composition, 76-79
    antibiotics, 78
    fatty acids, 78-79
    hormones, 77-78
      DES, 77-78
      growth promotants, 78
    plane of nutrition, 76-77
    vitamin A, 77
  freeze-preservation, nutrient loss, 269-278
  genetic factors, 58-60
  grades, 64-66
  grading and composition, 64-66, 128-130
    nutrients, 129
    vitamin content, 132
  insecticide residues, 75
  minerals in, cadmium, 73
    copper, 70
    mercury, 73

molybdenum, 70
selenium, 72
refrigeration, 134
salable yield, 126
seasonal effects, 64
sex, 64-66
Beer, 344-345
  composition, 335-345
Beer-like beverages, 345-346
Benzo/a/pyrene, 371-372
BET monolayer, 291-292
Beverages, alcoholic, 343-346
  beer, 335, 344-345
  beer-like, 345-346
  kaffir beer, 346
  sake, 345-346
  wine, 343-344
Biochemistry cycle, 5-6
Biotin, in, grape musts, 344
  milk, 325
Blanching, food storage, 224
  freeze-preservation, vegetables, 245-248
  loss factors, 220
  methods, 220-224
  hot gas, 221
  hot water, 220
  microwaves, 221
  steam, 221
  IQB, 221
  superheated steam, 224
Boiling and steaming, mineral losses, 505-507
  calcium, 505-507
  chlorine, 506
  iron, 506-507
  magnesium, 506-507
  phosphorus, 507
  potassium, 506
Boiling foods, ratio water to solids, 497-499, 540, 542-543
  nutrient retention, 497
  vitamin retention, 499
Braising, meat and poultry, nutrient retention, 481-483
Bread, 433-435
  amino acids, 624
  packaging, 433
  riboflavin retention, 434-435
Broasting foods, 488
Broiling, meats and poultry, nutrient retention, 478-481
Buttermilk, 329

Cabbage, fermented, sauerkraut, 335-336
  fresh, for sauerkraut, 333-334
Cadmium, beef, 73
Calcium, in, cheese, 326-328, 330
  fermented milk products, 325, 329, 331-332
  fermented vegetable products, 334-336
  fish paste, 342
  milk, 325, 335
  tofu, 341-342
  light, effect on, 40
  location, effect on, 44
  soil fertilization, 45-49
Canned foods, nutrient stability, 228-234
  vitamin retention, 230-233
Carbohydrates, in, beer and beer-like beverages, 335, 345-346
  fermented milk products, 325, 329-332
  miso, 340
  shoyu, 340
  sufu, 341
  tempeh, 342
  vegetables for pickle and non-pickle use, 332-337
  ionizing radiation, 401
Carcinogenicity, in smoking foods, 371-373
Carotene, genetic effect on, 36
  sweet potato, 36
  tomato, 36
  in, pickled and salted vegetables, 333-335, 337
  light, effect on, 39
  tomato, 39
  location, 43
  maturity, 50
  nitrogen fertilization, 47
  season, 42
  synthesis by industrial fermentation, 348
  temperature, 41
  carrots, 41
Carrots, carotene, 41
  genetic manipulation, 23-27
Carving, juice, 494
Cauliflower, for pickling, 333-334
Cereal grains, See also Barley; Corn; Oats; Rice; Rye; Wheat
  composition, 139
  amino acid, 139, 141
  chemical, 139-140
  protein, 139-140

history, 139
products, 141
per capita consumption, 142
Cheese, composition, 326-330
making, nutrient changes, 324, 326-329
Chopping, vegetables, nutrient losses, 476-477
Citric acid, synthesis by industrial fermentation, 348
Cobalt, livestock products, 69-70
Computer-assisted menu planning, 467
Concentration. *See* Dehydration
Controlled atmosphere storage, 115
Cooking, and protein quality of meat, 493
  cabbage, vitamin C loss, 519-521
  drippings, protein loss from addition of water, 493-494
  vitamin loss, 493-494
  electronic, 491
  electronic and conventional, meat patties, 490
Cooking, home, 468, 532, 539
  baking, 551-555
    baking powder, 553
    bread, 552-555
    vegetables, 551
  boiling, 540
    additives, 541-543
    nutrient retention, ascorbic acid, 541, 545, 548
    carotene, 541-542
    effect of equipment, 540-541, 544
    minerals, 541
    niacin, 542
    riboflavin, 542
    thiamin, 541-542
    time of cooking, 540
    water: food ratio, 540, 542-543
  frozen vegetables, 539
  frying, 550-551
  preparation of food for cooking, 539
  pressure cooking, 544
  reheating, 534, 555
  steam cooking, 545-547
  vitamin losses, ascorbic acid, 541, 545, 548, 555
    biotin, 537
    carotene, 541-542
    choline, 537
    niacin, 541
    pantothenic acid, 535-536
    pyridoxine (B-6), 536-537
    S. carlsbergenesis assay, 537
    riboflavin, 535, 541
    thiamin, 534-535, 541-542
      effect of cooking method, 535
  large piece vegetables, 499
  large-scale, 468
  losses of food of animal origin, 478, 532-537
  pressure, nutrient loss in turkeys, 488
  pressure versus steam, vitamin retention, 499, 502-504
  temperature, 468
  thawed versus frozen, 474
  time, 468, 540
  water, vitamin losses, 502-503, 505
Copper, beef liver, 70
  eggs, 70
  milk, 70
Corn, dry-milled products, 145-147
  analyses, 146
  corn germ, 148
  minerals, 146-147
  opaque-2, 146, 148
  vitamins, 146-147
  yields, 146
  genetic manipulation, 29-30
  niacin, 36-37
  protein, 46
  sweet, nonpickle use, 335
  salted, 336-337
Cottonseed supplementation, 588
Cucumbers, brined and pickled, 332-334
Curing, agents, reaction with meat proteins, 358
  and nitrosamines, 361-363
    blocking action of ascorbic acid, 362
    carcinogenicity, 361-362
    dimethylnitrosamine, 362
    effect of smoking, 362
    formation in humans, 362
    role of vitamin A, 362-363
    sodium nitrite, allowable level, 363
  and smoking meats, composition, 361
  meat, amino acids, 358
    composition, 361
    methods, 359
    protein quality and loss, 359
    sodium nitrite, 358
    vitamins and minerals, 360-361
  meat and fish, organoleptic, 355
    preservation, 355

INDEX

Dairy products, freeze-preservation, 279
  packaging, 435-437
    butter, nutrient stability, 437
    cheeses, 437
    fluid milk, 435-437
      ascorbic acid, 436-437
      light transmission, 435-437
      riboflavin, 436-437
Dehydrated foods, packaging, 437-440
  fruits and vegetables, 439-440
    vitamin retention, 439-440
  high-protein, 440
    fish meal, 440
    meat, 440
    skim milk, 440
  nonenzymatic browning, 437-438
  oxidation, water content effects, 438-439
Dehydration, 11-12
  chemical kinetics, 290
  concentration processes, 295-299
    evaporation, 295-298
      economics, 295, 297-298
      equipment classes, 296-297
      "multieffect evaporation," 298
      steam-jacketed kettle, 295-296
    freezing, 298
    membrane, 298-299
      reverse osmosis, 298
      ultrafiltration, 298-299
  nutrient losses, 304-306, See also Proteins; Vitamins
    browning reactions, 304-306
    drying model, 304
    time-temperature-moisture factors, 304
  process, 289, 299-304
    drier designs, 300
    drying rate, 300
    drying techniques, 300-304
      drum, 302-303
      freeze, 303
      new processes, 303-304
      spray, 302
      sun, 300-301
      tunnel, 301-302
    nutrient retention, 300
    wet- and dry-bulb properties, 299-300
  temperature, 290-291
    activation energies, 291
    range, 290-291
  water, 291-295
    activity, 291-292
      decrease, effects, 292
      BET monolayer, 291-292
      chemical reactions, 292-294
      nutrient loss, 292-295
        oil-soluble nutrients, 293
        rate, effects, 294
        water-soluble, 293
      sorption, isotherm, 291-292
Diet, effect on animal products, beef, 76-79
  eggs, 85-88
  lamb, 79
  milk, 88-90
  pork, 80-81
  poultry, 81-85
Drying. See Dehydration

Eggs, composition, 85-88
  cholesterol, 86
  fatty acids, 85-86
  protein, 85
  sitosterol, 86-87
  vitamins, 87-88
  xanthophyll, 87
  grading and composition, 130-131
  vitamin content, 133
    retention, 136-137
  insecticide residues, 74-75
  minerals in, copper, 70
    iodine, 71
    manganese, 70
    molybdenum, 70
    zinc, 71
  seasonal effects, 67-68
  vitamin D, 68

Fabricated foods, 465
Fast food establishments, 529
Fat, in, cheese, 326-328, 330
  fermented milk products, 325
  milk, 325, 335
  miso, 340
  olives, 334
  shoyu, 340
  tempeh, 339
  light, effect on, 40
  synthesis by industrial fermentation, 347
Fats and oils, oxidation, 420

packaging techniques, 441
Fermented foods, 324-354
  aflatoxin in, 343
  alcoholic beverages, 343-346
  fungal fermentation products, 337-342
  growth stimulants in, 342
  milk products, 325-332
  Oriental, 324, 337-343
  salt in, 332, 336, 338-341
Fiber, in diet, 630-631
Fish, curing, 355
  freeze-preservation, nutrient loss, 270-277
  grading and composition, 129
    vitamin, content, 133
      retention, 135-136
  mercury, 73-74
  pastes, 357-358
  salting, 355-358
  sauces, 357
    amino acid composition, 357-358
Fluorine, poultry, 74
Folic acid, in milk, 325
Food additives. See Additives
Food processing, methods, 11-13
  acidity control, 12-13
  additives, 12-13
  heat, 11-12
    Maillard reaction, 12
  irradiation, 12-14
    thiamine stability, 14
  low temperatures, 12-13
  moisture removal, 11-13
  water activity, 12
Food service, FDA nutritional labeling, 463
  manufacturing kitchens, 469
  nutrient retention, 463
  systems, 469-470
  U.S. market, 464-467
Food storage, See also Canned foods
  blanching, 224
  pasteurization, 226
  sterile foods, 229
Foods, animal origin, 58-90
  high-sugar, 382
    nutrients, 382
  plant origin, 33-51
Freeze-preservation, animal tissues, cooking effects, 278-279
  entire freeze process, 278

B-vitamins loss, 277-278
freezing, 269-271
  B-vitamins, loss, 269-271
  time-temperature conditions, 271
frozen storage, 271-272
  B-vitamins loss, 271-272
prefreezing, 269
thaw-exudate, 273-276
  composition, mammalian meat, 273
  nutrient loss, 273-274
  processing variables, 275-276
    time-temperature conditions, 275-276
  product properties, 275
  product type, 274
thawing, 271-273
  thiamin loss, 271-273
dairy products, 279
data interpretation, 244-245
fruits, 262-269
  entire freeze process, 267-268
    vitamin C loss, 267-268
  freezing, 262
  frozen storage, 262-266
    vitamin C loss, 262-266
  prefreezing, 262
  preservation methods, comparative vitamin losses, 268-269
  thawing, 266-267
    vitamin C loss, 266-267
process, 244
proteins, 280
vegetables, 245-262
  blanching and cooling, leaching, 246
    steam versus water, 246-247
    vitamin loss, 245-248
  canning versus freezing, vitamin loss, 261-262
  cooking effects, 255-261
    mineral loss, 255, 258
    vitamin C loss, 255-257, 260-261
    vitamin losses, 255, 259-260
  entire freeze process, 254-255
    vitamin C loss, 254
  freezing, vitamin loss, 248-249
  frozen storage, 250-253
    mineral loss, 250
    time-temperature condition, 252
    vitamin C stability, 252-253
    vitamin loss, 250-253
  thawing, 254

vitamin loss, comparative (fruits, vegetables and animal tissues), 280-282
vitamin stability, 244-245
Frozen foods, packaging, 446-448
  flesh foods, 447
    oxidation, 447
  fruits and vegetables, 446-447
    ascorbic acid, 446-447
  requirements, 446
  thaw indicators, 447-448
Fruits. *See* Freeze-preservation
Fruits and vegetables, ascorbic acid, 98-99
  bananas, 103
  tomato juice, 104
  tomatoes, 103
  cellular organization, 105-108
    ascorbic acid, stability, 105
      oxidative destruction, 106
    carotenoid pigments, 106
    mechanical injury, 105
  chilling injury, 112-113
  controlled atmosphere (CA) storage, 115
    apples, 115
    fumigants, 115-116
    grapes, 115-116
    sulfur dioxide, 115
  market quality, 98
    and economic loss, 98
    and nutritional value, 98
  maturation, corn, 105
    peas, 104
  mechanical harvesting, 106-108
    cherries, 108
    damage, 107
    tomatoes, 107-108
  packaging, dehydrated, 437-440
    fresh, 441-443
      factors, 441
      nutrient retention, 442-443
      permeability, 442
      respiration rates, 442
  precooling, 113-115
    hydrocooling, 113
    refrigeration, 114-115
    vacuum, 114
  respiration, fruits, 101-102
    climacteric, 101-102
    nonclimacteric, 101-102
    vegetables, 104
  ripening, 100-104
    apples, 100-101
    bananas, 102-103
    ethylene gas, 101
    pears, 100-101
    pineapples, 103-104
    tomatoes, 103
  stages of growth, 99-100
  temperature effects, 110-112
    fruits, 110-111
    potatoes, 111
  wilting, 108-109
    and ascorbic acid, 109
    and carotene, 109-110
Frying, nutrient retention, 507-508
  liver, 481
Fumigants, 115-116
  storage life, 115
  sulfur dioxide, 115
Fungi, as food, 347
  fermentation products from, 337-343
    aflatoxin formation in, 342
    antibiotic formation in, 343
    nutritive value, 343

Genetic factors, beef, 58-60
  composition, 58-59
  cutability scores, 58, 60
  heritability estimates, 60
  quality, 58-60
  eggs, 63
  lamb, 60-61
  milk, 63-64
    color, 63
    composition, 63-64
    heritability estimates, 63-64
  pork, 61-62
    composition, 61
    heritability estimates, 62
    quality, 62
    USDA grades, 61-62
  poultry, 62-63
    composition, 62-63
    heritability estimates, 63
    yield, 63
Genetic manipulation, nutritional quality, 19-30
  beans, 28-29
    flavor, 28
    protein, 28-29
      methionine, 28-29

carrots, 23-27
  carotene biosynthesis, 24-26
  dry matter, 26
  hybrids, 26-27
  isocoumarin, 26
corn, 29-30
  protein, 30
  sugar, 29-30
effect of environment, 20
peas, 29
  protein, 29
tomato, 21-23
  carotene biosynthesis, 21-22
  flavor, 22-23
  solids/acid ratio, 23
  provitamin A content, 22
Glass. See Packaging
Goitrogenic agents, 190-191
Gossypol, 193-195
Grading. See Animal processing
Grapes, for wines, composition, 335, 343-344

Heat-processed foods, packaging, 443-446
  ascorbic acid retention, 443-446
  bacon, 445
  cranberry sauce, 444
  vegetables, 445
  cans versus glass, 443-446
  iron retention, 446
  thiamin retention, pork, 445
Heat processing, See also Blanching; Pasteurization; Sterilization
  enzymes, 215
  high temperature-short time (HTST) method, 216-220
  microorganisms, 215
  nutrient degradation, 205, 207-215
    kinetic parameters, 210-215
    pigments, 214
    time-temperature effects, 207, 210-215
    vitamins, 214
  nutrient retention, 215-220
    conduction-heating foods, 218
    enzymes versus microorganisms, 216-217
    food particulates, 217
    optimization, 215-220
  objectives, 205
  quality attributes, 215
  reviews, general, 207-209
  storage life, 205
Heat sterilization, 11-12, 18
Holding foods, dry heat, 495-496
  refrigerated before reheating, 513-515
  steam table, 515-518
    comparison of microwave and conventional, 516
    nutrient retention, 494, 515-518
Holding meals, before freezing, 515
  insulated containers, 518-519
  vitamin C retention, 518-519
HTST process, 216-220, 227-228

Idli, 338, 341
Infrared cooking, meats, 488
Insecticide residues, eggs, 74-75
  meat, 75
  milk, 75
  poultry, 74-75
  DDT, 74-75
  Dieldrin, 74-75
Iodine, 635-636
  in eggs, 71
  organic compound, 636
  plant sanitation, 636
  thyroid, 636
Ionizing radiation, amino acids, 395-399
  beef, 395, 397
  fish, 395, 398
  fruit, 396
  rat diet, 396
  vegetables, 399
  wheat, 398-399
  carbohydrates, 401
  composite diets, 394-395
    radappertization versus thermal processing, 394-395
    radappertized versus nonirradiated, 394
  lipids, 399-400
    chemical reactions, 400
    digestibility, 400
  literature, 407
  proteins, 395-399
    fruit, 396
    rat diet, 395-396
    vegetables, 399
    wheat, 398-399
  types, 393
  vitamins, 401-407

retention, animal diets, 402-403
   fat-versus water-soluble, 405-407
   fish, 403-404
   fruit, 405
   pork, 402
   vegetables, 405-406
   wheat, 403-404
Iron, fortification, 618-620
  light, effect on, 40
  nitrogen fertilization, effect on, 47

Kaffir beer, 346
Kefir, 329, 331-332
Kinetics, chemical, 290
  parameters, 210-213
  thermal destruction, 210
Kumiss, 329, 331-332

Lactic acid, in cheese and cheese making, 326, 328
  fermented milk preparations, 329-332
  fermented vegetable products, 332-333, 336
  idli, 338
  kefir and kumiss, 331
  yoghurt, 331
Lactose, changes in cheese making, 327-328
  in fermented milk preparations, 329, 332
  yoghurt, 331
Lamb, age, 66-67
  composition, 79
    plane of nutrition, 79
    vitamin A, 79
  genetic factors, 60-61
  grading and composition, 130
    vitamin content, 132-133
  salable yield, 126
  seasonal effects, 66-67
  selenium, 71-72
  sex, 66-67
Leaching, nutrients during cooking, 503
Lead, livestock, 72-73
Legumes, seed supplementation, 587. *See also* Beans; Peas; Soybeans
Light, packaging, 419-424
Lima beans, salted, 335-337
Lipids, *See also* Fat
  from microorganisms, 324
  ionizing radiation, 399-400
Lysine, in, cheese, 328
  oilseed meals, 195-196
  tempeh, 339
  synthesis by industrial fermentations, 348-349

Macronutrients, 5
Magnesium, season, effect on, 42
  soil fertilization, 48
Maize supplementation, 581-582
Manganese, eggs, 70
Meat, curing. *See* Curing
Mechanical harvesting, fruits and vegetables, 106-108
Mercury, beef, 73
  fish, 73-74
  pork, 73
  poultry, 73
Metabolism, nutrients, animal experimentation, 621-626
  application to humans, 621, 623-625
  dietary habits, 650
Methionine, in, idli, 341
  milk and cheese, 328
  tempeh, 339
  synthesis by industrial fermentations, 349
Microbial food, 347
Micronutrients, 5, 636, 649
  minerals, 646-649
  vitamins, 636-646
Microwave cooking, 491-493, 508-513
  nutrient retention, frozen precooked foods, 491
  meats, 491-493
  turkey, 491
  vegetables, 508-513
Milk, composition, 88-90
  carotene, 89-90
  fatty acids, 88-89
    linoleic, 89
    oleic, 88
  vitamins, A, 89
    D, 89
  fermented, acidophilus, 329
    buttermilk, 329
    cream, sour, 329
    kefir and kumiss, 331, 332
    yoghurt, 329, 331
  genetic factors, 63-64
  grading and composition, 131
    vitamin, content, 133

insecticide residues, 75
minerals in, 325, 335
  copper, 70
  molybdenum, 70
  zinc, 71
nutrients, 325, 335
  pasteurization effects on, 325-326
pasteurization, 225-226
packaging, 226
seasonal effects, 68
vitamin content, 133
sterilization, 225-226
Millet supplementation, 583-584
Minerals, calcium, 647-649
  absorption, 648-649
  carbonated beverages, 648
  osteoporosis, 648
fortification, 612-620
iron, deficiency, 646-647
livestock products, 69-74
phosphorus, 649
sodium, 646
Miso, 338-340, 342
  vitamin content, 340
Molybdenum, beef liver, 70
  eggs, 70
  milk, 70
Monosodium glutamate, 634-635
  effect on newborn animals, 634-635
    application to humans, 634-635
  food additive, 634-635

Natto, 338-340
Niacin, cholesterol level, 644
fortification, effects, 643-644
genetics, 36-37
  corn, 36-37
in, cheese, 328-330
  fermented milk preparations, 332
  milk, 325-335
    pasteurization effects, 326
  tempeh, 339
  tofu, 342
  yoghurt, 331
requirement, 643
retention in cheese making, 327
Nitrogen, soil fertilization, 46-47
Nitrosamines. *See* Curing
Nutrient standard menu planning, 463-467
Nutrients, available versus RDA, 9

components, 5
  macronutrients, 5
  micronutrients, 5
essential, source, 5
foods, biochemistry cycle, 5
composition, 6-7
per capita consumption, 6-9
function, 5
limiting in diet, 496
loss, before cooking, 470
  holding after cooking, 469
  lack of information, 469
Recommended Dietary Allowances (RDA), 9
  versus available nutrients, 9
restoration, 612-620
retention after thawing and cooking, 471
Nutrition labeling, 626-627, 650
  RDA, 627
  shortcomings, 626

Oats, milling, 151-152
  products, 152-154
  composition, 153
  vitamins, 154
supplementation, 583
Oilseed, flour supplementation, 588-589
processing, 197-200
Oilseed meals, food use, 187
  heat effects, 195-197
    amino acids, 196
    lysine, 195-196
      levels, 195-196
      processing methods, 195-196
    methionine, 196
    vitamin requirement, B-12, 196-197
    pyridoxine, 197
  physiologically active substances, 188-195
    allergens, 191
    anticoagulants, 189
    cyaniferous glycosides, 195
    cyclopropenoid fatty acids, 192-193
      eggs, 192-193
      growth rate, 193
      plant species, 192
    flatus factors, 192
    goitrogenic agents, 190-191
      glycosides, 190-191
      processing methods, 191
      thyroid hypertrophy, 190

gossypol, 193-195
  and available lysine, 194
  broilers, growth rate, 194-195
  hemagglutins, 191
  hemostatic agents, 189
  richetogenic agents, 189
  saponins, 190
  thiocyanates, 191
  trypsin inhibitors, 188
    growth rate, 188
  solvent processes, 197-200
    alcohol, 198
    azeotrope versus commercial hexane, 199
      effects, 199-200
    chlorinated hydrocarbons, 198-199
    mixtures, 198
    petroleum naptha, 197-198
    water, 198
  toxic substances, 188
    safety levels, 188
Okra, nonpickle use, 335
  salted, 336
Olives, pickled or salted, 333-334
Onions, pickling, 333-334

Packaging, factors affecting nutrients, 419-429
  light, 419-424
    absorption, 421
    amino acids, 420
      cellophanes, 422, 424
      flexible materials, 422
      glass, 422-423
    oxidation, fats and oils, 420
    protection, 421
    proteins, 420
    transmission curves, 421-424
    vitamins, water-soluble, 420
  moisture, 429-431
  oxygen, 426-429
    oxidation, fats and oils, 426-427
    permeability, 428-429
    protein loss, 428
    vitamin stability, 428
  temperature, 424-426
    control factors, 425
    eutectic substances, 425-426
    freezing rate, 425
    heat transfer, 424
    insulating containers, 424-426

  interaction with product, 429-432
    food additives, 430
    food-simulating solvents, 431
    free oil, 432
    glass, 432
    metal containers, 432
    toxicological implications, 431-432
  materials, 412-418
    combinations, 418
    glass, 412-413
      container types, 412-413
      design problems, 412
      properties, 412
    metals, 413-414
      foils, 414
      rigid containers, 413-414
        aluminum, 413
        tinplate, 413
    paper, 414-415
      advantages, 414
      applications, 414-415
        flexible, 414-415
        rigid, 415
      properties, 414
    plastics, 415-418
      cellophanes, 415
      cellulosics, 415
      films, 417-418
        water-soluble and edible, 417-418
      pliofilm, 417
      polyamides, 417
      polyesters, 417
      polyfluorocarbons, 417
      polyolefins, 416
      vinyl derivatives, 416-417
        formulae, 416-417
        properties, 416
    wood, 418
pasteurized foods, 226
quantitative analysis of nutritive retention, 448-453
  accelerated testing, 452-453
    storage life, prediction, 452
  moisture-sensitive foods, 450-452
    nutrient deterioration reactions, 450-452
      cabbage, 451-452
  oxygen-sensitive foods, 453
  shelf-life, prediction, 448-450
  storage, 448
sociological impact, 454

666   NUTRITIONAL EVALUATION OF FOOD PROCESSING

"natural foods," 454
specialized food delivery systems, 455
  space programs, 455
Pasteurization, 11-12
  definition, 206
  food storage, 226
    packaging, 226
  methods, 224-225
  nutrient retention, 216, 220, 225-226
  milk, 225-226, 325-326
Peanut supplementation, 588-589
Peas, composition, 177-180
  amino acids, 177-178
  carbohydrates, 177
  fatty acids, 177, 180
  minerals, 177, 179
  proximate analyses, 177
  vitamins, 177, 179
  genetic manipulation, 29
  green, nonpickle use, 335
  salted, 336-337
  nutritional properties, 180-181
  heat, effects, 181
  oligosaccharides, 180-181
  processing, 181-183
  products, 182
  protein concentrates, 182-183
  protein isolates, 183
  structure, 176
  varieties, 176
Peppers, pickling, 333-334
Pesticides, livestock products, 74-75
Phosphorus, in, cheese, 326, 328, 330
  fermented milk preparations, 325, 329-332
  fermented vegetable products, 334-335
  livestock products, 69-70
  milk, 325-335
  location, effect on, 44
  soil fertilization, 48
Phytates, and zinc, 629-630
  maturity, effect on, 51
  grain, 51
Phytic acid, genetics, 37
Pickles, 332-333
  low-sodium, 333
Plastics. See Packaging
Pork, age, 67
  antemortem stress, 69
  composition, 80-81

    antibiotics, 80
    fat, 80
    hormones, 80-81
    plane of nutrition, 80
    vitamins, thiamin, 80-81
  freeze-preservation, nutrient loss, 269-279
  genetic factors, 61-62
  grading and composition, 130
    vitamin, content, 132
    retention, 134-135
  salable yield, 126
  seasonal effects, 67
  sex, 67
Potatoes, mashed, ascorbic acid, 477
Poultry, age, 67
  antemortem stress, 69
  composition, 81-85
    antibiotics, 85
    calorie-protein ratio, 82-83
    fatty acids, 83-84
      linoleic, 83-84
      oleic, 84
    hormones, 84-85
      DES, 84
    plane of nutrition, 81-82
      force-feeding, 82
    vitamin content, 83
  freeze-preservation, nutrient loss, 270-277
  genetic factors, 62-63
  grading and composition, 130
    nutrients, 131
    vitamin content, 133
  insecticide residues, 74-75
  minerals in, fluorine, 74
    mercury, 73
    selenium, 72
  salable yield, 126
  seasonal effects, 67
  sex, 67
Protein, content and amino acid needs, 574-579
  from microorganisms, 324
  in, brewing industry by-product, 345
    cheese, 327-328, 330
    fungal fermentation products, 339-342
    milk, 325-326, 335
  ionizing radiation, 395-399
  losses in, brewing of green beans, 337
    cheese making, 326

nitrogen fertilization, effect on, 46
nutrient loss from dehydration, 306-311
  alfalfa, amino acids, 308-309
  browning reactions, 306
  fish meal, 309
  lysine and reducing sugars, 306
  milk, 310-311
    drying methods, 310-311
    evaporated, 311
    lysine, 310-311
  soy products, 306-308
    lysine, 307
    PER, 307-308
    trypsin inhibitor, 306-307
nutrient metabolism, 622-626
  amino acids, bread, 624
  flour, wheat, 622-624
    white, 625
synthesis, 569
  by industrial fermentations, 347, 350

Rabbit, antemortem stress, 69
Radiation. *See* Ionizing radiation
RDA, nutrients, versus available, 9
  U.S., 467
  Nutrimeter, 467
Riboflavin, absorption, 642
  deficiency, 642-643
  in cheese, 327-328, 330
    fermented milk preparations, 325, 329
    fungal fermentation products, 339-342
    idli, 341
    milk, 325-326, 335
  light, 39
  location, 43
  maturity, 50-51
  season, 42
  stability, 643
  temperature, 40
Rice, constituents, 148
  amino acid, 149
  protein, 148-149
  consumption, 148
  in fermented foods, 338-339, 341-342
  milled, 148, 150-151
    chemical composition, 150
    vitamins, 151
  production, 148
  Sierra, 342
  supplementation, 582-583

Roasting, meats, nutrient retention, 483-489
  addition of water, 483
  comparison conventional versus electronic, 488
  oven temperature, 485
Rye, grades, 156
  milling, 155-156
  products, 155
  supplementation, 584

Sake, 345-346
Salting, fish, and smoking, composition, 357
  brine method, 356
  composition, 357
  dry method, 356
  effect on vitamins, 358
  prooxidant activity, 358
  protein, denaturation, 356-357
    loss from, 356
    solubility, 356
  flesh foods, 355
    effect on flavor, 355
    preservation, 355-356
  vegetables, 324, 332-337, *See also* Miso; Shoyu; Sufu
Sauerkraut and sauerkraut-type vegetable blends, 333-336
Seed storage, damage, control factors, 119-120
  environmental conditions, 119
  mechanical harvesters, 119
  moisture levels, 120
  mycological aspects, 121-124
    fungi, 121-124
  nutritional quality, factors, 120
  respiration rate, 120
  seed coat, 118-119
Selenium, beef, 72
  lambs, 71-72
  poultry, 72
Sesame seed supplementation, 589
Shelf-life, retail, 558-559, *See also* Packaging
Shoyu, 338, 340
Single-cell protein, 347
Smoking, and polycyclic aromatic hydrocarbons, 371-373
  benzo/a/pyrene, 371-372
  carcinogenicity, 371-373

factors, 372
formation from organic substances, 372-373
reduction of, 372
antioxidant effects, 364
particle smoke phase, 364
phenols, 364
vapor phase, 364
bactericidal effects, 364-365
compounds in, 364
effect on nutrients, 363-364
fish, 357
flavor improvement, 355
flesh foods, effect of, flow of gases, 365
heat, 365
interaction with smoke components, 365-366
effect on vitamins and minerals, 367-370
formaldehyde, 366
protein quality and loss, 365
methods, 363
preservation effects, 363
Sodium chloride, 633-634
baby food, 634
hypertension, 633-634
Soil fertilization, calcium, 47-48
mineral elements, 44-46, 48-49
nitrogen, 46-47
phosphorus, 48
Sorghum supplementation, 583-584
Soybeans, curd or tofu, 338, 341-342
full-fat flour, 164, 166-167
composition, 166
nutritional properties, 166-167
heat treatment, 166-167
in foods produced by fungal fermentations, 337-342
in sauerkraut-like product, 336
milk, 341
oil and flakes, 160, 167-168
edible products, 167-168
flours, 169
grits, 169
nutrient composition, 170
raw meal, 169
Oriental foods, 175-176
composition, 175
soy sauce, 176
protein concentrates, 170-172
composition, 171

heat treatment, 172
processes, 170-171
solubility, 171-172
protein isolates, 172-175
heat treatment, 174
nutrient composition, 173-174
pH effects, 174-175
process, 172-173
supplementation, 588
whole, 159-164
composition, 159-164
amino acids, 159-161
fatty acids, 160, 163
minerals, 159, 161
minor constituents, 164
polysaccharides, 161, 163
sugars, 161, 163
U.S. varieties, 159-160
vitamins, 164-165
consumption, 164
products, 164
seed structure, 159
Stability of nutrients, essential amino acids, 1-3
essential fatty acids, 1, 3
factors, 16-18
coenzyme forms, 16
heat, 17-18
light, 17-18
packaging, 16
pH, 16-17
Steam table. See Holding foods
Steamers, cooking, 499, 502
Sterilization, definition, 206-207
methods, 227-229
nutrient retention, 216, 220, 226-229
Storage, See also Packaging
home, 530
butter, 531
freezing, 556-557
rate, 530
margarine, 531
precooked frozen foods, 532
refrigeration, 530-535
vitamin losses, 556-557
shelf-life, 558-559
shell eggs, 532
soy products, 532
thawing, 531
in refrigerator, 531
in water, 531

INDEX 669

vitamin losses, frozen storage, 556-557
  refrigeration, 556-557
Sucrose, 632-633
  cardiovascular disease, 633
  dental caries, 632-633
  hexose, 633
  regulations, 633
Sufu, 338, 341-342
Sugar, high concentration foods, 382
Sulfites. See Curing

Tempeh, 337-339, 342
Thawing, juices, 472-474
  loss, fat, 473
    minerals, 473
    protein, 473
    use, 473
    weight, 472
Thermal processing foods, 468
  correlation with nutrient retention, 468
Thiamin, 640-642
  destruction of, milk pasteurization, 326
  heat processing effects, 214
  in, cheese, 327-328, 330
    fermented milk products, 325
    idli, 341
    milk, 325, 335
    natto, 340
    tempeh, 339
    tofu, 342
  irradiation sensitivity, 14
  light, effect on, 39-40
  location, 43-44
    beans, 43-44
    cereals, 43-44
  loss in cheese making, 327
  nitrogen fertilization, 47
  season, 42
  size, 51
  synthesis in beer-like beverages, 346
  temperature, 41
Tofu, 338, 341-342
Tomato, ascorbic acid, 35, 38, 49-50
  in pulp, 477
  carotene, 36, 39
  genetic manipulation, 21-23
Trimming, fat, 470
  composition, 472
  specifications, 470
    USDA, 470
  foods of plant origin, 473-478

factors, 474-475
  nutrient losses, 475
  waste, 475
Triticale supplementation, 584
Trypsin inhibitors, oilseed meals, 188-189
Tryptophan, in tempeh, 339
  synthesis by industrial fermentation, 349

USDA Agricultural Handbook 8, 228, 467

Veal, grading and composition, 130
  vitamin content, 132
Vegetables. See Freeze-preservation;
  Fruits and vegetables
Vitamin A, absorption, 636-637
  additive, 637
  effect of milk pasteurization, 326
  fortification, 637-638
  in, cheese, 326-328, 330
    fermented milk preparations, 329, 332
    milk, 325, 335
  nutrition labeling, 637-638
  RDA, 637-638
  synthesis by industrial fermentation, 348
  toxicity, 638
Vitamin B complex, See also Biotin;
  Niacin; Riboflavin; Thiamin
  in beer and beer-like beverages, 345-346
  by-products of brewing industry, 345
  cheese, 328
  fermented milk preparations, 329
  idli, 341
  milk, 325
  pickled and salted vegetables, 332
  wine, 344
  synthesis by industrial fermentations, 347
Vitamin B-6, 644-645
  in milk, 325
  stability, 645
Vitamin B-12, absorption, 645
  in, milk, 325
    natto, 340
    tempeh, 339
  pernicious anemia, 645-646
  stability, 646
  synthesis by industrial fermentations, 347-348
  tapeworm, 645

Vitamin C. See Ascorbic acid
Vitamin D, absorption, 638
  fortification, 639
  in milk, 325
  metabolic function, 639
  synthesis by industrial fermentation, 348
Vitamin E, aging process, 640
  dietary alterations, 640
  function, 639
  hepatic reactions, 640
  in milk, 325
  requirement, 640
  synthesis by industrial fermentation, 348
Vitamin K, 641
  in milk, 325
  synthesis by industrial fermentation, 348
Vitamins, See also Freeze-preservation
  fat-soluble, nutrient loss from dehydration, 318-319
    carotene, carrots, 318-319
    milk, 318
    vitamin E, 318-319
  fortification, 612-620
  heat processing effects, 214
  ionizing radiation, 401-407
  retention, canned foods, 229-233
  water-soluble, 420
    nutrient loss from dehydration, 311-318
      ascorbic acid, 311-315
      water activity, reaction rate constant, 312-313
      B vitamins, 316
      concentration processes, 317-318
        reverse osmosis, 318
        ultrafiltration, 317-318
      thiamin, 315-317
        meats, 315
        vegetables, 315
        wheat toasting, 316-317

Washing and soaking, nutrient losses, apples, 476
  potatoes, 476
  rice, 476
Water. See Dehydration
Wheat, in foods from fungal fermentations, 337-338, 340, 342
  milling, 142-143
  products, 142
    composition, 143
    protein, 143
    supplementation, 579-580
    vitamins, 143-145
Wine and wine-like beverages, 335, 343-344

Yeast, as food, 347
Yoghurt, 329, 331
  and lactose-intolerance, 331

Zinc, eggs, 71
  milk, 71

NOTES

NOTES

NOTES

NOTES